# Lecture Notes in Computer Science 4310

Commenced Publication in 1973
Founding and Former Series Editors:
Gerhard Goos, Juris Hartmanis, and Jan van Leeuwen

T0189788

Todor Boyanov   Stefka Dimova
Krassimir Georgiev   Geno Nikolov (Eds.)

# Numerical Methods and Applications

6th International Conference, NMA 2006
Borovets, Bulgaria, August 20-24, 2006
Revised Papers

 Springer

Volume Editors

Todor Boyanov
Stefka Dimova
Geno Nikolov
St. Kl. Ohridski University of Sofia
Faculty of Mathematics & Informatics
5 J. Bourchier Blvd., 1164 Sofia, Bulgaria
E-mail: {boyanovt, dimova, geno}@fmi.uni-sofia.bg

Krassimir Georgiev
Bulgarian Academy of Sciences
Institute for Parallel Processing
Acad. G. Bonchev str., bl. 25A, 1113 Sofia, Bulgaria
E-mail: georgiev@parallel.bas.bg

Library of Congress Control Number: 2007920821


CR Subject Classification (1998): G.1, F.2.1, G.4, J.2, J.6

LNCS Sublibrary: SL 1 – Theoretical Computer Science and General Issues

ISSN        0302-9743
ISBN-10     3-540-70940-1 Springer Berlin Heidelberg New York
ISBN-13     978-3-540-70940-4 Springer Berlin Heidelberg New York

Springer is a part of Springer Science+Business Media

springer.com

© Springer-Verlag Berlin Heidelberg 2007
Printed in Germany

Typesetting: Camera-ready by author, data conversion by Scientific Publishing Services, Chennai, India
Printed on acid-free paper      SPIN: 12021581      06/3142      5 4 3 2 1 0

# Preface

The international conference Numerical Methods and Applications has been a traditional forum for scientists of well-known research groups from various countries providing an opportunity for sharing ideas and establishing fruitful scientific cooperation.

The papers in this volume were presented at its sixth issue: International Conference on Numerical Methods and Applications (ICNM&A 2006) held in Borovets, Bulgaria, August 20–24, 2006. The conference was organized by the Faculty of Mathematics and Informatics at "St. Kliment Ohridski" University of Sofia, in cooperation with GAMM and IMACS. The Institute of Mathematics and Informatics and the Institute for Parallel Processing, Bulgarian Academy of Sciences, were co-organizers of this traditional scientific meeting.

In total, 119 participants from 27 countries all over the world attended the conference and 111 talks, including nine invited talks, were delivered. This volume contains 87 papers submitted by authors from 24 countries.

During ICNM&A 2006 a wide range of problems concerning recent theoretical achievements in numerical methods and their applications in mathematical modeling were discussed. Specific topics of interest were the following: *finite difference and finite volume methods; finite element and boundary element methods; multigrid and domain decomposition; level set and phase field methods; Monte Carlo methods; numerical linear algebra; parallel algorithms; computational mechanics; engineering applications.* The keynote lectures reviewed some of the advanced achievements in these fields. The ICNM&A 2006 talks were delivered by researchers representing some of the strongest research teams in the field of numerical methods and their application for solving wide range of practical problems.

The success of the conference and the present volume are due to the joint efforts of many colleagues from various institutions and organizations. We express our deep gratitude to all the members of the Scientific Committee for their valuable contribution to the scientific spirit of the conference, as well as for their help in reviewing the submitted papers. The special sessions represented the combined efforts of organizers whose contributions deserve to be recognized: Enrique Alba, Rene Alt, Radim Blaheta, Stefka Fidanova, Krasimir Georgiev, Todor Gurov, Aneta Karaivanova, Johannes Kraus, Svetozar Margenov, Svetoslav Markov, Gradimir Milovanović, Bojan Popov, Per Grove Thomsen, and Zahari Zlatev. We are also grateful to the staff involved in the local organization.

The conference was partly supported by project BIS-21++ funded by the European Commission in FP6 INCO via grant 016639/2005.

We hope that this meeting among scientists who develop and study numerical methods, on one hand, and researchers who use them for solving real-life problems, on the other, has broadened their horizons and has contributed to their mutual enrichment.

December 2006

Todor Boyanov
Stefka Dimova
Krasimir Georgiev
Geno Nikolov

# Organization

## International Scientific Committee

A. Abramov (Russia)
A. Andreev (Bulgaria)
P. Binev (USA)
R. Blaheta (Czech
   Republic)
P. Bochev (USA)
T. Boyadjiev (Bulgaria)
B. Boyanov (Bulgaria)
C. Budd (UK)
B. Chetverushkin
   (Russia)
C. Christov (USA)
I. Dimov (Bulgaria)
I. Farago (Hungary)
K. Georgiev (Bulgaria)

A. Gulin (Russia)
M. Gunzburger (USA)
R. Herbin (France)
O. Iliev (Germany)
R. Jeltsch (CH)
A. Karaivanova
   (Bulgaria)
M. Kaschiev (Bulgaria)
O. Kunchev (Bulgaria)
R. Lazarov (USA)
I. Lirkov (Bulgaria)
V. Makarov (Ukraine)
S. Margenov (Bulgaria)
S. Markov (Bulgaria)
P. Matus (Belarus)

G. Milovanovic (Serbia)
P. Minev (Canada)
M. Neytcheva (Sweden)
G. Nikolov (Bulgaria)
J. Pasciak (USA)
Y. Popov (Russia)
I. Puzynin (Russia)
S. Radev (Bulgaria)
V. Thomee (Sweden)
P. Vabishchevich
   (Russia)
E. Varbanova (Bulgaria)
P. Vassilevski (USA)
L. Zikatanov (USA)
Z. Zlatev (Denmark)

## Organizing Committee

Chairperson: S. Dimova

T. Boyanov
T. Chernogorova
I. Dimov
N. Kolkovska

S. Margenov
L. Milev
N. Naidenov
V. Rakidzi

S. Stoilova
D. Vasileva

# Table of Contents

## IV  Monte Carlo and Quasi-Monte Carlo for Diverse Applications

# V  Metaheuristics for Optimization Problems

# VI  Uncertain/Control Systems and Reliable Numerics

## VII  Interpolation and Quadrature Processes

# VIII  Large-Scale Computations in Environmental Modelling

# IX  Contributed Talks

# On the Discretization of the Coupled Heat and Electrical Diffusion Problems

Abdallah Bradji[1] and Raphaèle Herbin[2]

[1] Weierstrass Institute for Applied Analysis and Stochastics,
Mohrenstr. 39 10117 Berlin Germany
bradji@wias-berlin.de
http://www.wias-berlin.de/~bradji
[2] Laboratoire d'Analyse, Topologie et Probabilités,
Université Aix–Marseille 1, 39 rue Joliot Curie 13453 Marseille, France
raphaele.herbin@latp.univ-mrs.fr
http://www.cmi.univ-mrs.fr/~herbin

**Abstract.** We consider a nonlinear system of elliptic equations, which arises when modelling the heat diffusion problem coupled with the electrical diffusion problem. The ohmic losses which appear as a source term in the heat diffusion equation yield a nonlinear term which lies in $L^1$. A finite volume scheme is proposed for the discretization of the system; we show that the approximate solution obtained with the scheme converges, up to a subsequence, to a solution of the coupled elliptic system.

**Keywords:** Nonlinear elliptic system, Diffusion equation, Finite volume scheme, $L^1$-data, Ohmic losses.

## 1 Introduction

It is well known that the diffusion of electricity in a resistive medium induces some heating, known as ohmic losses. Such a situation arises for instance in the modelling of fuel cells, see e.g. [16], [17] and references therein. Let $\phi$ denote the electric potential, and let $\kappa$ denote the electrical conductivity. Then the ohmic losses may be written as $\kappa\nabla\phi\cdot\nabla\phi$. Since $\phi$ is the solution of a diffusion equation, it is reasonable to seek $\phi$ in the space $H^1(\Omega)$, so that $\nabla\phi\cdot\nabla\phi\in L^1$. Hence the heat diffusion equation has a right–hand–side in $L^1$, and its analysis falls out of the usual variational framework. Our aim in this paper is to study the convergence of approximate solutions to the resulting coupled problem obtained with a cell centred finite volume scheme.

The theory of elliptic and parabolic equations with irregular right–hand–side goes back to the pioneering work of G. Stampacchia [23], where solutions to the linear problem are defined by duality. Later on, L. Boccardo, T. Gallouët and co-authors (see [3] and references therein) introduced the tools and setting in which one may define solutions to such problems: these so-called entropy solutions [4] were found to be equivalent to the so-called renormalized solutions of P.-L. Lions and F. Murat [22], as well as to the solutions obtained by approximation, as defined by [10]. In the linear case, all these solutions are also equivalent to those of [23].

T. Boyanov et al. (Eds.): NMA 2006, LNCS 4310, pp. 1–15, 2007.

Other solutions obtained by approximation were defined thanks to numerical schemes. They also lead to the existence existence of a solution, but more importantly, they yield a constructive way to compute approximate solutions of the problem. The convergence of the finite volume scheme was proven in [20] for the Laplace equation with right–hand–side measure; the proof was generalized in [11] to noncoercive convection diffusion problems. The convergence of the finite element scheme, with irregular data, on bi-dimensional polygonal domains was proven for Delaunay triangular meshes in [18] and in [6] for three–dimensional tetrahedral meshes under geometrical conditions. Under regularity assumptions on the solutions, error estimates may be obtained by interpolation [8], [6]. The convergence order of finite element solutions is also studied in [24] for elliptic boundary value problems when the second member is piecewise smooth but discontinuous along some curve.

In the present paper, we recall some of the properties which have been established for the discretization of elliptic problems by finite volumes. We then show how the techniques introduced in the above references may be used to prove the convergence of a discretization scheme for the approximation of the above mentioned heat and electricity diffusion problem, since the resulting system of semilinear elliptic partial differential equations is such that the right-hand-side of the second equation depends on the solution of the first one and is in $L^1$.

The paper is organized as follows: in Section 2 we recall the main principle and properties of finite volume schemes for elliptic problems; in Section 3, we present the continuous problem, its weak form and the known result about existence [19]. In section 4, we describe the finite volume method for the approximation of the system and prove the existence of a solution to the resulting discrete system for both cases. The convergence of the finite volume cases in section 5. The proof of convergence is based on a priori estimates, compactness result and a passage to the limit in the scheme. Some conclusions are drawn in the last section.

## 2     The Cell Centered Finite Volume Method

Finite volume methods are known to be well suited for the discretization of conservation laws; these conservation laws may yield partial differential equations of different nature (elliptic, parabolic or hyperbolic) and also to coupled systems of equations of different nature.

Let $\Omega$ be a polygonal open subset of $\mathbb{R}^d$, $T \in \mathbb{R}$, and let us consider a balance law written under the general form:

$$u_t + \mathrm{div}(F(u, \nabla u)) + s(u) = 0 \text{ on } \Omega \times (0, T), \tag{1}$$

where $F \in C^1(\mathbb{R} \times \mathbb{R}^d, \mathbb{R}^d)$ and $s \in C(\mathbb{R}, \mathbb{R})$. Let $\mathcal{T}$ be a finite volume mesh of $\Omega$. For the time being, we shall only assume that $\mathcal{T}$ is a collection of convex polygonal control volumes $K$, disjoint one to another, and such that: $\bar{\Omega} = \cup_{K \in \mathcal{T}} \bar{K}$. The balance equation is obtained from the above conservation law by integrating it over a control volume $K$ and applying the Stokes formula:

$$\int_K u_t \, dx + \int_{\partial K} F(u, \nabla u) \cdot \mathbf{n}_K \, d\gamma(x) + \int_K s(u) \, dx = 0,$$

where $\mathbf{n}_K$ stands for the unit normal vector to the boundary $\partial K$ outward to $K$ and $d\gamma$ denotes the integration with respect to the $(d-1)$–dimensional Lebesgue measure. Let us denote by $\mathcal{E}$ the set of edges (faces in 3D) of the mesh, and $\mathcal{E}_K$ the set of edges which form the boundary $\partial K$ of the control volume $K$. With these notations, the above equation reads:

$$\int_K u_t \, dx + \sum_{\sigma \in \mathcal{E}_K} \int_\sigma F(u, \nabla u) \cdot \mathbf{n}_K \, d\gamma(x) + \int_K s(u) \, dx = 0.$$

Let $k = T/M$, where $M \in \mathbb{N}, M \geq 1$, and let us perform an explicit Euler discretization of the above equation (an implicit or semi-implicit discretization could also be performed, and is sometimes preferable, depending on the type of equation). We then get:

$$\int_K \frac{u^{(n+1)} - u^{(n)}}{k} \, dx + \sum_{\sigma \in \mathcal{E}_K} \int_\sigma F(u^{(n)}, \nabla u^{(n)}) \cdot \mathbf{n}_K \, d\gamma(x) + \int_K s(u^{(n)}) \, dx = 0,$$

where $u^{(n)}$ denotes an approximation of $u(\cdot, t^{(n)})$, with $t^{(n)} = nk$. Let us then introduce the discrete unknowns $(u_K^{(n)})_{K \in \mathcal{T}, \, n \in \mathbb{N}}$ (one per control volume and time step); assuming the existence of such a set of real values, we may define a piecewise constant function by:

$$u_{\mathcal{T}}^{(n)} \in X(\mathcal{T}) : u_{\mathcal{T}}^{(n)} = \sum_{K \in \mathcal{T}} u_K^{(n)} 1_K,$$

where $X(\mathcal{T})$ denotes the space of functions from $\Omega$ to $\mathbb{R}$ which are constant on each control volume of the mesh $\mathcal{T}$, and $1_K$ is the characteristic function of the control volume $K$, where $1_K(x) - 1$ if $x \in K$ and $1_K(x) = 0$ otherwise. In order to define the scheme, the fluxes $\int_\sigma F(u^{(n)}, \nabla u^{(n)}) \cdot \mathbf{n}_K \, d\gamma(x)$ need to be approximated as a function of the discrete unknowns. We denote by $F_{K,\sigma}(u_{\mathcal{T}}^{(n)})$ the resulting numerical flux, the expression of which depends on the type of flux to be approximated.

The coupled system which is the aim of our study is a system of diffusion equations. Let us now consider a linear diffusion reaction equation, that is equation (1) with $F(u, \nabla u) = -\nabla u$, and $s(u) = bu, b \in \mathbb{R}_+$:

$$u_t - \Delta u + bu = 0 \text{ on } \Omega, \tag{2}$$

the flux through a given edge then reads:

$$\int_\sigma F(u) \cdot \mathbf{n}_{K,\sigma} = \int_\sigma -\nabla u \cdot \mathbf{n}_{K,\sigma},$$

so that we need to discretize the term $\int_\sigma -\nabla u \cdot \mathbf{n}_{K,\sigma}$; this diffusion flux involves the normal derivative to the boundary, for which a possible discretization is obtained by considering the differential quotient between the value of $u_{\mathcal{T}}$ in $K$ and in the neighbouring control volume, let's say $L$:

$$F_{K,\sigma}(u_{\mathcal{T}}) = -\frac{\mathrm{m}(\sigma)}{d_{KL}}(u_L - u_K). \tag{3}$$

where $\mathrm{m}(\sigma)$ stands for the $(d-1)$–dimensional Lebesgue measure of $\sigma$ and $d_{KL}$ is the distance between some points of $K$ and $L$, which will be defined further. We then obtain the following numerical flux:

$$F_{K,\sigma}(u_{\mathcal{T}}) = -\frac{\mathrm{m}(\sigma)}{d_{KL}}(u_L - u_K).$$

However, we are able to prove that this choice for the discretization of the diffusion flux yields accurate results only if the mesh satisfies the so-called orthogonality condition, that is, there exists a family of points $(x_K)_{K \in \mathcal{T}}$, such that for a given edge $\sigma_{KL}$, the line segment $x_K x_L$ is orthogonal to this edge (see figure 1). The length $d_{KL}$ is then defined as the distance between $x_K$ and $x_L$. This geometrical feature of the mesh will be exploited to prove the consistency of the flux. Of course, this orthogonality condition is not satisfied for any mesh. Such a family of points exists for instance in the case of triangles, rectangles or Voronoï meshes. We refer to [12] for more details.

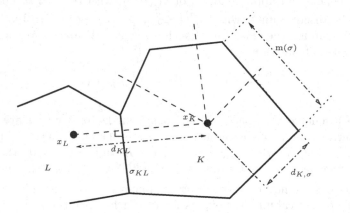

**Fig. 1.** Notations for a control volume

## 3   The Continuous Coupled Problem

We wish to find some numerical approximation of solutions to the following nonlinear coupled elliptic system, which models the thermal and electrical diffusion in a material subject to ohmic losses:

$$- \nabla \cdot (\kappa(x, u(x)) \nabla \phi(x)) = f(x, u(x)), \; x \in \Omega \tag{4}$$

$$\phi(x) = 0, \; x \in \partial\Omega, \tag{5}$$

$$-\nabla \cdot (\lambda(x, u(x)) \nabla u(x)) = \kappa(x, u(x)) |\nabla \phi|^2 (x), \; x \in \Omega, \tag{6}$$

$$u(x) = 0, \; x \in \partial\Omega, \tag{7}$$

where $\Omega$ is a convex polygonal open subset of $\mathbb{R}^d$, $d = 2$ or $3$, with boundary $\partial\Omega$, $\phi$ denotes the electrical potential and $u$ the temperature; the electrical conductivity $\kappa$, the thermal conductivity $\lambda$ and the source term $f$ are functions from $\Omega \times \mathbb{R}$ to $\mathbb{R}$ satisfying the following Assumptions:

**Assumption 1.** *The functions $\kappa, \lambda$ and $f$, defined from $\Omega \times \mathbb{R}$ to $\mathbb{R}$, are bounded and continuous with respect to $y \in \mathbb{R}$ for a.e. $x \in \Omega$, and measurable with respect to $x \in \Omega$ for any $y \in \Omega$, and such that:*

$$\exists \alpha > 0; \ \alpha \le \kappa(x, y) \text{ and } \alpha \le \lambda(x, y), \ \forall y \in \mathbb{R}, \text{ for a.e. } x \in \Omega. \tag{8}$$

The following existence result was proven in [19]:

**Theorem 1.** *Under Assumption 1, there exists a solution to the following weak form of Problem (4)– (7):*

$$\begin{cases} (\phi, u) \in H_0^1(\Omega) \times \cap_{p < \frac{d}{d-1}} W_0^{1,p}(\Omega), \\ \int_\Omega \kappa(\cdot, u)\nabla\phi \cdot \nabla\psi \, dx = \int_\Omega f(\cdot, u)\psi \, dx, \ \forall \psi \in H_0^1(\Omega) \\ \int_\Omega \lambda(\cdot, u)\nabla u \cdot \nabla v \, dx = \int_\Omega \kappa(\cdot, u)|\nabla\phi|^2 v \, dx, \ \forall v \in \cup_{r > d} W_0^{1,r}(\Omega), \end{cases} \tag{9}$$

*where here and in the sequel, $dx$ denotes the integration symbol with respect to the Lebesgue measure in $\mathbb{R}^d$ or $\mathbb{R}^{d-1}$.*

Note that the exponents $\frac{d}{d-1}$ and $d$ are conjugate, ant that, for $r > d$ the space $W_0^{1,r}(\Omega)$ is continuously imbedded in the space $C(\overline{\Omega}, \mathbb{R})$; therefore all terms in (9) make sense. In the case $d = 2$, we have $u \in W_0^{1,p}$ for all $p < 2$, but in general, $u \notin H_0^1(\Omega)$. Similarly, if $d = 3$, $u \in W_0^{1,p}$ for all $p < \frac{3}{2}$.

The proof of this theorem relies mainly on the analysis tools which were developed for the analysis of elliptic equations with irregular right–hand–side, see for instance [3] and references therein. We shall not need to assume this existence result for our present analysis. Indeed, the existence of a solution to (4)–(7) is obtained as a by–product of the convergence of the scheme. Nevertheless, a large part of the convergence analysis of the schemes is inspired from the ideas developed in [19] for the existence result, and we shall again use the ideas of [3] in our proofs.

## 4   The Discretization Schemes

In [17], the numerical simulation of solid oxide fuel cells relies on a mathematical model involving a set of semilinear partial differential equations, the unknowns of which are the temperature, the electrical potential and the concentrations of various chemical species in the porous media of the cell. System (4)–(7) is a sub–problem of this latter model, obtained by leaving out the chemical species diffusion equations. In [17], three different discretization schemes were implemented and compared, namely the linear finite element method, the mixed finite element method, and the cell centred finite volume method. Because of interface conditions involving the electrical current, a precise approximation of the electrical flux is needed at the interfaces, the linear finite element method was found to be less adapted than the two latter methods, so that finally the mixed finite element method and the cell centered methods were numerically compared. The cell centred finite volume method was found to be easier to implement and comparable to the mixed finite element method as to the ratio precision *vs.* computing time, so that it was finally chosen for the simulations of different geometries of fuel cells [16]. Here we shall give a theoretical justification of the convergence of and the cell centred finite volume method for the discretization of system (4)–(7). The convergence of a linear finite element scheme is proven in a forthcoming paper [5].

To define a finite volume approximation, we introduce an admissible mesh $\mathcal{T}$ in the sense of [12, Definition 9.1 page 762], which we recall here for the sake of completeness:

**Definition 1 (Admissible meshes).** *Let $\Omega$ be an open bounded polygonal subset of $\mathbb{R}^d$, $d = 2$ or 3. An admissible finite volume mesh of $\Omega$, denoted by $\mathcal{T}$, is given by a family of "control volumes", which are open polygonal convex subsets of $\Omega$ , a family of subsets of $\overline{\Omega}$ contained in hyperplanes of $\mathbb{R}^d$, denoted by $\mathcal{E}$ (these are the edges in two space dimensions, or faces in three space dimensions, of the control volumes), with strictly positive $(d-1)$-dimensional measure, and a family of points of $\Omega$ satisfying the following properties:*

(i) *The closure of the union of all the control volumes is $\overline{\Omega}$.*
(ii) *For any $K \in \mathcal{T}$, there exists a subset $\mathcal{E}_K$ of $\mathcal{E}$ such that $\partial K = \overline{K} \setminus K = \cup_{\sigma \in \mathcal{E}_K} \overline{\sigma}$. Furthermore, $\mathcal{E} = \cup_{K \in \mathcal{T}} \mathcal{E}_K$.*
(iii) *For any $(K, L) \in \mathcal{T}^2$ with $K \neq L$, either the $(d-1)$-dimensional Lebesgue measure of $\overline{K} \cap \overline{L}$ is 0 or $\overline{K} \cap \overline{L} = \overline{\sigma}$ for some $\sigma \in \mathcal{E}$, which will then be denoted by $\sigma_{KL}$.*
(iv) *The family of points $(x_K)_{K \in \mathcal{T}}$ is such that $x_K \in \overline{K}$ (for all $K \in \mathcal{T}$) and, if $\sigma = \sigma_{KL}$, it is assumed that $x_K \neq x_L$, and that the straight line going through $x_K$ and $x_L$ is orthogonal to $\sigma_{KL}$.*
(v) *For any $\sigma \in \mathcal{E}$ such that $\sigma \subset \partial\Omega$, let $K$ be the control volume such that $\sigma \in \mathcal{E}_K$. We assume that if $x_K \notin \sigma$, then the straight line going through $x_K$ and orthogonal to $\sigma$ intersects $\sigma$.*

An example of two cells of such a mesh is given in Figure 1, along with some notations. Item $(iv)$ of the above definition was referred to in the previous section as the "orthogonality property":

We refer to [12] for a description of such admissible meshes, which include triangular meshes, rectangular meshes, or Voronoï meshes. Here, for the sake of simplicity, we assume that the points $x_K \in K$. The finite volume approximations $\phi_\mathcal{T}$ and $u_\mathcal{T}$ of $\phi$ and $u$ solution to (9) are sought in the space $X(\mathcal{T})$ of functions from $\Omega$ to $\mathbb{R}$ which are constant over each control volume of the mesh, that is:

$$X(\mathcal{T}) = \{u \in C(\Omega, \mathbb{R}); u|_K \in \mathcal{P}_0 \text{ for all } K \in \mathcal{T}\}, \tag{10}$$

where $\mathcal{P}_0$ denotes the set of constant functions. The finite volume scheme is classically obtained from the balance form of Equations (4) and (6) on a control volume $K$, that is:

$$- \int_{\partial K} (\kappa(\cdot, u) \nabla \phi) \cdot \mathbf{n}_K dx = \int_K f(\cdot, u) dx \qquad (11)$$

$$- \int_{\partial K} (\lambda(\cdot, u) \cdot \nabla u) \cdot \mathbf{n}_K dx = \int_K \kappa(\cdot, u) |\nabla \phi|^2 dx, \qquad (12)$$

where $\mathbf{n}_K$ denotes the unit normal vector to $\partial K$ outward to $K$ and $dx$ denotes the integration symbol on $d$ dimensional domain $\Omega$ or the $d-1$ dimensional boundary, with respect to the Lebesgue measure. Let $\mathcal{E}_K$ denote the set of edges or faces of $\partial K$, decomposing the boundary of $K$ into edges or faces, $\partial K = \cup_{\sigma \in \mathcal{E}_K} \sigma$, we may rewrite (11)-(12) as:

$$- \sum_{\sigma \in \mathcal{E}_K} \int_\sigma (\kappa(\cdot, u) \nabla \phi) \cdot \mathbf{n}_{K,\sigma} dx = \int_K f(\cdot, u) dx \qquad (13)$$

$$- \sum_{\sigma \in \mathcal{E}_K} \int_\sigma (\lambda(\cdot, u) \cdot \nabla u) \cdot \mathbf{n}_{K,\sigma} dx = \int_K \kappa(\cdot, u) |\nabla \phi|^2 dx. \qquad (14)$$

Let us write the sought approximations as $\phi_T = \sum_{K \in T} \phi_K 1_K$, $u_T = \sum_{K \in T} u_K 1_K$; we then set

$$f_K(u_K) = \frac{1}{m(K)} \int_K f(x, u_K) dx. \qquad (15)$$

Let $\mathcal{E}$ denote the set of edges (or faces in 3D) of the mesh, and $\mathcal{E}_{int}$ (resp. $\mathcal{E}_{ext}$) the set of edges laying in $\Omega$ (resp. on $\partial\Omega$). For $\sigma \in \mathcal{E}$, let $F_{K,\sigma}^\kappa$ (resp. $F_{K,\sigma}^\lambda$) be an approximation of the flux $\int_\sigma (\kappa(x, u(x)) \nabla \phi(x)) \cdot \mathbf{n}_{K,\sigma} d\gamma(x)$ (resp. $\int_\sigma (\lambda(x, u(x)) \cdot \nabla u(x)) \cdot \mathbf{n}_{K,\sigma} d\gamma(x))$, and let $\mathcal{J}_K(u_T, \phi_T)$ denote an approximation of the nonlinear right-hand-side $\frac{1}{m(K)} \int_K \kappa(x, u(x)) |\nabla \phi|^2(x) dx$, with the notation $u_T = (u_K)_{K \in T}$ and $\phi_T = (\phi_K)_{K \in T}$. With these notations, a finite volume approximation may then be written under the form:

$$\begin{cases} \sum_{\sigma \in \mathcal{E}_K} F_{K,\sigma}^\kappa = m(K) f_K(u_K), \ \forall K \in T, \\ \sum_{\sigma \in \mathcal{E}_K} F_{K,\sigma}^\lambda = m(K) \mathcal{J}_K(u_T, \phi_T), \ \forall K \in T, \end{cases} \qquad (16)$$

provided one defines the expressions $F_{K,\sigma}^{\kappa,\lambda}$ and $\mathcal{J}_K(u_T, \phi_T)$ with respect to the discrete unknowns $(\phi_K)_{K \in T}$ and $(u_K)_{K \in T}$.

The discrete fluxes are consistent and conservative, and are given by the classical two–points formula:

$$F_{K,\sigma}^\kappa = \begin{cases} m(\sigma) \tau_\sigma^\kappa(u_T)(\phi_K - \phi_L), \text{ if } \sigma = K|L \in \mathcal{E}_{int}, \\ m(\sigma) \tau_\sigma^\kappa(u_T) \phi_K \text{ if } \sigma \in \mathcal{E}_K \cap \mathcal{E}_{ext}, \end{cases} \qquad (17)$$

$$F_{K,\sigma}^\lambda = \begin{cases} m(\sigma) \tau_\sigma^\lambda(u_T)(u_K - u_L) \text{ if } \sigma = K|L \in \mathcal{E}_{int}, \\ m(\sigma) \tau_\sigma^\lambda(u_T)(u_K), \text{ if } \sigma \in \mathcal{E}_K \cap \mathcal{E}_{ext}, \end{cases} \qquad (18)$$

where $\tau_\sigma^\kappa$ (and, similarly $\tau_\sigma^\lambda$) is defined through a harmonic average, that is:

$$\tau_\sigma^\kappa(u_{\mathcal{T}}) = \begin{cases} \dfrac{\kappa_K(u_K)\kappa_L(u_L)}{d_{K,\sigma}\kappa_L(u_L) + d_{L,\sigma}\kappa_K(u_K)} & \text{if } \sigma = K|L \in \mathcal{E}_{\text{int}}, \\ \dfrac{\kappa_K(u_K)}{d_{K,\sigma}}, & \text{if } \sigma \in \mathcal{E}_{\text{ext}} \cap \mathcal{E}_K, \end{cases} \tag{19}$$

where the values $\kappa_K(u_K)$ and $\lambda_K(u_K)$ are defined by (15), replacing $f$ by $\kappa$ or $\lambda$.
The term $\mathcal{J}_K(u_{\mathcal{T}}, \phi_{\mathcal{T}})$ is defined as:

$$\mathcal{J}_K(u_{\mathcal{T}}, \phi_{\mathcal{T}}) = \frac{1}{\text{m}(K)} \sum_{\sigma \in \mathcal{E}_K} \text{m}(\mathcal{D}_{K,\sigma}) \mathcal{J}_\sigma(u_{\mathcal{T}}, \phi_{\mathcal{T}}), \tag{20}$$

where, for $K \in \mathcal{T}$ and $\sigma \in \mathcal{E}_K$, we define the half dual cell $\mathcal{D}_{K,\sigma}$ delimited by $x_K$ and $\sigma$ (see Figure 1) by

$$\mathcal{D}_{K,\sigma} = \{tx_K + (1-t)x, \ (x,t) \in \sigma \times (0,1)\},$$

and

$$\mathcal{J}_\sigma(u_{\mathcal{T}}, \phi_{\mathcal{T}}) = \frac{\tau_\sigma^\kappa(u_{\mathcal{T}})}{d_\sigma}(D_\sigma\phi)^2 d \tag{21}$$

with

$$D_\sigma\phi = \begin{cases} |\phi_K - \phi_L| \text{ if } \sigma \in \mathcal{E}_{\text{int}}, \\ |\phi_K| \text{ if } \sigma \in \mathcal{E}_{\text{ext}} \end{cases}$$

We show in Theorem 2 below the existence of $(\phi_K)_{K \in \mathcal{T}}$ and $(u_K)_{K \in \mathcal{T}}$ solution to (16)–(21). This entitles us to define the functions $\phi_{\mathcal{T}}$ and $u_{\mathcal{T}} \in X(\mathcal{T})$ with respective values $\phi_K$ and $u_K$ on cell $K$, along with the function $\mathcal{J}_{\mathcal{T}}(u_{\mathcal{T}}, \phi_{\mathcal{T}}) \in X(\mathcal{T})$ with value $\mathcal{J}_K(u_{\mathcal{T}}, \phi_{\mathcal{T}})$ on cell $K$. To prove the existence of a finite volume solution $(\phi_{\mathcal{T}}, u_{\mathcal{T}})$ to the problem (16)-(21), we use Brouwer's theorem. To this end we combine Lebesgue's theorem with the fact that $X(\mathcal{T})$ is a finite dimensional space.

**Theorem 2.** *Let $(\kappa, \lambda, f)$ be three functions satisfying the Assumption 1. Let $X(\mathcal{T})$ be the finite volume space defined in Definition 2. Then there exists at least a solution $(u_{\mathcal{T}}, \phi_{\mathcal{T}}) \in (X(\mathcal{T}))^2$ to the problem (16)-(21).*

The proof is based on the fixed point theorem. In fact, the existence of a solution to (9) was proven in [19] using Schauder's fixed point theorem; here, since the space is finite–dimensional, we need only use Brouwer's theorem. The proof is an easy adaptation of that of [19] so we omit it.

## 5    Convergence of the Finite Volume Approximation

### 5.1    The Convergence Result

In this section, we shall prove that a solution of (16)-(21) converges, as $h_{\mathcal{T}} = \sup\{diam(K), K \in \mathcal{T}\}$ tends to 0, towards a solution of (9), as stated in the following theorem:

**Theorem 3.** *Under Assumption 1, let $(\mathcal{T}_n)_{n\in\mathbb{N}}$ be a sequence of admissible meshes in the sense of Definition 1. Let $(\phi^n, u^n)$ be a solution of the system (16)-(20) for $\mathcal{T} = \mathcal{T}_n$, and let $\mathcal{J}^n(u^n, \phi^n)$ be defined by (20). Assume that $h_n = \sup\{diam(K), K \in \mathcal{T}_n\} \to 0$, as $n \to \infty$, and that there exists $\zeta > 0$ (not depending on $n$), such that:*

$$d_\sigma \le \zeta d_{K,\sigma}, \forall \sigma \in \mathcal{E}_n, \forall K \in \mathcal{T}_n. \tag{22}$$

*Then, there exists a subsequence of $(\mathcal{T}_n)_{n\in\mathbb{N}}$, still denoted by $(\mathcal{T}_n)_{n\in\mathbb{N}}$, such that $(\phi^n, u^n)$ converges to a solution $(\phi, u) \in H_0^1(\Omega) \times \cap_{q < \frac{d}{d-1}} W_0^{1,q}$ of (9), as $n \to \infty$, in the following sense:*

$$\|\phi^n - \phi\|_{L^2(\Omega)} \to 0, \text{ as } n \to +\infty, \tag{23}$$

$$\|u^n - u\|_{L^p(\Omega)} \to 0, \text{ as } n \to +\infty, \text{ for all } p < \frac{d}{d-2}. \tag{24}$$

*Moreover,*

$$\int_\Omega \mathcal{J}^n(u^n, \phi^n)(x)dx \to \int_\Omega \kappa(x, u(x))|\nabla\phi|^2(x)dx \text{ as } n \to +\infty. \tag{25}$$

**Proof.** For the sake of clarity, we only list here the main ingredients of the proof and refer to lemmata proven below for the details.

Let $(\phi^n)_{n\in\mathbb{N}} \subset L^2(\Omega)$ and $(u^n)_{n\in\mathbb{N}} \subset L^2(\Omega)$ be such that, for any $n \in \mathbb{N}$, the pair $(\phi^n, u^n)$ is a solution of (16)-(20), with $\mathcal{T} = \mathcal{T}_n$ (recall that this solution exists by Theorem 2).

From Lemma 1 below, we know that the sequences $(\phi^n)_{n\in\mathbb{N}}$ and $(u^n)_{n\in\mathbb{N}}$ are bounded for respectively, the $L^2$ norm and the $L^q$ norm, with $q < \frac{d}{d-2}$. Note that condition (22) is needed here since it is required when using the discrete Sobolev inequality (see e.g. [9]) to obtain the uniform bound of $(u^n)_{n\in\mathbb{N}}$ in an $L^q$ norm from a discrete $W_0^{1,p}$ estimate. Then, following [12] Lemma 9.3 page 770 or [13], one may easily get some uniform estimates on the translates of $\phi^n$ in the $L^2$ norm and of $u^n$ in the $L^p$ norm.

We may therefore use a discrete Rellich theorem (see e.g. [11], Proposition 2.3) to obtain that the sequences $(\phi^n)_{n\in\mathbb{N}}$ and $(u^n)_{n\in\mathbb{N}}$ are relatively compact in, respectively, $L^2(\Omega)$ and $L^p(\Omega)$, for $p < \frac{d}{d-2}$. The estimates on the tranlations also yield the regularity of the limit (Proposition 2.4 of [11]), that is, if $\phi$ is a limit of the sequence $(\phi^n)_{n\in\mathbb{N}}$ in $L^2(\Omega)$, then $\phi \in H_0^1(\Omega)$; similarly, if $u$ is a limit of the sequence $(u^n)_{n\in\mathbb{N}}$ in $L^p(\Omega)$, then $u \in \cap_{q < \frac{d}{d-1}} W_0^{1,q}(\Omega)$.

Hence, for any sequence $(\mathcal{T}_n)_{n\in\mathbb{N}}$ of admissible meshes satisfying (22) and such that size$(\mathcal{T}_n) \to 0$, as $n \to \infty$, there exists a subsequence, still denoted by $(\mathcal{T}_n)_{n\in\mathbb{N}}$, such that:

1. $u^n$ converges to some $u \in \cap_{q < \frac{d}{d-1}} W_0^{1,q}(\Omega)$ in $L^p(\Omega)$, for all $p < \frac{d}{d-2}$, and therefore in $L^2(\Omega)$, as $n \to \infty$.
2. $\phi^n$ converges to some $\phi \in H_0^1(\Omega)$, in $L^2(\Omega)$, as $n \to \infty$.

Thanks to the Lebesgue dominated theorem, we also get that $\kappa(\cdot, u^n) \to \kappa(\cdot, u)$ and $f(\cdot, u^n) \to f(\cdot, u)$ as $n \to \infty$, in the $L^p$ norm for any $p < \frac{d}{d-2}$.

One then obtains by a straightforward adaptation of the proof of Theorem 2 in [13], that the function $\phi \in H_0^1(\Omega)$ is the (unique, for the considered function $u$) weak solution of the first equation of (9), that is:

$$\int_\Omega \kappa(x, u(x))\nabla\phi(x) \cdot \nabla\psi(x)\,dx = \int_\Omega f(x, u(x))\psi(x)\,dx, \ \forall \psi \in H_0^1(\Omega).$$

A straightforward adaptation of the proof of the convergence of the discrete $H_0^1$ norm in [12] (Theorem 9.1, proof page 776) yields that the discrete ohmic losses converge to the continuous ones, that is:

$$\sum_{\sigma\in\mathcal{E}} \mathrm{m}(\sigma)\tau_{K|L}^\kappa(u_K^n)(\phi_L^n - \phi_K^n)^2 \to \int_\Omega \kappa(x, u(x))|\nabla\phi|^2(x)dx \text{ as } n \to +\infty,$$

which proves (25).

In order to prove (23) and (24), there now only remains to show that $u$ satisfies the second equation of (9). In order to do so, we proceed in a now classical way, that is, we multiply the second equation of the scheme (16) by $\psi(x_K)$ where $\psi$ is a smooth function with compact support in $\Omega$, we sum over $K \in \mathcal{T}_n$, and obtain:

$$\sum_{K\in\mathcal{T}} \sum_{\sigma\in\mathcal{E}} \mathrm{m}(\sigma)\tau_\sigma^\lambda(u_K^n)(u_K^n - u_L^n)\psi(x_K) = \sum_{K\in\mathcal{T}} \mathrm{m}(K)\mathcal{J}_K^n(u^n, \phi^n)\psi(x_K) \quad (26)$$

Let us now pass to the limit as $n \to +\infty$. A straightforward adaptation of the proof of e.g. Theorem 2 in [13] yields that the left hand side of (26) tends to $-\int_\Omega u\nabla \cdot (\lambda(\cdot, u)\nabla\psi)(x)dx$, as $n \to +\infty$. Moreover, we show in Lemma 2 below that the right hand side of (26) tends to $\int_\Omega \kappa(x, u(x))|\nabla\phi|^2(x)\psi(x)\,dx$, so that, by density of $C^\infty(\overline{\Omega})$ we get that $u$ satisfies

$$-\int_\Omega \lambda(x, u(x))\nabla u(x) \cdot \nabla\psi(x)\,dx = \int_\Omega \kappa(x, u(x))|\nabla\phi|^2(x)\psi(x)\,dx$$

for all $\psi \in \cup_{q>d} W_0^{1,q}(\Omega)$. This concludes the proof of the theorem. ∎

In the following sections, we shall derive the estimates and the intermediate convergence results which were used in the above proof.

## 5.2   Estimate on the Approximate Solutions and Compactness

Recall that the approximate finite volume solutions are piecewise constant; hence they are not in the spaces $W^{1,p}$, and we need therefore to define a discrete $W^{1,p}$ norm (see also [9,12]) in order to obtain some compactness results.

**Definition 2 (Discrete $W^{1,p}$ norm).** *Let $\Omega$ be an open bounded subset of $\mathbb{R}^d$, $d = 1, 2$ or $3$, and let $\mathcal{T}$ be an admissible finite volume mesh in the sense of*

*Definition 1.* For $u_T \in X(T)$, $u_T = \sum_{K \in T} u_K 1_K$, and $p \in [1, +\infty)$,

$$\|u_T\|_{1,p,T} = \left( \sum_{\sigma \in \mathcal{E}} m(\sigma) d_\sigma \left( \frac{D_\sigma u}{d_\sigma} \right)^p \right)^{\frac{1}{p}},$$

*with the notation*

$$D_\sigma u = \begin{cases} |u_K - u_L| & \text{if } \sigma \in \mathcal{E}_{\text{int}}, \\ |u_K| & \text{if } \sigma \in \mathcal{E}_{\text{ext}} \end{cases}$$

To prove the convergence of $(\phi_T, u_T)$, we prove at first some estimates on $\phi_T$ and $u_T$.

**Lemma 1.** *Under Assumption 1, let $T$ be an admissible mesh in the sense of Definition 1, and let $\zeta_T > 0$ be such that:*

$$d_\sigma \leq \zeta_T \, d_{K,\sigma}, \ \forall \sigma \in \mathcal{E}, \text{ and for any } K \in T. \tag{27}$$

*Let $(\phi_T, u_T)$ be a solution of (16)-(20). Then there exists $C_1 \in \mathbb{R}_+^\star$, only depending on $\Omega$, $d$, $\|f\|_{L^\infty(\Omega \times \mathbb{R}, \mathbb{R})}$, $\|\kappa\|_{L^\infty(\Omega \times \mathbb{R}, \mathbb{R})}$, and $\alpha$ such that*

$$\|\phi_T\|_{1,2,X(T)} \leq C_1 \tag{28}$$

$$\|\phi_T\|_{L^2(\Omega)} \leq C_1, \tag{29}$$

*and*

$$\|\mathcal{J}_T(u_T, \phi_T)\|_{L^1(\Omega)} \leq C_1. \tag{30}$$

*Moreover, for all $q \in [1, \frac{d}{d-1})$, there exists $C_2 \in \mathbb{R}_+^\star$ only depending on $\Omega$, $d$, $\|f\|_{L^\infty(\Omega \times \mathbb{R}, \mathbb{R})}, \|\kappa\|_{L^\infty(\Omega \times \mathbb{R}, \mathbb{R})}, \|\lambda\|_{L^\infty(\Omega \times \mathbb{R}, \mathbb{R})}, \zeta_T, q, \alpha$ such that*

$$\|u_T\|_{1,q,X(T)} \leq C_2, \tag{31}$$

*and*

$$\|u_T\|_{L^{p^*}} \leq C_2. \tag{32}$$

*Proof.* The proof of (28) follows [12], Lemma 9.2 page 768 and the estimate (29) is then obtained by the discrete Poincaré inequality [12] Lemma 9.1 page 765. Let us then prove the $L^1$ estimate (30) on the right hand side $\mathcal{J}_T(u_T, \phi_T)$. Indeed, by definition of $\mathcal{J}_T(u_T, \phi_T)$,

$$\|\mathcal{J}_K(u_T, \phi_T)\|_{L^1(\Omega)} = \sum_{K \in T} \sum_{\sigma \in \mathcal{E}_K} m(\mathcal{D}_{K,\sigma}) \mathcal{J}_\sigma(u_T, \phi_T) = \sum_{\sigma \in \mathcal{E}} m(\mathcal{D}_\sigma) \mathcal{J}_\sigma(u_T, \phi_T),$$

where $\mathcal{D}_\sigma$ denotes the "diamond cell" around $\sigma$, that is $\mathcal{D}_\sigma = \mathcal{D}_{K,\sigma} \cup \mathcal{D}_{L,\sigma}$ if $\sigma = K|L \in \mathcal{E}_{\text{int}}$, and $\mathcal{D}_\sigma = \mathcal{D}_{K,\sigma}$ if $\sigma \in \mathcal{E}_{\text{ext}} \cap \mathcal{E}_K$. From the definition of $\mathcal{J}_\sigma(u_T, \phi_T)$, noting that $m(\mathcal{D}_\sigma) = \frac{1}{d} m(\sigma) d_\sigma$, and using Assumption 1, one then obtains that:

$$\|\mathcal{J}_K(u_T, \phi_T)\|_{L^1(\Omega)} \leq \frac{\|\kappa\|}{\alpha} \|\phi_T\|_{1,2,T}.$$

which proves (30). Since $\mathcal{J}_T(u_T, \phi_T) \in L^1(\Omega)$, one obtains (31) by a straightforward adaptation of [20], Lemma 1 (see also [11], Theorem 2.2). The estimate (32) follows from a discrete Sobolev inequality [9].

## 5.3    Passage to the Limit on the $L^1$ Term

**Lemma 2.** *Under Assumption 1, let $(\mathcal{T}_n)_{n\in\mathbb{N}}$ be a sequence of admissible meshes in the sense of Definition 1. Let $(\phi^n, u^n)$ be a solution of the system (16)-(20) for $\mathcal{T} = \mathcal{T}_n$, and let $\mathcal{J}^n(u^n, \phi^n) \in X(\mathcal{T}_n)$ be defined by (20). Assume that $h_n = \max\{\mathrm{diam}(K), K \in \mathcal{T}_n\} \to 0$, as $n \to \infty$, and that there exists $\zeta > 0$, not depending on $n$, such that (22) holds. Assume that*

1. *$u^n$ converges to some $u \in \cap_{q < \frac{d}{d-1}} W_0^{1,q}(\Omega)$ in $L^2(\Omega)$, as $n \to \infty$.*
2. *$\phi^n$ converges to some $\phi \in H_0^1(\Omega)$, in $L^2(\Omega)$, as $n \to \infty$.*

*Let $\psi \in C_c^1(\Omega, \mathbb{R})$, and let $\psi^n \in X(\mathcal{T}_n)$ be defined by:*

$$\psi^n(x) = \psi(x_K), \ for \ a.e. \ x \in K, \ \forall K \in \mathcal{T}_n.$$

*Then:*

$$\int_\Omega \mathcal{J}^n(u^n, \phi^n)(x)\psi^n(x)\,dx \ \to \ \int_\Omega \kappa(x, u(x))|\nabla\phi|^2(x)\psi(x)\,dx \ as \ n \to +\infty. \quad (33)$$

*Thus, $\mathcal{J}^n(u^n, \phi^n) \to |\nabla\phi|^2$ as $n \to +\infty$ for the weak $\star$ topology of $(C(\overline{\Omega}))'$, where $C(\overline{\Omega}))$ denotes the set of continuous functions with compact support on $\overline{\Omega}$.*

*Proof.* Assume that $n$ is large enough so that the function $\psi$ (which has a compact support in $\Omega$) vanishes on the cells neighbouring the boundary $\partial\Omega$. Noting that $\mathrm{m}(\mathcal{D}_{K,\sigma}) = \frac{1}{d}\mathrm{m}(\sigma)d_{K,\sigma}$, one has:

$$\int_\Omega \mathcal{J}^n(u^n, \phi^n)(x)\psi^n(x)\,dx = \sum_{K\in\mathcal{T}^n}\sum_{\sigma\in\mathcal{E}_K} \mathrm{m}(\sigma)d_{K,\sigma}\tau_\sigma^\kappa(u^n)\frac{(D_\sigma\phi^n)^2}{d_\sigma}\psi(x_K)$$
$$= \mathbb{T}_7^n + \mathbb{T}_8^n,$$

where

$$\mathbb{T}_7^n = \sum_{K\in\mathcal{T}_n}\sum_{\sigma\in\mathcal{E}_K} \mathrm{m}(\sigma)d_{K,\sigma}\tau_\sigma^\kappa(u_n)\frac{(D_\sigma\phi^n)^2}{d_\sigma}\psi(x_L), \quad (34)$$

and

$$\mathbb{T}_8^n = \sum_{K\in\mathcal{T}_n}\sum_{\sigma=K|L\in\mathcal{E}_K} \mathrm{m}(\sigma)d_{K,\sigma}\tau_\sigma^\kappa(u^n)\frac{(D_\sigma\phi^n)^2}{d_\sigma}(\psi(x_K) - \psi(x_L)).$$

Since $d_{K,\sigma} \leq d_\sigma$ and $|\psi(x_K) - \psi(x_L)| \leq 2h_n\|\nabla\psi\|_{(L^\infty(\overline{\Omega}))^d}$ and $\tau_\sigma^\kappa(u^n) \leq \frac{\|\kappa\|_{L^\infty(\Omega\times\mathbb{R},\mathbb{R})}^2}{\alpha d_\sigma}$, we have

$$|\mathbb{T}_8^n| \leq 2\frac{\|\kappa\|_{L^\infty(\Omega\times\mathbb{R},\mathbb{R})}^2}{\alpha} \ h_n\|\nabla\psi\|_{(L^\infty(\overline{\Omega}))^d}\|\phi^n\|_{1,2,X(\mathcal{T})}^2.$$

Using (28) we then obtain that:

$$|\mathbb{T}_8^n| \to 0, \ as \ n \to +\infty.$$

We turn now to the term $\mathbb{T}_7^n$, reordering the sum on the edges in the right hand side of (34), we get

$$\mathbb{T}_7^n = \sum_{\sigma=K|L \,\in\mathcal{E}_{\mathrm{int}}} \mathrm{m}(\sigma)\tau_\sigma^\kappa(u^n)\frac{(D_\sigma\phi^n)^2}{d_\sigma}\psi_\sigma$$

where

$$\psi_\sigma = \frac{d_{K,\sigma}\psi(\mathrm{x}_L) + d_{L,\sigma}\psi(\mathrm{x}_K)}{d_\sigma}, \quad \sigma \in \mathcal{E}_{\mathrm{int}} \text{ and } \sigma = K|L \tag{35}$$

We may then decompose $\mathbb{T}_7^n = \mathbb{T}_9^n + \mathbb{T}_{10}^n$, with

$$\mathbb{T}_9^n = - \sum_{\sigma=K|L \,\in\mathcal{E}_{\mathrm{int}}} \mathrm{m}(\sigma)\tau_\sigma^\kappa(u^n)(\phi_L^n - \phi_K^n)(\phi_K^n\psi_K^n - \phi_L^n\psi_L^n), \tag{36}$$

and

$$\mathbb{T}_{10}^n = - \sum_{\sigma=K|L \,\in\mathcal{E}_{\mathrm{int}}} \mathrm{m}(\sigma)\tau_\sigma^\kappa(u^n)\left((\phi_L^n - \phi_K^n)\phi_K^n(\psi_\sigma - \psi_K) - (\phi_L^n - \phi_K^n)\phi_L^n(\psi_\sigma - \psi_L)\right).$$

Reordering the sum of the right hand side of (36) on the control volumes, using the fact that $\phi^n$ is the solution of the first equation of the finite volume scheme (16), and that $u^n$ (resp.$\phi^n$) converges to $u$ (resp. $\phi$) in $L^2(\Omega)$ as $n \to \infty$, we get that:

$$\mathbb{T}_9^n \to \int_\Omega \kappa(x, u(x))|\nabla\phi|^2(x)\psi(x)dx + \int_\Omega \kappa(x, u(x))\nabla\phi(x) \cdot \nabla\psi(x)\phi(x)dx \tag{37}$$

as $n \to +\infty$. Reordering the sum of $\mathbb{T}_{10}^n$ on the edges of the control volumes and using Lemma 2 in [14] (see [5] for details), we get

$$\mathbb{T}_{10}^n \to - \int_\Omega \nabla\phi(x) \cdot \nabla\psi(x)\phi(x)\kappa(x, u(x))\, dx \text{ as } n \to +\infty, \tag{38}$$

from which it is easy to see that

$$\int_\Omega \mathcal{J}^n(u^n, \phi^n)(x)\psi^n(x) \to \int_\Omega \kappa(x, u(x))|\nabla\phi|^2(x)\psi(x)dx,$$

which proves (33).

## 6   Conclusion and Perspectives

We proved here the convergence of a cell centred finite volume method for the coupled heat and potential equation; the condition on the considered meshes is such that the discrete maximum principle holds. Indeed, the technique of proof mimics the tools used for the existence the continuous case, which requires the monotonicity of the operator. In the case of the cell centred finite volume, the scheme satisfies the maximum principle for any admissible mesh. These include triangles and rectangles in two space dimensions, and Voronoï meshes in any dimension.

In two space dimensions, the piecewise linear finite element method satisfies the discrete maximum principle for triangular meshes under the Delaunay condition. It is easy to show that under this condition, the matrix of the scheme is identical to that of the cell-centred finite volume on the dual Voronoï mesh. Therefore, the convergence of the finite element scheme may be obtained from that of the finite volume scheme, as explained in [18].

In three space dimensions, there is no known way to build a Voronoï mesh from a tetrahedral one, and therefore one must proceed directly with the finite element interpolation operator, as in [6] in the case of a linear diffusion operator, and [5] in the case of the present coupled problem. In the three–dimensional case, a known sufficient condition for the maximum principle to hold on a tetrahedral meshes is that all angles of all the faces be strictly acute. Unfortunately, there does not seem to be an easy way to construct such meshes in practise [2], so that convergence results for the finite element scheme in 3D remain quite academic.

# References

1. Eymard, R., Gallouët T. , Herbin R.: A hybrid finite volume schemes for the discretization of anistropic diffusion operator on general meshes. submitted
2. Bern M., Eppstein D. : Mesh generation and optimal triangulation. Computing in Euclidean geometry, 23-90, Lecture Notes Ser. Comput., 1, World Sci. Publ., River Edge, NJ (1992)
3. Boccardo,L., Gallouët, T. , Vazquez, J.-L.: Nonlinear Elliptic Equations in $\mathbb{R}^n$ without Growth Restrictions on the Data. Journal of Differential Equations **105** n° 2 (1993) 334–363
4. Bénilan, Ph., Boccardo, L., Gallouët, T., Gariepy, R., Pierre, M., Vazquez, J.-L.: An $L^1$ theory of existence and uniqueness of solutions of nonlinear elliptic equations. Ann. Scuola. Norm. Sup. Pisa **22** (1995) 241–273
5. Bradji, A., Herbin, R.: Discretization of ohmic losses by the finite element and the finite volume method. (in preparation)
6. Casado-Diaz, J., Chacon Rebollo, T., Girault, V., Gomez, M., Murat, F.: Finite elements approximation of second order linear elliptic equations in divergence form with right-hand side in $L^1$. ( Submitted)
7. Ciarlet, P.A.: Basic error estimates for elliptic problems. Handbook of Numerical Analysis., **II**, (North-Holland, Amsterdam) (1991) 17–352
8. Clain, S.: Finite element approximations for Laplace operator with right hand side measure. Math. Models Meth. App. Sci. **6** n° 5 (1996) 713–719
9. Coudière, Y., Gallouët, T., Herbin, R.: Discrete Sobolev inequalities and $L^p$ error estimates for approximate finite volume solutions of convection diffusion equations. M2AN **35** n° 4 (2001) 767–778
10. Dall'aglio, A.: Approximated solutions of equations with $L^1$ solutions wiht $L^1$ data. Application to the $H$–convergence of quasi–linear parabolic equations. Ann. Mat. Pura Appl. **170** (1996) 207–240
11. Droniou, J., Gallouët, T., Herbin, R.: A finite volumes schemes for a noncoercive elliptic equation with measure data. SIAM J. Numer. Anal. **41** n° 6 (2003) 1997-2031

12. Eymard, R., Gallouët, T., Herbin, R.:  Finite volume methods. *Handbook of Numerical Analysis. P. G. Ciarlet and J. L. Lions (eds.)* **vol. VII** (2000) 723–1020
13. Eymard, R., Gallouët, T., Herbin, R.:  Convergence of finite volume schemes for semilinear convection diffusion equations. Numer. Math. **82** (1999) 91–116
14. Eymard, R., Gallouët, T.:  H-convergence and numerical schemes for elliptic problems. SIAM J. Numer. Anal. **41** (2003) 539–562
15. Eymard, R., Gallouët, T., Herbin, R.:  A cell-centered finite-volume approximation for anisotropic diffusion operators on unstructured meshes in any space dimension. IMA J. of Numer. Anal. **26** (2006) 326–353
16. Ferguson, J.R., Fiard , J. M., Herbin, R.:  A mathematical model of solid oxide fuel cells. Journal of Power Sources **58** (1996) 109–122
17. Fiard, J. M., Herbin, R.:  Comparison between finite volume and finite element methods for an elliptic system arising in electrochemical engineerings.  Comput. Methods App. Mech. Engrg. **115** (1994) 315–338
18. Gallouët, T. , Herbin, R.:  Convergence of linear finite element for diffusion equations with measure data. CRAS . **338** (2004) 81–84
19. Gallouët, T. , Herbin, R.:  Existence of a solution to a coupled elliptic system. Appl. Math. Lett.  **7** (1994) 49–55
20. Gallouët, T. , Herbin, R.:  Finite volume approximation of elliptic problems with irregular data. In Finite Volumes for Complex Applications, II, F. Benkhaldoun, M. Hänel and R. Vilsmeier eds, Hermes (1999) 155–162
21. Gallouët, T. :  Measure data and numerical schemes for elliptic problems. Progress in Nonlinear Differential Equations and their Applications.  **63** (2005) 1–13
22. Murat, F. : Equations elliptiques non linéaires avec second membre $L^1$ ou mesure. In Actes du 26-ème Congrès d'Analyse Numérique, Les Karellis, Université de Lyon (1994)
23. Stampacchia G. : Le problème de Dirichlet pour les équations elliptiques du second orders à coefficients discontinus. Ann. Inst. Fourier **15** (1965) 189 258
24. Scott, R. : Finite element convergence for singular data. Numer. Math. **21** (1973) 317-327

# The Vector Analysis Grid Operators for Applied Problems

Petr Vabishchevich

Institute for Mathematical Modeling, RAS
4-A Miusskaya Square, 125047 Moscow, Russia
vab@imamod.ru

**Abstract.** Mathematical physics problems are often formulated by means of the vector analysis differential operators: divergence, gradient and rotor. For approximate solutions of such problems it is natural to use the corresponding operator statements for the grid problems, i.e., to use the so-called VAGO (**V**ector **A**nalys **G**rid **O**perators) method. We discuss the possibilities of such an approach in using general irregular grids. The vector analysis difference operators are constructed using the Delaunay triangulation and the Voronoi diagrams. The truncation error and the consistency property of the difference operators constructed on two types of grids are investigated.

## 1 Introduction

Applied problems are mostly defined in the form of systems of partial differential equations supplemented by the corresponding boundary and initial conditions. To solve approximately the boundary-value problems, the finite-difference and finite-element methods are used.

To formulate the mathematical physics problems, one can use the differential vector analysis operators: the invariant first-order divergence, gradient and rotor operators. For approximate solution of such problems, it is natural to use the corresponding operator statements for the grid problems. In this case [3], one speaks of Mimetic Finite Difference Operators. Mimetic discretization methods for the numerical solution of continuum mechanics problems use analogs of identities from vector calculus or differential forms to both derive and analyze the discretization.

The advantages of such an approach are that we do not have to define a concrete coordinate system. This is especially interesting for considering irregular computational grids. In using vector analysis grid operators it is natural to use the approach with direct substitution of the vector analysis differential operators by the grid operators.

The universal balance method (integro-interpolational method, finite-volume method) is mainly used to construct discrete problems [4]. In this case, the difference scheme is constructed by integrating the input equation with respect to the control volume (part of the computational domain adjoining to given computational node). For the Delauney triangulation it is natural to use the

T. Boyanov et al. (Eds.): NMA 2006, LNCS 4310, pp. 16–27, 2007.

Voronoi polygons (the set of points lying closer to this node than to the others) as the control volume.

In this paper the questions concerning the construction of computational grids are considered. For the given set of nodes the Delauney triangulation ($D$-grid) is constructed. The dual grid ($V$-grid) constructed on the vertices of the Voronoi polyhedrons linked to the $D$-grid. On these grids the scalar and vector grid functions are defined. On the introduced grids the vector analysis grid operators of the gradient (grad), the divergence (div) and the rotor (rot, curl) are constructed. Primary consideration is given to the investigation of the truncation error of the vector analysis grid operators. The important properties of the consistency of the approximations of individual operators are determined. The paper [5] gives examples of applying the introduced vector analysis grid operators to the approximate solution of some classes of the mathematical physics problems.

## 2    Grids and Grid Functions

Consider the computational grids and their corresponding scalar and vector functions. The general unstructured grid is constructed on the basis of the Delaunay triangulation and the Voronoi diagrams.

### 2.1    Delaunay Triangulation and Voronoi Diagrams

For the given points the triangulation can be performed in different ways. Note also that for a given set of nodes we obtain the same number of triangles by any triangulation method. So we need to optimize the triangulation by some criteria. The main optimization criterion consists in the following: the obtained triangles should be close to equilateral ones (there should be no too sharp angles). This is a local criterion belonging to one triangle. The second (global) criterion consists in that adjacent triangles do not differ too widely in area — the criterion of grid uniformity.

There is a special triangulation — the Delaunay triangulation [1], which has a number of optimum properties. One of them is the tendency of obtained triangles to equiangular ones. The above mentioned property can be formulated more exactly in the following way: in the Delaunay triangulation the minimum value of inner angles of triangles is maximized. The formal definition of the Delaunay triangulation is associated with the property that for each triangle all the other nodes are situated outside the circumcircle. For our further presentation the relation between the Delaunay triangulation and the Voronoi diagram is very important.

The Voronoi polygon for a separate node is a set of points lying closer to this node than to all the other nodes. For two points the sets are defined by the half-plane bounded by a perpendicular to the middle of the segment connecting these two points. The Voronoy polygon thereby will be the intersection of such half-planes for all pairs of nodes created by this node and all the other nodes. Note that this polygon is always convex.

Each vertex of the Voronoi polygon is a point of contact of three Voronoi polygons. The triangle constructed by the corresponding nodes of contacting Voronoi polygons is associated with each of these vertices. This is exactly the Delaunay triangulation. Thus between the Voronoi diagram and the Delaunay triangulation a unique correspondence is established.

## 2.2   General Notations

Assume that the computational domain is a convex polyhedron $\Omega$ with the boundary $\partial\Omega$. In the domain $\overline{\Omega} = \Omega \cup \partial\Omega$ we consider the grid $\overline{\omega}$, which consists of nodes $\mathbf{x}_i^D$, $i = 1, 2, \ldots, M_D$, and the angles of the polyhedron $\Omega$ are nodes. Let $\omega$ be a set of inner nodes and $\partial\omega$ is a set of boundary nodes, i.e., $\omega = \overline{\omega} \cap \Omega$, $\partial\omega = \overline{\omega} \cap \partial\Omega$.

Each node $\mathbf{x}_i^D$, $i = 1, 2, \ldots, M_D$, connect a certain part of the computational domain, namely, the Voronoi polyhedron or its part belonging to $\Omega$. The Voronoi polyhedron (polygon) for a separate node is a set of points lying closer to this node than to all the other ones:

$$V_i = \{\, \mathbf{x} \mid \mathbf{x} \in \Omega, \ |\mathbf{x} - \mathbf{x}_i^D| < |\mathbf{x} - \mathbf{x}_j^D|, \ j = 1, 2, \ldots, M_D \,\}, \quad i = 1, 2, \ldots, M_D,$$

where $|\cdot|$ is the Euclidean distance. Each vertex $\mathbf{x}_k^V$, $k = 1, 2, \ldots, M_V$ of the Voronoi polyhedron is associated with the tetrahedron constructed by the appropriate nodes contacting the Voronoi polyhedrons. We will assume that all vertices of the Voronoi polyhedrons lie either inside the computational domain $\Omega$ or on its boundary $\partial\Omega$. These tetrahedrons determine the Delaunay triangulation — a dual triangulation to the Voronoi diagram. The $D$-grid in the domain $\Omega$ is determined by the set of nodes (vertices of tetrahedrons of the Delaunay triangulation) $\mathbf{x}_i^D$, $i = 1, 2, \ldots, M_D$, the $V$-grid is defined by the set of nodes (vertices of polyhedron of the Voronoi diagram) $\mathbf{x}_k^V$, $k = 1, 2, \ldots, M_V$.

We mark a separate tetrahedron $D_k$ of the Delaunay triangulation. This tetrahedron is identified by the number $k$ of the Voronoi polyhedron vertex, $k = 1, 2, \ldots, M_V$. The tetrahedrons $D_k, k = 1, 2, \ldots, M_V$ cover the entire computational domain, so

$$\overline{\Omega} = \overset{M_V}{\underset{k=1}{\cup}} \overline{D}_k, \quad \overline{D}_k = D_k \cup \partial D_k, \quad D_k \cap D_m = \emptyset, \quad k \neq m, \quad k, m = 1, 2, \ldots, M_V.$$

For common planes of the tetrahedron we use the notations

$$\partial D_{km} = \partial D_k \cap \partial D_m, \quad k \neq m, \quad k, m = 1, 2, \ldots, M_V.$$

The boundary of the computational domain $\partial\Omega$ consists of the planes of Delaunay tetrahedrons. Let

$$\partial D_0 = \partial\Omega, \quad \partial D_{k0} = \partial D_k \cap \partial D_0, \quad k = 1, 2, \ldots, M_V.$$

We associate the Voronoi polyhedron $V_i$, $i = 1, 2, \ldots, M_D$ with the node of the main grid $i$. Thus, we have

$$\overline{\Omega} = \overset{M_D}{\underset{i=1}{\cup}} \overline{V}_i, \quad \overline{V}_i = V_i \cup \partial V_i, \quad V_i \cap V_j = \emptyset, \quad i \neq j, \quad i, j = 1, 2, \ldots, M_D$$

and
$$\partial V_{ij} = \partial V_i \cap \partial V_j, \quad i \neq j, \quad i, j = 1, 2, \ldots, M_D.$$

For each tetrahedron $D_k$, $k = 1, 2, \ldots, M_V$, we define the set of neighbors $\mathcal{W}^D(k)$, having common planes with $D_k$, i.e.,

$$\mathcal{W}^D(k) = \{m \mid \partial D_k \cap \partial D_m \neq \emptyset, \; m = 0, 1, \ldots, M_V\}, \quad k = 1, 2, \ldots, M_V.$$

In this case, $m = 0$ means that the tetrahedron $D_k$ contacts the boundary. We define also the neighbors for each Voronoi polyhedron $V_i$, $i = 1, 2, \ldots, M_D$:

$$\mathcal{W}^V(i) = \{j \mid \partial V_i \cap \partial V_j \neq \emptyset, \; j = 1, 2, \ldots, M_D\}, \quad i = 1, 2, \ldots, M_D.$$

We assume that the introduced Delaunay triangulation and the Voronoi diagram are regular [2]. For the notations

$$h_k^D = \mathrm{diam}(D_k) \; - \; \text{diameter } D_k,$$

$$\varrho_k^D = \sup\{\mathrm{diam}(S) \mid S \; - \; \text{sphere in } D_k\}, \quad k = 1, 2, \ldots, M_V$$

the regularity condition of the Delaunay triangulation is

$$\frac{h_k^D}{\varrho_k^D} \leq \sigma > 0, \quad k = 1, 2, \ldots, M_V.$$

Likewise, for the Voronoi diagram we have

$$h_i^V = \mathrm{diam}(V_i) \; - \; \text{diameter } V_i,$$

$$\varrho_i^V = \sup\{\mathrm{diam}(S) \mid S \; - \; \text{sphere in } V_i\},$$

$$\frac{h_i^V}{\varrho_i^V} \leq \sigma > 0, \quad i = 1, 2, \ldots, M_D,$$

and
$$h = \max_{i,k}\{h_i^V, h_k^D\},$$

$$\mathrm{meas}(D_k) = \int_{D_k} d\mathbf{x}, \quad k = 1, 2, \ldots, M_V,$$

$$\mathrm{meas}(V_i) = \int_{V_i} d\mathbf{x}, \quad i = 1, 2, \ldots, M_D.$$

The other notations for the Voronoi diagram and Delaunay triangulation will be considered later.

## 2.3    Scalar and Vector Grid Functions

We will approximate the scalar functions of the continuous argument by the scalar grid functions that are defined in the nodes of the $D$-grid or in the nodes of the $V$-grid. We denote by $H_D$ the set of grid functions defined on the $D$-grid

$$H_D = \{\, y(\mathbf{x}) \mid y(\mathbf{x}) = y(\mathbf{x}_i^D) = y_i^D, \ i = 1, 2, \ldots, M_D \,\}.$$

For the functions $y(\mathbf{x}) \in H_D$, vanishing on the boundary $\partial \omega$, we define

$$H_D^0 = \{\, y(\mathbf{x}) \mid y(\mathbf{x}) \in H_D, \ y(\mathbf{x}) = 0, \ \mathbf{x} \in \partial \omega \,\}.$$

We consider the scalar product and the norm for the scalar grid functions from $H_D$ by

$$(y, v)_D = \sum_{i=1}^{M_D} y_i^D \, v_i^D \ \mathrm{meas}(V_i), \quad \|y\|_D = (y, y)_D^{1/2}.$$

This scalar product and the norm are grid analogs of the scalar product and the $L_2(\Omega)$-norm for the scalar functions of the continuous argument.

Likewise for the grid functions defined on the $V$-grid we define the space

$$H_V = \{\, y(\mathbf{x}) \mid y(\mathbf{x}) = y(\mathbf{x}_k^V) = y_k^V, \ k = 1, 2, \ldots, M_V \,\}.$$

For the functions $y(\mathbf{x}) \in H_V$ we have

$$(y, v)_V = \sum_{k=1}^{M_V} y_k^V \, v_k^V \ \mathrm{meas}(D_k), \quad \|y\|_V = (y, y)_V^{1/2}.$$

The approximation of continuous argument vector functions is a more difficult task. The simplest approach is connected with the assignment of three Cartesian components of the vector function in the nodes of the $D$-grid or in the nodes of the $V$-grid for approximating the vector field in the appropriate control volume — the Voronoi polygon for the nodes of the $D$-grid or the Delaunay tetrahedron for the $V$-grid nodes. The use of such an approximation for general unstructured grids is unjustified and connected with technical difficulties.

To determine the vector field in the control volume, it is natural to use the components of the sought function normal to the corresponding planes of the control volume. Choosing the initial and the final node, we connect with each tetrahedron edge $D_k$, $k = 1, 2, \ldots, M_V$ or polyhedron edge $V_i$, $i = 1, 2, \ldots, M_D$, a vector — the directed edge. For Delaunay triangulation the normals to the planes are the directed edges of Voronoi diagram and vice versa. For the approximation of the vector functions thereby we can use projections of the vectors on the directed edges. We will further use exactly this variant with the description of the vector field in the control volume by means of vector projections on the edges of the control volume.

We will orient the Delaunay triangulation edges by the unit vector

$$\mathbf{e}_{ij}^D = \mathbf{e}_{ji}^D, \quad i = 1, 2, \ldots, M_D, \quad j \in \mathcal{W}^V(i),$$

directed from the node with a smaller number to the node of a larger number. Likewise, we define the unit vectors

$$\mathbf{e}_{km}^V = \mathbf{e}_{mk}^V, \quad k = 1, 2, \ldots, M_V, \quad m \in \mathcal{W}^D(k)$$

for the directed edges of the Voronoi diagram.

The vector function $\mathbf{y}(\mathbf{x})$ on the Delaunay triangulation is defined by the components

$$y_{ij}^D = \mathbf{y}\, \mathbf{e}_{ij}^D, \quad i = 1, 2, \ldots, M_D, \quad j \in \mathcal{W}^V(i),$$

that are given in the middle of the edges

$$\mathbf{x}_{ij}^D = \frac{1}{2}(\mathbf{x}_i^D + \mathbf{x}_j^D).$$

Using the Voronoi diagram, the components are given at the point on the edge $\mathbf{x}_{km}^V$ defined as a point of intersection of the edge and the corresponding plane of the Delaunay triangulation, thereby

$$y_{km}^V = \mathbf{y}\, \mathbf{e}_{km}^V, \quad k = 1, 2, \ldots, M_V, \quad m \in \mathcal{W}^D(k).$$

Taking into consideration the introduced notations, the points $\mathbf{x}_{k0}^V$ lie on the boundary $\partial \Omega$.

Note that in the two-dimensional case the nodes of the vector function assignment in the Delaunay triangulation and the Voronoi diagram coincide. In this case, we have the local orthogonal coordinate system and this property can be used for constructing grid problems. In general three-dimensional case this is not true.

For the Delaunay triangulation used and the Voronoi diagram we define the length of the edges in the following way:

$$l_{ij}^D = |\mathbf{x}_i^D - \mathbf{x}_j^D|, \quad i = 1, 2, \ldots, M_D, \quad j \in \mathcal{W}^V(i),$$

$$l_{km}^V = |\mathbf{x}_k^V - \mathbf{x}_m^V|, \quad k = 1, 2, \ldots, M_V, \quad m \in \mathcal{W}^D(k).$$

We denote by $\mathbf{H}_D$ the set of grid vector functions determined by the components $y_{ij}^D$, $i = 1, 2, \ldots, M_D$, $j \in \mathcal{W}^V(i)$ that are given in the middle of the edges. In a similar way we denote by $\mathbf{H}_V$ the set of grid vector functions defined by the components $y_{km}^V$, $k = 1, 2, \ldots, M_V$, $m \in \mathcal{W}^D(k)$. If the tangential components of the grid vector functions $\mathbf{y} \in \mathbf{H}_D$ vanish on the boundary, we define

$$\mathbf{H}_D^0 = \{\, \mathbf{y} \mid \mathbf{y} \in \mathbf{H}_D, \quad \mathbf{y}(\mathbf{x})\, \mathbf{e}_{ij}^D = 0, \quad \mathbf{x} = \mathbf{x}_{ij}^D \in \partial\omega, = 1, 2, \ldots, M_D, \, j \in \mathcal{W}^V(i)\,\},$$

$$\mathbf{H}_V^0 = \{\, \mathbf{y} \mid \mathbf{y} \in \mathbf{H}_V, \quad \mathbf{y}(\mathbf{x})\, \mathbf{e}_{k0}^V = 0, \quad k = 1, 2, \ldots, M_V \,\}.$$

Consider the scalar product and the norm in $\mathbf{H}_D$:

$$(\mathbf{y}, \mathbf{v})_D = \frac{1}{2} \sum_{k=1}^{M_V} \sum_{m \in \mathcal{W}^D(k)} \sum_{(i,j)\, \in \mathcal{Q}^D(k,m)} y_{ij}^D\, v_{ij}^D\, |\mathbf{x}_m^V - \mathbf{x}_{km}^V|\, \operatorname{meas}(\partial D_{km}),$$

$$\|\mathbf{y}\|_D = (\mathbf{y}, \mathbf{y})_D^{1/2},$$

where

$$\mathcal{Q}^D(k, m) = \{(i, j) \mid \mathbf{x}_i^D, \mathbf{x}_j^D \in \partial D_{km}, \; i = 1, 2, \ldots, M_D, \; j \in \mathcal{W}^V(i)\}$$

is the set of vertices of the plane $\partial D_{km}$. Likewise, we define

$$(\mathbf{y}, \mathbf{v})_V = \frac{1}{2} \sum_{i=1}^{M_D} \sum_{j \in \mathcal{W}^V(i)} \sum_{(k,m) \in \mathcal{Q}^V(i,j)} y_{km}^V v_{km}^V \, |\mathbf{x}_i^D - \mathbf{x}_{ij}^D| \, \mathrm{meas}(\partial V_{ij}),$$

$$\|\mathbf{y}\|_V = (\mathbf{y}, \mathbf{y})_V^{1/2},$$

$$\mathcal{Q}^V(i, j) = \{(k, m) \mid \mathbf{x}_k^V, \mathbf{x}_m^V \in \partial V_{ij}, \; k = 1, 2, \ldots, M_V, \; m \in \mathcal{W}^D(k)\}$$

for the scalar product and the norm in $\mathbf{H}_V$.

## 3    Vector Analysis Grid Operators

We consider the problems of mathematical physics defined in terms of the vector analysis operators: divergence, gradient and rotor operators. Turning to the discrete problem, we should have the grid analogs of these operators. On the other hand we cannot always use the standard finite-element approximation. In particular, this also concerns the construction of the grid analogs of the vector analysis operator.

### 3.1    Grid Gradient and Divergence Operators

The set of grid functions $H_D$ or $H_V$ can be the domain of definition of the grid gradient operator. In the first case, we denote the grid gradient operator by $\mathrm{grad}_D$, and in the second case, $\mathrm{grad}_V$. Taking into account the chosen edge orientation, at the points $\mathbf{x}_{ij}^D$ we set

$$(\mathrm{grad}_D y)_{ij}^D = \eta(i, j) \, \frac{y_j^D - y_i^D}{l_{ij}}, \quad i = 1, 2, \ldots, M_D, \quad j \in \mathcal{W}^V(i), \qquad (1)$$

i.e., the range of values of the operator $\mathrm{grad}_D : H_D \to \mathbf{H}_D$ is the set of vector grid functions $\mathbf{H}_D$. In (1), we use the following notation:

$$\eta(i, j) = \begin{cases} 1, & j > i, \\ -1, & j < i. \end{cases}$$

For the truncation error of the grid operator $\mathrm{grad}_D$ we have

$$(\mathrm{grad}_D u)(\mathbf{x}) = (\mathrm{grad}\, u)(\mathbf{x}) + \mathbf{g}(\mathbf{x}), \quad \mathbf{g} = \mathcal{O}(h^2), \quad \mathbf{x} = \mathbf{x}_{ij}^D, \qquad (2)$$

$$i = 1, 2, \ldots, M_D, \quad j \in \mathcal{W}^V(i)$$

provided that $u(\mathbf{x})$ is a sufficiently smooth functions. Here and below we shall not give an accurate formulation of the requirements for the smoothness of the functions used (solution, coefficients etc.).

For the values of the grid operator $\mathrm{grad}_V : H_V \to \mathbf{H}_V$ at the points $\mathbf{x}_{km}^V \in \Omega$ we set

$$(\mathrm{grad}_V y)_{km}^V = \eta(k, m) \, \frac{y_m^V - y_k^V}{l_{km}}, \tag{3}$$

$$k = 1, 2, \ldots, M_V, \quad m \in \mathcal{W}^D(k), \quad m \neq 0.$$

In this case, the truncation error has only the first order because the first derivatives are approximated at the point $\mathbf{x}_{km}^V$ that is not situated in the middle of the edge of the Voronoi polyhedron

$$(\mathrm{grad}_V u)(\mathbf{x}) = (\mathrm{grad}\, u)(\mathbf{x}) + \mathbf{g}(\mathbf{x}), \quad \mathbf{g} = \mathcal{O}(h), \quad \mathbf{x} = \mathbf{x}_{km}^V, \tag{4}$$

$$k = 1, 2, \ldots, M_V, \quad m \in \mathcal{W}^D(k), \quad m \neq 0.$$

Now construct the grid analogs of operator div on the set of the vector grid functions $\mathbf{y} \in \mathbf{H}_D$ and $\mathbf{y} \in \mathbf{H}_V$. Start from the divergence equality. For the Voronoi polyhedron this equality is written in the following form:

$$\int_{V_i} \mathrm{div}\, \mathbf{u} \, d\mathbf{x} = \sum_{j \in \mathcal{W}^V(i)} \int_{\partial V_{ij}} (\mathbf{u} \cdot \mathbf{n}_{ij}^V) \, d\mathbf{x}, \tag{5}$$

where $\mathbf{n}_{ij}^V$ is the normal to the edge $\partial V_{ij}$ outside with respect to $V_i$. To construct the grid operator $\mathrm{div}_D : \mathbf{H}_D \to H_D$ we use the elementary formulas of integration for the left- and right-hand sides (5). This leads to the grid analogs of the operator div in the form of

$$(\mathrm{div}_D \mathbf{y})_i^D = \frac{1}{\mathrm{meas}(V_i)} \sum_{j \in \mathcal{W}^V(i)} (\mathbf{n}_{ij}^V \cdot \mathbf{e}_{ij}^D) \, y_{ij}^D \, \mathrm{meas}(\partial V_{ij}), \quad i = 1, 2, \ldots, M_D. \tag{6}$$

Similarly, on the nodes of the $V$-grid we define the values of the grid analogs of the divergence operator on the set of vector grid functions $\mathbf{y} \in \mathbf{H}_V$. Start from the divergence theorem, rewriting for the Delauney tetrahedron:

$$\int_{D_k} \mathrm{div}\, \mathbf{u} \, d\mathbf{x} = \sum_{m \in \mathcal{W}^D(k)} \int_{\partial D_{km}} (\mathbf{u} \cdot \mathbf{n}_{km}^D) \, d\mathbf{x}, \tag{7}$$

where $\mathbf{n}_{km}^D$ is the normal to the edge $\partial D_{km}$ outside with respect to $D_k$. Approximation (7) leads to the following presentation:

$$(\mathrm{div}_V \mathbf{y})_k^V = \frac{1}{\mathrm{meas}(D_k)} \sum_{m \in \mathcal{W}^D(k)} (\mathbf{n}_{km}^D \cdot \mathbf{e}_{km}^V) \, y_{km}^V \, \mathrm{meas}(\partial D_{km}) \tag{8}$$

for the grid operator $\mathrm{div}_V : \mathbf{H}_V \to H_V$.

To estimate the error of the constructed approximations of the divergence operator, we take into account the fact that the introduced $D$-grid and $V$-grid are regular. For the left-hand side of (7) we have

$$\int_{D_k} \operatorname{div} \mathbf{u} \, dx = \operatorname{meas}(D_k) \left( (\operatorname{div}_V \mathbf{u})_k^V + g(\mathbf{x}_k^V) \right),$$

where $g(\mathbf{x}_k^V) = \mathcal{O}(h)$. Similarly, for the right-hand side we get

$$\sum_{m \, \in \mathcal{W}^D(k)} \int_{\partial D_{km}} (\mathbf{u} \cdot \mathbf{n}_{km}^D) \, dx = \sum_{m \, \in \mathcal{W}^D(k)} (\mathbf{n}_{km}^D \cdot \mathbf{e}_{km}^V) \, (u(\mathbf{x}_{km}^V) + q(\mathbf{x}_{km}^V)) \operatorname{meas}(\partial D_{km}),$$

where $q(\mathbf{x}_{km}^V) = \mathcal{O}(h)$ in the three-dimensional case. For two-dimensional problems our approximation of the right-hand side (7) corresponds to the quadrangle formula and so $q(\mathbf{x}_{km}^V) = \mathcal{O}(h^2)$.

Taking into account the last fact in the general three-dimensional case for the error of formula (8), we obtain

$$(\operatorname{div}_V \mathbf{u})(\mathbf{x}) = (\operatorname{div} \mathbf{u})(\mathbf{x}) + g(\mathbf{x}) + (\operatorname{div}_V \mathbf{q})(\mathbf{x}), \quad g = \mathcal{O}(h), \quad \mathbf{q} = \mathcal{O}(h), \quad (9)$$

$$\mathbf{x} = \mathbf{x}_k^V, \quad k = 1, 2, \ldots, M_V.$$

For two-dimensional problems we have

$$(\operatorname{div}_V \mathbf{u})(\mathbf{x}) = (\operatorname{div} \mathbf{u})(\mathbf{x}) + g(\mathbf{x}), \quad g = \mathcal{O}(h), \quad (10)$$

$$\mathbf{x} = \mathbf{x}_k^V, \quad k = 1, 2, \ldots, M_V.$$

Similarly to (9), we get

$$(\operatorname{div}_D \mathbf{u})(\mathbf{x}) = (\operatorname{div} \mathbf{u})(\mathbf{x}) + g(\mathbf{x}) + (\operatorname{div}_D \mathbf{q})(\mathbf{x}), \quad g = \mathcal{O}(h), \quad \mathbf{q} = \mathcal{O}(h), \quad (11)$$

$$\mathbf{x} = \mathbf{x}_i^D, \quad i = 1, 2, \ldots, M_D,$$

for the truncation error.

Thus, in general, the truncation error for the grid divergence operators $\operatorname{div}_D$ and $\operatorname{div}_V$ equals to $\mathcal{O}(1)$. However, there exists a special divergence expression of the truncation error saving the situation in the case of approximation of problems of mathematical physics.

## 3.2   Grid Rotor Operators

Now we construct grid analogs of the operator rot noticing, as for the other vector analysis operators, two modifications of such operators. To construct the grids operators on the derived grid, we use the following integral equality (the Stokes theorem):

$$\int_S (\operatorname{rot} \mathbf{u} \cdot \mathbf{n}) dx = \oint_{\partial S} (\mathbf{u} \cdot \mathbf{l}) \, dx, \quad (12)$$

where $S$ is a simply connected surface spanned on the contour $\partial S$, l thereby is a tangent vector. And the direction of the normal vector $\mathbf{n}$ coordinated with the orientation of the contour $\partial S$ according to the right-hand screw rule.

Construct the grid operator $\mathbf{rot}_D : \mathbf{H}_D \to \mathbf{H}_V$ on the set of vector grid functions $\mathbf{y} \in \mathbf{H}_D$. We use (12) for the edge $\partial D_{km}$. It follows that

$$
\int_{\partial D_{km}} (\mathrm{rot}\, \mathbf{u} \cdot \mathbf{n}_{km}^D)\, d\mathbf{x} = \sum_{(i,j)\,\in\mathcal{Q}^D(km)} \int_{\mathbf{x}_i^D}^{\mathbf{x}_j^D} \chi(\mathbf{n}_{km}^D, \mathbf{e}_{ij}^D)(\mathbf{u} \cdot \mathbf{e}_{ij}^D)\, d\mathbf{x}. \tag{13}
$$

The outside normal $\mathbf{n}_{km}^D$ to the side $\partial D_{km}$ defines the orientation of the tracing edges of the side $\partial D_{km}$. The value of $\chi(\mathbf{n}_{km}^D, \mathbf{e}_{ij}^D)$ depends on the orientation of the edges of the side $\partial D_{km}$:

$$
\chi(\mathbf{n}_{km}^D, \mathbf{e}_{ij}^D) = \begin{cases} 1, & \text{if the vector } \mathbf{e}_{ij}^D \text{ corresponds to positive tracing,} \\ -1, & \text{otherwise.} \end{cases}
$$

for the left-hand side of (13) we set

$$
\int_{\partial D_{km}} (\mathrm{rot}\, \mathbf{u} \cdot \mathbf{n}_{km}^D)\, d\mathbf{x} = \mathrm{meas}(\partial D_{km})\, (\mathbf{n}_{km}^D \cdot \mathbf{e}_{km}^V)\, ((\mathbf{rot}_D \mathbf{u})_{km}^V + \mathbf{g}_1(\mathbf{x}_{km}^V)), \tag{14}
$$

where $\mathbf{g}_1(\mathbf{x}_{km}^V) = \mathcal{O}(h)$ for a sufficiently smooth vector function $\mathbf{u}$. The nodes $\mathbf{x}_{ij}^D$ are situated in the middle of the edges, and using the quadrangle formula for the approximation of right-hand side of (13), we obtain

$$
\sum_{(i,j)\,\in\mathcal{Q}^D(km)} \int_{\mathbf{x}_i^D}^{\mathbf{x}_j^D} \chi(\mathbf{n}_{km}^D, \mathbf{e}_{ij}^D)(\mathbf{u} \cdot \mathbf{e}_{ij}^D)\, d\mathbf{x} =
$$

$$
\sum_{(i,j)\,\in\mathcal{Q}^D(km)} \chi(\mathbf{n}_{km}^D, \mathbf{e}_{ij}^D) l_{ij}^D\, (\mathbf{u}(\mathbf{x}_{ij}^D) + \mathbf{g}_2(\mathbf{x}_{ij}^D)), \tag{15}
$$

where $\mathbf{q}_2(\mathbf{x}_{ij}^D) = \mathcal{O}(h^2)$. Taking into account (13)–(15), we define the values of the grid operator $\mathbf{rot}_D : \mathbf{H}_D \to \mathbf{H}_V$ by the rule

$$
(\mathrm{rot}_D \mathbf{y})_{km}^V = (\mathbf{n}_{km}^D \cdot \mathbf{e}_{km}^V)\, \frac{1}{\mathrm{meas}(\partial D_{km})} \sum_{(i,j)\,\in\mathcal{Q}^D(km)} \chi(\mathbf{n}_{km}^D, \mathbf{e}_{ij}^D)\, y_{ij}^D\, l_{ij}^D, \tag{16}
$$

$$
k = 1, 2, \ldots, M_V, \quad m \in \mathcal{W}^D(k).
$$

For the truncation error we have the following representation:

$$
(\mathrm{rot}_D \mathbf{u})(\mathbf{x}) = (\mathrm{rot}\, \mathbf{u})(\mathbf{x}) + \mathbf{g}(\mathbf{x}), \quad \mathbf{g} = \mathcal{O}(h), \quad \mathbf{x} = \mathbf{x}_{km}^V \tag{17}
$$

$$
k = 1, 2, \ldots, M_V, \quad m \in \mathcal{W}^D(k).
$$

In the same way we construct the grid operator $\mathbf{rot}_V : \mathbf{H}_V \rightarrow \mathbf{H}_D$:

$$(\mathbf{rot}_V \mathbf{y})_{ij}^D = (\mathbf{n}_{ij}^V \cdot \mathbf{e}_{ij}^D) \frac{1}{\text{meas}(\partial V_{ij})} \sum_{(k,m) \in \mathcal{Q}^V(ij)} \chi(\mathbf{n}_{ij}^V, \mathbf{e}_{km}^V) \, y_{km}^V \, l_{km}^V, \qquad (18)$$

$$i = 1, 2, \ldots, M_D, \quad j \in \mathcal{W}^V(i).$$

Similarly to (17), we can get the following special representation of the truncation error for the grid operator $\mathbf{rot}_V$

$$(\mathbf{rot}_V \mathbf{u})(\mathbf{x}) = (\mathbf{rot}\,\mathbf{u})(\mathbf{x}) + \mathbf{g}(\mathbf{x}) + (\mathbf{rot}_V \mathbf{q})(\mathbf{x}), \quad \mathbf{g} = \mathcal{O}(h), \quad \mathbf{q} = \mathcal{O}(h),$$
$$\mathbf{x} = \mathbf{x}_{ij}^D, \quad i = 1, 2, \ldots, M_D, \quad j \in \mathcal{W}^V(i). \tag{19}$$

The last component of the error appears due to the fact that the contour integral is approximated by the edges the point $\mathbf{x}_{km}^V$, in general, is not situated in the middle of the edges.

## 3.3    Consistency of the Vector Analysis Grid Operators

We note the following most important properties of the vector analysis operators

$$\text{rot}\,\text{grad} = 0, \tag{20}$$

$$\text{div}\,\text{rot} = 0. \tag{21}$$

If relation (20) (or (21)) holds for the grid analogs, then we say that such grid operators are consistent as to property (20) (or (21)).

Taking into account (1) and (16), we get

$$\text{rot}_D\,\text{grad}_D\, y = 0, \quad y(\mathbf{x}) \in H_D, \tag{22}$$

i.e., the grid operators $\text{rot}_D$ and $\text{grad}_D$ are consistent as to property (20). Likewise, we determine the consistency of the operators $\text{rot}_V$ and $\text{grad}_V$

$$\text{rot}_V\,\text{grad}_V\, y = 0, \quad y(\mathbf{x}) \in H_V. \tag{23}$$

The consistency as to property (21) holds for pairs of the operators $\text{div}_D$, $\text{rot}_V$ and $\text{div}_V$, $\text{rot}_D$. For the grid divergence and rotor operators we have

$$\text{div}_V\,\mathbf{rot}_D\,\mathbf{y} = 0, \quad \mathbf{y}(\mathbf{x}) \in \mathbf{H}_V, \tag{24}$$

$$\text{div}_D\,\mathbf{rot}_V\,\mathbf{y} = 0, \quad \mathbf{y}(\mathbf{x}) \in \mathbf{H}_D. \tag{25}$$

We consider that the self-adjointness and antisymmetry to be the most important properties of the grid operators of mathematical physics. These properties arise from the corresponding properties of the vector analysis grid operators. In order to clarify this question, we first recall the following facts concerning the vector analysis differential operators.

We have
$$\text{div}_D^* = -\,\text{grad}_D \tag{26}$$
on the set of grid functions $v \in H_D^0$ and $\mathbf{y} \in \mathbf{H}_D$. Similarly, using (8), we obtain

$$(v, \text{div}_D\,\mathbf{y})_D + (\text{grad}_D\,v, \mathbf{y})_D = 0$$

on the set of functions $v(\mathbf{x}_{km}^V) = 0$, $\mathbf{x}_{km}^V \in \partial\Omega$, i.e.,

$$\text{div}_V^* = -\,\text{grad}_V. \tag{27}$$

For the grid vector functions $\mathbf{v} \in \mathbf{H}_D^0$, $\mathbf{u} \in \mathbf{H}_V$, using (16), (18), we obtain

$$(\mathbf{v}, \text{rot}_V\,\mathbf{u})_D = (\text{rot}_D\,\mathbf{v}, \mathbf{u})_V.$$

Therefore,
$$\text{rot}_V^* = \text{rot}_D \tag{28}$$
holds. This equality can be considered as the grid analog of self-adjointness of the differential rotor operator.

## Acknowledgements

This work was supported by the INTAS under grant 03-50-4395.

## References

1. Aurenhammer, F., Klein, R.: Voronoi diagrams. In Sack, J., Urrutia, G., eds.: Handbook of Computational Geometry, Elsevier Science Publishing (2000) 201–290
2. Ciarlet, P.G.: The Finite Element Method for Elliptic Problems. North–Holland, Amsterdam, New York (1978)
3. Robidoux, N., Steinberg, S.: A discrete vector calculus in tensor grids. (2002)
4. Samarskii, A.A.: The theory of difference schemes. Volume 240 of Monographs and Textbooks in Pure and Applied Mathematics. Marcel Dekker Inc., New York (2001)
5. Vabishchevich, P.: Finite-difference approximation of mathematical physics problems on irregular grids. Computational Methods in Applied Mathematics **5**(3) (2005) 294–330

# On Some Computational Aspects of the Variational Data Assimilation Techniques

Zahari Zlatev

National Environmental Research Institute
Frederiksborgvej 399, P. O. Box 358, DK-4000 Roskilde, Denmark

**Abstract.** It is important to incorporate all available observations when large-scale mathematical models arising in different fields of science and engineering are used to study various physical and chemical processes. Variational data assimilation techniques can be used in the attempts to utilize efficiently observations in a large-scale model. Variational data assimilation techniques are based on a combination of three very important components

- numerical methods for solving differential equations,
- splitting procedures

and

- optimization algorithms.

It is crucial to select an optimal (or, at least, a good) combination of these three components, because models which are very expensive computationally become much more expensive (the computing time being often increased by a factor greater than 100) when a variational data assimilation technique is applied. Therefore, it is important to study the interplay between the three components of the variational data assimilation techniques as well as to apply powerful parallel computers in the computations. Some results obtained in the search for a good combination will be reported. Parallel techniques described in [1] are used in the runs related to this paper.

Modules from a particular large-scale mathematical model, the Unified Danish Eulerian Model (UNI-DEM), are used in the experiments. The mathematical background of UNI-DEM is discussed in [1], [24] The ideas are rather general and can easily be applied in connection with other mathematical models.

## 1 Statement of the Problem

It is important to start the computations with a good set of initial values of the concentrations when short-time air pollution forecasts are to be calculated by applying large-scale air pollution models. Good initial values are normally not available and, therefore, one has to find a way to produce good initial values. Different data assimilation techniques and available observations can be used in the efforts to resolve this task. It should be noted here that data assimilation techniques are very useful also in the efforts to solve many other important tasks

T. Boyanov et al. (Eds.): NMA 2006, LNCS 4310, pp. 28–39, 2007.
© Springer-Verlag Berlin Heidelberg 2007

related to large-scale computations with environmental models (see [6], [7] and [20]). Data assimilation techniques can also be applied in many other fields of science and engineering (see [15]).

Assume that observations are available at every $t_p$, $p \in \{0, 1, 2, \ldots, P\}$. These observations can be taken into account in an attempt to improve the results obtained by a given model. This can be done by minimizing the value of the following functional (see, for example, [16]):

$$J\{\bar{c}_0\} = \frac{1}{2} \sum_{p=0}^{P} < W(t_p)\,(\bar{c}_p - \bar{c}_p^{obs})\,,\, \bar{c}_p - \bar{c}_p^{obs} >, \tag{1}$$

where

- the functional $J\{\bar{c}_0\}$ is depending on the initial value $\bar{c}_0$ of the vector of the concentrations at time $t_0$ (because the model results $\bar{c}_p$ depend on $\bar{c}_0$),
- $W(t_p)$ is a matrix containing some weights, and
- $<,>$ is an inner product in an appropriately defined Hilbert space (it will be assumed here that the usual vector space is used, i.e. it is assumed that $\bar{c} \in \mathbf{R}^q$ where $q$ is the number of chemical species which are involved in the model).

It is seen that the functional $J\{\bar{c}_0\}$ depends on both the weights and the differences between calculated by the model concentrations $\bar{c}_p$ and observations $\bar{c}_p^{obs}$ at the time-levels $t_p$ at which observations are available. $W(t_p)$ will be assumed to be the identity matrix $I$ in this paper, but in general weights are to be defined in some way and used in the computations.

The task is to find an improved initial field $\bar{c}_0$, which minimizes the functional $J\{\bar{c}_0\}$. This can be achieved by using some optimization algorithm. Most of the optimization algorithms are based on the application of the gradient of $J\{\bar{c}_0\}$. The adjoint equation has to be defined and used in the calculation of the gradient of the functional $J\{\bar{c}_0\}$.

## 2   Algorithmic Representation

A data assimilation algorithm can be represented by applying the procedure described in Fig. 1. An optimization procedure is needed for the calculations that are to be carried out in the loop "DO ITERATIONS". In many optimization subroutines, the direction of the steepest descent is to be found and then the value of parameter $\rho$ that gives the largest decrease in the direction found is to be used in an attempt to improve the current solution. In practice, however, it is only necessary here to find a good standard minimization subroutine. In our experiments we used the subroutine E04DGF from the NAG Numerical Library [18]. This subroutine uses a preconditioned conjugate gradient methods and is based on the algorithm PLMA, described in [9] and in Section 4.8.3 of [10].

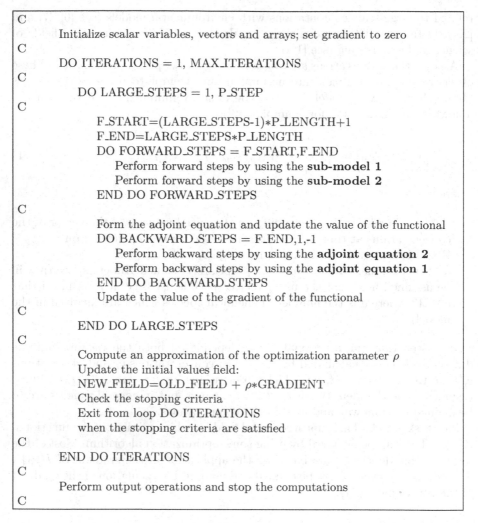

**Fig. 1.** An algorithm for performing variational data assimilation by performing multiple backward calculations. P_STEP is equal to $P$. P_LENGTH is equal to the number of time-steps that are to be carried out between two time-points $t_p$ and $t_{p+1}$ at which observations are available.

It is assumed in the derivation of the algorithm presented in Fig. 1 that a simple sequential splitting procedure is used and that the model under consideration is split into two sub-models. The results can easily been extended for the case where the model is split into more than two sub-models and/or where some other splitting procedures are applied (different splitting procedures are discussed in detail in Chapter 2 of [25]). Note that the sub-models in the loop related to the adjoint equations are used in reverse order (see Fig. 1).

The algorithm sketched in Fig. 1 is designed under the assumption that the backward calculations (with the adjoint equations) are carried out every time when the forward computations are performed to a time-point at which observations are available. It can be proved that this is not necessary. One can perform the forward calculations from the starting time-point to the end time-point and than carry out the backward mode only once from the end time-point to the starting time-point (see Chapter 10 in [25]).

## 3  Major Components of the Data Assimilation Algorithm

The data assimilation algorithm presented in Fig. 1 consists of three major computational parts:

- At every time-step (both a forward time-step and a backward time-step) we need a **numerical algorithm for solving differential equations**, because both the model and the adjoint equation are normally represented by systems of differential equations that are to be handled numerically.
- The discretization of the systems of differential equations leads to very time-consuming numerical tasks (these tasks are described by huge systems of ordinary differential equations). This is why some kind of **splitting** is normally used when large models are to be handled numerically. As mentioned before, a simple sequential splitting procedure is used in Fig. 1. This splitting procedure consists of two sub-models only. In practice, the original model is normally split into more than two sub-models.
- **An optimization algorithm** has to be used in order to minimize the functional (1). Most of the optimization algorithms are using the gradient of the functional in the computational process. The sub-models obtained by the selected splitting procedure and the corresponding adjoint equations are used to calculate the gradient.

From the algorithm given in Fig. 1, it can clearly be seen that the computations are carried out on three levels:

- **On the lowest level**, numerical methods for solving differential equations are to be used.
- **On the middle level**, some splitting procedure is to be applied in order (i) to split the model, (ii) to construct the corresponding adjoint equations and (iii) to treat the sub-models and the corresponding equations in some prescribed order.
- **On the highest level**, the optimization procedure is to be used.

The interplay of the methods used on these three levels must be studied carefully in the efforts to design a good data assimilation algorithm which is robust, fast and sufficiently accurate.

## 4  Treatment of a Simple Transport-Chemistry Problem

The following simple transport-chemistry problem is used in this sections in order to illustrate how the data assimilation techniques discussed in the previous section can be implemented. The problem is described mathematically by a system of partial differential equations (PDEs), in which it is assumed that $V$ is a positive constant:

$$\frac{\partial \bar{c}}{\partial t} = -V \frac{\partial \bar{c}}{\partial x} + f(t, x, \bar{c}), \quad x \in [a, b], \quad t \in [0, T], \quad \bar{c} \in \mathbf{R^n}, \quad c(0, x) = \bar{c}_0. \quad (2)$$

The system of PDEs given by (2) can be considered as a mathematical description of a very simple environmental transport-chemistry model. Normally, three-dimensional transport is used in the modern model. Moreover several other physical processes (diffusion, deposition, emissions, etc.) have to be included (see, for example, [3], [4], [5], [8], [13], [19], [21] and [24]). However, some important topics related to the implementation of data assimilation techniques can easily be explained by using (2). Note also that in the actual computations some modules from the Unified Danish Eulerian Model (UNI-DEM), ([24], [25]) have been implemented and used.

The model described by the system of PDEs (2) is split into two sub-models: a pure transport sub-model and a pure chemistry sub-model:

$$\frac{\partial \bar{g}}{\partial t} = -V \frac{\partial \bar{g}}{\partial x}, \quad x \in [a, b], \quad t \in [0, T], \quad \bar{g} \in \mathbf{R^q}, \quad (3)$$

$$\frac{\partial \bar{h}}{\partial t} = f(t, x, \bar{h}), \quad x \in [a, b], \quad t \in [0, T], \quad \bar{h} \in \mathbf{R^q}. \quad (4)$$

The calculations with the two sub-models (3) and (4) are carried out as follows. Assume that the calculations from $t = t_0$ to $t = t_n$ have been completed and an approximation $\bar{c}_n$ to $\bar{c}(t_{n+1}, x)$ has to be computed. We set $\bar{g}_n = \bar{c}_n$ and calculate with the selected numerical method for the treatment of (3) $\bar{g}_{n+1}$. We proceed by setting $\bar{h}_n = \bar{g}_{n+1}$ and by calculating $\bar{h}_{n+1}$ using the selected numerical method for the treatment of (4). Then we set $\bar{c}_{n+1} = \bar{h}_{n+1}$. In this way, the computations needed to calculate an approximation at $t = t_{n+1}$ are completed and we can proceed in the same way to compute an approximation at the next time-point $t = t_{n+2}$. It is still necessary to explain how to start the computations, i.e. how to calculate an approximation $\bar{c}_1$ to $\bar{c}(t_1, x)$. However, this is not a problem, because $\bar{c}_0$ is given, see (2).

## 4.1   The Transport Sub-model and Its Adjoint Equation

It is important to emphasize here the fact that (3) is a system of $q$ independent scalar PDEs, where $q$ is the number of the chemical species involved in the model. Therefore it is quite sufficient to show how any of these scalar equations can be solved. Assume that the number of grid-points on the $Ox$ axis is $N_x + 1$ and let

$$\frac{\partial g}{\partial t} = -V \frac{\partial g}{\partial x}, \quad x \in [a, b], \quad t \in [0, T], \quad g \in \mathbf{R}, \quad (5)$$

be any of the scalar equations forming the system of PDEs (3), i.e. $g$ is equal to some component $g_i$ of $\bar{g}$ where $i = 1, 2, \ldots, q$. Consider the grids:

$$\mathbf{G_x} = \{x_i | i = 0, 1, \ldots, N_x, \ x_0 = a, \ x_{N_x} = b\} \tag{6}$$

and

$$\mathbf{G_t} = \{t_n | n = 0, 1, \ldots, N_t, \ t_0 = 0, \ t_{N_t} = T\} \tag{7}$$

Let $x_i \in \mathbf{G_x}$ and $t_n \in \mathbf{G_t}$. Different numerical methods can be used to calculate approximations

$$g_{i,n} \approx g(t_n, x_i) \tag{8}$$

of the exact solution at point $(t_n, x_i)$.

Formula (9), which is given below, can be obtained by using the notation $w_i = (Vx_i \triangle t)/4\triangle x)$ and simple finite differences:

$$- w_i g_{i-1,n+1} + g_{i,n+1} + w_i g_{i+1,n+1} = w_i g_{i-1,n} + g_{i,n} - w_i g_{i+1,n}. \tag{9}$$

where $i = 2, 3, \ldots, N_x - 2$.

Let us assume, for the sake of simplicity, that periodic boundary conditions are given by the following formulae (in fact, only one boundary condition is quite sufficient, but considering two boundary condition will be necessary if diffusion terms are added):

$$c_{0,n} = 0, \quad c_{N_x,n} = 0, \quad n = 0, 1, \ldots, N_t. \tag{10}$$

By applying (10), formula (9) can be rewritten for $i = 1$ as

$$g_{1,n+1} + w_1 g_{2,n+1} = g_{1,n} - w_1 g_{2,n} \tag{11}$$

and for $i = N_x - 1$ as

$$- w_{N_x-1} g_{N_x-2,n+1} + g_{N_x-1,n+1} = w_{N_x-1} g_{N_x-2,n} + g_{N_x-1,n}. \tag{12}$$

Denote

$$\tilde{g}_n = (g_{1,n}, g_{2,n}, \ldots, g_{N_x-1,n})^T. \tag{13}$$

By using the notation introduced by (13), it is possible to rewrite (9), (11) and (12) in a matrix form:

$$(I - A)\tilde{g}_{n+1} = (I + A)\tilde{g}_n \quad \Rightarrow \quad \tilde{g}_{n+1} = D\tilde{g}_n, \tag{14}$$

$$D \overset{\text{def}}{=} (I - A)^{-1}(I + A), \tag{15}$$

where $I$ is the identity matrix and $A$ is a matrix which has non-zero elements only on the two diagonals that are adjacent to the main diagonal (and, more precisely, $-w_i$ on the diagonal below the main diagonal and $w_i$ on the diagonal above the main diagonal). Formula (15) has to be modified, by adding an appropriate vector $\alpha_n$ in its right-hand-side, when some other definition, different from that given in (10), is to be used in connection with the boundary conditions.

If $\tilde{g}_n$ has already been calculated, then (14) can be used to proceed with the calculation of $\tilde{g}_{n+1}$ . Thus, if an initial field, $\tilde{g}_0$, is given, then (14) can be used to calculate successively, step-by-step, approximations of the exact solution.

As mentioned above, it is necessary to calculate the gradient $Grad\{J\}$ of the functional $J\{\bar{c}_0\}$ from (1) in order to find an improved initial field $\bar{c}_0$. Assume as in the first section of this chapter that observations are available at times $\{t_p \mid p = 0, 1, \ldots, P\}$. Assume also that the calculations by formula (14) for some $n + 1 = p$ have been completed. It is then necessary to form the adjoint variable

$$\tilde{q}_p = \tilde{W}_p \left( \tilde{c}_p - \tilde{c}_p^{obs} \right) \tag{16}$$

and to use it as a starting value in the integration of the adjoint equation backward from $t = t_p$ to $t = t_0$ (let us reiterate that we have assumed that $\bar{W}_p = I$ for all values of $p$). The backward calculations can be carried out by using the following formula, which is the discrete adjoint equation corresponding to (14):

$$\tilde{q}_n = -D^T \tilde{q}_{n+1}, \tag{17}$$

where $D$ is defined in (15).

## 4.2   The Chemistry Sub-model and Its Adjoint Equation

There are no spatial derivatives in the chemical sub-model. If we consider this sub-model at the spatial grid-points used in the transport model, then a system of ordinary differential equations (ODEs) can be obtained. It is easily seen that in fact this systems of ODEs consists of several independent and smaller systems. Each of the smaller systems contains $q$ equations and the number of such systems is equal to the number $N_x + 1$ of spatial grid-points. It is quite sufficient to illustrate how any of the small systems can be handled numerically.

Denote by $\tilde{h}(t)$ and and $\tilde{f}(t, \tilde{h})$ the values $h(t, x_i)$ and $f(t, x_i, h)$ for any $i$, $i = 0, 1, \ldots, N_x$. Then the system of ODEs corresponding to the chosen index $i$ can be written as

$$\frac{d\tilde{h}}{dt} = \tilde{f}(t, \tilde{h}) \tag{18}$$

The system (18) contains $q$ equations and one such system has to be treated at every time-step and at every spatial grid-point. The system is stiff (discussion about different aspects of the concept of stiffness for systems of ODEs can be found, for example, in [11], [12], [14] and [23]) and therefore it has to be solved by using implicit numerical methods for stiff systems of ODEs. The use of the simple

Backward Euler Method, which has in some sense best stability properties, is based on the following formula:

$$\tilde{h}_{n+1} = \tilde{h}_n + \triangle t \tilde{f}(t_n, \tilde{h}_{n+1}), \tag{19}$$

where $\triangle t$ is the time-stepsize used and $\tilde{h}_n$ is an approximation of $\tilde{h}(t_n)$.

The adjoint equation of (19), which has to be handled in the backward mode, can be written in the following form:

$$\tilde{q}_n = \tilde{q}_{n+1} - \triangle t \left[ -\frac{\partial f(c_n)}{\partial c} \right]^T \tilde{q}_n, \tag{20}$$

where the adjoint variable is denoted by $\tilde{q}$.

### 4.3   Algorithmic Representation of the Calculations

The algorithm given in Fig. 1 can easily be modified for the particular problem described by (2). The resulting algorithm is given in Fig. 2.

Comparing the algorithm given in Fig. 1 with the algorithm given in Fig. 2, it is seen that there are four differences:

- The action "Perform forward steps by using the **sub-model 1**" from Fig. 1 is replaced in Fig. 2 by a loop carried out over the chemical species. This loop can be executed in parallel.
- The action "Perform forward steps by using the **sub-model 2**" from Fig. 1 is replaced in Fig. 2 by a loop carried out over the spatial grid-points. This loop can be executed in parallel.
- The action "Perform backward steps by using the **adjoint equation 2**" from Fig. 1 is replaced in Fig. 2 by a loop carried out over the spatial grid-points. This loop can be executed in parallel.
- The action "Perform backward steps by using the **adjoint equation 1**" from Fig. 1 is replaced in Fig. 2 by a loop carried out over the chemical species. This loop can be executed in parallel.

The algorithm shown in Fig. 2 illustrates an important fact: not only is the splitting procedure used leading to the treatment of smaller problems, but additionally **any parallel tasks appear in a very natural way**. This can be used to solve the problem more efficiently if a parallel computer is available.

### 4.4   Need of Synchronization of the Numerical Methods

It is desirable to synchronize the numerical methods that are used in the different parts of the data assimilation algorithm in order to get better performance. It is easy to see that the numerical methods used in the data assimilation algorithm presented in this section are not synchronized. Consider the order of accuracy of the numerical methods used in the different parts:

- The numerical method used in the treatment of the transport sub-model is of second order.

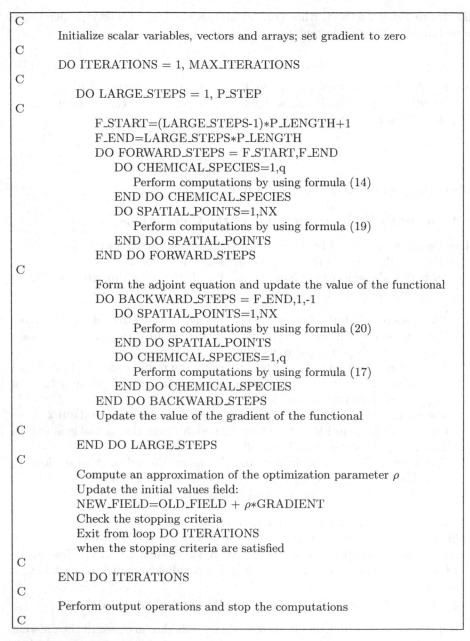

```
C
          Initialize scalar variables, vectors and arrays; set gradient to zero
C
          DO ITERATIONS = 1, MAX_ITERATIONS
C
             DO LARGE_STEPS = 1, P_STEP
C
                F_START=(LARGE_STEPS-1)*P_LENGTH+1
                F_END=LARGE_STEPS*P_LENGTH
                DO FORWARD_STEPS = F_START,F_END
                   DO CHEMICAL_SPECIES=1,q
                      Perform computations by using formula (14)
                   END DO CHEMICAL_SPECIES
                   DO SPATIAL_POINTS=1,NX
                      Perform computations by using formula (19)
                   END DO SPATIAL_POINTS
                END DO FORWARD_STEPS
C
                Form the adjoint equation and update the value of the functional
                DO BACKWARD_STEPS = F_END,1,-1
                   DO SPATIAL_POINTS=1,NX
                      Perform computations by using formula (20)
                   END DO SPATIAL_POINTS
                   DO CHEMICAL_SPECIES=1,q
                      Perform computations by using formula (17)
                   END DO CHEMICAL_SPECIES
                END DO BACKWARD_STEPS
                Update the value of the gradient of the functional
C
             END DO LARGE_STEPS
C
             Compute an approximation of the optimization parameter ρ
             Update the initial values field:
             NEW_FIELD=OLD_FIELD + ρ*GRADIENT
             Check the stopping criteria
             Exit from loop DO ITERATIONS
             when the stopping criteria are satisfied
C
          END DO ITERATIONS
C
          Perform output operations and stop the computations
C
```

**Fig. 2.** Adapting the algorithm presented in Fig. 1 to the simple transport-chemistry example studied in this section

- The Backward Euler Method, which is used in the treatment of the chemistry sub-model, is of first order.
- The sequential splitting procedure is of order one.

The order of accuracy $s$ of the combined method (methods for solving differential equation + splitting procedure) is usually given by the formula:

$$s = \min\{p_1, p_2, r\} \tag{21}$$

where (i) $p_1$ is the order of the numerical methods used in the treatment of the first sub-model, (ii) $p_2$ is the order of the numerical methods used in the treatment of the first sub-model and (iii) $r$ is the order of the splitting procedure.

Formula (21) shows that the order of the combined method is one. This indicates that if we are interested to have a combined method of order two, then it is necessary to use the following rules in the selection of numerical methods for solving differential equations and splitting procedures:

- A numerical method of order two has to be chosen for the treatment of the chemistry sub-model. Both the Trapezoidal Rule and the Implicit Mid-point Rule are obvious candidates. However, the stability properties of these two numerical methods are poorer than the stability properties of the used in this paper Backward Euler Method (see, for example, [11] and [12]). This means that if the chemical sub-model leads to system of very stiff ODEs (which is often the case), then it is not advisable to use these two methods. One can apply some Runge-Kutta method of order two with good stability properties, but these methods are as a rule much more expensive than the three methods mentioned above. This short discussion shows very clearly that the choice is, at least, not very easy.
- Also a splitting procedure of order two has to be chosen. A possible choice is the symmetric splitting procedure introduced by Marchuk and Strang (see, [17], and [22]). It should be mentioned here, however, that the symmetric splitting is more expensive computationally than the simpler sequential splitting procedure. Thus, also in this part the decision is not an easy task.

## 5   Concluding Remarks

A data assimilation technique consists of three computational parts:

- numerical method for solving differential equations,
- splitting procedures

and

- optimization algorithms.

These three parts were discussed in this paper. Many conditions are to be satisfied if one is interested in achieving an accurate, robust and fast algorithm. It is not very easy to satisfy all these conditions. On the other hand, however, the data assimilation techniques lead to huge computational tasks. Therefore, efficiency is highly desirable.

The stability properties of the data assimilation techniques are an important issue. Only a few results related to the stability are obtained until now. More research is needed in this direction. Some stability results are given in [2].

Better optimization algorithms are also highly desirable. The big problem is that most of the optimization subroutines are using vector norms in the stopping criteria. The problems arising in the the transport-chemical models are extremely badly scaled (see [24] and [25]). Therefore it is desirable to use relative component-wise checks in the stopping criteria.

# Acknowledgements

A grant (CPU-1101-17) from the Danish Centre for Scientific Computing (DCSC) gave us access to the Sun computers at the Technical University of Denmark. The members of the staff of DCSC helped us to resolve some difficult problems related to the efficient exploitation of the grid of Sun computers.

This project was also supported by the NATO Scientific Programme (Grant No. 980505).

# References

1. Alexandrov, V. N., Owczarz, W., Thomsen, P. G. and Zlatev, Z.: Parallel runs of a large air pollution model on a grid of Sun computers, Mathematics and Computers in Simulation, **65** (2004) 557-577.
2. Cao, Y., Shengtai, L., Petzold, L. and Serban, R.: Adjoint sensitivity anslysis for differential-algebraic equations: The adjoint DAE system and its numerical solution, SIAM J. Sci. Comput., **24** (2003) 1076-1089.
3. Djouad, R. and Sportisse, B.: Solving reduced chemical models in air pollution modelling, Applied Numerical Mathematics, **44** (2003) 49-61.
4. Ebel, A.: Chemical Transfer and Transport Modelling, In: Transport and Chemical Transformation of Pollutants in the Troposphere (Borrell, P. and Borrell, P. M., eds.), pp. 85-128, Springer, Berlin, 2000.
5. Ebel, A., Memmesheimer, M. and Jacobs, H.: Regional modelling of tropospheric ozone distributions and budgets, In: Atmospheric Ozone Dynamics: I. Global Environmental Change, (Varotsos, C., ed.), pp. 37-57. NATO ASI Series, Vol. 53, Springer, Berlin, 1997.
6. Elbern, H., Schmidt, H.: A four-dimensional variational chemistry data assimilation scheme for Eulerian chemistry transport modelling. Journal of Geophysical Research, **104** (1999) 18583–18598
7. Elbern, H., Schmidt, H., Talagrand, O., Ebel, A.: 4D-variational data assimilation with an adjoint air quality model for emission analysis. Environmental Modelling & Software, **15** (2000) 539–548.
8. Gery, M. W., Whitten, G. Z., Killus, J. P. and Dodge, M. C.: A photochemical kinetics mechanism for urban and regional modeling, Journal of Geophysical Research, **94** (1989) 12925-12956.
9. Gill, P. E. and Murray. W.: Conjugate-gradient methods for large-scale nonlinear optimization. Technical Report SOL 79-15. Department of Operations Research, Stanford University, Stanford, California, USA, 1979.

10. Gill, P. E., Murray. W. and Wright, M. H.: Practical Optimization. Academic Press, New York, 1981.
11. Hairer, E., Wanner, G.: Solving Ordinary Differential Equations, II: Stiff and Differential-algebraic Problems. Springer, Berlin-Heidelberg-New York-London, 1991.
12. Hundsdorfer, W., Verwer, J. G.: Numerical solution of time-dependent advection-diffusion-reaction equations. Springer, Berlin, 2003.
13. Jonson, J. E., Sundet, J. and Tarrason, L.: Model calculations of present and future levels of ozone and ozone precursors with a global and a regional model. Atmospheric Environment, **35** (2001) 525-537.
14. Lambert, J. D.: Numerical Methods for Ordinary Differential Equations. Wiley, Chichester-New York-Brisbane-Toronto-Singapore, 1991.
15. Le Dimet, F.-X., Navon, I. M., Daescu, D. N.: Second order information in data assimilation. Monthly Weather Review, **130** (2002) 629–648.
16. Lewis, J. M., Derber, J. C.: The use of adjoint equations to solve a variational adjustment problem with advective constraints. Tellus, **37A** (1985) 309–322
17. Marchuk, G. I.: Some application of splitting-up methods to the solution of mathematical physics problems. Applik. Mat., —bf 13 (1968) No. 2.
18. NAG Library Fortran Manual: E04 - minimizing and maximizing a function, http://www.nag.co.uk. Numerical Algorithms Group (NAG), Banbury Road 7, Oxford, England, 2004.
19. Peters, L. K., Berkowitz, C. M., Carmichael, G. R., Easter, R. C., Fairweather, G., Ghan, S. J., Hales, J. M., Leung, L. R., Pennell, W. R., Potra, F. A., Saylor, R. D. and Tsang, T. T.: The current state and future direction of Eulerian models in simulating tropospheric chemistry and transport of trace species: A review, Atmospheric Environment, **29** (1995) 189-222.
20. Sandu, A., Daescu, D. N., Carmichael, G. R., Chai, T.: Adjoint sensitivity analysis of regional air quality models. Journal of Computational Physics, 2005; to appear.
21. Simpson, D., Fagerli, H., Jonson, J. E., Tsyro, S. G., Wind, P., Tuovinen, J-P.: Transboundary Acidification, Eutrophication and Ground Level Ozone in Europe, Part I. Unified EMEP Model Description. EMEP/MSC-W Status Report 1/2003. Norwegian Meteorological Institute, Oslo, Norway, 2003.
22. Strang, G.: On the construction and comparison of difference schemes, SIAM J. Numer. Anal., **5** (1968) 505-517.
23. Zlatev, Z.: Modified diagonally implicit Runge-Kutta methods. SIAM Journal on Scientific and Statistical Computing, **2** (1981) 321-334.
24. Zlatev, Z.: Computer treatment of large air pollution models. Kluwer Academic Publishers, Dordrecht-Boston-London, 1995.
25. Zlatev, Z. and Dimov, I.: Computational and Environmental Challenges in Environmental Modelling", Studies in Computational Mathematics, Vol. 13, Elsevier Science, Amsterdam, 2006

# Weighted Iterative Operator-Splitting Methods: Stability-Theory

Jürgen Geiser

Humboldt-Universität zu Berlin
Department of Mathematics
Unter den Linden 6
D-10099 Berlin, Germany
geiser@mathematik.hu-berlin.de

**Abstract.** In the last years the need to solve complex physical models increased. Because of this motivation to solve complex models with efficient methods, we deal with advanced operator-splitting methods. They are based on weighted iterative operator-splitting methods and decouple complicate problems in simpler problems. The stability of the weighted splitting method is discussed and the efficiency of such methods. For the stiff-problems we present the A-stability property and the choice of the weighted parameters. The theory for the semi-discretized equations is introduced with respect to the gained ODE's. A general stability-theory for linearized operators is proposed and discussed for stiff-problems. Finally we concern the weighted operator-splitting methods for multi-dimensional and multi-physical problems.

## 1 Introduction

We motivate our studying on multi physics problems with decomposable processes, e.g. flow- and reaction parts. Decomposition methods provide an efficient methodology for solving PDE's, because of the possibility of adapting componentwise the discretization and solver methods to the local behavior of the solution. One could adapt the methods to different scales in time and space with respect to the physical behavior. Based on these contributions we present a flexible iterative operator-splitting method for applications on stiff partial differential equations. The possibility to decouple the full problem in simpler problems reduce the amount of computations. The discussion of the methods with respect to the stability and consistency is done to develop a flexible solver method. The stability is considered for commutative operators. We obtain stable methods in the sense of A-stability and could apply our method to well-known linear test-examples.

The paper is organized as follows. A mathematical model based on the convection-reaction equations is introduced in section 2. The iterative operator-splitting methods and the modifications to weighted methods are described in section 3. In section 4 we introduce the stability analysis of the methods and derive the A-stability. The numerical experiments are discussed in section 5. Finally we discuss our future works in the area of splitting and decomposition methods.

T. Boyanov et al. (Eds.): NMA 2006, LNCS 4310, pp. 40–47, 2007.

## 2   Mathematical Model

The motivation for the study presented below is coming from a computational simulation of time-dependent processes, for example in bio-remediation [1] or radioactive contaminants [4].

The mathematical equations are given by

$$\partial_t\, R\, c + \nabla \cdot \mathbf{v}c = f(c,t), \tag{1}$$

$$f(c,t) = \lambda(t)\, c\,,\ \text{time-dependent reactions}, \tag{2}$$

$$c_0 = c(x,0)\,,\ \text{initial condition}, \tag{3}$$

$$\mathbf{n} \cdot \mathbf{v}\, c = 0\,,\ \text{boundary condition}. \tag{4}$$

The unknown $c = c(x,t)$ is considered in $\Omega \times (0,T) \subset \mathbb{R}^d \times \mathbb{R}^+$, the space-dimension is given by $d$ . The parameter $R \in \mathbb{R}^+$ is constant and is named as retardation factor. The reaction $\lambda(t)$ is a nonlinear time-dependent functions. $\mathbf{v}$ is the divergence-free velocity.

In the following we describe the Operator-Splitting method for decoupling in two equation-parts in 2 operators as a basic tool for solving our equations.

## 3   Unsymmetric Weighted Iterative Splitting Method

The proposed unsymmetric weighted iterative splitting method is a combination between a sequential splitting method, see [6], and an iterative operator splitting method, see [2]. The weighting factor $\omega$ is used as an adaptive switch between lower and higher order splitting methods, see [3]. The following algorithm is based on the iteration with fixed splitting discretization step-size $\tau$. On the time interval $[t^n, t^{n+1}]$ we solve the following sub-problems consecutively for $i = 0, 2, \ldots 2m$.

$$\frac{dc_i(t)}{dt} = Ac_i(t) + \omega\, Bc_{i-1}(t),\ \text{with}\ c_i(t^n) = c^n \tag{5}$$

and $c_0(t^n) = c^n$ , $c_{-1} = 0.0$,

$$\frac{dc_{i+1}(t)}{dt} = \omega\, Ac_i(t) + Bc_{i+1}(t), \tag{6}$$

with $c_{i+1}(t^n) = \omega\, c^n + (1 - \omega)\, c_i(t^{n+1})$,

where $c^n$ is the known split approximation at the time level $t = t^n$. The split approximation at the time-level $t = t^{n+1}$ is defined as $c^{n+1} = c_{2m+1}(t^{n+1})$. Our parameter $\omega \in [0, 1]$. For $\omega = 0$ we have the sequential-splitting and for $\omega = 1$ we have the iterative splitting method, cf. [2].

Because of the weighting between the sequential splitting and iterative splitting method, also the initial-conditions are weighted. So, we have the final results of the first equation (5) appearing in the initial condition for the second (6).

## 4  Stability Theory

We concentrate on the stability theory for the linear ordinary differential equations with commutative operators. First we apply the recursion for the general case and obtain the commutative case.

### 4.1  Recursion

We study the stability for the linear system (5) and (6). We treat the special case for the initial-values with $c_i(t^n) = c^n$ and $c_{i+1}(t^n) = c^n$ for an overview. The general case $c_{i+1}(t^n) = \omega c^n + (1 - \omega)c_i(t^{n+1})$ could be treated in the same manner.

We consider the suitable vector norm $|| \cdot ||$ on $\mathbb{R}^M$, together with its induced operator norm. The matrix exponential of $Z \in \mathbb{R}^{M \times M}$ is denoted by $\exp(Z)$. We assume that

$$|| \exp(\tau A)|| \leq 1 \quad \text{and} \quad || \exp(\tau B)|| \leq 1 \quad \text{for all} \quad \tau > 0.$$

It can be shown that the system (5)–(6) implies $|| \exp(\tau (A + B))|| \leq 1$ and is itself stable.

For the linear problem (5) and (6) it follows by integration that

$$c_i(t) = \exp((t - t^n)A)c^n + \int_{t^n}^{t} \exp((t - s)A) \, \omega \, Bc_{i-1}(s) \, ds, \tag{7}$$

$$c_{i+1}(t) = \exp((t - t^n)B)c^n + \int_{t^n}^{t} \exp((t - s)B) \, \omega \, Ac_i(s) \, ds. \tag{8}$$

With elimination of $c_i$ we get

$$c_{i+1}(t) = \exp((t - t^n)B)c^n + \omega \int_{t^n}^{t} \exp((t - s)B) \, A \, \exp((s - t^n)A) \, c^n \, ds$$
$$+\omega^2 \int_{s=t^n}^{t} \int_{s'=t^n}^{s} \exp((t - s)B) \, A \, \exp((s - s')A) \, B \, c_{i-1}(s') \, ds' \, ds. \tag{9}$$

For the following commuting case we could evaluate the double integral $\int_{s=t^n}^{t} \int_{s'=t^n}^{s}$ as $\int_{s'=t^n}^{t} \int_{s=s'}^{t}$ and could derive the weighted stability-theory.

### 4.2  Commuting Operators

For more transparency of the formula (9) we consider a well-conditioned system of eigenvectors and the eigenvalues $\lambda_1$ of $A$ and $\lambda_2$ of $B$ instead of the operators $A, B$ themselves. Replacing the operators $A$ and $B$ by $\lambda_1$ and $\lambda_2$ respectively, we obtain after some calculations

$$c_{i+1}(t) = c^n \frac{1}{\lambda_1 - \lambda_2} \left( \omega \lambda_1 \exp((t - t^n)\lambda_1) + ((1 - \omega)\lambda_1 - \lambda_2) \exp((t - t^n)\lambda_2) \right)$$

$$+ c^n \, \omega^2 \, \frac{\lambda_1 \lambda_2}{\lambda_1 - \lambda_2} \int_{s=t^n}^{t} (\exp((t - s)\lambda_1) - \exp((t - s)\lambda_2)) \, ds. \tag{10}$$

Note that this relation is symmetric in $\lambda_1$ and $\lambda_2$.

**A-Stability.** We define $z_k = \tau\lambda_k$, $k = 1, 2$. We start with $c_0(t) = u^n$ and we obtain

$$c_{2m}(t^{n+1}) = S_m(z_1, z_2)\, c^n, \tag{11}$$

where $S_m$ is the stability function of the scheme with $m$-iterations. We use (10) and obtain after some calculations

$$S_1(z_1, z_2) = \omega^2\, c^n + \frac{\omega\, z_1 + \omega^2\, z_2}{z_1 - z_2}\, \exp(z_1)\, c^n \tag{12}$$
$$+ \frac{(1 - \omega - \omega^2)\, z_1 - z_2}{z_1 - z_2}\, \exp(z_2)\, c^n,$$

$$S_2(z_1, z_2) = \omega^4\, c^n + \frac{\omega\, z_1 + \omega^4\, z_2}{z_1 - z_2}\, \exp(z_1)\, c^n \tag{13}$$
$$+ \frac{(1 - \omega - \omega^4)\, z_1 - z_2}{z_1 - z_2}\, \exp(z_2)\, c^n$$
$$+ \frac{\omega^2\, z_1\, z_2}{(z_1 - z_2)^2}\, ((\omega z_1 + \omega^2 z_2)\exp(z_1)$$
$$+ (-(1 - \omega - \omega^2)z_1 + z_2)\exp(z_2))\, c^n$$
$$+ \frac{\omega^2\, z_1\, z_2}{(z_1 - z_2)^3}\, ((-\omega z_1 - \omega^2 z_2)(\exp(z_1) - \exp(z_2))$$
$$+ ((1 - \omega - \omega^2)z_1 - z_2)(\exp(z_1) - \exp(z_2)))\, c^n.$$

Let us consider the stability given by the following eigenvalues in a wedge

$$\mathcal{W} = \{\zeta \in \mathbb{C} : |\arg(\zeta)| \le \alpha\}.$$

For stability we have $|S_m(z_1, z_2)| \le 1$ whenever $z_1, z_2 \in \mathcal{W}_{\pi/2}$.

In the following theorem the stability is given for the first two iteration steps of the unsymmetric weighted iterative splitting method.

**Theorem 1.** *We have the stability given as:*

*For $S_1$ we have A-stability with*
$\max_{z_1 \le 0, z_2 \in W_\alpha} |S_1(z_1, z_2)| \le 1$, $\forall\, \alpha \in [0, \pi/2]$ *with* $\omega = \frac{1}{\sqrt[4]{3}}$.

*For $S_2$ we have A-stability with*
$\max_{z_1 \le 0, z_2 \in W_\alpha} |S_2(z_1, z_2)| \le 1$, $\forall\, \alpha \in [0, \pi/2]$ *with* $\omega \le \left(\frac{1}{8\,\tan^2(\alpha)+1}\right)^{1/8}$.

*Proof.* We consider a fixed $z_1 = z$, $Re(z) < 0$ and $z_2 \to -\infty$. Then we obtain

$$S_1(z, \infty) = \omega^2(1 - e^z) \tag{14}$$
$$\text{and}\ \ S_2(z, \infty) = \omega^4(1 - (1 - z)e^z). \tag{15}$$

If $z = x + iy$, $x < 0$ then we obtain:

1.) For $S_1$ we get

$$|S_1(z,\infty)|^2 = \omega^4(1 - 2\exp(x)\cos(y) + \exp(2x)), \tag{16}$$

and hence $|S_1(z,\infty)| \leq 1 \Leftrightarrow \omega^4 \leq \dfrac{1}{1 - 2\exp(x)\cos(y) + \exp(2x)}.$ (17)

Because of $x < 0$ and $y \in \mathbb{R}$ we could estimate $-2 \leq 2\exp(x)\cos(y)$ and $\exp(2x) \geq 0$. From (17) we obtain $\omega \leq \frac{1}{\sqrt[4]{3}}$.

2.) For $S_2$ we get

$$|S_2(z,\infty)|^2 = \omega^8\{1 - 2\exp(x)[(1-x)\cos(y) + y\sin(y)] \tag{18}$$
$$+ \exp(2x)[(1-x)^2 + y^2]\}.$$

After some calculations we could obtain

$$|S_2(z,\infty)| \leq 1 \Leftrightarrow \exp(x) \leq (\frac{1}{\omega^8} - 1)\frac{\exp(-x)}{(1-x)^2 + y^2} + 2\frac{|1-x| + |y|}{(1-x)^2 + y^2}. \tag{19}$$

We could estimate for $x < 0$ and $y \in \mathbb{R}$ : $\frac{|1-x|+|y|}{(1-x)^2+y^2} \leq 3/2$ and $\frac{1}{2\tan^2(\alpha)} < \frac{\exp(-x)}{(1-x)^2+y^2}$ where $\tan(\alpha) = y/x$. Finally, we get the bound $\omega \leq \left(\frac{1}{8\tan^2(\alpha)+1}\right)^{1/8}$.

*Remark 1.* The stability is derived for ordinary differential equations with linear operators. For applications in linear partial differential equations we assume a discretization of the spatial operators, so that we obtain a system of linear ordinary differential equations. These equations can be treated as described below.

# 5   Numerical Results

## 5.1   First Example

We deal with a first order partial differential-equation given as a transport equation in the following example

$$\partial_t u_1 = -v_1\partial_x u_1 - \lambda_1 u_1, \tag{20}$$

$$\partial_t u_2 = -v_2\partial_x u_2 + \lambda_1 u_1 - \lambda_2 u_2, \tag{21}$$

$$u_1(x,0) = \begin{cases} 1 \text{ , for } 0.1 \leq x \leq 0.3, \\ 0 \text{ , otherwise,} \end{cases} \tag{22}$$

$$u_2(x,0) = 0 \text{ , for } x \in [0,X], \tag{23}$$

where $\lambda_1, \lambda_2 \in \mathbb{R}^+$ and $v_1, v_2 \in \mathbb{R}^+$. We have the time-interval $t \in [0,T]$ and the space-interval $x \in [0,X]$.

The analytical solutions are given in [5].

We rewrite the equation-system (20)–(23) in operator notation, and end up with the following equations

$$\partial_t u = Au + Bu, \tag{24}$$

$$u(x,0) = \begin{cases} (1,0)^T \text{ , for } 0.1 \leq x \leq 0.3, \\ (0,0)^T \text{ , otherwise,} \end{cases} \tag{25}$$

where $u = (u_1, u_2)^T$. Our splitted operators are

$$A = \begin{pmatrix} -v_1 \partial_x & 0 \\ 0 & -v_2 \partial_x \end{pmatrix}, \quad B = \begin{pmatrix} -\lambda_1 & 0 \\ \lambda_1 & -\lambda_2 \end{pmatrix}. \tag{26}$$

We use the finite difference method as spatial discretization method and solve the time-discretization analytically.

The spatial discretization is done as follows, we concentrate on the interval $x \in [0, 1.5]$ and we consider a uniform partition of it with step $\Delta x = 0.1$. For the transport-term we use an upwind finite difference discretization given as :

$$\partial_x u_i = \frac{u_i - u_{i-1}}{\Delta x}. \tag{27}$$

For initial-values we use the given impulses

$$u_1(x) = \begin{cases} 1 & , \; 0.1 \le x \le 0.3 \\ 0 & , \; \text{otherwise} \end{cases} \tag{28}$$

$$\text{and} \;\; u_2(x) = 0 \, , \; x \in [0, 1.5]. \tag{29}$$

In the following equations we write the iterative operator splitting algorithm taking into account the discretization in space. The time-discretization is solved analytically. For time-integration we apply implicit Euler methods for the semi-discretized equations.

For the parameters of equations (20)–(23) we use $\lambda_1 = 0.01$, $\lambda_2 = 10^4$, $v_1 = 1.0$ and $v_2 = 0.5$. For the time-interval $t \in [0, 1]$ we apply the time-steps $\Delta t = 1.0$ and $\Delta t = 0.2$.

For the end-time $t_{end} = 1$, we check the results for the end-point $x_1 = 1.0$. We get the exact solution of our equation, see [5]

$$u_1(x_1, t_{end}) = 9.9004 \times 10^{-1} \, , \; u_2(x_1, t_{end}) = 9.901 \times 10^{-7}.$$

We present the computed results in Table 1. The results are improved with the selected weighting factors $\omega$. For weighting factors around 0.3 we can stabilize

**Table 1.** Numerical results for the first example with the weighted method

| Number of time-part. | Iterative Steps | approx$_1$ | approx$_2$ | error$_1$ | error$_2$ | $\omega$ |
|---|---|---|---|---|---|---|
| 1 | 2 | 0.9010315283 | 0.0000179191 | $8.900847 \times 10^{-2}$ | $1.692903 \times 10^{-5}$ | 0.3 |
| 1 | 2 | 0.9028283841 | 0.0000494056 | $8.721162 \times 10^{-2}$ | $4.841555 \times 10^{-5}$ | 0.5 |
| 1 | 2 | 0.9040595631 | 0.0000710110 | $8.598044 \times 10^{-2}$ | $7.002093 \times 10^{-5}$ | 0.6 |
| 1 | 2 | 0.9055125761 | 0.0000965245 | $8.452742 \times 10^{-2}$ | $9.553440 \times 10^{-5}$ | 0.7 |
| 1 | 4 | 0.9000930952 | $-0.0018767031$ | $8.994690 \times 10^{-2}$ | $1.877693 \times 10^{-3}$ | 0.3 |
| 1 | 4 | 0.9006999095 | $-0.0142537213$ | $8.934009 \times 10^{-2}$ | $1.425471 \times 10^{-2}$ | 0.5 |
| 5 | 2 | 0.9227053726 | 0.0505299263 | $6.733463 \times 10^{-2}$ | $5.052894 \times 10^{-2}$ | 0.3 |
| 5 | 2 | 1.2627573122 | 0.0009437033 | $2.727173 \times 10^{-1}$ | $9.427132 \times 10^{-4}$ | 0.5 |

our method for small time-steps, but we have problems with large time-steps. A better stabilization for larger time-steps is obtained with values of $\omega$ around 0.5. A balance between the number of time-partitions and iterative steps can lead to an optimal weighting factor.

## 5.2   Second Test-Example

We deal with a nonlinear ordinary differential equation given as

$$\frac{du(t)}{dt} = -(1 + t + t^2/2)u(t) , \tag{30}$$

$$u(0) = 1 , \text{ (initial conditions)} , \tag{31}$$

where we have the time-interval $t \in [0, 2]$. The analytical solution is:

$$u(t) = \exp(-(t + t^2/2 + t^3/6)), \text{ for } 0 \leq t \leq 2 , \tag{32}$$

We rewrite the equation-system (30)–(31) in operator notation, and end up with the following equations :

$$\frac{du(t)}{dt} = A(t)u(t) + B(t)u(t) , \quad u(0) = 1, \tag{33}$$

Our splitted operators are

$$B(t) = -t^2/2 \text{ quadratic part.} \tag{34}$$

According to the weighted splitting method, we divide our system of ODE's in step $i$ and $i + 1$ as following

$$\frac{du_i(t)}{dt} = -(1 + t)u_i(t) - \omega(t^2/2)u_{i-1}(t) ,$$

$$\frac{du_{i+1}(t)}{dt} = -\omega(1 + t)u_i(t) - (t^2/2)u_{i+1}(t) ,$$

with $u_i(0) = 1$ and $u_{i+1}(0) = \omega + (1 - \omega)u_i(t^{n+1})$.

For the steps $i$ and $i + 1$ we can derive analytical solutions and apply them in our numerical scheme.

The numerical results are compared for the end-time $t = 2$ with the exact solution given as 0.0048279500. Our results are presented in Table 2.

Because of the time-dependent problem one of the best weighting-factor $\omega$ is around 0.3 and one shift to a lower order method. In table 2 we can see that the computed results are positive with more time-partitions and more iteration-steps and we obtain more the analytical results. It can also be found an optimal weighting factor is a balance between the time-steps and the iterative steps.

**Table 2.** Numerical results for the second example with the weighted method

| Number of time-part. | Iterative Steps | approx | error | $\omega$ |
|---|---|---|---|---|
| 5 | 2 | $-0.0081201319$ | $1.294808 \times 10^{-2}$ | 0.3 |
| 5 | 4 | $-0.0081201324$ | $1.294808 \times 10^{-2}$ | 0.3 |
| 10 | 2 | $0.0683816109$ | $6.355366 \times 10^{-2}$ | 0.3 |
| 10 | 2 | $0.0208173736$ | $1.598942 \times 10^{-2}$ | 0.5 |
| 10 | 4 | $0.0683816109$ | $6.355366 \times 10^{-2}$ | 0.3 |
| 10 | 4 | $0.0208173736$ | $1.598942 \times 10^{-2}$ | 0.5 |

# 6    Conclusion and Discussions

We present a modified iterative Operator-Splitting method and we could study the behavior for the stiff case. Because of the weighting factor we obtain stabilized results and a good choice between iteration-steps and weighting factors is possible. As in the analysis presented, we can see an good result with $2 - 4$ iteration-steps and weighting factors of $0.5 - 0.6$. In the next step the discussion for nonlinear operators with respect to applications in fluid-dynamics is proposed.

# References

1. R.E. Ewing. Up-scaling of biological processes and multiphase flow in porous media. *IIMA Volumes in Mathematics and its Applications*, Springer-Verlag, 295 (2002), 195-215.
2. I.Farago, J.Geiser, "Iterative Operator-Splitting Methods for Linear Problems", *IJCS, International Journal of Computational Sciences*, Vol. 1, Nos. 1/2/3, pp. 64-74, October 2005.
3. J. Geiser, C. Kravvaritis, "Weighted Iterative Operator-Splitting Methods for stiff problems in complex applications", *Preprint at Humboldt-University*, April 2006.
4. J. Geiser. *Numerical Simulation of a Model for Transport and Reaction of Radionuclides*. Proceedings of the Large Scale Scientific Computations of Engineering and Environmental Problems, Sozopol, Bulgaria, 2001.
5. J. Geiser. *Gekoppelte Diskretisierungsverfahren für Systeme von Konvektions-Dispersions-Diffusions-Reaktionsgleichungen*. Doktor-Arbeit, Universität Heidelberg, 2003.
6. D. Lanser and J.G. Verwer. *Analysis of Operator Splitting for advection-diffusion-reaction problems from air pollution modelling*. Journal of Computational Applied Mathematics, 111(1-2):201–216, 1999.

# Weighted Iterative Operator-Splitting Methods and Applications

Jürgen Geiser[1] and Christos Kravvaritis[2]

[1] Humboldt-Universität zu Berlin
Department of Mathematics
Unter den Linden 6
D-10099 Berlin, Germany
geiser@mathematik.hu-berlin.de
[2] Department of Mathematics, University of Athens,
Panepistimiopolis 15784, Athens, Greece
ckrav@math.uoa.gr

**Abstract.** The subject of our research is to solve accurately ODEs, which appear in mathematical models arising from several physical processes. For this purpose we develop a new class of weighted iterative operator splitting methods. We present applications to systems of linear ODEs, which might contain also stiff parameters. The benefit of the proposed method is demonstrated with regard to convergence results and comparison to analytical solutions. We provide improved results and convergence rates in comparison with classical operator splitting methods.

## 1 Introduction

Mathematical equations representing complex models with coupled processes often involve transport and reaction equations. An important design principle for many successful numerical methods for such systems is *operator splitting* (OS). The efficiency of OS lies in the possibility to deal with simpler problems and solve them with adapted methods. We present a new weighted iterative OS method and confirm its effectiveness with two applications on systems of ODEs. Weighted OS schemes are based on the idea that the methods can be stabilized with some weighting factor, which actually weights between first order and higher order OS methods, and lead to better approximations with few iterative steps. The numerical results show that the proposed scheme offers higher accuracy compared with already known OS methods.

## 2 Mathematical Model

The motivation for our research originates from a computational simulation of bio-remediation or radioactive contaminants [1]. The mathematical model is illustrated by following equations

$$R \frac{\partial c}{\partial t} + \nabla \cdot (\mathbf{v}c - \mathbf{D}\nabla c) = f(c) ,$$

T. Boyanov et al. (Eds.): NMA 2006, LNCS 4310, pp. 48–55, 2007.

$$f(c) = c^p, \ p > 0, \ \text{chemical-reaction}$$
$$f(c) = \frac{c}{1-c}, \ \text{bio-remediation}$$

The unknown function $c = c(x,t)$ is considered in $\Omega \times (0,T) \subset \mathrm{IR}^d \times \mathrm{IR}$. The parameter $R \in \mathrm{IR}^+$ is constant and is named retardation factor. The functions $f(c)$ are nonlinear, for example bio-remediation or chemical reaction. $\mathbf{D}$ is the Scheidegger diffusion-dispersion tensor and $\mathbf{v}$ is the velocity. In this work we deal only with linear systems of ODEs in order to verify the effectiveness of our approach and it is a subject currently under research how to apply these ideas to more complicated nonlinear problems, as the one mentioned above.

## 3   The New Weighted Iterative Splitting Method

Our goal is to improve the convergence of the results for the iterative OS method [2], which is a traditional, powerful concept used in many diverse fields of applied mathematics for the design of effective numerical schemes. We focus our study on Cauchy problems of the form

$$\frac{dc(t)}{dt} = A\,c(t) + B\,c(t), \ \ t \in (0,T), \ c(0) = c_0,$$

where $A$ and $B$ are linear operators represented by matrices.

The following algorithm is based on the iteration with fixed splitting discretization step-size $\Delta t = t^{n+1} - t^n$, which is actually the step-size of a uniform partition $0 = t^0 < t^1 < \ldots < t^{N-1} < t^N = T$ of the time interval $[0,T]$. On the intervals $[t^n, t^{n+1}]$, $n = 0, 1, \ldots N - 1$, we solve the following sub-problems consecutively for $i = 0, 2, \ldots 2m$.

The Initial idea is the unsymmetric weighted iterative splitting method:

$$\frac{dc_i(t)}{dt} = Ac_i(t) + \omega\, Bc_{i-1}(t), \ \text{with} \ c_i(t^n) = c^n \tag{1}$$
and $c_0(t^n) = c^n$ , $c_{-1} = 0$,
$$\frac{dc_{i+1}(t)}{dt} = \omega\, Ac_i(t) + Bc_{i+1}(t), \tag{2}$$
with $c_{i+1}(t^n) = \omega\, c^n + (1 - \omega)\, c_i(t^{n+1})$ ,

where $c^n$ is the known split approximation at the time level $t = t^n$. The split approximation at the time-level $t = t^{n+1}$ is defined as $c^{n+1} = c_{2m+1}(t^{n+1})$. The parameter $\omega \in [0,1]$. For $\omega = 0$ we have the sequential splitting and for $\omega = 1$ we have the iterative splitting method, cf. [2].

Because of the weighting between the sequential splitting and iterative splitting method, also the initial-conditions are weighted. So, we have the final results of the first equation (1) appearing in the initial condition for the second (2). This initial weighting idea faces problems in the convergence analysis, because of the unsymmetry. We are led to construct a new weighted splitting method, according to the following scheme:

The new idea is the symmetric weighted iterative splitting method:

$$\frac{dc_i(t)}{dt} = 2\omega A c_i(t) + (1 - 2\omega) B c_{i-1}(t), \tag{3}$$

$$\text{with } c_i(t^n) = c^n \text{ and } c_0(t^n) = c^n , \; c_{-1} = 0,$$

$$\frac{dc_{i+1}(t)}{dt} = (1 - 2\omega) A c_i(t) + 2\omega B c_{i+1}(t), \tag{4}$$

$$\text{with } c_{i+1}(t^n) = 2\omega \, c^n + (1 - 2\omega) \, c_i(t^{n+1}) ,$$

where $c^n$ is the known split approximation at the time level $t = t^n$. The split approximation at the time-level $t = t^{n+1}$ is defined as $c^{n+1} = c_{2m+1}(t^{n+1})$. The parameter $\omega \in [0,1]$. For $\omega = 0$ we have the sequential splitting and for $\omega = 0.5$ we have the iterative splitting method, cf. [2].

From a software development point of view, the above described numerical scheme can be realized in a stepwise manner, starting with a simple solver for each subproblem and then replacing each solver independently of the other by a more advanced solver, until a desired level of sophistication is reached.

## 4   Numerical Results

### 4.1   First Test-Example of an ODE

In order to verify the efficiency of our proposed scheme we concentrate on a simple system of ODEs. We could then study the behavior for stiff-problems when $\lambda_2 >> \lambda_1 \approx 0$.

$$\frac{du_1(t)}{dt} = -\lambda_1 u_1 + \lambda_2 u_2 , \tag{5}$$

$$\frac{du_2(t)}{dt} = \lambda_1 u_1 - \lambda_2 u_2 , \tag{6}$$

$$u_1(0) = 1 , \; u_2(0) = 1 \text{ (initial conditions) }, \tag{7}$$

where $\lambda_1 = 0.04$ and $\lambda_2 = 1 \; 10^4$ are the decay factors. The time-interval is $t \in [0,1]$.

We rewrite the equation-system (5)–(7) in operator notation, and end up with the following equations :

$$\frac{du}{dt} = Au + Bu , \quad u(0) = (1,1)^T ,$$

where $u(t) = (u_1(t), u_2(t))^T$ for $t \in [0,1]$ and our splitted operators are

$$A = \begin{pmatrix} -\lambda_1 & \lambda_2 \\ 0 & 0 \end{pmatrix} , \; B = \begin{pmatrix} 0 & 0 \\ \lambda_1 & -\lambda_2 \end{pmatrix} .$$

For the system of ODEs (5)–(7) we can derive the analytical solution by integrating it :

$$u_1(t) = u_{10} + u_{20} \exp(-(\lambda_1 + \lambda_2)t) ,$$

$$u_2(t) = \frac{\lambda_1}{\lambda_2} u_{10} - u_{20} \exp(-(\lambda_1 + \lambda_2)t) ,$$

According to the new second weighted splitting method, we divide our system of ODEs in step $i$ and $i+1$ as follows

Step $i$

$$\frac{du_1^i}{dt} = -2\omega\lambda_1 u_1^i + 2\omega\lambda_2 u_2^i\ ,$$

$$\frac{du_2^i}{dt} = (1-2\omega)\lambda_1 u_1^{i-1} - (1-2\omega)\lambda_2 u_2^{i-1}\ ,$$

$$u_1^i(0) = u_{10}\ ,\ u_2^i(0) = u_{20}\ ,$$

Step $i+1$

$$\frac{du_1^{i+1}}{dt} = -(1-2\omega)\lambda_1 u_1^i + (1-2\omega)\lambda_2 u_2^i\ ,$$

$$\frac{du_2^{i+1}}{dt} = 2\omega\lambda_1 u_1^{i+1} - 2\omega\lambda_2 u_2^{i+1}\ ,$$

$$u_1^{i+1}(0) = u_{10}'\ ,\ u_2^{i+1}(0) = u_{20}'\ ,$$

where $t \in [0, \Delta t]$, $u_{10}' = 2\omega u_{10} + (1-2\omega)u_1^i(t^{n+1})$ and $u_{20}' = 2\omega u_{20} + (1-2\omega)u_2^i(t^{n+1})$.

For steps $i$ and $i+1$ we can derive analytical solutions and apply them in our numerical scheme. The analytical solutions are given as

$$u_1^i(t) = u_{10}\exp(-2\omega\lambda_1 t) + u_{20}\frac{\lambda_2}{\lambda_1} + u_1^{i-1}(t)\left[(1-2\omega)\lambda_2 t - \frac{1-2\omega}{2\omega}\frac{\lambda_2}{\lambda_1}\right]$$

$$+ u_2^{i-1}(t)\left[-(1-2\omega)\frac{\lambda_2^2}{\lambda_1}t + \frac{1-2\omega}{2\omega}\frac{\lambda_2^2}{\lambda_1^2}\right]\ ,$$

$$u_2^i(t) = (1-2\omega)(\lambda_1 u_1^{i-1}(t) - \lambda_2 u_2^{i-1}(t))t + u_{20},$$

and

$$u_1^{i+1}(t) = (2\omega-1)(\lambda_1 u_1^i(t) - \lambda_2 u_2^i(t))t + u_{20}',$$

$$u_2^{i+1}(t) = u_{20}'\exp(-2\omega\lambda_2 t) + u_{10}'\frac{\lambda_1}{\lambda_2} + u_2^i(t)\left[(1-2\omega)\lambda_1 t - \frac{1-2\omega}{2\omega}\frac{\lambda_1}{\lambda_2}\right]$$

$$+ u_1^i(t)\left[-(1-2\omega)\frac{\lambda_1^2}{\lambda_2}t + \frac{1-2\omega}{2\omega}\frac{\lambda_1^2}{\lambda_2^2}\right]\ ,$$

We compute with our given scheme and compare with the values of the analytical solutions at the end-time $t = 1$, which are $u_{1,exact} = 1$ and $u_{2,exact} = 4 \cdot 10^{-6}$. The numerical results are presented in Table 1, for $\omega = 0.6$ and $0.9$. Table 2 shows the results of the traditional iterative method for comparison. Figure 1 shows the behavior of the error for the solution $u_2$ as a function of the number of iterations, for several values of $\omega$. We see clearly that if we have 10 or more time partitions, the method provides convergence for all values of $\omega$. A closer examination informs that $\omega = 0.6$ is the optimal value of $\omega$ for this specific example.

**Table 1.** Numerical results for the first example with the new weighted method

| Number of time-partitions | Iterative Steps | $err_1$ | $err_2$ | $\omega$ |
|---|---|---|---|---|
| 1 | 2 | $1.000299 \times 10^0$ | $1.195833 \times 10^{-9}$ | 0.6 |
| 1 | 2 | $1.001168 \times 10^0$ | $4.670030 \times 10^{-9}$ | 0.9 |
| 1 | 10 | $1.000308 \times 10^0$ | $1.230005 \times 10^{-9}$ | 0.6 |
| 1 | 10 | $1.001455 \times 10^0$ | $5.817809 \times 10^{-9}$ | 0.9 |
| 1 | 200 | $1.000308 \times 10^0$ | $1.230005 \times 10^{-9}$ | 0.6 |
| 1 | 200 | $1.001455 \times 10^0$ | $5.819514 \times 10^{-9}$ | 0.9 |
| 10 | 2 | $3.428597 \times 10^1$ | $1.330296 \times 10^{-4}$ | 0.6 |
| 10 | 2 | $4.430855 \times 10^2$ | $1.767357 \times 10^{-3}$ | 0.9 |
| 10 | 10 | $2.311510 \times 10^{-5}$ | $3.999908 \times 10^{-6}$ | 0.6 |
| 10 | 10 | $5.415010 \times 10^{-1}$ | $1.835200 \times 10^{-6}$ | 0.9 |
| 10 | 200 | $3.269149 \times 10^{-6}$ | $3.999987 \times 10^{-6}$ | 0.6 |
| 10 | 200 | $1.580502 \times 10^{-5}$ | $3.999937 \times 10^{-6}$ | 0.9 |
| 150 | 2 | $2.285723 \times 10^0$ | $5.028608 \times 10^{-6}$ | 0.6 |
| 150 | 2 | $2.953858 \times 10^1$ | $1.131697 \times 10^{-4}$ | 0.9 |
| 150 | 10 | $1.337681 \times 10^{-6}$ | $3.999995 \times 10^{-6}$ | 0.6 |
| 150 | 10 | $3.609851 \times 10^{-2}$ | $3.856809 \times 10^{-6}$ | 0.9 |
| 150 | 200 | $1.462200 \times 10^{-8}$ | $4.000000 \times 10^{-6}$ | 0.6 |
| 150 | 200 | $7.084913 \times 10^{-8}$ | $4.000000 \times 10^{-6}$ | 0.9 |

**Table 2.** Numerical results for the first example with the iterative method

| Number of time-partitions | Iterative Steps | $err_1$ | $err_2$ |
|---|---|---|---|
| 1 | 2 | $9.607895 \times 10^3$ | $3.842758 \times 10^{-2}$ |
| 1 | 10 | $9.607894 \times 10^3$ | $3.842757 \times 10^{-2}$ |
| 1 | 200 | $9.607894 \times 10^3$ | $3.842757 \times 10^{-2}$ |
| 10 | 2 | $9.896297 \times 10^3$ | $1.027203 \times 10^0$ |
| 10 | 10 | $2.548589 \times 10^8$ | $7.638750 \times 10^2$ |
| 10 | 200 | $2.548589 \times 10^8$ | $7.638750 \times 10^2$ |
| 150 | 2 | $4.000800 \times 10^{-4}$ | $1.600320 \times 10^{-9}$ |
| 150 | 10 | $3.809891 \times 10^{-2}$ | $3.047906 \times 10^{-7}$ |
| 150 | 200 | $3.809891 \times 10^{-2}$ | $3.047906 \times 10^{-7}$ |

## 4.2  Second Test-Example of an ODE

We study another ODE and separate the complex operator in two simpler operators.

$$\frac{du_1(t)}{dt} = -16u_1 + 12u_2 + 16\cos(t) - 13\sin(t) , \tag{8}$$

$$\frac{du_2(t)}{dt} = 12u_1 - 9u_2 - 11\cos(t) + 9\sin(t) , \tag{9}$$

$$u_1(0) = 1 , \ u_2(0) = 0 \ \text{(initial conditions)} , \tag{10}$$

where the time-interval is $t \in [0, \pi/4]$.

For the equation-system (8)–(10) we can derive the analytical solution:

$$u_1(t) = \cos(t), \quad u_2(t) = \sin(t)$$

At the end-point $t = \pi/4$ we have $u_{1,exact} = \frac{\sqrt{2}}{2}$, $u_{2,exact} = \frac{\sqrt{2}}{2}$.

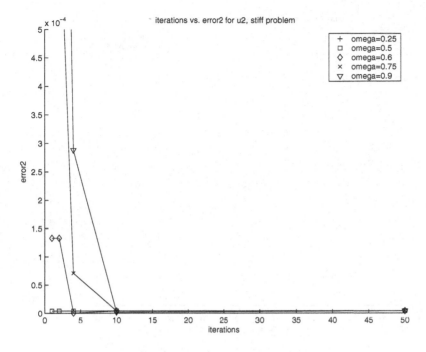

**Fig. 1.**

We rewrite the equation-system (8)–(10) in operator notation, and end up with the following equations :

$$\frac{du}{dt} = A(u) + B(u) , \quad u(0) = (1,0)^T,$$

where $u(t) = (u_1(t), u_2(t))^T$ for $t \in [0, \pi/4]$. Due to the singularity of this example, we must choose among several possibilities the optimal assignment for the splitted operators. We select

$$A(u) = \begin{pmatrix} -16u_1 - 13\sin(t) \\ 12u_1 + 9\sin(t) \end{pmatrix} , \quad B(u) = \begin{pmatrix} 12u_2 + 16\cos(t) \\ -9u_2 - 11\cos(t) \end{pmatrix} .$$

For the sake of simplicity and for economy of space we write here the operator-splitting scheme and the solutions for every step only for the iterative method. The new weighted method is applied absolutely similarly, according to the scheme (3)–(4).

Step $i$

$$\frac{du_{1,i}}{dt} = -16u_{1,i} + 12u_{2,i-1} + 16\cos(t) - 13\sin(t) ,$$

$$\frac{du_{2,i}}{dt} = 12u_{1,i} - 9u_{2,i-1} - 11\cos(t) + 9\sin(t) ,$$

$$u_1^i(t^n) = 1 , \quad u_2^i(t^n) = 0 ,$$

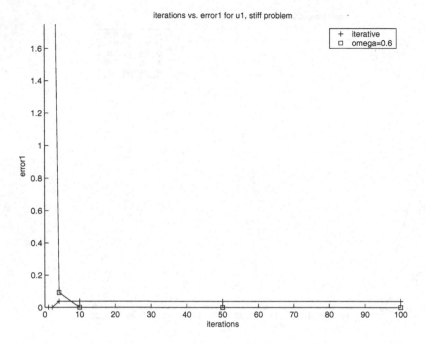

**Fig. 2.**

Step $i + 1$

$$\frac{du_{1,i+1}}{dt} = -16u_{1,i} + 12u_{2,i+1} + 16\cos(t) - 13\sin(t) \,,$$

$$\frac{du_{2,i+1}}{dt} = 12u_{1,i} - 9u_{2,i+1} - 11\cos(t) + 9\sin(t) \,,$$

$$u_1^{i+1}(t^n) = 1 \,, \ u_2^{i+1}(t^n) = 0 \,,$$

where $t \in [0, \Delta t]$, and $\Delta t = \pi/4$.

For the steps $i$ and $i + 1$ we can derive analytical solutions and apply them in our numerical scheme. The analytical solutions are given as

$$u_1^i(t) = \frac{3}{4}u_{2,i-1} + \frac{269}{257}\cos(t) - \frac{192}{257}\sin(t) - \frac{12}{257}\exp(-16t)$$

$$u_2^i(t) = \frac{401}{257}\sin(t) - \frac{9}{257}\cos(t) + \frac{3}{4}\exp(-16t)\frac{12}{257},$$

and

$$u_1^{i+1}(t) = \frac{8}{41}\sin(t) + \frac{113}{41}\cos(t) + \frac{4}{3}\exp(-9t)\frac{2}{123} + \frac{26896}{15129}$$

$$u_2^{i+1}(t) = \frac{4}{3}u_{1,i} - \frac{54}{41}\cos(t) + \frac{35}{41}\sin(t) - \exp(-9t)\frac{2}{123}$$

**Table 3.** Numerical results for the second example with the new weighted method, $\omega = 0.9$, in comparison with the iterative method

| Number of time-part. | Iterative Steps | $err_1$(weighted) | $err_2$(weighted) | $err_1$(iter) | $err_2$(iter) |
|---|---|---|---|---|---|
| 1 | 4 | $3.53579 \times 10^2$ | $1.56711 \times 10^0$ | $6.90483 \times 10^{-1}$ | $7.52328 \times 10^{-1}$ |
| 1 | 10 | $3.53579 \times 10^2$ | $1.53395 \times 10^0$ | $6.90483 \times 10^{-1}$ | $9.33215 \times 10^{-1}$ |
| 1 | 50 | $3.53579 \times 10^2$ | $1.53370 \times 10^0$ | $6.90483 \times 10^{-1}$ | $5.22929 \times 10^0$ |
| 10 | 4 | $3.40879 \times 10^2$ | $7.75019 \times 10^{-1}$ | $3.72598 \times 10^{-3}$ | $2.93505 \times 10^{-1}$ |
| 10 | 10 | $3.40846 \times 10^2$ | $7.75013 \times 10^{-1}$ | $3.72598 \times 10^{-3}$ | $1.36090 \times 10^0$ |
| 10 | 50 | $3.40846 \times 10^2$ | $7.75013 \times 10^{-1}$ | $3.72598 \times 10^{-3}$ | $4.06530 \times 10^1$ |
| 100 | 4 | $2.31925 \times 10^{-1}$ | $2.68301 \times 10^{-3}$ | $4.40903 \times 10^{-4}$ | $5.20919 \times 10^{-2}$ |
| 100 | 10 | $1.67892 \times 10^{-3}$ | $2.68301 \times 10^{-3}$ | $4.40903 \times 10^{-4}$ | $3.08888 \times 10^0$ |
| 100 | 50 | $3.96724 \times 10^{-5}$ | $2.68301 \times 10^{-3}$ | $4.40903 \times 10^{-4}$ | $7.52127 \times 10^1$ |

The results are presented in Table 3 in comparison with the results of the new proposed weighted method.

# 5    Conclusion and Discussions

The intention of this work is to introduce a modified weighted iterative Operator-Splitting method and to confirm the accuracy of the proposed scheme through application on two systems of ODEs. We obtain better convergence results in comparison with traditional iterative splitting, even for a stiff case of parameters. With appropriate assignment of the weighting factor we can stabilize the method, and actually with less iterations. The suitable modifications of the ideas presented here for applying them on PDEs with respect to convection-diffusion-reaction equations are issues currently under research.

# References

1. R.E. Ewing. Up-scaling of biological processes and multiphase flow in porous media. *IIMA Volumes in Mathematics and its Applications*, Springer-Verlag, **295** (2002), pp. 195-215.
2. I.Farago, J.Geiser, Iterative Operator-Splitting Methods for Linear Problems, *IJCS, International Journal of Computational Sciences*, **1**, Nos. 1/2/3, pp. 64-74, October 2005.

# Multilevel Preconditioning of 2D Rannacher-Turek FE Problems; Additive and Multiplicative Methods

Ivan Georgiev[1], Johannes Kraus[2], and Svetozar Margenov[3]

[1] Institute of Mathematics and Informatics and Institute for Parallel Processing, Bulgarian Academy of Sciences, Acad. G. Bonchev Bl. 25A, 1113 Sofia, Bulgaria
john@parallel.bas.bg

[2] Johann Radon Institute for Computational and Applied Mathematics, Altenbergerstraße 69, A-4040 Linz, Austria
johannes.kraus@oeaw.ac.at

[3] Institute for Parallel Processing, Bulgarian Academy of Sciences, Acad. G. Bonchev Bl. 25A, 1113 Sofia, Bulgaria
margenov@parallel.bas.bg

**Abstract.** In the present paper we concentrate on algebraic two-level and multilevel preconditioners for symmetric positive definite problems arising from discretization by Rannacher-Turek non-conforming rotated bilinear finite elements on quadrilaterals. An important point to make is that in this case the finite element spaces corresponding to two successive levels of mesh refinement are not nested (in general). To handle this, a proper two-level basis is required in order to fit the general framework for the construction of two-level preconditioners for conforming finite elements and to generalize the methods to the multilevel case.

The proposed variants of hierarchical two-level basis are first introduced in a rather general setting. Then, the involved parameters are studied and optimized. As will be shown, the obtained bounds – in particular – give rise to optimal order AMLI methods of additive type. The presented numerical tests fully confirm the theoretical estimates.

## 1   Introduction

In this paper we consider the elliptic boundary value problem

$$Lu \equiv -\nabla \cdot (a(\mathbf{x})\nabla u(\mathbf{x})) = f(\mathbf{x}) \text{ in } \quad \Omega,$$
$$u = 0 \quad \text{on} \quad \Gamma_D, \qquad (1)$$
$$(a(\mathbf{x})\nabla u(\mathbf{x})) \cdot \mathbf{n} = 0 \quad \text{on} \quad \Gamma_N,$$

where $\Omega$ is a polygonal domain in $\mathbb{R}^2$, $f(\mathbf{x})$ is a given function in $L^2(\Omega)$, the diffusion coefficient $a(\mathbf{x})$ is a piece-wise smooth and strictly positive function, uniformly bounded in $\Omega$, $\mathbf{n}$ is the outward unit vector normal to the boundary $\Gamma = \partial\Omega$, and $\Gamma = \bar{\Gamma}_D \cup \bar{\Gamma}_N$. The weak formulation of problem (1) reads as follows:

T. Boyanov et al. (Eds.): NMA 2006, LNCS 4310, pp. 56–64, 2007.

Given $f \in L^2(\Omega)$ find $u \in V \equiv H^1_D(\Omega) = \{v \in H^1(\Omega) : v = 0 \text{ on } \Gamma_D\}$, satisfying

$$\mathcal{A}(u, v) = (f, v) \quad \forall v \in H^1_D(\Omega), \text{ where } \mathcal{A}(u, v) = \int_{\Omega} a(\mathbf{x})\nabla u(\mathbf{x}) \cdot \nabla v(\mathbf{x})d\mathbf{x}. \quad (2)$$

The variational problem (2) is discretized using the finite element method. That is, the continuous space $V$ is replaced by a finite dimensional subspace $V_h$ which corresponds to a given partition $\mathcal{T}_h$ of the domain $\Omega$. Moreover, we assume that $\mathcal{T}_h$ is obtained by a proper refinement of a coarser partition $\mathcal{T}_H$. Then the problem is: find $u_h \in V_h$, such that

$$\mathcal{A}_h(u_h, v_h) = (f, v_h) \quad \forall v_h \in V_h, \quad \text{where} \quad \mathcal{A}_h(u_h, v_h) = \sum_{e \in \mathcal{T}_h} \int_e a(e)\nabla u_h \cdot \nabla v_h d\mathbf{x}.$$

Here $a(e)$ is a constant defined by the integral averaged value of $a(\mathbf{x})$ over each element from the coarser partition $\mathcal{T}_H$.

## 1.1 The Two-Level Setting

We are concerned with the construction of a two-level preconditioner $M$ for $A_h$, such that the spectral condition number $\varkappa(M^{-1}A_h)$ of the preconditioned matrix $M^{-1}A_h$ is uniformly bounded with respect to the meshsize parameter $h$ and possible coefficient jumps. The classical theory for constructing optimal order two-level preconditioners was first developed in [3,8], see also [2], for the case of linear conforming finite elements. The general framework requires to define two nested finite element spaces $V_H \subset V_h$ that correspond to two consecutive (regular) mesh refinements $\mathcal{T}_H$ and $\mathcal{T}_h$ of the domain $\Omega$. Let $\{\phi_H^{(k)}, k = 1, 2, \cdots, |\mathcal{N}_H|\}$ and $\{\phi_h^{(k)}, k = 1, 2, \cdots, |\mathcal{N}_h|\}$ denote the set of standard finite element nodal basis functions for the spaces $V_H$ and $V_h$, respectively. We split the set of meshpoints (nodes) $\mathcal{N}_h$ from $\mathcal{T}_h$ into two groups: the first group contains the nodes $\mathcal{N}_H$ from $\mathcal{T}_H$ and the second one consists of the rest, where the latter are the newly added node-points $\mathcal{N}_{h \backslash H}$ from $\mathcal{T}_h \backslash \mathcal{T}_H$. Next we define the so-called hierarchical basis functions

$$\{\widetilde{\phi}_h^{(k)}, k = 1, 2, \cdots, |\mathcal{N}_h|\} = \{\phi_H^{(l)} \text{ on } \mathcal{T}_H\} \cup \{\phi_h^{(m)} \text{ on } \mathcal{T}_h \backslash \mathcal{T}_H\}. \quad (3)$$

Let then $\widetilde{A}_h$ be the corresponding hierarchical stiffness matrix. Under the splitting (3) both matrices $A_h$ and $\widetilde{A}_h$ admit in a natural way a two-by-two block structure

$$A_h = \begin{bmatrix} A_{11} & A_{12} \\ A_{21} & A_{22} \end{bmatrix} \begin{matrix} \}\mathcal{N}_{h \backslash H} \\ \}\mathcal{N}_H \end{matrix}, \qquad \widetilde{A}_h = \begin{bmatrix} A_{11} & \widetilde{A}_{12} \\ \widetilde{A}_{21} & A_H \end{bmatrix} \begin{matrix} \}\mathcal{N}_{h \backslash H} \\ \}\mathcal{N}_H \end{matrix}. \quad (4)$$

As is well-known, there exists a transformation matrix $J = \begin{bmatrix} I_1 & 0 \\ J_{21} & I_2 \end{bmatrix}$, which relates the nodal point vectors for the standard and the hierarchical basis functions as follows,

$$\widetilde{\mathbf{v}} = \begin{bmatrix} \widetilde{\mathbf{v}}_1 \\ \widetilde{\mathbf{v}}_2 \end{bmatrix} = J \begin{bmatrix} \mathbf{v}_1 \\ \mathbf{v}_2 \end{bmatrix}, \quad \begin{matrix} \widetilde{\mathbf{v}}_1 = \mathbf{v}_1 \\ \widetilde{\mathbf{v}}_2 = J_{21}\mathbf{v}_1 + \mathbf{v}_2 \end{matrix}.$$

## 1.2   Two-Level Preconditioners of Additive and Multiplicative Type

Consider a general matrix $A$, which is assumed to be symmetric positive definite and partitioned as in (4). The quality of this partitioning is characterized by the corresponding CBS inequality constant:

$$\gamma = \sup_{\mathbf{v}_1 \in \mathbb{R}^{n_1},\, \mathbf{v}_2 \in \mathbb{R}^{n_2}} \frac{\mathbf{v}_1^T A_{12} \mathbf{v}_2}{\left(\mathbf{v}_1^T A_{11} \mathbf{v}_1\right)^{1/2} \left(\mathbf{v}_2^T A_{22} \mathbf{v}_2\right)^{1/2}}, \tag{5}$$

where $n_1 = |\mathcal{N}_{h \backslash H}|$ and $n_2 = |\mathcal{N}_H|$ denote the cardinality of the sets $\mathcal{N}_{h \backslash H}$ and $\mathcal{N}_H$, respectively.

Consider now two preconditioners to $A$ under the assumptions

$$A_{11} \leq C_{11} \leq (1 + \delta_1) A_{11} \quad \text{and} \quad A_{22} \leq C_{22} \leq (1 + \delta_2) A_{22}. \tag{6}$$

The inequalities (6) are in a positive semidefinite sense where $C_{11}$ and $C_{22}$ are symmetric and positive definite matrices for some positive constants $\delta_i$, $i = 1, 2$. The additive preconditioner $M_A$ and the multiplicative preconditioner $M_F$ are then introduced as

$$M_A = \begin{bmatrix} C_{11} & 0 \\ 0 & C_{22} \end{bmatrix}, \quad \text{and} \quad M_F = \begin{bmatrix} C_{11} & 0 \\ A_{21} & C_{22} \end{bmatrix} \begin{bmatrix} I_1 & C_{11}^{-1} A_{12} \\ 0 & I_2 \end{bmatrix}, \tag{7}$$

respectively. When $C_{11} = A_{11}$ and $C_{22} = A_{22}$, then the following estimates hold (see, e.g., [2]):

$$\varkappa(M_A^{-1} A) \leq \frac{1 + \gamma}{1 - \gamma}, \quad \text{and} \quad \varkappa(M_F^{-1} A) \leq \frac{1}{1 - \gamma^2}.$$

## 2   Rannacher-Turek Finite Elements

Nonconforming finite elements based on *rotated* multilinear shape functions were introduced by Rannacher and Turek [12] as a class of simple elements for the Stokes problem. More generally, the recent activities in the development of efficient solution methods for non-conforming finite element systems are inspired by their attractive properties as a stable discretization tool for illconditioned problems.

The unit square $[-1, 1]^2$ is used as a reference element $\hat{e}$ to define the isoparametric rotated bilinear element $e \in \mathcal{T}_h$. Let $\psi_e : \hat{e} \rightarrow e$ be the corresponding bilinear one-to-one transformation, and let the nodal basis functions be determined by the relation

$$\{\phi_i\}_{i=1}^4 = \{\hat{\phi}_i \circ \psi_e^{-1}\}_{i=1}^4, \qquad \{\hat{\phi}_i\} \in \mathrm{span}\{1, x, y, x^2 - y^2\}.$$

For the variant MP (mid point), $\{\hat{\phi}_i\}_{i=1}^4$ are found by the point-wise interpolation condition $\hat{\phi}_i(b_\Gamma^j) = \delta_{ij}$, where $b_\Gamma^j, j = 1, 4$ are the midpoints of the edges of the quadrilateral $\hat{e}$. Then,

$$\hat{\phi}_1(x, y) = (1 - 2x + (x^2 - y^2))/4, \quad \hat{\phi}_2(x, y) = (1 + 2x + (x^2 - y^2))/4,$$
$$\hat{\phi}_3(x, y) = (1 - 2y - (x^2 - y^2))/4, \quad \hat{\phi}_4(x, y) = (1 + 2y - (x^2 - y^2))/4.$$

The variant MV (mean value) corresponds to integral midvalue interpolation conditions. Let $\Gamma_{\hat{e}} = \bigcup_{j=1}^{4} \Gamma_{\hat{e}}^{j}$. Then $\{\hat{\phi}_i\}_{i=1}^{4}$ are determined by the equality $|\Gamma_{\hat{e}}^{j}|^{-1} \int_{\Gamma_{\hat{e}}^{j}} \hat{\phi}_i d\Gamma_{\hat{e}}^{j} = \delta_{ij}$, which leads to

$$\hat{\phi}_1(x,y) = (2 - 4x + 3(x^2 - y^2))/8, \quad \hat{\phi}_2(x,y) = (2 + 4x + 3(x^2 - y^2))/8,$$
$$\hat{\phi}_3(x,y) = (2 - 4y - 3(x^2 - y^2))/8, \quad \hat{\phi}_4(x,y) = (2 + 4y - 3(x^2 - y^2))/8.$$

## 3  Hierarchical Two-Level Splittings

Let us consider two consecutive discretizations $\mathcal{T}_H$ and $\mathcal{T}_h$. Figure 1 illustrates a macro-element obtained after one regular mesh-refinement step. We see that in this case $\mathcal{V}_H$ and $\mathcal{V}_h$ are not nested. As shown in [3], the constant $\gamma$ can be estimated locally over each finite element (macro-element) $E \in \mathcal{T}_H$, which means that $\gamma = \max_{E} \gamma_E$, where

$$\gamma_E = \sup_{u \in \mathcal{V}_1(E), \ v \in \mathcal{V}_2(E)} \frac{\mathcal{A}_E(u,v)}{\sqrt{\mathcal{A}_E(u,u)\mathcal{A}_E(v,v)}}, \quad v \neq const.$$

The spaces $\mathcal{V}_k(E)$ above contain the functions from $\mathcal{V}_k$ restricted to $E$ and $\mathcal{A}_E(u,v)$ corresponds to $\mathcal{A}(u,v)$ restricted over the element $E$ of $\mathcal{T}_H$ (see also [10]). Let us introduce the following macro-element level transformation matrix $J_E$:

$$J_E = \frac{1}{2} \begin{bmatrix} 2 & & & & & & & & \\ & 2 & & & & & & & \\ & & 2 & & & & & & \\ & & & 2 & & & & & \\ & & & & 1 & -1 & & & \\ & & & & & & 1 & -1 & \\ & & & & & 1 & -1 & & \\ & & & & & & & & 1 & -1 \\ \alpha_{11} & \alpha_{12} & \alpha_{13} & \alpha_{14} & 1 & 1 & & & \\ \alpha_{21} & \alpha_{22} & \alpha_{23} & \alpha_{24} & & & 1 & 1 & \\ \alpha_{31} & \alpha_{32} & \alpha_{33} & \alpha_{34} & & 1 & 1 & & \\ \alpha_{41} & \alpha_{42} & \alpha_{43} & \alpha_{44} & & & & 1 & 1 \end{bmatrix} \tag{8}$$

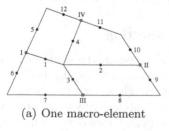

(a) One macro-element

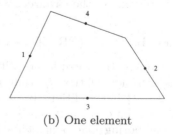

(b) One element

**Fig. 1.** Uniform refinement on a general mesh

## 3.1   Two-Level Splitting by Differences and Aggregates (DA)

Let $\phi_1, \ldots, \phi_{12}$ denote the standard nodal basis functions for the macro-element $E$ depicted in Figure 1. Then we define

$$
\begin{aligned}
\mathcal{V}(E) &= \operatorname{span}\{\phi_1, \ldots, \phi_{12}\} = \mathcal{V}_1(E) \oplus \mathcal{V}_2(E)\,, \\
\mathcal{V}_1(E) &= \operatorname{span}\{\phi_1,\, \phi_2,\, \phi_3,\, \phi_4,\, \phi_5 - \phi_6,\, \phi_9 - \phi_{10},\, \phi_7 - \phi_8,\, \phi_{11} - \phi_{12}\} \\
\mathcal{V}_2(E) &= \operatorname{span}\{\phi_5 + \phi_6 + \sum_{j=1,4} \alpha_{1j}\phi_j,\, \phi_9 + \phi_{10} + \sum_{j=1,4} \alpha_{2j}\phi_j, \\
&\qquad \phi_7 + \phi_8 + \sum_{j=1,4} \alpha_{3j}\phi_j,\, \phi_{11} + \phi_{12} + \sum_{j=1,4} \alpha_{4j}\phi_j\}\,.
\end{aligned}
$$

Using the related transformation matrix (8), the macro-element stiffness matrix is transformed into a hierarchical form

$$
\tilde{A}_E = J_E A_E J_E^T = \begin{bmatrix} \tilde{A}_{E,11} & \tilde{A}_{E,12} \\ \tilde{A}_{E,21} & \tilde{A}_{E,22} \end{bmatrix} \begin{matrix} \tilde{\phi}_i \in \mathcal{V}_1(E) \\ \tilde{\phi}_i \in \mathcal{V}_2(E) \end{matrix}\,. \tag{9}
$$

Following the local definitions, we can similarly construct the new hierarchical basis $\tilde{\varphi} = \{\tilde{\varphi}_h^{(i)}\}_{i=1}^{|\mathcal{N}_h|}$ and the corresponding splitting

$$
\mathcal{V}_h = \mathcal{V}_1 \oplus \mathcal{V}_2\,. \tag{10}
$$

Our aim is to analyze the constant $\gamma = \cos(\mathcal{V}_1, \mathcal{V}_2)$ for the splitting (10) locally. If $A_e$ denotes the element matrix (with constant coefficients on macro-element level) then a necessary condition serving this purpose is

$(i)$ $\ker(\tilde{A}_{E,22}) = \ker(A_e)$ .

In view of (8) and (9), condition $(i)$ holds if and only if

$$
\sum_{i=1}^{4} \alpha_{ij} = 1\,, \qquad \forall j \in \{1,2,3,4\}. \tag{11}
$$

When the two-level algorithm is recursively generalized to the multilevel case, it is further useful to have

$(ii)$ $\tilde{A}_{E,22} = pA_e$

for some positive $p$. This proportionality can be met in a very general setting for the DA splitting of the Crouzeix-Raviart finite element space, see [9].

## 3.2   "First Reduce" (FR) Two-Level Splitting

The "first reduce" (FR) two-level splitting is based on the simplified local two-level transformation matrix $J_E$ that is obtained from (8) by taking $\alpha_{ij} = 0$ $\forall i, j \in \{1, 2, 3, 4\}$ but additionally passing through an exact elimination of the degrees of freedom corresponding to "interior" nodes, which are not shared by any two neighboring macro-elements in the global mesh. For our analysis we proceed as follows:

Step 1: We observe that the upper left block of

$$\widetilde{A}_h = \sum_{E \in \mathcal{T}_h} \widetilde{A}_E = \begin{bmatrix} \widetilde{A}_{11} & \widetilde{A}_{12} \\ \widetilde{A}_{21} & \widetilde{A}_{22} \end{bmatrix}$$

is a block-diagonal matrix. The diagonal entries of $\widetilde{A}_{11}$ are $4 \times 4$ blocks, corresponding to the interior points $\{1, 2, 3, 4\}$, cf. Figure 1, which are not connected to nodes in other macro-elements. Thus, the corresponding unknowns can be eliminated exactly, i.e., to be done locally. Therefore, we first compute the local Schur complements arising from static condensation of the "interior degrees of freedom" in $\widetilde{A}_E$ and obtain the $(8 \times 8)$ matrix $B_E$. Next we split $B_E$ as

$$B_E = \begin{bmatrix} B_{E,11} & B_{E,12} \\ B_{E,21} & B_{E,22} \end{bmatrix} \begin{array}{l} \}\text{two-level half-difference basis functions} \\ \}\text{two-level half-sum basis functions} \end{array}$$

written again in two-by-two block form with blocks of order $(4 \times 4)$.

Step 2: We are now in a position to estimate the CBS constant corresponding to the $2 \times 2$ splitting of $B$. Following the general theory, it suffices to compute the minimal eigenvalue of the generalized eigenproblem

$$S_E \mathbf{v}_E = \lambda_E^{(1)} B_{E,22} \mathbf{v}_E, \quad \mathbf{v}_E \perp (1, 1, \ldots, 1)^T,$$

where $S_E = B_{E,22} - B_{E,21} B_{E,11}^{-1} B_{E,12}$, and then

$$\gamma^2 \leq \max_{E \in \mathcal{T}_h} \gamma_E^2 = \max_{E \in \mathcal{T}_h} (1 - \lambda_E^{(1)}). \tag{12}$$

# 4   Uniform Estimates of the CBS Constant

## 4.1   DA Algorithm

The following results can be verified using a computer algebra program such as MATHEMATICA. A more detailed discussion, including numerical experiments for anisotropic problems, can be found in Ref. [11].

**Variant MP:**

**Lemma 1.** *There exists a DA two-level splitting satisfying the condition (ii) if and only if $p \geq 3/7$. Then, the obtained solutions for $\alpha_{ij}$ are invariant with respect to the local CBS constant $\gamma_E^2 = 1 - 1/(4p)$, and for the related optimal splitting we have $\gamma_{MP}^2 \leq 5/12$.*

**Variant MV:**

**Lemma 2.** *There exists a DA two-level splitting satisfying the condition (ii) if and only if $p \geq 2/5$. Then, the obtained solutions for $\alpha_{ij}$ are invariant with respect to the local CBS constant $\gamma_E^2 = 1 - 1/(4p)$, and for the related optimal splitting we have $\gamma_{MV}^2 \leq 3/8$.*

**Table 1.** Linear AMLI V-cycle: number of PCG iterations

| $1/h$ | 32 | 64 | 128 | 256 | 512 | 1024 |
|---|---|---|---|---|---|---|
| DA/MP $\varepsilon = 1$ | 15 (8) | 21 (10) | 29 (13) | 39 (16) | 49 (19) | 61 (22) |
| $\varepsilon = 0.01$ | 15 (8) | 22 (10) | 29 (13) | 39 (16) | 50 (19) | 63 (22) |
| FR/MP $\varepsilon = 1$ | 11 (6) | 15 (8) | 19 (9) | 24 (11) | 28 (12) | 34 (14) |
| $\varepsilon = 0.01$ | 11 (6) | 15 (8) | 20 (9) | 24 (11) | 30 (12) | 36 (14) |
| DA/MV $\varepsilon = 1$ | 14 (8) | 20 (10) | 28 (13) | 36 (16) | 45 (18) | 56 (21) |
| $\varepsilon = 0.01$ | 14 (8) | 20 (11) | 28 (13) | 37 (16) | 47 (18) | 59 (22) |
| FR/MV $\varepsilon = 1$ | 12 (7) | 16 (9) | 21 (10) | 26 (12) | 31 (14) | 37 (16) |
| $\varepsilon = 0.01$ | 13 (7) | 17 (9) | 22 (10) | 27 (12) | 33 (14) | 39 (16) |

## 4.2 FR Algorithm

For the two-level FR splitting we get the following uniform bounds with respect to the size of the discrete problem and any possible jumps of the (piece-wise constant) diffusion coefficient $a(e)$ between macro-elements $E \in \mathcal{T}_H$:

**Variant MP:**

$$\lambda_E^{(1)} = \frac{5}{7}, \qquad \gamma_E^2 = 1 - \lambda_E^{(1)} = \frac{2}{7}, \quad \text{and therefore} \quad \gamma_{MP}^2 \le \frac{2}{7}.$$

**Variant MV:**

$$\lambda_E^{(1)} = \frac{5}{8}, \qquad \gamma_E^2 = 1 - \lambda_E^{(1)} = \frac{3}{8}, \quad \text{and therefore} \quad \gamma_{MV}^2 \le \frac{3}{8}.$$

Let us remind that the obtained estimates hold theoretically for the two-level algorithm only. This is because the matrix $B_{E,22}$ is only associated with the coarse discretization $\mathcal{T}_H$ and is not proportional to the related element stiffness matrix $A_e$.

## 5 Numerical Results

We compare the convergence properties of the preconditioned conjugate gradient (PCG) method using either the additive or the multiplicative variant of the multilevel preconditioner based on either DA or FR splitting for MP and MV discretization of the model problem (1). We subdivide the square domain $\Omega = [0,1]^2$ into four subdomains of equal shape and size, i.e., $\Omega = \Omega_1 \cup \ldots \cup \Omega_4$, where $\Omega_1 = [0, 1/2]^2$, $\Omega_2 = [1/2, 1] \times [0, 1/2]$, $\Omega_3 = [0, 1/2] \times [1/2, 1]$, and $\Omega_4 = [1/2, 1]^2$. The diffusion coefficient is given by $a(e) = 1$ on subdomains $\Omega_1$ and $\Omega_4$, $a(e) = \varepsilon$ on $\Omega_2$, and $a(e) = \varepsilon^{-1}$ on $\Omega_3$. The first Table 1 summarizes the number PCG iterations that reduce the residual norm by a factor $10^6$ when performing a single V-cycle of linear algebraic multilevel iteration (AMLI). In the second Table 2 we list the corresponding results for the linear AMLI W-cycle (employing properly shifted second-order Chebyshev polynomials for stabilizing the condition number, see [5,6]). Finally, the third

**Table 2.** Linear AMLI W-cycle: number of PCG iterations

| $1/h$ | 32 | 64 | 128 | 256 | 512 | 1024 |
|---|---|---|---|---|---|---|
| DA/MP $\varepsilon = 1$ | 15 (8) | 16 (8) | 17 (8) | 18 (8) | 18 (8) | 18 (8) |
| $\varepsilon = 0.01$ | 15 (8) | 17 (8) | 17 (8) | 18 (8) | 18 (8) | 19 (8) |
| FR/MP $\varepsilon = 1$ | 11 (6) | 12 (6) | 12 (6) | 13 (6) | 13 (6) | 13 (6) |
| $\varepsilon = 0.01$ | 11 (6) | 12 (6) | 13 (6) | 13 (6) | 13 (6) | 13 (6) |
| DA/MV $\varepsilon = 1$ | 14 (8) | 15 (9) | 16 (9) | 16 (9) | 16 (9) | 16 (9) |
| $\varepsilon = 0.01$ | 14 (8) | 16 (10) | 16 (9) | 16 (9) | 17 (9) | 17 (9) |
| FR/MV $\varepsilon = 1$ | 12 (7) | 14 (7) | 14 (7) | 14 (7) | 14 (7) | 14 (7) |
| $\varepsilon = 0.01$ | 13 (7) | 14 (7) | 15 (7) | 15 (7) | 15 (7) | 15 (7) |

**Table 3.** Non-linear AMLI W-cycle: number of (outer) GCG iterations

| $1/h$ | 32 | 64 | 128 | 256 | 512 | 1024 |
|---|---|---|---|---|---|---|
| DA/MP $\varepsilon = 1$ | 15 (8) | 17 (8) | 17 (8) | 17 (8) | 17 (8) | 17 (8) |
| $\varepsilon = 0.01$ | 15 (8) | 17 (8) | 17 (8) | 17 (8) | 17 (8) | 17 (8) |
| FR/MP $\varepsilon = 1$ | 11 (6) | 12 (6) | 12 (6) | 13 (6) | 13 (6) | 13 (6) |
| $\varepsilon = 0.01$ | 11 (6) | 12 (6) | 13 (6) | 13 (6) | 13 (6) | 13 (6) |
| DA/MV $\varepsilon = 1$ | 14 (8) | 15 (8) | 15 (8) | 16 (8) | 16 (8) | 16 (8) |
| $\varepsilon = 0.01$ | 14 (8) | 15 (8) | 16 (8) | 16 (8) | 16 (8) | 16 (8) |
| FR/MV $\varepsilon = 1$ | 13 (7) | 14 (7) | 14 (7) | 14 (7) | 14 (7) | 14 (7) |
| $\varepsilon = 0.01$ | 13 (7) | 14 (7) | 15 (7) | 15 (7) | 15 (7) | 15 (7) |

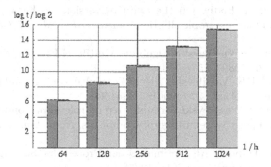

**Fig. 2.** CPU-time for additive (dark) and multiplicative (light) preconditioning (logarithmic scale)

Table 3 refers to the (variable-step) non-linear AMLI method stabilized by two inner generalized conjugate gradient iterations at every coarse level, cf., [7], and using a direct solve on the coarsest mesh with mesh-size $1/h = 16$ as in the other tests. The results for the multiplicative variant are put in parentheses in each case. Though the number of iterations approximately doubles in most cases when switching from multiplicative to additive preconditioning the CPU-time (in most situations) increases only by 10 to 50 per cent, which is due to the lower operation count per application of the additive preconditioner, see [1,4]. This effect is illustrated in Figure 5, which depicts the logarithm (to the basis 2) of the CPU-time in milliseconds measured on a 2 GHz Linux-PC for the case of the DA splitting and MV discretization. Moreover, (in accordance with the analysis) both preconditioners, the additive as well as the multiplicative method, are perfectly robust with respect to jump discontinuities in the coefficient $a(e)$ as can be seen from the almost identical results shown in the respective first ($\varepsilon = 1$) and second ($\varepsilon = 0.01$) column of Tables 1–3. The solution of the largest problem with approximately 2 million degrees of freedom took around $2^{15}$ milliseconds, which is approximately 30 seconds on a single processor! Finally, we want to stress that the additive method has also excellent parallelization properties.

**Acknowledgments.** Parts of this work have been conducted during the Special Radon Semester on Computational Mechanics, held at RICAM, Linz, Oct. 3rd - Dec. 16th 2005. The authors gratefully acknowledge the support by the Austrian Academy of Sciences. The authors have been also partially supported by EC INCO Grant BIS-21++ 016639/2005.

# References

1. O. Axelsson, Stabilization of algebraic multilevel iteration methods; additive methods, *Numerical Algorithms*, 21(1999), 23-47.
2. O. Axelsson, *Iterative solution methods*. Cambridge University Press, 1994.
3. O. Axelsson and I. Gustafsson, Preconditioning and two-level multigrid methods of arbitrary degree of approximations, *Math. Comp.*, 40(1983), 219-242.
4. O. Axelsson and A. Padiy, On the additive version of the algebraic multilevel iteration method for anisotropic elliptic problems, *SIAM J. Sci. Comput.*, 20(1999), 1807-1830.
5. O. Axelsson and P.S. Vassilevski, Algebraic Multilevel Preconditioning Methods I, *Numer. Math.*, 56 (1989), 157-177.
6. O. Axelsson and P.S. Vassilevski, Algebraic Multilevel Preconditioning Methods II, *SIAM J. Numer. Anal.*, 27 (1990), 1569-1590.
7. O. Axelsson and P.S. Vassilevski, Variable-step multilevel preconditioning methods, I: self-adjoint and positive definite elliptic problems, *Num. Lin. Alg. Appl.*, 1 (1994), 75-101.
8. R. Bank and T. Dupont, An Optimal Order Process for Solving Finite Element Equations, *Math. Comp.*, 36 (1981), 427-458.
9. R. Blaheta, S. Margenov, M. Neytcheva, Uniform estimate of the constant in the strengthened CBS inequality for anisotropic non-conforming FEM systems, *Numerical Linear Algebra with Applications*, Vol. 11 (4) (2004), 309-326.
10. V. Eijkhout and P.S. Vassilevski, The Role of the Strengthened Cauchy-Bunyakowski-Schwarz Inequality in Multilevel Methods, *SIAM Review*, 33 (1991), 405-419.
11. I. Georgiev, J. Kraus, S. Margenov, Multilevel preconditioning of rotated bilinear non-conforming FEM problems, submitted. Also available as RICAM-Report 2006-3, RICAM, Linz, Austria, 2006.
12. R. Rannacher, S. Turek, *Simple non-conforming quadrilateral Stokes Element*, Numerical Methods for Partial Differential Equations, 8(2) (1992), 97-112.

# A Parallel Algorithm for Systems of Convection-Diffusion Equations

János Karátson[1], Tamás Kurics[1], and Ivan Lirkov[2]

[1] Department of Applied Analysis and Computational Mathematics,
ELTE University, H-1117 Budapest, Hungary
karatson@cs.elte.hu, fantom@cs.elte.hu
[2] Institute for Parallel Processing, Bulgarian Academy of Sciences,
Acad. G. Bonchev, Bl. 25A, 1113 Sofia, Bulgaria
ivan@parallel.bas.bg

**Abstract.** The numerical solution of systems of convection-diffusion equations is considered. The problem is described by a system of second order partial differential equations (PDEs). This system is discretized by Courant-elements. The preconditioned conjugate gradient method is used for the iterative solution of the large-scale linear algebraic systems arising after the finite element discretization of the problem. Discrete Helmholtz preconditioners are applied to obtain a mesh independent superlinear convergence of the iterative method. A parallel algorithm is derived for the proposed preconditioner. A portable parallel code using Message Passing Interface (MPI) is developed. Numerical tests well illustrate the performance of the proposed method on a parallel computer architecture.

**2000 Mathematics Subject Classification:** 65N12, 68W10, 65F10, 74S05.

## 1  Introduction

The generalized conjugate gradient (GCG) method has become the most widespread way of solving nonsymmetric linear algebraic systems arising from discretized elliptic problems, see [3] where an extensive summary is given on the convergence of the CGM. For discretized elliptic problems, the CGM is mostly used with suitable preconditioning (cf. [3]), which sometimes relies on Hilbert space theory (cf. [6]) and then provides mesh independent convergence. Moreover, it has been shown in [6] that the GCG method can be competitive with multigrid methods.

The CGM for nonsymmetric equations in Hilbert space has been studied in [4,5]: in the latter superlinear convergence has been proved in Hilbert space and, based on this, mesh independence of the superlinear estimate has been derived for FEM discretizations of elliptic Dirichlet problems. The mesh independent superlinear convergence results have been extended from a single equation to systems of PDEs in a recent paper [7] in the framework of normal operators in Hilbert space. An important advantage of the obtained preconditioning method

T. Boyanov et al. (Eds.): NMA 2006, LNCS 4310, pp. 65–73, 2007.
© Springer-Verlag Berlin Heidelberg 2007

for systems is that one can define decoupled preconditioners, hence the size of the auxiliary systems remains as small as for a single equation, moreover, parallelization of the auxiliary systems is available. The main goal of this paper is to develop an efficient MPI parallel code using multiple processors, based on a proper summary of the theoretical result for systems of PDEs.

We consider systems of the form

$$\left.\begin{array}{l} - \operatorname{div}(K_i \nabla u_i) + \mathbf{b}_i \cdot \nabla u_i + \sum_{j=1}^{l} V_{ij} u_j = g_i \\ u_{i_{|\partial\Omega}} = 0 \end{array}\right\} \qquad (i = 1, \ldots, l) \qquad (1)$$

under the following

ASSUMPTIONS BVP.

(i) the bounded domain $\Omega \subset \mathbb{R}^N$ is $C^2$-diffeomorphic to a convex domain;
(ii) for all $i, j = 1, \ldots, l$, $K_i \in C^1(\overline{\Omega})$, $V_{ij} \in L^\infty(\Omega)$ and $\mathbf{b}_i \in C^1(\overline{\Omega})^N$;
(iii) there is $m > 0$ such that $K_i \geq m$ holds for all $i = 1, \ldots, l$;
(iv) letting $V = \{V_{ij}\}_{i,j=1}^{l}$, the coercivity property

$$\lambda_{\min}\left(V + V^T\right) - \max_i \operatorname{div} \mathbf{b}_i \geq 0 \qquad (2)$$

holds pointwise on $\Omega$, where $\lambda_{\min}$ denotes the smallest eigenvalue;
(v) $g_i \in L^2(\Omega)$.

Items (iii) and (iv) ensure the coercivity property (6) which is a crucial assumption for Theorem 1.

Systems of the form (1) arise, e.g., from the time discretization and Newton linearization of nonlinear reaction-convection-diffusion systems which occur frequently in meteorological air-pollution models [12].

We write the considered system in a short vector form using the corresponding $n$-tuples:

$$\left.\begin{array}{l} L\mathbf{u} \equiv - \operatorname{div}(\mathbf{K}\nabla\mathbf{u}) + \mathbf{b} \cdot \nabla\mathbf{u} + V\mathbf{u} = \mathbf{g} \\ \mathbf{u}_{|\partial\Omega} = \mathbf{0} \end{array}\right\}, \qquad (3)$$

where

$$\mathbf{u} = \begin{pmatrix} u_1 \\ \vdots \\ u_l \end{pmatrix}, \ \mathbf{g} = \begin{pmatrix} g_1 \\ \vdots \\ g_l \end{pmatrix}, \ \operatorname{div}(\mathbf{K}\nabla\mathbf{u}) = \begin{pmatrix} \operatorname{div}(K_1 \nabla u_1) \\ \vdots \\ \operatorname{div}(K_l \nabla u_l) \end{pmatrix}, \ \mathbf{b}\cdot\nabla\mathbf{u} = \begin{pmatrix} \mathbf{b}_1 \cdot \nabla u_1 \\ \vdots \\ \mathbf{b}_l \cdot \nabla u_l \end{pmatrix}.$$

The FEM discretization of (3) leads to a linear algebraic system $\mathbf{L}_h \mathbf{c} = \mathbf{g}_h$. This can be solved by the GCG method using a preconditioner. In this paper we consider decoupled symmetric Helmholtz preconditioners

$$S_i u_i := - \operatorname{div}(K_i \nabla u_i) + \eta_i u_i \qquad (i = 1, \ldots, l) \qquad (4)$$

where $\eta_i \in C(\overline{\Omega})$, $\eta_i \geq 0$ are suitable functions. The $n$-tuple $S$ of the elliptic operators $S_i$ and the corresponding matrix $\mathbf{S}_h$ can be defined in the same way as previously, hence the preconditioned form of the discretized equation is

$$\mathbf{S}_h^{-1} \mathbf{L}_h \mathbf{c} = \mathbf{f}_h \equiv \mathbf{S}_h^{-1} \mathbf{g}_h. \qquad (5)$$

## 2   The Preconditioned Generalized Conjugate Gradient Method

Now let us consider the operator equation $Lu = g$ with an unbounded linear operator $L : D \to H$ defined on a dense domain $D$, and with some $g \in H$, where $H$ is an infinite dimensional complex separable Hilbert space. We have the following
ASSUMPTIONS A.

(i) The operator $L$ is decomposed in $L = S + Q$ on its domain $D$ where $S$ is a self-adjoint operator in $H$.
(ii) $S$ is a strongly positive operator, i.e., there exists $p > 0$ such that

$$\langle Su, u \rangle \geq p\|u\|^2 \quad (u \in D). \tag{6}$$

(iii) There exists $\varrho > 0$ such that $\Re \langle Lu, u \rangle \geq \varrho \langle Su, u \rangle$ $(u \in D)$.
(iv) The operator $Q$ can be extended to the energy space $H_S$, and then $S^{-1}Q$ is assumed to be a compact normal operator on $H_S$.

The generalized conjugate gradient, least square (GCG-LS) method is defined in [2]. The full version of the GCG-LS method constructs a sequence of search directions $d_k$ and simultaneously a sequence of approximate solutions $u_k$. Following the terminology of [2,4], the definition also involves an integer $s \in \mathbb{N}$, further, we let $s_k = \min\{k, s\}$ $(k \geq 0)$. The full version of the algorithm for the solution of the preconditioned operator equation

$$S^{-1}Lu = f = S^{-1}g. \tag{7}$$

in $H_S$ is as follows:

$$
\begin{cases}
(1) & \text{Let } u_0 \in D \text{ be arbitrary, let } r_0 \text{ be the solution of } Sr_0 = Lu_0 - g; \\
& d_0 = -r_0; \text{ and } z_0 \text{ be the solution of } Sz_0 = Ld_0; \\[4pt]
& \text{for any } k \in \mathbb{N}: \text{ when } u_k, d_k, r_k, z_k \text{ are obtained, let} \\[4pt]
(2a) & \text{the numbers } \alpha^{(k)}_{k-j} \quad (j = 0, \ldots, k) \quad \text{be the solution of} \\
& \displaystyle\sum_{j=0}^{k} \alpha^{(k)}_{k-j} \langle Sz_{k-j}, z_{k-l} \rangle = -\langle r_k, Sz_{k-l} \rangle \quad (0 \leq l \leq k); \\[4pt]
(2b) & u_{k+1} = u_k + \displaystyle\sum_{j=0}^{k} \alpha^{(k)}_{k-j} d_{k-j}; \\[4pt]
(2c) & r_{k+1} = r_k + \displaystyle\sum_{j=0}^{k} \alpha^{(k)}_{k-j} z_{k-j}; \\[4pt]
(2d) & \beta^{(k)}_{k-j} = \langle Lr_{k+1}, z_{k-j} \rangle / \|z_{k-j}\|^2_S \quad (j = 0, \ldots, s_k); \\[4pt]
(2e) & d_{k+1} = -r_{k+1} + \displaystyle\sum_{j=0}^{s_k} \beta^{(k)}_{k-j} d_{k-j}; \\[4pt]
(2f) & z_{k+1} \text{ be the solution of } Sz_{k+1} = Ld_{k+1}.
\end{cases} \tag{8}
$$

When symmetric part preconditioning is used, a more simple truncated algorithm is applicable, namely the so-called GCG-LS(0) (see [4] for details), where only the previous search direction $d_k$ and the auxiliary vector $z_k$ are used, so the previous ones do not have to be stored. Assumptions A imply that the operator of the preconditioned equation $S^{-1}L$ has the form $I + S^{-1}Q$, which is a compact perturbation of the identity operator, hence the following convergence result (cf. [5,7]) is applicable. Recall that a compact operator has countably many eigenvalues (with multiplicity), clustering at zero.

**Theorem 1.** *Let Assumptions A hold. Denoting the unique solution by $u^*$, the generalized conjugate gradient method applied for equation (7) yields for all $k \in \mathbb{N}$*

$$Q_k := \left( \frac{\|e_k\|_L}{\|e_0\|_L} \right)^{1/k} \leq \frac{2}{\varrho} \left( \frac{1}{k} \sum_{i=1}^{k} |\lambda_i(S^{-1}Q)| \right) \to 0 \quad as \; k \to \infty \qquad (9)$$

*where $e_k = u_k - u^*$ is the error vector and $\lambda_i = \lambda_i(S^{-1}Q)$ $(i \in \mathbb{N})$ are the ordered eigenvalues of the operator $S^{-1}Q$ $(|\lambda_i| \geq |\lambda_{i+1}|)$.*

## 3    Superlinear Convergence for Elliptic Systems

Let us consider the Hilbert space $H = L^2(\Omega)^l$ with the inner product $\langle \mathbf{u}, \mathbf{v} \rangle = \int_\Omega \sum_{i=1}^{l} u_i \bar{v}_i$ and define the operators $L$ and $S$ according to (3) and (4) on the dense domain

$$D(L) = D(S) = D := \left( H^2(\Omega) \cap H_0^1(\Omega) \right)^l .$$

Now we can use the convergence theorem for this problem in the space $L^2(\Omega)^l$ by verifying that $L$ and $S$ satisfy Assumptions A. First, we apply Theorem 1 using the truncated algorithm when $S$ is the symmetric part of $L$. Then we consider the full version (8) and use Theorem 1 for problems with constant coefficients when the normality of the preconditioned operator in the corresponding Sobolev space can be ensured.

First symmetric part preconditioning is considered, that is $S = (L + L^*)/2$. Since $Q = L - S$ is antisymmetric, it can be shown easily, that the operator $S^{-1}Q$ is antisymmetric in $H_S$, therefore it is normal automatically. We have for $\mathbf{u}, \mathbf{v} \in D$

$$\langle L\mathbf{u}, \mathbf{v} \rangle = \int_\Omega \left( \sum_{i=1}^{l} (K_i \nabla u_i \cdot \overline{\nabla} v_i + (\mathbf{b}_i \cdot \nabla u_i) \bar{v}_i) + \sum_{i,j=1}^{l} V_{ij} u_j \bar{v}_i \right) . \qquad (10)$$

The divergence theorem and the boundary conditions imply (see [4]) that

$$\langle S\mathbf{u}, \mathbf{v} \rangle = \int_\Omega \left( \sum_{i=1}^{l} \left( K_i \nabla u_i \cdot \overline{\nabla} v_i - \frac{1}{2}(\operatorname{div} \mathbf{b}_i) u_i \bar{v}_i \right) + \frac{1}{2} \sum_{i,j=1}^{l} (V_{ij} + V_{ji}) u_j \bar{v}_i \right) .$$

The operator $S$ itself falls into the type (4) if and only if

$$V_{ij} = -V_{ji} \quad (i \neq j) \quad \text{and} \quad \eta_i = V_{ii} - \frac{1}{2}(\operatorname{div} \mathbf{b}_i) . \qquad (11)$$

**Proposition 1.** (cf. [7]). *Under Assumptions BVP and condition* (11), *Assumptions A are satisfied and therefore the truncated GCG-LS algorithm for system* (1) *converges superlinearly in the space* $H_0^1(\Omega)^l$ *according to the estimate* (9) *with the parameter* $\varrho = 1$.

Using the truncated algorithm can be beneficial, but it is a significant restriction not to have the freedom to choose the coefficients $\eta_i$ of $S$ in (4). For convection-dominated problems, large values of $\eta_i$ might compensate the large **b** [8]. Now let us consider the preconditioner operator (4) with arbitrary nonnegative parameters $\eta_i$.

**Proposition 2.** (cf. [7]). *Assume that* $K_i \equiv K \in \mathbb{R}$, $\eta_i \equiv \eta \in \mathbb{R}$ *and* $\mathbf{b}_i \equiv \mathbf{b} \in \mathbb{R}^N$ *are constants,* $V \in \mathbb{R}^{l \times l}$ *is a normal matrix and suppose that Assumptions BVP hold. Then the full version of the preconditioned GCG-LS algorithm* (8) *for system* (1) *with the preconditioning operator* (4) *converges superlinearly in the space* $H_0^1(\Omega)^l$ *according to the estimate* (9).

Now let us consider the discretized problem (5). Then as shown in [7], the GCG method can be defined similarly as in (8), simply replacing $L$ and $S$ by $\mathbf{L}_h$ and $\mathbf{S}_h$, in particular, in step (2f) $z_{k+1}$ is defined as the FEM solution of the problem $S z_{k+1} = L d_{k+1}$ in the considered subspace $V_h$. Then the right-hand side of (9) provides a mesh independent superlinear convergence estimate for the discretized problem. Besides the superlinear convergence result, the advantage of the preconditioning method (4) is that the elliptic operators are decoupled, i. e. the corresponding matrix $\mathbf{S}_h$ is symmetric block-diagonal, hence auxiliary equations for the discretized system like $\mathbf{S}_h \mathbf{z}_h = \mathbf{L}_h \mathbf{d}_h$ (step (2f) in algorithm (8)) can be divided into $l$ parts and they can be solved simultaneously.

## 4   Parallelization of the GCG-LS Algorithm

The basic advantage of the proposed preconditioner is its inherent parallelism. The $k$th iteration of the full version of GCG-LS algorithm consists of two matrix-vector multiplications with matrix $\mathbf{L}_h$, one preconditioning step (solving a system of equations with the preconditioner), solving a system of $k$ equations, $2k + s + 2$ inner products, and $s + 2$ linked triads (a vector updated by a vector multiplied by a scalar).

Let us consider a parallel system with $p$ processors. We divide the vectors $u_k, d_k, r_k, z_k$ (defined in (8)) in such a way that first $\left\lceil \frac{l}{p} \right\rceil$ blocks are stored in the first processor, blocks for $i = \left\lceil \frac{l}{p} \right\rceil + 1, \ldots, 2 \left\lceil \frac{l}{p} \right\rceil$ in the second processor and so on. Then the preconditioning step and linked triads do not need any communication between processors. The computation of inner products requires one global communication to accumulate the local inner products computed on each processor. Communication time for computing inner products increases with the number of processors but in general it is small. The matrix-vector multiplication requires exchanging of data between all processors. Communication time for matrix-vector multiplication depends on the size of the matrix and on the number of processors.

## 5   Numerical Experiments

In this section we report the results of the experiments executed on a Linux cluster consisting of 4 dual processor PowerPCs with G4 450 MHz processors, 512 MB memory per node. The developed parallel code has been implemented in C and the parallelization has been facilitated using the MPI library [10,11]. We use the LAPACK library [1] for computing the Cholesky factorization of the preconditioner and for solving the linear systems arising in GCG-LS. The optimization options of the compiler have been tuned to achieve the best performance. Times have been collected using the MPI provided timer. In this paper we report the best results from multiple runs.

The first test problem is a class of systems (1) with $l = 2, 3, \ldots, 10$ equations, where $\mathbf{b}_i = \begin{pmatrix} 1 \\ 0 \end{pmatrix}$ and the matrix $V$ is skew-symmetric with elements which are randomly generated constants. Our second test problem comes from the time discretization and Newton linearization of a nonlinear reaction-convection-diffusion system of 10 equations, used in meteorological air-pollution models [12]. Since the run times here have proved to be very similar to the case of a random $10 \times 10$ matrix in the first test problem, we will only present the test results for the first problem.

In what follows, we analyze the obtained parallel time $T_p$ on $p$ processors, relative parallel speed-up $S_p = \frac{T_1}{T_p} \leq p$ and relative efficiency $E_p = \frac{S_p}{p} \leq 1$.

In our experiments we used a stopping criterion $\|r_k\| \leq 10^{-14}$. Table 1 shows the required number of iterations. The obtained parallel time $T_p$ on $p$ processors is presented in Tables 2 and 3. Here $l$ denotes the number of equations. The first column consists of the number of processors. The execution time for problems with $h^{-1} = 32, 64, 128, 192, 256$ in seconds is shown in the next columns. The execution times of the full and truncated version of the algorithm are similar. Because of that we put in Table 3 execution times only for systems of 8 and 10 equations. One can see that for relatively small problems, the execution time on

**Table 1.** Number of iterations

| 1/h | $l$ | | | | | | | | | |
|---|---|---|---|---|---|---|---|---|---|---|
|  | 1 | 2 | 3 | 4 | 5 | 6 | 7 | 8 | 9 | 10 |
| 8 | 9 | 10 | 11 | 12 | 12 | 12 | 13 | 13 | 14 | 14 |
| 16 | 9 | 10 | 12 | 12 | 13 | 13 | 13 | 14 | 14 | 14 |
| 32 | 9 | 10 | 12 | 12 | 13 | 13 | 14 | 14 | 14 | 14 |
| 64 | 9 | 10 | 12 | 12 | 13 | 13 | 14 | 14 | 14 | 14 |
| 128 | 9 | 10 | 12 | 12 | 13 | 13 | 14 | 14 | 14 | 14 |

**Table 2.** Execution time for full version of GCG-LS

| $p$ | $h^{-1}$ | | | |
|---|---|---|---|---|
| | 32 | 64 | 128 | 256 |
| $l = 2$ | | | | |
| 1 | 0.13 | 1.06 | 11.30 | 130.06 |
| 2 | 0.46 | 0.99 | 6.50 | 69.31 |
| $l = 3$ | | | | |
| 1 | 0.22 | 1.91 | 19.05 | 207.86 |
| 2 | 0.55 | 1.47 | 13.24 | 143.40 |
| 3 | 0.60 | 1.39 | 8.41 | 79.30 |
| $l = 4$ | | | | |
| 1 | 0.32 | 2.64 | 25.62 | 648.18 |
| 2 | 0.63 | 1.86 | 14.43 | 332.55 |
| 3 | 0.62 | 1.67 | 14.58 | 149.23 |
| 4 | 0.65 | 1.66 | 10.05 | 84.37 |
| $l = 5$ | | | | |
| 1 | 0.43 | 3.44 | 32.73 | 912.90 |
| 2 | 0.66 | 2.26 | 20.79 | 216.12 |
| 3 | 0.68 | 2.10 | 16.25 | 153.08 |
| 4 | 0.69 | 1.95 | 16.31 | 155.75 |
| 5 | 0.76 | 2.06 | 12.38 | 94.59 |
| $l = 6$ | | | | |
| 1 | 0.54 | 3.96 | 39.92 | 1237.71 |
| 2 | 0.74 | 2.59 | 22.10 | 219.50 |
| 3 | 0.75 | 2.22 | 17.15 | 156.95 |
| 4 | 0.76 | 2.24 | 18.09 | 161.69 |
| 5 | 0.82 | 2.19 | 19.06 | 165.57 |
| 6 | 0.86 | 2.27 | 14.98 | 105.21 |

| $p$ | $h^{-1}$ | | | | |
|---|---|---|---|---|---|
| | 32 | 64 | 128 | 192 | 256 |
| $l = 7$ | | | | | |
| 1 | 0.66 | 5.13 | 47.11 | 171.49 | 1479.28 |
| 2 | 0.79 | 3.17 | 28.60 | 103.44 | 667.80 |
| 3 | 0.77 | 2.74 | 23.54 | 82.53 | 227.45 |
| 4 | 0.82 | 2.70 | 19.14 | 62.73 | 166.62 |
| 5 | 0.88 | 3.55 | 20.95 | 66.59 | 361.98 |
| 6 | 0.94 | 2.80 | 21.71 | 68.22 | 176.53 |
| 7 | 0.97 | 2.78 | 18.56 | 51.21 | 119.14 |
| $l = 8$ | | | | | |
| 1 | 0.79 | 5.96 | 54.17 | 306.79 | 1725.53 |
| 2 | 0.86 | 3.74 | 29.99 | 104.48 | 771.83 |
| 3 | 0.84 | 3.30 | 25.52 | 86.95 | 233.69 |
| 4 | 0.86 | 3.08 | 19.95 | 64.44 | 170.92 |
| 5 | 0.94 | 3.55 | 22.14 | 69.20 | 178.03 |
| 6 | 1.02 | 3.62 | 24.37 | 73.58 | 183.49 |
| 7 | 1.07 | 3.78 | 25.52 | 76.36 | 190.79 |
| 8 | 1.08 | 4.67 | 22.30 | 59.38 | 132.55 |
| $l = 10$ | | | | | |
| 1 | 1.08 | 7.97 | 70.15 | 688.04 | |
| 2 | 0.97 | 4.89 | 38.64 | 132.98 | 1111.04 |
| 3 | 0.95 | 4.16 | 32.82 | 113.15 | 685.93 |
| 4 | 0.99 | 4.43 | 28.75 | 94.33 | 248.61 |
| 5 | 1.12 | 4.13 | 25.35 | 76.26 | 434.87 |
| 6 | 1.18 | 4.50 | 27.88 | 81.52 | 197.62 |
| 7 | 1.22 | 4.69 | 29.99 | 86.40 | 205.91 |
| 8 | 1.30 | 5.49 | 32.45 | 92.05 | 212.42 |

one processor is less than one second and parallelization is not necessary. For medium size problems the parallel efficiency on two processors is close to 90% but on three and more processors it decreases. The reason is that communication between two processors in one node is much faster than communication between nodes. For the largest problems ($h^{-1} = 256$) the available physical memory was not enough to solve the problem on one processor. The corresponding numbers in boxes show an atypical progression which is due to the usage of swap memory. The numerical results show that the main advantage of the parallel algorithm is that we can easily solve large problems using a parallel system with distributed memory.

Figure 1 shows the speed-up $S_p$ of the full version of the algorithm obtained for $h^{-1} = 128$ and $l = 3, 4, \ldots 10$. As it was expected when the number of equations $l$ is divisible by the number of processors $p$ the parallel efficiency of the parallel algorithm is higher. The reason is the partitioning of the vectors $u_k, d_k, r_k, z_k$ onto the processors described in previous section.

**Table 3.** Execution time for GCG-LS(0) for $s = 5$

| $p$ | $h^{-1}$ | | | | $p$ | $h^{-1}$ | | | |
|---|---|---|---|---|---|---|---|---|---|
| | 32 | 64 | 128 | 256 | | 32 | 64 | 128 | 256 |
| | $l = 8$ | | | | | $l = 10$ | | | |
| 1 | 0.84 | 6.07 | 57.02 | 2046.74 | 1 | 1.16 | 8.51 | 76.50 | |
| 2 | 0.48 | 3.46 | 31.01 | 935.01 | 2 | 0.65 | 4.87 | 41.57 | 1335.88 |
| 3 | 0.51 | 3.16 | 26.69 | 255.81 | 3 | 0.67 | 4.55 | 36.44 | 817.74 |
| 4 | 0.59 | 2.99 | 21.45 | 189.93 | 4 | 0.71 | 4.46 | 32.03 | 275.20 |
| 5 | 0.67 | 3.52 | 23.86 | 428.05 | 5 | 0.86 | 4.72 | 29.53 | 522.18 |
| 6 | 0.76 | 3.62 | 26.81 | 437.50 | 6 | 0.96 | 5.14 | 32.62 | 533.91 |
| 7 | 0.82 | 4.15 | 29.04 | 215.17 | 7 | 1.06 | 5.77 | 35.31 | 471.83 |
| 8 | 0.85 | 5.38 | 26.00 | 155.73 | 8 | 1.09 | 6.60 | 38.63 | 482.45 |

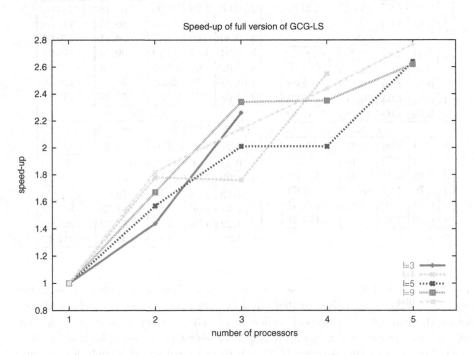

**Fig. 1.** Speed-up of the full version of GCG-LS algorithm

## 6    Concluding Remarks and Future Work

In this paper we have reported on the parallel performance of a new preconditioner applied to the generalized conjugate gradient method used to solve a
sparse linear system arising from systems of convection-diffusion equations. The
proposed preconditioner has inherent parallelism — the preconditioning step is
implemented without any communications between processors. We have shown

that the code parallelizes well, resulting in a highly efficient treatment of large-scale problems.

The next step in development of the parallel code will be the implementation of matrix vector products using nonblocking MPI_Isend functions and avoiding communications for zero elements of the matrix $V$. In this way we can overlap the computation of part of the product and communication between processors. Our future plans include an approximation of the blocks of the preconditioner in order to implement a parallel preconditioning step on multiprocessor systems with more processors.

## Acknowledgments

This research was supported by grant I-1402/2004 from the Bulgarian NSF and by the project BIS-21++ funded by FP6 INCO grant 016639/2005, and by Hungrarian Research Grant OTKA No. T 043765.

## References

1. Anderson, E., Bai, Z., Bischof, C., Blackford, S., Demmel, J., Dongarra, J., Du Croz, J., Greenbaum, A., Hammarling, S., McKenney, A., Sorensen, D.: *LAPACK Users' Guide*, third edition, SIAM, Philadelphia, 1999, http://www.netlib.org/lapack/.
2. Axelsson, O.: A generalized conjugate gradient least square method. Numer. Math. **51** (1987), 209–227.
3. Axelsson, O.: Iterative Solution Methods. Cambridge University Press, 1994.
4. Axelsson, O., Karátson J.: Symmetric part preconditioning for the conjugate gradient method in Hilbert space. Numer. Funct. Anal. **24** (2003), No. 5-6, 455–474.
5. Axelsson, O., Karátson J.: Superlinearly convergent CG methods via equivalent preconditioning for nonsymmetric elliptic operators. Numer. Math. **99** (2004), No. 2, 197-223.
6. Faber, V., Manteuffel, T., Parter, S.V.: On the theory of equivalent operators and application to the numerical solution of uniformly elliptic partial differential equations. Adv. in Appl. Math., **11** (1990), 109–163.
7. Karátson J., Kurics T.: Superlinearly convergent PCG algorithms for some nonsymmetric elliptic systems. J. Comp. Appl. Math., to appear Preprint: http://www.cs.elte.hu/applanal/eng/preprint_eng.html **2006-09**.
8. Manteuffel, T., Otto, J.: Optimal equivalent preconditioners. SIAM J. Numer. Anal. **30** (1993), 790–812.
9. Manteuffel, T., Parter, S. V.: Preconditioning and boundary conditions. SIAM J. Numer. Anal. **27** (1990), no. 3, 656–694.
10. Snir, M., Otto, St., Huss-Lederman, St., Walker, D., Dongara, J.: *MPI: The Complete Reference*. Scientific and engineering computation series (The MIT Press, Cambridge, Massachusetts, 1997) Second printing.
11. Walker, D., Dongara, J.: MPI: a standard Message Passing Interface, *Supercomputer* **63** (1996) 56–68.
12. Zlatev, Z.: Computer treatment of large air pollution models. Kluwer Academic Publishers, Dordrecht-Boston-London, 1995.

# Comparative Analysis of Mesh Generators and MIC(0) Preconditioning of FEM Elasticity Systems

Nikola Kosturski and Svetozar Margenov

Institute for Parallel Processing, Bulgarian Academy of Sciences

**Abstract.** In this study, the topics of grid generation and FEM applications are studied together following their natural synergy. We consider the following three grid generators: Triangle, NETGEN and Gmsh. The quantitative analysis is based on the number of elements/nodes needed to obtain a triangulation of a given domain, satisfying a certain minimal angle condition. After that, the performance of two displacement decomposition (DD) preconditioners that exploit modified incomplete Cholesky factorization MIC(0) is studied in the case of FEM matrices arising from the discretization of the two-dimensional equations of elasticity on non-structured grids.

**Keywords:** FEM, PCG, MIC(0), displacement decomposition.

## 1   Introduction

Mesh generation techniques are now widely employed in various scientific and engineering fields that make use of physical models based on partial differential equations. While there are a lot of works devoted to finite element methods (FEM) and their applications, it appears that the issues of meshing technologies in this context are less investigated. Thus, in the best cases, this aspect is briefly mentioned as a technical point that is possibly non-trivial.

In this paper we consider the problem of linear elasticity with isotropic materials. Let $\Omega \subset \mathbb{R}^2$ be a bounded domain with boundary $\Gamma = \partial \Omega$ and $\mathbf{u} = (u_1, u_2)$ the *displacement* in $\Omega$. The components of the *small strain tensor* are

$$\varepsilon_{ij} = \frac{1}{2} \left( \frac{\partial u_i}{\partial x_j} + \frac{\partial u_j}{\partial x_i} \right), \quad 1 \le i, j \le 2 \tag{1}$$

and the components of the *Cauchy stress tensor* are

$$\tau_{ij} = \sum_{k,l=1}^{2} c_{ijkl} \varepsilon_{kl}(\mathbf{u}), \quad 1 \le i, j \le 2 , \tag{2}$$

where the coefficients $c_{ijkl}$ describe the behavior of the material. In the case of isotropic material the only non-zero coefficients are

$$c_{iiii} = \lambda + 2\mu, \quad c_{iijj} = \lambda, \quad c_{ijij} = c_{ijji} = \mu . \tag{3}$$

T. Boyanov et al. (Eds.): NMA 2006, LNCS 4310, pp. 74–81, 2007.

Now, we can introduce the *Lamé's* system of linear elasticity (see, e.g., [2])

$$(\lambda + \mu)\, \nabla(\nabla \cdot \mathbf{u})_i + \mu \Delta u_i + F_i = 0, \quad 1 \le i \le 2 \tag{4}$$

equipped with boundary conditions

$$u_i(\mathbf{x}) = g_i(\mathbf{x}), \qquad \qquad \mathbf{x} \in \Gamma_D \subset \partial\Omega,$$

$$\sum_{j=1}^{2} \tau_{ij}(\mathbf{x}) n_j(\mathbf{x}) = h_i(\mathbf{x}), \quad \mathbf{x} \in \Gamma_N \subset \partial\Omega ,$$

where $n_j(\mathbf{x})$ denotes the components of the outward unit normal vector $\mathbf{n}$ onto the boundary $\mathbf{x} \in \Gamma_N$. The finite element method (FEM) is applied for discretization of (4) where linear finite elements on a triangulation $\mathcal{T}$ are used. The preconditioned conjugate gradient (PCG) [1] method will be used for the solution of the arising linear algebraic system $K\mathbf{u_h} = \mathbf{f_h}$.

## 2   MIC(0) DD Preconditioning

We first recall some known facts about the modified incomplete Cholesky factorization MIC(0), see, e.g. [4,6]. Let $A = (a_{ij})$ be a symmetric $n \times n$ matrix and let

$$A = D - L - L^T , \tag{5}$$

where $D$ is the diagonal and $-L$ is the strictly lower triangular part of $A$. Then we consider the factorization

$$C_{\text{MIC}(0)} = (X - L)X^{-1}(X - L)^T , \tag{6}$$

where $X = \text{diag}(x_1, \ldots, x_n)$ is a diagonal matrix, such that the sums of the rows of $C_{\text{MIC}(0)}$ and $A$ are equal

$$C_{\text{MIC}(0)}\mathbf{e} = A\mathbf{e}, \quad \mathbf{e} = (1, \ldots, 1) \in \mathbb{R}^n . \tag{7}$$

**Theorem 1.** *Let $A = (a_{ij})$ be a symmetric $n \times n$ matrix and let*

$$\begin{aligned} L &\ge 0 \\ A\mathbf{e} &\ge 0 \\ A\mathbf{e} + L^T\mathbf{e} &> 0 \quad where \quad \mathbf{e} = (1, \ldots, 1)^T . \end{aligned} \tag{8}$$

*Then there exists a stable MIC(0) factorization of $A$, defined by the diagonal matrix $X = \text{diag}(x_1, \ldots, x_n)$, where*

$$x_i = a_{ii} - \sum_{k=1}^{i-1} \frac{a_{ik}}{x_k} \sum_{j=k+1}^{n} a_{kj} > 0 . \tag{9}$$

It is known, that due to the positive offdiagonal entries of the coupled stiffness matrix $K$, the MIC(0) factorization is not directly applicable to precondition the FEM elasticity system. Here we consider a composed algorithm based on a separable displacement two-by-two block representation

$$\begin{pmatrix} K_{11} & K_{12} \\ K_{21} & K_{22} \end{pmatrix} \mathbf{u_h} = \mathbf{f_h} \ . \tag{10}$$

In this setting, the stiffness matrix $K$ is spectrally equivalent to the block-diagonal approximations $C_{\text{SDC}}$ and $C_{\text{ISO}}$

$$C_{\text{SDC}} = \begin{pmatrix} K_{11} & \\ & K_{22} \end{pmatrix}, \quad C_{\text{ISO}} = \begin{pmatrix} A & \\ & A \end{pmatrix}, \tag{11}$$

where $A = \dfrac{1}{2}(K_{11} + K_{22})$. This theoretical background of this displacement decomposition (DD) step is provided by the second Korn's inequality [2]. Now the MIC(0) factorization is applied to the blocks of (11). In what follows, the related preconditioners will be referred to as $C_{\text{SDC-MIC(0)}}$ and $C_{\text{ISO-MIC(0)}}$, cf., References [2,4,5].

## 3   Diagonal Compensation: Condition Number Estimate

The blocks $K_{11}$, $K_{22}$, and $A$ correspond to a certain FEM elliptic problem on the triangulation $\mathcal{T}$. When there are some positive off-diagonal entries in the matrix, the stability conditions for MIC(0) factorization are not satisfied. The diagonal compensation is a general procedure to substitute $A$ by a proper $M$-matrix $\bar{A}$, to which the MIC(0) factorization is then applied. Here, we will restrict our analysis to the case of isotropic DD, i.e., we will consider the piece-wise Laplacian matrix

$$A = \sum_{e \in \mathcal{T}} A_e \tag{12}$$

where the summation sign stands for the standard FEM assembling procedure. The following important geometric interpretation of the current element stiffness matrix holds (see, e.g., in [3])

$$A_e = t_e \begin{pmatrix} \alpha + \beta & -\alpha & -\beta \\ -\alpha & \alpha + 1 & -1 \\ -\beta & -1 & \beta + 1 \end{pmatrix}, \tag{13}$$

where $\theta_1 \geq \theta_2 \geq \theta_3 \geq \tau > 0$ are the angles of the triangle $e \in \mathcal{T}$, $a = \cot\theta_1$, $b = \cot\theta_2$, $c = \cot\theta_3$, $\alpha = \dfrac{a}{c}$ and $\beta = \dfrac{b}{c}$. Since $|a| \leq b \leq c$, the element-by-element diagonal compensation is mandatory applied if and only if $a < 0$. Then, the modified element and global stiffness matrices read respectively as follows

$$\bar{A}_e = t_e \begin{pmatrix} \beta & 0 & -\beta \\ 0 & 1 & -1 \\ -\beta & -1 & \beta + 1 \end{pmatrix}, \quad \bar{A} = \sum_{e \in \mathcal{T}} \bar{A}_e. \tag{14}$$

Note that $\bar{A}_e \equiv A_e$ if $a \geq 0$.

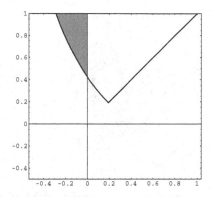

**Fig. 1.** $\tilde{D} = \left\{ \dfrac{1-t^2}{2t^2} \le \alpha < 0, \quad \dfrac{1-\alpha t^2}{t^2(1+\alpha)} \le \beta \le 1 \right\}$

**Theorem 2.** *The relative condition number $\kappa(\bar{A}^{-1}A)$ is uniformly bounded by a constant, depending only on the minimal angle $\tau$. More precisely*

$$\kappa = \kappa(\bar{A}^{-1}A) \le c(\tau) = t^2 \tag{15}$$

*where $t = \cot \tau$.*

*Proof.* We consider the generalized eigenvalue problem $\bar{A}_e \mathbf{u} = \lambda A_e \mathbf{u}$, $\mathbf{u} \ne c\mathbf{e}$. The case $a < 0$ corresponds to $(\alpha, \beta) \in \tilde{D}$, see Fig. 1. Straightforward computations lead to $\lambda_1 = 1 + \alpha + \dfrac{\alpha}{\beta}$, $\lambda_2 = 1$, and therefore

$$\kappa \le \kappa_e = \frac{\lambda_2}{\lambda_1} = \frac{\beta}{\alpha + \beta + \alpha\beta} \le t^2 \, .$$

In the final estimate we have used that the maximal value of $\kappa_e$ is achieved at the corner point of $\tilde{D}$, $(\alpha, \beta) = \left( \dfrac{1-t^2}{2t^2}, 1 \right)$, which completes the proof. $\square$

## 4   Comparison of Mesh Generators

In this section, we compare the following three mesh generators:

- Triangle (http://www.cs.cmu.edu/~quake/triangle.html);
- NETGEN (http://www.hpfem.jku.at/netgen/);
- Gmsh (http://geuz.org/gmsh/).

In the previous section we have seen the impact of the minimal angle on the preconditioning. Let us remind also that the minimal angle directly reflects on the accuracy of the FEM approximation as well as on the condition number of the related stiffness matrix. Since a larger minimal angle usually leads to a larger

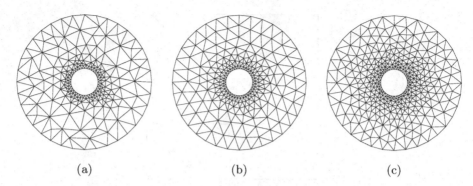

(a)                          (b)                          (c)

**Fig. 2.** Meshes generated by: (a) Triangle; (b) NETGEN; (c) Gmsh

number of elements and nodes in the resulting mesh it is natural to compare the generators based on the numbers of elements and nodes, needed to obtain a mesh with a certain minimal angle.

The domain we chose for this comparison is a disc. The generated meshes, related minimal angles, and numbers of elements and nodes are shown in Fig. 2 and Table 1 respectively.

The results clearly show that Triangle achieved the biggest minimal angle, while also the smallest number of elements and nodes. Note that various parameters may influence the quality of the resulting mesh. Triangle is also the only one of the compared generators that has the minimal angle as a parameter. Triangle's documentation states that the algorithm often succeeds for minimum angles up to $33°$ and usually doesn't terminate for larger angles.

## 5    Numerical Experiments

The presented numerical test illustrate the PCG convergence rate of the two studied displacement decomposition algorithms. For a better comparison, the number of iterations for the CG method are also given. Starting with a given coarse mesh, we refine it uniformly connecting the midpoints of each element. Obviously, such a refinement preserves the minimal angle.

*Remark 1.* The experiments are performed using the perturbed version of the MIC(0) algorithm, where the incomplete factorization is applied to the matrix

**Table 1.** Resulting Mesh Properties

| Generator | Minimal angle | Elements | Nodes |
|-----------|---------------|----------|-------|
| Triangle  | $33.122°$     | 386      | 229   |
| NETGEN    | $27.4256°$    | 440      | 256   |
| Gmsh      | $31.8092°$    | 688      | 380   |

$\tilde{A} = A + \tilde{D}$. The diagonal perturbation $\tilde{D} = \tilde{D}(\xi) = \mathrm{diag}(\tilde{d}_1, \ldots, \tilde{d}_n)$ is defined as follows:

$$\tilde{d}_i = \begin{cases} \xi a_{ii} & \text{if } a_{ii} \geq 2w_i \\ \xi^{1/2} a_{ii} & \text{if } a_{ii} < 2w_i \end{cases} , \tag{16}$$

where $0 < \xi < 1$ is a constant and $w_i = -\sum_{j>i} a_{ij}$.

*Remark 2.* A generalized coordinate-wise ordering is used to ensure the conditions for a stable MIC(0) factorization.

## 5.1   Model Problem in the Unit Square

We consider first a model pure displacement problem in the unit square $\Omega = [0,1] \times [0,1]$ and $\Gamma_D = \partial\Omega$. The material is homogeneous with $\lambda = 1$ and $\mu = 1.5$, and the right-hand side corresponds to the given solution $u_1 = x^2 + \cos(x+y)$, $u_2 = x^3 + y^4 + \sin(y-x)$. A uniform initial (coarsest) triangulation with a mesh size $h = 1/8$ is used. The stopping criteria used is $\|\mathbf{r}^{(k)}\|_\infty \leq 10^{-10}\|\mathbf{r}^{(0)}\|_\infty$.

## 5.2   Model Problem in a Disc

The pure displacement problem with the same given solution and the same stopping criteria as in the the unit square is considered. The computational domain $\Omega$ is a disc with outer radius 1 and inner radius 0.2. The unstructured initial mesh is shown in Fig. 2(a).

## 5.3   Computer Simulation of a Pile Foundation System

We consider the simulation of a foundation system in multi-layer soil media is considered. The system consists of two piles with a linking plate. Fig. 3 (a) shows the geometry of $\Omega$ and the related weak soil layers. The mesh is locally refined in areas with expected concentration of stresses, see Fig. 3 (b). The material characteristics of the concrete (piles) are $\lambda_p = 7666.67\,\mathrm{MPa}$, $\mu_p = 11500\,\mathrm{MPa}$. The related parameters for the soil layers are as follows: $\lambda_{L_1} = 28.58\,\mathrm{MPa}$, $\mu_{L_1} = 7.14\,\mathrm{MPa}$, $\lambda_{L_2} = 9.51\,\mathrm{MPa}$, $\mu_{L_2} = 4.07\,\mathrm{MPa}$, $\lambda_{L_3} = 2.8\,\mathrm{MPa}$, $\mu_{L_3} = 2.8\,\mathrm{MPa}$, $\lambda_{L_4} = 1.28\,\mathrm{MPa}$, $\mu_{L_4} = 1.92\,\mathrm{MPa}$. The forces, acting on the top cross-sections of the piles are $F_1 = (150\,\mathrm{kN}, 2000\,\mathrm{kN})$ and $F_2 = (150\,\mathrm{kN}, 4000\,\mathrm{kN})$.

**Table 2.** Model Problem in the Unit Square

| Mesh | N | CG | ISO-MIC(0) | SDC-MIC(0) |
|------|-------|-----|------------|------------|
| 1 | 81 | 67 | 24 | 19 |
| 2 | 289 | 129 | 30 | 24 |
| 3 | 1089 | 246 | 41 | 32 |
| 4 | 4225 | 445 | 59 | 44 |
| 5 | 16641 | 853 | 81 | 61 |

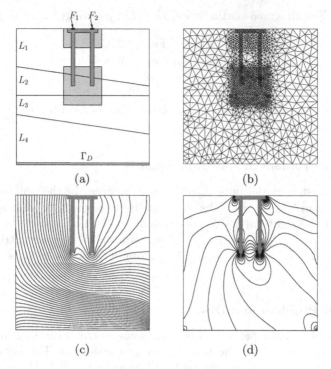

**Fig. 3.** Pile Foundation. (a) Geometry of the pile system and the soil layers; (b) The initial mesh with a local refinement; (c) vertical displacements; (d) vertical stresses.

Dirichlet boundary conditions are applied on the bottom side. Table 4 contains the PCG convergence rate for Jacobi[1] and the two MIC(0) DD preconditioners. The stopping criteria here is $\|\mathbf{r}^{(k)}\|_\infty \leq 10^{-3}\|\mathbf{r}^{(0)}\|_\infty$. Fig. 3 (c) and (d) show the vertical displacements and vertical stresses respectively.

**Table 3.** Model Problem in the Disc

| Mesh | Nodes | CG | ISO-MIC(0) | SDC-MIC(0) |
|------|-------|-----|-----------|-----------|
| 1 | 229 | 68 | 28 | 27 |
| 2 | 844 | 150 | 42 | 40 |
| 3 | 3232 | 335 | 62 | 61 |
| 4 | 12640 | 712 | 97 | 98 |
| 5 | 49984 | 1448 | 159 | 161 |

## 5.4    Concluding Remarks

The rigorous theory of MIC(0) preconditioning is applicable to the first test problem only. For a structured grid with a mesh size $h$ and smoothly varying

---

[1] The diagonal of the original matrix is used as a preconditioner.

**Table 4.** Pile Foundation System

| Mesh | Nodes | Jacobi | ISO-MIC(0) | SDC-MIC(0) |
|------|-------|--------|-----------|-----------|
| 1 | 1449 | 725 | 291 | 227 |
| 2 | 5710 | 1604 | 412 | 349 |
| 3 | 22671 | 3523 | 585 | 527 |
| 4 | 90349 | 7468 | 848 | 811 |
| 5 | 360729 | 15370 | 1274 | 1334 |

material coefficents, the estimate $\kappa(C_h^{-1}A_h) = O(h^{-1}) = O(N^{1/2})$ holds, where $C_h$ is the SDC-MIC(0) or ISO-MIC(0) preconditioner. The number of PCG iterations in this case is $n_{it} = O(N^{1/4})$. The reported number of iterations fully confirm this estimate. Moreover, we observe the same asymptotics of the PCG iterations for the next two problems, which is not supported by the theory up to now. As we see, the considered algorithms have a stable behaviour for unstructured meshes in a curvilinear domain (see Fig. 2(a)). The robustness in the case of local refinement and strong jumps of the coefficients is well illustrated by the last test problem.

## Acknowledgment

The authors gratefully acknowledge the support provided via EC INCO Grant BIS-21++ 016639/2005. The second author has been also partially supported by the Bulgarian NSF Grant I1402.

## References

1. O. Axelsson: *Iterative solution methods*, Cambridge University Press, Cambridge, MA, 1994.
2. O. Axelsson, I. Gustafsson: *Iterative methods for the Navier equations of elasticity*, Comp. Meth. Appl. Mech. Engin., **15** (1978), 241–258.
3. O. Axelsson, S. Margenov: *An optimal order multilevel preconditioner with respect to problem and discretization parameters*, In Advances in Computations, Theory and Practice, Minev, Wong, Lin (eds.), Vol. **7** (2001), Nova Science: New York, 2–18.
4. R. Blaheta: *Displacement Decomposition - incomplete factorization preconditioning techniques for linear elasticity problems*, Numer. Lin. Alg. Appl., **1** (1994), 107–126
5. I. Georgiev, S. Margenov: *DD-MIC(0) preconditioning of rotated trilinear FEM elasticity systems*, Journal of Computer Assisted Mechanics and Engineering Sciences, **11** (2004), 197–209.
6. I. Gustafsson: *An incomplete factorization preconditioning method based on modification of element matrices*, BIT **36:1** (1996), 86–100.

# Multigrid–Based Optimal Shape and Topology Design in Magnetostatics*

Dalibor Lukáš

Department of Applied Mathematics, VŠB–Technical University of Ostrava,
17. listopadu 15, 708 33 Ostrava–Poruba, Czech Republic

**Abstract.** The paper deals with an efficient solution technique to large–scale discretized shape and topology optimization problems. The efficiency relies on multigrid preconditioning. In case of shape optimization, we apply a geometric multigrid preconditioner to eliminate the underlying state equation while the outer optimization loop is the sequential quadratic programming, which is done in the multilevel fashion as well. In case of topology optimization, we can only use the steepest–descent optimization method, since the topology Hessian is dense and large–scale. We also discuss a Newton–Lagrange technique, which leads to a sequential solution of large–scale, but sparse saddle–point systems, that are solved by an augmented Lagrangian method with a multigrid preconditioning. At the end, we present a sequential coupling of the topology and shape optimization. Numerical results are given for a geometry optimization in 2–dimensional nonlinear magnetostatics.

## 1 Introduction

The process of engineering design involves proposing a new prototype, testing it and improvements towards another prototype. This loop can be simulated on a computer if we can exactly determine the design space of improvements, the objective function which evaluates the tests, and the state constraints that model the underlying physical laws. Nowadays, with the rapid progress in computing facilities, the computer aided design requires software tools that are able to solve problems which still fit to the memory (milions of unknowns) and their solution times are in terms of hours. In this spirit, multigrid techniques [2,7] proved to be the relevant methods for partial differential equations. Recently, they have been extended to multigrid–based optimization in optimal control [1,11], in parameter identification [4] or in topology optimization [6]. This paper is a summary of our recent contributions within the multigrid optimization context [8,9]. Moreover, it extends our latest development in multigrid preconditioning of mixed systems [10] towards the topology optimization.

---

* This research has been supported by the Czech Ministry of Education under the grant AVČR 1ET400300415, by the Czech Grant Agency under the grant GAČR 201/05/P008 and by the Austrian Science Fund FWF within the SFB "Numerical and Symbolic Scientific Computing" under the grant SFB F013, subproject F1309.

T. Boyanov et al. (Eds.): NMA 2006, LNCS 4310, pp. 82–90, 2007.

We consider a sufficiently regular fixed computational domain $\Omega \in \mathbb{R}^2$ and the weak formulation of the following 2–dimensional nonlinear magnetostatical state problem:

$$\begin{cases} -\mathrm{div}\left(\nu(x, \|\mathrm{grad}(u(x))\|^2, q(x))\,\mathrm{grad}(u(x))\right) = J(x) \text{ for } x \in \Omega, \\ \qquad\qquad\qquad\qquad\qquad\qquad u(x) = 0 \quad \text{ for } x \in \partial\Omega, \end{cases} \quad (1)$$

where $u$ denotes a scalar magnetic potential so that $\mathrm{curl}(u):=(\partial u/\partial x_2, -\partial u/\partial x_1)$ is a magnetic flux density, $J$ denotes an electric current density and $\nu$ is the following nonlinear magnetic reluctivity:

$$\nu(x,\eta,q(x)) := \nu_0 + (\nu(\eta) - \nu_0)q(x), \quad \nu(\eta) := \nu_1 + (\nu_0 - \nu_1)\frac{\eta^4}{\eta^4 + \nu_0^{-1}},$$

where $\nu_0 := 1/(4\pi 10^{-7})$ [mH$^{-1}$], $\nu_1 := 5100\,\nu_0$ is the air and ferromagnetic reluctivity, respectively, and where $q : \Omega \to \{0,1\}$ denotes a topology design, which tells us whether the point $x$ belongs to the air or the ferromagnetics.

As a typical example we consider a direct electric current (DC) electromagnet, see Fig. 1. It is used for measurements of Kerr magnetooptic effects with

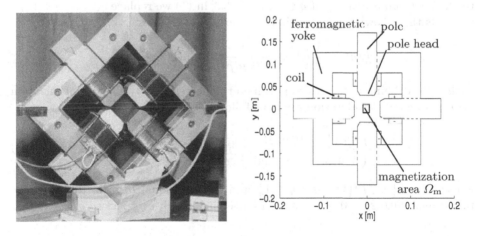

**Fig. 1.** Benchmark problem

applications in development of high–density magnetic or optic data recording media. The measurements require the magnetic field among the pole heads to match a prescribed constant field. Our aim is to design a geometry of the ferromagnetic yoke and pole heads that preserves the requirement. We will discuss topology optimization, where only the current sources are fixed, as well as shape optimization, where, additionally, some initial topology is fixed, which makes the computation less expensive due to the less freedom in the design.

In the shape optimization case, we fix the topology, which means that the shape design splits the domain $\Omega$ into two distinct subdomains as follows:

$$q(x, p) := \begin{cases} 0 \text{ for } x \in \Omega_0(p), \\ 1 \text{ for } x \in \Omega_1(p), \end{cases}$$

where $p \in P$ denotes a parametrization, e.g. Béziér, of the shape of the splitting interface $\Gamma(p) := \partial\Omega_0(p) \cap \partial\Omega_1(p)$. Then, we consider the following shape optimization problem:

$$\begin{cases} \min_{p \in P} \mathcal{I}(u(p)) \\ \text{subject to (1) and } |\Omega_1(p)| \leq V_{\max}, \end{cases} \tag{2}$$

where $V_{\max} > 0$ denotes a maximal admissible volume of the ferromagnetic parts. We assume $\mathcal{I}$ to be twice differentiable and coercive and $P$ to be a compact set of sufficiently regular shapes. In our experiments $\mathcal{I}$ will measure inhomogeneities of the magnetic field density in the following manner:

$$\mathcal{I}(u) := \frac{1}{2|\Omega_{\mathrm{m}}|} \int_{\Omega_{\mathrm{m}}} \|\mathrm{curl}(u(x)) - B_{\mathrm{given}}\|^2 \, dx + \frac{\varepsilon_u}{2|\Omega|} \int_{\Omega} \|\nabla u\|^2 \, dx,$$

where the term with $\varepsilon_u > 0$ regularizes the coercitivity and $\Omega_{\mathrm{m}}$ is as in Fig. 1.

In the topology optimization case, we relax the integer constraint $q(x) \in \{0,1\}$ to the continuous constraint $q \in [0,1]$ so that in (1) we replace $q(x)$ by $q_\rho(q(x))$, which is the following penalization of intermediate values:

$$q_\rho(q) := \frac{1}{2}\left(1 + \frac{1}{\arctan(\rho)}\arctan(\rho(2q - 1))\right)$$

with $\rho \gg 0$ being the penalty parameter. The relevant topology optimization problem under consideration reads as follows:

$$\begin{cases} \min_{q \in Q} \left\{\mathcal{I}(u(q_\rho(q))) + \frac{\varepsilon_q}{2|\Omega|} \int_\Omega q^2 \, dx\right\} \\ \text{subject to (1) and } \int_\Omega q_\rho(q(x)) \, dx \leq V_{\max}, \end{cases} \tag{3}$$

where $Q := \{q \in L^2(\Omega) : 0 \leq q \leq 1\}$ and the additional term in the objective functional, with $\varepsilon_q > 0$, regularizes its coercitivity with respect to $q$.

## 2  Multigrid Nested Shape and Topology Optimization

We aim at an efficient numerical solution to large–scale discretized shape optimization problems arising from (2). In this case the number of design variables is one–order less than the number of state variables, thus the shape Hessian is dense, but rather small, and the overall computational work is performed in the state elimination. After a discretization, the state equation (1) leads to the following nonlinear system of equations:

$$A(u, q(p))u = J.$$

The latter is solved by the nested approach, which means that $u$ in the equation above is eliminated for each shape design $p$ using a nested Newton method with multigrid preconditioned conjugate gradients (MCG) method in the most inner iterations. We propose to couple this nested Newton method with the most outer quasi–Newton optimization method as depicted in Fig. 2.

Given $p_{\text{init}}$, discretize at the first level: $h^{(1)} \longrightarrow p_{\text{init}}^{(1)}, A^{(1)}, J^{(1)}$.
Solve by a quasi-Newton method coupled with a nested Newton method, while using a nested direct solver: $p_{\text{init}}^{(1)} \longrightarrow p_{\text{opt}}^{(1)}$.
Store the first level preconditioner $C^{(1)} := \left[ A_{\text{lin}}^{(1)}(p_{\text{opt}}^{(1)}) \right]^{-1}$.
FOR $l = 2, \ldots$ DO
    Refine: $h^{(l-1)} \longrightarrow h^{(l)}, p_{\text{init}}^{(l)}, A^{(l)}, J^{(l)}$.
    Prolong: $p_{\text{opt}}^{(l-1)} \longrightarrow p_{\text{init}}^{(l)}$.
    Solve by a quasi-Newton method coupled with a nested Newton method, while using the nested MCG method preconditioned with $C^{(l-1)}$: $p_{\text{init}}^{(l)} \longrightarrow p_{\text{opt}}^{(l)}$.
    Store the $l$–th level multigrid preconditioner $C^{(l)} \approx \left[ A_{\text{lin}}^{(l)}(p_{\text{opt}}^{(l)}) \right]^{-1}$.
END FOR

**Fig. 2.** Multigrid shape optimization: the algorithm

For numerical experiments, which were computed using the software Netgen/NgSolve developed by Joachim Schöberl et al. at the University Linz, Austria, see Tab. 1. In the table the second and the fourth column respectively depict the numbers of shape design variables and the numbers of nodes in the discretizations. We can observe an optimal behaviour in terms of the CG iterations, see the numbers before the slash in the sixth column, that are preconditioned using the same geometric multigrid preconditioner throughout the whole algorithm, thus it effectively acts for changing designs $p$. The multigrid preconditioner is built from the linearizations $A_{\text{lin}}^{(l)}(p_{\text{opt}}^{(l)}) := \mathrm{d}A^{(l)}(0, p_{\text{opt}}^{(l)})/\mathrm{d}u$, therefore, the numbers of iterations within the nonlinear steps decay, see the numbers after the slash in the sixth column. Note also that the sensitivity analysis for the shape optimization is performed via an adjoint nonlinear equation, the solution

**Table 1.** Multigrid shape optimization: numerical results

| level | design variables | outer iterations | state variables | maximal inner iterations | CG steps linear/nonlinear | time |
|-------|------------------|------------------|-----------------|--------------------------|---------------------------|------|
| 1 | 19 | 10 | 1098 | 3 | direct solver | 32s |
| 2 | 40 | 15 | 4240 | 3 | 3/14–25 | 2min 52s |
| 3 | 82 | 9 | 16659 | 4 | 4–5/9–48 | 9min 3s |
| 4 | 166 | 10 | 66037 | 4 | 4–6/13–88 | 49min 29s |
| 5 | 334 | 13 | 262953 | 5 | 3–6/20–80 | 6h 36min |

of which takes about the same computational work as the nested state elimination Newton method, which we have to differentiate in the usual adjoint sense,
see [9] for the details.

In case of nested topology optimization we avoid a Newton technique in the
outer optimization loop, as the topology Hessian is both dense and large–scale.
Therefore, we use a steepest–descent optimization method. With this only difference we can apply the algorithm from Fig. 2 to the problem (3). The topology
sensitivity analysis is again as expensive as the cost evaluation, when using the
adjoint method.

## 3   Multigrid All–at–Once Topology Optimization

Now we aim at developing a Newton method for large–scale discretized topology optimization problems arising from (3). For sake of clarity, let us assume
the ferromagnetic reluctivity to be constant $\nu(\eta) := \nu_1$, which is the case of
linear magnetostatics. Contrary to shape optimization, in topology optimization
the numbers of state and design variables are of the same order, therefore the
topology Hessian is both dense and large–scale, and the nested Newton approach
can not be applied. We rather prescribe the state constraint in terms of a Lagrange multiplier $\lambda \in H_0^1(\Omega)$ and we propose to use an active set strategy to
fulfill the other inequality constraints. This leads to a sequence of the following saddle–point systems, which are large–scale, but sparse and well–structured:
Find $(\delta u_k, \delta q_k, \delta \lambda_k) \in H_0^1(\Omega) \times L^2(\Omega) \times H_0^1(\Omega)$:

$$
\begin{pmatrix}
L & , & \text{sym.} & , \text{sym.} \\
B(\lambda_k, q_k) & , & I(u_k, q_k, \lambda_k) & , \text{sym.} \\
L(q_k) & , & B(u_k, q_k)^T & , 0
\end{pmatrix}
\begin{pmatrix}
\delta u_k \\
\delta q_k \\
\delta \lambda_k
\end{pmatrix}
= -
\begin{pmatrix}
f(u_k, q_k, \lambda_k) \\
g(u_k, q_k, \lambda_k) \\
c(u_k, q_k)
\end{pmatrix}
\qquad (4)
$$

where the entries are the following bilinear or linear forms:

$$
L(u, v) := \int_\Omega (1|_{\Omega_m} + \varepsilon_u) \nabla u \nabla v \, dx, \quad L(q_k)(u, v) := \int_\Omega \nu(q_\rho(q_k)) \nabla u \nabla v \, dx,
$$

$$
B(v_k, q_k)(p, v) := \int_\Omega (\nu_1 - \nu_0) \frac{d q_\rho}{d q}(q_k) \nabla v_k p \nabla v \, dx,
$$

$$
I(u_k, q_k, \lambda_k)(p, q) := \int_\Omega \left( \varepsilon_q + (\nu_1 - \nu_0) \frac{d^2 q_\rho}{d q^2}(q_k) \nabla u_k \nabla \lambda_k \right) p q \, dx,
$$

$$
f(u_k, q_k, \lambda_k)(v) := \int_{\Omega_m} (\operatorname{curl}(u_k) - B_{\text{given}}) \operatorname{curl}(v) \, dx
$$

$$
+ \int_\Omega (\varepsilon_u \nabla u_k + \nu(q_\rho(q_k)) \nabla \lambda_k) \nabla v \, dx,
$$

$$
g(u_k, q_k, \lambda_k)(p) := \int_\Omega \left( \varepsilon_q q_k + (\nu_1 - \nu_0) \frac{d q_\rho}{d q}(q_k) \nabla u_k \nabla \lambda_k \right) p \, dx,
$$

$$c(u_k, q_k)(v) := \int_\Omega \nu(q_\rho(q_k)) \nabla u_k \nabla v \, dx - \int_\Omega J \, v \, dx,$$

where $u, v \in H_0^1(\Omega)$, $p, q \in Q$. The update can be given by the following line–search:

$$u_{k+1} := u_k + t_k \, \delta u_k, \quad q_{k+1} := P_{\widetilde{Q}}(q_k + t_k \, \delta q_k), \quad \lambda_{k+1} := \lambda_k + t_k \, \delta \lambda_k,$$

where $t_k > 0$ and $P_{\widetilde{Q}} : L^2(\Omega) \to \widetilde{Q}$ is the projection onto $\widetilde{Q} := \{q \in Q : \int_\Omega q_\rho(q(x)) \, dx \leq V_{\max}\}$. Unfortunately, so far we have not found a successful globalization strategy to find a proper $t_k$, while the simple one based on minimization of the norm of the right–hand side of (4) failed.

### 3.1   Multigrid–Lagrange Method for the Stokes Problem

As a first step towards solution to (4), we focuse on solution to a linear system. We realize that (4) is rather similar to the 2–dimensional Stokes problem: Find $(u, q) \in [H_0^1(\Omega)]^2 \times L^2(\Omega)$:

$$\begin{pmatrix} A, & \text{sym.} \\ B, & 0 \end{pmatrix} \begin{pmatrix} u \\ q \end{pmatrix} = \begin{pmatrix} F \\ 0 \end{pmatrix} \tag{5}$$

where for $v, w \in [H_0^1(\Omega)]^2$ and for $p \in L^2(\Omega)$ the operators reads as follows: $A(v, w) := \int_\Omega \sum_{i=1}^2 \nabla v_i \nabla w_i \, dx$, $B(v, p) := \int_\Omega \text{div}(v) \, p \, dx$ and $F(v) := \int_\Omega f \, v \, dx$, where $f \in [L^2(\Omega)]^2$. Note that the solution is unique up to a constant hydrostatical pressure $q$.

Our algorithmic development is based on a variant of the augmented Lagrangian method proposed in [5], the convergence of which was proven to depend only on the smallest eigenvalue of $A$. Therefore, we build a multigrid preconditioner to $A$, denoted by $\widehat{A}^{-1}$ and the multigrid preconditioner to the $L^2(\Omega)$–inner product, denoted by $\widehat{M}^{-1}$, which leads to a method of linear computational complexity, see [10] for details. Denote the augmented Lagrange functional of (5) by

$$\mathcal{L}(u, q, \rho) := \frac{1}{2} A(u, u) - F(u) + B(q, u) + \frac{\rho}{2} \|Bu\|_{L^2(\Omega)'}^2,$$

where $\|.\|_{L^2(\Omega)'}$ denotes the norm in the dual space to $L^2(\Omega)$, which can be evaluated due to the Riesz theorem as follows: $\|Bu\|_{L^2(\Omega)'} \approx \|Bu\|_{\widehat{M}^{-1}} :=$ $\sqrt{(Bu)^T \widehat{M}^{-1}(Bu)}$, where $\widehat{M}$ is an approximation of the mass matrix. The algorithm based on a semi–monotonic augmented Lagrangian technique and multigrid preconditioning is described in Fig. 3. In the algorithm the inner minimization is realized via the conjugate gradients method preconditioned with $\widehat{A}^{-1}$, i.e. the inner loop is optimal. It is important that the inner loop is terminated with a precision proportional to the violence of the constraint $Bu = 0$. From the theory in [5] and the fact that we use optimal preconditioners, it follows that also the number of outer iterations is bounded by a constant independently from the fineness of discretization, i.e. the algorithm is of assymptoticaly linear complexity with respect to the number of unknowns. The key point for the optimality of the outer loop is that we preserve a kind of monotonicity of the augmented Lagrange functional, see Fig 3 for the condition for the increase of the penalty $\rho$. Under this condition we have also proven [10] a uniform upper bound on $\rho$.

---

Given $\eta > 0$, $\beta > 1$, $\nu > 0$, $\rho^{(0)} > 0$, $u^{(0)} \in V$, $p^{(0)} \in Z$, precision $\varepsilon > 0$
and feasibility precision $\varepsilon_{\text{feas}} > 0$
FOR $k := 0, 1, 2, \ldots$ DO
    Find $u^{(k+1)} : \|\nabla_u \mathcal{L}(u^{(k+1)}, p^{(k)}, \rho^{(k)})\|_{\widehat{A}^{-1}} \leq \min\left\{\nu\|Bu^{(k+1)}\|_{\widehat{M}^{-1}}, \eta\right\}$
    IF $\|\nabla_u \mathcal{L}(u^{(k+1)}, p^{(k)}, \rho^{(k)})\|_{\widehat{A}^{-1}} \leq \varepsilon$ and $\|Bu^{(k+1)}\|_{\widehat{M}^{-1}} \leq \varepsilon_{\text{feas}}$
        BREAK
    END IF
    $p^{(k+1)} := p^{(k)} + \rho^{(k)} \widehat{M}^{-1} B u^{(k+1)}$
    IF $k > 0$ and $\mathcal{L}(u^{(k+1)}, p^{(k+1)}, \rho^{(k)}) < \mathcal{L}(u^{(k)}, p^{(k)}, \rho^{(k-1)}) + \frac{\rho^{(k)}}{2}\|Bu^{(k+1)}\|_{\widehat{M}^{-1}}^2$
        $\rho^{(k+1)} := \beta\rho^{(k)}$
    ELSE
        $\rho^{(k+1)} := \rho^{(k)}$
    END IF
END FOR
$u^{(k+1)}$ is the solution.

**Fig. 3.** Multigrid preconditioned semi–monotonic augmented Lagrangian method

---

There is an advantage to the classical inexact Uzawa method [3], since in our algorithm we do not need to have independent constraints, i.e. $B$ does not need to be a full rank matrix. Nevertheless, in case of $B$ being full rank and with a special setup ($\beta\nu^2 \approx \rho^{(0)}$), the penalty $\rho$ is never updated and the algorithm in Fig. 3 becomes the inexact Uzawa method applied to the system

$$\begin{pmatrix} A + \rho^{(0)} B^T \widehat{M}^{-1} B \,, \text{sym.} \\ B \qquad\qquad , \quad 0 \end{pmatrix} \begin{pmatrix} u \\ q \end{pmatrix} = \begin{pmatrix} F \\ 0 \end{pmatrix}$$

with $(1/2)\widehat{A}^{-1}$ as a preconditioner for the (1,1) block and $(\lambda/\|B\|^2)\widehat{M}^{-1}$ as a preconditioner for the Schur complement, where $\lambda$ is the ellipticity constant of $A$. Thus, the theory of [3] applies, see [10].

The numerical results were obtained for the data $\Omega := (0,1)^2$, $f(x_1, x_2) := \text{sign}(x_1)\,\text{sign}(x_2)(1,1)$, $u^{(0)} := 0$, $p^{(0)} := 0$, $\varepsilon := 10^{-3}$, $\varepsilon_{\text{feas}} := 10^{-3}$, $\eta := 0.1$, $\rho^{(0)} := 1$, $\beta := 10$, $\nu := 1$. We employed Crouzeix–Raviart finite elements and a block multiplicative multigrid smoother with 3 pre– and 3 post–smoothing steps. The results are depicted in Tab. 2 and they were computed in Matlab. The columns in Tab. 2 respectively denote the level, the numbers of Crouzeix–Raviart nodes, the numbers of elements, the numbers of outer iterations (before the slash in the fourth column), the numbers of inner PCG iterations throughout all the outer iterations (after the slash in the fourth column), and the sum of all the

**Table 2.** Multigrid solution to the Stokes problem

| level $l$ | size($u_l$) | size($q_l$) | outer/PCG iterations | total PCG iterations |
|:---:|:---:|:---:|:---:|:---:|
| 1 | 56 | 32 | 6/1,0,1,2,4,8 | **16** |
| 2 | 208 | 128 | 6/1,0,1,2,5,13 | **22** |
| 3 | 800 | 512 | 6/1,0,1,2,5,14 | **23** |
| 4 | 3136 | 2048 | 6/1,0,1,2,6,14 | **24** |
| 5 | 12416 | 8192 | 6/1,0,1,2,6,15 | **25** |
| 6 | 49408 | 32768 | 6/1,0,1,2,6,16 | **26** |

inner PCG iterations per level. From the last column we can see that the total number of PCG iterations becomes at higher levels almost constant, which is the expected multigrid behaviour.

## 4    Coupling of Topology and Shape Optimization

The coarsely discretized optimal topology design serves as the initial guess for the shape optimization. The first step towards a fully automatic procedure is a shape identification. The second step, we are treating now, is a piecewise smooth approximation of the shape by a Bézier curve $\Gamma(p)$. Let $q_{\text{opt}} \in \mathcal{Q}$ be an optimized discretized material distribution. We solve the following least square fitting problem:

$$\min_{p \in P} \int_{\Omega} [q_{\text{opt}} - \chi \left( \Omega_1 \left( \Gamma(p) \right) \right)]^2 \, dx, \tag{6}$$

where $\chi(\Omega_1)$ is the characteristic function of $\Omega_1$.

When solving (6) numerically, one encounters a problem of intersection of the Bézier shapes with the mesh on which $q_{\text{opt}}$ is elementwise constant. In order to avoid it we use the property that the Bézier control polygon converges linearly to the curve under the procedure that adds control nodes so that the resulting Bézier shape remains unchanged. The integration in (6) is then replaced by a sum over the elements and we deal with intersecting of the mesh and a polygon.

Note that the least square functional in (6) becomes non-differentiable whenever a shape touches the grid. Nevertheless, we compute forward finite differences, which is still acceptable for the steepest-descent optimization method that we use. The smoothness can be achieved by smoothing the characteristic function $\chi(\Omega_1)$.

We consider the benchmark problem depicted in Fig. 1 and simplified as in Fig. 4 **(a)**. Given the initial design $q_{\text{init}} := 0.5$, we start with the topology optimization. The coarse topology optimization problem involves 861 design, 1105 state variables and the optimization runs in 7 steepest descent iterations taking 2.5 seconds, when using the adjoint method for the sensitivity analysis. The second part of the computation is the shape approximation. We are looking for three Bézier curves that fit the optimized topology. There are 19 design

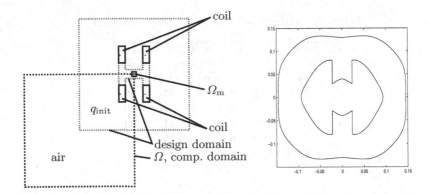

**Fig. 4.** Coupled topology and shape optimization: **(a)** initial topology design; **(b)** optimal shape design

parameters in total and solving the least square problem (6) runs in 8 steepest–descent iterations taking 26 seconds when using numerical differentiation. See Fig. 4 **(b)** for the resulting geometry.

# References

1. Borzi, A.: Multigrid Methods for Optimality Systems, habilitation thesis, TU Graz (2003)
2. Bramble, J. H.: Multigrid Methods. John Wiley & Sons (1993)
3. Bramble, J. H., Pasciak, J. E., Vassilev, A. T.: Analysis of the inexact Uzawa algorithm for saddle point problems, SIAM J. Numer. Anal. **34** (1997) 1072–1092
4. Burger, M., Mühlhuber, W.: Numerical approximation of an SQP–type method for parameter identification, SIAM J. Numer. Anal. **40** (2002) 1775–1797
5. Dostál, Z.: Semi-monotonic inexact augmented Lagrangians for quadratic programming with equality constraints. Optimization Methods and Software **20** (2005), 715–727
6. Dreyer, T., Maar, B., Schulz, V.: Multigrid optimization in applications, J. Comp. Appl. Math. **120** (2000) 67–84
7. Hackbusch, W.: Multi–grid Methods and Applications. Springer (1985)
8. Lukáš, D., Langer, U., Lindner, E., Stainko, R., Pištora, J.: Computational shape and topology optimization with applications to 3–dimensional magnetostatics, Oberwolfach Reports **1** (2004) 601–603
9. Lukáš, D., Chalmovianský, P.: A sequential coupling of optimal topology and multilevel shape design applied to 2–dimensional nonlinear magnetostatics, Comp. Vis. Sci. (to appear)
10. Lukáš, D., Dostál, Z.: Optimal multigrid preconditioned semi–monotonic augmented Lagrangians Applied to the Stokes Problem, Num. Lin. Alg. Appl. (submitted)
11. Schöberl, J., Zulehner, W.: Symmetric indefinite preconditioners for saddle point problems with applications to PDE–constrained optimization problems, SFB Report 2006-19, University Linz (2006)

# Generalized Aggregation-Based Multilevel Preconditioning of Crouzeix-Raviart FEM Elliptic Problems

Svetozar Margenov[1] and Josef Synka[2]

[1] Institute for Parallel Processing, Bulgarian Academy of Sciences,
Acad. G. Bonchev Str. Bl. 25-A, 1113 Sofia, Bulgaria
margenov@parallel.bas.bg
[2] Industrial Mathematics Institute, Johannes Kepler University,
Altenberger Str. 69, A-4040 Linz, Austria
josef.synka@jku.at

**Abstract.** It is well-known that iterative methods of optimal order complexity with respect to the size of the system can be set up by utilizing preconditioners based on various multilevel extensions of two-level finite element methods (FEM), as was first shown in [5]. Thereby, the constant $\gamma$ in the so-called Cauchy-Bunyakowski-Schwarz (CBS) inequality, which is associated with the angle between the two subspaces obtained from a (recursive) two-level splitting of the finite element space, plays a key role in the derivation of optimal convergence rate estimates. In this paper a generalization of an algebraic preconditioning algorithm for second-order elliptic boundary value problems is presented, where the domain is discretized using linear Crouzeix-Raviart finite elements and the two-level splitting is defined by differentiation and aggregation (DA). It is shown that the uniform estimate on the constant $\gamma$ (as presented in [6]) can be improved if a minimum angle condition, which is an integral part in any mesh generator, is assumed to hold in the triangulation. The improved values of $\gamma$ can then be exploited in the set up of more problem-adapted multilevel preconditioners with faster convergence rates.

**Keywords:** multilevel preconditioning, hierarchical basis, differentiation and aggregation.

## 1 Introduction

The target problems in our study are self-adjoint elliptic boundary value problems of the form

$$\mathcal{L}u \equiv -\nabla \cdot (a(\mathbf{x})\, \nabla u(\mathbf{x})) = f(x) \quad \text{in} \quad \Omega$$

$$u = 0 \qquad \text{on} \quad \Gamma_D \qquad (1)$$

$$(a(\mathbf{x})\, \nabla u(\mathbf{x})) \cdot \mathbf{n} = 0 \qquad \text{on} \quad \Gamma_N$$

T. Boyanov et al. (Eds.): NMA 2006, LNCS 4310, pp. 91–99, 2007.

where $\Omega \subset \mathbb{R}^2$ denotes a polygonal domain and $f(\mathbf{x})$ is a given function in $L^2(\Omega)$. The matrix $a(\mathbf{x}) := (a_{ij}(\mathbf{x}))_{i,j \in \{1,2\}}$ is assumed to be bounded, symmetric and uniformly positive definite (SPD) on $\Omega$ with piecewise smooth functions $a_{ij}(\mathbf{x})$ in $\overline{\Omega} := \Omega \cup \partial\Omega$, and $\mathbf{n}$ represents the outward unit normal vector onto the boundary $\Gamma := \partial\Omega$ with $\Gamma = \overline{\Gamma}_D \cup \overline{\Gamma}_N$.

Let us assume that the domain $\Omega$ is discretized using triangular elements and that the fine-grid partitioning, denoted by $\mathcal{T}_h$, is obtained by a uniform refinement of a given coarser triangulation $\mathcal{T}_H$. If the coefficient functions $a_{ij}(\mathbf{x})$ are discontinuous along some polygonal interfaces, we assume that the partitioning $\mathcal{T}_H$ is aligned with these lines to ensure that $a(\mathbf{x})$ is sufficiently smooth over each element $E \in \mathcal{T}_H$. The discrete weak problem of the above problem then reads as follows:

Given $f \in L^2(\Omega)$, find $u_h \in \mathcal{V}_h$ such that $\mathcal{A}_h(u_h, v_h) = (f, v_h) \ \forall v_h \in \mathcal{V}_h$ (2)

is satisfied, where

$$\mathcal{A}_h(u_h, v_h) := \sum_{e \in \mathcal{T}_h} \int_e a(e)\, \nabla u_h(\mathbf{x}) \cdot \nabla v_h(\mathbf{x})\, d\mathbf{x}, \quad (f, v_h) := \int_e f\, v_h\, d\mathbf{x}, \quad (3)$$

and $\mathcal{V}_h := \{v \in L^2(\Omega) : v|_e$ is linear on each $e \in \mathcal{T}_h$, $v$ is continuous at the midpoints of the edges of triangles from $\mathcal{T}_h$ and $v = 0$ at the midpoints on $\Gamma_D\}$.

In our study we will consider the case of piecewise linear Crouzeix-Raviart finite elements, where the nodal basis is associated with the midpoints of the edges rather than at the triangle vertices as used in standard conforming FEM. The construction of multilevel hierarchical preconditioner is based on a two-level framework (cf., e.g., [5,6]). The needed background for the estimates on the constant $\gamma$ in the CBS inequality can be found in [1]. It was shown in [3] for hierarchical basis splitting of the conforming finite elements that under certain assumptions $\gamma$ can be estimated locally by considering a single finite macro-element $E \in \mathcal{T}_H$, which means that $\gamma := \max_{E \in \mathcal{T}_H} \gamma_E$, where

$$\gamma_E = \sup_{u \in \mathcal{V}_{1,E}, v \in \mathcal{V}_{2,E}} \frac{\mathcal{A}_E(u, v)}{[\mathcal{A}_E(u, u)\, \mathcal{A}_E(v, v)]^{1/2}} \quad (4)$$

and $\mathcal{V}_{i,E} := \mathcal{V}_i|_E$ simply denotes the restriction of the functions of the vector space $\mathcal{V}_i$ ($i = 1, 2$) to the macro-element $E \in \mathcal{T}_H$ and, analogously, $\mathcal{A}_E(.,.) := \mathcal{A}(.,.)|_E$. With this local definition it can then be shown that $\gamma$ depends on the construction of the subspaces $\mathcal{V}_1$ and $\mathcal{V}_2$ (i.e. on the type of basis functions chosen), but is independent of the mesh-size parameter $h$ if, e.g., the refinement is performed by congruent triangles, of the geometry of the domain $\Omega$ and is also independent of any discontinuities of the coefficients involved in the bilinear form $\mathcal{A}_E(.,.)$ as long as they do not occur within the element $E$ itself.

In Section 2 we discuss the generalization of the two-level splitting by differences and aggregates (DA) without restrictions to mesh and coefficient anisotropy and show that it comprises the standard DA-splitting. In Section 3 the effect of this generalization on $\gamma$ w.r.t. a minimum angle condition, as commonly used in numerical implementations, is studied. We show that the application of the generalized DA-splitting in a multilevel approach yields the convergence of the underlying stiffness matrix to its counterpart for the equilateral triangle for which the best CBS constant holds. This means that the local CBS constant is continuously improved at each level when applied in a multilevel preconditioning framework based on the generalized DA-splitting. The paper concludes in Section 4 with a brief summary of the main results.

## 2    Generalization of the 2-Level Splitting by Differences and Aggregates (GDA-Splitting)

In order to derive estimates for the CBS-constant, it is known, that it suffices to consider an isotropic (Laplacian) problem in an arbitrarily shaped triangle $T$. Let us denote the angles in such a triangle, as illustrated in Fig. 1, by $\theta_1$, $\theta_2$ and $\theta_3 := \pi - \theta_1 - \theta_2$ and without loss of generality let us assume $\theta_1 \geq \theta_2 \geq \theta_3$. Then, with $a := \cot\theta_1$, $b := \cot\theta_2$ and $c := \cot\theta_3$, the condition $|a| \leq b \leq c$ holds in the triangle (cf. [4,2]).

A simple computation shows that the standard nodal basis element stiffness matrix for a non-conforming Crouzeix-Raviart (CR) linear finite element $A_e^{CR}$ coincides with that for the conforming (c) linear element $A_e^c$, up to a factor 4 and can be written as

$$A_e^{CR} = 2 \begin{pmatrix} b+c & -c & -b \\ -c & a+c & -a \\ -b & -a & a+b \end{pmatrix}. \tag{5}$$

The hierarchical stiffness matrix at macro-element level is then obtained by assembling four such matrices according to the numbering of the nodal points, as shown in Fig. 1(b). But, first we have to know on how to obtain a proper decomposition of the vector space $\mathcal{V}_h$, which is associated with the basis functions at the fine grid. This will be dealt with in the following.

For the non-conforming Crouzeix-Raviart finite element, where the nodal basis functions correspond to the midpoints along the edges of the triangle rather that

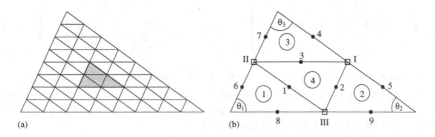

(a)                                                    (b)

**Fig. 1.** Crouzeix-Raviart finite element (a) Discretization (b) Macro-element in detail

at its vertices (cf. Fig. 1), the natural vector spaces $\mathcal{V}_H := \mathrm{span}\{\phi_I, \phi_{II}, \phi_{III}\}$ and $\mathcal{V}_h := \mathrm{span}\{\phi_i\}_{i=1}^9$ (cf. the macro-element in Fig.1(b)) are no longer nested. This renders a direct construction with $\mathcal{V}_2 := \mathcal{V}_H$, as used for conforming elements, impossible. Consequently, the hierarchical basis functions have to be chosen such that the vector space can be written as a direct sum of the resulting vector subspaces $\mathcal{V}_1$ and $\mathcal{V}_2$. That means, the decomposition by differences and aggregates (DA) on the macro-element level is to be based on the splitting $\mathcal{V}(E) := \mathrm{span}\{\Phi_E\} = \mathcal{V}_1(E) \oplus \mathcal{V}_2(E)$, where $\Phi_E := \{\phi_E^{(i)}\}_{i=1}^9$ denotes the set of the "midpoint" basis functions of the four congruent elements in the macro-element $E$, as depicted in Fig. 1(b).

A generalized form of the DA-decomposition can be obtained by enhancing the aggregates in the vector space $\mathcal{V}_2(E)$, while leaving $\mathcal{V}_1(E)$ unaltered: Instead of utilizing only the nodal basis function along the corresponding edge of the inner element, numbered as element 4 in the macro-element, as illustrated in Fig. 1, we now enhance the aggregates by using a linear combination of all three inner basis functions. The generalized DA-splitting (GDA) is thus defined by

$$\mathcal{V}_1(E) := \mathrm{span}\{\phi_1, \phi_2, \phi_3, \phi_4 - \phi_5, \phi_6 - \phi_7, \phi_8 - \phi_9\} \quad \text{and}$$

$$\mathcal{V}_2(E) := \mathrm{span}\{\phi_{123} + \phi_4 + \phi_5, \phi_{312} + \phi_6 + \phi_7, \phi_{231} + \phi_8 + \phi_9\}, \qquad (6)$$

where $\phi_{ijk} := c_i\phi_i + c_j\phi_j + c_k\phi_k$ and $i, j, k \in \{1, 2, 3\}$. The corresponding transformation matrix for this generalized DA-splitting reads

$$J_{DA} = \begin{pmatrix} I_3 & 0 \\ 0 & J^- \\ \frac{1}{2}I_3 + C & J^+ \end{pmatrix} \quad \text{with} \quad C := \begin{pmatrix} c_1 - \frac{1}{2} & c_2 & c_3 \\ c_3 & c_1 - \frac{1}{2} & c_2 \\ c_2 & c_3 & c_1 - \frac{1}{2} \end{pmatrix}. \qquad (7)$$

Compared to the standard DA-splitting, as introduced in [6], solely the lower-left block $\frac{1}{2}I_3$ is now enhanced by the non-zero matrix $C$, as given above, where $I_3$ denotes the $3 \times 3$ unity matrix. The coefficients $c_1, c_2$ and $c_3$ have yet to be determined such that the CBS constant is improved compared to the standard DA-algorithm, where the latter can be retrieved by the setting $c_1 = \frac{1}{2}$ and $c_2 = c_3 = 0$. The matrices $J^-$ and $J^+$ in Eq. (7) are given as

$$J^- := \frac{1}{2}\begin{pmatrix} 1 & -1 & \\ & 1 & -1 \\ & & 1 & -1 \end{pmatrix} \quad \text{and} \quad J^+ := \frac{1}{2}\begin{pmatrix} 1 & 1 & \\ & 1 & 1 \\ & & 1 & 1 \end{pmatrix}. \qquad (8)$$

The matrix $J_{DA}$ transforms a vector of macro-element basis functions $\varphi_E := (\phi_E^{(i)})_{i=1}^9$ to the hierarchical basis vector $\tilde{\varphi}_E := (\tilde{\varphi}_E^{(i)})_{i=1}^9 = J_{DA}\varphi_E$. The hierarchical stiffness matrix at macro-element level can then be computed in a cost-saving way from its standard counterpart as

$$\tilde{A}_E = J_{DA} A_E J_{DA}^T \qquad (9)$$

and can be decomposed into a $2 \times 2$ block-matrix according to the basis functions of $\mathcal{V}_1$ and $\mathcal{V}_2$. Similarly, a two-by-two block structure for the hierarchical global stiffness matrix

$$\sum_{E \in \mathcal{T}_H} \tilde{A}_E =: \tilde{A}_h = \begin{pmatrix} \tilde{A}_{11} & \tilde{A}_{12} \\ \tilde{A}_{21} & \tilde{A}_{22} \end{pmatrix} \tag{10}$$

can be obtained, whereby the upper-left block $\tilde{A}_{11}$ corresponds to the interior degrees of freedom (such as to element 4 in the macro-element, as depicted in Fig. 1(b)) plus the differences of the nodal unknowns along the edges of the macro-element $E \in \mathcal{T}_H$. The lower-right block on the other hand corresponds to the aggregates of nodal unknowns, as defined by $\mathcal{V}_2(E)$ for the macro-elements.

The analysis of the related two-level method can again be performed locally by considering the corresponding problems at macro-element level. In order to ensure the extension of the generalized two-level splitting to the multilevel case, we have to show that $\tilde{A}_{22}$ is properly related to the $A_H$, the stiffness matrix at the coarser level.

*Remark 1.* In [6] it was shown that for the standard DA-splitting the CBS constant is uniformly bounded w.r.t. both coefficient and mesh anisotropy by $\gamma^2 \leq \frac{3}{4}$. And, if $\tilde{A}_{22}$ denotes the stiffness matrix as defined by (10) and $A_H$ is the stiffness matrix corresponding to the finite element space $\mathcal{V}_H$ of the coarse grid discretization $\mathcal{T}_H$, equipped with the standard nodal basis $\{\phi_H^{(k)}\}_{k=1,\ldots,N_H}$, then these two matrices are related as $\tilde{A}_{22} = 4 A_H$.

For the generalized DA-splitting this latter relation takes the form

$$\tilde{A}_{22} = R \cdot A_H, \tag{11}$$

where $R$ is a non-diagonal matrix. As will be shown in the next section, a successive application of $R$ yields convergence of the stiffness matrix to its counterpart corresponding to the equilateral case with the smallest CBS-constant. Hence, this property can be exploited in a multilevel GDA-preconditioner, where the CBS-constant is continuously improved with each level of refinement.

*Remark 2.* From symmetry reasons (since the macro-element was obtained by uniform refinement of a triangle at the coarser level) one obtains $c_3 = c_2$. Note that $\ker(\tilde{A}_{E,22}) = \ker(\tilde{S}_E)$, where $\tilde{S}_E$ is the Schur complement in the generalized eigenvalue problem and subscript $E$ indicates that the evaluations are performed at macro-element level, is a necessary and sufficient condition to have $\gamma_E^2 < 1$. As a result and based on the construction of the hierarchical basis, on the relation (11) and on the $c_3$-setting given above, it follows that $c_1 = \frac{1}{2} - 2 c_2$ must hold.

Hence, the generalized DA-splitting only depends on $c_2 \in [0, \frac{1}{4}]$, which can now be used to optimize the CBS-constant. In the following section we will study the behaviour of $\gamma_E$ for varying values of $c_2$ and varying angles $\theta_i$, $i = 1, 2$, with $\theta_3 = \pi - (\theta_1 + \theta_2)$ in the triangle.

# 3    Study on the Local CBS Constant and on the Property of $A^{(k)}$ in a Multilevel Preconditoning Approach

The analysis in this section is performed locally, i.e., for a single but arbitrary macro-element $E \in \mathcal{T}^{(k)}$. Therefore the obtained results are local, but they straightforwardly provide the global estimates, such as $\gamma \leq \max_{E \in \mathcal{T}_H} \gamma_E$.

It can be shown that the function $\max_{c_2 \in [0,1/4]} \gamma_E^2(c_2, \theta_1, \theta_2, \theta_3)$ attains its minimum at $\theta_3 = \theta_2$ for arbitrary but fixed $\theta_1$ and its overall minimum of $\frac{3}{8}$ for the equilateral triangle. This means, that $\gamma_E^2$ can be improved compared to its value of $\frac{3}{4}$, as obtained for the standard DA-algorithm, where $c_2 = 0$, if the angles in the triangles are chosen sufficiently large. Hence, we want to investigate the effect of $c_2$ on the CBS constant subject to a minimum angle condition. Together with the assumptions of Section 2, we now use $\theta_1 \geq \theta_2 \geq \theta_3 \geq \theta_{min}$, which yields $|a| \leq b \leq c \leq d := \cot(\theta_{min})$.

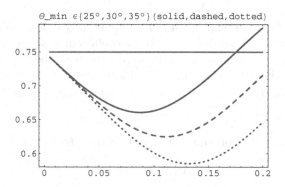

$\theta\_min \in \{25°, 30°, 35°\}$ (solid, dashed, dotted)

**Fig. 2.** Distribution of $\max_{\theta_k \geq \theta_{min}, \, k=1,2,3} \gamma_E^2(c_2, \theta_1, \theta_2, \theta_3)$ for different values of $\theta_{min}$ and comparison with the distribution for the standard DA-algorithm (solid straight line)

Figure 2 shows the distribution of the maximum values of $\gamma_E^2(c_2, \theta_1, \theta_2, \theta_3)$ w.r.t. $\theta_1$, $\theta_2$ and $\theta_3$ for different settings of $\theta_{min} \in \{20°, 25°, 30°\}$, as commonly used in commercial mesh generators. The three $\gamma_E^2$-curves are nearly identical for values of $c_2$ less than about 0.05, but spread out for larger values of $c_2$. It can also be seen that the CBS constant is significantly improved for larger $\theta_{min}$-values and properly chosen values of $c_2$.

Applied in a multilevel context, Eq. (11) (at elemental level) takes the form

$$A_e^{(k+1)} = R^{(k)} \cdot A_e^{(k)} , \tag{12}$$

where $A_e^{(k)}$ denotes the element stiffness matrix at level $k$. The stiffness matrix at the coarsest level is given by $A_e^{(0)}$, which is assembled by element stiffness matrices of the form (5) up to a certain constant factor, and $A_e^{(k+1)} := \tilde{A}_{E,22}^{(k)}$. Since we will normalize the matrix w.r.t. one of its entries in the following, the constant factor, however, will cancel out. $R^{(k)} := R^{(k)}(a, b, c)$ designates the relation matrix between the local stiffness matrices at levels $k + 1$ and $k$. Note that $A_e^{(k+1)}$ is used as the coarse grid matrix at the next finer level and can again be written in the form (5) with the only difference that it now corresponds to a triangle with modified angles (cf. [7]). Consequently, by the GDA construction the stiffness matrix $A_e^{(k)}$ is positive semi-definite at all levels. For all element stiffness matrices $A_e^{(k)}$ the following results can be shown:

**Lemma 1.** *Let $A_e^{(k)}$ be the stiffness matrix at level $k$ and let $A_e^{(k+1)} := \tilde{A}_{E,22}^{(k)}$ (with $\tilde{A}_{E,22}^{(k)}$ defined in analogy with (10)) be the corresponding element stiffness matrix at level $k + 1$, which is related to $A_e^{(k)}$ according to (12). Let*

$$B^{(k)} := c A_e^{(k)} / A_{e,12}^{(k)} \tag{13}$$

*be the normalized local stiffness matrix at level $k$, where $A_{e,12}^{(k)}$ denotes the $(1, 2)$-entry of the matrix $A_e^{(k)}$. Then, under the assumption of convergence and $c_2 > 0$, the limiting value of $B^{(k)}$ is given by the normalized stiffness matrix corresponding to the equilateral triangle, denoted as $B_{eq}$, with $a = b = c = 1/\sqrt{3}$.*

*Note that for $c_2 = 0$ (the standard DA-splitting) the triangle remains the same at all levels, which yields a constant sequence of the cotangens of the angles in the initial triangle.*

**Theorem 1.** *Let $B^{(k)}$ be the normalized element stiffness matrix at a given level $k \in \mathbb{N}_0$, as given by Eq. (13), and let $c_2$ be bounded away from zero, i.e., $c_2 \in [c_{2,min}, 1/4]$ with $c_{2,min} > 0$. Then,*

$$\|B^{(k+1)} - B_{eq}\|_F \leq q \|B^{(k)} - B_{eq}\|_F \tag{14}$$

*is satisfied for some positive $q < 1$, where $\|.\|_F$ denotes the Frobenius norm of a given matrix. Since $q \in [0, 1)$ this yields*

$$\|B^{(k)} - B_{eq}\|_F \longrightarrow 0 \quad if \quad k \to \infty. \tag{15}$$

*The limiting value is thus given by the stiffness element matrix $B_{eq}$, which corresponds to the equilateral triangle.*

*Remark 3.* Note that due to Eqs. (12) and (13) the matrices $A_e^{(k)}$ and $B^{(k)}$ only differ by the constant scaling factor $\alpha^{(k)} := c/A_{e,12}^{(k)}$ at each level. Hence, the convergence (to the minimal value of $\frac{3}{8}$) of the CBS-constants for the normalized matrices $B^{(k)}$ also holds for the original matrices $A_e^{(k)}$.

If $c_2 \in (0, 1/4]$, it can be shown that the upper bound value of the convergence factor, $\bar{q}$, lies in the interval $[4/5, 1)$ and is monotonically decreasing for increasing values of $c_2$.

The improvement of the convergence rate in Eq. (15) is confirmed in the following example, where the behaviour of the true convergence factor $q$ ($< \bar{q}$), as obtained

**Table 1.** Study of the convergence factor $q$ and of $\|B^{(k)} - B_{eq}\|_F$ for different $c_2$-values at different levels $k$

| $k$ | $c_2 = 0.12$ | | $c_2 = 0.20$ | |
|---|---|---|---|---|
| | $q$ | $\|B^{(k)} - B_{eq}\|_F$ | $q$ | $\|B^{(k)} - B_{eq}\|_F$ |
| 1 | 0.963 | 1.496 | 0.853 | 1.325 |
| 3 | 0.961 | 1.379 | 0.826 | 0.876 |
| 5 | 0.959 | 1.260 | 0.803 | 0.518 |
| 10 | 0.954 | 0.971 | 0.763 | 0.104 |
| 20 | 0.945 | 0.502 | 0.732 | 0.003 |

from Eq. (14), and of the Frobenius norm of the distance between $B^{(k)}$ and $B_{eq}$ is studied at different levels and for different settings of $c_2$ for the triangle with $(\theta_1, \theta_2, \theta_3) = (85^\circ, 70^\circ, 25^\circ)$ with an initial value of $\|B^{(0)} - B_{eq}\|_F = 1.554$ at level 0.

From the results in Table 1 it can be seen that the larger the value of $c_2$ the smaller the value of the convergence factor $q$ and thus the faster the convergence (w.r.t. level $k$) is towards the (normalized) equilateral case. Consequently, the CBS-constant decreases faster for larger values of $c_2$ since the stiffness matrix resembles more and more its equilateral counterpart.

For further details and proofs of the stated results, the reader is referred to [7].

## 4   Concluding Remarks

A generalized splitting based on differences and aggregates (GDA) was introduced by enhancing the aggregates in the decomposition of the vector spaces. For $c_2 = 0$ the standard DA-approach can be retrieved for which the CBS constant $\gamma^2$ is uniformly bounded by $3/4$ w.r.t. to both coefficient and mesh anisotropy. It was shown that for the GDA-splitting the local CBS constant can significantly and continuously be improved with each level of refinement if the mesh triangulation satisfies a reasonable minimum angle condition. An exploitation of this favourable behaviour for the set up of a self-adaptive multilevel GDA-preconditioner is subject to future research.

## Acknowledgements

This work was started during the Special Radon Semester on Computational Mechanics, held at RICAM, Linz, Oct. 3rd - Dec. 16th, 2005. The authors gratefully acknowledge the support by the Austrian Academy of Sciences. The first author has also partially been supported by the Bulgarian NSF Grant I1402.

# References

1. O. Axelsson: Iterative solution methods. Cambridge University Press, Cambridge, MA, 1994.
2. O. Axelsson and R. Blaheta: Two simple derivations of universal bounds for the C.B.S. inequality constant, Applications of Mathematics, **49(1)** (2004) 57–72.
3. O. Axelsson and I. Gustafsson: Preconditioning and two-level multigrid methods of arbitrary degree of approximations. Mathematics of Computation **40** (1983) 219–242.
4. O. Axelsson and S. Margenov: An optimal order multilevel preconditioner with respect to problem and discretization parameters. In Advances in Computations, Theory and Practice, Minev, Wong, Lin (eds.), Vol. **7** (2001), Nova Science: New York, 2–18.
5. O. Axelsson and P.S. Vassilevski: Algebraic multilevel preconditioning methods, I. Numer. Math., 56 (1989) 157–177.
6. R. Blaheta, S. Margenov, and M. Neytcheva: Uniform estimate of the constant in the strengthened CBS inequality for anisotropic non-conforming FEM systems. Numerical Linear Algebra with Applications, **11** (2004) 309–326.
7. S. Margenov and J. Synka: Generalized aggregation-based multilevel preconditioning of Crouzeix-Raviart FEM elliptic problems. RICAM-Report 2006-23 (2006), RICAM: Linz, Austria, 1–19.

# Solving Coupled Consolidation Equations

Felicja Okulicka-Dłuzewska

Faculty of Mathematics and Information Science,
Warsaw University of Technology,
Pl. Politechniki 1, 00-661 Warsaw Poland
okulicka@mini.pw.edu.pl

**Abstract.** The iterative method to solve the equations modelling the coupled consolidation problem is presented. The algorithm is tested in the finite element package Hydro-geo for the geotechnical constructions.

## 1  Introduction

Coupled problems appear in the real world everywhere. Description of different mechanical phenomena such as flow, mechanical behavior, thermal effects, leads to coupled systems of differential equations. The finite element method is widely used to solve such problems. The most important part of the finite element method algorithm is the procedure of solving the set of linear and non-linear equations. For big problems it is the most time consuming part of all calculations. In the case when the coupled problems are taken under consideration the structure of the system of equations leads to the application of the block solvers. Each block describes different phenomena or the coupled part, has different condition coefficient and can be calculated independently from others. As the example the consolidation problem is considered and calculated using Hydro-geo package. The large systems of linear equation, especially in the situation when matrices are sparse or bounded, can be easily solved by the parallel iterative method. This method with block structure of the matrix can also be very easy implemented on distributed memory system. The iterative methods allow splitting matrix between separated memory and require to send only the solution vector at each step of the calculation. The cost of the communication is small comparing the direct methods which require sending the parts of matrices. The application of the iterative methods unfortunately can lead to the problems with convergence. In the paper the conjugate gradient method is used to solve the set of equations.

In Section 2 numerical procedure implemented in the finite element package Hydro-geo is recall after [6]. The benchmark problem used for testing is presented in Section 3. The algorithm used to solve the set of equation by iterative method is described in Section 4.

The work is sponsored by the project of Polish Ministry of Science No 4 T07E 020 29 at Warsaw University of Technology, Faculty of Environmental Engineering.

T. Boyanov et al. (Eds.): NMA 2006, LNCS 4310, pp. 100–105, 2007.

## 2   Numerical Procedure

The finite element method package Hydro-Geo [7] oriented at hydro and geotechnical problems is developed at Warsaw University of Technology and next extended to allow the parallel calculations [9,10]. In the paper the new algorithm for solving the set of equations is presented.

In the finite element package Hydro-Geo the virtual work principle, continuity equation with boundary conditions is the starting points for numerical formulation. The finite element method is applied to solve initial boundary value problems. Several procedures stemming from elasto-plastic modelling can be coupled with the time stepping algorithm during the consolidation process. The elasto-plastic soil behavior is modelled by means of visco-plastic theory (Perzyna, 1966). The finite element formulation for the elasto-plastic consolidation combines overlapping numerical processes. The elasto pseudo-viscoplastic algorithm for numerical modelling of elasto-plastic behavior is used after Zienkiewicz and Cormeau (1974). The stability of the time marching scheme was proved by Cormeau (1975). The pseudo-viscous algorithm developed in finite element computer code Hydro-Geo is successfully applied to solve a number of boundary value problems, Długzewski (1993). The visco-plastic procedure was extended to cover the geometrically non-linear problems by Kanchi et al (1978) and also developed for large strains in consolidation, Długzewski (1997) [5]. The pseudo-viscous procedure is adopted herein for modelling elasto-plastic behavior in consolidation. In the procedure two times appear, the first is the real time of consolidation and the second time is only a parameter of the pseudo-relaxation process. The global set of equations for the consolidation process, called the Biot's equation, is derived as follows

$$
\begin{pmatrix} K_T & L \\ L^T & -(S + \Theta \Delta t H^i) \end{pmatrix} \bullet \begin{pmatrix} \Delta u^i \\ \Delta p^i \end{pmatrix} = \begin{pmatrix} 0 & 0 \\ 0 & -\Delta t H^i \end{pmatrix} \bullet \begin{pmatrix} u^i \\ p^i \end{pmatrix} + \begin{pmatrix} \Delta F^i \\ \Delta q \end{pmatrix} \quad (1)
$$

where $K_T$ is the tangent stiffness array, considering large strains effects, $L$ is the coupling array, $S$ is the array responsible for the compressibility of the fluid, $H$ is the flow array, $u$ are the nodal displacements, $p$ are the nodal excesses of the pore pressure, $\Delta F^i$ is the load nodal vector defined below

$$
\Delta F^i = \Delta F_L + \Delta R_I^i + \Delta R_{II}^i \quad (2)
$$

$\Delta F^i$ is the load increment, $\Delta R_I^i$ is the vector of nodal forces due to pseudo-visco iteration, $\Delta R_{II}^i$ is the unbalanced nodal vector due to geometrical nonlinearity. $\Delta R_I^i$ takes the following form

$$
\Delta R_I^i = \int_{{}_{t+\Delta t}^{(i-1)}V} B_{(i-1)}^T D({}_{t+\Delta t}^{t+\Delta t}\Delta \epsilon_i^{vp})_{i-1}^{t+\Delta t} dv \quad (3)
$$

and is defined in the current configuration of the body. The subscripts indicate the configuration of the body, and superscripts indicate time when the value is defined (notation after Bathe (1982)). $\Delta R_I^i$ stands for the nodal vector which results from the relaxation of the stresses. For each time step the iterative procedure is engaged to solve the material non-linear problem. The i-th indicates steps of iterations. Both local and global criterions for terminating the iterative process are used. The iterations are continued until the calculated stresses are acceptable close to the yield surface, $F \leq$ Tolerance at all checked points, where $F$ is the value of the yield function. At the same time the global criterion for this procedure is defined at the final configuration of the body. The global criterion takes its roots from the conjugated variables in the virtual work principle, where the Cauchy stress tensor is coupled with the linear part of the Almansi strain tensor. For two phase medium, the unbalanced nodal vector $\Delta R_{II}^i$ is calculated every iterative pseudo-time step.

$$\Delta R^{k-1} = \int_{t+\Delta t V} N^T f^{t+\Delta t} dV + \int_{t+\Delta t S} N^T t^{t+\Delta t} dS$$
$$- \int_{t+\Delta t}^{(k-1)} V B_{(k-1)}^T D\left(_{t+\Delta t}^{t+\Delta t} \sigma^{j(k-1)} + m_{t+\Delta t}^{t+\Delta t} p^{(k-1)}\right)_{t+\Delta t}^{(k-1)} dV \tag{4}$$

The square norm on the unbalanced nodal forces is used as the global criterion of equilibrium. The iterative process is continued until both criterions are fulfilled.

More details can be find in [8].

In the problem presented above the integration over time leads to more accurate results for the shorter time steps provided the time step is not shorter than the critical one [5,14]. There is a time step bellow which the results show the oscillation of the excess pore pressure in space. The shortest time step which still gives the smooth distribution of the excess pore pressure is called the critical time step. In the paper we do not discuss the problem. The appropriate condition of the solution convergence is presented in [5] and assumed to be fulfilled.

## 3   Benchmark Problem: Two-Dimensional Consolidation

The problem is presented precisely in [5]. The axisymmetrical problem solved analytically by Gibson et al (1970) is modelled. A layer of clay is analyzed. The clay layer resting in the rigid smooth and impermeable foundation is loaded by uniformly distributed traction within a circle. The time stepping process is started from very short time step and next continued until the complete decay of the express pore pressure is observed. The following material data are assumed: $G = 333.3kPa$, $\nu = 0.25$, permeability $k = 10^{-7}m/s$, unit weight of water $\gamma_w = 10kN/m^3$.

## 4   Parallel Numerical Algorithm for Solving the Set of Equations for Consolidation Problem

The block formulation of the coupled problems makes natural the application of the block methods for solving the sets of linear equations. The large matrixes can be split into blocks and put into separate memories of the distributed machine. The parallel calculations are reached due to the matrix operations on separate blocks. The standard numerical algorithms should be rebuilt for the block version or the standard libraries can be used. For big problems the matrix of the system of linear equations is put on distributed memories and iterative methods is used to obtain solution of the set of equations. For the consolidation problem the coefficient matrix is bad-conditioned [1,2,11,12,13].

The condition coefficient for whole matrix of the consolidation problem is of the range $10^6$ at least. At the same time the condition coefficients of blocks built for separate parts, responsible for different phenomena, are much smaller. For the iterative methods with the whole uncoupled system calculation we do not reach the convergence. The method which allows us to have solution is the uncoupled approach.

We use the following notation to simplify the Biot's equation:

$$\begin{pmatrix} K & L \\ L^T & W \end{pmatrix} \bullet \begin{pmatrix} \Delta u \\ \Delta p \end{pmatrix} = \begin{pmatrix} f_1 \\ f_2 \end{pmatrix} \tag{5}$$

where $K$ is the tangent stiffness array, considering large strains effects, $L$ is the coupling array, $W$ is the array responsible for the compressibility and flow of the fluid, $u$ are the nodal displacements, $p$ are the nodal excesses of the pore pressure, $\Delta x$ means the $x$'s increment.

*Remark.* When we consider the norms of the matrices in consolidation equation the following occurs:

$$||K|| >> ||L|| >> ||W||$$

where $||.||$ denotes the norm of matrix.

*Example.* The calculated matrix $K$ for the Gibson problem has the norm of the range $O(10^3)$, the matrix $W$ has the norm of the range $O(10^{-3})$ and the matrix $L$ has the norm of the range $0(1)$. The solution vector $\Delta u$ has the norm equal $O(10^{-3})$ and the vector $\Delta p$ has the norm equal $O(1)$. The differences of the range causes the problems with convergence of the iterative method.

## 4.1 Algorithm

The iterative methods for the set of equations for consolidation problem on the main level on matrix division leads to the following iterative formula:

$$K \bullet \Delta u^{i+1} = f_1 - L \bullet \Delta p^i \tag{6}$$

$$W \bullet \Delta p^{i+1} = f_2 - L^T \bullet \Delta u^i$$

Next, both above equations can be solved in parallel, possibly by block methods.

For each of the equations (6) the Conjugate Gradient algorithm is used. The algorithm is as follows:

*Algorithm*

1. Determine $\Delta u^0$ and $\Delta p^0$ as the solution of the block Gauss elimination.
2. For $i = 1, 2, \ldots$ until convergence do
   (i) solve
   $$K \bullet \Delta u^{i+1} = f_1 - L \bullet \Delta p^i$$
   and
   $$W \bullet \Delta p^{i+1} = f_2 - L^T \bullet \Delta u^i$$
   each by the Conjugate Gradient method
   (ii) $\Delta u^{i+1} = \Delta u^i$
   (iii) $\Delta p^{i+1} = \Delta p^i$
   End For

As the stop condition the error of both equations are used:

$$min(f_1 - L \bullet \Delta p^{i+1} - K \bullet \Delta u^{i+1}, f_2 - L^T \bullet \Delta u^{i+1} - W \bullet \Delta p^{i+1}).$$

and expected to be smaller than the assumed tolerance.

## 4.2    Convergence

The main problem to achieve convergence is to find the start vector for iterations. We use simple block Gaussian elimination. Such solution for large problems is not proper because of the numerical error but it is very good start point for iterative methods. In fact we improve the solution reached by direct method.

The block uncoupled iterative formula allows us to reach the convergence. The iterative methods used for the whole coupled set of equations do not converge for our test Gibson problem independently on the size of the matrix. We do not obtain the solution even for small problems. The speed of the convergence depends on the start point. It means that the change of the solutions in the successive time steps can not be too big.

## 5    Conclusion

The uncoupled iterative method allows us to calculate big problems in parallel. The most difficult part is the convergence of the method. In the paper the start point was chosen as the non-accurate solution from the direct method. In future by applying the preconditioning we can reach better convergence for the arbitrary start point for iteration.

## References

1. Axelsson, O.: *Iterative solution Methods*, Cambridge (1994)
2. Axelsson, O., Barker, V.A.: *Finite Element Solution of Boundary Value Problems*, Academic Press,Inc. (1984)
3. Barret, R., Berry, M., Chan, T., Demmel, J., Donato, J., Dongara, J., Eijkhout, V., Pozo, R., Romine, C., Van der Vost, H.: *Templates for the Solution of Linear Systems: Building Blocks for Iterative Methods*, SIAM, (1994)

4. Demmel, J.: *Applied numerical linear algebra*, (1997)
5. Dłuzewski, J. M.: *Non-linear consolidation in finite element modelling*, Proceedings of the Ninth International Conference on Computer Methods and Advances in Geomechanics, Wuhan, China, November (1997)
6. Dłuzewski, J. M.: *Nonlinear problems during consolidation process*, Advanced Numerical Applications and Plasticity in Geomechanics ed. D.V. Griffiths, and G. Gioda, Springer Verlag, Lecture Notes in Computer Science, (2001)
7. Dłuzewski, J. M.: *HYDRO-GEO - finite element package for geotechnics*, hydrotechnics and environmental engineering, Warsaw (1997) (in Polish)
8. Lewis, R.W., Schrefler, B.A.: *The Finite Element Method in the Static and Dynamic Deformation and Consolidation of Porous Media*, John Wiley and Sons, (1998)
9. Okulicka, F. : *High-Performance Computing in Geomechanics by a Parallel Finite Element Approach*, Lecture Notes in Computer Science 1947, "Applied parallel Computing", 5th International Workshop, PARA 2000, Bergen, Norway, June 2000, pp 391-398
10. Okulicka, F.: *Parallelization of Finite Element ement Package by MPI library*, International Conference of MPI/PVM Users, MPI/PVM 01, Santorini 2001, Lecture Notes in Computer Science 2131, pp 425-436
11. Parter, S.: *Preconditioning Legrendre spectral collocation methods for elliptic problems I: Finite element operators*, SIAM Journal on Numerical Analysis, Vol. 39, No 1, (2001), pp 348-362
12. Phoon, K.K., Toh, K.C., Chan, S.H., Lee, F.H.: *An Efficient Diagonal Preconditioner for Finite Element Solution of Biot's Consolidation Equations*, International Journal of Numerical Methods in Engineering.
13. Saad, Y.: *Iterative methods for Sparse linear systems*, SIAM (2003)
14. Wienands, R., Gaspar, F.J., Lisbona, F.J. , Oosterlee, C.W.: *An effcient multigrid solver based on distributive smoothing for poroelasticity equations.* Computing 73, (2004), pp 99-119

# Parallel Schwarz Methods for T-M Modelling

Jiří Starý, Ondřej Jakl, and Roman Kohut

Institute of Geonics, Academy of Sciences of the Czech Republic
{stary,jakl,kohut}@ugn.cas.cz

**Abstract.** The paper deals with a finite element solution of transient thermo-elasticity problems. In this context, it is especially devoted to the parallel computing of nonstationary heat equations, when the linear systems arising in each time step are solved by the overlapping domain decomposition method. The numerical tests are performed by OpenMP and/or MPI solvers on a large benchmark problem derived from geoenvironmental model KBS.

## 1 Introduction

The work is motivated by the model of the prototype nuclear waste repository located at Äspö in Sweden. The aim is mathematical modelling and computer simulation of complex phenomena in the studied domain.

In this paper, we consider the finite element solution of thermo-elasticity problems, which are not fully coupled. We suppose, that deformations are very slow and do not influence temperature fields. Thus, we can divide the problem in two parts. Firstly, we determine the temperature distribution by solving the nonstationary heat equation. Secondly, we solve the linear elasticity problem at given time levels.

The numerical solution of both problems leads to the repeated solution of large linear systems. For this purpose, we use iterative solvers based on the conjugate gradient method with Schwarz-type preconditioners. Such approach naturally allows the parallel implementation of the solvers and so consequently enables to process the original problem more efficiently. We shall investigate these facts by numerical tests on a large geoenvironmental benchmark problem.

The presented work is a continuation of [2]. By the numerical experiments, it shows the stability of the parallel solvers with the one-level Schwarz preconditioner, when the heat conduction problem is solved. Herewith, the efficiency of both the OpenMP and MPI solvers is demonstrated on a parallel computer up to 16 used processors.

## 2 From Thermo-Elasticity to Linear Equations

The thermo-elasticity problem is formulated to find the temperature $\tau = \tau(x,t)$ and the displacement $u = u(x,T)$,

$$\tau\colon\ \Omega \times (0,T) \to R, \qquad u\colon\ \Omega \times (0,T) \to R^3,$$

T. Boyanov et al. (Eds.): NMA 2006, LNCS 4310, pp. 106–113, 2007.

that fulfill the following equations

$$\kappa\rho\frac{\partial\tau}{\partial t} = k\sum_i \frac{\partial^2\tau}{\partial x_i{}^2} + Q(t) \qquad\qquad \text{in}\quad \Omega \times (0,T)\,,$$

$$-\sum_j \frac{\partial\sigma_{ij}}{\partial x_j} = f_i \quad (i=1,\dots,3) \qquad\qquad \text{in}\quad \Omega \times (0,T)\,,$$

$$\sigma_{ij} = \sum_{kl} c_{ijkl}\left[\varepsilon_{kl}(u) - \alpha_{kl}(\tau - \tau_0)\right] \qquad \text{in}\quad \Omega \times (0,T)\,,$$

$$\varepsilon_{kl}(u) = \frac{1}{2}\left(\frac{\partial u_k}{\partial x_l} + \frac{\partial u_l}{\partial x_k}\right) \qquad\qquad \text{in}\quad \Omega \times (0,T)$$

together with the corresponding boundary and initial conditions specified below.

The expressions represent the nonstationary heat conduction equation, the equilibrium equations, the Hook's law and the tensor of small deformations, respectively. The used symbols have the following meaning: $\kappa$ is the specific heat, $\rho$ is the density of material, $k$ is the coefficient of the heat conductivity, $Q$ is the density of the heat source, $f$ is the density of the volume (gravitational) forces, $c_{ijkl}$ are the components of the elasticity tensor, $\alpha_{kl}$ are the coefficients of the heat expansion and $\tau_0$ is the reference (initial) temperature.

For the heat conduction, we use the boundary conditions

$$\tau(x,t) = \hat\tau(x,t) \qquad\qquad \text{on}\quad \Gamma_0 \times (0,T)\,,$$

$$-k\sum_i \frac{\partial\tau}{\partial x_i}n_i = q \qquad\qquad \text{on}\quad \Gamma_1 \times (0,T)\,,$$

$$-k\sum_i \frac{\partial\tau}{\partial x_i}n_i = H(\tau - \hat\tau_{out}) \qquad\quad \text{on}\quad \Gamma_2 \times (0,T)\,,$$

where $\Gamma = \Gamma_0 \cup \Gamma_1 \cup \Gamma_2$. These conditions prescribe the temperature, the heat flow through the surface heat flux $q$ and the heat transfer to the surrounding medium with the temperature $\hat\tau_{out}$. The symbol $H$ denotes the heat transfer coefficient.

For the elasticity, we apply the boundary conditions

$$u_n = \sum_i u_i n_i = 0 \qquad\qquad \text{on}\quad \tilde\Gamma_0 \times (0,T)\,,$$

$$\sigma_t = 0 \qquad\qquad \text{on}\quad \tilde\Gamma_0 \times (0,T)\,,$$

$$\sum_j \sigma_{ij}n_j = g_i \quad (i=1,\dots,3) \qquad \text{on}\quad \tilde\Gamma_1 \times (0,T)\,,$$

setting the displacement, the stresses and the surface loading. Here, $\Gamma = \tilde\Gamma_0 \cup \tilde\Gamma_1$.

As the initial conditions, we specify the initial temperature,

$$\tau(x,0) = \hat\tau_0(x) \qquad\qquad \text{in}\quad \Omega\,.$$

After the variational formulation, the whole thermo-elasticity problem is discretized by the finite elements in space and the finite differences in time. Using the linear finite elements and the backward Euler time discretization, it leads to the solution of linear equations for vectors $\tau^j$, $u^j$ of nodal temperatures and displacements at the time levels $t_j$ $(j = 1, \ldots, N)$ with the time steps $\Delta t_j = t_j - t_{j-1}$. It gives the time stepping algorithm in Figure 1.

Here, $M_h$ is the capacitance matrix, $K_h$ is the conductivity matrix, $A_h$ is the stiffness matrix, $q_h$ comes from the heat sources and $b_h$ represents the volume and the surface forces including a thermal expansion term.

> find $\tau^0$:    $M_h \tau^0 = \tau_0$
> find $u^0$:    $A_h u^0 = b^0 = b_h(\tau^0)$
>
> for $j = 1, \ldots, N$:
>
> > compute    $d^j = M_h \tau^{j-1} + q_h^j$
> > find $\tau^j$:    $(M_h + \Delta t_j K_h)\tau^j = d^j$
> > find $u^j$:    $A_h u^j = b^j = b_h(\tau^j)$
>
> end for

**Fig. 1.** The time stepping algorithm for thermo-elasticity problems

To optimize the computational effort, we use the adaptive time steps. It means, that we can test the time change of the solution and change the time step size if the variation is too small or too large. The testing is based on a local comparison of the backward Euler and Crank-Nicholson steps [3].

## 3   Solution of Linear Systems

The most of the computational work is concentrated to the repeated numerical solution of two large systems of linear equations. In each time step, we must solve the linear system for the heat conduction,

$$(M + \Delta t K)\tau = d.$$

At several time levels, chosen according to the needs of geomechanical expertise (e.g. when the temperature reaches its maximum), we solve also the linear system for the elasticity,

$$Au = b,$$

as a post-processing step. We use iterative solution of both systems based on the well proven preconditioned conjugate gradient method. Whereas in the sequential case the preconditioning is based on the incomplete factorization, parallel solvers take advantage of the additive Schwarz method for the preconditioning step.

More precisely, in the parallel solution the domain is decomposed along the $\mathcal{Z}$ direction into $m$ non-overlapping subdomains $\tilde{\Omega}_k$, which are then extended so that adjacent subdomains $\Omega_k$ overlap themselves by two or more layers of elements, see Figure 2. In practice, the minimal overlap is usually applied. Using the one-level additive Schwarz method, the preconditioning step can be expressed as

$$g = Gr = \sum_{k=1}^{m} I_k B_k^{-1} R_k r \,,$$

where $B_k$ are the finite element matrices corresponding to subproblems on $\Omega_k$ and $I_k$, $R_k = I_k^T$ are the interpolation and restriction matrices, respectively. If $B$ denotes the finite element matrix of the whole problem, $B_k = R_k B I_k$. The local subproblems are solved inexactly, when the matrices $B_k$ are replaced by their incomplete factorizations $B_k^*$.

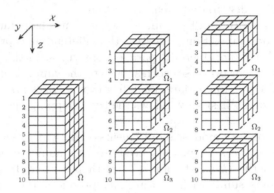

**Fig. 2.** 1-D domain decomposition along the vertical $\mathcal{Z}$ axes. Each hexahedron is further divided into six tetrahedral finite elements.

Moreover, in the preconditioner for the elliptic elasticity problems, we employ the coarse grid created algebraically by aggregation of the original fine grid nodes. Such improvement results in the two-level Schwarz method [5] and ensures the numerical scalability.

On the contrary, in the parabolic problem of the heat conduction, when reasonable assumptions hold [4], we can maintain numerical scalability without help of a coarse grid correction in the preconditioner, see [2] and Table 1. But here, another trouble can occure. For small values of $\Delta t$, the matrix $(M + \Delta t K)$ is not an M-matrix and its incomplete factorization fails for the preconditioning. However, we can apply the incomplete factorization to the matrix $(M^L + \Delta t K)$, where $M^L$ is the lumped matrix $M$, which has each diagonal element equal to the sum of the elements on the corresponding row.

## 4   Parallel Implementation of Solvers

The described thermo-elasticity solvers were integrated into the in-house finite element software package GEM, which serves both for experimental purposes and practical computations in geomechanics. Here, we shall describe only the new solvers for the nonstationary heat conduction part. The elasticity part, when needed, is solved separately by the original parallel conjugate gradient solver using the two-level additive Schwarz preconditioner, see [7].

We conceived the implementation of the solvers as an opportunity to make a practical comparison between the two main standards in parallel programming, message passing and shared memory, and their widely accepted representants, MPI and OpenMP standards. Accordingly, we implemented two parallel solvers.

Let us recall that OpenMP requires shared-memory parallel hardware and allows a simple and scalable directive-based programming, which exploits primarily loop-level parallelism based on local analysis of individual loops. Whereas message passing of MPI is supported and generally available on all parallel architectures including distributed-memory systems.

Both the solvers, written in Fortran, follow the same algorithm and work with the same parallel decomposition, reflecting the different conception, syntax and semantics of the MPI and OpenMP parallel constructs. In this decomposition, the $k$-th of $m$ concurrent processes corresponds to the subproblem $\Omega_k$ and works with a locally stored portion of data, including the matrices $M_k$, $K_k$ and the vectors $\tau_k$, $q_k$, for example. It simply follows the time stepping algorithm presented in Figure 1.

This approach has very modest requirements on data exchange. In fact during the iteration phase, the $k$-th process needs to communicate just locally with its neighbours, i.e. the $(k+1)$-th and $(k-1)$-th processes. In addition to few single value transfers summarizing local inner products, the communication is realized mainly when the matrix-by-vector multiplication, the preconditioning or the nonstationary loading are computed. The amount of the data transferred is quite small, proportional to the overlapped region. Thus, the parallelization has very good assumptions to be efficient and scalable.

# 5    Numerical Experiments

In the background of our interest in the modelling of the thermo-mechanical phenomena is its relevancy to the assessment of the underground repositories of the nuclear waste - a highly urgent topic worldwide, with a great impact on the future of nuclear power utilization. In this context, one of the most internationally recognised project is the Äspö Prototype Repository in Sweden, which is a full-scale experimental realization of the KBS-3 concept of spent nuclear fuel repository [6], where modelling of such phenomena as heat transfer, moisture migration, solute transport and stress/strain development can be verified. We have chosen this model in a simplified form for our numerical experiments.

## 5.1    Geoenvironmental Model KBS

Our benchmark problem is set up as a coupled thermo-elasticity problem with the thermal effects caused by the radioactive waste. The computational domain has dimensions $158\times57\times115$ m and lies 450 m below the ground surface. It closes about a 65 m long tunnel with two separate sections consisting of four and two deposition holes, respectively. Each of deposition holes, intended for a storage of canisters filled with spent nuclear fuel, is 8 m deep and has 1.75 m diameter.

The model is discretized by linear tetrahedral finite elements with 2 586 465 degrees of freedom for the heat conduction and 7 759 395 degrees of freedom for the elasticity computations. The time interval is 100 years, the adaptive time stepping begins with the time step $10^{-4}$ and requires totally 47 time steps.

## 5.2   Computer Systems

With courtesy of the UPPMAX [1], the solvers, originally developed on our small local computing facilities, were ported to the following parallel systems, where the experiments presented below were conducted:

Ra: A cluster delivered by Sun (2005) and based on the AMD Opteron CPUs. In total 99 (non-homogenous) nodes with 280 cores (the total peak performance 1.34 TFlops), 688 GB of (distributed) memory, low-latency InfiniBand (10Gbit/s) and Gigabit Ethernet interconnects, 12 TB of raw disk space. Employed computing nodes Sun V20z with two AMD Opteron 250 processors (2.4 GHz). For more than 4 processors of this new machine, we experience a strange behaviour with rather long computing times, which are still under investigation.

Simba: A shared-memory multiprocessor of the type Sun Fire E 15000 (2001), in total having 48 UltraSPARC-III/900 processors (the theoretical peak performance 86 GFlops), 48 GB of shared memory, Sun Fireplane system interconnect (9.6 GB/s) and 3.4 TB disk storage. Simba is a "virtual server" on this system with 36 CPUs and 36 GB of main memory assigned.

## 5.3   Results

We tested our parallel solvers on the benchmark problem derived from the geoenvironmental model KBS. With respect to our previous long-term experience with

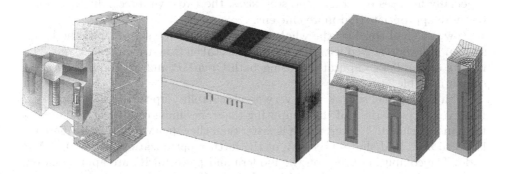

**Fig. 3.** KBS-3 concept and the finite element mesh of $391\times63\times105$ nodes

**Table 1.** The dependence of the number of PCG iterations on the time step size $\Delta t$ and various number of processors #P. Values of $\Delta t$ are presented in years.

| #P | $\Delta t$ | | | | | | | | | |
|---|---|---|---|---|---|---|---|---|---|---|
| | 0.0001 | 0.001 | 0.01 | 0.1 | 1.0 | 5.0 | 10.0 | 100.0 | 1000.0 | |
| 1 | 11 | 11 | 16 | 26 | 38 | 46 | 60 | 109 | 193 | |
| 2 | 12 | 12 | 16 | 26 | 38 | 49 | 64 | 118 | 222 | |
| 4 | 12 | 12 | 16 | 26 | 38 | 49 | 64 | 125 | 238 | |
| *4* | *18* | *17* | *17* | *27* | *41* | *50* | *53* | *83* | *142* | *aggr.* |
| 8 | 14 | 16 | 20 | 26 | 39 | 50 | 68 | 146 | 281 | |
| 12 | 14 | 16 | 20 | 25 | 42 | 54 | 78 | 183 | 328 | |
| 16 | 14 | 16 | 20 | 26 | 42 | 56 | 84 | 212 | 395 | |

sequential and parallel solvers for the elasticity problems [7], only the nonstationary heat conduction part was computed.

We shall begin with the solution of linear system only in the first time step, which started from the initial zero guess and continued up to the relative residual accuracy $10^{-6}$. The results, showing the number of preconditioned conjugate gradient iterations depending on the time step size $\Delta t$ and the number of processors #P, are collected in Table 1. The number of subproblems corresponds to the number of used processors. We tested the usage of the one-level additive Schwarz preconditioner. In case of #P $= 4$, the efficiency of the two-level method was also investigated. The local subproblems are solved inexactly by using an incomplete factorization as well as the subproblem on the coarse grid of $60 \times 10 \times 17$ nodes created by aggregation.

The results prove the numerical stability of the parallel solvers, when for the given time step size, the number of iterations is almost constant with the growing number of subproblems. It holds for sufficiently "small" time step sizes, say $\Delta t \leq 5$, which are acceptable for most of applications. However for the given number of subproblems, the number of iterations naturally grows with the increasing time step size. But such growth is again acceptable for even $\Delta t \leq 10$. This fact supports an idea to use only the one-level preconditioner. Otherwise especially in cases of larger time step sizes, the two-level preconditioner could be more appropriate and more efficient.

Now, we shall consider the whole sequence of 47 time steps, when the linear system is allways solved with the initial guess taken from the previous step. The results of parallel computations using both OpenMP and MPI approaches are shown in Table 2.

Both the parallel solvers show very good scalability up to 16 used processors. However, a comparison of the computing times on Simba confirms a higher performance of the MPI solver, which is faster than the OpenMP one. The difference is up to 36 % and shows necessity of the further optimization of the OpenMP code. The computing times on the modern and powerful Ra are approximately five times shorter than on the older and slower Simba.

**Table 2.** Parallel computations on Simba and Ra. The total number of iterations # It and the computing time T in dependence on the number of processors # P. The relative speed-up S of each parallel solver is related to the sequential run of the same code.

| # P | Simba - OpenMP | | | Simba - MPI | | | Ra - MPI | | |
|---|---|---|---|---|---|---|---|---|---|
| | # It | T [s] | S | # It | T [s] | S | # It | T [s] | S |
| 1 | 1341 | 6292 | | 1344 | 5931 | | 1344 | 1144 | |
| 2 | 1421 | 4101 | 1.63 | 1424 | 3169 | 1.87 | 1421 | 643 | 1.78 |
| 4 | 1425 | 2082 | 3.44 | 1428 | 1577 | 3.76 | 1426 | 314 | 3.64 |
| 8 | 1514 | 1120 | 6.34 | 1514 | 833 | 7.12 | | | |
| 12 | 1578 | 872 | 8.48 | 1581 | 596 | 9.95 | | | |
| 16 | 1614 | 751 | 10.09 | 1618 | 483 | 12.28 | | | |

# 6   Conclusion

This work outlines applications of parallel computing in geotechnics. The experiments with the solution of the nonstationary heat conduction part of the model KBS confirm a good efficiency of the used parallel solvers based on the conjugate gradient method, domain decomposition technique. Especially in the case of the MPI code, the parallel solver shows to be much more efficient than the sequential one for the solution of such kind of large practical problems.

**Acknowledgement.** This work has been supported by the Grant Agency of the Czech Republic, contract No. 105/04/P036.

# References

1. *UPPMAX home page.* Uppsala Multidisciplinary Center for Advanced Computational Science, http://www.uppmax.uu.se (December 15, 2005).
2. R. Kohut, J. Starý, R. Blaheta, K. Krečmer: *Parallel Computing of Thermoelasticity Problems.* In: I. Lirkov, S. Margenov, J. Wasniewski (eds.): Proceedings of the Fifth International Conference on Large-Scale Scientific Computing LSSC'05 held in Sozopol, Springer Verlag, Berlin, 2006, pp. 671–678.
3. R. Blaheta, P. Byczanski, R. Kohut, A. Kolcun, R. Šňupárek: *Large-Scale Modelling of T-M Phenomena from Underground Reposition of the Spent Nuclear Fuel.* In: P. Konečný et al (eds.): EUROCK 2005, Impact of Human Activity on Geological Environment. A.A.Balkema, Leiden, 2005, pp. 49–55.
4. X.-C. Cai: *Multiplicative Schwarz methods for parabolic problems.* SIAM Journal on Scientific Computing 15, 1994, pp. 587–603.
5. B. Smith, P. Bjørstad, W. Gropp: *Domain decomposition. Parallel multilevel methods for Elliptic Partial Differential Equations.* Cambridge University Press, New York, 1996.
6. C. Svemar, R. Pusch: *Prototype Repository - Project description.* IPR-00-30, SKB, Stockholm, 2000.
7. R. Blaheta, P. Byczanski, O. Jakl, R. Kohut, A. Kolcun, K. Krečmer, J. Starý: *Large-scale parallel FEM computations of far/near stress field changes in rocks.* Future Generation Computer Systems, International Journal of Grid Computing: Theory, Methods and Applications - special issue Numerical Modelling in Geomechanics and Geodynamics. Elsevier, volume 22, issue 4, 2006, pp. 449–459.

# Parallel Incomplete Factorization of 3D NC FEM Elliptic Systems

Yavor Vutov

Institute for Parallel Processing, Bulgarian Academy of Sciences

**Abstract.** A new parallel preconditioner for solution of large scale second order 3D FEM elliptic systems is presented. The problem is discretized by rotated trilinear non-conforming finite elements. The algorithm is based on application of modified incomplete Cholesky factorisation (MIC(0)) to a locally constructed modification $B$ of the original stiffness matrix $A$. The matrix $B$ preserves the robustness of the point-wise factorisation and has a special block structure allowing parallelization. The performed numerical tests are in agreement with the derived estimates for the parallel times.

**Keywords:** FEM, PCG, MIC(0), Parallel Algorithms.

## 1 Introduction

We consider the model elliptic boundary value problem:

$$Lu \equiv -\nabla \cdot (a(x)\nabla u(x)) = f(x) \text{ in } \Omega,$$
$$u = 0 \text{ on } \Gamma_D, \tag{1}$$
$$(a(x)\nabla u(x)) \cdot n = 0 \text{ on } \Gamma_N,$$

where $\Omega = [0,1]^3 \subset \mathbb{R}^3$, $\Gamma_D \cup \Gamma_N = \partial\Omega$ and $a(x)$ is a symmetric and positive definite coefficient matrix. The problem is discretized using non-conforming finite elements method (FEM). The resulting linear algebraic system is assumed to be large. The stiffness matrix $A$ is symmetric and positive definite. For large scale problems, the preconditioned conjugate gradient (PCG) method is known to be the best solution method [1].

The recent efforts in development of efficient solution methods for non-conforming finite element systems is inspired by their importance for various applications in scientific computations and engineering [10,3,9]. The goal of this study is to develop a new parallel PCG solver for the arising 3D FEM elliptic systems. A locally modified approximation of the global stiffness matrix is proposed allowing for: a) a stable MIC(0) (modified incomplete Cholesky) factorization; and b) a scalable parallel implementation. The considered non-conforming FEM and MIC(0) factorization are robust for problems with possible jumps of the coefficients.

The algorithm is based on the experience in developing such kind of algorithms for 2D problems using conforming FEM elements on skewed meshes [8] and

T. Boyanov et al. (Eds.): NMA 2006, LNCS 4310, pp. 114–121, 2007.

non-conforming rotated bilinear FEM elements [5,9,6]. The rotated trilinear non-conforming finite elements on hexahedrons are used for the numerical solution of (1).

We assume that $\Omega^h = w_1^{h_1} \times w_2^{h_2} \times w_3^{h_3}$ is a decomposition of the computational domain $\Omega \subset \mathbb{R}^3$ into hexahedrons. The degrees of freedom are associated with the midpoints of the sides. The standard computational procedure leads to the linear system of equations $Ax = b$, where the stiffness matrix $A$ is sparse, symmetric and positive definite.

The rest of this paper is organised as follows. Section 2 describes the element-by-element construction of the preconditioner. Section 3 contains the parallel implementation details and estimates of parallel times. Some results from numerical experiments, are presented in Section 4. Short concluding remarks are given at the end.

## 2    Preconditioning Strategy

We use PCG algorithm with preconditioner based on MIC(0) factorization. The MIC(0) factorization of a real sparse symmetric matrix $A$ has the form:

$$C_{MIC(0)}(A) = (X - L) X^{-1} \left(X - L^T\right). \tag{2}$$

Here $(-L)$ is the strictly lower triangular part of $A$, and $X$ is a diagonal matrix. Since we are going to use $C_{MIC(0)}$ as a preconditioner, we are interested in the case when $X > 0$. This holds when $A$ is a M matrix. More details about MIC(0) factorization can be found in [7,4].

To solve the system with preconditioner (2) one have to solve one system with lower triangular matrix, one with upper triangular and one with diagonal matrix. The solution of the systems with triangular matrices is based on recursive computations. That is why the PCG algorithm with MIC(0) factorization as a preconditioner is inherently sequential. To construct a parallel MIC(0) solver, we introduce a locally constructed approximation $B$ of the original stiffness matrix $A$.

Following the standard FEM assembling procedure we write $A$ in the form $A = \sum_{e \in \omega_h} L_e^T A_e L_e$, where $A_e$ is the element stiffness matrix, $L_e$ stands for the restriction mapping of the global vector of unknowns to the local one corresponding to the current element $e$. Let us consider the following approximation $B_e$ of $A_e$.

$$
A_e = \begin{bmatrix}
a_{11} & a_{12} & a_{13} & a_{14} & a_{15} & a_{16} \\
a_{21} & a_{22} & a_{23} & a_{24} & a_{25} & a_{26} \\
a_{31} & a_{32} & a_{33} & a_{34} & a_{35} & a_{36} \\
a_{41} & a_{42} & a_{43} & a_{44} & a_{45} & a_{46} \\
a_{51} & a_{52} & a_{53} & a_{54} & a_{55} & a_{56} \\
a_{61} & a_{62} & a_{63} & a_{64} & a_{65} & a_{66}
\end{bmatrix}
\quad
B_e = \begin{bmatrix}
b_{11} & 0 & a_{13} & a_{14} & a_{15} & a_{16} \\
0 & b_{22} & a_{23} & a_{24} & a_{25} & a_{26} \\
a_{31} & a_{32} & b_{33} & 0 & a_{35} & a_{36} \\
a_{41} & a_{42} & 0 & b_{44} & a_{45} & a_{46} \\
a_{51} & a_{52} & a_{53} & a_{54} & b_{55} & 0 \\
a_{61} & a_{62} & a_{63} & a_{64} & 0 & b_{66}
\end{bmatrix}
$$

The local numbering follows the pairs of the opposite nodes of the reference element. On fig. 1 (a) and (b) are shown the connectivity patterns of the matrices

$A_e$ and $B_e$. The modification actually removes the links between the degrees of freedom on pairs of opposite sides. The diagonal entries of $B_e$ are modified to hold the rowsum criteria. Assembling the locally defined matrices $B_e$ we get the global one $B = \sum_{e \in \omega_h} L_e^T B_e L_e,$. The sparsity structure of the matrices $A$

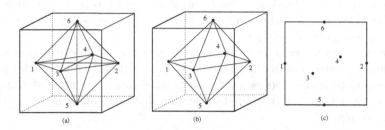

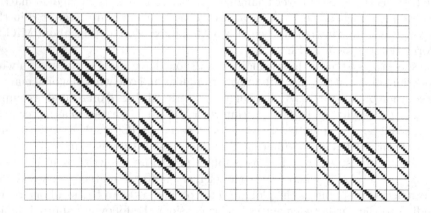

**Fig. 1.** (a) Connectivity pattern of $A_e$;(b) Connectivity pattern of $B_e$; (c) 2D projection of the finite element used in further figures

and $B$ is illustrated by Figure 2. Lexicographic node numbering is used. The important property of the matrix $B$ is that its diagonal blocks are diagonal matrices.

**Fig. 2.** Sparsity pattern of the matrices $A$ (on the left) and $B$ (on the right), for the division of $\Omega$ into 2x2x6 hexahedrons. Non-zero elements are drawn as black dots.

It can be shown that matrices $A$ and $B$ are spectrally equivalent with an uniform estimate for the condition number $\kappa(B^{-1}A)$ . We can now introduce the preconditioner $\mathcal{C}$ for $A$ which is defined as a MIC(0) factorization of $B$, that is, $\mathcal{C} = \mathcal{C}_{MIC(0)}(B)$. This needs of course $B$ to allow for a stable MIC(0) factorization which has been analysed in [5] for the 2D case. The diagonal blocks of $B$ allow a parallel implementation of the resulting PCG algorithm.

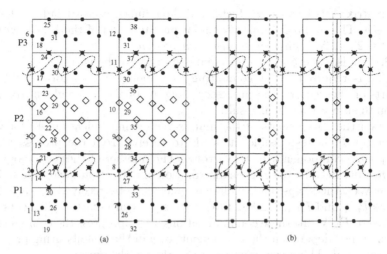

**Fig. 3.** Data distribution: $n_p = 3$, $n_1 = 2$, $n_2 = 2$, $n_3 = 6$. Communication scheme for matrix-vector multiplication (a), and for the solution of systems with lower triangular matrices (b).

# 3   Parallel Implementation

Let us have $n_p \leq n_3 + 1$ processors denoted by $P_1$, $P_2$, ..., $P_{n_p}$. We assume that the domain is decomposed into $n_1 \times n_2 \times n_3$ non-conforming hexahedral elements. On fig. 3 is illustrated the partitioning of the domain into finite elements. A 2D projection on each "slice" of finite elements is used (see fig. 2 (c)). Since the "slices" have common sides, the nodes belonging to those sides appear twice (once on each 2D projection). Each vertical line of nodes corresponds to one of the diagonal blocks of the matrices $A$ and $B$. The corresponding blocks have a varying size of $n_3$ or $n_3 + 1$. The total number of these blocks is $k = n_1(3n_2 + 1) + n_2$. To handle the system with the preconditioner $C_{MIC(0)}(B)w \equiv (X - L) X^{-1} \left( X - L^T \right) w = v$ one has to solve systems $\tilde{L}y \equiv (X - L)y = v$, $X^{-1}z = y$ and $\tilde{L}^T w = z$, $L$ is the strictly lower triangular part of the matrix $B$. The triangular systems are solved using standard forward or backward recurrences. This can be done in $k$ stages. Within stage $i$ the block $y_i$ is computed. Since the blocks $\tilde{L}_{ii}$, are diagonal, computations of each component of $y_i$ are independent of each other and can be performed in parallel. That is the reason to distribute the entries of $y_i$ among all the processors as shown on fig. 3. Therefore each processor $P_j$ receives a strip of the computational domain. These strips are almost equally sized. Elements of all vectors and rows of all matrices which participate in PCG algorithm are distributed in the same manner. The processor $P_i$ is responsible for the local computations on its strip. Let us see now how are performed the operations in the PCG algorithm, and what kind of communications are required.

Each PCG iteration consists of one solution of a system with the preconditioning matrix $C_{MIC(0)}(B)$, one matrix vector multiplication with the original matrix $A$, two inner products, and three linked vector triads of the form $v := \alpha v + u$. The number of operations for one iteration of the PCG algorithm is $\mathcal{N}_{it}^{PCG} \approx 27N$, $N = 3n_1n_2n_3 + n_1n_2 + n_1n_3 + n_2n_3$.

For the triads, each processor calculates its part of the vector $v$ and no communication is required. After computing the inner products corresponding to their parts of the vectors, the processors have to perform one global reduction operation to sum up the final result.

To obtain the components of the matrix-vector multiplication $Av$ for which the processor $P_i$ is responsible, it needs to receive from the processors $P_{i-1}$ and $P_{i+1}$ some components of the vector $v$. The number of these components is $4n_1n_2 + n_1 + n_2$. Because of the even distribution of nodes that lay on the splitting planes between the strips, half of that number is to be received by each of the processors $P_{i-1}$ and $P_{i+1}$. On fig. 3 (a), with sign $\times$ are marked ellements to be transferred to $P_2$. While these communications are in progress the components of $Av$ which do not depend on the components of $v$ in the neighbouring processors can be computed. These components are marked with sign $\diamond$.

Let us go back to the solution of the preconditioner system (see (2)). The solution of a system with a diagonal matrix is trivial and does not require any communications. As we saw, the solution of the triangular systems can be done in $k$ stages. On each stage, the part of the solution, corresponding to one vertical line of nodes, is computed. After each stage, the processors have to exchange some components in order the computations in the next stage to be performed. Three different patterns of transfers are required. They are illustrated with differently dashed lines and arrows. Transfer of one or two components between each pair of nodes in both directions is required per each vertical line. Again, computations for the inner components, marked with sign $\diamond$, can be overlapped with the communications.

Estimation of the parallel times is derived under the following assumptions: a) the execution of $M$ arithmetic operations on one processor takes time $T = Mt_a$, where $t_a$ is the average unit time to perform one arithmetic operation on a single processor, b) the time to transfer $M$ data elements between two neighbouring processors can be approximated by $T^{comm} = t_s + Mt_c$, where $t_s$ is the start-up time and $t_c$ is the incremental time for each of the $M$ elements to be transferred, and c) send and receive operations between each pair of neighbouring processors can be done in parallel. We get the following expressions for the communication times:

$$T^{comm}(C^{-1}v) \approx 6n_1n_2(t_s + 2t_c), \quad T^{comm}(Av) \approx t_s + 2n_1n_2t_c.$$

The above communications are completely local. The inner product needs one broadcasting and one gathering global communication but they do not contribute to the leading terms of the total parallel time. The parallel properties of the algorithm do not depend on the number of iterations, so it is enough to evaluate the parallel time per iteration, and use it in the speedup and efficiency analysis. As the computations are almost equally distributed among the processors, assuming there is no overlapping of the communications and computations one can write for the total time on $n_p$ processors:

$$T_{n_p}^{it} \approx \frac{27Nt_a}{n_p} + T^{comm}(C^{-1}v) + T^{comm}(Av)$$

$$= \frac{27(n_1n_2 + n_3(3n_1n_2 + n_1 + n_2))t_a}{n_p} + t_s + 2n_1n_2(3t_s + 7t_c)$$

What we can observe is that the communication time is practically independent of the number of processors. The situation changes if we have overlapping of computations and communications. One can expect in that case some reduction of the time $T_{n_p}^{it}$. However, with increasing of $n_p$ this effect will weaken. The overlapping can notably reduce the influences of networks' latencies and slow speeds. This is true only if the amount of the computations overlapped is big enough. Of course, there is always an overhead for the communications related to the communication calls – buffering, addresses computations, etc. The speedup $S_{n_p} = T_1/T_{n_p}$ will grow with $n_3$ and it will grow even if $n_1$ and $n_2$ grow as $n_3$ up to the theoretical limit $S_{n_p} = n_p$. However, typically, for the real life parallel systems $t_s \gg t_c$ and $t_s \gg t_a$, and we could expect good speedups and efficiencies $E_{n_p} = S_{n_p}/n_p$ only when $n_3 \gg n_p t_s/t_a$.

## 4  Numerical Tests

The presented algorithm is coded in C++ using the MPI library. Two different reorderings of the calculations are performed to improve the overlapping of the computations and communications: 1) when computing $Ax$ and solving the triangular systems, first are performed the computations which do not depend on the values stored in neighbouring processors. 2) some of the vector operations of the PCG algorithm are performed simultaneously with the solution of the preconditioner. It is observed that the later optimisation also improves the cache utilisation.

The experiments are performed on two parallel platforms. These platforms are referenced further as "A" and "B". Platform "A" is an "IBM SP Cluster 1600" made of 64 nodes p5-575 interconnected with a pair of connections to the Federation HPS (High Performance Switch). Each p5-575 node contains 8 SMP processors Power5 at 1.9GHz and 16GB of RAM. Platform "B" is a "Cray XD1" cabinet, fully populated with 72 2-way nodes, totally 144 AMD Opteron processors at 2.4GHz. Each node has 4GB of memory. The CPUs are interconnected with the Cray RaidArray network.

We consider the model Poisson equation in a unit cube with homogeneous Dirichlet boundary conditions assumed on the right side of the domain. The partitioning is uniform, let $n_1 = n_2 = n_3 = n$. The size of the discrete problem is $N = 3(n^3 + n^2)$. A relative stopping criterion $\frac{(C^{-1}r^i, r^i)}{(C^{-1}r^0, r^0)} < 10^{-9}$ is used in the PCG algorithm, where $r^i$ stands for the residual at the i-th iteration step.

**Table 1.** Number of iterations and sequential times in seconds

| $n$ | $N$ | $n_{it}$ | $T_A$ | $T_B$ |
|------|-----------|----|--------|-------|
| 31 | 92 256 | 22 | 1.2 | 0.7 |
| 63 | 762 048 | 31 | 10.7 | 7.4 |
| 127 | 6 193 536 | 44 | 105.6 | 78.4 |
| 255 | 49 939 200 | 63 | 1098.6 | 834.0 |

The mesh size parameter $n$, the total number of unknowns $N$, the iteration count $n_{it}$ and the sequential times $T_A$ and $T_B$ on platforms "A" and "B" respectively are presented in Table 1. The experiment with $n = 255$ was not run on platform "B" due to lack of enough physical memory available on a single node. The corresponding time in the table is obtained by a simple extrapolation. It is used later to estimate the speedups and the efficiencies of the tests on 4 and 8 processors. One can observe that: 1) the iteration count grows as $O(n^{1/2}) = O(N^{1/6})$; 2) the time per iteration is proportional to $N$.

**Table 2.** Parallel speedups and efficiencies

| $n$ | $n_p$ | $S_{n_p}^A$ | $S_{n_p}^B$ | $E_{n_p}^A$ | $E_{n_p}^B$ |
|------|----|------|------|------|------|
| 31 | 2 | 1.18 | 1.39 | 0.59 | 0.70 |
| | 4 | 1.00 | 1.61 | 0.25 | 0.40 |
| | 8 | 0.96 | 1.98 | 0.12 | 0.25 |
| 63 | 2 | 1.30 | 1.59 | 0.65 | 0.79 |
| | 4 | 1.83 | 2.00 | 0.44 | 0.50 |
| | 8 | 2.15 | 2.46 | 0.27 | 0.33 |
| 127 | 2 | 1.60 | 1.74 | 0.80 | 0.87 |
| | 4 | 2.30 | 2.70 | 0.57 | 0.67 |
| | 8 | 2.22 | 3.79 | 0.28 | 0.47 |
| 255 | 2 | 1.64 | - | 0.82 | - |
| | 4 | 2.63 | 2.91 | 0.65 | 0.72 |
| | 8 | 4.06 | 4.09 | 0.51 | 0.51 |

In the Table 2. are presented the speedups and efficiencies of the performed parallel tests, $n_p$ is the number of processors used. With $S_{n_p}^X$ and $E_{n_p}^X$ are denoted speedup and efficiency on platform "X" using $n_p$ processors. One can see that for a given number of processors the speedup and efficiency grow with the problem size. Conversely for fixed $n$, the speedup and the efficiency decrease with the number of processors. For small ratios $n/n_p$ they are still far from the theoretical upper bounds $S_{n_p} \leq n_p$ and $E_{n_p} \leq 1$. Unfortunately the ratio $n/n_p$ is strongly bounded in real-life computations. As $n$ increases, the total memory requirements increase as $n^3$. Therefore one has to choose $n_p$ sufficiently large in order to fit the required for the computations vectors in the RAM. The applicability of the proposed algorithm will increase in the future when larger amount of memory will be available per CPU.

# 5    Conclusions

A new parallel MIC(0) preconditioner for 3d elliptic problems is proposed and studied. Estimates for the parallel times have been derived. The presented numerical tests are in a good agreement with the theoretical results. The future plans are addressed to improvement of the parallel efficiency based on a more appropriate modification $B$ of the original stiffness matrix $A$ [2]. Its block structure will be with larger diagonal blocks on the diagonal corresponding to mesh nodes of the entire plane. As a result, the number of communication steps in the solution of the preconditioner will decrease and more computations could be overlapped with the communications.

# Acknowledgements

This work has been performed under the EC Project HPC-EUROPA RII3-CT-2003-506079. The author has also been supported by the EC INCO Grant BIS-21++ 016639/2005.

# References

1. O. Axelsson, Iterative Solution Methods, *Cambridge University Press* (1995)
2. P. Arbenz, S. Margenov, Parallel MIC(0) preconditioning of 3D nonconforming FEM systems, Iterative Methods, Preconditioning and Numerical PDEs, Processings (2004), 12-15.
3. D.N. Arnold and F. Brezzi, Mixed and nonconforming finite element methods: implementation, postprocessing and error estimates, *RAIRO, Model. Math. Anal. Numer.*, **19** (1985), 7–32.
4. R. Blaheta, Displacement decomposition - incomplete factorization preconditioning techniques for linear elasticity problems, *Numer. Lin. Alg. Appl.*, **1** (1994), 107–126.
5. G. Bencheva, S. Margenov, Parallel incomplete factorization preconditioning of rotated bilinear FEM systems, *J. Comp. Appl. Mech.*, **4(2)** (2003), 105-117.
6. G. Bencheva, S. Margenov, J. Star, MPI Implementation of a PCG Solver for Nonconforming FEM Problems: Overlapping of Communications and Computations, *Technical Report 2006-023, Uppsala Univerty*
7. I. Gustafsson, Stability and rate of convergence of modified incomplete Cholesky factorization methods, *Report 79.02R. Dept. of Comp. Sci., Chalmers University of Technology*, Goteborg, Sweden, 1979
8. I. Gustafsson, G. Lindskog, On parallel solution of linear elasticity problems: Part I: Theory, *Numer. Lin. Alg. Appl.*, **5** (1998), 123–139.
9. R.D. Lazarov, S.D. Margenov, On a two-level parallel MIC(0) preconditioning of Crouzeuz-Raviart non-conforming FEM systems, *Springer Lecture Notes in Computer Science*, Vol. 2542 (2003), 192-201.
10. R. Rannacher, S. Turek, Simple nonconforming quadrilateral Stokes Element, *Numerical Methods for Partial Differential Equations*, 8(2) (1992), pp. 97–112.

# Extended Object Tracking
# Using Mixture Kalman Filtering*

Donka Angelova[1] and Lyudmila Mihaylova[2]

[1] Institute for Parallel Processing, Bulgarian Academy of Sciences, 25A Acad. G. Bonchev St,
1113 Sofia, Bulgaria
donka@bas.bg
[2] Department of Communication Systems, Lancaster University, South Drive,
Lancaster LA1 4WA, UK
mila.mihaylova@lancaster.ac.uk

**Abstract.** This paper addresses the problem of tracking extended objects. Examples of extended objects are ships and a convoy of vehicles. Such kind of objects have particularities which pose challenges in front of methods considering the extended object as a single point. Measurements of the object extent can be used for estimating size parameters of the object, whose shape is modeled by an ellipse. This paper proposes a solution to the extended object tracking problem by mixture Kalman filtering. The system model is formulated in a conditional dynamic linear (CDL) form. Based on the specifics of the task, two latent indicator variables are proposed, characterising the mode of maneuvering and size type, respectively. The developed Mixture Kalman filter is validated and evaluated by computer simulation.

## 1 Introduction

Most of the target tracking algorithms available in the literature consider the moving object as a single point and estimate its center of mass based on the incoming sensor data, e.g., range and bearing. However, recent high-resolution sensor systems are able to resolve individual features or measurement sources on an extended object. The possibility to additionally make use of this measurements is referred to extended target tracking.

There exist several ways for modelling object extent parameters. A simple ellipsoidal object model is proposed in [1,2] and adopted in our work. The lengths of the major and minor axes of the ellipse have to be calculated, based on the measurements of down-range object extent. Shape parameters are included in [1] in the state vector together with kinematic parameters and are estimated by Extended and Unscented Kalman Filters (EKFs and UKFs) and particle filtering. However, it is pointed out in [1,2] that the EKF implementation is prone to divergence due to high nonlinearity conditions and a particle filtering approach can avoid this problem.

Having in mind the inferences in [1,2] and suggestions in [3], we developed in [4] a joint state and parameter estimation algorithm. It combines the advantages of a *suboptimal Bayesian* interacting multiple model (IMM) filter and of the *Markov Chain Monte*

---

* Research supported in part by the Bulgarian Foundation for Scientific Investigations: MI-1506/05 and by Center of Excellence BIS21++, 016639.

T. Boyanov et al. (Eds.): NMA 2006, LNCS 4310, pp. 122–130, 2007.

*Carlo* (MCMC) approach. Based on the assumption that the unknown shape parameters are defined over a discrete set of values, a *data augmentation* (DA) algorithm for finite mixture estimation [5] is proposed for shape parameters estimation. On the other hand, an IMM filter estimates the kinematic states of a maneuvering ship. The DA procedure provides a precise and convergent size estimate, but it is achieved at the cost of increased computational complexity, since DA is basically an iterative procedure.

In this paper we propose an alternative solution, using a mixture Kalman filter (MKF). The MKF is a sequential Monte Carlo technique for state estimation of conditional dynamic linear systems [6,7]. It recursively generates samples of some indicator variables based on sequential importance sampling and integrates out the linear and Gaussian state variables conditioned on these indicators. Due to marginalisation, the MKF is more accurate than the conventional particle filters. In addition, the features of the task, connected with the motion regimes and size type can be modelled by the indicator variables. The algorithm implemented here is inspired by the ideas in [8], addressing the problem of tracking maneuvering and bending extended target in cluttered environment. Kinematics and shape modes are included in a common modal state vector. In contrast to [8], we consider an indicator vector, comprising two indicator variables characterising the motion regime and the object size. An algorithm with two indicator variables is proposed in [9] (Ch.11) relying on different models and aimed at another application. In this paper we explore the capabilities of mixture Kalman filtering to accomplish both on-line tracking and size type determination of the extended object. The developed MKF is compared with an IMM-DA algorithm.

The paper is organised as follows. Section 2 describes the system dynamics and the measurement model. Section 3 presents the general MKF framework and the designed MKF algorithm for ship tracking. Section 4 illustrates and compares the proposed algorithm performance with an IMM-DA algorithm.Conclusions are given in Section 5.

## 2   System Dynamics and Measurement Models

**System Model.** Consider the following model of a discrete-time jump Markov system, describing the object dynamics and sensor measurements

$$x_k = F(\lambda_{k,1}) x_{k-1} + G(\lambda_{k,1}) w_k(\lambda_{k,1}), \tag{1}$$

$$z_k^1 = H(\lambda_{k,1}) x_k + v_k^1(\lambda_{k,1}), \tag{2}$$

$$z_k^2 = L(\theta_{\lambda_{k,2}}, x_k) + v_k^2(\lambda_{k,2}), \quad k = 1, 2, \ldots, \tag{3}$$

where $x_k \in \mathbb{R}^{n_x}$ is the *base (continuous) state* vector with the transition matrix $F(\lambda_{k,1})$, $\lambda_{k,1}$ and $\lambda_{k,2}$ are modal states, $z_k^1 \in \mathbb{R}^{n_z}$ is the measurement vector with the measurement matrix $H(\lambda_{k,1})$, $z_k^2$ is the scalar measurement of the object extent and $k = 1, 2, \ldots$ is a discrete time. The noises are Gaussian distributed processes having characteristics: $w_k(\lambda_{k,1}) \sim N(0, Q(\lambda_{k,1}))$, $v_k^1(\lambda_{k,1}) \sim N(0, R(\lambda_{k,1}))$ and $v_k^2(\lambda_{k,2}) \sim N(0, R_L(\lambda_{k,2}))$. The matrices are known, assuming that $\Lambda_k = \{\lambda_{k,1}, \lambda_{k,2}\}$ is known. Consider a base state vector in the form $x_k = (x_k, \dot{x}_k, y_k, \dot{y}_k)'$, where $x$ and $y$ specify the object position with respect to known observer position and $(\dot{x}, \dot{y})$ is the object

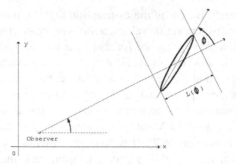

**Fig. 1.** Position of the ship versus the position of the observer

velocity in the Cartesian plane, centered at the observer location (Fig. 1). In particular, we focus on *ship* tracking.

The *modal (discrete) state* $\lambda_{k,1} \in \mathbb{S} \triangleq \{1, 2, \ldots, s\}$ is a first-order Markov chain with transition probabilities $p_{ij} \triangleq Pr\{\lambda_{k,1} = j \mid \lambda_{k-1,1} = i\}$, $(i, j \in \mathbb{S})$ and known initial probability distribution. All possible motion regimes of the maneuvering ship are modelled by the modal state $\lambda_{k,1}$. Denote by $\boldsymbol{\theta} = (\ell, \gamma)'$ a parameter vector, containing the unknown shape parameters: the length of the major axis of the ship ellipse $\ell$ and the ratio of the lengths of the minor and major axes $\gamma$ (aspect ratio). Based on a priori information about ship types, we assume that $\boldsymbol{\theta}$ takes values from a known discrete (size type) set $\boldsymbol{\theta} \in \mathbb{T} \triangleq \{\boldsymbol{\theta}_1, \boldsymbol{\theta}_2, \ldots, \boldsymbol{\theta}_t\}$ with known prior distribution: $P_{\boldsymbol{\theta}_0}(i) \triangleq Pr\{\boldsymbol{\theta} = \boldsymbol{\theta}_i\}, i \in \{1, \ldots, t\}$, such that $P_{\boldsymbol{\theta}_0}(i) \geq 0$, and $\sum_{i=1}^{t} P_{\boldsymbol{\theta}_0}(i) = 1$. Consider the mapping $\varphi : \{1, 2, \ldots, t\} \rightarrow \{\boldsymbol{\theta}_1, \boldsymbol{\theta}_2, \ldots, \boldsymbol{\theta}_t\}$.

The *modal* state $\lambda_{k,2}$ takes its values from the set $\mathbb{N}_t \triangleq \{1, 2, \ldots, t\}$ with the probability $P(\lambda_{k,2} = i \mid \lambda_{k-1,2}) = P(\lambda_{k,2} = i) = 1/t$ and represents the *size type*. In the MKF framework, the *modal* states $\lambda_{k,1}$ and $\lambda_{k,2}$ are referred to *indicator* variables. The index $k$ of $\lambda_{k,2}$ indicates that the size parameters are calculated at time instant $k$, not that the size type is time-varying. Note also that $\mathbb{S}$ is the set of maneuvering modes, whereas $\mathbb{N}_t$ is the set of integers, corresponding to the ship types.

**Measurement Equation.** Similarly to [1,2], we assume, that a high-resolution radar provides measurements of range $r$ and bearing $\beta$ to the object centroid, as well as the object down-range extent $L$ along the observer-object line-of-sight (LOS), Fig. 1. The relationship between $L$ and the angle $\phi$ between the major axis of the ellipse and the target-observer LOS is given by $L(\phi) = \ell\sqrt{\cos^2\phi + \gamma^2\sin^2(\phi)}$. If it is assumed that the target ellipse is oriented so that its major axis is parallel to the velocity vector $(\dot{x}, \dot{y})$, the along-range target extent can be written in terms of the state vector and $\boldsymbol{\theta}$

$$L(\phi(\boldsymbol{x}_k)) = \boldsymbol{\theta}(1)\sqrt{\cos^2\phi(\boldsymbol{x}_k) + \boldsymbol{\theta}(2)^2\sin^2\phi(\boldsymbol{x}_k)}, \qquad (4)$$

where $\phi(\boldsymbol{x}_k) = \arctan((x_k\dot{y}_k - \dot{x}_k y_k)/(x_k\dot{x}_k + y_k\dot{y}_k))$.

Let us denote $\boldsymbol{z}_k^1 = (r_k, \beta_k)'$, $\boldsymbol{z}_k^2 = L_k$ and $\boldsymbol{z}_k = ((\boldsymbol{z}_k^1)', \boldsymbol{z}_k^2)'$. There exist a non-linear relationship between the state and measurements: $\boldsymbol{z}_k^1 = \boldsymbol{h}(\boldsymbol{x}_k) + \boldsymbol{v}_k^1$, where

$$\boldsymbol{h}(\boldsymbol{x}_k) = \left(\sqrt{(x_k - x_o)^2 + (y_k - y_o)^2}, \quad \arctan((y_k - y_o)/(x_k - x_o))\right)' \qquad (5)$$

and $(x_o, y_o)$ is the location of the observer. It is common practice in tracking applications, to perform measurement conversion from polar $(r_k, \beta_k)$ to Cartesian $(x_k, y_k)$ coordinates: $z_k^1 = (r_k \cos(\beta_k), r_k \sin(\beta_k))'$. Thus, the measurement equation becomes linear with a simple measurement matrix $\boldsymbol{H}$. The components of the converted measurement noise covariance matrix $\boldsymbol{R}_c$ can be found in [12].

The problem that we consider has own particularities: the measurements of $L$ are not used for the base state estimation. The kinematic states are estimated through $r$ and $\beta$. The estimated kinematic states are, however, used for the estimation of $\ell$ and $\gamma$. This is the motivation for presenting the system under consideration in the form (1)-(3).

**The goal** is to estimate the *state* vector $\boldsymbol{x}_k$ and the *extent parameter* vector $\boldsymbol{\theta}$, based on all available measurement information $\boldsymbol{Z}^k = \{\boldsymbol{z}_1, \boldsymbol{z}_2, \ldots, \boldsymbol{z}_k\}$.

## 3   Extended Object Tracking by MKF

**MKF summary.** The object-measurement equations (1)-(3) describe a special state-space model, namely the conditional dynamic linear model (CDLM). When conditioning on the indicator vectors $\Lambda_i = \{\lambda_{i,1}, \lambda_{i,2}\}, i = 1, \ldots, k$, the CDLM becomes a dynamic linear model (DLM), and all the $\boldsymbol{x}_s, s = 1, \ldots, k$ can be integrated out recursively by using a standard Kalman filter [6,9,10]. If the Monte Carlo (MC) sampling is working in the space of indicator variables instead of in the space of the state variables, we obtain the MKF, which gives more accurate results than the MC filters dealing with $\boldsymbol{x}_k$ directly. The MKF uses a Gaussian mixture distribution to represent the posterior distribution of $\boldsymbol{x}_k$, which distinguishes it from the MC filters where the posterior state distribution is approximated by a set of samples. When the model is CDLM the "true" target distribution is indeed a mixture of Gaussians, although the number of mixture components increases exponentially with $k$.

Suppose we have a collection of $N$ Kalman filters, $KF_{k-1}^{(1)}, \ldots, KF_{k-1}^{(j)}, \ldots, KF_{k-1}^{(N)}$ at time $k-1$. Each $KF_{k-1}^{(j)}$ is characterisied by the mean vector $\boldsymbol{\mu}_{k-1}^{(j)}$, its state covariance $\boldsymbol{P}_{k-1}^{(j)}$, and the set of indicator vectors $\widetilde{\boldsymbol{\Lambda}}_{k-1}^{(j)} = \left(\Lambda_1^{(j)}, \Lambda_2^{(j)}, \ldots, \Lambda_{k-1}^{(j)}\right)$ up to time $k - 1$, i.e., with $\left(\boldsymbol{\mu}_{k-1}^{(j)}, \boldsymbol{P}_{k-1}^{(j)}, \widetilde{\boldsymbol{\Lambda}}_{k-1}\right)$. Since the CDLM is reduced to a DLM when conditioning on $\widetilde{\boldsymbol{\Lambda}}_{k-1}^{(j)}$, the mean vector $\boldsymbol{\mu}_{k-1}^{(j)}$ and covariance matrix $\boldsymbol{P}_{k-1}^{(j)}$, constitute a sufficient statistics at time $k - 1$. Each filter is associated with a weight $w_{k-1}^{(j)}$. The guideline on how to update the filter $KF_{k-1}^{(j)} \rightarrow KF_k^{(j)}$ at time $k$ can be described as follows [11]:

*The first* step is connected with the computation of a trial sampling density for $\Lambda_k = \{i_1, i_2\}, i_1 \in \mathbb{S}, i_2 \in \mathbb{N}_t$:

$$\mathcal{L}_{k,\{i_1,i_2\}}^{(j)} \triangleq P\left(\Lambda_k = \{i_1, i_2\} \,|\, \widetilde{\boldsymbol{\Lambda}}_{k-1}^{(j)}, \boldsymbol{Z}^k\right) \propto$$
$$p\left(\boldsymbol{z}_k | \Lambda_k = \{i_1, i_2\}, \widetilde{\boldsymbol{\Lambda}}_{k-1}^{(j)}, \boldsymbol{Z}^{k-1}\right) P\left(\Lambda_k = \{i_1, i_2\} \,|\, \widetilde{\boldsymbol{\Lambda}}_{k-1}^{(j)}\right), \quad (6)$$

where (6) holds, since $\Lambda_k$ is independent on $\boldsymbol{Z}^{k-1}$. It can be seen that the measurement $\boldsymbol{z}_k$ has a Gaussian conditional density

$$p(z_k|\Lambda_k = \{i_1, i_2\}, \widetilde{\Lambda}_{k-1}^{(j)}, Z^{k-1}) = \tag{7}$$

$$p(z_k^1|\lambda_{k,1} = i_1, KF_{k-1}^{(j)}) \, p(z_k^2|\lambda_{k,2} = i_2, \lambda_{k,1} = i_1, KF_{k-1}^{(j)})$$

$$p(z_k^1|\lambda_{k,1} = i_1, KF_{k-1}^{(j)}) \sim \mathcal{N}\left(z_k^1; H\boldsymbol{\mu}_{k|k-1}^{(j)}, S_k^{(j)}\right),$$

$$p(z_k^2|\lambda_{k,2} = i_2, \lambda_{k,1} = i_1, KF_{k-1}^{(j)}) \sim \mathcal{N}\left(L_k; L\left(\theta_{i_2}, \boldsymbol{\mu}_{k|k-1}^{(j)}\right), R_L\right), \text{ where } \hat{\boldsymbol{\mu}}_{k|k-1}^{(j)} \text{ is}$$

the predicted state and $S_k^{(j)}$ is the measurement prediction covariance, calculated by a Kalman filter, adjusted for $\Lambda_k = \{i_1, i_2\}$.

*At the second* step, the indicator vector $\Lambda_k = \{i_1, i_2\}$ is imputed with probability, proportional to $\mathcal{L}_{k,\{i_1,i_2\}}^{(j)}$. Then the filter mean $\hat{\boldsymbol{\mu}}_{k|k}$ and covariance $P_{k|k}$ are updated for the sampled indices $\{i_1, i_2\}$. Thus, the required quantities at time instant $k$: $KF_k^{(j)}$ and $\widetilde{\Lambda}_k^{(j)} = \left(\widetilde{\Lambda}_{k-1}^{(j)}, \Lambda_k^{(j)}\right)$ are obtained.

*Finally*, the weight for this updated filter estimate is calculated as

$$w_k^{(j)} = w_{k-1}^{(j)} p\left(z_k|\widetilde{\Lambda}_{k-1}^{(j)}, Z^{k-1}\right) = w_{k-1}^{(j)} \times \sum_{i_1, i_2} \mathcal{L}_{k,\{i_1,i_2\}}^{(j)}. \text{ The on-line estimate}$$

of the state vector is: $\hat{x}_k = \frac{1}{W_k}\sum_{j=1}^{N} w_k^{(j)} \hat{\boldsymbol{\mu}}_{k|k}^{(j)}$, where $W_k = \sum_{j=1}^{N} w_k^{(j)}$.

**MKF for state and size parameters estimation.** The detailed scheme of the algorithm, designed for the model (1)-(3) is given below:

---

### Algorithm Outline

---

1. Initialisation, $k = 0$
   * For $j = 1, \ldots, N$,
   sample $\lambda_{0,1}^{(j)}$ from the set $\mathbb{S}$ with probability $P(\lambda_{0,1}^{(j)} = i_1) \propto P_0(i_1), i_1 \in \mathbb{S}$ and $\lambda_{0,2}^{(j)}$ from the set $\mathbb{N}_t$ with probability $P(\lambda_{0,2}^{(j)} = i_2) \propto P_{\theta_0}(i_2), i_2 \in \mathbb{N}_t$. Form $\widetilde{\Lambda}_0^{(j)} = \left\{\lambda_{0,1}^{(j)}, \lambda_{0,2}^{(j)}\right\}$. Set $KF_0^{(j)} = \left\{\boldsymbol{\mu}_0^{(j)}, P_0^{(j)}, \widetilde{\Lambda}_0^{(j)}\right\}$, where $\boldsymbol{\mu}_0^{(j)} = \hat{\boldsymbol{\mu}}_0$ and $P_0^{(j)} = P_0$ are the mean and covariance of the initial state $x_0 \sim \mathcal{N}(\hat{\boldsymbol{\mu}}_0, P_0)$. Set the initial weights $w_0^{(j)} = 1/N$.
   * end for j ;    Set $k = 1$.

2. For $j = 1, \ldots, N$ complete:

   – For each $\Lambda_k = \{i_1, i_2\}$, $i_1 \in \mathbb{S}$ and $i_2 \in \mathbb{N}_t$ compute
   - one step prediction for each Kalman filter $KF_{k-1}^{(j)}$

   $$(\boldsymbol{\mu}_{k|k-1}^{(j)})^{(i_1)} = F(i_1)\boldsymbol{\mu}_{k-1|k-1}^{(j)},$$
   $$(P_{k|k-1}^{(j)})^{(i_1)} = F(i_1)P_{k-1|k-1}^{(j)}F(i_1)' + GQ(i_1)G',$$
   $$(z_{k|k-1}^{(j)})^{(i_1)} = H(\boldsymbol{\mu}_{k|k-1}^{(j)})^{(i_1)},$$
   $$(S_k^{(j)})^{(i_1)} = H(P_{k|k-1}^{(j)})^{(i_1)}H' + R_c(i_1).$$

   - on receipt of a measurement $z_k$ calculate the likelihood
   $$\mathcal{L}_{k,\{i_1,i_2\}}^{(j)} = p(z_k^1|\lambda_{k,1} = i_1, KF_{k-1}^{(j)}) \, p(z_k^2|\lambda_{k,2} = i_2, \lambda_{k,1} = i_1, KF_{k-1}^{(j)}) \times$$

$$p(\Lambda_k = \{i_1, i_2\} \,|\, \widetilde{\Lambda}_{k-1}^{(j)}), \text{ where}$$

$$p(z_k^1 | \lambda_{k,1} = i_1, KF_{k-1}^{(j)}) = \mathcal{N}\left(z_k^1; (z_{k|k-1}^{(j)})^{(i_1)}, (S_k^{(j)})^{(i_1)}\right),$$

$$p(z_k^2 | \lambda_{k,2} = i_2, \lambda_{k,1} = i_1, KF_{k-1}^{(j)}) = \mathcal{N}\left(L_k; L\left(\theta_{i_2}, (\mu_{k|k-1}^{(j)})^{(i_1)}\right), R_L\right),$$

$$p(\Lambda_k = \{i_1, i_2\} | \widetilde{\Lambda}_{k-1}^{(j)}) = p(\lambda_{k,1} = i_1 | \widetilde{\Lambda}_{k-1}^{(j)}) p(\lambda_{k,2} = i_2);$$

$$p(\lambda_{k,1} = i_1 | \lambda_{k-1,1}^{(j)}) = p_{\lambda_{k-1,1}^{(j)}, i_1} \text{ and } p(\lambda_{k,2} = i_2) = 1/t.$$

- sample $\Lambda_k^{(j)} = \{i_1, i_2\}$ with a probability, proportional to $\mathcal{L}_{k,\{i_1,i_2\}}^{(j)}$;

  suppose that $\Lambda_k^{(j)} = \{\ell_1, \ell_2\}$. Append $\Lambda_k^{(j)}$ to $\widetilde{\Lambda}_{k-1}^{(j)}$ and obtain $\widetilde{\Lambda}_k^{(j)}$.

- perform $KF_k^{(j)}$ update step for $\Lambda_k^{(j)} = \{\ell_1, \ell_2\}$:

$$K_{k|k}^{(j)} = (P_{k|k-1}^{(j)})^{(\ell_1)}(H)'[(S_k^{(j)})^{(\ell_1)}]^{-1},$$

$$\mu_{k|k}^{(j)} = (\mu_{k|k-1}^{(j)})^{(\ell_1)} + K_{k|k}^{(j)}[z_k^1 - (z_{k|k-1}^{(j)})^{(\ell_1)}],$$

$$P_{k|k}^{(j)} = (P_{k|k-1}^{(j)})^{(\ell_1)} - K_{k|k}^{(j)}(S_k^{(j)})^{(\ell_1)}(K_{k|k}^{(j)})',$$

- update the importance weights: $w_k^{(j)} = w_{k-1}^{(j)} \sum_{i_1,i_2} \mathcal{L}_{k,\{i_1,i_2\}}^{(j)}$;

  normalise the weights $w_k^{(j)} = w_k^{(j)} / \sum_{j=1}^N w_k^{(j)}$.

3. *Compute* the output estimates and posterior probabilities of indicator variables

$$\hat{x}_k = \sum_{j=1}^N \mu_{k|k}^{(j)} w_k^{(j)}, \quad P(\lambda_{k,1} = i_1) = \sum_{j=1}^N 1(\lambda_{k,1}^{(j)} = i_1) w_k^{(j)}, \; i_1 \in \mathbb{S}$$

$$P(\lambda_{k,2} = i_2) = \sum_{j=1}^N 1(\lambda_{k,2}^{(j)} = i_2) w_k^{(j)}, \; i_2 \in \mathbb{N}_t, \quad \hat{\theta}_k = \sum_{i=1}^t P(\lambda_{k,2} = i)\,\theta_i$$

$1(\cdot)$ is an indicator function such that

$1(\lambda_k = \ell) = 1$, if $\lambda_k = \ell$ and $1(\lambda_k = \ell) = 0$ otherwise;

*Compute* the effective sample size: $N_{eff} = 1/ \sum_{j=1}^N \left(w_k^{(j)}\right)^2$

4. If $N_{eff} < N_{thres}$, resample with replacement $N$ particles :

$$\left(\mu_{k|k}^{(j)}, P_{k|k}^{(j)}, \widetilde{\Lambda}_k^{(j)}\right), j = 1, \ldots, N, \text{ according to the weights; set } w_k^{(j)} = 1/N.$$

5. Set $k \longleftarrow k + 1$ and go to step 2.

## 4   Simulation Results

The algorithm performance is evaluated by simulations over trajectories, including consecutive segments of uniform motion and maneuvers (a typical scenario is shown in Fig.2(a), [4]). The observer is static, located at the origin of $x - y$ plane. The initial target state is $x_0 = (18000, -14, 90000, 5)'$. The object performs two turn maneuvers with a normal acceleration of $\pm 1.4 \; [m/s^2]$. Its length is $\ell = 50 \; [m]$ and the aspect ratio is $\gamma = 0.2$. The sensor parameters are as follows [2]: sampling interval $T = 0.2 \; [s]$; the noise covariances of measurement errors along range, azimuth and along-range target extent are respectively: $R = \text{diag}\{5^2 \; [m]^2, 0.2^2 \; [\deg]^2\}$ and $R_L = 5^2 \; [m]^2$.

*Root-Mean Squared Errors* (RMSEs) [12] are selected as a quantitative measure for the algorithm performance evaluation. The results presented below are based on 100 Monte Carlo runs.

The set $\mathbb{S}$ of the indicator variable $\lambda_{k,1}$ contains $s = 3$ elements, corresponding to three motion models. The first model corresponds to the nearly constant velocity motion. The next two models are matched to the nearly coordinated turn maneuvers with a turn rate of $\omega = \pm 0.1\ [rad/s]$. The form of the transition matrices in (1) for these models can be found in [12].

We assume that $\theta$ takes values from a set $\mathbb{T} = \{(30, 0.15), (50, 0.2), (70, 0.25), (100, 0.3)\}$ with equal initial probabilities. Therefore, the set $\mathbb{N}_t$ of the indicator variable $\lambda_{k,2}$ contains $t = 4$ elements. Note that $\theta_2$ corresponds to the true $\theta$.

The scenario and initial conditions are selected the same as in [4], in order to compare the MKF with the combined IMM-DA algorithm, developed in [4] in terms of accuracy and complexity. The DA is implemented in a sliding window mode, with a window size of 160 scans. The number of iterations is $M = 180$ and the "warming up" initial interval of the Markov chain, producing the $\theta$ estimate, is $m_0 = 100$.

The posterior probability of indicator variable $\lambda_{k,1}, k = 1, \ldots, 200$, estimated by the MKF procedure, is given in Fig.2(b). The switches between maneuvering modes ($\lambda_1 = 2$ and $\lambda_1 = 3$) reproduce well the left and right turns performed by the extended object.

Comparative plots of the true and estimated ship parameters, $\ell$ and $\gamma$, obtained by MKF and IMM-DA, are presented in Fig.(3). Both algorithms - MKF and DA provide

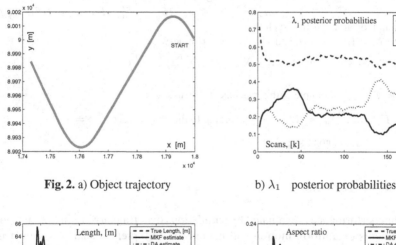

**Fig. 2.** a) Object trajectory          b) $\lambda_1$   posterior probabilities

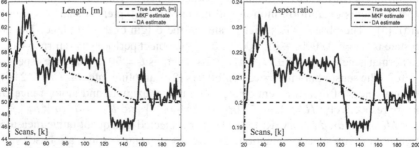

**Fig. 3.** MKF and IMM-DA comparison: True and estimated $\ell$ and $\gamma$

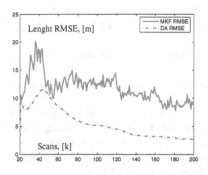
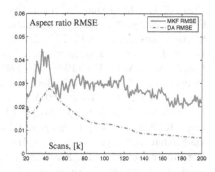

**Fig. 4.** MKF and IMM-DA comparison: RMSE of $\ell$ and $\gamma$

size estimates converging to the true parameters. It can be seen from Fig.(4), that MKF produces larger RMSEs than the IMM-DA procedure. This has an explanation with the fact that the DA is an off-line procedure and the extent parameters are estimated based on collected measurements in a window. The size estimates are computed along with tracking, but with a certain delay. The MKF gives on-line estimates. In some applications, this is important for rapid classification and decision making.

The better IMM-DA performance is achieved at the increased computational load. The IMM-DA execution time is approximately 1.7 times larger than the MKF computational time. The MKF is implemented with a sample size $N = 300$.

## 5   Conclusions

A MKF for extended object tracking is developed and studied in this paper. The algorithm is designed for a ship, whose shape is modelled by an ellipse. Simulation results for an object having three dynamic models and four size types are given to illustrate the ability of MKF to track a maneuvering ship and estimate its length and aspect ratio. Comparative results with the IMM-DA algorithm show, that the algorithm offers a good trade-off between accuracy and computational time.

## References

1. Salmond, D., Parr, M.: Track maintenance using measurements of target extent. IEE Proc.-Radar Sonar Navig., Vol. **150**, No. 6, (2003) 389–395.
2. Ristic, B., Salmond, D.: A study of a nonlinear filtering problem for tracking an extended target. Proc. Seventh Intl. Conf. on Information Fusion, (2004) 503–509.
3. Jilkov, V., Li, X. Rong, Angelova, D.: Estimation of Markovian Jump Systems with Unknown Transition Probabilities through Bayesian Sampling. LNCS, Vol. 2542 (2003), Springer-Verlag, Berlin, pp. 307–315.
4. Angelova, D., Mihaylova, L.: A Monte Carlo Algorithm for State and Parameter Estimation of Extended Targets. ICCS 2006, Part III, LNCS 3993, Springer-Verlag, 2006, pp. 624–631.
5. Diebolt, J., Robert, C.P.: Estimation of Finite Mixture Distributions through Bayesian Sampling. J. of Royal Statist. Soc. B **56**, No. 4, (1994) 363–375.

6. Chen, R., Liu, J.: Mixture Kalman filters, J. Roy. Statist. Soc. B 62 (2000) 493-508.
7. Guo, D., Wang, X., Chen, R.: Multilevel Mixture Kalman Filter. EURASIP J.on Applied Signal Processing, 2004:15, 2255-2266.
8. Dezert, J.: Tracking manoeuvring and bending extended target in cluttered environment. In Proc. of SPIE, Vol. 3373, (1998) 283-294.
9. Doucet, A., de Freitas, N., Gordon, N. (ed.): Sequential Monte Carlo Methods in Practice. Springer-Verlag, New York, (2001).
10. Liu, J.: Monte Carlo Strategies in Scientific Computing. Springer-Verlag, New York, (2003).
11. Chen, R., Wang, X., Liu, J.: Adaptive Joint Detection and Decoding in Flat-Fading Channels via Mixture Kalman Filtering. IEEE Trans. Inform. Theory, Vol. 46, No. 6, (2000) 493–508.
12. Bar-Shalom, Y., Li, X.R., Kirubarajan, T.: Estimation with Applications to Tracking and Navigation: Theory, Algorithms, and Software. Wiley, New York (2001).

# Exact Error Estimates and Optimal Randomized Algorithms for Integration*

Ivan T. Dimov[1] and Emanouil Atanassov[2]

[1] Institute for Parallel Processing, Bulgarian Academy of Sciences Acad. G. Bonchev Str., bl. 25 A, 1113 Sofia, Bulgaria and ACET Centre, University of Reading Whiteknights, PO Box 217, Reading, RG6 6AH, UK
I.T.Dimov@reading.ac.uk
[2] Institute for Parallel Processing, Bulgarian Academy of Sciences, Acad. G. Bonchev Str., bl. 25 A, 1113 Sofia, Bulgaria
atanasov@parallel.bas.bg

**Abstract.** Exact error estimates for evaluating multi-dimensional integrals are considered. An estimate is called *exact* if the rates of convergence for the low- and upper-bound estimate coincide. The algorithm with such an exact rate is called *optimal*. Such an algorithm has an *unimprovable* rate of convergence.

The problem of existing exact estimates and optimal algorithms is discussed for some functional spaces that define the regularity of the integrand. Important for practical computations data classes are considered: classes of functions with bounded derivatives and Hölder type conditions.

The aim of the paper is to analyze the performance of two optimal classes of algorithms: deterministic and randomized for computing multi-dimensional integrals. It is also shown how the smoothness of the integrand can be exploited to construct better randomized algorithms.

## 1 Introduction: Definitions and Basic Notations

The problem of evaluating integrals of high dimension is an important task since it appears in many important scientific applications of financial mathematics, economics, environmental mathematics and statistical physics. Randomized (Monte Carlo) algorithms have proved to be very efficient in solving multidimensional integrals in composite domains [16], [6].

In this paper we are interested in exact error estimates for evaluating multi-dimensional integrals. An estimate is called *exact* if the rates of convergence for the low- and upper-bound estimate coincide. An algorithm which reaches such an *unimprovable* rate of convergence is called *optimal*. The class of functions with bounded derivatives and Hölder type conditions are considered. We discuss the

---

* Partially supported by the NSF of Bulgaria through grant number I-1405/04 and by the Bulgarian IST Centre of Competence in 21 Century – BIS-21++ (contract # INCO-CT-2005-016639).

T. Boyanov et al. (Eds.): NMA 2006, LNCS 4310, pp. 131–139, 2007.

unimprovable limits of complexity for two classes of algorithms: *deterministic–* $\mathcal{A}$ and *randomized–* $\mathcal{A}^{\mathcal{R}}$. Having these unimprovable rates an important question arises: which one of the existing algorithms reaches these unimprovable rates? We analyze the *performance*, i.e., number of operations (or the computational cost) of $\mathcal{A}$ and $\mathcal{A}^{\mathcal{R}}$ classes of algorithms. It should be mentioned here that the performance analysis is connected with complexity that will be defined in Section 2. The complexity characterizes the problem for a given class of algorithms (not the algorithm itself). In Section 2 we present the computational model and show how the computational cost is connected with the complexity. In Section 3 we prove some error estimates for Hölder functions. Performance analysis of algorithms with unimprovable convergence rate is given in Section 4. Complexity of the integration problem for Hölder functions are considered in Section 5. In Section 6 we present some concluding remarks.

Let us introduce some basic notations used in the paper. By $x = (x_1, \ldots, x_d)$ we denote a point in a closed domain $G \subset \mathbb{R}^d$, where $\mathbb{R}^d$ is $d$-dimensional Euclidean space. The $d$-dimensional unite cube is denoted by $E^d = [0,1]^d$.

**Definition 1.** *Let $d, p$ be integers, and $d, p \geq 1$. Consider the class $\mathbf{W}^p(\alpha; G)$ of real functions $f$ defined over $G$, possessing all the partial derivatives: $D^r f = \frac{\partial^r f(x)}{\partial x_1^{r_1} \ldots \partial x_d^{r_d}}$, $r_1 + \ldots + r_d = r \leq p$, which are continuous when $r < p$ and bounded in sub norm when $r = p$. The semi-norm $\| \, . \, \|$ on $\mathbf{W}^p(\alpha; G)$ is defined as $\alpha = \| f \| = \sup \left\{ |D^p f|, |r_1, \ldots, r_d| = p, \quad x \equiv (x_1, \ldots, x_d) \in E^d \right\}$.*

**Definition 2.** *Define the class $H_\lambda^p(\alpha, G), (0 < \lambda \leq 1)$ of functions from $W^p$, which derivatives of order $p$ satisfy the Hölder condition with a parameter $\lambda$:*

$$H_\lambda^p(\alpha, G) \equiv \left\{ f \in W^p : |D^p f(y_1, \ldots, y_d) - D^p f(z_1, \ldots, z_d)| \leq \alpha \sum_{j=1}^d |y_j - z_j|^\lambda \right\}.$$

Usually randomized algorithms reduce problems to the approximate calculation of mathematical expectations. The mathematical expectation of the random variable $\theta$ is denoted by $E_\mu(\theta)$, where $\mu$ denotes some probability measure.(The definition of probability measure is given in [11].) Sometimes $E_\mu(\theta)$ is abbreviated to $E\theta$. We shall further denote the values (realizations) of a random point $\xi$ or random variable $\theta$ by $\xi^{(i)}$ and $\theta^{(i)} (i = 1, 2, \ldots, n)$ respectively. If $\xi^{(i)}$ is a $d$-dimensional random point, then usually it is constructed using $d$ random numbers $\gamma$, i.e., $\xi^{(i)} \equiv (\gamma_1^{(i)}, \ldots, \gamma_d^{(i)})$. Let $I$ be the desired value of the integral. Assume for a given random variable $\theta$ one can prove that $E\theta = I$. Suppose the mean value of $n$ realizations of $\theta$: $\theta^{(i)}$, $i = 1, \ldots, n$ is considered as a Monte Carlo approximation to the solution: $\bar{\theta}_n = 1/n \sum_{i=1}^n \theta^{(i)} \approx I$. One can only state that a certain randomized algorithm can produce the result with a given probability error.

**Definition 3.** *If $I$ is the exact solution of the problem, then the probability error is the least possible real number $R_n$, for which $P = Pr \left\{ |\bar{\xi}_n - I| \leq R_n \right\}$, where $0 < P < 1$. If $P = 1/2$, then the probability error is called probable error.*

So, dealing with randomized algorithms one has to accept that the result of the computation can be true only with a certain (even high) probability. In most cases of practical computations it is reasonable to accept an error estimate with a given probability.

## 2    Computational Model

Consider the following problem of integration:

$$I = \int_{E^d} f(x)dx, \tag{1}$$

where $E^d \equiv [0,1]^d$, $x \equiv (x_1, \ldots, x_d) \in E^d \subset \mathbb{R}^d$ and $f \in C(E^d)$ is an integrable function on $E^d$. The computational problem can be considered as a mapping of function $f : \{[0,1]^d \to \mathbb{R}^d\}$ to $\mathbb{R}$ [10]: $S(f) : f \to \mathbb{R}$, where $S(f) = \int_{E^d} f(x)dx$ and $f \in F_0 \subset C(E^d)$. We will call $S$ the solution operator. The elements of $F_0$ are the data, for which the problem has to be solved; and for $f \in F_0$, $S(f)$ is the exact solution. For a given $f$ we want to compute (or approximate) $S(f)$. We will be interested to consider subsets $F_0$ of $C(E^d)$ and try to study how the smoothness of $F_0$ can be exploited. A similar approach (which is in fact included in the above mentioned consideration) is presented in [18].

We will call a *quadrature formula* any expression $A = \sum_{i=1}^n c_i f(x^{(i)})$, which approximates the value of the integral $S(f)$. The real numbers $c_i \in \mathbb{R}$ are called weights and $d$ dimensional points $x^{(i)} \in E^d$ are called nodes. It is clear that for fixed weights $c_i$ and nodes $x_i$ the quadrature formula $A$ may be used to define an algorithm. The algorithm $A$ belongs to the class of deterministic algorithms $\mathcal{A}$. We call a *randomized quadrature formula* any formula of the following kind: $A^R = \sum_{i=1}^n \sigma_i f(\xi^{(i)})$, where $\sigma_i$ and $\xi^{(i)}$ are random weights and nodes.

The computational cost of a deterministic algorithm $A$ will be defined as a supremum (over all integrands $f$ from $F_0$) of the time (number of operations) needed to perform the algorithm $A$: $\tau(A) = \sup_{f \in F_0} \tau(A, f)$. For a randomized algorithm $A^R \in \mathcal{A}^R$ we will have: $\tau(A^R) = \sup_{f \in F_0} E_\mu\{\tau(A^R, f, \omega)\}$. As a good measure of the cost can be considered

$$\tau(A, f) = kn + c \quad \text{and} \quad \tau(A^R, f, \omega) = k^R n + c^R,$$

where $n$ is the number of nodes and $k, k^R$ are constants depending on the function $f$, dimensionality $d$ and on the domain of integration (in our case on $E^d$) and constants $c$ and $c^R$ depend only on $d$ and on the regularity parameter of the problem (in the case of $H_\lambda^p(\alpha, G)$ - on $p + \lambda$). These constants describe the so-called preprocessing operations, i.e., operations that are needed to be performed beforehand.

We assume that one is happy to obtain an $\varepsilon$-approximation to the solution with a probability $0 < P < 1$. For a given positive $\varepsilon$ the $\varepsilon$-complexity of the integration problems $S$ and $S^R$ are defined as follows: $C_\varepsilon(S) = \inf_{A \in \mathcal{A}}\{\tau(A) : r(A) \leq \varepsilon\}$ and $C_\varepsilon(S^R) = \inf_{A^R \in \mathcal{A}^R}\{\tau(A^R) : r(A^R) \leq \varepsilon\}$, where the errors $r(A)$

and $r(A^R)$ are defined in the Section 3. One can see that in our consideration $\varepsilon$-complexity characterizes the problem for a given class of algorithms (not the algorithms itself).

# 3    Exact Error Estimates in Functional Spaces

Generally, we assume that the problem of integration is not solved exactly, that is $S(f)$ differs from $A(f)$. We define the error as

$$r(A) = \sup_{f \in F_0} |S(f) - A(f)|$$

in the deterministic case and as

$$r(A^R) = \sup_{f \in F_0} E_\mu \left| S(f) - A^R(f, \omega) \right| = \sup_{f \in F_0} \int_{E^d} \left| S(f) - A^R(f, \omega) \right| d\mu(\omega),$$

where $A(f, \omega)$ is $\Sigma$-measurable in $\omega$ for each $f$ in the randomized case.

Let us now define the subset $F_0 \equiv H_\lambda^p(\alpha, E^d)$. In [2] Bakhvalov proved the following theorem:

**Theorem 1.** *(Bakhvalov [2]) For any deterministic way of evaluating the integral (1), i.e., for any algorithm from $\mathcal{A}$*

$$\sup_{f \in H_\lambda^p(\alpha, E^d)} r(A) \geq c'(d, p + \lambda)\alpha n^{-\frac{p+\lambda}{d}} \tag{2}$$

*and for any randomized way of evaluating the integral (1), i.e., for any algorithm from $\mathcal{A}^R$*

$$\sup_{f \in H_\lambda^p(\alpha, E^d)} r(A^R) \geq c''(d, p + \lambda)\alpha n^{-\frac{p+\lambda}{d} - \frac{1}{2}}. \tag{3}$$

The constants $c'(d, p + \lambda)$ and $c''(d, p + \lambda)$ depend only on $d$ and $p + \lambda$. This theorem gives the best possible order for both algorithmic classes $\mathcal{A}$ and $\mathcal{A}^R$.

In our work [1] we construct two randomized algorithms $A_1^R$ and $A_2^R$, and prove that both have the best possible rate (3) for integrands from $W^p(\alpha, E^d)$. The proposed algorithms allow to extend the estimates for the functional class $H_\lambda^p(\alpha, E^d)$, where $0 < \lambda \leq 1$. Here we give the essential idea of the algorithms (for more details we refer to [1]). In algorithm $A_1^R$ we divide the unit cube $\mathbf{E}^d$ into $n = q^d$ disjoint cubes: $\mathbf{E}^d = \bigcup_{j=1}^{q^d} K_j$. Then we select $m$ random points $\xi(j, s) = (\xi_1(j, s), ..., \xi_d(j, s))$ from each cube $K_j$, such that all $\xi_i(j, s)$ are uniformly distributed and mutually independent. We consider the Lagrange interpolation polynomial of the function $f$ at the point $z$: $L_p(f, z)$, which uses the information from the function values at exactly $\binom{p + d - 1}{d}$ points satisfying a special property [1]. The second algorithm $A_2^R$ is a modification which calculates the Newton interpolation polynomial. $A_2^R$ involves less operations for the same number of random nodes. Finally, we use the following randomized quadrature formula:

$$I(f) \approx A_1^R = \frac{1}{q^d m} \sum_{j=1}^{q^d} \sum_{s=1}^{m} (f(\xi(j,s)) - L_p(f, \xi(j,s))) + \int_{K_j} L_p(f,x)dx. \quad (4)$$

Now, for functions from $H_\lambda^p(\alpha, E^d)$ we can prove the following theorem:

**Theorem 2.** *The quadrature formula (4) satisfies*

$$R_n \leq c'(d, p+\lambda) \frac{1}{m} \alpha n^{-\frac{1}{2} - \frac{p+\lambda}{d}} \quad and$$

$$\left( E \left( \int_{E^d} f(x)dx - I(f) \right)^2 \right)^{1/2} \leq c''(d, p+\lambda) \frac{1}{m} \alpha n^{-\frac{1}{2} - \frac{p+\lambda}{d}},$$

*where the constants $c'(d, p+\lambda)$ and $c''(d, p+\lambda)$ depend implicitly on the points $x^{(r)}$, but not on $n$.*

*Proof.* The proof is a modification of the proof given in [1]. Indeed, taking into account that $f$ belongs to the space $H_\lambda^p(\alpha, E^d)$ one can use the following inequality: $|f(\xi(s,t) - L_p(f, \xi(j,s))| \leq c_{d,p+\lambda} \alpha n^{-p-\lambda}$. Using the above inequality and applying similar technique used in the proof of Theorem 2.1 from [1] we prove the theorem.

Both algorithms $A_1^R$ and $A_2^R$ are unimprovable by rate for all functions from $H_\lambda^p(\alpha, E^d)$. Indeed, $r(A_{I_1}^R) \leq c_1''(d, p+\lambda)\alpha n^{-\frac{p+\lambda}{d} - \frac{1}{2}}$ for the algorithm $A_1^R$ and $r(A_{I_2}^R) \leq c_2''(d, p+\lambda)\alpha n^{-\frac{p+\lambda}{d} - \frac{1}{2}}$ for the algorithm $A_2^R$.

# 4    Performance Analysis of Algorithms with Unimprovable Convergence Rate

In this subsection the computational cost of both algorithms $A_1^R$ and $A_2^R$ are presented. The following theorem can be proved:

**Theorem 3.** *[1] The computational cost of the numerical integration of a function from $H_\lambda^p(\alpha, E^d)$ using randomized algorithm $A_i^R$ $(i = 1,2)$ can be presented in the following form:*

$$\tau(A_i^R, x, \omega) = k_i^R n + c_i^R,$$

$$k_1^R \leq \left[ m + \binom{d+p-1}{d} \right] a_f + m[d(b_r + 2) + 1] \quad (5)$$

$$+ 2 \binom{d+p-1}{d} \left[ m + 1 + d + \binom{d+p-1}{d} \right], \quad (6)$$

$$k_2^R \le \left[ m + \binom{d+p-1}{d} \right] a_f + m[d(b_r + 2 + k) + 1] \qquad (7)$$

$$+2 \binom{d+p-1}{d} (d+1+m), \qquad (8)$$

*where $b_r$ denotes the number of operations used to produce a uniformly distributed random number in $[0,1)$, $a_f$ stands for the number of operations needed for each calculation of a function value, and $c_i^R = c_i^R(d, p+\lambda)$ depends only on $d$ and $p+\lambda$.*

*Remark 1.* The performance analysis results of Theorem 3 shows that the computational cost of both algorithms is linear with the number of nodes $n$. With such a cost an error of order $n^{-\frac{p+\lambda}{d} - \frac{1}{2}}$ is reached. Such an order is unimprovable in $H_\lambda^p(\alpha, E^d)$.

Optimal algorithms for functions from $W^p(\alpha, E^d)$ are also proposed in [7,12,14,4,15,17,9]. It is not an easy task to construct a unified algorithm with unimprovable rate of convergence for any dimension $d$ and any value of $p$. Various methods for Monte Carlo integration that achieve the order $O\left(N^{-\frac{1}{2} - \frac{p}{d}}\right)$ are known. While in the case of $p=1$ and $p=2$ these methods are fairly simple and are widely used (see, for example, [17,14,13]), when $p \ge 3$ such methods become much more sophisticated.

Using the same construction as in [1] it is easy to show that for the deterministic case there exists an algorithm for which $r(A) \le c_A'(d, p+\lambda)\alpha n^{-\frac{p+\lambda}{d}}$. As an example of such an algorithm could be considered the algorithm $A_1^R$ proposed in [1] in which the nodes are fixed points.

# 5    Complexity of the Integration Problem for Functional Spaces

## 5.1    Complexity for Hölder Spaces

Now we are ready to formulate a theorem given the estimates of the $\varepsilon$-complexity of the problem.

**Theorem 4.** *For $F_0 \equiv H_\lambda^p(\alpha, E^d)$ the $\varepsilon$-complexity of the problem of integration $S$ is $C_\varepsilon(S) = k \left(c_A(d, p+\lambda)\alpha\right)^{\frac{d}{p+\lambda}} \left(\frac{1}{\varepsilon}\right)^{\frac{d}{p+\lambda}}$ for the class of deterministic algorithms $\mathcal{A}$, and $C_\varepsilon(S) = k^R \left(c_{A^R}(d, p+\lambda)\alpha\right)^{\frac{d}{p+\lambda+d/2}} \left(\frac{1}{\varepsilon}\right)^{\frac{d}{p+\lambda+d/2}}$ for the class of randomized algorithms $\mathcal{A}^R$.*

*Proof.* According to the definition of the cost of the algorithm we should take the worst algorithm in sense of $\tau(A, f)$ corresponding to $f \in H_\lambda^p(\alpha, E^d)$. According to the Bakhvalov's theorem [2] one can write:

$$\sup_{f \in H_\lambda^p(\alpha, E^d)} \tau(A, f) = kn + c = k \left(c_A'(d, p+\lambda)\alpha\right)^{\frac{d}{p+\lambda}} \left(\frac{1}{r(A)}\right)^{\frac{d}{p+\lambda}} + c.$$

Now, for a given $\varepsilon > 0$ we should take $\inf\left\{k\left(c'_A(d, p+\lambda)\alpha\right)^{\frac{d}{p+\lambda}}\left(\frac{1}{r(A)}\right)^{\frac{d}{p+\lambda}}\right.$ :

$\left. r(A) \leq \varepsilon\right\}$. Let us note, that this is a non-uniform complexity notion: for each $\varepsilon > 0$ a separate $A$ can be designed. However, following the remark that the algorithms $A$ are uniform over the set of problems, and the fact that the infimum of the number of preprocessing operations described by $c$ is zero, one can get:

$$C_\varepsilon(S) = k\left(c'_A(d, p+\lambda)\alpha\right)^{\frac{d}{p+\lambda}}\left(\frac{1}{\varepsilon}\right)^{\frac{d}{p+\lambda}},$$

which proves the first part of the theorem concerning the deterministic algorithms. The result for the randomized algorithms can be proved similarly.

**Corollary 1.** *The $\varepsilon$-complexity of the problem of integration strongly depends on the dimension of the problem for the class of deterministic algorithms. With the increasing of dimensionality the $\varepsilon$-complexity goes **exponentially** to infinity for the class $F_0 = H^p_\alpha(\lambda, E^d)$.*

**Corollary 2.** *In the case of randomized algorithms the $\varepsilon$-complexity of the integration problem for functions from $F_0 = H^p_\alpha(\lambda, E^d)$ goes asymptotically to $\left(\frac{1}{\varepsilon}\right)^2$.*

*Remark 2.* The fact that the $\varepsilon$-complexity exponentially depends on $d$ makes the class of deterministic algorithms infeasible for large dimensions.

*Remark 3.* In the last case the $\varepsilon$-complexity does not increase exponentially with $d$. This is why for high-dimensional integration Monte Carlo is a right choice. Nevertheless, the results presented here demonstrate that the smoothness can be exploited to improve the rate of convergence by a factor of $n^{-\frac{p+\lambda}{d}}$ over the rate of standard randomized algorithms $n^{-\frac{1}{2}}$. This fact allows to decrease the $\varepsilon$-complexity from $(1/\varepsilon)^2$ by a factor of $\left(\frac{1}{\varepsilon}\right)^{-\frac{4(p+\lambda)}{2(p+\lambda)+d}}$.

## 6    Concluding Remarks

As a general remark, one can conclude that as smaller is the order of regularity as simpler randomized algorithm should be used. Even for low dimensions ($d = 1, 2$) Monte Carlo is a right choice if the functional class has no smoothness. It is important to note that the level of confidence $P$ ($0 < P < 1$) does not reflect on the rate of convergence of the probability error $R_n$. It reflects only on the constant $k^R$. That's why the choice of the value of $P$ is not important for the convergence rate (respectively, for the rate of algorithmic complexity). Nevertheless, for practical computations it may be of great importance to have the value of the constant in order to get the number of operations for a given algorithm (as we have done in Section 4).

In case of non-regular input data (discontinues functions and/or singularities) there are special techniques well developed in Monte Carlo algorithms [8,16,3,1]. These allow to *include* the singularity into the density function of special choice (see, for instance [4,5]).

As a general remark it should be emphasized that the randomized algorithms have better convergence rate for the same regularity of the input data. The results can be extended to optimal algorithms for solving integral equations. An important obvious advantage of randomized algorithms is the case of *bad functions*, i.e., functions that do not satisfy some additional conditions of regularity. The main problem with the deterministic algorithms is that normally they need some additional approximation procedure that require additional regularity. The randomized algorithms do not need such procedures. But one should be careful because

- the better convergence rate for randomized algorithms is reached with a given probability less than 1, so the advantage of Monte Carlo algorithms is a matter of definition of the probability error. Such a setting of the problem of error estimation may not be acceptable if one needs a guaranteed accuracy or strictly reliable results. In fact, we see that this is a price paid by randomized algorithms to increase their convergence rate.
- If the nature if the problem under consideration do not allow to use the probability error for estimates or the answer should be given with a guaranteed error then the higher convergence order randomized algorithms are not acceptable.

# References

1. E. Atanassov, I. Dimov, *A new Monte Carlo method for calculating integrals of smooth functions*, Monte Carlo Methods and Applications, **Vol. 5**, No 2 (1999), pp. 149-167.
2. N.S. Bachvalov, *On the approximate computation of multiple integrals*, Vestnik Moscow State University, Ser. Mat., Mech., **Vol. 4** (1959), pp. 3–18.
3. Dimov I.T., *Minimization of the probable error for some Monte Carlo methods*, Proc. of the Summer School on Mathematical Modelling and Scientific Computations, Bulgaria, Sofia, Publ. House of the Bulg. Acad. Sci. (1991), pp. 159–170.
4. I. Dimov, *Efficient and Overconvergent Monte Carlo Methods. Parallel algorithms.*, Advances in Parallel Algorithms (I. Dimov, O. Tonev, Eds.), Amsterdam, IOS Press (1994), pp. 100–111.
5. I. Dimov, A. Karaivanova, H. Kuchen, H. Stoltze, *Monte Carlo Algorithms for Elliptic Differential Equations. Data Parallel Functional Approach*, Journal of Parallel Algorithms and Applications, **Vol. 9** (1996), pp. 39–65.
6. I. Dimov, O. Tonev, *Monte Carlo algorithms: performance analysis for some computer architectures*, J. of Computational and Applied Mathematics, **Vol. 48** (1993), pp. 253–277.
7. J.H. Halton, D.C. Handscome, *A method for increasing the efficiency of Monte Carlo integrations*, J. Assoc. comput. machinery, **Vol. 4, No. 3** (1957), pp. 329–340.
8. M.H. Kalos, *Importance sampling in Monte Carlo calculations of thick shield penetration*, Nuclear Sci. and Eng., **2, No. 1** (1959), pp. 34–35.
9. A. Karaivanova, I. Dimov, *Error analysis of an Adaptive Monte Carlo Method for Numerical Integration*, Mathematics and Computers in Simulation, **Vol. 47** (1998), pp. 201-213.

10. Ker-I Ko, *Computational Complexity of Real Functions*, Birkhauser Boston, Boston, MA, 1991
11. A.N. Kolmogorov, *Foundations of the Theory of Probability*, Second English Edition, Chelsea Publishing Company, New York, 1956.
12. E. Novak, K. Ritter, *Optimal stochstic quadrature farmulas for convex functions*, BIT, **Vol. 34** (1994), pp. 288–294.
13. E. Novak, K. Ritter, *High Dimensional Integration of Smooth Functions over cubes*, Numerishche Mathematik (1996), pp. 1–19.
14. Bl. Sendov, A. Andreev, N. Kjurkchiev, *Numerical Solution of Polynomial Equations, Handbook of Numerical Analysis (General Editors: P.G. Ciarlet and J. L. Lions), Vol. 3)*, **Solution of Equations in $R^n$ (Part 2), North-Holland**, Amsterdam, N.Y., 1994.
15. S.A. Smolyak, *Quadrature and inperpolation formulas for tensor products of certain classes of functions*, Soviet Math. Dokl., **Vol. 4** (1963), pp. 240–243.
16. I.M. Sobol, *Monte Carlo numerical methods*, Nauka, Moscow, 1973.
17. I.M. Sobol, *On Quadratic Formulas for Functions of Several Variables Satisfying a General Lipschitz Condition*, USSR Comput. Math. and Math. Phys., **Vol. 29(6)** (1989), pp. 936 – 941.
18. Traub, J.F., Wasilkowski, G.W. and Wozniakowski, H., *Information-Based Complexity*, Acad. Press, INC., New York, 1988.

# Parallel Monte Carlo Approach for Integration of the Rendering Equation⋆

Ivan T. Dimov[1], Anton A. Penzov[2], and Stanislava S. Stoilova[3]

[1] Institute for Parallel Processing, Bulgarian Academy of Sciences,
Acad. G. Bonchev Str., bl. 25 A, 1113 Sofia, Bulgaria and ACET Centre, University
of Reading, Whiteknights, PO Box 217, Reading, RG6 6AH, UK
`I.T.Dimov@reading.ac.uk`
[2] Institute for Parallel Processing, Bulgarian Academy of Sciences,
Acad. G. Bonchev Str., bl. 25 A, 1113 Sofia, Bulgaria
`apenzov@parallel.bas.bg`
[3] Institute of Mathematics and Informatics, Bulgarian Academy of Sciences,
Acad. G. Bonchev Str., bl. 8, 1113 Sofia, Bulgaria
`stoilova@math.bas.bg`

**Abstract.** This paper is addressed to the numerical solving of the rendering equation in realistic image creation. The rendering equation is integral equation describing the light propagation in a scene accordingly to a given illumination model. The used illumination model determines the kernel of the equation under consideration. Nowadays, widely used are the Monte Carlo methods for solving the rendering equation in order to create photorealistic images.

In this work we consider the Monte Carlo solving of the rendering equation in the context of the parallel sampling scheme for hemisphere. Our aim is to apply this sampling scheme to stratified Monte Carlo integration method for parallel solving of the rendering equation. The domain for integration of the rendering equation is a hemisphere. We divide the hemispherical domain into a number of equal sub-domains of orthogonal spherical triangles. This domain partitioning allows to solve the rendering equation in parallel. It is known that the Neumann series represent the solution of the integral equation as a infinity sum of integrals. We approximate this sum with a desired truncation error (systematic error) receiving the fixed number of iteration. Then the rendering equation is solved iteratively using Monte Carlo approach. At each iteration we solve multi-dimensional integrals using uniform hemisphere partitioning scheme. An estimate of the rate of convergence is obtained using the stratified Monte Carlo method.

This domain partitioning allows easy parallel realization and leads to convergence improvement of the Monte Carlo method. The high performance and Grid computing of the corresponding Monte Carlo scheme are discussed.

---

⋆ Supported by the Ministry of Education and Science of Bulgaria under Grand No.
I-1405/04 and by FP6 INCO Grand 016639/2005 Project BIS-21++.

T. Boyanov et al. (Eds.): NMA 2006, LNCS 4310, pp. 140–147, 2007.

# 1 Introduction

The main task in the area of computer graphics is photorealistic image creation. From mathematical point of view, photorealistic image synthesis is equivalent to the solution of the rendering equation [9]. The rendering equation is a Fredholm type integral equation of second kind. It describes the light propagation in closed domains called scenes. The kernel of the rendering equation is determined by the used illumination model. The illumination model (see [11] for a survey of illumination models) describes the interaction of the light with a point on the surface in the scene. Each illumination model approximates the BRDF (bidirectional reflectance distribution function), taking into account the material surface characteristics. The physical properties like reflectivity, roughness, and colour of the surface material are characterized by the BRDF. This function describes the light reflection from a surface point as a ratio of outgoing to incoming light. It depends on the wavelength of the light, incoming, outgoing light directions and location of the reflection point. The BRDF expression receives various initial values for the objects with different material properties. Philip Dutré in [4] presents a good survey of the different BRDF models for realistic image synthesis.

One possible approach for the solution of rendering equation is the Monte Carlo methods, which has been in the focus of mathematical research for several decades. Frequently the Monte Carlo methods for numerical integration of the rendering equation are the only practical method for multi-dimensional integrals. The convergence rate of conventional Monte Carlo method is $O(N^{-\frac{1}{2}})$ which gives relatively slow performance at realistic image synthesis of complex scenes and physical phenomena simulation. In order to improve Monte Carlo method and speed up the computation much of the efforts are directed to the variance reduction techniques. The separation of the integration domain [12] is widely used Monte Carlo variance reduction method. Monte Carlo algorithms using importance separation of the integration domain are presented in [8], [7], [2], [5] and [6]. The method of importance separation uses a special partition of the domain and computes the given integral as a sum of the integrals on the sub-domains. An adaptive sub-division technique of spherical triangle domains is proposed by Urena in [14]. Keller [10] suggests the usage of low discrepancy sequences for solving the rendering equation and proposes Quasi Monte Carlo approach. The idea is to distribute the samples into the domain of integration, as uniformly as possible in order to improve the convergence rate.

Further in this paper we consider the Monte Carlo solving of rendering equation with uniform hemisphere separation. The technique of uniform hemisphere partition was introduced by us and described in [3]. The uniform separation of the integration domain into *uniformly small by probability as well as by size* subdomains fulfills the conditions of the *Theorem for super convergence* presented in [12]. We show that the variance is bounded for numerical solving of the multidimensional integrals. Due to uniform separation of integration domain, this approach has hierarchical parallelism which is suitable for Grid implementations.

## 2   Rendering Equation for Photorealistic Image Creation

The light propagation in a scene is described by rendering equation [9], which is a second kind Fredholm integral equation. The radiance $L$, leaving from a point $x$ on the surface of the scene in direction $\omega \in \Omega_x$ (see Fig. 1), where $\Omega_x$ is the hemisphere in point $x$, is the sum of the self radiating light source radiance $L^e$ and all reflected radiance:

$$L(x,\omega) = L^e(x,\omega) + \int_{\Omega_x} L(h(x,\omega'), -\omega') f_r(-\omega', x, \omega) \cos\theta' d\omega'.$$

The point $y = h(x,\omega')$ indicates the first point that is hit when shooting a ray from $x$ into direction $\omega'$. The radiance $L^e$ has non-zero value if the considered point $x$ is a point from solid light source. Therefore, the reflected radiance in direction $\omega$ is an integral of the radiance incoming from all points, which can be seen through the hemisphere $\Omega_x$ in point $x$ attenuated by the surface BRDF $f_r(-\omega', x, \omega)$ and the projection $\cos\theta'$, which puts the surface perpendicular to the ray $(x, \omega')$. The angle $\theta'$ is the angle between surface normal in $x$ and the direction $\omega'$. The law for energy conservation holds, because a real scene always reflects less light than it receives from the light sources due to light absorption of the objects, i.e.: $\int_{\Omega_x} f_r(-\omega', x, \omega) \cos\theta' d\omega' < 1$. That means the incoming photon is reflected with a probability less than 1, because the selected energy is less than the total incoming energy. Another important property of the BDRF is the Helmholtz principle: the value of the BRDF will not change if the incident and reflected directions are interchanged, $f_r(-\omega', x, \omega) = f_r(-\omega, x, \omega')$.

Many BRDF for realistic image synthesis are based on surface microfacet theory. They are considered as function defined over all directions $\omega' \in \Omega_x$ (see [15]). For example, the BRDF function of Cook-Torrance (see [1], [4] and [11]) depends on the product of three components: Fresnel term - $F$, microfacets distribution function - $D$ and geometrical attenuation factor - $G$; all depending on $\omega'$. More detailed look at those functions gives us the assumption that Cook-Torrance BRDF has continuous first derivative.

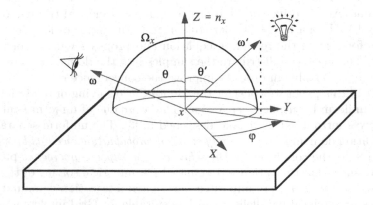

**Fig. 1.** The geometry for the rendering equation

## 3 Parallel Monte Carlo Approach for the Rendering Equation

In order to solve the rendering equation by classical Monte Carlo approach we estimate the integral over the domain $\Omega_x$. This is done by independently sampling $N$ points according to some convenient probability density function $p(\omega')$, and then computing the Monte Carlo estimator $\xi_N$. Let us consider the sampling of the hemisphere $\Omega_x$ with $p(\omega') = \frac{1}{\Omega_x} = \frac{1}{2\pi}$, where $p = \int\limits_{\Omega_x} p(\omega')d\omega' = 1$. It is known that the estimator $\xi_N$ has the following form:

$$\xi_N = \frac{2\pi}{N} \sum_{i=1}^{N} L^e(h(x,\omega_i'), -\omega_i')f_r(-\omega_i', x, \omega) \cos\theta_i'.$$

The parallel Monte Carlo approach for solving the rendering equation is based on the strategy for separation of the integration domain $\Omega_x$ into non-overlapping sub-domains, as described in [3]. We apply the symmetry property for partitioning of the hemisphere $\Omega_x$. The coordinate planes partition the hemisphere into 4 equal areas. The partitioning of each one area into 6 equal sub-domains is continued by the three bisector planes. In Fig. 2 is shown the partitioning of the area with positive coordinate values of $X, Y$ and $Z$ into 6 equal sub-domains.

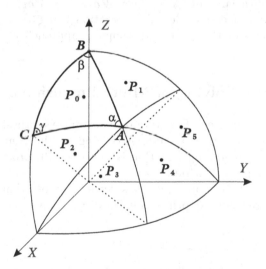

**Fig. 2.** Partitioning of the domain of integration

Let us now apply the partitioning of the hemisphere $\Omega_x$ into 24 non-overlapping equal size sub-domains of orthogonal spherical triangles $\Omega_{i_x}$, where $\Omega_{\triangle ABC} = \Omega_{i_x} = \frac{1}{24}\Omega_x = \frac{\pi}{12}$ for $i_x = 1, 2, \ldots, 24$.

We can rewrite the rendering equation as:

$$L(x,\omega) = L^e(x,\omega) + \sum_{i_x=1}^{24} \int\limits_{\Omega_{i_x}} L(h(x,\omega'), -\omega')f_r(-\omega', x, \omega) \cos\theta' d\omega',$$

where $\Omega_x = \bigcup_{i_x=1}^{24} \Omega_{i_x}$. Therefore, the solution of rendering equation can be find as a sum of integrals over equal size non-overlapping sub-domains $\Omega_{i_x}$.

Consider the probability:

$$p = \int_{\Omega_x} p(\omega')d\omega' = \sum_{i_x=1}^{24} \int_{\Omega_{i_x}} p(\omega')d\omega' = \sum_{i_x=1}^{24} p_{i_x} = 1,$$

it is obvious that $p_{i_x} = \int_{\Omega_{i_x}} p(\omega')d\omega' = \frac{1}{24}$ for $i_x = 1, 2, \ldots, 24$. Each sub-domain is sampled by random points $N_{i_x} \in \Omega_{i_x}$ with a density function $p(\omega')/p_{i_x}$. For all sub-domains $N$ independent sampling points are generated in parallel using the sampling scheme for hemisphere from [3], where $N = 24N_{i_x}$ for $i_x = 1, 2, \ldots, 24$, t.e. the random sampling point are equal number in each sub-domain. In this case the sum of integrals for solving the separate rendering equation is estimated (see [12]) by:

$$\xi_N^* = \sum_{i_x=1}^{24} \frac{\pi}{12N_{i_x}} \sum_{s=1}^{N_{i_x}} L^e(h(x, \omega_{i_x,s}'), -\omega_{i_x,s}') f_r(-\omega_{i_x,s}', x, \omega) \cos\theta_{i_x,s}'.$$

Comparing the two approach it is known (see in [12]) that the variance of $\xi_N^*$ is not bigger to the variance of $\xi_N$ or always $Var[\xi_N^*] \leq Var[\xi_N]$. Somting more, one can see that in our case of domain separation $Var[\xi_N^*] = \frac{1}{24}Var[\xi_N]$. Also, the main advantages of this stratified sampling approach is the easy parallel realization.

## 4    Monte Carlo Solving of Multi-dimensional Integrals

The global illumination (see in Fig. 3) in realistic image synthesis can be modeled as stationary linear iterative process. According to the Neumann series, the numerical solving of rendering equation is iterative [13] process, where multi-dimensional integrals are considered.

For solving multi-dimensional integrals, let us suppose that $k_\varepsilon$ is maximum level of recursion (recursion depth or number of iterations, see [1]) sufficient for

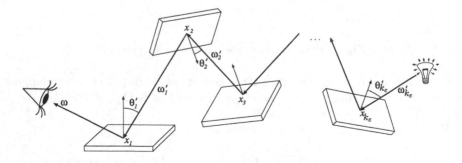

**Fig. 3.** Global illumination as iterative process

numerical solving of the integral with a desired truncation error $\varepsilon$. In this case on each iteration we have to solve the multi-dimensional integrals of the following type:

$$L^{(j)} = L_j - L_{j-1} = \int_{\Omega_{x_1}} \ldots \int_{\Omega_{x_j}} K_1(x_1, \omega_1') \ldots K_j(x_j, \omega_j') L^e(x_{j+1}, \omega_j') d\omega_1' \ldots d\omega_j',$$

where $K_j(x_j, \omega_j') = f_r(-\omega_j', x_j, \omega_{j-1}') \cos\theta_j'$ for $j = 1, \ldots, k_\varepsilon$ and $L_0 = L^e(x_1, \omega)$ (note that $L_0 = L^e(x_1, \omega) = 0$ if the point $x_1$ is not a point from solid light source). The total domain of integration $\Omega_x$ can be represented as:

$$\Omega_x = \Omega_{x_1} \times \Omega_{x_2} \times \ldots \times \Omega_{x_{k_\varepsilon}} = \prod_{j=1}^{k_\varepsilon} \left( \bigcup_{i_{x_j}=1}^{24} \Omega_{i_{x_j}} \right) = 24^{k_\varepsilon} \left( \frac{\pi}{12} \right)^{k_\varepsilon}.$$

Let us consider the integral $L^{(j)}$ in the case when $j = k_\varepsilon$ or $L^{(j)} = L^{(k_\varepsilon)}$. Using the partitioning of each domain $\Omega_{x_j}$ (for $j = 1, 2, \ldots, k_\varepsilon$) of non-overlap equal size spherical triangle sub-domains $\Omega_{x_j} = \bigcup_{i_{x_j}=1}^{24} \Omega_{i_{x_j}}$ with size $\Omega_{i_{x_j}} = \left( \frac{\pi}{12} \right)$ for $i_{x_j} = 1, 2, \ldots, 24$; we can rewrite the multi-dimensional integral $L^{(k_\varepsilon)}$ as:

$$L^{(k_\varepsilon)} = \sum_{i_{x_1}=1}^{24} \ldots \sum_{i_{x_{k_\varepsilon}}=1}^{24} \int_{\Omega_{i_{x_1}}} \ldots \int_{\Omega_{i_{x_{k_\varepsilon}}}} L^e(x_{k_\varepsilon+1}, \omega_{k_\varepsilon}') F(\omega_1', \ldots, \omega_{k_\varepsilon}') d\omega_1' \ldots d\omega_{k_\varepsilon}',$$

where $F(\omega_1', \ldots, \omega_{k_\varepsilon}') = \prod_{j=1}^{k_\varepsilon} K_j(x_j, \omega_j')$. For numerical solving of integral $L^{(k_\varepsilon)}$, we use $N$ realization of random samples and $N = 24^{k_\varepsilon}$. It means that only one random sample is generated in each sub-domain $\Omega_s$ for $s = 1, 2, \ldots, N$, received after partitioning of $\Omega_x$. Then approximate the integral $L^{(k_\varepsilon)}$ with $\xi_N^{*(k_\varepsilon)}$:

$$\xi_N^{*(k_\varepsilon)} = \left( \frac{\pi}{12} \right)^{k_\varepsilon} \sum_{s=1}^{N} L_s^e(x_{k_\varepsilon+1}, \omega_{k_\varepsilon}') F_s(\omega_1', \ldots, \omega_{k_\varepsilon}')$$

with the integral approximation error $\varepsilon_N = \left| \xi_N^{*(k_\varepsilon)} - L^{(k_\varepsilon)} \right| = \sqrt{\frac{Var\left[ \xi_N^{*(k_\varepsilon)} \right]}{N}}$.

According to the statements proofed in [12], the variance $Var\left[ \xi_N^{*(k_\varepsilon)} \right]$ can be estimated as:

$$Var\left[ \xi_N^{*(k_\varepsilon)} \right] \le c^2 L^2 N^{-1-\frac{2}{k_\varepsilon}},$$

where the first partial derivatives of $F(\omega_1', \ldots, \omega_{k_\varepsilon}')$ are limited by an existing constant $L$, $\left| \frac{\partial F}{\partial \omega_j'} \right| \le L$ for $j = 1, 2, \ldots, k_\varepsilon$ and the constant $c = k_\varepsilon c_1 c_2$. Also, there exist constants $c_1$ and $c_2$ such that:

$$p_s \le \frac{c_1}{N} \quad \text{and} \quad d_s \le \frac{c_2}{N^{\frac{1}{k_\varepsilon}}},$$

where $p_s$ is the probability and $d_s$ is diameter of the domain $\Omega_s$ for each $s = 1, 2, \ldots, N$. Since all $\Omega_s$ for each $s = 1, 2, \ldots, N$ are of equal size, it is obvious that $p = 24^{k_\varepsilon} p_s = 1$ or $p_s = \frac{1}{24^{k_\varepsilon}}$. The diameter $d_s$ for each $s = 1, 2, \ldots, N$ can be calculated as $d_s = \sqrt{k_\varepsilon} \left| \max \left( \widehat{AC}, \widehat{AB}, \widehat{BC} \right) \right| = \sqrt{k_\varepsilon} \left| \widehat{AB} \right|$, where $\left| \widehat{AB} \right|$ is the length of arc $\widehat{AB}$ in the spherical triangle $\triangle ABC$ shown in Fig. 2. We recall the derived in [3] transformations, where $\tan \widehat{AB} = \frac{1}{\cos \varphi}$ at $\varphi = \frac{\pi}{4}$ and therefore the length of arc $\widehat{AB}$ is $\arctan \left( \sqrt{2} \right)$. Therefore, $d_s = \sqrt{k_\varepsilon} \arctan \left( \sqrt{2} \right)$ for $s = 1, 2, \ldots, N$.

Now we estimates the constants $c_1$ and $c_2$ by the inequalities:

$$\frac{N}{24^{k_\varepsilon}} \le c_1 \implies 1 \le c_1 \qquad \text{and} \qquad d_s N^{\frac{1}{k_\varepsilon}} \le c_2 \implies \sqrt{k_\varepsilon} \arctan \left( \sqrt{2} \right) N^{\frac{1}{k_\varepsilon}} \le c_2.$$

Therefore we can write:

$$Var \left[ \xi_N^{*(k_\varepsilon)} \right] \le c^2 L^2 N^{-1 - \frac{2}{k_\varepsilon}} = k_\varepsilon^2 c_1^2 c_2^2 L^2 N^{-1 - \frac{2}{k_\varepsilon}}$$

which is equivalent to:

$$Var \left[ \xi_N^{*(k_\varepsilon)} \right] \le k_\varepsilon^3 \arctan^2 \left( \sqrt{2} \right) L^2 N^{-1} \Rightarrow Var \left[ \xi_N^{*(k_\varepsilon)} \right] \le \arctan^2 \left( \sqrt{2} \right) L^2 \frac{k_\varepsilon^3}{24^{k_\varepsilon}}.$$

The variance is bounded if we solve multi-dimensional integrals with uniform hemisphere separation approach. The last inequality shows us that the convergence rate for iterative solution of rendering equation with a desired truncation error $\varepsilon$ depends on the sufficient recursion depth $k_\varepsilon$. Also, the multi-dimensional integrals are numerically solved with a rate of convergence $O(N^{-1})$. This is through the uniform separation of the integration domain into *uniformly small by probability as well by size* sub-domains, all of them matching the conditions of the *Theorem for super convergence* (see the proof in [12]). Summing the variance for all $k_\varepsilon$ iterations we obtain: $Var \left[ \xi_N^* \right] = \sum_{j=1}^{k_\varepsilon} Var \left[ \xi_N^{*(j)} \right] = \left( \frac{N-1}{23N} \right) Var \left[ \xi_N \right]$, where the variance $Var \left[ \xi_N \right]$ indicates the variance for solving the rendering equation by $N$ independent random sampling points without uniform separation of integration domain. Therefore, the total variance for solving of the rendering equation with uniform hemisphere separation is reduced.

## 5   Conclusion

The parallel Monte Carlo approach for solving of the rendering equation presented in this paper is based on partitioning of the hemispherical domain of integration by a way introduced by us in [3]. Essentially, this approach accumulates the stratified sampling by uniform separation of the integration domain. In fact, the uniform separation scheme is variance reduction approach and speed up the computations. The uniform separation of the integration domain hints for the applying of low discrepancy sequences as shown in [10]. The combination of uniform separation with the usage of low discrepancy sequences for numerical solving of the rendering equation could improve the uniformity of sampling points distribution and more so to reduce the variance. On the other hand this

Monte Carlo approach includes hierarchical parallelism. Therefore, it is suitable for implementation in algorithms with parallel realization of computations and completely can utilize the power of Grid computations. Thus, the main advantages of this approach lie in the efficiency of parallel computations. The future research of the parallel Monte Carlo approach under consideration for rendering equation could be developed in the following directions: 1) Investigation of the utilization of low discrepancy sequences with the uniform separation of integration domain. 2) Development of computational parallel Monte Carlo algorithms for creation of photorealistic images. 3) Creation of parallel Monte Carlo and Quasi Monte Carlo algorithms for high performance and Grid computing.

# References

1. Dimov, I. T., Gurov, T. V. and Penzov, A. A., A Monte Carlo Approach for the Cook-Torrance Model, NAA 2004, LNCS 3401, pp. 257–265, Springer-Verlag Berlin Heidelberg, (2005).
2. Dimov, I., Karaivanova A., Georgieva, R. and Ivanovska, S., Parallel Importance Separation and Adaptive Monte Carlo Algorithms for Multiple Integrals, NMA 2002, LNCS 2542, pp. 99–107, Springer-Verlag Berlin Heidelberg, (2003).
3. Dimov, I. T., Penzov, A. A. and Stoilova, S. S., Parallel Monte Carlo Sampling Scheme for Sphere and Hemisphere, NMA 2006, LNCS 4310, pp. 148–155, Springer-Verlag Berlin Heidelberg, (2007).
4. Dutré, Philip, Global Illumination Compendium, Script of September 29 2003, http://www.cs.kuleuven.ac.be/~phil/GI/TotalCompendium.pdf
5. Georgieva, Rayna and Ivanovska, Sofiya, Importance Separation for Solving Integral Equations, LSSC 2003, LNCS 2907, pp. 144–152, Springer-Verlag Berlin Heidelberg, (2004).
6. Ivanovska, Sofiya and Karaivanova, Aneta, Parallel Importance Separation for Multiple Integrals and Integral Equations, ICCS 2004, LNCS 3039, pp. 499–506, Springer-Verlag Berlin Heidelberg, (2004).
7. Karaivanova, Aneta, Adaptive Monte Carlo methods for numerical integration, Mathematica Balkanica, vol. 11, pp. 201 213, (1997).
8. Karaivanova, A. and Dimov, I., Error analysis of an adaptive Monte Carlo method for numerical integration.Mathematics and Computers in Simulation, vol. 47, pp. 391 406, (1998).
9. Kajiya, J. T., The Rendering Equation, Computer Graphics, vol. 20, No. 4, pp. 143–150, Proceedings of SIGGRAPH'86, (1986).
10. Keller, A., Quasi-Monte Carlo Methods in Computer Graphics: The Global Illumination Problem, Lectures in Applied Mathematics, vol. 32, pp. 455–469, (1996).
11. Penzov, Anton A., Shading and Illumination Models in Computer Graphics - a literature survey, MTA SZTAKI, Research Report CG-4, Budapest, (1992).
12. Sobol, I., Monte Carlo Numerical Methods, (in Russian), Nauka, Moscow, (1975).
13. Szirmay-Kalos, Laszlo, Monte-Carlo Methods in Global Illumination, Script in WS of 1999/2000, http://www.fsz.bme.hu/ szirmay/script.pdf
14. Urena, Carlos, Computation of Irradiance from Triangles by Adaptive Sampling, Computer Graphics Forum, Vol. 19, No. 2, pp. 165–171, (2000).
15. Veach, Eric, Robust Monte Carlo Methods for Light Transport Simulation, Ph.D. Dissertation, Stanford University, December (1997).

# Parallel Monte Carlo Sampling Scheme for Sphere and Hemisphere[*]

I.T. Dimov[1], A.A. Penzov[2], and S.S. Stoilova[3]

[1] Institute for Parallel Processing, Bulgarian Academy of Sciences
Acad. G. Bonchev Str., bl. 25 A, 1113 Sofia, Bulgaria and ACET Centre, University
of Reading Whiteknights, PO Box 217, Reading, RG6 6AH, UK
I.T.Dimov@reading.ac.uk
[2] Institute for Parallel Processing, Bulgarian Academy of Sciences,
Acad. G. Bonchev Str., bl. 25 A, 1113 Sofia, Bulgaria
apenzov@parallel.bas.bg
[3] Institute of Mathematics and Informatics, Bulgarian Academy of Sciences,
Acad. G. Bonchev Str., bl. 8, 1113 Sofia, Bulgaria
stoilova@math.bas.bg

**Abstract.** The sampling of certain solid angle is a fundamental operation in realistic image synthesis, where the rendering equation describing the light propagation in closed domains is solved. Monte Carlo methods for solving the rendering equation use sampling of the solid angle subtended by unit hemisphere or unit sphere in order to perform the numerical integration of the rendering equation.

In this work we consider the problem for generation of uniformly distributed random samples over hemisphere and sphere. Our aim is to construct and study the parallel sampling scheme for hemisphere and sphere. First we apply the symmetry property for partitioning of hemisphere and sphere. The domain of solid angle subtended by a hemisphere is divided into a number of equal sub-domains. Each sub-domain represents solid angle subtended by orthogonal spherical triangle with fixed vertices and computable parameters. Then we introduce two new algorithms for sampling of orthogonal spherical triangles.

Both algorithms are based on a transformation of the unit square. Similarly to the Arvo's algorithm for sampling of arbitrary spherical triangle the suggested algorithms accommodate the stratified sampling. We derive the necessary transformations for the algorithms. The first sampling algorithm generates a sample by mapping of the unit square onto orthogonal spherical triangle. The second algorithm directly compute the unit radius vector of a sampling point inside to the orthogonal spherical triangle. The sampling of total hemisphere and sphere is performed in parallel for all sub-domains simultaneously by using the symmetry property of partitioning. The applicability of the corresponding parallel sampling scheme for Monte Carlo and Quasi-Monte Carlo solving of rendering equation is discussed.

---

[*] Supported by the Ministry of Education and Science of Bulgaria under Grand No. I-1405/04 and by FP6 INCO Grand 016639/2005 Project BIS-21++.

T. Boyanov et al. (Eds.): NMA 2006, LNCS 4310, pp. 148–155, 2007.

# 1    Introduction

The main task in the area of computer graphics is realistic image synthesis. For creation of photorealistic images the solution of a Fredholm type integral equation must be found. This integral equation is called rendering equation and it is formulate first by Kajiya in [3]. The rendering equation describes the light propagation in closed domains called frequently scenes (see Fig. 1).

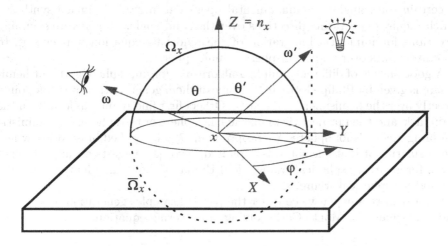

**Fig. 1.** The geometry for the rendering equation

The radiance $L$, leaving from a point $x$ on the surface of the scene in direction $\omega \in \Omega_x$, where $\Omega_x$ is the hemisphere in point $x$, is the sum of the self radiating light source radiance $L^e$ and all reflected radiance:

$$L(x,\omega) = L^e(x,\omega) + \int\limits_{\Omega_x} L(h(x,\omega'),-\omega')f_r(-\omega',x,\omega)\cos\theta' d\omega'.$$

Here $h(x,\omega')$ is the first point that is hit when shooting a ray from $x$ into direction $\omega'$. The radiance $L^e$ has non-zero value if the considered point $x$ is a point from solid light source. Therefore, the reflected radiance in direction $\omega$ is an integral of the radiance incoming from all points, which can be seen through the hemisphere $\Omega_x$ in point $x$ attenuated by the surface BRDF (Bidirectional Reflectance Distribution Function) $f_r(-\omega',x,\omega)$ and the projection $\cos\theta'$. The angle $\theta'$ is the angle between surface normal in $x$ and the direction $\omega'$. The law for energy conservation holds, i.e.: $\int_{\Omega_x} f_r(-\omega',x,\omega)\cos\theta' d\omega' < 1$, because a real scene always reflects less light than it receives from the light sources due to light absorption of the objects.

When the point $x$ is a point from a transparent object the transmitted light component must be added to the rendering equation. This component estimates the total light transmitted trough the object and incoming to the point $x$ from all directions opposite to the hemisphere $\Omega_x$. The transmitted light in direction

$\omega$ is an integral similar to the the the reflected radiance integral where the domain of integration is the hemisphere $\overline{\Omega}_x$ in point $x$ and BRDF is substituted by the surface BTDF (Bidirectional Transmittance Distribution Function) [2]. In this case the integration domain for solving the rendering equation is a sphere $\Omega^{(x)}$ in point $x$, where $\Omega^{(x)} = \Omega_x \bigcup \overline{\Omega}_x$.

Applying Monte Carlo methods for solving the rendering equation, we must sample the solid angle subtended by unit hemisphere or unit sphere in order to perform the numerical integration of the rendering equation. The sampling of certain solid angle is a fundamental operation in realistic image synthesis, which requires generating directions over the solid angle. To generate sampling directions for numerical integration of the rendering equation it is enough to generate points over unit hemisphere or unit sphere.

A good survey of different sampling algorithms for unit sphere and unit hemisphere is given by Philip Dutré in [2]. Some of them generate the sampling points directly over the hemisphere and sphere. Others first find points uniformly on the main disk, and then project them on the hemisphere or the sphere. Deterministic sampling methods for spheres are proposed in [7] and applied in robotics, where the regularity of Platonic solids is exploited. Arvo [1] suggests a sampling algorithm for arbitrary spherical triangle and Urena [6] shows an adaptive sampling method for spherical triangles.

Further in this paper we consider the parallel samples generation over sphere and hemisphere for Monte Carlo solving of rendering equation.

## 2   Partitioning of Sphere and Hemisphere

Consider hemisphere and sphere with center in the origin of a Descartes coordinate system. Similarly to the Bresenham algorithm [5] for raster display of circle we apply the symmetry property for partitioning of hemisphere and sphere. It is obvious that the coordinate planes partition the hemisphere into 4 equal areas and the sphere into 8 equal areas. The partitioning of each one area into subdomains can be continued by the three bisector planes. One can see that the bisector planes to the dihedral angles $(\vec{X}, \vec{Y})$, $(\vec{X}, \vec{Z})$ and $(\vec{Z}, \vec{Y})$, partition each area into 6 equal sub-domains. In Fig. 2 we show the partitioning of the area with positive coordinate values of $X, Y$ and $Z$ into 6 equal sub-domains.

As described above we can partition the hemisphere into 24 and respectively the sphere into 48 equal sub-domains. Something more, due to the planes of partitioning each sub-domain is symmetric to all others. The symmetric property allows us to calculate in parallel the coordinates of the symmetric points. For example to calculate the coordinates of the point $P_3$ we consecutively multiply the coordinates of the point $P_0(x_0, y_0, z_0)$ by two matrix of symmetry:

$$P_0(x_0, y_0, z_0) * \begin{pmatrix} 0\,0\,1 \\ 0\,1\,0 \\ 1\,0\,0 \end{pmatrix} * \begin{pmatrix} 1\,0\,0 \\ 0\,0\,1 \\ 0\,1\,0 \end{pmatrix} = P_2(z_0, y_0, x_0) * \begin{pmatrix} 1\,0\,0 \\ 0\,0\,1 \\ 0\,1\,0 \end{pmatrix} = P_3(z_0, x_0, y_0).$$

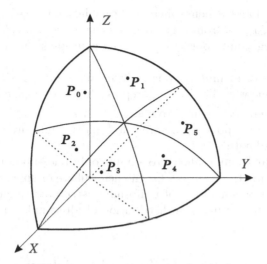

**Fig. 2.** Partition of a spherical area into 6 sub-domains

**Table 1.** Parallel Sample Coordinates Calculation of Hemisphere

| | | | |
|---|---|---|---|
| $P_0(x_0, y_0, z_0)$ | $P_0'(-x_0, y_0, z_0)$ | $P_0''(-x_0, -y_0, z_0)$ | $P_0'''(x_0, -y_0, z_0)$ |
| $P_1(y_0, x_0, z_0)$ | $P_1'(-y_0, x_0, z_0)$ | $P_1''(-y_0, -x_0, z_0)$ | $P_1'''(y_0, -x_0, z_0)$ |
| $P_2(z_0, y_0, x_0)$ | $P_2'(-z_0, y_0, x_0)$ | $P_2''(-z_0, -y_0, x_0)$ | $P_2'''(z_0, -y_0, x_0)$ |
| $P_3(z_0, x_0, y_0)$ | $P_3'(-z_0, x_0, y_0)$ | $P_3''(-z_0, -x_0, y_0)$ | $P_3'''(z_0, -x_0, y_0)$ |
| $P_4(x_0, z_0, y_0)$ | $P_4'(-x_0, z_0, y_0)$ | $P_4''(-x_0, -z_0, y_0)$ | $P_4'''(x_0, -z_0, y_0)$ |
| $P_5(y_0, z_0, x_0)$ | $P_5'(-y_0, z_0, x_0)$ | $P_5''(-y_0, -z_0, x_0)$ | $P_5'''(y_0, -z_0, x_0)$ |

**Table 2.** Parallel Sample Coordinates Calculation of Sphere

| | | | |
|---|---|---|---|
| $P_0(x_0, y_0, z_0)$ | $P_0'(-x_0, y_0, z_0)$ | $P_0''(-x_0, -y_0, z_0)$ | $P_0'''(x_0, -y_0, z_0)$ |
| $P_1(y_0, x_0, z_0)$ | $P_1'(-y_0, x_0, z_0)$ | $P_1''(-y_0, -x_0, z_0)$ | $P_1'''(y_0, -x_0, z_0)$ |
| $P_2(z_0, y_0, x_0)$ | $P_2'(-z_0, y_0, x_0)$ | $P_2''(-z_0, -y_0, x_0)$ | $P_2'''(z_0, -y_0, x_0)$ |
| $P_3(z_0, x_0, y_0)$ | $P_3'(-z_0, x_0, y_0)$ | $P_3''(-z_0, -x_0, y_0)$ | $P_3'''(z_0, -x_0, y_0)$ |
| $P_4(x_0, z_0, y_0)$ | $P_4'(-x_0, z_0, y_0)$ | $P_4''(-x_0, -z_0, y_0)$ | $P_4'''(x_0, -z_0, y_0)$ |
| $P_5(y_0, z_0, x_0)$ | $P_5'(-y_0, z_0, x_0)$ | $P_5''(-y_0, -z_0, x_0)$ | $P_5'''(y_0, -z_0, x_0)$ |
| $\overline{P}_0(x_0, y_0, -z_0)$ | $\overline{P}_0'(-x_0, y_0, -z_0)$ | $\overline{P}_0''(-x_0, -y_0, -z_0)$ | $\overline{P}_0'''(x_0, -y_0, -z_0)$ |
| $\overline{P}_1(y_0, x_0, -z_0)$ | $\overline{P}_1'(-y_0, x_0, -z_0)$ | $\overline{P}_1''(-y_0, -x_0, -z_0)$ | $\overline{P}_1'''(y_0, -x_0, -z_0)$ |
| $\overline{P}_2(z_0, y_0, -x_0)$ | $\overline{P}_2'(-z_0, y_0, -x_0)$ | $\overline{P}_2''(-z_0, -y_0, -x_0)$ | $\overline{P}_2'''(z_0, -y_0, -x_0)$ |
| $\overline{P}_3(z_0, x_0, -y_0)$ | $\overline{P}_3'(-z_0, x_0, -y_0)$ | $\overline{P}_3''(-z_0, -x_0, -y_0)$ | $\overline{P}_3'''(z_0, -x_0, -y_0)$ |
| $\overline{P}_4(x_0, z_0, -y_0)$ | $\overline{P}_4'(-x_0, z_0, -y_0)$ | $\overline{P}_4''(-x_0, -z_0, -y_0)$ | $\overline{P}_4'''(x_0, -z_0, -y_0)$ |
| $\overline{P}_5(y_0, z_0, -x_0)$ | $\overline{P}_5'(-y_0, z_0, -x_0)$ | $\overline{P}_5''(-y_0, -z_0, -x_0)$ | $\overline{P}_5'''(y_0, -z_0, -x_0)$ |

Therefore, sampling the hemisphere and calculating the coordinates of a sampling point $P_0(x_0, y_0, z_0)$ from a given sub-domain, we can calculate in parallel the other sampling point coordinates for the hemisphere in accordance to the Table 1.

The coordinate of symmetric points when we sampling the sphere can be calculated in a same way. The Table 2. represents the parallel coordinates calculations for the sphere. Note that the marked with $\overline{P}_{(\cdot)}^{(\cdot)}$ points in Table 2. are the same as the respective points presented in Table 1. and only differ in negative sign of the $Z$ coordinate.

This kind of partitioning allows to sample only one sub-domain and to calculate in parallel all other samples for the hemisphere or sphere. Since the symmetry is identity the generation of uniformly distributed random samples in a sub-domain leads to the uniform distribution of all samples in the hemisphere and sphere.

## 3    Algorithms for Parallel Sampling Scheme

In this section we consider the problem for sampling a sub-domain in the terms of hemisphere and sphere partitioning, described in the previous section. Each sub-domain represents solid angle subtended by orthogonal spherical triangle with fixed vertices and computable parameters. In order to generate uniformly distributed random samples of the sub-domain we propose two algorithms.

### 3.1    Algorithm 1

This algorithm is very similar to the Arvo's [1] algorithm for sampling of arbitrary spherical triangle. Let us consider the solid angle subtended by the spherical triangle $\triangle ABC$ shown in Fig. 3.

Due to the partitioning planes, we can observe, that the arcs $\widehat{AB}$, $\widehat{BC}$ and $\widehat{AC}$ are the arcs of the main (central) circles. The angle $\gamma$ is equal to $\frac{\pi}{2}$ and the length of the arc $\widehat{BC}$ is $\frac{\pi}{4}$. One can write the following identities:

$$\cos \widehat{AB} = \cos \widehat{BC} \cdot \cos \widehat{AC} + \sin \widehat{BC} \cdot \sin \widehat{AC} \cdot \cos \gamma$$

$$\cos \beta = -\cos \gamma \cdot \cos \alpha + \sin \gamma \cdot \sin \alpha \cdot \cos \widehat{AC}$$

$$\frac{\sin \widehat{BC}}{\sin \alpha} = \frac{\sin \widehat{AB}}{\sin \gamma}.$$

The first two expressions are spherical cosine law for the arc $\widehat{AB}$ and for the angle $\beta$, as well the third is spherical sine law. By definition the angle $\beta$ is the angle between tangents to the spherical arcs in the point $B$ which is equal to the dihedral angle between the partitioning planes. Therefore, the angle $\beta$ is

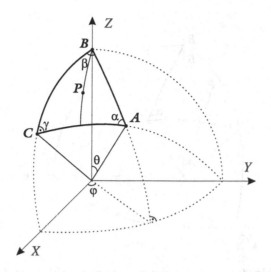

**Fig. 3.** Sampling point generation by Algorithm 1

always equal to the angle $\varphi$. Since the arc $\widehat{AB}$ is arc from the central circle in the orthogonal spherical triangle, it equals to $\theta$, we can write the following equality:

$$\tan \widehat{AB} = \frac{1}{\cos \beta} \Rightarrow \tan \theta = \frac{1}{\cos \varphi}.$$

Similarly to the strategy presented in [1] we attempt to generate a sample by mapping of the unit square onto orthogonal spherical triangle. With other words we seek a bijection $F(u, v) : [0, 1]^2 \rightarrow \triangle ABC$, where $u$ and $v$ are random variables uniformly distributed in $[0, 1]$. Now we introduce the following transformation:

$$\varphi = \frac{u\pi}{4} \quad \text{and} \quad \theta = \arctan \frac{v}{\cos \varphi} = \arctan \frac{v}{\cos \frac{u\pi}{4}},$$

where $u, v \in [0, 1]$; $\varphi \in [0, \frac{\pi}{4}]$ and $\theta \in [0, \arctan \frac{1}{\cos \varphi}]$. The algorithm can be described as:

▷ Generate Random Variables:( real $u$ , real $v$)
▷ Calculate angles : $\varphi = \dfrac{u\pi}{4}$    and    $\theta = \arctan \dfrac{v}{\cos \frac{u\pi}{4}}$
▷ Calculate the sampling point coordinates:
   $P_x = \cos \varphi \cdot \sin \theta$,    $P_y = \sin \varphi \cdot \sin \theta$    and    $P_z = \cos \theta$
▷ Return Sampling Point:    $P(P_x, P_y, P_z)$.

### 3.2    Algorithm 2

This algorithm tries to compute directly the unit radius vector of a sampling point inside to the orthogonal spherical triangle. Consider a point $P$ inside

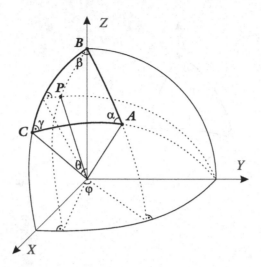

**Fig. 4.** Sampling point generation by Algorithm 2

for the spherical triangle $\triangle ABC$ shown in Fig. 4. The coordinates of an arbitrary sampling point $P$ could be calculated by finding the intersection point of spherical triangle $\triangle ABC$ with the two coordinate planes having normal vectors $N_x(1,0,0)$ and $N_y(0,1,0)$, and rotated respectively on angle $\varphi$ and angle $\theta$, where $\theta, \varphi \in [0, \frac{\pi}{4}]$. It is clear that when $\theta = \varphi = \frac{\pi}{4}$, the point $P \equiv A$ and when $\theta = \varphi = 0$, the point $P \equiv B$.

The rotations of the coordinate planes are defined by the matrices of rotation. Let, $R_{y\theta}$ is the matrix of rotation on angle $\theta$ around the axis $Y$ and $R_{z\varphi}$ is the matrix of rotation on angle $\theta$ around the axis $Z$:

$$R_{y\theta} = \begin{pmatrix} \cos\theta & 0 & -\sin\theta \\ 0 & 1 & 0 \\ \sin\theta & 0 & \cos\theta \end{pmatrix} \quad and \quad R_{z\varphi} = \begin{pmatrix} \cos\varphi & \sin\varphi & 0 \\ -\sin\varphi & \cos\varphi & 0 \\ 0 & 0 & 1 \end{pmatrix}.$$

Applying the rotations to the normal vectors $N_x(1,0,0)$ and $N_y(0,1,0)$ we calculate the normal vectors $N_x'$ and $N_y'$ to the rotated coordinate planes:

$$N_x' = N_x R_{y\theta} = (\cos\theta, 0, -\sin\theta) \quad and \quad N_y' = N_y R_{z\varphi} = (-\sin\varphi, \cos\varphi, 0).$$

The normalized vector product of the vectors $N_x'$ and $N_y'$ compute the coordinate of point $P$ as:

$$P = \frac{N_x' \times N_y'}{|N_x' \times N_y'|} \quad with \quad |P| = 1$$

where $N_x' \times N_y' = (\sin\theta\cos\varphi, \sin\theta\sin\varphi, \cos\theta\cos\varphi)$ and

$$|N_x' \times N_y'| = \sqrt{\sin^2\theta + \cos^2\theta\cos^2\varphi} = \sqrt{\cos^2\varphi + \sin^2\theta\sin^2\varphi}.$$

The algorithm can be described as:

▷ Generate Random Variables:( real $u$ , real $v$)
▷ Calculate angles : $\varphi = \dfrac{u\pi}{4}$    and    $\theta = \dfrac{v\pi}{4}$
▷ Calculate the sampling point coordinates:

$$P_x = \frac{\sin\theta \cdot \cos\varphi}{\sqrt{\sin^2\theta + \cos^2\theta\cos^2\varphi}}, \quad P_y = \frac{\sin\theta \cdot \sin\varphi}{\sqrt{\sin^2\theta + \cos^2\theta\cos^2\varphi}}$$

and  $P_z = \dfrac{\cos\theta \cdot \cos\varphi}{\sqrt{\sin^2\theta + \cos^2\theta\cos^2\varphi}}$

▷ Return Sampling Point:    $P(P_x, P_y, P_z)$.

## 4  Conclusion

The presented parallel sampling scheme for Monte Carlo solving of the rendering equation uses partitioning of the hemisphere or sphere by a natural way. The sphere or hemisphere is divided into equal sub-domains of orthogonal spherical triangles by applying the symmetry property. The advantages of this approach lie in the parallel computations. Sampling only one sub-domain, the sampling points over hemisphere or sphere are calculated in parallel. This approach is suitable for realization over parallel (MIMD, multiple instruction - multiple data) architectures and implementation on Grid infrastructures. The proposed algorithms for sampling of orthogonal spherical triangles accommodate the stratified sampling. Instead of using random variables $u$ and $v$ for sampling point generation we can apply low discrepancy sequences as shown in [4]. This fact leads to parallel Quasi-Monte Carlo approach for solving of the rendering equation, which is a subject of future study and research.

## References

1. Arvo, James, Stratifed sampling of spherical triangles. In: Computer Graphics Proceedings, Annual Conference Series, ACM Siggraph, pp. 437–438, (1995).
2. Dutré, Philip, Global Illumination Compendium, Script of September 29 2003, http://www.cs.kuleuven.ac.be/~phil/GI/TotalCompendium.pdf
3. Kajiya, J. T., The Rendering Equation, Computer Graphics, **vol. 20**, No. 4, pp. 143–150, Proceedings of SIGGRAPH'86, (1986).
4. Keller, Alexander, Quasi-Monte Carlo Methods in Computer Graphics: The Global Illumination Problem, Lectures in Applied Mathematics, **vol. 32**, pp. 455–469, (1996).
5. Rogers, D. F., Procedural Elements for Computer Graphics. McGraw-Hill Inc, (1985).
6. Urena, Carlos, Computation of Irradiance from Triangles by Adaptive Sampling, Computer Graphics Forum, **Vol. 19, No. 2**, pp. 165–171, (2000).
7. Yershova, Anna and LaValle, Steven M., Deterministic Sampling Methods for Spheres and $SO(3)$, Robotics and Automation, 2004. IEEE Proceedings of ICRA'04, **vol.4**, pp. 3974–3980, (2004).

# A Hybrid Monte Carlo Method for Simulation of Quantum Transport*

Todor Gurov, Emanouil Atanassov, and Sofiya Ivanovska

Institute for Parallel Processing - Bulgarian Academy of Sciences,
Acad. G. Bonchev St., Bl.25A, 1113 Sofia, Bulgaria
gurov@parallel.bas.bg, emanouil@parallel.bas.bg, sofia@parallel.bas.bg

**Abstract.** In this work we propose a hybrid Monte Carlo method for solving the Levinson equation. This equation describes the electron-phonon interaction on a quantum-kinetic level in a wire. The evolution problem becomes inhomogeneous due to the spatial dependence of the initial condition. The properties of the presented algorithm, such as computational complexity and accuracy, are investigated on the Grid by mixing quasi-random numbers and pseudo-random numbers. The numerical results are obtain for a physical model with GaAs material parameters in the case of zero electrical field.

## 1 Introduction

The Levinson equation [1] describes a femtosecond relaxation process of optically excited electrons which interact with phonons in an one-band semiconductor. The equation is an important tool for studying quantum effects in semiconductor devices [2]. Monte Carlo (MC) algorithms for solving such equations require large amounts of CPU time [3]. The spread of computational grids makes the parallelization properties of these algorithms important for the practical implementation. On the other hand quasi-Monte Carlo (QMC) methods [4] can offer higher precision and/or faster convergence for some kinds of integral equation problems, that are usually solved by MC methods. Careful study of the equation under consideration is needed in order to justify their usage.

In this paper we study a hybrid Monte Carlo method by mixing quasi-random numbers and pseudo-random numbers for solving the Levinson equation. After reduction of the dimension [5], the quantum-kinetic equation has the following integral form:

$$f_w(z, k_z, t) = f_{w,0}(z - \tfrac{\hbar k_z}{m}, t, k_z) + \tag{1}$$

$$+ \int_0^t dt'' \int_{t''}^t dt' \int_G d^3\mathbf{k}' \{K_1(k_z, \mathbf{k}', t', t'')f_w(z + h(k_z, q_z', t, t', t''), k_z', t'')\}$$

$$+ \int_0^t dt'' \int_{t''}^t dt' \int_G d^3\mathbf{k}' \{K_2(k_z, \mathbf{k}', t', t'')f_w(z + h(k_z, q_z', t, t', t''), k_z, t'')\},$$

* Supported by the Ministery of Education and Science of Bulgaria under Grant No. I1405/04 and by the EC FP6 under Grant No: INCO-CT-2005-016639 of the project BIS-21++.

T. Boyanov et al. (Eds.): NMA 2006, LNCS 4310, pp. 156–164, 2007.

where

$$h(k_z, q'_z, t, t', t'') = -\frac{\hbar k_z}{m}(t - t'') + \frac{\hbar q'_z}{2m}(t' - t'')$$

$$K_1(k_z, \mathbf{k}', t', t'') = S(k'_z, k_z, t', t'', \mathbf{q}'_\perp) = -K_2(\mathbf{k}', k_z, t', t'')$$

$$S(k'_z, k_z, t', t'', \mathbf{q}'_\perp) = \frac{2V}{(2\pi)^3} |G(\mathbf{q}'_\perp)\mathcal{F}(\mathbf{q}'_\perp, k_z - k'_z)|^2 \times$$

$$\left[ (n(\mathbf{q}') + 1) \cos\left( \frac{\epsilon(k_z) - \epsilon(k'_z) + \hbar\omega_{\mathbf{q}'}}{\hbar}(t' - t'') \right) \right.$$

$$\left. +n(\mathbf{q}') \cos\left( \frac{\epsilon(k_z) - \epsilon(k'_z) - \hbar\omega_{\mathbf{q}'}}{\hbar}(t' - t'') \right) \right],$$

and $\int_G d^3\mathbf{k}' = \int d\mathbf{q}'_\perp \int_{-Q_2}^{Q_2} dk_z$. The domain $G$ is specified in the next section. The Bose function, $n_{\mathbf{q}'} = 1/(\exp(\hbar\omega_{\mathbf{q}'}/\mathcal{K}T) - 1)$, describes the phonon distribution, where $\mathcal{K}$ is the Boltzmann constant and $T$ is the temperature of the crystal. $\hbar\omega_{\mathbf{q}'}$ is the phonon energy which generally depends on $\mathbf{q}' = \mathbf{q}'_\perp + q'_z = \mathbf{q}'_\perp + (k_z - k'_z)$, and $\varepsilon(k_z) = (\hbar^2 k_z^2)/2m$ is the electron energy.

The electron-phonon coupling constant $\mathcal{F}$ is chosen according to a Fröhlich polar optical interaction:

$$\mathcal{F}(\mathbf{q}'_\perp, k_z - k'_z) = -\left[ \frac{2\pi e^2 \omega_{\mathbf{q}'}}{\hbar V} \left( \frac{1}{\varepsilon_\infty} - \frac{1}{\varepsilon_s} \right) \frac{1}{(\mathbf{q}')^2} \right]^{\frac{1}{2}}, \tag{2}$$

$(\varepsilon_\infty)$ and $(\varepsilon_s)$ are the optical and static dielectric constants. $G(\mathbf{q}'_\perp)$ is the Fourier transform of the square of the ground state wave function $|\Psi|^2$.

The equation describes electron evolution which is quantum in both, the real space due to the confinements of the wire, and the momentum space due to the early stage of the electron-phonon kinetics. The kinetics resembles the memory character of the homogeneous Levinson or Barker-Ferry models [9], but the evolution problem becomes inhomogeneous due to the spatial dependence of the initial condition $f_{w,0}$. The cross-section of the wire is chosen to be a square with side $a$ so that:

$$|G(\mathbf{q}'_\perp)|^2 = |G(q'_x)G(q'_y)|^2 =$$

$$\left( \frac{4\pi^2}{q'_x a ((q'_x a)^2 - 4\pi^2)} \right)^2 4\sin^2(aq'_x/2) \left( \frac{4\pi^2}{q'_y a ((q'_y a)^2 - 4\pi^2)} \right)^2 4\sin^2(aq'_y/2)$$

We note that the Neumann series of integral equations of type (1) converges, [3]. Thus, we can construct a MC estimator to evaluate suitable functionals of the solution.

## 2    The Monte Carlo Method

The values of the physical quantities are expressed by the following general functional of the solution of (1):

$$J_g(f) \equiv (g, f) = \int_0^T \int_D g(z, k_z, t) f_w(z, k_z, t) dz dk_z dt. \tag{3}$$

Here we specify that the phase space point $(z, k_z)$ belongs to a rectangular domain $D = (-Q_1, Q_1) \times (-Q_2, Q_2)$, and $t \in (0, T)$. The function $g(z, k_z, t)$ depends on the quantity of interest. In particular, we focus on the Wigner function $f_w(z, k_z, t)$, the wave vector (and respectively the energy) $f(k_z, t)$, and the density distribution $n(z, t)$. The latter two functions are given by the integrals

$$f(k_z, t) = \int \frac{dz}{2\pi} f_w(z, k_z, t); \qquad n(z, t) = \int \frac{dk_z}{2\pi} f_w(z, k_z, t). \tag{4}$$

The evaluation is performed in fixed points by choosing $g(z, k_z, t)$ as follows:

$$
\begin{aligned}
(i) \quad & g(z, k_z, t) = \delta(z - z_0)\delta(k_z - k_{z,0})\delta(t - t_0), \\
(ii) \quad & g(z, k_z, t) = \frac{1}{2\pi}\delta(k_z - k_{z,0})\delta(t - t_0), \\
(iii) \quad & g(z, k_z, t) = \frac{1}{2\pi}\delta(z - z_0)\delta(t - t_0).
\end{aligned}
\tag{5}
$$

We construct a biased MC estimator for evaluating the functional (3) using backward time evolution of the numerical trajectories in the following way:

$$\xi_s[J_g(f)] = \frac{g(z, k_z, t)}{p_{in}(z, k_z, t)} W_0 f_{w,0}(., k_z, 0) + \frac{g(z, k_z, t)}{p_{in}(z, k_z, t)} \sum_{j=1}^{s} W_j^\alpha f_{w,0}\left(., k_{z,j}^\alpha, t_j\right),$$

$$\tag{6}$$

where

$$f_{w,0}\left(., k_{z,j}^\alpha, t_j\right) = \begin{cases} f_{w,0}\left(z + h(k_{z,j-1}, q'_{z,j}, t_{j-1}, t'_j, t_j), k_{z,j}, t_j\right), & \text{if } \alpha = 1, \\ f_{w,0}\left(z + h(k_{z,j-1}, q'_{z,j}, t_{j-1}, t'_j, t_j), k_{z,j-1}, t_j\right), & \text{if } \alpha = 2, \end{cases}$$

$$W_j^\alpha = W_{j-1}^\alpha \frac{K_\alpha(k_{zj-1}, k_j, t'_j, t_j)}{p_\alpha p_{tr}(k_{j-1}, k_j, t'_j, t_j)}, \quad W_0^\alpha = W_0 = 1, \quad \alpha = 1, 2, \quad j = 1, \ldots, s.$$

The probabilities $p_\alpha, (\alpha = 1, 2)$ are chosen to be proportional to the absolute value of the kernels in (1). The initial density $p_{in}(z, k_z, t)$ and the transition density $p_{tr}(\mathbf{k}, \mathbf{k}', t', t'')$ are chosen to be tolerant[1] to the given function $g(z, k_z, t)$ and the kernels, respectively. The first point $(z, k_{z0}, t_0)$ in the Markov chain is chosen using the initial density, where $k_{z0}$ is the third coordinate of the wave vector $\mathbf{k}_0$. Next points $(k_{zj}, t'_j, t_j) \in (-Q_2, Q_2) \times (t_j, t_{j-1}) \times (0, t_{j-1})$ of the Markov chain:

$$(k_{z0}, t_0) \to (k_{z1}, t'_1, t_1) \to \ldots \to (k_{zj}, t'_j, t_j) \to \ldots \to (k_{zs}, t'_s, t_s), \quad j = 1, 2, \ldots, s$$

---

[1] $r(x)$ is tolerant of $g(x)$ if $r(x) > 0$ when $g(x) \neq 0$ and $r(x) \geq 0$ when $g(x) = 0$.

do not depend on the position $z$ of the electrons. They are sampled using the transition density $p_{tr}(\mathbf{k}, \mathbf{k}', t', t'')$ as we take only the $k_z$-coordinate of the wave vector $\mathbf{k}$. Note the time $t'_j$ conditionally depends on the selected time $t_j$. The Markov chain terminates in time $t_s < \varepsilon_1$, where $\varepsilon_1$ is a fixed small positive number called a truncation parameter. In order to evaluate the functional (3) by $N$ independent samples of the estimator (6), we define a Monte Carlo method

$$\frac{1}{N} \sum_{i=1}^{N} (\xi_s[J_g(f)])_i \xrightarrow{P} J_g(f^{(s)}) \approx J_g(f). \tag{7}$$

$f^{(s)}$ is the iterative solution obtained by the Neumann series of (1), and $s$ is the number of iterations. In order to obtain a MC computational algorithm, we have to specify the initial and transition densities. Also, we have to describe the sampling rule needed to calculate the states of the Markov chain by using SPRNG library [8].

We note the MC estimator is constructed using the kernels of the equation (1). That is why we suggest the transition density function to be proportional of the term (2) that contains the singularity, namely: $p_{tr}(\mathbf{k}, \mathbf{k}', t', t'') = p(\mathbf{k}'/\mathbf{k})p(t, t', t'')$, where

$$p(t, t', t'') = p(t, t'')p(t'/t'') = \frac{1}{t} \frac{1}{(t - t'')}, \quad p(\mathbf{k}'/\mathbf{k}) = c_1/(\mathbf{k}' - \mathbf{k})^2.$$

$c_1$ is the normalized constant. Thus, if we know $t$, the next times $t''$ and $t'$ are computed by using the inverse-transformation rule. The function $p(\mathbf{k}'/\mathbf{k})$ is chosen in spherical coordinates $(\rho, \theta, \varphi)$, and the wave vector $\mathbf{k}'$ are sampled in the same way as it is described in [7].

The choice of $p_{in}(z, k_z, t)$ depends on the choice of the function $g(z, k_z, t)$ in (5). Thus, by using one and the same Markov chains the desired physical quantities (values of the Wigner function, the energy and the density distributions) can be evaluated simultaneously.

## 3   Background on Quasi-Monte Carlo Methods and Quasirandom Numbers

Quasi-Monte Carlo methods and algorithms are popular in many areas from physics to economy. The constructive dimension of the algorithms can be several hundreds or even more, but fewer dimensions offer decreased error. Monte Carlo and quasi-Monte Carlo algorithms are frequently executed on parallel supercomputers or clusters. Scrambling (see [6] provides a way of combining the advantages of Monte Carlo and quasi-Monte Carlo methods, offering authomatic error estimation.

**Definition 1.** *Discrepancy of a sequence* $\sigma = \{x_i\}_{i=1}^{N}$ *in* $E^s$:

$$D_N^*(\sigma) = \sup_{I \subset E^s} \left| \frac{A_N(\tau, I)}{N} - \mu(E) \right|.$$

*For low-discrepancy sequences $D_N^*(\sigma) = \mathcal{O}(N^{-1}\log^s N)$.*

The Koksma-Hlawka inequality relates integration error with discrepancy:

$$\left| \int_{\mathbb{E}^s} f(x)\, dx - \frac{1}{N}\sum_{i=1}^{N} f(x_i) \right| \le V(f) D_N^*(\sigma).$$

Improved convergence - in practice the log factor is not observed. Can be either completely deterministic or include some randomness. Some families of sequences can be tuned to the particular application. Faster generation than pseudo-random numbers. Authomatic error estimation can be achieved. Amenable to parallel implementation.

**Definition 2.** *Let $A_1, \ldots, A_s$ be infinite matrices,*

$$A_k = \left\{ a_{ij}^{(k)} \right\}, \quad i,j = 0, 1, \ldots,$$

*with $a_{ij}^{(k)} \in \{0,1\}$, such that $a_{ii}^{(k)} = 1$ for all $i$ and $k$, $a_{ij}^{(k)} = 0$ if $i < j$, and let $\tau^{(1)}, \ldots, \tau^{(s)}$ be sequences of permutations of the set $\{0,1\}$. The nth term of the low-discrepancy sequence $\sigma$ is obtained by representing $n$ in binary number system:*

$$n = \sum_{i=0}^{r} b_i 2^i, \quad \text{and setting:} \quad x_n^{(k)} = \sum_{j=0}^{r} 2^{-j-1} \tau_j^{(k)} \left( \bigoplus_{i=0}^{j} b_i a_{ij}^{(k)} \right).$$

**Definition 3.** *Let $p_1, \ldots, p_s$ be distinct prime numbers, and let the integer modifiers $k_1, \ldots, k_s$ be given, $k_i \in [0, p_i - 1]$. Fix also some scrambling terms $b_j^{(i)} \in [0, p_i - 1]$. If the representation of $n$ in $p_i$-adic number system is:*

$$n = \sum_{j=0}^{m} a_j^{(i)} p_i^j,$$

*then the ith coordinate of the modified Halton sequence $\sigma(p_1, \ldots, p_s; k_1, \ldots, k_s)$ is defined as*

$$x_n^{(i)} = \sum_{j=0}^{m} \mathrm{imod}\left( a_j^{(i)} k_i^{j+1} + b_j^{(i)}, p_i \right) p_i^{-j-1}.$$

*The original Halton sequence is obtained when all the modifiers are one, and all the scrambling terms are zero.*

## 4    The Hybrid Monte Carlo Algorithm

The idea of the hybrid Monte Carlo algorithm consists of using pseudo-random numbers for some dimensions and quasi-random numbers for the other dimensions. Error estimation is achieved by means of scrambling of the quasi-random sequence. The use of scrambling corrects the correlation problem found in the previous purely quasi-Monte Carlo algorithm. We decided to sample the times using the Halton quasi-random sequence and to sample the other quantities, related to spacial dimensions, using pseudo-random numbers. A schematic description of the algorithm is given below, assuming that we only need to compute the Wigner function at one point $(k_1, z_1)$. In the algorithm, $\varepsilon_1$ is the truncation parameter.

- Input number of trajectories to be used $N$, relaxation time $T$, other parameters, describing the initial condition.
- For $i$ from 1 to $N$ sample a trajectory as follows:
  - set time $t := T$, weight $W := 1, k = k_1, z := z_1$
  - prepare the next point of the Halton sequence to be used $(x_1, x_2, \ldots, x_n)$, with $n$ sufficiently big ($n = 100$ in our case), and set $j = 1$
  - repeat until $t > \varepsilon_1$:
    * $k$ is simulated using pseudorandom numbers
    * $t'$, $t$ are simulated using consecutive dimensions of the Halton sequence, i.e. the points $x_{2j-1}, x_{2j}$, by the formulae

    $$t_2 := tx_{2j-1}, t_1 := t_2 + x_{2j}(t - t_2), t' := t_1, t = t_2$$

    * multiply the weight: $W := W * t(t - t_2)$
    * compute the two kernels $K_1$ and $K_2$
    * select which one to use with probability proportional to their absolute values.
    * multiply the weight: $W := W * (|K_1| + |K_2|) \, sgn \, (K_m)$ if $K_m$ is the kernel selected
    * sample $q$ using a spline approximation of the inverse function
    * multiply the weight by the appropriate integral: $W := W * I$
    * modify $k$, depending on the kernel and the electric field applied: $k_{new} = k - c_3 * (t - t2)$ if $K_1$ was chosen or $k_{n}ew = k - c_3 * (t - t2)$ if $K_2$ was chosen
    * modify $z$: $z_{n}ew = z - c_1 * k * (t - t_2) - c_2 * (t - t_2) * (t + t_2)$
    * compute the contribution of this iteration to the Wigner function: add $W * \psi(z, k)$ to the estimator, where $\psi(z, k)$ is the value of the initial condition
    * increment $j := j + 1$

The constructive dimensionality of the algorithm is $4n$, where $n$ is the maximal length of the trajectory. We use $2n$ pseudorandom numbers for each trajectory, and the dimensionality of the Halton sequence is $2n$.

# 5   Parallel Implementation for Grids

Monte Carlo methods are inherently parallelizable. Parallelization over Grid is done via pre-processing and post-processing phase. The Monte Carlo and quasi-Monte Carlo simulations that we need to perform require hours of CPU time, due to the exponential dependence on the time $T$ [3]. Naturally we come to the idea of using computational Grids in order to perform the computations in a feasible timeframe (a day or two for obtaining a set of results). We divided the work into chunks and we used as a package of parallel pseudo-random generators the SPRNG [8]. For generating the scrambled Halton sequences, we used our ultra-fast generators, which provide the necessary functionality:

- portable, multi-platform
- using assembly language where possible for best performance
- providing a fast-forward operation, which permits jumping ahead in the sequence without unnecessary overhead

We used our framework for massive job submission, which allows submission and control of hundreds, even thousands of jobs on the Grid, with redundancy and failover. Due to the use of quasi-random numbers, we had to achieve fault tolerant operation of the framework. The inputs are entered into a MySQL database. The jobs that we submit to the Grid are *placeholder* jobs, which obtain their input from a web service, offering authentication and authorization using Grid certificates and connecting to the database. The jobs are monitored throughout their lifetime and their status is available from the database. The failover procedures allowed us to discard lost/aborted jobs without problems. In this way we can see when all the work is completed. The jobs were submitted to several clusters from the EGEE production service[2], and we strove to achieve the maximal throughput that we could. During our work on this paper, we submitted several thousands of jobs in total, each requiring between 1 and 20 hours of CPU time.

## 6    Numerical Experiments

The numerical results presented in Figures 1-2 are obtained for the following GaAs material parameters: the electron effective mass is 0.063, the optimal phonon energy is 36 meV, the static and optical dielectric constants are $\varepsilon_s = 12.9$ and $\varepsilon_\infty = 10.92$. The initial condition is a product of two Gaussian distributions of the energy and space. The $k_z^2$ distribution corresponds to a generating laser pulse with an excess energy of about 150 meV. The $z$ distribution is centered around zero. The side of the wire is chosen to be 10 nanometers. Hybrid MC solutions for the energy distribution, $f(k_z, t)$, are presented on Fig.1, where $t = 120, 140, 160, 180$ femtoseconds. The solutions are estimated for 130 points in

---

[2] The Enabling Grids for E-sciencE (EGEE) project is funded by the European Commission and aims to build on recent advances in grid technology and develop a service grid infrastructure which is available to scientists 24 hours-a-day. The project aims to provide researchers in both academia and industry with access to major computing resources, independent of their geographic location. The EGEE project identifies a wide-range of scientific disciplines and their applications and supports a number of them for deployment. To date there are five different scientific applications running on the EGEE Grid infrastructure. For more information see http://public.eu-egee.org/

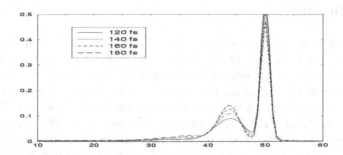

**Fig. 1.** Energy distribution at different evolution times. The quantum solution shows broadening of the replicas.

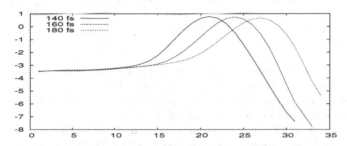

**Fig. 2.** Electron density along the wire at different evolution times

the interval $(0, Q_2)$, where $Q_2 = 60 \times 10^7 m^{-1}$. Hybrid MC solutions for the electron density, $n(z, t0$, are shown on Fig.2 at fixed evolution times. Here, solutions are estimated for 200 points in the interval $(0, Q_1)$, where $Q_1 = 35 \times 10^9 m^{-1}$. The number of Markov chain realizations (stochastic trajectories) is $N = 2.5$ millions, when $t = 140 fs$. The number of trajectories increases up to 10 millions with increasing the evolution time. The results for the computational complexity (CPU time) are shown in Table 1. It is clear that the use of the hybrid algorithm is preferable than the MC algorithm. The QMC algorithm has the least complexity. However, noise appears in the QMS solutions. That is why they are not shown in this paper.

**Table 1.** The computational complexity (CPU time) and RMS error of all algorithms for 2 500 000 trajectories

|  | $t$ | CPU time | RMS error |
|---|---|---|---|
| MC | $120 fs$ | 242m5.220s | 0.038623 |
| algorithm | $140 fs$ | 251m0.690s | 0.146971 |
| Hybrid | $120 fs$ | 233m8.530s | 0.037271 |
| algorithm | $140 fs$ | 234m9.980s | 0.150905 |
| QMC | $120 fs$ | 201m0.870s | 0.028626 |
| algorithm | $140 fs$ | 223m3.050s | 0.136403 |

# 7    Conclusions

We developed a hybrid Quasi-Monte Carlo algorithm, which combines the advantages of Monte Carlo and Quasi-Monte Carlo algorithms. The new algorithm does not exhibit the bias, observed in the previous Quasi-Monte Carlo algorithm for high values of the evolution time. The statistical error of the hybrid algorithm is of the same magnitude as that of the Monte Carlo algorithm. The results obtained by the hybrid algorithm gave results, consistent with the results obtained by using pseudo-random numbers. The generation of the low-discrepancy sequences is faster than that of pseudo-random numbers, which makes the new method faster overall. The algorithm was implemented in a Grid environment, which enabled us to obtain results much faster than we could using our local cluster, because we utilized several Grid clusters at once.

# References

1. I. Levinson, Translational invariance in uniform fields and the equation for the density matrix in the Wigner representation, Sov. Phys. JETP, 30, 362–367, 1970.
2. M. Herbst et al. Electron-phonon quantum kinetics for spatially inhomogeneous excitations, Physical Review B, 67, 195305:1–18, 2003.
3. T.V. Gurov, P.A. Whitlock. An efficient backward Monte Carlo estimator for solving of a quantum kinetic equation with memory kernel, Mathematics and Computers in Simulation, Vol. 60(1-2), 85-105, 2002.
4. H. Niederreiter. Random Number Generations and Quasi-Monte Carlo Methods. SIAM, Philadelphia, 1992.
5. M. Nedjalkov et al. Femtosecond Evolution of Spatially Inhomogeneous Carrier Excitatons - Part I: Kinetic Approach., LSSC 2005, LNCS, Vol. 3743, Springer-Verlag, 149-155, 2006.
6. E. Atanassov. A New Efficient Algorithm for Generating the Scrambled Sobol Sequence, Numerical Methods and Applications (I. Dimov, I. Lirkov, S. Margenov, Z. Zlatev - Eds.), LNCS 2542, 83–90, Springer, 2003.
7. T.V. Gurov, I.T. Dimov. A Parallel Monte Carlo Method For Electron Quantum Kinetic Equation, LNCS, Vol. 2907, Springer-Verlag, 151-161, 2004.
8. M. Mascagni. SPRNG: A Scalable Library for Pseudorandom Number Generation. Recent Advances in Numerical Methods and Applications II (O. Iliev, M. Kaschiev, Bl. Sendov, P.S. Vassilevski eds.), Proceeding of NMA 1998, World Scientific, Singapore, 284–295, 1999.
9. M. Nedjalkov et al. Unified Particle Approach to Wigner-Boltzmann Transport in Small Semiconductor Devices. Phys. Rev. B, Vol. 70, 115319–115335, 2004.

# Quasi-random Walks on Balls Using C.U.D. Sequences*

Aneta Karaivanova[1], Hongmei Chi[2], and Todor Gurov[1]

[1] IPP - Bulgarian Academy of Sciences,
Acad. G. Bonchev St., Bl.25A, 1113 Sofia, Bulgaria
{anet,gurov}@parallel.bas.bg
[2] Dept. of Computer & Information Sciences Florida A&M University, Tallahassee,
FL 32307-5100, USA
hchi@cis.famu.edu

**Abstract.** This paper presents work on solving elliptic BVPs problems based on quasi-random walks, by using a subset of uniformly distributed sequences—completely uniformly distributed (c.u.d.) sequences. This approach is novel for solving elliptic boundary value problems. The enhanced uniformity of c.u.d. sequences leads to faster convergence. We demonstrate that c.u.d. sequences can be a viable alternative to pseudorandom numbers when solving elliptic boundary value problems. Analysis of a simple problem in this paper showed that c.u.d. sequences achieve better numerical results than pseudorandom numbers, but also have the potential to converge faster and so reduce the computational burden.

**Keywords:** completely uniformly distributed sequences, Monte Carlo methods, acceptance-rejection, Markov chains, BVPs.

## 1  Introduction

Monte Carlo (MC) methods are based on the simulation of stochastic processes whose expected values are equal to computationally interesting quantities. Standard MC simulation using pseudorandom sequences can be quite slow due to its convergence rate is only $O(N^{-1/2})$ [21] for $N$ sample paths. Quasi-Monte Carlo (QMC) simulation, using deterministic sequences that are more uniform than random ones, holds out the promise of much greater accuracy, close to $O(N^{-1})$ [15] in optimal cases. Randomized versions of QMC simulation can in some cases bring a typical error close to $O(N^{-3/2})$ [22]. QMC simulation is not a magic bullet, however, as shown in the works of Morokoff [18]. The asymptotic error magnitudes are the ones "close to" above, multiplied by $(logN)^k$, where $k$ depends on the dimension of the simulation. In high dimensions these powers of $logN$ do not become negligible at any computationally possible sample size. This loss of effectiveness has been documented for a series of test problems.

* Supported by the Ministry of Education and Science of Bulgaria under Grant No. I1405/04 and by the EC FP6 under Grant No: INCO-CT-2005-016639 of the project BIS-21++.

T. Boyanov et al. (Eds.): NMA 2006, LNCS 4310, pp. 165–172, 2007.

Perhaps more startling is the fact that a considerable fraction of the enhanced convergence of the quasi-Monte Carlo integration is lost when the integrand is discontinuous. In fact, even in two dimensions one can loose the approximately $O(N^{-1})$ QMC convergence for an integrand that is discontinuous on a curve such as a circle. In the best cases the convergence drops to $O(N^{-2/3})$, which is only slightly better than regular MC integration.

There are some very efficient MC methods for the solution of elliptic boundary value problems (BVPs), see for example [4,24,16]. While it is often preferable to solve a partial differential equation with a deterministic numerical method, there are some circumstances where MC methods have a distinct advantage. For example, when the geometry of a problem is complex, when the required accuracy is moderate, when a geometry is defined only statistically, or when a linear functional of the solution, such as a solution at a point, is desired, MC methods are often the most efficient method of solution.

The use of QMC methods for BVPs is studied in [20,18,13,17,7] etc. Here we refer to [17] where quasi-Monte Carlo variants of three Monte Carlo algorithms: Grid-Walk, Walk-on-Spheres and Walk-on-Balls, are considered. The first one uses a discretization of the problem on a mesh and solves the linear algebraic system which approximates the original problem. The second two methods use an integral representation of the problem which leads to a random walk on spheres or to a random walk on balls method. Different strategies for using quasirandom sequences are proposed and tested in order to generate quasirandom walks on grids, spheres, and balls. Theoretical error bounds are established, numerical experiments with model elliptic BVPs in two and three dimensions are also solved. The rate of convergence and the computational complexity of the QMC methods and the corresponding MC methods are compared both theoretically and numerically. The QMC methods preserve the advantages of the Monte Carlo for solving problems in complicated domains, and show better rates of convergence for QMC methods with direct simulation (Grid-Walk and Walk-on-Spheres) and slightly better for Walk-on-Balls (WoB) only for some parameters of the differential equation.

Recently, Owen and Tribble [23] showed that a subset of quasi-random numbers (QRNs) called, completely uniformly distributed (c.u.d.) sequences, can replace pseudorandom numbers in Markov chain Monte Carlo methods. Their results motivated us to apply c.u.d. sequences in the WoB method.

## 2    Background(Random Walks on Balls)

Let $G \subset \mathbb{R}^3$ be a bounded domain with boundary $\partial G$. Consider the following elliptic BVP:

$$Mu \equiv \sum_{i=1}^{3}(\frac{\partial^2}{\partial x_i^2} + b_i(x)\frac{\partial}{\partial x_i})u(x) + c(x)u(x) = -\phi(x), \qquad x \in G \qquad (1)$$

$$u(x) = \psi(x), \qquad x \in \partial G. \qquad (2)$$

Assume that the boundary $\partial G$, and the given function $b_i(x)$, $c(x) \leq 0$, $\phi(x)$, and $\psi(x)$, satisfy conditions ensuring that the solution of the problem (1), (2) exists and is unique, [4,19]. In addition, assume that $\nabla \cdot b(x) = 0$. The solution, $u(x)$, has a local integral representation for any standard domain, $T$, lying completely inside the domain $G$, [19]. To derive this representation we proceed as follows. The adjoint operator of $M$ has the form:

$$M^* = \sum_{i=1}^{3} \left( \frac{\partial^2}{\partial x_i^2} - b_i(x) \frac{\partial}{\partial x_i} \right) + c(x)$$

In the integral representation we use Levy's function defined as:

$$L_p(y,x) = \mu_p(R) \int_r^R (1/r - 1/\rho) p(\rho) d\rho, \quad r < R,$$

where

$$\mu_p(R) = (4\pi q_p)^{-1}, \quad q_p(R) = \int_0^R p(\rho) d\rho,$$

$p(r)$ is a density function, and $r = |x - y| = \left( \sum_{i=1}^{3} (x_i - y_j)^2 \right)^{1/2}$. Then the following integral representation holds, [4]:

$$u(x) = \int_T [u(y) M_y^* L_p(y,x) + L_p(y,x)\phi(y)] dy$$

$$+ \int_{\partial T} \sum_{j=1}^{3} \nu_j \left[ \left( L_p(y,x) \frac{\partial u(y)}{\partial y_j} - u(y) \frac{\partial L(y,x)}{\partial y_i} \right) - b_j(y)u(y)L_p(y,x) \right] d_y S,$$

where $\nu = (\nu_1, \nu_2, \nu_3)$ is the exterior normal to the boundary $\partial G$, and $T$ is any closed domain in $G$.

For the special case where the domain is a ball $T = B(x) = \{y : |y-x| \leq R(x)\}$ with center $x$ and radius $R(x)$, $B(x) \subset \overline{G}$, and for $p(r) = e^{-kr}$, $k \geq b^* + Rc^*$ ($b^* = \max_{x \in G} |b(x)|$, $c^* = \max_{x \in G} |c(x)|$), the above representation can be simplified, [3,4]:

$$u(x) = \int_{B(x)} M_y^* L_p(y,x) u(y) dy + \int_{B(x)} L_p(y,x)\phi(y) dy. \tag{3}$$

Moreover, $M_y^* L_p(y,x) \geq 0$, $y \in B(x)$ for the above parameters $k, b^*$ and $c^*$, and so it can be used as a transition density in a Markov process.

We can consider our problem (1, 2) as an integral equation (3) with kernel:

$$k(x,y) = \begin{cases} M_y^* L_p(y,x) & \text{,when } x \notin \partial G \\ 0 & \text{,when } x \in \partial G, \end{cases}$$

and the right-hand side given by

$$f(x) = \begin{cases} \int_{B(x)} L_p(y,x)\phi(y) dy & \text{, when } x \notin \partial G \\ \psi(x) & \text{, when } x \in \partial G. \end{cases}$$

However, for the above choice of $k(x, y)$, $||K|| < 1$, and the corresponding Neumann series of the integral equation (3) converges. Thus, the last representation allows to construct the WoB procedure for solving the problem. To ensure the convergence of the MC method, a biased estimate is constructed (see [4]) by introducing an $\varepsilon$-strip, $\partial G_\varepsilon$, of the boundary, $\partial G$, $\partial G_\varepsilon = \{x \in G : d(x) < \varepsilon\}$ where $d(x)$ is the distance from $x$ to the closest point of the boundary, $\partial G$. The transition density function of the Markov chain is chosen as follow:

$$p(x, y) = k(x, y) = M_y^* L_p(y, x) \geq 0, \tag{4}$$

$|x - y| \leq R$, where $R$ is the radius of the maximal ball with center $x$, and lying completely in $G$. Now, we construct a random walk $\xi_1, \xi_2, \ldots, \xi_{k_\varepsilon}$ such that every point $\xi_j$, $j = 1, \ldots, k_\varepsilon - 1$ is chosen in the maximal ball $B(x_{j-1})$, lying in $G$, in accordance with density (4). The Markov chain terminates when it reaches $\partial G_\varepsilon$, so that $\xi_{k_\varepsilon} \in \partial G_\varepsilon$.

If we are interested to estimate the solution at the point $\xi_0$, we choose the initial density function of the Markov chain to be $\delta(x - \xi_0)$. Then $u(\xi_0) = E[\theta(\xi_0)]$, where

$$\theta(\xi_0) = \sum_{j=0}^{k_\varepsilon - 1} \int_{B(\xi_j)} L_p(y, \xi_j)\phi(y)dy + \psi(\xi_{k_\varepsilon}), \tag{5}$$

is the biased MC estimator for the solution of the (3) and $E[\theta(\xi_0)]$ is its mathematical expectation.

## 2.1  The Monte Carlo Simulation

The Monte Carlo algorithm for estimating $u(\xi_0)$ consists of simulating $N$ Markov chains with a transition density given in (4), scoring the corresponding realizations of $\theta[\xi_0]$ and accumulating them. Direct simulation of the points in the Markov chain using WoB procedure by the density function (4) is problematic due to the complexity of the expression for $M_y^* L(y, x)$. It is computationally easier to represent $p(x, y)$ in spherical coordinates as $p_1(r)p_2(\mathbf{w}|r)$. Thus, given $\xi_{j-1}$, the next point is $\xi_j = \xi_{j-1} + r\mathbf{w}$, where the distance, $r$, is chosen with density

$$p_1(r) = (ke^{-kr})/(1 - e^{-kR}), \tag{6}$$

and the direction, $\mathbf{w}$, is chosen according to:

$$p_2(\mathbf{w}/r) =$$
$$1 + \frac{\sum_{i=1}^{3} b_i(x + r\mathbf{w})w_i + c(x + r\mathbf{w})r}{e^{-kr}} \int_r^R e^{-k\rho}d\rho - \frac{c(x + r\mathbf{w})r^2}{e^{-kr}} \int_r^R \frac{e^{-k\rho}}{\rho}d\rho. \tag{7}$$

To simulate the direction, $\mathbf{w}$, the **acceptance-rejection method** (ARM) is used with the following majorant:

$$h(r) = 1 + \frac{b^*}{e^{-kr}} \int_r^R e^{-k\rho}d\rho. \tag{8}$$

# 3   Completely Uniformly Distributed (C.U.D.) Sequences

Quasirandom (Uniformly distributed mod 1) sequences are constructed to minimize the *discrepancy*, a measure of the deviation from uniformity and therefore
quasirandom sequences are sometimes described as low-discrepancy sequences.
The discrepancy quantifies the lack of uniformity or equidistribution of points
placed in a set, usually in the unit hypercube, $[0,1)^s$. The most widely studied
discrepancy measures are based on the star discrepancy [20]. For any sequence
$\{x_n\} \in [0,1)^s$ with $N$ points, define $E$ as a subset of the unit cube, $vol(E)$
as the volume of $E$, and $\#\{x_i \in E\}$ as the number of points in $E$. Then the
star-discrepancy of this sequence is given by

$$D_N^* = \sup_{E \subseteq [0.1)^s} \left| \frac{\#\{x_i \in E\}}{N} - vol(E) \right|. \tag{9}$$

For a one-dimensional point set, the star-discrepancy is the Kolmogorov-Smirnov
statistic [8] based on the uniform distribution. Quasirandom sequences aim to
have the fraction of their points within any subset $E$ as close as possible to the
subset's volume fraction. A sequence $\{x_i\}_{1 \le i \le N}$ is a low-discrepancy sequence
if its star-discrepancy satisfies

$$D_N^* \le C_s \frac{(\log N)^s}{N}, \tag{10}$$

where the constant $C_s$ depends only on the dimension $s$. The accuracy of a
QMC estimate strongly depends on the discrepancy of the sequence. Therefore,
the convergence rate of QMC is close to $O(N^{-1})$ [21,1] instead of $O(N^{-1/2})$.

An infinite sequence $\{x_i\}$ is uniformly distributed mod 1, if we have

$$\lim_{n \longrightarrow \infty} D_n^*(x_1, x_2, \ldots, x_n) = 0, \tag{11}$$

An infinite sequence $\{x_i\}$ is completely uniformly distributed if $z_i^k$ is uniformly
distributed mod 1 for any positive integer $k$, where $z_i^k = (x_i, x_{i+1}, \ldots, x_{i+k-1}) \in$
$(0,1]^k$. Therefore, if a sequence is completely uniformly distributed, it is a uniformly distributed sequence. Those c.u.d. sequences have a slightly stronger property than uniformly distributed sequences, namely, c.u.d. sequences are a subset
of uniformly distributed sequences. The idea of c.u.d. sequences was introduced
by Korabov in [11]. The description of the construction of c.u.d. sequences can
be found in [10,12,25,14]. An example showing the difference between c.u.d. sequences and quasirandom sequences in MC simulation was given by [6]. A proved
statement that guarantees the existence of c.u.d. sequences can be found in [9].
The c.u.d. sequence is the universal sequence for computing multi-dimensional
integrals, modeling Markov chains and pseudorandom numbers [8].

## 3.1   Quasirandom Walks on Balls

The use of quasirandom numbers for integration of high-dimensional problems
improved the results over standard MC techniques. Loh (see [15]) theoretically

showed that QMC can converge faster than MC, and QMC has been successfully used to approximate high-dimension integrals [2] in situations where MC
has difficulties in achieving good results. One would think that improvements
in MCMC should easily translate to QMCMC, but difficulties arise because of
the sequential nature of Markov Chain. Despite these difficulties, Owen [23]
extended work of [2,6] and showed that in principle, QMC can deliver superior results over Markov Chain runs, although many problems still need to
be explored.

According to the two papers [6,23], any c.u.d. sequence can replace pseudo-
random numbers in sequential Monte Carlo methods, such as Metropolis-Hasting
algorithm. Random walk is a special case of Metroplolis algorithm. In the paper
[17], we proposed an algorithm for quasirandom walk on balls, where we used a
QRN-PRN hybrid to generate the walks on balls or spheres. We have to restrict
to use quasirandom numbers in the algorithm of quasirandom walks. Here, we
use the same algorithms, but we replace all PRNs with c.u.d. sequences.

## 3.2   Constructions of C.U.D. Sequences

The original construction of completely uniformly distributed sequences is related to the $\{\theta^n\}$ sequence [12], where $\theta > 1$. A simple and practical construction of c.u.d. sequence is to use the full period of a linear congruential generator [5]. This helps us to explain why the only successful sequence in Hofmann
and Mathe' paper[6] is the pseudorandom number sequence. Recently, Levin
[14] modified low-discrepancy sequences and constructed c.u.d. sequences with
lower discrepancy. The low-discrepancy sequences are based on digital inversion
method, which is the central idea behind the construction of current low-discrepancy sequences, such as Halton, Faure and Sobol sequences. Niederreiter [21]
extended these methods in arbitrary bases and dimensions.

We use the construction of c.u.d. sequences [10,25]

**Theorem 1.** *Let* $\mu_n = k \log p_i$, *with* $1 \leq k \leq n_r$, *and* $1 \leq i \leq r$, *where* $p_1, p_2, \dots$.
*are prime,* $n_r = [e^{(\ln r)^3}] + 1$, *and* $n = m_r + (i-1)n_r + k$ *with* $m_1 = 0, m_r = \sum_{j=1}^{r-1} n_j$. *Then* $\mu_n$ *is completely uniformly distributed.*

Here $[x]$ is the operation of keeping the integer part of any real number, for
example $[2.345] = 2$. This c.u.d. sequence is easy to implement. Shparlinskii (see
[25]) modified the coefficients of $k$, $\log p_i$ and obtained a c.u.d. sequence with
discrepancy close to $O(N^{-1})$.

## 4   Numerical Tests

We use the same test problem as in [17] to test our new scheme.

The results are shown on Figure 1. Here we use LCG as pseudorandom number
generator and the experiments are done for different number of trajectories. The
advantage of using c.u.d. sequences is obvious.

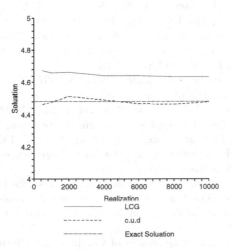

**Fig. 1.** Walk-on-Balls. Relative errors in the solution at the center point (0.5, 0.5, 0.5) using PRNs and c.u.d. sequences.

## 5   Conclusions

A new scheme for solving BVPs via completely uniformly distributed sequences is proposed. The advantage of this scheme is that it provides faster convergent rate based on c.u.d. sequences. This scheme is an alternative for solving BVPs using pseudorandom numbers. However, our numerical results are very preliminary results. In future, more numerical tests have to be done using a set of c.u.d. generators and different transition probabilities. Also, problems with more complicated differential operators and domains have to be considered. We will provide a library of various c.u.d. sequences.

## References

1. R. E. CAFLISCH, "Monte Carlo and quasi-Monte Carlo methods," *Acta Numerica*, **7**: 1–49, 1998.
2. S. CHAUDHARY, *Acceleration of Monte Carlo Methods using Low Discrepancy Sequences*, Dissertation, University of California, Los Angeles, 2004.
3. I.T. DIMOV, T.V. GUROV, Estimates of the computational complexity of iterative Monte Carlo algorithm based on Green's function approach, *Mathematics and Computers in Simulation*, **47**: 183–199, 1998.
4. S. ERMAKOV, V. NEKRUTKIN, V. SIPIN, *Random Processes for solving classical equations of the mathematical physics*, Nauka, Moscow, 1984.
5. K. ENTACHER, P. HELLEKALEK, P. L'ECUYER, "Quasi-Monte Carlo Node Sets from Linear Congruential Generators", *Monte Carlo and Quasi-Monte Carlo methods 1998, H. Niederreiter and J. Spanier Eds.,* , Springer, Berlin, pp.188–198, 2000,
6. N. HOFMANN AND P. MATHE, "On Quasi-Monte Carlo Simulation of stochastic Differential equations ," *Mathematics of Computation*, Vol. **66**, pp. 573–589, 1997.

7. KARAIVANOVA, A., MASCAGNI, M. AND SIMONOV, N.A.: Solving BVPs using quasirandom walks on the boundary. In: *Lecture Notes in Computer Science*, Vol. 2907. Springer-Verlag, Berlin Heidelberg New York (2004) 162–169.

8. D. E. KNUTH. *The Art of Computer Programming, vol. 2: Seminumerical Algorithms*. Addison-Wesley, Massachusets, 1997.

9. D. E. KNUTH,"Construction of a random sequence," *BIT* Vol. **4**, pp. 264-250, 1965.

10. N. M. KOROBOV, "Bounds of trigonometric sums involving completely uniformly distributed functions, " *Soviet. Math. Dokl.* Vol. **1**, pp. 923-926, 1964.

11. N. M. KOROBOV, "On some topics of uniform distribution, " *Izvestia Akademii Nauk SSSR, Seria Matematicheskaja* Vol. **14**, pp. 215-238, 1950.

12. L. KUIPERS AND H. NIEDERREITER , *Uniform Distribution of Sequences*, John Wiley and Sons- - New York, 1974.

13. C. LECOT AND B. TUFFIN. "Quasi-Monte Carlo Methods for Estimating Transient Measures of Discrete Time Markov Chains". In *Fifth International Conference on Monte Carlo and Quasi- Monte Carlo Methods in Scientific Computing*, Springer, pages 329-344, 2002.

14. M. B. LEVIN, "Discrepancy Estimates of Completely Uniformly Distributed and Pseudorandom Number Sequences ," *International Mathematics Research Notices*, Vol. **22**, pp. 1231-1251, 1999.

15. W. L. LOH. "On the asymptotic distribution of scrambled net quadrature." *Annals of Statistics*, **31**: 1282-1324, 2003.

16. G. A. MIKHAILOV, *New Monte Carlo Methods with Estimating Derivatives*, Utrecht, The Netherlands, 1995.

17. M. MASCAGNI, A. KARAIVANOVA, C. HWANG, "Monte Carlo Methods for elliptic boundary Value Problems," *Proceedings of MCQMC 2002: Monte Carlo and Quasi-Monte Carlo Methods 2002, Springer Verlag*, 345–355, 2004.

18. W. MOROKOFF, "Generating Quasi-Random Paths for Stochastic Processes," *SIAM Rev.*, Vol. **40**, No.4, pp. 765–788, 1998.

19. C. MIRANDA, *Equasioni alle dirivate parziali di tipo ellittico*, Springer Verlag, Berlin, 1955.

20. W. MOROKOFF AND R. E. CAFLISCH, "A quasi-Monte Carlo approach to particle simulation of the heat equation," *SIAM J. Numer. Anal.*, **30**: 1558–1573, 1993.

21. H. NIEDERREITER. *Random Number Generations and Quasi-Monte Carlo Methods*. SIAM, Philadelphia, 1992.

22. A. B. OWEN, " Scrambled Net Variance for Integrals of Smooth Functions," *Annals of Statistics*, **25**(4): 1541–1562, 1997.

23. A. B. OWEN AND S. TRIBBLE, "A quasi-Monte Carlo Metroplis Algorithm," *Proceedings of the National Academy of Sciences of the United States of America*, **102**: 8844–8849, 2005.

24. K. SABELFELD, *Monte Carlo Methods in Boundary Value Problems*, Springer Verlag, Berlin - Heidelberg - New York - London, 1991.

25. I. E. SHPARLINSKII , " On a Completely Uniform Distribution," *USSR Computational Mathematics and Mathematical Physics* , **19**: 249-253, 1979.

# On the Exams of a Multi-Attribute Decision Making Electronic Course

Cornel Resteanu[1], Marius Somodi[2], and Marin Andreica[3]

[1] National Institute for Research and Development in Informatics,
8-10 Averescu Avenue, 011455, Bucharest 1, Romania
resteanu@ici.ro
[2] Department of Mathematics, University of Northern Iowa, Cedar Falls,
IA 50614, USA
somodi@uni.edu
[3] Economic Studies Academy, 6 Romana Square, 010572, Bucharest 1, Romania
marinandreica@yahoo.com

**Abstract.** This paper gives brief information about a new electronic course for enhanced learning in making optimal decisions using the Multi-Attribute Decision Making paradigm. Emphasis is put on the construction of the exams of this electronic course. In order to provide ready to use tests for students' exams, an algorithm based on a Monte Carlo method was conceived. This algorithm and its benefits are presented.

**Keywords:** Knowledge Based Society, Creativity and Personal Development, LifeLong Learning, Enhanced Learning Technology, Multi-Attribute Decision Making, Monte Carlo Method.

## 1 Introduction

Any modern hands-on approach in education is based on the idea that, in the learning process, the students should be active participants instead of passive learners. In order to improve the outcomes of the teaching act, the educational system must involve professors who are experts in the course area (Multi-Attribute Decision Making (MADM) [1], [2], in this case), understand students' psychology, master the pedagogical art, and are familiar with the modern Information and Communication Technology (ICT) tools [3], [4], [5]. As a part of the project *Excellency Level Tools for Multi-Attribute Decision Making Field's Promotion*, belonging to *The Romanian National Excellency Research Program*, an electronic course designed to teach students how to make optimal decisions is being prepared. This electronic course benefits from the work of a large team of researchers that includes several international collaborators.

In this paper we will focus on the MADM electronic course's testing component. In order to generate exams from a questions and problems bank for a multitude of students in a continuous world wide open session, a Monte Carlo procedure has been defined. Despite the fact that the equitable selection of random questions and problems, with collisions avoided, even when

T. Boyanov et al. (Eds.): NMA 2006, LNCS 4310, pp. 173–180, 2007.
© Springer-Verlag Berlin Heidelberg 2007

considering exams given (a defined) long time ago is time consuming, this procedure ensures that the *exam's generation time* appears insignificant to each student. The Monte Carlo procedure is not a simulation, but it is running *at warm*, because it generates exams and shows when the questions and problems bank needs to be enriched if one desires to maintain the initial exigency in producing exams.

## 2    The MADM Electronic Course and Its Exams

The electronic course will provide knowledge regarding a basic problem of Decision Theory: the *Optimal Choice Problem* (OCP) [6]. The course is structured into five modules:

**a. Defining the mathematical model.** This module presents the first part of the MADM mathematical model, which includes *structured information* and defines the common knowledge about that.

**b. Treating the model's inconsistency.** This module presents the second part of the MADM mathematical model, which includes *unstructured information*, and defines the expert knowledge about that. The expert knowledge [7] is expressed through a set of production rules that deal with the potential drawbacks of a MADM model, i.e. *syntactic/semantic errors, incorrectness, incompleteness* or *incredibleness*.

**c. Normalization and solving methods for OCPs.** This module presents different procedures of reducing multiple decision makers, multiple states of nature problems to single decision maker, single state of nature problems, as well as a large set of solving methods for this kind of problems.

**d. IT for Design of MADM software applications.** This module covers three levels of technology for rapid design of MADM software applications.

**e. OPTCHOICE - MADM modeling and solving software.** The last module is a tutorial on the MADM modeling and solving pervasive service (available to anyone on the Internet, free of charge, from any place and at any time). The service uses the software named *OPTCHOICE*, which will cover the needs to define and solve optimal choice problems in the MADM paradigm.

The five modules of the MADM electronic course can only be approached sequentially, following the order established by design, so that a higher-order module becomes available only after passing the exam given at the end of the preceding module. The course is successfully completed when passing a graduation exam, which can only be taken after passing five end-of-module exams.

At the end of each module, 100 questions, divided into 10 classes of 10 questions each, are made available but no answers are provided. Throughout module #4, 100 problems with solutions are included, and the students are expected to learn how to solve them or similar problems.

An end-of-module exam consists of 10 randomly selected questions, one from each of the equivalence classes. A graduation exam consists of 5 questions, one from each module, plus one of the available problems but with modified data. The problem may be solved either by hand or by using the *OPTCHOICE* software. However, an indication of the success of the authors' efforts in this course is when the students are capable of using proficiently the *OPTCHOICE* software.

Being administered on the Internet, in order to avoid certain possible fraud methods specific to electronic-format exams, it is essential that each exam to be randomly generated from the questions and problems bank in such a way that elementary-level collisions in a long period of time in the past are avoided. A period of time is defined by two dates: one in the past and the other one given by the current date of the course's server. A direct approach of this problem, without any precaution, has a serious drawback: when a large number of students located in different parts of the world take exams, the response time to the query of generating new exams can be large enough to become frustrating for some of the students. The algorithm that will provide a solution to this problem is based on a Monte Carlo method. Using mechanisms of planning the inputs crowding in the system (i.e. the queries of generating exams) and releasing the outputs from the system (i.e. the exams completion by students), the algorithm prompts the administrators when to enrich the questions and problems bank, and also indirectly shows the necessary volume in each category.

In this setting, there are two random variables to consider: the *generating time* of a randomly selected exam and the *response time* (or the exam completion time) of a randomly selected student. We will use the real exam generating time instead of the theoretical exam generating time. The response time, on the other hand, needs to be modeled carefully as described in the next section.

## 2.1   Modeling the Response Time

Let $X$ be the response time of a randomly selected individual that takes the interactive, computer-based exam. Then $X$ is a continuous random variable with probability density function $f(x)$. If $n$ randomly selected individuals take the exam, then their response times $X_1$, ..., $X_n$ represent independent (as there is no communication between students and their exams are different), identically distributed random variables with probability density function $f(x)$. The function $f$ can be either completely specified (when both the type of the distribution of $X$ and the corresponding parameters are identified) or partially specified (the type is determined, but the corresponding parameters are unknown). This section presents a method of approximating the function $f$:

**a.** One uses the sample $X_1$, ..., $X_n$ to construct a relative frequency histogram for the variable $X$. Using a sample of size equal to 90, provided by a real two-hour exam from an Operations Research electronic course, one observes the shape of the histogram.

**b.** One uses the relative frequency histogram to hypothesize the type of the distribution of $X$. One anticipates, based on the mound-shaped histogram, that the response time $X$ has a Weibull, gamma, or normal distribution. Because the possible distributions of $X$ are now partially specified, one of the methods of estimating parameters, for instance the *method of moments*, must be used before going to the next step (see for instance [8] for details).

**c.** One formulates the null and the alternative hypotheses of a goodness-of-fit statistical test for each of the three possible distributions. The hypotheses are: i) $H_0^1$: $X$ has a Weibull distribution (with specified parameters) and $H_a^1$: $X$ does not have a Weibull distribution, ii) $H_0^2$: $X$ has a Gamma distribution and $H_a^2$: $X$ does not have a Gamma distribution, iii) $H_0^3$: $X$ has a normal distribution and $H_a^3$: $X$ does not have a normal distribution.

**d.** One conducts the goodness-of-fit statistical tests (see [9]). The sample of size 90 indicates that $X$ has approximately a normal distribution.

When the MADM course will run and the first data about its exams duration will become available, the entire procedure for modeling the response time as a random variable will be applied again for more accurate results.

The following algorithm can be used to simulate the response time of a randomly selected student under the assumption that if there are no time constraints then the response time is normally distributed with mean $\mu$ and standard deviation $\sigma$. The algorithm makes use of a random number generator capable of generating random numbers uniformly distributed on the interval (0,1). In C, C++, or Pascal, the function RANDOM uses such a random number generator.

```
EXITSTUD
START
z = 0
FOR i = 1, 12
    u = RANDOM
    z = z + u
ENDFOR
z = z - 6
x = miu + z * sigma
STORE exam_duration FROM MIN(x, 120)
STOP
```

The variable exam_duration stores the simulated response time of a randomly selected student for a 120-minute exam.

## 2.2   A Monte Carlo Type Algorithm for Exams Generation

At the beginning one mentions that the authors have constructed a bank of questions and problems that is large enough to generate a large number of exams. As indicated before, the Monte Carlo algorithm runs at warm primarily on this bank but also on the other electronic course data / knowledge components. What this algorithm does is more than just a simulation. When an exam is generated, it is stored in the section of generated exams. When a student requests an exam, that exam is typically already generated, and if not, then the exam is generated momentarily by the Monte Carlo procedure which starts automatically. If the electronic course administrator follows the generation process and the statistical information provided by the Monte Carlo procedure, it is impossible to end up in this situation. The exams which have been given to students are kept for a while in the active zone of the data base, and any newly generated exam is compared to the existing exams to avoid collisions on any components.

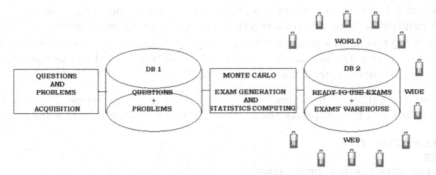

**Fig. 1.** From questions and problems to ready to use exams

When the administrator decides that the validation base is too large and contains too many old exams, these exams are transferred from the active zone into the statistical component of the data base.

Before presenting the algorithm, we will make several preliminary remarks:

**a.** The processes of generating exams and administering them run concurrently. At the beginning, if the Monte Carlo procedure has not been launched before the exams start, no problems are anticipated, as most exams are typically generated in less than 10 seconds, including the time necessary for tele-transmission. This time is considered good in today's technological conditions.

**b.** If an exam is interrupted by a student and continued within 24 hours (which is a facility for students) the difference between the exam duration time and the exam generation time is even larger than in the previous case, which allows the process of generating new exams to run and produce a reserve of new exams. Therefore, in this case the condition that the students only wait during a part of the exam generation time (i.e. while the exams are transferred on the Internet) is easily satisfied.

**c.** The concurrent algorithm stops when no new questions are left in the questions bank. No stop criterion is used for problems, as they are selected from the problems bank for which the data are randomly modified.

**d.** In order to avoid stopping the algorithm, we have two possible solutions: first, enrich the bank of questions and problems; second, release the questions from exams given prior to a certain date selected by the administrator to be used when generating new exams; in this way, the algorithm used to generate exams can be revived automatically.

**e.** The exam generation time can exceed the duration of the exam itself if the following conditions are satisfied simultaneously:

i) a large number of students are taking exams and the criterion of avoid collisions is too strong (there is a high sensitivity for collisions);

ii) a large number of students request new exams and the request rate (i.e. number of new exams to be generated in one hour) remains constant or increases.

If these conditions are satisfied then the number of collisions when generating new questions can be very large, leading to an unusual increase in exam generation time, and therefore blocking the process of generating exams.

**f.** Depending on hardware configuration and power, one must take into consideration a limitation for exams concurrency. At the beginning, when all the course software (administration, modules and exams) will be installed on one computer, the rational concurrency limit can be set to 25.

The true values of the terms "rich", "high sensitivity", or "large" can only be determined by an appropriate Monte Carlo simulation handled by this algorithm:

```
MC-EXAM-GENERATION
START
DISPLAY WITH THEIR NATURAL NAMES:
nr_of_ready_to_use_exams, nr_of_exams_in_statistics,
last_limit_date, last_crowding_rate, last_nr_of_histories,
last_acceleration_of_crowding_rate, last_nr_of_cycles
ACCEPT AND VALIDATE:
limit_date (last_limit_date < limit_date  < current_date),
crowding_rate (integer, > 0), nr_of_histories (integer, > 99),
acceleration_of_crowding_rate (integer, > 0), nr_of_cycles (integer, > 0)
FOR last_limit_date, 1, limit_date
    REMOVE FROM GENERATED-EXAMS-Q&P-BASE
       Questions AND Problems WITH date = limit_date
    IN STATISTICS-EXAMS-Q&P-BASE
    nr_of_exams_in_statistics = nr_of_exams_in_statistics +1
ENDFOR
FOR k = 1, 1, nr_of_cycles
    nr_of_exams_to_be_generated = crowding_rate +
                         (k-1)*acceleration_of_crowding_rate
    DISPLAY "Simulation number" k
    "at crowding rate = " nr_of_exams_to_be_generated
    DISPLAY current_date AND time FROM SYSTEM
    cycle_exam_generation_time = 0, generated_exams_per_cycle = 0
    FOR l = 1, 1, nr_of_histories
       FOR m = 1, 1, nr_of_exams_to_be_generated
          IF GENERATED-EXAMS-Q&P-BASE IS FULL THEN
             DISPLAY "GENERATED-EXAMS_Q&P-BASE IS FULL"
             GO TO FINAL
          ENDIF
          GET current_time FROM SYSTEM INTO initial_exam_generation_time
          GENERATE exam FROM Q&P-BASE
          VERIFY WITH GENERATED-EXAMS_Q&P-BASE
          IF exam IS NOT IN GENERATED-EXAMS_Q&P-BASE THEN
             MOVE exam IN GENERATED-EXAMS-Q&P-BASE
             generated_exams_per_cycle = generated_exams_per_cycle + 1
             GET current_time FROM SYSTEM INTO final_exam_generation_time
             exam_generation_time = final_exam_generation_time -
                            initial_exam_generation_time
          IF exam_generation_time * crowding_rate >= 60   THEN
             DISPLAY
                "The generation is impossible at this rate" crowding_rate
```

```
              GO TO FINAL
         ENDIF
         cycle_exam_generation_time = cycle_exam_generation_time +
                              exam_generation_time
       ENDIF
     ENDFOR
   ENDFOR
   exam_generation_mean_time = cycle_exam_generation_time /
                         generated_exams_per_cycle
   nr_of_ready_to_use_exams = nr_of_ready_to_use_exams +
                         generated_exams_per_cycle
   DISPLAY current_date AND time FROM SYSTEM / "Cycle number" k /
   "Number of generated exams" generated_exams_per_cycle
   "Exam generation mean time" exam_generation_mean_time
ENDFOR
STORE UP-TO-DATE data AND DISPLAY them WITH their NATURAL NAMES:
nr_of_exams_in_statistics, nr_of_ready_to_use_exams,
last_limit_date FROM limit_date,
last_crowding_rate FROM nr_of_exams_to_be_generated,
last_acceleration_of_crowding_rate FROM acceleration_of_crowding_rate,
last_nr_of_histories FROM nr_of_histories,
last_nr_of_cycles FROM nr_of_cycles
USING cycles, numbers of generated exams, exam generation mean time
GIVE a graphical image of generation process evolution
FINAL
concurrency_power=25
GET current date AND time FROM SYSTEM INTO exams_initial_time
FOR i =1, 1, concurrency_power
   exam_current_time (i) = exams_initial_time
ENDFOR
FOR n = 1, 1, nr_of_ready_to_use_exams
   i = INDEX OF ELEMENT exam_current_time (i)
       WHICH IS MIN(exam_current_time (1:concurrency_power))
   CALL EXITSTUD GIVING exam_duration
   exam_current_time (i) = exam_current_time (i) + 2*exam_duration
ENDFOR
DISPLAY "Generated exams are enough till" exam_current_time (i)
"at full concurrency power and the exams sustaining only in the active
period of the day"
STOP
```

In this way one can determine the size of the reserve of exams and the mean exam generation time in the conditions described above, and one can decide whether or not it is necessary to enrich the bank of questions and problems.

## 3   Conclusions

The MADM electronic course presented in this paper is an active project whose goal is to allow easy and efficient access to everyone in the world to the MADM methods. The course incorporates the fundamental MADM techniques, which are presented in an attractive and easy to assimilate format. Designed to fully take advantage of the advanced features of the OPTCHOICE software, this electronic course has the potential to meet the learning needs of diverse groups of individuals. One of the critical components of any electronic course is the capability of administering exams tailored for this new form of learning. Because at the course's graduation a certificate recognized by the Romanian Education and Research Ministry is awarded, the course exams must take place in optimal conditions. An algorithm based on a Monte Carlo method is used to solve certain problems that occur when many students take exams at the same time. This algorithm is capable of generating new and equitable exams for all the students so that duplications of questions or problems are avoided and, at the same time, it is a support for taking decision in administering the whole system. Multiple simulations have been run and the conclusion is that the algorithm is efficient and may be used with confidence.

# References

1. Hwang, C-L., Yoon, K.: Multiple Attribute Decision Making. Springer-Verlag, Berlin Heidelberg New York (1981).
2. Hwang, C-L., Lin, M.J.: Group Decision Making under Multiple Criteria. Springer-Verlag, Berlin Heidelberg New York (1997).
3. Anderson, T.: Modes of Interaction in Distance Education: Recent Developments and Research Questions. In Moore, M., Anderson, T., (eds.): Handbook of Distance Education, Mahwah, New Jersey: Lawrence Erlbaum Associates, Inc. (2003) 129–144.
4. Gibson, C.: Learners and Learning: The Need for Theory. In Moore, M., Anderson, T., (eds.): Handbook of Distance Education, Mahwah, New Jersey: Lawrence Erlbaum Associates, Inc. (2003) 147–160.
5. Rumble, G.: Reinventing Distance Education, 1971 - 2001. Int. J. Lifelong Educ. **20** (2001) 31–43.
6. Resteanu, C., Filip, F.G., Ionescu, C., Somodi, M.: On Optimal Choice Problem Solving. In Sage, A.P., Zheng, W., (eds.): Proceedings of SMC '96 Congress (Beijing, October 14-17). IEEE Publishing House, Piscataway NJ (1996) 1864–1869.
7. Giarratano, J.C., Riley, G.D.: Expert Systems: Principles and Programming. 3rd edition. PWS Publishing Company, Boston, (1999).
8. Wackerly, D.D., Mendenhall III, W., Scheaffer, R.L.: Mathematical Statistics with Applications, Duxbury (2002).
9. D'Agostino, R.B., Stephens, M.A., Eds.: Goodness-of-Fit Techniques. Marcel Dekker (1996).

# Random Walks for Solving Boundary-Value Problems with Flux Conditions[*]

Nikolai A. Simonov[1,2]

[1] Institute of Molecular Biophysics, Florida State University, Tallahassee, FL, USA
[2] Institute of Computational Mathematics and Mathematical Geophysics SB RAS, Novosibirsk, Russia
simonov@scs.fsu.edu
http://osmf.sscc.ru/LabSab/nas

**Abstract.** We consider boundary-value problems for elliptic equations with constant coefficients and apply Monte Carlo methods to solving these equations. To take into account boundary conditions involving solution's normal derivative, we apply the new mean-value relation written down at boundary point. This integral relation is exact and provides a possibility to get rid of the bias caused by usually used finite-difference approximation. We consider Neumann and mixed boundary-value problems, and also the problem with continuity boundary conditions, which involve fluxes. Randomization of the mean-value relation makes it possible to continue simulating walk-on-spheres trajectory after it hits the boundary. We prove the convergence of the algorithm and determine its rate. In conclusion, we present the results of some model computations.

## 1 Introduction

Different stochastic algorithms can be applied for treating numerically boundary-value problems. For elliptic equations, the most efficient and commonly used Monte Carlo methods are the walk-on-spheres (WOS) algorithm [1,2,3], and the random walk on the boundary algorithm [4]. The WOS algorithm provides the tool for efficient simulation of exit points of the diffusion process to the domain's boundary. Averaging of known values at these points provides an estimate for the solution. This algorithm works well for the Dirichlet boundary conditions. Common way of treating flux conditions is to simulate reflection from the boundary in accordance with the finite-difference approximation of normal derivative [5,6,7,8]. Such approach introduces bias into the estimate.

Application of walk on the boundary algorithms makes it possible to solve not only Dirichlet, but also Neumann and third boundary-value problems. The same approach can also be used to solving problems with continuity boundary conditions [9,10]. However, this class of algorithms has its limitations.

Recently, we proposed new approach to constructing Monte Carlo algorithms for solving elliptic boundary-value problems with flux conditions. This approach

[*] Supported in part by NATO Linkage Grant and Grant for Leading Scientific Schools.

T. Boyanov et al. (Eds.): NMA 2006, LNCS 4310, pp. 181–188, 2007.
© Springer-Verlag Berlin Heidelberg 2007

is based on the mean-value relation written for the value of solution at boundary point. It provides a possibility to get rid of the bias when treating algorithmically boundary conditions that involve normal derivative.

## 2 Formulation of the Problem and Walk-on-Spheres Algorithm

Consider the linearized Poisson-Boltzmann (Helmholtz) equation

$$\Delta u - \kappa^2 u = 0 \ , \ \kappa = \text{const} \geq 0 \tag{1}$$

in a bounded domain $G \subset \mathbb{R}^m$, and mixed boundary-value problem for it:

$$\alpha(y)\frac{\partial u}{\partial n}(y) + \beta(y)u(y) = g(y) \ , \ y \in \Gamma = \partial G \ . \tag{2}$$

Here, $\alpha = 1$, $\beta = 0$ on $\Gamma_0$, and vice versa: $\alpha = 0$, $\beta = 1$ on $\Gamma_1 = \Gamma \setminus \Gamma_0$. We suppose the boundary to be piece-wise smooth and the parameters of the problem to guarantee that the unique solution exists [11]. To simplify the inference and formulas involved, in the rest of the paper we consider only the three-dimensional Euclidean space. It is clear, however, that the integral formula derived stays valid for an arbitrary $m \geq 2$.

To find the solution at a fixed point, $x_0$, we use the walk-on-spheres algorithm. Let $x \in G$. We define $d(x)$ as the distance from this point to the boundary $\Gamma$. Next we consider the ball $B(x, d(x))$ and write down the integral Green's formula for solution, $u$, and the Green's function for this ball, $\Phi_{\kappa,d}$:

$$u(x) = \int_{S(x,d(x))} \frac{\partial \Phi_{\kappa,d}}{\partial n} u \ ds \ . \tag{3}$$

Here $\Phi_{\kappa,d}(x, y) = -\dfrac{1}{4\pi} \dfrac{\sinh(\kappa(d - |y - x|))}{|y - x|\sinh(\kappa d)}$, and $S(x, d(x))$ denotes the sphere of radius $d(x)$ centered at the point $x$.

Randomization of (3) leads to the estimate based on simulation of the walk-on-spheres Markov chain. The chain is defined by the recursive relation: $x_{i+1} = x_i + d(x_i) \ \omega_i$, where $\{\omega_0, \omega_1, \ldots\}$ is a sequence of independent isotropic unit vectors.

From (3) we have $u(x_i) = \mathbb{E}(q(\kappa, d(x_i))u(x_{i+1})|x_i)$, and we treat the factor, $q(\kappa, d(x_i)) = \dfrac{\kappa d}{\sinh(\kappa d)}$, as the survival probability. For $\kappa = 0$ (or if we use the weight estimate [3]), WOS with probability one converges to the boundary. Let $x_k$ be the first point of the Markov chain that hit $\Gamma_\varepsilon$, the $\varepsilon$-strip near the boundary, and denote by $x_k^* \in \Gamma$ the point nearest to $x_k$. Then, $u(x_0) = \mathbb{E}(u(x_k)\chi)$, where $\chi = 0$, if the trajectory was terminated inside the domain, $\chi = 1$, if Markov chain reached $\Gamma_\varepsilon$, and $\chi = \prod q(\kappa, d(x_i))$, if we use the weight estimate.

For $x_k^* \in \Gamma_1$, we have

$$u(x_k) = g(x_k^*) + \phi_1(x_k, x_k^*) , \tag{4}$$

where $\phi_1(x_k, x_k^*)$ is $O(\varepsilon)$ for elliptic $x_k^*$.

For $x_k^* \in \Gamma_0$, the boundary conditions give

$$u(x_k) = u(x_k^*) - g(x_k^*)d(x_k) + \phi_0(x_k, x_k^*) , \tag{5}$$

where $\phi_0(x_k, x_k^*) = O(\varepsilon^2)$ as $\varepsilon \to 0$.

To be sure that the nearest point on the boundary, $x_k^*$, is elliptic, we use the walk-in-subdomains technique [12].

Note that for $x_k^* \in \Gamma_0$ the value of $u(x_k^*)$ is not known. To estimate it, we will use the integral formula derived in the next section.

## 3   Mean-Value Relation at a Point on the Boundary

For an elliptic point on the boundary, $x \in \Gamma_0$, consider the ball $B(x, a)$ and the Green's function $\Phi_{\kappa,a}$ for this ball taken at its center. For every $y \neq x$, we have $\Delta_y \Phi_{\kappa,a} - \kappa^2 \Phi_{\kappa,a} = 0$.

Denote by $B_i(x, a) = B(x, a) \bigcap G$ the interior part of the ball that lies inside the domain, and let $S_i(x, a)$ be the interior part of its surface. We apply the Green's formula to the pair of functions, $u$ and $\Phi_{\kappa,a}$, in $B_i(x, a) \setminus B(x, \varepsilon)$, excluding small neighborhood of the point. Both functions satisfy the Poisson-Boltzmann equation in this domain, $\Phi_{\kappa,a} = 0$ on $S(x, a)$, and we suppose that everywhere on $\Gamma \bigcap B(x, a)$, $u$ satisfies the Neumann boundary conditions. Next we take into account smoothness of $u$ and that for elliptic points the surface area of $S_i(x, \varepsilon)$ is $2\pi\varepsilon^2(1 + O(\varepsilon))$ when $\varepsilon \to 0$. As a consequence, we obtain in the limit the following integral formula:

$$u(x) = \int_{\partial B_i(x,a)\setminus\{x\}} 2 \frac{\partial \Phi_{\kappa,a}}{\partial n} u \, ds$$
$$- \int_{\Gamma \bigcap B(x,a)\setminus\{x\}} 2 \Phi_{\kappa,a} \, g \, ds . \tag{6}$$

Clearly, this formula stays valid when $\kappa = 0$. In this case the Green's function is $\Phi_{0,a}(x, y) = -\dfrac{1}{4\pi}\left(\dfrac{1}{|y - x|} - \dfrac{1}{a}\right)$.

For convenience, we explicitly isolate singularities in the kernels of integral operators and rewrite (6) in the following form:

$$u(x) = \int_{\partial B_i(x,a)\setminus\{x\}} \frac{1}{2\pi} \frac{\cos \varphi_{yx}}{|y - x|^2} W_{\kappa,a} \, u(y) \, ds(y)$$
$$+ \int_{\Gamma \bigcap B(x,a)\setminus\{x\}} \frac{1}{2\pi|y - x|} \left(1 - \frac{|y - x|}{a}\right) W_{\kappa,a}^1 g(y) \, ds(y) . \tag{7}$$

Here, $\cos\varphi_{yx}$ is the angle between the external (with respect to $B_i(x,a)$) normal vector at a point, $y$, and vector, $y - x$. The weight function,

$$W_{\kappa,a}(|y - x|) = \frac{\sinh(\kappa(a - |y - x|)) + \kappa|y - x|\cosh(\kappa(a - |y - x|))}{\sinh(\kappa a)} ,$$

is smooth, and $W_{\kappa,a} = q(\kappa,a) = \dfrac{\kappa a}{\sinh(\kappa a)}$ on the surface of auxiliary sphere, $S(x,a)$. Clearly, everywhere in $\overline{B}(x,a) \setminus \{x\}$, this weight function is positive and its value is less than one. For $\kappa = 0$ it equals one identically.

The second weight function, $W_{\kappa,a}^1 = \dfrac{\sinh(\kappa(a - |y - x|))}{a - |y - x|}\dfrac{a}{\sinh(\kappa a)}$, is also smooth. It is less than or equal to one, and $W_{\kappa,a}^1 \geq \dfrac{\kappa a}{\sinh(\kappa a)}$. Obviously, for $\kappa = 0$ it equals one identically.

## 4    Construction of the Algorithm and Its Convergence

Suppose first that the part of the domain's boundary with Neumann conditions, $\Gamma_0$, is convex. In this case, the kernel of the integral operator in (7) is sub-stochastic, which means that it is non-negative and its integral is less than or equal to one. Therefore, we can use this kernel as the transition density. The term $\dfrac{1}{2\pi}\dfrac{\cos\varphi_{x_{i+1}x_i^*}}{|x_{i+1} - x_i^*|^2}$ corresponds to isotropic distribution of $x_{i+1} \in \partial B_i(x_i^*,a)$ in solid angle with its vertex at $x_i^*$. The weight, $W_{\kappa,a}(|x_{i+1} - x_i^*|)$, is treated as the survival probability on this transition. (Note that for the plane boundary, with probability one $x_{i+1}$ is distributed isotropically on the half-sphere $S_i(x_i^*,a)$, and survival probability equals $q(\kappa,a)$.)

The described construction of Markov chain corresponds to so-called direct simulation of integral equation [2]. This means that the resulting estimate for the solution's value at a point, $x = x_0 \in \overline{G}$, is

$$\xi[u](x) = \sum_{i=0}^{N} \xi[F](x_i) . \tag{8}$$

Here, $N$ is the random length of Markov chain, and $\xi[F]$ are estimates for the right-hand side of the integral equation. This function is defined by (4), (5), (7), which gives

$$F(x) = 0 , \text{ when } x \in G \setminus \overline{\Gamma}_\varepsilon ;$$

$$= \int_{\Gamma \cap B(x^*,a)\setminus\{x^*\}} \frac{1}{2\pi|y - x^*|}\left(1 - \frac{|y - x^*|}{a}\right) W_{\kappa,a}^1 \, g(y) \, ds(y)$$

$$- g(x^*) \, d(x) + \phi_0(x,x^*) , \text{ when } x \in \overline{\Gamma}_\varepsilon \text{ and } x^* \in \Gamma_0 ;$$

$$= g(x^*) + \phi_1(x,x^*) , \text{ when } x \in \overline{\Gamma}_\varepsilon \text{ and } x^* \in \Gamma_1 . \tag{9}$$

For Markov chains based on direct simulation, the finiteness of mean number of steps, $\mathbb{E}N < \infty$, is equivalent to convergence of Neumann series for the correspondent integral operator, which kernel coincides with the transition density of this Markov chain. Besides that, the kernel of integral operator that defines the second moment of the estimate is also equal to this density [2]. This means that for exactly known free term, $F$, the estimate (8) is unbiased and has finite variance. The same is true if estimates, $\xi[F](x_i)$, are unbiased and have uniformly in $x_i$ bounded second moments. It is clear that we can easily choose such density that estimate for the integral in (9) will have the requested properties.

To prove that the mean number of steps is finite, we consider the auxiliary boundary-value problem:

$$\Delta p_0 = 0 , \quad p_0|_{\Gamma_1} = 0 , \quad \frac{\partial p_0}{\partial n}\big|_{\Gamma_0} = 1 . \tag{10}$$

From (9) it follows that for this problem $F(x_i) = \dfrac{a}{2}\left(1 + O\left(\dfrac{a}{2R}\right)\right)$ as $a/R \to 0$ in such a way that $a/R > c_0\varepsilon^{1/2}$ for some constant $c_0$. Here, $R$ is the smallest curvature radius at the elliptic point, $x_i^* \in \Gamma_0$. Consider the estimate (8) for $p_0$. Note that the mean-value relation (7) is valid only when $\Gamma \cap B(x_i^*, a) \subset \Gamma_0$. From here it follows that radii of auxiliary spheres, $a_i$, can be arbitrary small. Nevertheless, it can be easily shown that the probability of reflected diffusion coming to $\Gamma_1$ and thus being terminated remains to be separated from zero. Therefore, the mean number of reflections near the line that separates $\Gamma_0$ and $\Gamma_1$ is bounded by some constant, $c^*$. Next we fix some value, $a^*$, and divide the set of all reflections into two classes: in the first one we put reflections with radii, $a_i$, which are greater than $a^*$, and the second class includes the reflections with $a_i < a^*$. Let $N_{0,1}$ and $N_{0,2}$ be, respectively, mean numbers of reflections in the correspondent classes. Then we have $p_0 > N_{0,1}a^*c_1$ for some constant $c_1 < 1$, and, hence, the overall number of reflections, $N_0 = N_{0,1} + N_{0,2}$ is less than $C_a = p_0/(a^*c_1)+c^*$. From here it follows that the overall mean number of steps of the described version of the walk-on-spheres algorithm is of the same order as for the case of the Dirichlet boundary conditions, which means $\mathbb{E}N < \text{const}|\log(\varepsilon)|$.

In the previous reasoning we supposed error functions $\phi_0$ and $\phi_1$ to be known exactly. This presupposition provided us with a possibility of proving that the estimate (8) is unbiased. To obtain a functioning estimate we are forced to get rid of these unknowns. The resulting bias in the estimate is $\phi_1 + N_0\phi_0$, which is $O(\varepsilon)$ when $\varepsilon \to 0$. This means that we proved the following proposition.

**Proposition 1.** *The new version of the walk-on-spheres algorithm provides the estimate, $\xi[u](x)$, for the mixed boundary-value problem (1), (2). Variance of this estimate is bounded, and its bias is $O(\varepsilon)$, where $\varepsilon$ is the width of the strip near the boundary of the domain. For the required accuracy of solution, $\delta = \varepsilon$, computational cost of the algorithm is $O(\log(\delta)\, \delta^{-2})$.*

*For $\kappa > 0$ and for pure Neumann boundary-value problem, the algorithm has the same properties.* □

Note that with the finite-difference approximation for the solution's normal derivative, every hit of the $\varepsilon$-strip near the boundary, $\Gamma_0$, introduces $O(\varepsilon + h^2)$ bias into the estimate ($h$ is the step value in approximation) [7,6]. Mean number of reflections in this case is $O(h^{-1})$. It means that if $\varepsilon \sim h^2$, then the resulting bias is $O(h)$, and the overall mean number of steps is $O(\log(h)\,h^{-1})$. Thus, for the required accuracy $\delta$, we have to take $\varepsilon \sim \delta^2$ and $h \sim \delta$. The computational cost of this algorithm is $O(\log(\delta)\,\delta^{-3})$.

So, as we see, the approach we described here makes it possible to substantially improve the efficiency of the walk-on-spheres algorithm when applied to solving mixed and Neumann boundary-value problems.

The algorithm described above works for convex $\Gamma_0$. If it is not the case, the kernel of the integral operator in (7) can be negative, and direct simulation is not possible. This also means that changing the kernel of the integral operator to its modulus can lead to non-convergent Neumann series, and, thus, to nonoperable Monte Carlo estimates.

To solve this problem, we propose using simple approximation to the integral relation. Consider the tangent plane at the point, $x_i^* \in \Gamma_0$, and choose the next point of Markov chain isotropically on the half-sphere, $S^-(x_i^*, a)$, which lies inside the domain. It can be easily shown that for small $a/R$ such algorithm introduces an $O(a/2R)^3$ error. Therefore, the resulting bias is $O(a/2R)^2$, and the computational cost of such biased algorithm is $O(\log(\delta)\,\delta^{-5/2})$.

## 5   Other Boundary-Value Problems and Applications

The approach proposed in this paper works also for other boundary-value problems with flux conditions on the boundary. Consider, in particular, the problem of finding the electrostatic potential in a system of two dielectrics, $G_i$ and $G_e$, with different permittivities, $\epsilon_i$ and $\epsilon_e$, respectively. Both solution and flux are required to be continuous on the surface between two bodies:

$$u_i(y) = u_e(y)\,, \ \epsilon_i \frac{\partial u_i}{\partial n}(y) = \epsilon_e \frac{\partial u_e}{\partial n}(y)\,, \ y \in \Gamma\,, \tag{11}$$

where $u_i$ is defined in $G_i$ and $u_e$ is defined in $G_e$, $\Gamma = \overline{G}_i \bigcap \overline{G}_e$. For elliptic point, $x \in \Gamma$, and no charges present in its vicinity, we consider the ball, $B(x, a)$, and apply the Green's formula separately in two domains: to the pair of functions, $u_i$ and $\Phi_{0,a}$, in $(G_i \bigcap B(x, a)) \setminus B(x, \varepsilon)$, and to the pair of functions, $u_e$ and $\Phi_{0,a}$, in $(G_e \bigcap B(x, a)) \setminus B(x, \varepsilon)$. Taking into account boundary conditions (11) we have in the limit as $\varepsilon \to 0$:

$$u(x) = \frac{\epsilon_e}{\epsilon_e + \epsilon_i} \int_{S_e(x,a)} \frac{1}{2\pi a^2} u_e \, ds$$

$$+ \frac{\epsilon_i}{\epsilon_e + \epsilon_i} \int_{S_i(x,a)} \frac{1}{2\pi a^2} u_i \, ds$$

$$- \frac{\epsilon_e - \epsilon_i}{\epsilon_e + \epsilon_i} \int_{\Gamma \bigcap B(x,a) \setminus \{x\}} \frac{1}{2\pi} \frac{\cos \varphi_{yx}}{|y - x|^2} u \, dy\,. \tag{12}$$

Here, $S_i(x, a) = S(x, a) \bigcap G_i$ and $S_e(x, a) = S(x, a) \bigcap G_e$. Randomization of this integral relation provides a possibility to apply our revised version of the walk-on-spheres algorithm to this class of problems.

It is essential to note that solving such electrostatic problems is an extremely important task for molecular biophysics. In this case, compact $G_i$ can be considered as a molecule (e.g. protein) immersed into aqueous solution.

Another problem coming from molecular physics is the problem of computing diffusive and reactive properties of a molecule. If one part of the molecule's surface is reactive and the other part is reflective, then the problem can be reduced to calculating a functional of the solution to mixed boundary-value problem considered in this paper.

# 6   Results of Computations

To show the efficiency of the proposed walk-on-spheres algorithm in use, we consider the model problem of solving (10) in the unit cube. We consider the Neumann conditions to be given only on one side, whereas on all other sides of the cube the solution equals zero.

We took the center of the cube as the initial point of the walk-on-spheres Markov chain. The computed mean number of reflections in our new algorithm does not depend on $\varepsilon$ when $\varepsilon \to 0$ and tends to the constant value, which equals approximately 0.365. The mean value of radius of auxiliary sphere is 0.21. The results in Table 1 clearly show that the mean number of steps in the whole trajectory linearly depend on $\log(\varepsilon)$. This means that, asymptotically, number of steps in this algorithm is of the same order as in the case of ordinary walk-on-spheres algorithm applied to solving Dirichlet boundary-value problem.

For comparison, the same model problem was solved using the walk-on-spheres algorithm in combination with the finite-difference approximation of the normal derivative. For this method, mean number of reflections linearly depends on $h^{-1} = \varepsilon^{-1/2}$, and mean number of steps in the whole trajectory obeys power law and can be approximated by $O(h^{-1.059})$ dependence when $h \to 0$.

**Table 1.** Mean number of steps in the walk-on-spheres trajectory

| $\varepsilon$ | Our algorithm | With FD approximation | Mean number of reflections with FD |
|---|---|---|---|
| $10^{-2}$ | 10.99 | 12.82 | 0.49 |
| $10^{-3}$ | 20.97 | 33.98 | 1.31 |
| $10^{-4}$ | 31.06 | 94.28 | 3.92 |
| $10^{-5}$ | 41.39 | 292.61 | 11.93 |
| $10^{-6}$ | 51.52 | 1028.60 | 38.45 |
| $10^{-7}$ | 61.76 | 3614.34 | 120.86 |
| $10^{-8}$ | 71.96 | 13223.18 | 385.67 |

# References

1. Müller, M.E.: Some continuous Monte Carlo methods for the Dirichlet problem. Ann. Math. Statistics **27**(3) (1956) 569–589
2. Ermakov, S.M., Mikhailov, G.A.: Statisticheskoe modelirovanie [Statistical simulation]. Nauka, Moscow (1982) (Russian).
3. Elepov, B.S., Kronberg, A.A., Mikhailov, G.A., Sabelfeld, K.K.: Reshenie kraevyh zadach metodom Monte-Carlo [Solution of boundary value problems by the Monte Carlo method]. Nauka, Novosibirsk (1980) (Russian).
4. Sabelfeld, K.K., Simonov, N.A.: Random Walks on Boundary for solving PDEs. VSP, Utrecht, The Netherlands (1994)
5. Haji-Sheikh, A., Sparrow, E.M.: The floating random walk and its application to Monte Carlo solutions of heat equations. SIAM J. Appl. Math. **14**(2) (1966) 570–589
6. Kronberg, A.A.: On algorithms for statistical simulation of the solution of boundary value problems of elliptic type. Zh. Vychisl. Mat. i Mat. Phyz. **84**(10) (1984) 1531–1537 (Russian).
7. Mikhailov, G.A., Makarov, R.N.: Solution of boundary value problems of the second and third kind by Monte Carlo methods. Sibirsk. Mat. Zh. **38**(3) (1997) 603–614 (Russian).
8. Mascagni, M., Simonov, N.A.: Monte Carlo methods for calculating some physical properties of large molecules. SIAM J. Sci. Comp. **26**(1) (2004) 339–357
9. Karaivanova, A., Mascagni, M., Simonov, N.A.: Solving BVPs using quasirandom walks on the boundary. Lecture Notes in Computer Science **2907** (2004) 162–169
10. Karaivanova, A., Mascagni, M., Simonov, N.A.: Parallel quasi-random walks on the boundary. Monte Carlo Methods and Applications **10**(3-4) (2004) 311–320
11. Miranda, C.: Partial differential equations of elliptic type. Springer-Verlag, New York (1970)
12. Simonov, N.A.: A random walk algorithm for solving boundary value problems with partition into subdomains. In: Metody i algoritmy statisticheskogo modelirovanija [Methods and algorithms for statistical modelling]. Akad. Nauk SSSR Sibirsk. Otdel., Vychisl. Tsentr, Novosibirsk (1983) 48–58 (Russian).

# Examining Performance Enhancement of p-Channel Strained-SiGe MOSFET Devices

D. Vasileska[1], S. Krishnan[2], and M. Fischetti[3]

[1] Arizona State University, Tempe, AZ, USA
vasileska@asu.edu
[2] Intel Corp., Chandler, AZ, USA
santhosh.krishnan@intel.com
[3] University of Massachusetts, Amherst, MA, USA
fischett@umass.edu

**Abstract.** We examine performance enhancement of p-channel SiGe devices using our particle-based device simulator that takes into account self-consistently the bandstructure and the quantum mechanical space-quantization and mobility enhancement effects. We find surface roughness to be the dominant factor for the bad performance of p-channel SiGe devices when compared to conventional bulk p-MOSFETs at high bias conditions. At low and moderate bias conditions, when surface-roughness does not dominate the carrier transport, we observe performance enhancement in the operation of p-channel SiGe MOSFETs versus their conventional Si counterparts.

**Keywords:** p-channel MOSFETs, bandstructure effects, quantum confinement, strain, performance enhancement.

## 1 Introduction

For the past several decades now, electron transport in Si inversion layers has remained the favorite area of active research in semiconductors in the experimental as well as the theoretical field. On the other hand, while there have been several experimental efforts to understand hole transport in inversion layers and quantum wells, similar significant theoretical studies on hole transport have been few and far between. These studies can be categorized in many different ways but they can be classified broadly into those dealing with:

1 Bulk (3D) transport. This basically deals with transport in Si or SiGe material systems that are under strain or relaxed. These studies investigated drift velocity, bulk (3D) mobility, and phenomena such as velocity overshoot, impact ionization etc.

2 Transport of confined (2D) carriers for example in inversion layers or quantum wells. This can be further subdivided into
(*i*) Low field Mobility Calculations; (*ii*) MOSFET Device Simulations.

The starting point for any transport calculation is the underlying band-structure providing the energy dispersion for that system. In the past all these band-structure

T. Boyanov et al. (Eds.): NMA 2006, LNCS 4310, pp. 189–196, 2007.

calculations served as the basis for transport calculations either in category 1 or 2 defined earlier. Early work performed in this area concentrated on hole transport in bulk. Earlier band-structure calculations based on self-consistent or empirical pseudopotential approaches of strained Si and SiGe alloys produced little information, if any, on transport parameters such as effective masses and the different energy splitting near the symmetry points. Nonetheless there are certain exceptions where effective masses were studied and furthermore, in this regard, the work of Krishnamurthy et al. [1] must be mentioned as one of the early works on the effect of band-structure, although not comprehensive, on carrier mobility. Other more involved and comprehensive calculations are by Hinckley and Singh [2],[3], [4]. The main drawback of these calculations was that they relied on scattered and inconsistent data then available on band-structures and deformation potential parameters. In addition to these sophisticated calculations, there existed other more simple approaches [5].[6] which relied heavily on crude approximations for the band-structure and the simple scattering models. Manku et al. [7] used the effective mass approximation to calculate the hole drift mobility in strained and unstrained SiGe alloys. Yamada and Ferry [8] employed a simple two band-model with an energy dependent effective masses to define the dispersions for the heavy and light hole bands and ignored warping as well as the contribution of the split off band. Using non-local pseudopotentials, Fischetti and Laux [9] unified the band-structure and mobility calculations for SiGe alloys. This study also helped reduce the uncertainty in available empirical data in particular effective masses and energy splittings near band extrema needed to fit mobility curves. Results from this work also included valence band parameters extracted from pseudopotential calculations. Deformation potentials that control the phonon scattering were also extracted. In addition to these conventional approaches, Watling, Asenov and Barker from the Glasgow group proposed a geometro-analytical model [10] for the valence band in strained and relaxed SiGe which shows good agreement with a six band model for the valence band. This method allows for the extraction of an effective mass tensor for the warped valence bands. This method has been used successfully by Watling and Barker [11] to fit simulated drift velocities and mobilities to experiment. This prescription is essentially 3D in nature and the Glasgow group further proposed the use of quantum potentials to model transport in inversion layers or quantum wells.

Now we shift our focus of attention to hole transport calculations for confined carriers. Just as calculation of the band-structure to model transport is imperative for the bulk case, accurate calculation of the subband structure is essential for confined carriers in the 2D case. For the case of Si, the small split off separation between the two fold degenerate heavy-light hole bands and the split-off band means that at the very least six bands must be considered. This is typically done within the framework first suggested by Dresselhaus, Luttinger and Kohn; the exception being recent calculations by Nakatsuji et al. [12] who used self-consistent pseudopotentials. Most of the theoretical studies that have been conducted have mostly focused on evaluating the low field mobility of holes in Si inversion layers, the enhancement thereof in strained Si by Nakatsuji [12]

Oberhuber and Vogl [13], Fischetti et al. [14] or modeling hole transport in Si/SiGe/Si quantum wells [15].

Even as it is generally acknowledged that accurate hole mobility calculation in inversion layers is painfully expensive and daunting, it should be mentioned that even now, there are, to the best of our knowledge, no device simulators that can model hole carrier transport in p-channel Si MOSFETs, by properly addressing both the issues of hole band-structure and quantum confinement effects. The only approaches that are self-consistent employ a full band Monte Carlo approach but even they fall short by neglecting quantum effects on carrier transport. The importance of accurate subband structure calculation in 2D simulations cannot be overemphasized. In the aforementioned approaches, the density of states and subband mixing has been neglected at the cost of including band-structure effects accurately within a 3D framework for a 2D problem. The purpose of this work is to remedy this situation by presenting a new approach to modeling p-channel devices using a 2D Monte Carlo transport kernel that is coupled self-consistently to a 2D Poisson equation solver and to a six-band band-structure module. The need for full band solvers for hole transport is especially true in case of strained layer MOSFETs- buried channel strained SiGe p-MOSFETs (this work) and surface channel strained Si, for instance, which will be investigated in the future.

## 2    Monte Carlo Device Simulator Description

Having calculated the hole band-structure in the contacts (3D) and the subband structure (2D) in the active device region, i.e. under the gate, self-consistently with the 2D Poisson equation, the quantum mechanical hole density in the channel is then calculated by weighing the sheet density of each subband with the probability density corresponding to that subband along the device depth and then summing over all subbands of each of the six bands. The initialization of carriers in real space is based on the local 3D carrier density for holes in the reservoirs and the self-consistent quantum-mechanical hole density in the channel region. In the channel region the carriers are assigned to subbands in a probabilistic manner so that it reflects the contribution to the hole sheet density from the different subbands. The kinetic energy of the particles is initialized by assuming a Boltzmann distribution and their wave-vectors are then assigned by randomly choosing the azimuth and polar (in case of 3D) angles on the constant energy surface.

Now, in order to deal with inhomogeneous transport at high longitudinal fields, a transport kernel that handles the transport of holes through the device is needed. In our approach, we use an Ensemble Monte Carlo (EMC) particle based simulator that handles the transport of holes. After the carriers are initialized, a bias is applied on the drain contact and the Monte Carlo algorithm takes over the hole transport, performing the drift and scattering of carriers. As the simulation time evolves in steps of 0.1 fs, and the carriers drift under the influence of the electric field, the confining potential changes along the channel from the source end to the drain end, and this in turn, changes the hole subband structure in the channel region. As a result, the hole subband structure and subsequently the scattering rates must be updated frequently during the simulation to reflect the changed subband structure.

## 2.1    2D↔3D Transitions

In our method, we have assumed the holes to be quasi-3D-like particles in the source and drain regions. This frequently gives rise to situations where the particle energy and momentum are not conserved across the boundaries and one needs appropriate models to treat these transitions properly. While the carrier energy and momentum can be conserved in a 2D→3D transition, the same is not true for a 3D→2D transition. When converting a bulk (3D) Monte Carlo particle into a low-dimensionality (2D) particle occupying a subband in the inversion layer, the difference between the carrier energy $e_{3D}$ and the in-plane kinetic energy $e_{2D}$ gives the subband energy $e_{\nu n}$. The carrier subband is then determined by choosing a subband with the minimum error in subband energy and $e_{\nu n}$, the calculated energy. In the opposite case of converting a 2D-particle into a bulk carrier, the 3D carrier energy is given by $e_{3D} = e_{2D} + e_{\nu n}$. By scanning the polar angle $\theta$ from the tabulated values of the 3D $\mathbf{K}$-vector and preserving the in plane azimuth $\phi$, the $\mathbf{K}_{3D}$ vector which minimizes the error in the magnitude of the in-plane $\mathbf{K}_{2D}$ vector is chosen as the 3D carrier momentum of the bulk particle.

## 2.2    Band-Structure

Figure 1 shows the valence band-structure of strained SiGe on relaxed Si for Ge concentrations increased in steps of 5%. The uppermost bold lines correspond to the HH, LH and SO bands in Si while the lowermost bold lines correspond to that of Ge. This figure shows the effect of strain and alloying. It is seen that strain has two-fold effect:

1 It lifts the degeneracy of the HH and LH band at the gamma point. Further, as strain increases, it also causes the SO band to move further away from the HH minima at the gamma point. The bands too are sharper than in the case of relaxed SiGe and it can be inferred that the density of states in the strained case would be lower than in the case of relaxed SiGe.

2 Transport properties at moderate electric fields would be determined mostly by inter-band scattering events involving transitions between the HH and LH bands because of the large energy difference between the split off and the HH and LH bands.

The density of states for 2D confined carriers in the channel for the case of the triangular test potential is shown in Figure 2; the left panel for Si inversion layer, while the right panel is indicative of the same for strained SiGe inversion layer.

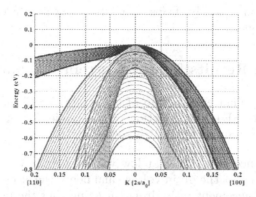

**Fig. 1.** Valence band-structure of strained SiGe for different Ge concentrations

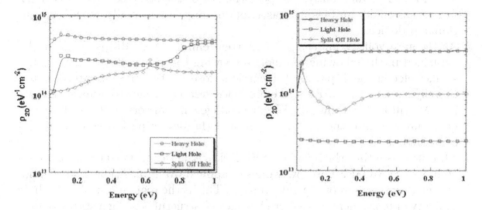

**Fig. 2.** Density of states for 2D carriers for the case of a triangular test potential: Left Panel: Si, right panel: strained SiGe

1 Considering the left panel,

    a The deviation of the 2D density of states obtained by a full band calculation from a regular step-like profile expected out of an effective-mass type approximation is clearly seen in the case of the light hole and split off bands.

    b Furthermore, subband crossings are seen in the case of light hole and split off subbands, where these subbands cross into higher lying heavy hole subbands, resulting in spikes in the density of states.

2 For the case of the right panel,

    a The heavy and light hole subbands have a clear density of states with no subband crossings.

    b The split off band actually follows the heavy hole subband density of states, where there is a subband crossing from split off subband into the heavy hole subband. The crossover then changes shape and the density of states consequently drops and settles down to a constant value at higher energies.

# 3    Simulation Results for the Current Enhancement

The drain current enhancement ratio of the strained SiGe MOSFET over the conventional Si MOSFET as a function of the applied drain bias for different gate voltages is shown in Figure 3. It is seen that:

1 The peak enhancement comes at low values of drain bias, in the low-field transport regime. As the drain voltage and hence the electric field increases, the current enhancement ratio drops: meaning that the performance of the Si p-channel MOSFET device is comparable to that of the strained SiGe MOSFET. Put differently, the performance of the strained SiGe MOSFET worsens as the drain bias increases, performing just as badly as the conventional Si device.

2 As the gate voltage increases, the current enhancement drops. This can be explained in the following manner: Increasing the gate voltage increases the surface electric field, pulling the carriers closer to the Si-SiO2 interface and thereby causing the carriers to experience greater surface roughness scattering. At still higher values of the gate voltage, the carriers spill over from the quantum well into the Si cap region and the device performance degrades even further.

3 Thus it is seen that the SiGe MOSFET clearly performs better than the conventional Si MOSFET at low values of applied drain bias (low field regime) and moderate values of the gate voltage. This is the regime in which the hole mobility enhancement is predicted for device structures using a strained SiGe layer as the active layer for carrier transport.

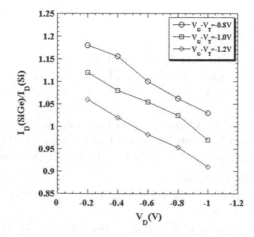

**Fig. 3.** Drain current enhancement of the strained SiGe over the conventional Si MOSFET

# 4 Conclusions

In summary, this work has presented a novel way of incorporating band-structure and quantum effects on hole transport in p-channel MOSFETs. In this approach, a full band Monte Carlo technique has been employed to investigate hole transport and the effect of valence band-structure on transport probed using a six band k.p model giving an accurate picture of the coupling between the heavy-hole, light-hole and the split-off bands. Further, within the scope of this approach, for lack of an accurate and computationally feasible and a reasonably fast method to incorporate open boundary conditions to model the contacts, carriers in the source and drain regions are treated as quasi-3D like particles with the band-structure information obtained by solving for the eigenstates of the more compact six band k.p Hamiltonian proposed initially by Dresselhaus, Kip and Kittel. The effect of carrier spatial confinement in the channel (along the depth direction) due to the confining potential under the gate is included by self-consistently coupling the Poisson, the discretized six band k.p solver and the Monte Carlo transport kernel in the device simulator. All relevant scattering mechanisms were included in the model: acoustic and optical phonon scattering (within the isotropic approximation), surface roughness scattering as well as Coulomb scattering. For the case of the strained SiGe MOSFET, alloy scattering was included in the transport model. Self-consistent device simulations of a 25 nm gate length conventional Si p-channel MOSFET were performed and the performance compared against similar self-consistent results of device simulations of a 25 nm gate length strained SiGe p-channel MOSFET. The results showed that the drive current performance of the strained SiGe MOSFET and the conventional Si MOSFET were comparable at the same normalized gate voltages ($V_G - V_T$) at moderate and high values of drain bias. In fact the drive current enhancement ratio $I_{D(SiGe)}/I_{D(Si)}$ is lower than 1.0 at high values of $V_D$ 0.8-1.0V, meaning that the expectation of performance enhancement expected out of the strained SiGe MOSFET is misplaced. On the other hand, the performance of the strained SiGe MOSFET is found to be better at moderate values of drain bias in what is known as the low-field regime. This is the region in which low-field mobility enhancement is expected out of using a strained active region of carrier transport. To conclude, transistor scaling in the decanano regime has reduced the performance gap between conventional Si MOSFETs and strained layer heterojunction devices.

# References

1. S. Krishnamurthy, A. Sher, and A.-B. Chen, Phys. Rev. B, **vol. 33**, pp. 1026, (1986).
2. K. Takeda, A. Taguchi, and M. Sakata, J. Phys. C, **vol. 16**, pp. 2237, (1983).
3. J. M. Hinckley and J. Singh, J. Appl. Phys., **vol. 76**, pp. 4192, (1994).
4. D. K. Nayak and S. K. Chun, Appl. Phys. Lett., **vol. 64**, pp. 2514, (1994).
5. A Abramo et al., Semicond. Sci. Technol., **vol. 7B**, pp. 597, (1992).
6. F. M. Bufler et al., J. Vac. Sci. Technol. B, **vol. 16**, pp. 1667, (1998).

7. T. Manku et al., IEEE Trans. Electron Devices, **vol. 40**, pp. 1990, (1993).
8. T. Yamada and D. K. Ferry, Solid State Electron., **vol. 38**, pp. 881, (1995).
9. M.V. Fischetti and S.E. Laux, J. Appl. Phys., **vol. 80**, pp. 2234, (1996).
10. J. R. Watling, A. Asenov and J. R. Barker, Proceedings of IWCE 98, IEEE **Cat. NO. 98EX116**, pp. 96, (2004).
11. J.R. Barker, J.R. Watling, Micoelectronic Engineering, **vol. 47**, pp. 369, (1999).
12. H. Nakatsuji, Y. Kamakura, and K. Taniguchi, IEDM Tech. Dig., pp. 727, (2002).
13. R. Oberhuber, G. Zandler, and P. Vogl, Phys. Rev. B, **vol. 58**, pp. 9941, (1998).
14. M. V. Fischetti et al., J. Appl. Phys., **vol. 94**, pp. 1079, (2003).
15. Z. Ikonic, P. Harrison, and R. W. Kelsall, Phys. Rev. B, **vol 64**, pp. 245311 (2001).

# A Monte Carlo Model of Piezoelectric Scattering in GaN

S. Vitanov, M. Nedjalkov, and V. Palankovski

Advanced Materials and Device Analysis Group, Inst. for Microelectronics, TU Wien,
Gusshausstrasse 27-29, A-1040 Vienna, Austria

**Abstract.** A non-parabolic piezoelectric model of electron-phonon interaction in Gallium Nitride is discussed. The Monte Carlo aspects of the model, needed for the simulation tools which provide the characteristics of GaN-based devices are analyzed in details. The piezo-scattering rate is derived by using quantum-mechanical considerations. The angular dependence is avoided by a proper spherical averaging and the non-parabolicity of the bands is accounted for. For the selection of the after-scattering state we deploy the rejection technique. The model is implemented in a simulation software. We employ a calibrated experimentally verified set of input material parameters to obtain valuable data for the transport characteristics of GaN. The simulation results are in good agreement with experimental data available for different physical conditions.

## 1 Introduction

Gallium Nitride (GaN) based devices demonstrate impressive power capabilities in radio-frequency range which recently became of interest for applications in state-of-the-art mobile communication technology, e.g. base stations amplifiers. The physical model of GaN, needed for the Monte Carlo (MC) simulation tools to describe the electronic and optical behavior of this material, is subject of an active research and development [1], [2], [3]. The model provides information about the band structure (analytical or full-band), the scattering mechanisms (caused by impurities, acoustic and optical phonons) and other microscopic characteristics which govern the carrier transport in the semiconductor.

There are two types of GaN crystal lattice structures: wurtzite or zink blende. Due to the the lack of inversion symmetry, elastic strain gives rise to macroscopic electric fields. These fields cause additional coupling between the acoustic waves and the free carriers, known as piezoelectric scattering. Nitrides are characterized by the largest piezoelectric constants among the III-V semiconductors so that this scattering must be taken into account in the MC simulations. The papers related to this subject stress on the simulation results and merely formulate the out-scattering rate of the utilized piezo-model. The next section of this work focuses on the MC aspects and peculiarities of a non-parabolic piezo-scattering model. The simulation results obtained by the proposed MC approach are discussed in the last section.

T. Boyanov et al. (Eds.): NMA 2006, LNCS 4310, pp. 197–204, 2007.
© Springer-Verlag Berlin Heidelberg 2007

## 2    The Model

According the Golden rule the probability for scattering from electron state $\mathbf{k}$ to state $\mathbf{k}'$ by phonons with wave vector $\mathbf{q}$ in branch $j$ is determined with the help of the matrix element $|\langle \mathbf{k}', \hat{n}'_{\mathbf{q},j}|H_{e-p}|\hat{n}_{\mathbf{q},j}, \mathbf{k}\rangle|^2 \delta(E - E')$. Standard notations are used, where $H_{e-p}$ is the interaction Hamiltonian $|\hat{n}, \mathbf{k}\rangle$ denotes the electron-phonon state, $\hat{n}$ and $\hat{n}'$, and $E$ and $E'$, refer to the initial and final phonon number and energies respectively. Phonons are described by waves $\mathbf{s} = \mathbf{e} \exp(i\mathbf{q}.\mathbf{r} - \omega_q t)$ where $\mathbf{r}$ is the position, $\omega_q$ the energy and $\mathbf{e}$ the unit vector of the polarization[1]. The basic piezo-interaction energy is proportional to the integral of the electric displacement $\mathbf{D}(\mathbf{r})$ associated with the electron, multiplied by the lattice polarization $\mathbf{P}(\mathbf{r})$. The screening is accounted via the Thomas-Fermi model, which introduces in $\mathbf{D}$ the reciprocal Debye screening length $q_0$. The polarization is proportional to the strain $\mathbf{S}$ caused by the propagating acoustic waves: $\mathbf{P}_i = \sum_{ik} e_{ik}\mathbf{S}_k/\epsilon_r$ where, in reduced notations $i, k$ run from 1 to 6, $e_{ik}$ denote the piezo coefficients, and $\epsilon_r$ is the dielectric constant. For zinc blende crystals $e_{14} = e_{25} = e_{36}$ and all other components are zero. For wurtzite only $e_{15} = e_{24}$, $e_{31} = e_{32}$ and $e_{33}$ are non-zero. The matrix element of $H_{e-p}$ gives rise to conservation rules for the phonon numbers: $n' = n \pm 1$ and the electron wave vector $\mathbf{k}' = \mathbf{k} \pm \mathbf{q}$. The Bloch assumption [4] allows to replace the phonon degrees of freedom with their mean equilibrium number $n_{\mathbf{q}}$ given by the Bose-Einstein distribution. The factor $H'(e_{kl}, \mathbf{e}, \alpha, \beta, \gamma, q_0, q)$ summarizes the complicated dependence of the matrix element on the polarization $\mathbf{e}$ and direction cosines $\alpha$, $\beta$, $\gamma$ of the direction of propagation of $\mathbf{q}$ with respect to the crystal axes. A simplification is certainly desirable and is achieved by a spherical averaging. The averaged scattering rate $W$ can be written explicitly as:

$$W = W_a + W_e = \sum_{\pm} \frac{2\pi}{\hbar} |F(q)|^2 (n_q + \frac{1}{2} \mp \frac{1}{2}) \delta\left(\epsilon(\mathbf{k} \pm \mathbf{q}) \mp \epsilon(\mathbf{k}) \mp \hbar\omega_q\right) \qquad (1)$$

where $W_a$ corresponds to absorption ($\mathbf{k}' = \mathbf{k}+\mathbf{q}$) and $W_e$ to emission ($\mathbf{k}' = \mathbf{k}-\mathbf{q}$) of a phonon with wave vector $\mathbf{q}$. The averaged isotropic coupling constant depends on $q$ as $|F(q)|^2 = Cf(q)$, $f(q) = \frac{q^3}{(q^2+q_0^2)^2}$. The constant $C$ will be introduced later.

We consider a three valley ($\Gamma$, U, and L) spherical non-parabolic energy dispersion model with $m$ the effective electron mass for the corresponding valley:

$$\frac{\hbar^2 k^2}{2m} = \epsilon(k)(1 + \alpha\epsilon(k)) = \gamma(k); \qquad k = \frac{1}{\hbar}\sqrt{2m\gamma}; \qquad \mathbf{v}(\mathbf{k}) = \frac{\hbar\mathbf{k}}{m(1 + 2\alpha\epsilon(k))}$$

### 2.1    Absorption

The absorption out-scattering rate $\lambda_a = \int W_a d\mathbf{k}'$ is calculated by using spherical coordinates $(q, \theta, \phi)$, where the $z$ axis is chosen along $\mathbf{k}$ so that $\theta$ becomes the angle between $\mathbf{k}$ and $\mathbf{q}$:

---

[1] In an isotropic media there are one longitudinal, $L$, $(\mathbf{e}\|\mathbf{q})$ and two transverse, $T$, $\mathbf{e} \perp \mathbf{q}$ branches. In crystals $L$ and $T$ exist for special directions only.

$$\lambda_a = \frac{V2\pi C}{(2\pi)^3 \hbar} \int_0^{2\pi} d\phi \int_{-1}^1 d\cos\theta \int_0^\infty dq q^2 f(q) n_q \delta(\epsilon(k,q,\theta) + \epsilon(k) - \hbar\omega_q)$$

where $\frac{V}{(2\pi)^3}$ is the density of states in the $\mathbf{q}$ space. The acoustic phonon energy $\hbar\omega_{\mathbf{q}} = \hbar v_s q$ introduces the sound velocity $v_s$ which is anisotropic. The following consideration can be applied if a particular direction of $\mathbf{q}$ is considered, or if a spherical average is taken for $v_s$. The argument of the delta function becomes zero if

$$\cos\theta = \frac{2v_s}{v(k)} - \frac{q}{2k}(1 - 4\alpha\epsilon_s); \qquad \epsilon_s = \frac{mv_s^2}{2}$$

which, furthermore, gives rise to the following condition for $q_1 \le q \le q_2$:

(i) if $\frac{v_s}{v} < 1$ then $-1 \le \cos\theta \le \frac{v_s}{v}$ and $q_1 = 0$, $q_2 = \frac{2k(v_s/v+1)}{1-4\alpha\epsilon_s}$;

(ii) else $-1 \le \cos\theta \le 1$ and $q_1 = \frac{2k(v_s/v-1)}{1-4\alpha\epsilon_s}$, $q_2 = \frac{2k(v_s/v+1)}{1-4\alpha\epsilon_s}$.

By using the equipartition approximation: $n_q = kT/\hbar\omega_q = kT/\hbar v_s q$ and introducing the dimensionless variable $x = q/q_0$ ($x_i = q_i/q_0$), the scattering rate is obtained:

$$\lambda_a = \underbrace{\frac{e^2 K_{av}^2 \sqrt{m}kT}{8\pi\epsilon_0\epsilon_r\hbar^2\sqrt{2\gamma(k)}}(1 + 2\alpha\epsilon(k))}_{C_1(k)} I_1(x_1, x_2) + \underbrace{\frac{e^2 v_s K_{av}^2 \sqrt{m}2\alpha kT q_0}{8\pi\epsilon_0\epsilon_r\hbar\sqrt{2\gamma(k)}}}_{C_2(k)} I_2(x_1, x_2)$$

where $e$ is the electric charge and the integrals $I_1$ and $I_2$ are evaluated as follows:

$$I_1 = \int_{x_1}^{x_2} dx \frac{x^3}{(x^2+1)^2} = \int_{x_1}^{x_2} J_1(x) dx; \qquad I_2 = \int_{x_1}^{x_2} dx \frac{x^4}{(x^2+1)^2} = \int_{x_1}^{x_2} J_2(x) dx$$

The coefficients $C_1(k)$ and $C_2(k)$ in front of the integrals are expressed in terms of the dimensionless quantity $K_{av}^2$. For zinc blende and wurtzite structures we have respectively

$$K_{av}^2 = \frac{e_{14}^2}{\epsilon_0\epsilon_r}\left(\frac{12}{35c_L} + \frac{16}{35c_T}\right) \qquad K_{av}^2 = \frac{e_L^2}{c_L\epsilon_0\epsilon_r} + \frac{e_T^2}{c_T\epsilon_0\epsilon_r}.$$

The longitudinal and transverse elastic constants $c_L$ and $c_T$ can be obtained from the elastic coefficients $c_{11}$, $c_{12}$, and $c_{44}$ or from the longitudinal and transverse sound velocities $v_{sL}$ and $v_{sT}$, if known.

$$c_L = 0.6 \cdot c_{11} + 0.4 \cdot c_{12} + 0.8 \cdot c_{44}; \qquad v_{sL} = \sqrt{c_L/\rho}$$

$$c_T = 0.2 \cdot c_{11} - 0.2 \cdot c_{12} + 0.6 \cdot c_{44}; \qquad v_{sT} = \sqrt{c_T/\rho}$$

The piezo coefficients $e_{15}$, $e_{31}$, and $e_{33}$ are used to calculate the corresponding $e_L^2$ and $e_T^2$, which are necessary to obtain the coupling coefficient $K_{av}$ taking into account the wurtzite structure.

$$e_L^2 = \frac{e_{33}^2}{7} + \frac{4e_{33}(e_{31} + 2e_{15})}{35} + \frac{8(e_{31} + 2e_{15})^2}{105};$$

$$e_T^2 = \frac{16e_{15}^2}{35} + \frac{16e_{15}(e_{33} - e_{31} - e_{15})}{105} + \frac{2(e_{33} - e_{31} - e_{15})^2}{35}.$$

**Selection of the after-scattering state.** Since $\cos\theta$ is uniquely determined by the value of $q$, the main task is to derive algorithm for selection of $q$. The angle $\phi$ is then selected randomly. $\phi$, $\cos\theta$ and $q$ determine $\mathbf{q}$, and the after-scattering state is given by $\mathbf{k'} = \mathbf{k} + \mathbf{q}$. The probabilities $P_1$ and $P_2 = 1 - P_1$ for the after scattering state to be selected by the corresponding terms which comprise $\lambda_e$ are

$$\lambda_a = C_1 I_1 + C_2 I_2; \qquad P_1 = \frac{C_1 I_1}{C_1 I_1 + C_2 I_2}; \qquad P_2 = \frac{C_2 I_2}{C_1 I_1 + C_2 I_2}$$

Furthermore, to select $q$ by either term we have to solve the equality $I_i(x_r, x_1) = r I_i(x_2, x_1)$; where $r$ is a random number and $q$ is determined from $q = x_r q_0$. Neither of these equations can be solved for $x_r$ in a simple way. The problem can be overcome by application of a rejection technique: The value of $x_r$ is generated by using a function $\xi_i(x)$ greater than the corresponding integrand $J_i$. Then, depending on the non equality $\xi_i(x_r)r' < J_i(x_r)$, where $r'$ is a second random number, the value of $x_r$ is accepted or rejected.

For the first case we choose $\xi_1(x) = \frac{x}{x^2+2}$, $\xi_1(x) > J_1(x)\forall x$ (Fig. 1). This choice gives the following expression for $x_r$:

$$x_r^2 = (x_2^2 + 2)^r (x_1^2 + 2)^{1-r} - 2$$

In the second case we choose $\xi_2(x) = \frac{x}{\sqrt{x^2+4}}$, $\xi_2(x) > J_2(x)\forall x$ (Fig. 2) so that:

$$x_r^2 = \left(r\sqrt{x_2^2 + 4} + (1 - r)\sqrt{x_1^2 + 4}\right)^2 - 4$$

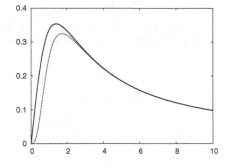     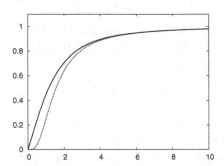

**Fig. 1.** The function $\xi_1$ (solid line) as compared to $J_1$ (dots)

**Fig. 2.** The function $\xi_2$ (solid line) as compared to $J_2$ (dots)

## 2.2   Emission

The necessary condition for emission is the initial electron energy to be greater than the phonon energy: $\epsilon(k) > \hbar\omega_q$. The out-scattering rate is calculated in the same way as in the case of absorption. In particular the delta function gives rise to the relation:

$$\cos\theta = \frac{2v_s}{v(k)} + \frac{q}{2k}(1 - 4\alpha\epsilon_s); \tag{2}$$

giving rise to the condition $q_1 \leq q \leq q_3$, where

(i) if $\frac{v_s}{v} < 1$ then $\frac{v_s}{v} \leq \cos\theta \leq 1$ and $q_1 = 0$, $q_3 = \frac{2k(1-v_s/v)}{1-4\alpha\epsilon_s}$;

(ii) else there is no solution.

Thus, the out-scattering rate is evaluated as:

$$\lambda_e = C_1(k)I_1(0, x_3) - C_2(k)I_2(0, x_3)$$

**Selection of the after-scattering state.** We utilize the condition $C_1 I_1(0, x_3) > \lambda_e$ to develop a rejection technique. Since $\xi_1$ is a majorant function for $J_1$, the value of $x_r$ is obtained according to:

$$x_r^2 = (x_3^2 + 2)^r 2^{1-r} - 2$$

A second random number $r'$ is used to accept or reject $x_r$ in the inequality:

$$C_1(k)\xi_1(x_r)r' < C_1(k)J_1(x_r) - C_2(k)J_2(x_r)$$

The functions $\xi_1$ and $\xi_2$ are compared on Figs. 1 and 2 with the corresponding counterparts $J_1$ and $J_2$. In both cases the difference is negligible for $x = q/q_0 > 3$. At room temperatures the average electron wave vector $q$ is of order of $10^7$ [1/cm], while $q_0$ is usually an order of magnitude smaller. Hence the region of significant rejection, below $x = 3$, is relatively rarely visited during the simulations.

## 3   Simulation Results

In order to establish a rigorous MC simulation, parameters from various publications have been collected and analyzed [5]. Table 1 provides a summary of bulk material parameters for GaN, necessary for analytical band-structure MC simulations, such as energies of lowest conduction bands, effective electron masses, non-parabolicity factors, and model parameters for the acoustic deformation potential (ADP) scattering, inter-valley scattering (iv), and polar optical phonon scattering (LO). $\varepsilon_\infty$ and $\varepsilon_s$ are the optical and static dielectric constants, $\rho$ is the mass density.

Table 2 summarizes the values for GaN of the elastic constants $c_{11}$, $c_{12}$, and $c_{44}$ together with the piezo coefficients $e_{15}$, $e_{31}$, and $e_{33}$ adopted in our MC simulation. From them, the corresponding $c_L$, $c_T$, $v_{sl}$, $v_{st}$, $e_L^2$, $e_T^2$, and $K_{av}$ are obtained.

Using the established setup of models and model parameters, we obtained MC simulation results for different physical conditions (doping, temperature, field, etc.) for bulk GaN. Fig. 3 shows the low-field electron mobility in hexagonal

**Table 1.** Summary of material parameters of wurtzite GaN for Monte Carlo simulation

| Bandgap energy | | | Electron mass | | | Non-parabolicity | | | Scattering models | | | | | |
|---|---|---|---|---|---|---|---|---|---|---|---|---|---|---|
| $\Gamma_1$ | U | $\Gamma_3$ | $m_{\Gamma1}$ | $m_U$ | $m_{\Gamma3}$ | $\alpha_{\Gamma1}$ | $\alpha_U$ | $\alpha_{\Gamma3}$ | ADP | hf$_{iv}$ | hf$_{LO}$ | $\rho$ | $\varepsilon_s$ | $\varepsilon_\infty$ |
| [eV] | [eV] | [eV] | [m$_0$] | [m$_0$] | [m$_0$] | [1/eV] | [1/eV] | [1/eV] | [eV] | [meV] | [meV] | [g/cm$^3$] | [-] | [-] |
| 3.39 | 5.29 | 5.59 | 0.21 | 0.25 | 0.40 | 0.189 | 0.065 | 0.029 | 8.3 | 91.0 | 92.0 | 6.07 | 8.9 | 5.35 |

**Table 2.** Summary of elastic constants of GaN and the resulting longitudinal and transverse elastic constants and sound velocities

| $c_{11}$ [GPa] | $c_{12}$ [GPa] | $c_{44}$ [GPa] | $c_L$ [GPa] | $c_T$ [GPa] | $v_{sl}$ [m/s] | $v_{st}$ [m/s] | $e_{15}$ [C/m$^2$] | $e_{31}$ [C/m$^2$] | $e_{33}$ [C/m$^2$] | $e_L^2$ [C$^2$/m$^4$] | $e_T^2$ [C$^2$/m$^4$] | $K_{av}$ [-] |
|---|---|---|---|---|---|---|---|---|---|---|---|---|
| 373 | 141 | 94 | 355 | 103 | 7641 | 4110 | -0.30 | -0.36 | 1.0 | 0.106 | 0.452 | 0.137 |

GaN as a function of free carrier concentration. Two MC simulation curves are included to demonstrate the effect of the piezo-scattering model and its impact on the low-field mobility. Our MC simulation is in fairly good agreement with experimental data from collections or single point measurements from [6,7,8,9,10]. The electron mobilities, selected for comparisons in this work, consider bulk material and are measured using the Hall effect. The discrepancy between our simulation results and the measured data might be attributed to dislocation scattering which is not considered in our work. This mechanism is considered to be a source of mobility degradation for GaN samples.

Fig. 4 shows the corresponding scattering rates as a function of the doping concentration in hexagonal GaN. Note, that the piezoelectric scattering is the dominant mobility limitation factor at low concentrations even at room temperature, beside the commonly accepted importance at low temperatures.

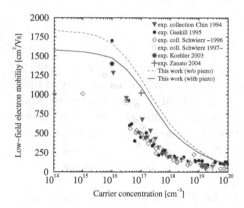
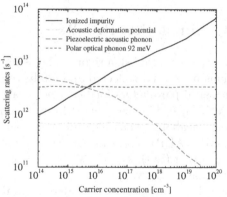

**Fig. 3.** Low-field electron mobility as a function of carrier concentration in GaN. Comparison of the MC simulation results and experimental data.

**Fig. 4.** Scattering rates utilized in our simulation model for wurtzite GaN as a function of carrier concentration at room temperature

Fig. 5 shows the low-field electron mobility as a function of lattice temperature in GaN at $10^{17}$ cm$^{-3}$ concentration. The experimental data are from [10,11,12]. Note, that mobility increases over the years because of the improved material quality (reduced dislocation density).

Fig. 6 provides the electron drift velocity versus the electric field. We compare our MC results with other simulations [3,13,14,15,16], and with the available experimental data [17,18]. The low field data points are in qualitatively good

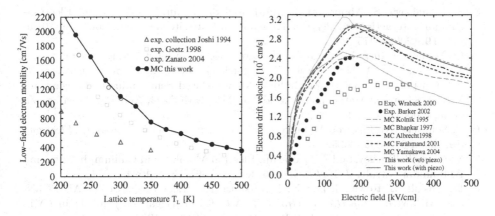

**Fig. 5.** Low-field electron mobility as a function of lattice temperature in GaN at carrier concentration of $10^{17}$ cm$^{-3}$

**Fig. 6.** Drift velocity vs. electric field in wurztite GaN: Comparison of MC simulation results and experimental data

agreement, at higher fields experimental values are significantly lower. Both experiments [17,18] of electron velocities in bulk GaN, employ pulsed voltage sources. The discrepancy in the MC results comes from differently chosen sets of parameter values and considerations of scattering mechanisms.

Our MC results prove that the piezoscattering mechanism has less influence at higher electric fields than other scattering mechanisms, such as polar optical scattering.

## 4    Conclusion

A non-parabolic piezoelectric model of electron-phonon interaction is derived. It is applied to materials with hexagonal crystal structure in a Monte Carlo simulator. The importance of the piezoelectric effect is illustrated by simulation results for different physical conditions.

## Acknowledgment

This work has been supported by the Austrian Science Funds, FWF Project START Y247-N13.

## References

1. B. Ridley, Quantum Processes in Semiconductors, Oxford University Press, third ed. (1993).
2. G. Kokolakis, F. Compagnone, A. Di Carlo, and P. Lugli, Exciton Relaxation in Bulk Wurtzite GaN: the role of piezoelectric interaction, Phys. Stat. Sol. (a) **195**, (2003), 618–627.

 3. S. Yamakawa, S. Aboud, M. Saraniti, and S. Goodnick, Influence of Electron-Phonon Interaction on Electron Transport in Wurtzite GaN, Semicond. Sci. Technol. **19**, (2004), 475–477
 4. O. Madelung, Introduction to Solid-State Theory, Springer Verlag, (1978).
 5. V. Palankovski, A. Marchlewski, E. Ungersböck, and S. Selberherr, Identification of Transport Parameters for Gallium Nitride Based Semiconductor Devices, 5th Vienna Symp. on Mathematical Modeling MATHMOD, **2**, AGRESIM-Verlag, Vienna, (2006), 14-1–14-9
 6. F. Schwierz, An Electron Mobility Model for Wurtzite GaN, *Solid-State Electron.*, **49**, no. 6, 889–895, 2005.
 7. V. Chin, T. Tansley, and T. Osotachn, Electron Mobilities in Gallium, Indium, and Aluminium Nitride, *J.Appl.Phys.*, **75**, no. 11, 7365–7372, 1994.
 8. D. Gaskill, L. Rowland, and K. Doverspike, Electrical Properties of AlN, GaN, and AlGaN, in *Properties of Group III Nitrides* (J. Edgar, ed.), no. 11 in EMIS Datareviews Series, section 3.2, 101–116, IEE INSPEC, 1994.
 9. K. Köhler, S. Müller, N. Rollbühler, R. Kiefer, R. Quay, and G. Weimann, Multiwafer Epitaxy of AlGaN/GaN Heterostructures for Power Applications, in *Proc. Intl. Symp. Compound Semiconductors*, Lausanne, 235–238, 2003.
10. D. Zanato, N. Balkan, G. Hill, and W. J. Schaff, Energy and Momentum Relaxation of Electrons in Bulk and 2D GaN, *Superlattices & Microstructures*, **36**, no. 4-6, 455–463, 2004.
11. R. Joshi, Temperature-dependent Electron Mobility in GaN: Effects of Space Charge and Interface Roughness Scattering, *Appl.Phys.Lett.*, **64**, no. 2, 223–225, 2004.
12. W. Götz, N. Johnson, C. Chen, H. Liu, C. Kuo, and W. Ilmler, Activation Energies of Si Donors in GaN, *Appl. Phys. Let.*, **6**, no. 22, 3144–3147, 1996.
13. J. Kolnik, I. Oguzman, K. Brennan, R. Wang, P. Ruden, and Y. Wang, Electronic Transport Studies of Bulk Zincblende Wurtzite Phases of GaN Based on an Ensemble Monte Carlo Calculation Including a Full Zone Band Structure, *J.Appl.Phys.*, **78**, no. 2, 1033–1038, 1995.
14. U. Bhapkar and M. Shur, Monte Carlo Calculation of Velocity-Field Characteristics of Wurtzite GaN, *J.Appl.Phys.*, **82**, no. 4, 1649–1655, 1997.
15. J. Albrecht, R. Wang, and P. Ruden, Electron Transport Characteristics of GaN for High Temperature Device Modeling, *J.Appl.Phys.*, **83**, no. 9, 4777–4781, 1998.
16. M. Farahmand, C. Garetto, E. Bellotti, K. Brennan, M. Goano, E. Ghillino, G. Ghione, J. Albrecht, and P. Ruden, Monte Carlo Simulation of Electron Transport in the III-Nitride Wurtzite Phase Materials System: Binaries and Ternaries, *IEEE Trans.Electron Devices*, **48**, no. 3, 535–542, 2001.
17. M. Wraback, H. Shen, J. Carrano, T. Li, J. Campbell, M.J.Schurman, and I. Ferguson, Time-Resolved Electroabsorption Measurement of the Electron Velocity-Field Characteristic in GaN, *Appl. Phys. Let.*, **76**, no. 9, 1154–1157, 2000.
18. J. Barker, R. Akis, D. Ferry, S. Goodnick, T. Thornton, D. Kolesk, A. Wickenden, and R. Henry, High-Field Transport Studies of GaN, *Physica B*, **314**, no. 1-4, 39–41, 2002.

# Solving the Illumination Problem with Heuristics[*]

Manuel Abellanas[1], Enrique Alba[2], Santiago Canales[3], and
Gregorio Hernández[1]

[1] Universidad Politécnica de Madrid. Facultad de Informática
Departamento de Matemática Aplicada. Spain
{mabellanas, gregorio}@fi.upm.es
http://www.dma.fi.upm.es
[2] Universidad de Málaga
Departamento de Lenguajes y CC. CC. Spain
eat@lcc.uma.es
http://www.lcc.uma.es/~eat
[3] Universidad Pontificia Comillas de Madrid
Departamento de Matemática Aplicada y Computación. Spain
scanales@icai.upcomillas.es
http://www.upcomillas.es/personal/scanales

**Abstract.** In this article we propose optimal and quasi optimal solutions
to the problem of searching for the *maximum lighting point* inside a polygon $P$ of $n$ vertices. This problem is solved by using three different techniques: *random search, simulated annealing* and *gradient*. Our comparative
study shows that simulated annealing is very competitive in this application. To accomplish the study, a new polygon generator has been implemented, which greatly helps in the general validation of our claims on the
illumination problem as a new class of optimization task.

## 1 Introduction

Illumination and visibility problems have been always an interesting topic of study
in Mathematics and Computer Science, and especially in the area of Computational Geometry. To summarize the problem, we can state that, given a set $D$ in
$\Re^2$, two points $x, y \in D$ are visible in $D$ (or $x$ illuminates $y$) if the segment $\overline{xy}$ is
completely contained in $D$. A classic problem is the *Art Gallery Problem* proposed
by V. Klee in 1973: *How many guards are needed to see every point in the interior
of an art gallery?* In 1975, Chvátal showed that $\lfloor \frac{n}{3} \rfloor$ guards are always sufficient
and occasionally necessary to guard a simple polygon with $n$ vertices.

For example, $\lfloor \frac{n}{3} \rfloor$ lights are always sufficient to illuminate any polygon with
$n$ vertices, but in many polygons this number of light sources is too large. Thus,
it makes sense to outline the following algorithmic problem: *given a polygon $P$
calculate the minimum number of light sources that illuminate $P$.* This problem
and many variants about guarding are $\mathcal{NP}$-hard [10,12].

---

[*] Partially supported by TIN 2005-08818-C04-01 and CAM S-0505/DPI/023.

The search for algorithms which obtain approximate solutions to these problems is reduced to [8]. In 2000 Eidenbenz [5] showed some results about approximability and inapproximability of these problems. In this article, which is a part of the doctoral dissertation of Canales [3], we present heuristic algorithms to find optimal and nearly-optimal solutions for the maximum lighting point-light in the interior of a polygon $P$. Our analysis includes three techniques: *Simulated Annealing (SA), Random Search (RS)* and *Gradient method (GRAD)*. Formally the problem can be stated in the following form:

SEARCH OF THE MAXIMUM INTERIOR ILLUMINATION POINT TO A POLYGON MAXA-P-PV1($P$):
INPUT: A polygon $P$ of $n$ vertices.
GOAL: Find an interior point in $P$ to locate a light source so that the area illuminated by such light is maximal.

In Sections 2, 3 and 4 we present the methods $SA$, $RS$ and $GRAD$, respectively. In Section 5 we discuss the experimental analysis over a set of polygons generated with the random generator $RPG$. In Section 6 we compare the three proposed techniques and finalize with some conclusions and future work.

## 2 Solving MAXA-P-PV1($P$) with *Simulated Annealing (SA)*

*Simulated Annealing (SA)* is an algorithm using local search to progress from a partial solution to another one of a higher quality [7]. The inconvenience of a local minimization algorithm is that it can be trapped in a local minimum and thus never converge to the global minimum. To overcome this drawback, $SA$ introduces a variable of control $T$ (temperature) which permits, with a certain probability, to temporary explore worse areas of the search space. This probability is designated *acceptance function*, and usually it is evaluated according to the function $e^{(\frac{-\delta}{T})}$, $\delta$ being the increase or deterioration of the *objective function* $O(x)$. If $x \in S$ is the initial configuration , $T$ the temperature in each iteration with $T_0 > 0$ the initial temperature and $N(T)$ the number of iterations for each temperature, the general plan of the *simulated annealing* is the following:

```
[01]   do
[02]     {do
[03]       {Generate solution y ∈ Neighborhood(x) ⊂ S;
[04]        Evaluate δ ← O(x) − O(y);
[05]        if (δ < 0) y ← x
[06]        else
[07]          if ((δ ≥ 0) ∧ (U(0,1) < e^(−δ/T))) y ← x;
[08]        n ← n + 1;
[09]        }while (n ≤ N(T));
[10]      T ← 0.99 · T;        //decrease of the temperature
[11]    }while (not stop);
```

The next section explains how $SA$ has been customized to deal with MAXA-P-PV1($P$).

## 2.1  Applying $SA$ to the Problem

The input of the problem is the $n$ vertices, (in positive sense), of the polygon $P$ and the output is the coordinates $(x, y)$ of a point interior to $P$.

**Set $S$ of configurations:** The set of configurations or feasible solutions of our problem will be all the points that are found in the interior of the polygon $P$. Thus, we will consider that the set of configurations is infinite and that each element (each point), comes determined by its coordinates $(x, y)$. $S = \{p_1 = (x_1, y_1), p_2 = (x_2, y_2), ..., p_n = (x_n, y_n), ...\}$.

**Objective function $O$:** The objective function $O : S \to \Re$ will assign a real value to each element of the configuration $S$. In our case for each $p_i \in S$ the objective function will produce a value which represents the area illuminated by a *light-point* located precisely in the point $p_i$, i.e., it will compute the area of the *visibility polygon* of the point $p_i$, $V(P, p_i) : O(p_i) = Area\ V(P, p_i)$.

**Neighborhood of each configuration:** As indicated in the general behavior of $SA$, for each point $p_i = (x_i, y_i)$ will find out a new point $p'_i = (x'_i, y'_i)$ to analyze. In our algorithm, the point $p'_i$ is calculated by adding to each coordinate of $p_i$ a random real value with normal distribution $N(0, 1)$.

**Initial configuration:** A initial point $p_0$ inner to $P$, will be considered as the first solution to analyze, created with uniform distribution $U(0, 1)$ for its two coordinates independently.

## 2.2  Annealing Strategies

The results of the $SA$ depend to a large extent on the conditions of temperature for each problem. That is, the initial temperature $T_0$ and the annealing plan of such temperature in each iteration are important.

**Initial temperature:** In this article we have used three criteria to define the initial temperature $T_0$: (a) an initial temperature according to $T_0 = O(S_0)$, where $S_0$ represents the initial configuration; (b) initial temperature depending of the number of vertices of the polygon $P$: $T_0 = f(n)$, concretely $f(n) = n$; and (c) a constant initial temperature: $T_0 = 100.0$.

**Decrease of the temperature in each iteration:** Numerous studies demonstrate that $SA$ can show a slow convergence [7]. Thereafter, Szu and Hartley [11] proposed a fast version ($FSA$), using the function $T(k) = \frac{T_0}{1+k}$   $k = 1, 2, ...,$ which was improved later by Ingber [9] ($VFSA$), where the function $T(k) = \frac{T_0}{e^k}$, $k = 1, 2, ...,$ was proposed. In Section 5 we evaluate both, $FSA$ and $VFSA$. All the obtained results have been compared with a base annealing recommended by Dowsland [4]: $T(k) = \alpha T(k-1)$, where $\alpha$ represents the cooling factor $0 \le \alpha \le 1$.

**Number of iterations for each temperature$N(T)$:** In our algorithm the number of iterations for each temperature value is $N(T) = 1/T$.

**Stop criterion:** In theory, the corresponding temperature for a "cold system" should be $T_f = 0$. However, much before zero the probability $e^{-\delta/T}$ is so smallthat no worse solution is ever accepted. Therefore, the stop condition in

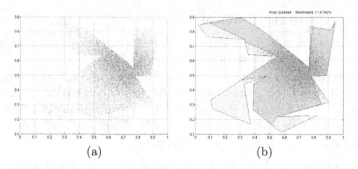

(a)                              (b)

**Fig. 1.** Solving MAxA-P-Pv1$(P)$ with SA using, $T_0 = 100$, $T(k) = \frac{T_0}{1+k}$, $T_f = 0.005$ and $N(T) = \frac{1}{T}$

our algorithm will be $T_f \leq 0.005$. We present in Figure 1 an example of execution of our *SA* for a polygon $P$ with 25 vertices.

## 3    Solving MAxA-P-Pv1$(P)$ with *Random Search* (*RS*)

In this section we propose what could be considered the simplest algorithm. The idea of the algorithm is to perform a *random search* on the inner points of the polygon $P$. For this, a set $N = \{p_1, ..., p_m\}$ of $m$ inner points of the polygon $P$ is generated, and a sequential search of the point $p_i \in N$ is computed to fulfill: $Area(V(P, p_i)) \geq Area(V(P, p_j))$  $\forall j \neq i$  $i, j \in \{1, ..., m\}$, with $V(P, p_i)$ being the visibility polygon of the point $p_i$ in $P$. Evidently, the size of the point set $N$ must be related to the number of vertices of the polygon, in such a form that for polygons with greater number of vertices will produce a larger value of $m$. We have considered a set size $m = 50\ n$. A pseudocode of *RS* follows:

---

**Algorithm. *RS*-MAxA-P-Pv1$(P)$**
INPUT: A polygon $P$ of $n$ vertices, $\{v_1, ..., v_n\}$
OUTPUT: A *light source* of maximum illumination in $P$.

```
[01]    area ← 0;
[02]    N ← Generate_Set(P, 50n);
[03]    for (p_i ∈ N)
[04]      {Calculate the visibility polygon V(P, p_i);
[05]        if (area < Area(V(P, p_i))
[06]          area ← Area(V(P, p_i));
[07]          p ← p_i;
[08]      }
[09]    return p;
```

---

We present in Figure 2 (a) an example of execution of this algorithm on a polygon $P$ with 25 vertices (generated by our generator *RPG*). We can see some differences between Figure 1(b) and Figure 2 (a), with respect to the uniformity of the points analyzed and the solution obtained.

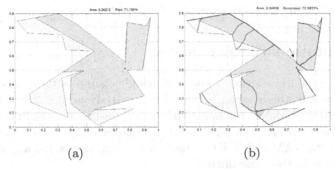

(a)                                    (b)

**Fig. 2.** (a) $RS$ ($m = 1250$).  (b) $GRAD$ ($v = 16$, $\beta = 0.001$).

## 4    Solving MAXA-P-PV1$(P)$ with *Gradient Method*($GRAD$)

We present in this section the third technique designed to solve the problem MAXA-P-PV1$(P)$. The fundamental idea consists of considering that every inner point $p$ to a polygon $P$ of $n$ vertices, not being a local maximum, can be displaced according to a "positive gradient" of illuminated area. Thus, given a point $p$ we consider a set $\{p_1, ..., p_v\}$ of $v$ neighboring points of $p$, at a distance $\beta$, such that the vectors defined by them $\overline{pp_i}$, will sequentially form a positive angle $2\pi/v$.

We propose a function **Generate-Candidate** which determines the best potential neighboring $v$. Evidently, this heuristic will frequently produce convergence to a local maximum. To avoid this undesired situation we initially set a light source in all the vertices of the polygon $P$, causing that each one of them converges to a local maximum. The pseudocode of $GRAD$ follows.

---

**Algorithm.** $GRAD$-MAXA-P-PV1$(P)$

INPUT: A polygon $P$ of $n$ vertices, $\{v_1, ..., v_n\}$
OUTPUT: A *light source* of maximum illumination in $P$.

```
[01]    changes ← 1;
[02]    max ← 1;
[04]    q ← v;              /*We make a copy of the vertices of P*/
[05]    while(changes ≠ 0)
[06]      {changes ← 0;
[07]       for (q_i)
[08]         {c_i ← Generate_Candidate q_i;
[09]          area ← Area(V(P,c_i));              /*Move each q_i */
[10]          if(area > Area(V(P,q_i))
[11]            {q_i ← c_i;
[12]             changes ←changes+1;
[13]             if(area > Area(V(P,q_max))
[14]                max ← i;}}}
[15]    return q_max;        /*Return the maximum illumination point*/
```

---

We show in Figure 2 (b) an example of a $GRAD$ execution.

## 5    Computational Experiments

The results exposed in Figures 1 and 2 are single examples for a concrete polygon. We present in this section the results of a study accomplished on a wide set of polygons with different number of vertices. All the algorithms have been implemented in C++. All the experiments have been accomplished on a Pentium IV 2.5 Ghz, with RAM of 512 KB. The Table 1 summarizes the algorithms and parameters used in the experiments.

**Table 1.** Parameters of the algorithms

| Problem | Algorithm | Parameters |
|---|---|---|
| | $Simulated\ Annealing - SA$ | $Case$ 1-9, ($see$ Table 2) |
| MAXA-P-Pv1$(P)$ | $Random\ Search - RS$ | $m = 50n$ |
| | $Gradient - GRAD$ | $v = 16\quad \beta = 0.001$ |

### 5.1    Results with $SA$

Let us start analyzing the influence of the initial temperature $T_0$ and its annealing plan. In this sense we have studied nine possible combinations. These combinations produce nine cases that we will compare using our random polygons generator $RPG$, a set of 50 polygons of 50, 100, 150 and 200 vertices. We report on the final illuminated area, the employed time, and the number of iterations performed by the algorithms. The cases are shown in Table 2.

**Table 2.** Cases studied for $SA$

| Case | Parameters |
|---|---|
| 1 | $T_0 = C(S_0)\quad T_k = \frac{T_0}{1+k}\ (FSA)$ |
| 2 | $T_0 = C(S_0)\quad T_k = \frac{T_0}{e^k}\ (VFSA)$ |
| 3 | $T_0 = C(S_0)\quad T_k = \alpha\ T_{k-1}\ (\alpha = 0.9)$ |
| 4 | $T_0 = n\quad T_k = \frac{T_0}{1+k}\ (FSA)$ |
| 5 | $T_0 = n\quad T_k = \frac{T_0}{e^k}\ (VFSA)$ |
| 6 | $T_0 = n\quad T_k = \alpha\ T_{k-1}\ (\alpha = 0.9)$ |
| 7 | $T_0 = 100.0\quad T_k = \frac{T_0}{1+k}\ (FSA)$ |
| 8 | $T_0 = 100.0\quad T_k = \frac{T_0}{e^k}\ (VFSA)$ |
| 9 | $T_0 = 100.0\quad T_k = \alpha\ T_{k-1}\ (\alpha = 0.9)$ |

The general conclusions that we can deduce from this study with for $SA$ are the following ones:

– A slow decrement of the temperature decreases improves the solution, though the time of response of the algorithm is large.

- The best solutions are procured taking an initial temperature depending on the number of vertices $n$ of the polygon $P$, though for polygons with a small number of vertices ($n \leq 200$) a constant initial temperature produces better results.
- The best results with respect to the percentage of illuminated area are obtained in Case 4: $T_0 = n$ $T_k = \frac{T_0}{1+k}$ ($FSA$). This will be selected for the forthcoming comparisons of $SA$ with $RS$ and $GRAD$ in the next sections.

## 5.2   Results with $RS$

We present in Figure 3 (b) the results obtained with $RS$, where we observe that $RS$ improves each one of the analyzed variables over $SA$: percentage of illuminated area, execution time, and number of iterations. If we reduce the final temperature in $SA$ from $x$ down to 0.005, $T_f < 0.005$, the results obtained by $SA$ then are superior to those obtained through $RS$, although the response time of $SA$ is always much larger.

## 5.3   Results with $GRAD$

The method converges faster if we use a smaller number of initial points, i.e., if we let some free vertices in $P$. However the number of local maxima can be enlarged to $O(n)$.

In Figure 3 (b) we show the data obtained for $SA$, $RS$ and $GRAD$, that relate the number of vertices of $P$ with the middle, (in sets of 50 polygons), percentage of illuminated area. We conclude that $GRAD$ obtains the best results for all the analyzed problem instances.

# 6   Analysis and Conclusions

For the conclusions, let us begin with the statistical confidence of our results. For this, we have performed $T$ hypothesis contrast, (using the mathematical software MatLab), with a meaning level of the 95%, for all instances with respect to the final illumination area found by the algorithms. From this statistical analysis we can deduce that $RS$ and $GRAD$ do not produce significantly different results, though in mean we observe that $GRAD$ gives better solution in percentage of illuminated area. The rest of heuristics generate significantly different results.

## 6.1   Performance of the Algorithms as the Dimension of the Problem Grows Up

To better understand the scalability features of the three algorithms we show here the *mean curve of growth* for each technique heuristic with polygons of 100 vertices. For this, each heuristic has been applied to 50 polygons with 100 vertices, obtaining the *growth curve* for each polygon and calculating finally the mean of all them in each iteration. In Figure 3 (a) we plot the resulting curves. Thus, we can draw the following conclusions:

- **With respect to the percentage of illuminated area** by the solution point, $GRAD$ provides better results in all the tests. $RS$ shows good results in percentage of illuminated area. Though sensibly better with respect of $SA$, it is very similar to the results provided by $GRAD$.
- **With respect to the fast convergence of the heuristics**, we can observe in Figure 3 (a), that the best method is $RS$, followed by $GRAD$ and $SA$.
- **With respect to the ratio iterations/time**, (see Figure 3 (b)). In this sense the best method is $SA$, followed by $RS$, and the $GRAD$. This means that $SA$ is the lightest algorithm from a computational point of view.

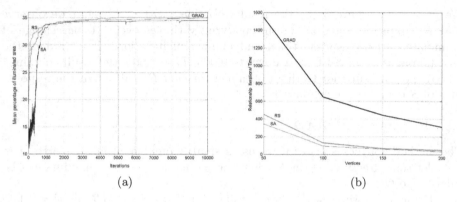

(a)                                            (b)

**Fig. 3.** (a) Comparison in scalability     (b) Relation $Iterations/Time$

## 6.2   Future Work

A future work will analyze some more parameters which influence on the heuristics. Also an important restriction that can be imposed to the illumination is the *limitation in the scope*. An interesting study for future investigations is the behavior of these methods when the lighting is constrained to cover only nearby points. It would be interesting to accomplish a similar study taking the concept of $t-$good illumination analyzed in [3].

# References

1. Auer T., Held, M.: Heuristics for the Generation of Random Polygons. Proc. 8th Canad. Conf. Comput. Geom. 38–44. Ottawa, Canada, Aug. 1996
2. Back, T.: Evolutionary Algorithms in Theory and Practice. Oxford Press (1996)
3. Canales, S.: Métodos Heurísticos en Problemas Geométricos. Visibilidad, Iluminación y Vigilancia. Ph. D. Thesis, UPM, Spain (2004)
4. Dowsland, K. A.: Simulated Annealing. In: Modern Heuristic Techniques for Combinatorial Problems (C. R. Reeves, ed.). Blackwell Scientific Pub. Oxford (1993)
5. Eidenbenz, S.: (In)-Approximability of Visibility Problems on Polygons and Terrains. Ph. D. Thesis, Swiss Federal Institute of Tecnology Zurich (2000)
6. Fogel, D.: Evolutionay Computation. IEEE Press (1995)

7. Gelatt, C. D., Kirkpatrick, S., Vecchi, M. P.: Optimazation by simulated annealing. Science **220** (1983) 671–680
8. Ghosh, S. K.: Approximation algorithms for Art Gallery Problems. Proceedings of the Canadian Information Processing Society Congress (1987)
9. Ingber, L.: Very fast simulated re-annealing. Math. Comput. Modelling **12(8)** (1989) 967–973
10. Lee D. T., Lin A. K.: Computational complexity of art gallery problem. IEEE Trans. Info. Th. IT-**32** (1979) 415–421
11. Szu, H. H., Hartley, R. L.: Fast simulated annealig. Physic Letters A **122**(1987) 157–162
12. Urrutia, J.: Art Gallery and Illumination Problems. Handbook on Computational Geometry (J. R. Sack and J. Urrutia ed.). Elsevier (1999)

# Optimal Placement of Antennae Using Metaheuristics

Enrique Alba, Guillermo Molina, and Francisco Chicano

Departamento de Lenguajes y Ciencias de la Computación
University of Málaga, 29071 Málaga, Spain
eat@lcc.uma.es, donguille125@hotmail.com, chicano@lcc.uma.es

**Abstract.** In this article we solve the radio network design problem (RND). This NP-hard combinatorial problem consist of determining a set of locations for placing radio antennae in a geographical area in order to offer high radio coverage using the smallest number of antennae. This problem is originally found in mobile telecommunications (such as mobile telephony), and is also relevant in the rising area of sensor networks. In this work we propose an evolutionary algorithm called CHC as the state of the art technique for solving RND problems and determine its expected performance for different instances of the RND problem.

## 1 Introduction

An important symbol of our present information society are telecommunications. With a rapidly growing number of user services, telecommunications is a field in which many open research lines are challenging the research community. Many of the problems found in this area can be formulated as optimization tasks. Some examples are assigning frequencies to cells in mobile communication systems [1], building multicast routing trees for alternate path computation in large networks [2], developing error correcting codes for the transmission of messages [3], and designing the telecommunication network [4,5]. The problem tackled in this paper belongs to this last broad class of network design tasks. When a geographically dispersed set of terminals needs to be covered by transmission antennae a key issue is to minimize the number and locations of these antennae and cover a large area at the same time. This is the central idea of the *radio network design problem* (RND).

In order to solve RND, metaheuristic techniques are used to overcome the large dimension and complexity of the problem, often unaffordable for exact algorithms. In the associated literature the problem has been solved with genetic algorithms [6,7]. In this article, our goal is to improve existing results and propose a state-of-the-art optimization method to solve the RND problem. In particular, we will compare the CHC algorithm against three other techniques: a simulated annealing (SA), a steady state genetic algorithm (ssGA), and a generational genetic algorithm (genGA). Another objective of this work is to extend the basic formulation of the problem to include more realistic kinds of antenna.

T. Boyanov et al. (Eds.): NMA 2006, LNCS 4310, pp. 214–222, 2007.

In summary, the contribution of this paper consists of: the application of a an algorithm not previously used, CHC, that improves all the results in the literature, the optimization of the algorithm parameters, the analysis of the scaling properties of the RND problem, and the extension of the basic problem to include more than one type of antenna.

The paper is organized as follows. In the next section we define and characterize the radio network design problem. Section 3 briefly describes the CHC algorithm. Section 4 provides the results of the tests performed either to compare algorithms or to analyze different types of antenna. Finally, some concluding remarks and future research lines are drawn in Section 5.

## 2    The Radio Network Design Problem

The radio coverage problem amounts to covering an area with a set of antennae. The part of an area that is covered by an antenna is called *a cell*. In the following we will assume that the cells and the area considered are discretized, that is, they can be described as a finite collection of geographical locations (taken from a geo-referenced grid).

Let us consider the set $L$ of all potentially covered locations and the set $M$ of all potential antenna locations. Let $G$ be the graph, $(M \cup L, E)$, where $E$ is a set of edges such that each antenna location is linked to the locations it covers, and let the vector $x$ be a solution to the problem, where $x_i$ with $i \in [1, |M|]$ indicates whether an antenna is being used or not at the $i$th available location.

Throughout this work we will consider different versions of the RND problem, which will differ in the type of antennae that might be placed in each location. There are simple versions using antennae that have no parameters, and more complex versions where antennae have parameters (i.e. azimuth) that determine the area they cover. In the last case, any solution $x$ must also indicate which values the parameters of the antennae have for each antenna used.

Searching for the minimum subset of antennae that covers a maximum surface of an area comes to searching for a subset $M' \subseteq M$ such that $|M'|$ is minimum and such that $|Neighbors(M', E)|$ is maximum, where

$$Neighbors(M', E) = \{u \in L \mid \exists v \in M', (u, v) \in E\}. \tag{1}$$

The problem we consider recalls the Unicost Set Covering Problem (USCP) that is known to be NP-hard. An objective function to combine the two goals has been proposed in [6]:

$$f(x) = \frac{Coverage(x)^\alpha}{Nb.\ of\ antennae(x)} \ , \ Coverage(x) = \frac{100 \cdot Neighbors(M', E)}{Neighbors(M, E)}, \tag{2}$$

where the parameter $\alpha$ can be tuned to favor the cover rate factor with respect to the number of antennae. Just like Calégari et al. did [6], we will use $\alpha = 2$, and a $287 \times 287$ point grid representing an open-air flat area.

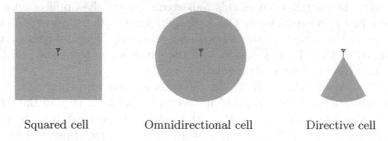

Squared cell          Omnidirectional cell          Directive cell

**Fig. 1.** Terrain coverages with different types of antenna

Three different antenna types will be used in this work: a square shaped cell antenna that covers a $41 \times 41$ point cell as used in [6,7], an omnidirectional antenna that covers a 22 point radius circular cell (new contribution here), and a directive antenna that covers one sixth of the omnidirectional cell (new contribution here). When directive antennae are employed, three of them are placed in the location site. Fig. 1 illustrates the terrain coverages obtained with the different kinds of antenna.

## 3   The CHC Algorithm

The algorithm we propose for solving the RND problem is Eshelman's CHC, a kind of Evolutionary Algorithm (EA) surprisingly not used in many studies despite it has unique operations usually leading to very efficient and accurate results [8]. Like all EAs, it works with a set of solutions (*population*) at any time. The algorithm works iteratively, producing new solutions at each iteration, some of which will be placed into the population instead of others that were previously included. The pseudocode for this algorithm is shown in Fig. 2.

The algorithm CHC works with a population of individuals (solutions) that we will refer to as $P_a$. In every step, a new set of solutions is produced by selecting pairs of solutions from the population (the parents) and recombining them. This selection is made in such a way that individuals that are too similar can not mate each other, and recombination is made using a special procedure known as HUX. This procedure copies first the common information for both parents into both offspring, then translates half the diverging information from each parent to each of the offspring. This is done in order to preserve the maximum amount of diversity in the population, as no new diversity is introduced during the iteration (there is no mutation operator). The next population is formed by selecting the best individuals among the old population and the new set of solutions (elitist criterion).

As a result of this, at some point of the execution population convergence is achieved, so the normal behavior of the algorithm should be to stall on it. A special mechanism is used to generate new diversity when this happens: the *restart* mechanism. When restarting, all of the solutions except the very best ones are significantly modified (*cataclysmically*). This way, the best results of the previous phase of evolution are maintained and the algorithm can proceed again.

```
t:=0;
Initialize(Pa,convergence_count);
while not ending_condition(t,Pa) do
        Parents := Selection_parents(Pa);
        Offspring := HUX(Parents);
        Evaluate(Offspring);
        Pn := Elitist_selection(Offspring,Pa);
        if not modified(Pa,Pn) then
                convergence_count := convergence_count-1;
                if (convergence_count == 0) then
                        Pn := Restart(Pa);
                        Initialize(convergence_count);
                end if
        end if
        t := t+1;
        Pa := Pn;
end while
```

**Fig. 2.** Pseudocode for CHC

## 4    Experiments

In this section we briefly present the results of performing an assorted set of experiments to solve the different RND problems using CHC. First we solve RND problems where antennae have no parameters. In this part, CHC will be faced against three other algorithms: SA, ssGA, and genGA, and the results will be compared to the best results of the literature [7] (dssGA8). Afterwards, we tackle the problem using antennae with parameters that shape the coverage cell. Only CHC will be employed in this part. Its behavior when facing different problem types will be studied here.

For each experiment, we will analyze the number of evaluations required to solve the problem if the execution is performed until an optimal solution is found (whenever possible). We perform 50 independent runs of each experiment. A statistical analysis is driven to validate the results obtained during the tests. The values of the parameters employed for CHC are shown in Table 1. When a range of values is shown instead of a single value, it means either that the parameter is tuned (population size) or that the value is selected to be adequate for each problem instance (maximum evaluations).

**Table 1.** Parameters of the CHC algorithm

| | |
|---|---|
| Maximum evaluations | 2,500,000−50,000,000 |
| Crossover probability | 0.8 |
| Restarting mutation probability(%) | 35 |
| Size of population | 50−10,000 |

## 4.1 RND with Squared and Circular Cell Antennae

In squared and circular cell antennae instances a solution is encoded with a bit string, where each bit relates to an available location site and determines whether an antenna is placed there (1) or not (0). Let $L$ be the problem size (the number of available location sites), the size of the solution space for these instances is $2^L$. For each instance the optimal solution is known beforehand.

The scalability of the problem is also studied by solving instances of sizes ranging from 149 to 349 available locations. Every time an algorithm is applied to solve an instance, we perform a parameter tuning in order to obtain the best possible performance from that algorithm. For the CHC algorithm the parameter tuned is the population size.

The results of the experiments are shown in Table 2 for square shaped cell antennae and Table 3 for omnidirectional antennae. All the algorithms were able to solve the problem with a very high hit ratio (percentage of executions where the optimal solution is found), with a few exceptions (highlighted in italics), therefore only the number of evaluations is shown. The best results obtained are highlighted in boldface. A Student t-test shows that all differences between CHC and the rest of algorithms are statistically significant with 95% of confidence.

**Table 2.** Comparison of the number of evaluations required by the different algorithms in RND with square shaped coverage antennae

| Algorithm | Size | | | | |
|---|---|---|---|---|---|
| | 149 | 199 | 249 | 299 | 349 |
| CHC | **30,319** | **78,624** | **148,595** | **228,851** | **380,183** |
| SA | 86,761 | 196,961 | 334,087 | 637,954 | 810,755 |
| ssGA | 239,305 | 519,518 | 978,573 | 1,872,463 | 3,460,110 |
| genGA | 141,946 | 410,531 | 987,074 | 1,891,768 | 3,611,802 |
| dssGA8 [7] | 785,893 | 1,467,050 | 2,480,883 | 2,997,987 | 4,710,304 |

**Table 3.** Comparison of the number of evaluations required by the different algorithms in RND with omnidirectional antennae

| Algorithm | Size | | | | |
|---|---|---|---|---|---|
| | 149 | 199 | 249 | 299 | 349 |
| CHC | **45,163** | **344,343** | **817,038** | **2,055,358** | **3,532,316** |
| SA | 83,175 | _262,282_ | 913,642 | 2,945,626 | _6,136,288_ |
| ssGA | 365,186 | 1,322,388 | 2,878,931 | _9,369,809_ | _9,556,983_ |
| gGA | 206,581 | 1,151,825 | 3,353,641 | _8,080,804_ | _19,990,340_ |

CHC proves to be the best technique among the four: it gets the lowest solving costs for all instances. In the first case (square shaped coverage) it improves the second best technique, SA, by costing less than 50%. In the second case (omnidirectional), the cost reduction regarding the second best technique (SA) is comprised between 10% and 40% (in the 199-size instance SA has a lower solving cost, but gets a low hit ratio). In both cases the increase of the number of evaluations is clearly superlineal, however, numeric approximations have returned subexponential models.

If we compare the two variants of RND (differing on the kind of antenna employed), we observe that the one using omnidirectional antennae seems to be more difficult to solve, since for the same instance size the required number of evaluations is higher. Furthermore, the problem becomes less tractable when its size grows, and the gap between efforts for solving the two kinds of problem increases.

In summary, CHC is better suited for solving RND than SA or any of the GAs. It is the best for the basic instance and allows a better scalability than the other two. The change of the antenna cell shape modifies the complexity of the optimization problem, but does not change the fact that the best results are obtained with the CHC algorithm. Therefore, from this point we will only employ CHC to solve the new instances of RND.

## 4.2   Complex RND Variants

Two variants of the RND are solved in this section: RND using directive antennae and RND using all kinds of antenna. When directive antennae are used, either three of them or none are placed in each available location. When three of them are placed, they are subject to one of the following restrictions: all antennae of the same location site must point in consecutive directions (case 1) or in different directions (case 2). When all kinds of antenna are employed, the restriction over the directive antennae is the second one (case 2).

The number of available locations of the instances considered is limited in both cases to only 149 as a base line for future research. For practical means, we will use the binary equivalent length (minimum length of a binary string that can store all the possible values of the solution space) as the instance size measure. Table 4 shows CHC's performance for all the problem instances solved in this work.

Fig. 3 illustrates the cost and size of all the different problem instances solved in this work: those using squared cells (unlabelled squared points), those using circular cells (unlabelled circular points), the ones using directional antennae under the first restriction (RND-3) and the second restriction (RND-4), and the variant using all antenna kinds (RND-5). Minimal mean square error approximations for the problems using squared cells and circular cells are also shown.

The problem variant using directive antennae seems to have a cost-size relation comprised between those of the variants using squared cell antennae and omnidirectional antennae. However, problem instances using only directive antennae do not have one single optimal solution (as the previous variants do), but a set of optimal solutions instead: $6^{52}$ and $20^{52}$ for the instances under the first and the second restriction respectively. Therefore the complexity reduction of this RND variant regarding the omnidirectional antennae variant might be due to the existence of many optimal solutions.

The variant of the problem using all antenna kinds simultaneously seems to have a cost-size relation lower than any of the other variants: for a

**Table 4.** Comparison of CHC's best performances for all the problem instances

| Problem Instance | | Binary Size | Fitness Evaluations | Optimal Population | Running Time(sec) | Hit Rate(%) |
|---|---|---|---|---|---|---|
| Square | 149 | 149 | 30,319 | 400 | 25.59 | 100 |
| | 199 | 199 | 78,624 | 1,200 | 76.66 | 100 |
| | 249 | 249 | 148,595 | 1,400 | 146.21 | 100 |
| | 299 | 299 | 228,851 | 1,800 | 237.38 | 100 |
| | 349 | 349 | 380,183 | 2,800 | 427.81 | 100 |
| Omnidirectional | 149 | 149 | 45,163 | 700 | 43.71 | 100 |
| | 199 | 199 | 344,343 | 2,800 | 374.01 | 100 |
| | 249 | 249 | 817,038 | 4,000 | 870.82 | 100 |
| | 299 | 299 | 2,055,358 | 8,000 | 2437.51 | 100 |
| | 349 | 349 | 3,532,316 | 10,000 | 4009.85 | 100 |
| Directive | case 1 | 419 | 2,383,757 | 4,000 | 4186.38 | 96 |
| | case 2 | 655 | 4,736,637 | 8,000 | 9827.60 | 88 |
| All antennae | | 675 | 829,333 | 10,000 | 1284.05 | 100 |

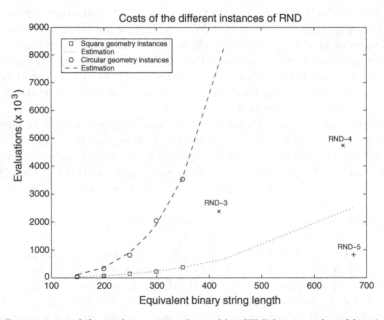

**Fig. 3.** Comparison of the evaluations performed by CHC for several problem instances

binary length of 675 (93% higher than the 349 squared coverage instance) its solving cost is only 829, 333 (118% higher). This would approximately correspond to a lineal growth, yet the measured growth has been estimated to be superlineal.

Therefore, the studied RND problems can be classified into two main different categories depending on their cost-size relation: a low complexity kind ($x^3$ law), and a high complexity kind ($x^4$ law). The geometry of the cell shape seems to be the decisive factor: both directive and omnidirectional antennae share the circular geometry so the two belong to the high complexity kind. The square shaped cells problem variant belongs to the low complexity kind. The variant of the problem where all antenna kinds are used simultaneously takes advantage of the possibility of using both geometries and achieves a complexity lower than any of the other variants.

## 5 Conclusions

We have established CHC as the best technique so far for solving the RND problem. This has been proven empirically by comparison with SA, ssGA, and genGA in two different scenarios: use of square shaped cell antennae and use of omnidirectional antennae. The cost of solving the problem has been estimated to grow in a subexponential manner as the size of the problem increases. The nature of that increase is mainly determined by the geometrical features of the antennae, being $x^3$ for square shaped cell and $x^4$ for circular shaped cell antennae. When directive antennae are placed, the fact of having many optimal solutions results in a cost reduction with respect to the RND using omnidirectional antennae. When several antennae are offered, the algorithm takes advantage of it and is able to solve the problem at a lower cost.

## Acknowledgements

This paper has been partially funded by the Spanish ministry of education and European FEDER under contract TIN2005-08818-C04-01 (The OPLINK project, http://oplink.lcc.uma.es). Guillermo Molina is supported by grant AP-2005-0914 from the spanish government. Francisco Chicano is supported by a grant (BOJA 68/2003) from the Junta de Andalucía (Spain).

## References

1. Matsui, S., Watanabe, I., Tokoro, K.: Application of the parameter-free genetic algorithm to the fixed channel assignment problem. Systems and Computers in Japan **36**(4) (2005) 71–81
2. Zappala, D.: Alternate Path Routing for Multicast. IEEE/ACM Transactions on Networking **12**(1) (2004) 30–43
3. Blum, C., Blesa, M.J., Roli, A.: Combining ILS with an effective constructive heuristic for the application to error correcting code design. In: Metaheuristics International Conference (MIC-2005), Viena, Austria (2005) 114–119
4. Maple, C., Guo, L., Zhang, J.: Parallel genetic algorithms for third generation mobile network planning. In: Proceedings of the International Conference on Parallel Computing in Electrical Engineering (PARELEC04). (2004) 229–236

5. Créput, J., Koukam, A., Lissajoux, T., Caminada, A.: Automatic mesh generation for mobile network dimensioning using evolutionary approach. IEEE Trans. Evolutionary Computation **9**(1) (2005) 18–30
6. Calégari, P., Guidec, F., Kuonen, P., Kobler, D.: Parallel island-based genetic algorithm for radio network design. Journal of Parallel and Distributed Computing (47) (1997) 86–90
7. Alba, E., Chicano, F.: On the behavior of parallel genetic algorithms for optimal placement of antennae in telecommunications. International Journal of Foundations of Computer Science **16**(2) (2005) 343–359
8. Eshelman, L.J.: The CHC Adaptive Search Algorithm: How to Have Safe Search When Engaging in Nontraditional Genetic Recombination. In: Foundations of Genetic Algorithms, Morgan Kaufmann (1991) 265–283

# Sparse Array Optimization by Using the Simulated Annealing Algorithm*

Vera Behar and Milen Nikolov

Institute for Parallel Processing, Bulgarian Academy of Sciences
"Acad. G. Bonchev" Str., bl. 25-A, 1113 Sofia, Bulgaria
behar@bas.bg, milenik@bas.bg

**Abstract.** Sparse synthetic transmit aperture (STA) imaging systems are a good alternative to the conventional phased array systems. Unfortunately, the sparse STA imaging systems suffer from some limitations, which can be overcome with a proper design. In order to do so, a simulated annealing algorithm, combined with an effective approach can used for optimization of a sparse STA ultrasound imaging system. In this paper, three two-stage algorithms for optimization of both the positions of the transmit sub-apertures and the weights of the receive elements are considered and studied. The first stage of the optimization employs a simulated annealing algorithm that optimizes the locations of the transmit sub-aperture centers for a set of weighting functions. Three optimization criteria used at this stage of optimization are studied and compared. The first two criteria are conventional. The third criterion, proposed in this paper, combines the first two criteria. At the second stage of optimization, an appropriate weighting function for the receive elements is selected.

The sparse STA system under study employs a 64-element array, where all elements are used in receive and six sub-apertures are used in transmit. Compared to a conventional phased array imaging system, this system acquires images of better quality 21 times faster than an equivalent phased array system.

## 1 Introduction

In conventional synthetic transmit aperture (STA) imaging systems only one transducer element is excited in transmit, while all the transducer elements receive the signals, reflected from the tissue. Each transducer element is fired consequently one after the other, and the signals received by each transducer element are recorded in the computer memory [1]. When the signals, received from each transmit/receive pair have been recorded, the synthetic beamforming is done by the appropriate algorithm. A disadvantage of STA imaging systems is the huge amount of the RF-data that must be stored in the computer memory and performed by the processor in order to reconstruct an image. For $N$-element

---

* This work is financially supported by the Bulgarian Foundation for Scientific Investigations: MI-1506/05 and by Center of Excellence BIS21++, 016639.

T. Boyanov et al. (Eds.): NMA 2006, LNCS 4310, pp. 223–230, 2007.
© Springer-Verlag Berlin Heidelberg 2007

array, $N$ RF-recordings are needed to form a conventional phased array image, while $(N \times N)$ RF-recordings are required to synthesize a STA image. The amount of data, used in the STA imaging, can be reduced to some extent, if only a small number of transducer elements $(M)$ are used as transmitters, where $M < N$ [2]. This is equivalent to using of a sparse array in transmit.

A disadvantage of a sparse STA imaging system is the low signal-to-noise ratio (SNR), that is caused by the use of a small number of transmit elements. A well-known approach to improve the SNR is to use *temporal encoding*, whereby linear frequency modulated (LFM) or phase shift key modulated (PSKM) signals excite the transducer elements [3]. As an alternative to temporal encoding, the *spatial encoding* can be used for improving the SNR, whereby all M transducer elements are fired simultaneously in each transmission [4]. Another approach is to use more than one elements in each transmission $(L)$, in order to create a spherical wave and, as a consequence, to improve the SNR proportionally to the number of elements used in the transmit sub-aperture [5].

The relation between the effective aperture function and the corresponding beam pattern of the imaging system can be used as a tool for analysis and for optimization of a sparse STA imaging system. In STA imaging the transmit aperture function depends on the number of transmit elements and their locations within the array. The receive aperture function depends on the length of a physical array and the weight coefficients applied to the array elements. Hence it appears that the shape of the effective aperture function of a system and, as consequence, the shape of the two-way beam pattern can be optimized depending on the element positions in transmit and the element weights in receive.

In this paper three two-stage algorithms for optimization of a sparse STA imaging system are studied. The first two of them are conventional. The third, proposed in the paper, combines the first two algorithms. These algorithms optimize both the positions of the transmit sub-aperture centers and the weights of the receive elements. At the first stage the simulated annealing algorithm optimizes the positions of the transmit sub-aperture centers for a set of weighting functions. At the second stage, an appropriate weighting function for the receive elements is selected.

## 2   Sparse STA Imaging

Consider a sparse STA imaging system that employs an array transducer with $N$ elements. In each transmission, a group of $L$ elements (transmit sub-aperture) are fired simultaneously to get higher transmit power. The spherical wave created by each transmit sub-aperture propagates in the whole region of interest and the echo signals received at $N$ transducer elements are recorded and stored in the computer memory. The process of data acquisition continues until all $M$ transmit sub-apertures are fired sequentially one by one as it is illustrated in Fig.1.

The back scattered echoes received after each transmission carry information from all directions, and a whole image can be formed by applying different delays on the received signals (*partial beamforming*). Since the image is focused only in receive, it has a low resolution. The low resolution images, although focused at the same points, have different phases due to the different locations of the transmit sub-apertures. The dynamic focusing in transmit is summing all low-resolution images. A high-resolution image is created as follows:

$$I_{High}(r, \theta) = \sum_{m=1}^{M} \sum_{n=1}^{N} a_n s_{m,n}(\tau_{m,n}(r, \theta)), \qquad (1)$$

where $a_n$ is weighting coefficient applied to the receive element $n$, $s_{m,n}$ is the echo signal received at the receive element $n$ after transmission $m$ and $\tau_{m,n}$ is the two-way propagation time from the location of the transmitter's group $m$ to the current focal point $(r, \theta)$ and back to the receive element $n$.

## 3   Description of the Optimization Problem

The image quality parameters, the lateral resolution and the contrast, are determined by the main lobe beam width ($W$) and the side lobe peak ($SL$) of the beam pattern of an imaging system. The two-way beam pattern of a sparse STA system that employs a transducer with $N$ elements is evaluated as the Fourier Transform of the effective aperture function $e_{STA}$, defined as:

$$e_{STA} = \sum_{m=1}^{M} a_m \otimes B, \quad a_m = [0, 0, \ldots, i_m, \ldots, 0], \quad i_m = 1, \qquad (2)$$

where $a_m$ is the full transmit aperture during the $m$-th firing, $i_m$ is the position of a transmit sub-aperture center within a full transmit aperture, $B$ is the weighting function applied to the receiver elements, and $\otimes$ is the convolution operator.

Since the positions of the transmit sub-apertures in a sparse array and the weighting applied to each receiver element impact the two-way beam pattern of a sparse STA system, the optimization of a sparse STA imaging system can be formulated as an optimization problem of both the location of the transmit sub-apertures within the sparse array, $(i_1, i_2, , i_M)$, and the weights assigned to the elements of the full array during receive ($B$).

In this paper, it is suggested to divide the process of optimization into two stages:

– **At the first stage,** a set of the optimal positions of transmit sub-aperture centers $(i_1, i_2, \ldots, i_M)_K$ are found, for a set of known weighting functions $\{B_k\}$, $k = 1, 2, \ldots, K$. Such a set of weighting functions may include several well-known window-functions (Hamming, Hann, Kaiser, Chebyshev and etc). At this stage, the optimization criterion can be written as follows:

$$\text{Given } M, \ N \text{ and } \{B\}_k, \text{ choose } (i_1, i_2, \ldots, i_M)_K \text{ to} \atop \text{minimize the cost } C(I, B_K), \text{ where } I = (i_1, i_2, \ldots, i_M). \qquad (3)$$

The following cost functions $C(I, B_K)$ are usually used:

$$C_1(I_K, B_K) = min \quad W(I, B_K), \qquad (4)$$

$$C_2(I_K, B_K) = min \quad SL(I, B_K).  \tag{5}$$

We propose using of the following cost function, which is a combination of the cost functions $C_1$ and $C_2$

$$C_3(I_K, B_K) = min \quad W(I, B_K) \quad subject \quad to \quad SL < Q,  \tag{6}$$

where $Q$ is the threshold of acceptable level of the side lobe peak.

- **At the second stage**, the final layout of transmit sub-apertures is chosen, which is a layout that corresponds to the most appropriate weighting function $B = (b_1, b_2, \ldots, b_N)$. This choice is a compromise between the minimal width of the main lobe and the acceptable level of the peak of the side lobes. Mathematically, it can be written as follows:

$$\begin{aligned} Given \quad M, \quad N, \quad \{B\}_k \quad and \quad \{i_1, i_2, \ldots, i_M\}_K, \quad choose \\ (b_1, b_2, \ldots, b_N) \quad to \quad minimize \quad W \quad subject \quad to \quad SL < Q, \end{aligned}  \tag{7}$$

where $\{i_1, i_2, \ldots, i_M\}_K$ are the selected positions of transmit elements, as found at the first stage of the optimization.

## 4   The Simulated Annealing Algorithm

One way of selecting the positions $\{i_1, i_2, \ldots, i_M\}_K$ is by using the simulated annealing algorithm suggested by Kirkpatrick et al. [6]. The simulated annealing algorithm realizes an iterative procedure that is determined by simulation of the arrays with variable positions of a transmit sub-aperture center. In order to maximize the lateral resolution of a system, it is assumed that the transmit sub-aperture 1 and the transmit sub-aperture $M$ are always located at the two outer elements of the physical array; their positions are not changed and are assigned numbers 1 and $N$. The positions of the other transmit sub-apertures are shifted randomly, where a shift in position to the left or to the right has equal probability (of 0.5). Once the process is initiated, with a random initial layout of transmit sub-aperture centers $I_0 = (i_1^0, i_2^0, \ldots, i_M^0)$, a neighbor layout $I_1 = (i_1^1, i_2^1, \ldots, i_M^1)$ is generated, and the algorithm accepts or rejects this layout according to a certain criterion. The corresponding simulated annealing algorithm is composed of two loops Fig.2:

The acceptance is described in terms of probability $p(T)$ that depends on the cost function $C(I, B_K)$. For the three cost functions defined by (4), (5) and (6), the expression for the acceptance probability $p(T)$ takes the form:

$$p_1(T) = \begin{cases} 1, & if \quad \Delta W \leq 0 \\ exp(-\Delta W/T_k), & otherwise \end{cases}, \quad p_2(T) = \begin{cases} 1, & if \quad \Delta SL \leq 0 \\ exp(-\Delta SL/T_k), & otherwise \end{cases}$$

$$p_3(T) = \begin{cases} 1, & if \quad \Delta W < 0 \ \& \ \Delta SL < 0 \\ exp(-\Delta W/T_k), & if \quad \Delta W > 0 \ \& \ \Delta SL < 0 \\ exp(-\Delta SL/T_k), & if \quad \Delta W < 0 \ \& \ \Delta SL > 0 \\ exp(-\Delta W/T_k - \Delta SL/T_k), & if \quad \Delta W > 0 \ \& \ \Delta SL > 0 \end{cases}$$

$$\tag{8}$$

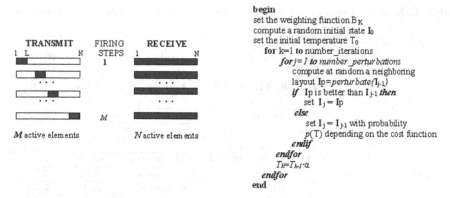

**Fig. 1.** Synthetic aperture data acquisition  **Fig. 2.** The simulated annealing algorithm

where $\Delta W$ is the difference of the width of the main lobe, $\Delta SL$ is the difference of the height of the peak of the side lobe between the current locations of transmit sub-apertures and the best one obtained at preceding steps (or Q for $C_3$). $T_k$ is the current value of the "temperature", where the current "temperature" is evaluated as $T_k = 0.95T_{k-1}$, and the algorithm proceeds until the number of iterations reaches the final value.

## 5  Simulation Results

### 5.1  Sparse Array Optimization

A physical array utilized in computer simulations is of 64 elements with a half wavelength spacing, where 64 active elements are used in the receive mode, and only six sub-apertures are used during six transmissions. The properties of the sparse STA system are optimized using the two-stage algorithm described in Section 3. First, the optimal positions of transmit sub-apertures are found for four weighting functions. For comparison, the optimization is done for the three optimization criterion. For each weighting function, the positions of transmit sub-apertures are shifted until optimal performance is obtained, as described earlier, using the simulated annealing algorithm presented in Fig.2. In order to obtain a beam pattern with a sharper main lobe, the optimization criteria $C_1$ and $C_3$ were formulated as the minimal width of the mainlobe at -20dB below the maximum.

The positions of transmit sub-apertures that were found to optimize the performance of the system, according the optimization criterion, together with the achieved widths of the main lobe (at -20 dB) and the levels of the peaks of the sidelobe, are presented in Table 1.

Both optimized functions, the effective aperture function and the corresponding two-way beam pattern, are plotted for each optimization criterion, for $C_1$ - Fig.3, for $C_2$ - Fig.4 and for $C_3$ - Fig.5. The two optimized functions plotted in Fig.3, 4 and 5, are obtained for the Hamming weighting function

**Table 1.** Numerical results obtained by employing the three-criterion optimization

| Optimization criterion | Transmitter Layout | Receiver Apodization | SLB [dB] | Beamwidth [-20dB][o] |
|---|---|---|---|---|
| | 1, 18, 27, 35, 49, 64 | – | -33.4 | 2.14 |
| | **1, 5, 8, 11, 16, 64** | **Hamming** | **-57.3** | **4.99** |
| $SLB_{[dB]}$=min | 1, 2, 7, 9, 10, 64 | Chebyshev | -58.2 | 5.18 |
| | 1, 15, 18, 20, 23, 64 | Blackman-Harris | -111.8 | 6.60 |
| | 1, 3, 5, 6, 19, 64 | Nattall-Harris | -103.3 | 7.37 |
| | 1, 2, 3, 62, 63, 64 | – | -28.3 | 1.28 |
| | **1, 11, 29, 38, 40, 64** | **Hamming** | **-30.6** | **2.21** |
| $Beamwidth_{[-20dB]}$ = min | 1, 11, 29, 38, 40, 64 | Chebyshev | -29.6 | 2.22 |
| | 1, 20, 24, 41, 45, 64 | Blackman-Harris | -29.06 | 2.30 |
| | 1, 17, 30, 35, 48, 64 | Nattall-Harris | -24.92 | 2.32 |
| | 1, 2, 3, 62, 63, 64 | – | -28.3 | 1.28 |
| $Beamwidth_{[-20dB]}$ = min | **1, 20, 24, 41, 45, 64** | **Hamming** | **-50.05** | **2.24** |
| subject to $SLB_{[dB]} \leq C$ | 1, 20, 24, 42, 44, 64 | Chebyshev | -51.03 | 2.32 |
| | 1, 18, 28, 41, 46, 64 | Blackman-Harris | -101.06 | 2.90 |
| | 1, 19, 25, 39, 49, 64 | Nattall-Harris | -99.3 | 2.88 |
| For comparison: | | | | |
| 64-element phased array: Transmit - no, Receive - Hamming | | | -85 | 4.04 |

(marked as bold in Table 1). It can be seen that the weighting applied to receiver elements reduces the peaks of the side lobes from -33 dB to -111 dB,
when the first optimization criterion $C_1$ is used to optimize the locations of the
transmit sub-apertures within a physical array. But it is done by Blackman-
Harris weighting at the cost of widening the main lobe of the beam pattern from
2° to 7°.

The minimal main lobe widths are provided by using of the optimization
criterion $C_2$ (no weighting). Unfortunately, the level of the peaks of the side lobes
is rather high and not acceptable (-28dB). The compromise solution is obtained
by using of the third optimization criterion $C_3$ that provides the minimal width
of the main lobe at -20dB below the maximum, where the condition that the
maximal level of the side lobe peak is below -50 dB. In that case, the beam
width is 2.2° and the peak of the side lobes equals -50dB. It is done by Hamming
weighting in receive.

For comparison, both parameters, the beam width and the side lobe peak,
evaluated for an equivalent phased array system, are presented in Table 1. Both
functions, the effective aperture function and the corresponding beam pattern,
are plotted for the conventional phased array system in Fig.6. Since the dynamic
range of a computer monitor is limited to about 50 dB, the comparison analysis
shows that the sparse array used with the Hamming weighting and the locations
of the transmit sub-aperture centers set at positions 1, 20, 24, 41, 45 and 64
within a physical array has a narrower beam width (2.2°) compared with the
equivalent phased array system (4°).

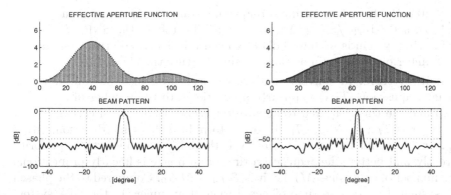

**Fig. 3.** The optimized effective aperture function and two-way array pattern $(C_1)$

**Fig. 4.** The optimized effective aperture function and two-way array pattern $(C_2)$

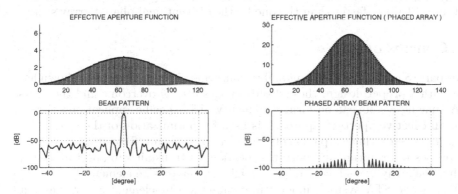

**Fig. 5.** The optimized effective aperture function and two-way array pattern $(C_0)$

**Fig. 6.** The phased array effective aperture function and two-way array pattern

## 5.2 Comparison Analysis

**Signal-to-Noise Ratio.** According to [2], the signal amplitude of a conventional phased array system is linearly proportional to the number of transmit elements $(N)$, i.e., $Signal \sim N$, while the noise is inversely proportional to the square root of the number of receive elements $(N)$, i.e., $Noise \sim 1/(\sqrt{N})$. Therefore, the SNR of a phased array system (that uses the same number of transmit and receive elements) can be expressed as $\text{SNR}_{PA} \sim N \cdot \sqrt{N}$. The number of signals added together during beamforming is determined as a product of the number of transmit sub-apertures $(M)$ and the number of receive elements $(N)$. Taking into account that each transmit sub-aperture includes $L$ array elements, the SNR is proportional to the product of the number of simultaneously excited transmit elements $(L)$, the square root of the number of transmit elements $(M)$, and the square root of the number of receive elements $(N)$: $\text{SNR}_{STA} \sim L \cdot \sqrt{M \cdot N}$. When compared to a phased array system, the relative SNR of a STA system with $M$ transmit sub-apertures can be written as $\text{SNR}_{STA}/\text{SNR}_{PA} \sim L \cdot \sqrt{M}/N$, which for $N = 64$, $M = 6$ and using decibels becomes: $(\text{SNR}_{STA}/\text{SNR}_{PA})_{dB} \sim$

($L_{dB} - 10.3dB$). If each transmit sub-aperture consists of 5 array elements, i.e, $L = 5$, then $(SNR_{STA}/SNR_{PA})_{dB} \sim -3.3dB$. Therefore, the SNR of a sparse STA imaging system is within -6 dB of an equivalent phased array system only if the number of elements in a transmit sub-aperture is at least 5.

**Acquisition time.** In conventional phased array systems, the time duration of image acquisition ($T_{PA}$) is linearly proportional to the number of scan lines ($N_{line}$) and the time required to acquire the echoes from one direction of view ($T_0$), i.e. it is $T_{PA} = N_{line} \cdot T_0$. In order to reduce this time duration, either $N_{line}$, or $T_0$ should be reduced. However, the STA imaging system performs differently, and better: the acquisition time of an image is linearly proportional to the number of emissions ($M$), i.e. it is $T_{STA} = M \cdot T_0$. Compared to the phased array system, the relative time of acquisition of an image in the STA system is $T_{STA}/T_{PA} = M/N_{line}$. For 64-element array $N_{line} > 127$, and, therefore $T_{STA}/T_{PA} < 1/21$ for $M = 6$. It means that the image formation of the STA system is 21 times faster than the one by the conventional phased array system.

# 6    Conclusions

The sparse STA imaging systems suffer from some disadvantages. It is shown here that with a proper design, these disadvantages can be overcome and the sparse STA imaging system can perform extremely well for specific applications. For this aim, an effective aperture approach is used for optimization of the sparse STA imaging system, which exploits sub-apertures in transmit. Two-stage algorithms are proposed for optimizing both the locations of transmit sub-aperture centers within the array transducer and the weights of the receive element.

The sparse STA system under study employs a 64-element array, where all elements are used in receive and six sub-apertures are used in transmit. Each transmit sub-aperture includes at least five transducer elements in order to create a spherical radiation wave. The analysis shows that a sparse STA system with 6 transmit sub-apertures obtains images of better quality and acquires data 21 times faster than a conventional phased array system.

# References

1. Ylitalo, J.: On the signal-to-noise ratio of a synthetic aperture ultrasound imaging method. European. J. Ultrasound **3** (1996) 277–281
2. Lockwood, G., Foster, F.: Design of sparse array imaging systems. IEEE Ultrasound Symposium 1995 1237–1243
3. Lokke, K., Jensen, J.: Multi-element synthetic transmit aperture imaging using temporal encoding. IEEE Trans. on Medical Imaging **22**(4) (2003) 552–563
4. Misaridis, T.: Ultrasound imaging using coded signals. PhD thesis, Ørsted•DTU, Technical University of Denmark, Lyngby, Denmark 2001
5. Frazier, C., O'Brien, W.: Synthetic aperture techniques with a virtual source element. IEEE Trans. Ultrason. Ferroelec. Freq. Contr. **45**(1) (1998) 196–207
6. Kirkpatrik, S., Gelatt, C., Vecchi, M.: Optimization by simulated annealing. Science, vol. 220, 4598 (1988) 671–680

# An Iterative Fixing Variable Heuristic for Solving a Combined Blending and Distribution Planning Problem

Bilge Bilgen

Dokuz Eylül University, Department of Industrial Engineering,
35100 Izmir, Turkey
bilge.bilgen@deu.edu.tr

**Abstract.** In this paper, we consider a combined blending and distribution planning problem faced by a company that manages wheat supply chain. The distribution network consists of loading ports, and customers. Products are loaded on bulk vessels of various capacity levels for delivery to overseas customers. The purpose of this model is simultaneous planning of the assignment of an appropriate type and number of vessels to each customer order, the planning of quantities blended at ports, loaded from ports, and transported from loading ports to customers. We develop a mixed integer programming (MIP) model and provide a heuristic solution procedure for this distribution planning problem. An iterative fixing variable heuristic algorithm is used to assure that acceptable solutions are obtained quickly. The effectiveness of the proposed heuristic algorithm is evaluated by computational experiment.

## 1 Introduction

Distribution network management is an area that remains critical to overall logistics and supply chain success. Proper planning of distribution network may give significant improvements in economic performance, which may be crucial for survival in an increasingly competitive market [6].

This research was motivated by a blending and shipment planning problem faced by a company that manages wheat distribution planning and theoretical research on the transportation planning within the supply chain management (SCM). This is an important problem involving high transportation costs, and even modest improvements in the performance may give significant savings. The problem concerns the assignment of an appropriate type and number of vessel to each customer order and the planning of the quantities transported from loading ports to customers. The planning is about making sure that the given demand is satisfied at the lowest possible cost. The large combinatorial size of this medium-term shipment planning problem necessitates the use of solution algorithms that do not assure an optimal solution. Hence, we have developed a heuristic algorithm in order to obtain an acceptable solution in a reasonable time.

T. Boyanov et al. (Eds.): NMA 2006, LNCS 4310, pp. 231–238, 2007.

Despite the fact that the majority of the challenging and important problems that arise in SCM, there has been relatively little effort devoted to the area of the use of maritime transportation with a focus on the whole supply chain [2,5]. Few studies that include decisions concerning maritime transportation can be found in [4,10,11,12].

The outline of the paper is as follows: The next section provides a description of the problem. Then in Section 3, the mathematical model for the problem is formulated. Section 4 describes the solution method and Section 5 presents computational results using data from a real-life case study. Finally, Section 6 presents some concluding remarks and suggests directions for future research.

## 2    Problem Description

Having given an overview of the problem and related literature, we now define precisely the problem that we address in this paper, its requirements and assumptions. The supply chain network considered in the model consists of loading ports, and customers. The planning horizon, in keeping with the midterm nature of the model, ranges from 1 month to 3 months. Original products are directly supplied to the loading ports. Different original products are blended at ports prior to outloading. Products are then loaded on bulk vessels of various capacity levels for delivery to overseas customers. It is assumed that a vessel can pick up products from at most two loading ports (1 or 2 loading ports) in a voyage. A vessel is discharged in a single destination port, where the customer is located. The size of each vessel type varies. Each vessel is hired for delivering one shipment of products to one customer, and is not allowed to discharge cargo at loading ports. It is assumed that there is unlimited number of vessels of each type available. The demand for both original and blended products exists at a set of customer locations. It is assumed that customers demand for a product and the shipment of that product must take place in the same period.

The model determines the quantities transported between sites, and assigns a set of vessel types to a set of routes that have to be satisfied demand at minimal cost while satisfying all operational requirements. In order to solve the complicated problems, the formulation of the problem is of vital importance. A huge amount of product is transported and so the savings generated by careful planning of transportation can be significant. The reader is referred to reference [3] for the detailed discussion on the problem and the model.

## 3    Model Formulation

Before presenting a MIP for the blending and shipment planning problem described in section 2, we first introduce the notation that will be used throughout the paper. Using this notation, a mathematical programming model can be formulated to solve the problem studied in this paper. Table 1 defines the notation for the shipment planning model.

**Table 1.** Notations

| Notation | Remark |
|---|---|
| **Index Sets** | |
| $J$ | set of original products $\{1, 2, \ldots, J\}$ |
| $I$ | set of blended products $\{1, 2, \ldots, I\}$ |
| $K$ | set of ports $\{1, 2, \ldots K\}$ |
| $L$ | set of customers $\{1, 2, \ldots, L\}$ |
| $V$ | set of vessel types $\{1, 2, \ldots, V\}$ |
| $T$ | set of time periods $\{1, 2, \ldots, T\}$ |
| **Input Parameters** | |
| $D_{ilt}$ | demand for product $i$ by customer $l$, in time period $t$. |
| $\text{Supply}_{jk}$ | the amount of original products $j$ supplied to port $k$. |
| $\text{BCap}_k$ | maximum blending capacity at port k (MT) |
| $\text{LCap}_k$ | maximum loading capacity at port $k$ (MT) |
| $\text{FCap}_v$ | maximum vessel capacity for each type $v$ (MT) |
| $\text{DCapp}_{kv}$ | draft capacity for vessel type $v$ at port $k$ (MT) |
| $\text{DCapc}_{lv}$ | draft capacity for vessel type $v$ at destination port $l$, where customers are located (MT) |
| $\text{TCost}_{kqlv}$ | total transportation cost for route from port $k$ to $q$ to customer $l$ on vessel type $v$ ($/Vessel) |
| $\text{LCost}_k$ | loading cost at port k ($/MT) |
| $\text{BCost}_k$ | Blending cost at port $k$ ($/MT) |
| $\text{MinBlnd}_{ij}$ | minimum ratio of original product $j$ in blended product $i$ |
| $\text{MaxBlnd}_{ij}$ | maximum ratio of original product $j$ in blended product $i$ |
| **Decision Variables** | |
| $x_{ikqlvt}^p$ | the amount of product $i$ loaded at port $p$ (either $k$ or $q$) on vessel type $v$ on route $k - q - l$, in time period $t$. |
| $n_{kqvlt}$ | the number of vessels of type $v$ on route $k - q - l$, in time period $t$. |
| $w_{ijkt}$ | the amount of original product $j$ used to make blended product $i$ at port $k$, in time period $t$. |
| $b_{ikt}$ | The amount of product $i$ blended at port $k$, in time period $t$. |
| $I_{jkt}$ | the amount of remaining stocks of original product $j$, at port $k$, at the end of period $t$. |
| $IB_{ikt}$ | the amount of remaining stocks of blended product $i$, at port $k$, at the end of period $t$. |

The MIP formulation of our model is as follows:
Minimize

$$\sum_{i,k,t} \text{BCost}_k b_{ikt} + \sum_{p \in \{k,q\}} \sum_{i,k,q,l,v,t} \text{LCost}_p x_{ikqlvt}^p + \sum_{k,q,l,v,t} \text{TCost}_{kqlv} n_{kqlvt} \qquad (1)$$

subject to

$$\sum_j w_{ijkt} = b_{ikt} \qquad \forall i, j, k, t, \qquad (2)$$

$$\begin{aligned} w_{ijkt} - \text{MaxBlnd}_{ij} b_{ikt} \leq 0 \qquad \forall i, j, k, t, \\ w_{ijkt} - \text{MinBlnd}_{ij} b_{ikt} \geq 0 \qquad \forall i, j, k, t, \end{aligned} \qquad (3)$$

$$\sum_i b_{ikt} \leq \text{BCap}_k \qquad \forall k, t, \qquad (4)$$

$$\text{Supply}_{jk} = I_{jk0} \qquad \forall j, k, \qquad (5)$$

$$\sum_{q,l,v} x_{ikqlvt}^k + \sum_{q,l,v} x_{iqklvt}^k + \sum_{i \in I} w_{ijkt} - I_{j,k,t-1} + I_{j,k,t} = 0 \qquad \forall i, j \in J, \quad \forall k, t, \quad (6)$$

$$\mathrm{IB}_{ik0} = 0 \qquad \forall i, k, \tag{7}$$

$$\sum_{q,l,v} \mathrm{x}^k_{ikqlvt} + \sum_{q,l,v} \mathrm{x}^k_{iqklvt} - \mathrm{IB}_{i,k,t-1} + \mathrm{IB}_{i,k,t} - \mathrm{b}_{ikt} = 0 \qquad \forall i, k, t, \tag{8}$$

$$\sum_{i \in \{I,J\}} \sum_{q,l,v} \mathrm{x}^k_{ikqlvt} + \sum_{i \in \{I,J\}} \sum_{q,l,v} \mathrm{x}^k_{iqklvt} \le \mathrm{LCap}_k \qquad \forall k, t, \tag{9}$$

$$\sum_{i \in \{I,J\}} \mathrm{x}^k_{ikqlvt} - \mathrm{DCap}_{kv} \mathrm{n}_{kqlvt} \le 0 \qquad \forall k, q, l, v, t, \tag{10}$$

$$\sum_{i \in \{I,J\}} \mathrm{x}^k_{ikqlvt} + \sum_{i \in \{I,J\}} \mathrm{x}^q_{ikqlvt} - \mathrm{DCap}_{qv} \mathrm{n}_{kqlvt} \le 0 \qquad \forall k, q, l, v, t, \tag{11}$$

$$\sum_{p \in \{k,q\}} \sum_{i \in \{I,J\}} \sum_{k,q} \mathrm{x}^p_{ikqlvt} - \sum_{k,q} \mathrm{DCap}_{lv} \mathrm{n}_{kqlvt} \le 0 \qquad \forall l, v, t, \tag{12}$$

$$\sum_{i \in \{I,J\}} \sum_{p \in \{k,q\}} \mathrm{x}^p_{ikqlvt} - \mathrm{FCap}_v \mathrm{n}_{kqlvt} \le 0 \qquad \forall k, q, l, v, t, \tag{13}$$

$$\sum_{p \in \{k,q\}} \sum_{k,q,v} = \mathrm{D}_{ilt} \qquad \forall i, l, t. \tag{14}$$

In the above formulation, the objective function of the problem seeks to minimize the total cost comprised of the blending, loading, and marine transportation costs. Constraint sets (2) and (3) are required to meet the blended product specifications. Constraints (4) ensure that the amount of blended products must be smaller than the blending capacity at each port. Constraints (5) guarantee that the initial amount of original product at each port is equal to the inventory for that original product at each port, at the end of the time period zero. Constraint sets (6) and (8) assure the availability of the original and blended products at each port, and in each time period, respectively. Constraints (7) ensure that there is no initial amount of blended products at ports at the end of time period 0. Constraints in set (9) assure that the amount of products that are loaded from the ports can not exceed the loading capacity for each port, and in each time period. Constraints (10), (11) and (12) assure the draft capacity for the first, second loading port and destination port and vessel type, respectively. Constraints (13) enforce the vessel capacity restriction. Finally, demand constraints are imposed by equations (14).

## 4    An Iterative Fixing Variable Heuristic Algorithm

In this section a solution procedure is developed that employs heuristic rules in conjunction with branch and bound methods. The size of the problem instance increases as with an increase in the number of ports, number of customers, vessel types, and products. Hence it is impractical to obtain exact solution for the model in a reasonable computation time. For large-sized test problems, we encountered out-of-memory difficulties before reaching any solution that was within 1% optimality. We have thus developed a heuristic algorithm to obtain near optimal solution in short computation time. Decomposition algorithms and heuristic algorithms are also of considerable computational interest, since they offer a possibility of avoiding the large amount of time needed to solve large models to be structured.

The out of memory difficulties associated with solving the model primarily stems from the large number of integer variables. This motivates the development of a heuristic procedure that generates a solution in an iterative fashion, in which each iteration involves a modified version of the model, where the integrality of only a subset of the integer variables is enforced. This heuristic operated by finding that integer variable n with the largest fractional value and then setting this to nearest integer. The problem was then resolved, and the process repeated. Data derived from an industrial operation was used to assemble a suit of ten problems of different sizes. The characteristics of the various test problems are provided in Table 2. We have generated these problems to test the model and efficiency of the solution procedures.

**Table 2.** The multi-period blending and shipping problems tested

| Problem No | Number of products $(j + i)$ | Number of ports $(k)$ | Number of customers $(l)$ | Number of vessel type $(v)$ | Number of viable routes $(k - q - l)$ | Number of variables | Number of constraints |
|---|---|---|---|---|---|---|---|
| 1 | 4 + 2 | 4 | 2 | 2 | 26 | 324 | 385 |
| 2 | 4 + 2 | 4 | 3 | 2 | 39 | 486 | 539 |
| 3 | 4 + 2 | 4 | 2 | 3 | 22 | 492 | 570 |
| 4 | 5 + 2 | 5 | 2 | 3 | 36 | 598 | 585 |
| 5 | 5 + 2 | 5 | 3 | 3 | 54 | 1006 | 1015 |
| 6 | 5 + 2 | 5 | 3 | 4 | 54 | 1243 | 1304 |
| 7 | 6 + 2 | 6 | 3 | 4 | 72 | 1604 | 1705 |
| 8 | 6 + 2 | 6 | 4 | 4 | 96 | 1725 | 1878 |
| 9 | 6 + 2 | 6 | 5 | 4 | 114 | 1504 | 1720 |
| 10 | 6 + 2 | 7 | 5 | 5 | 160 | 3085 | 3148 |

We have tested two approaches to solve the problem. The first approach is to use the integer programming solver CPLEX 8.0 [8] directly, and the second approach is to use the fixing variable heuristic. Below, we present the heuristic solution procedure that is designed to find good quality feasible solutions in practice.

The main steps of the heuristic algorithm can be described as follows:

1. Solve linear relaxation of the problem.
2. If there are no variables with fractional values in the solution, then stop.
3. Select a fractional variable according to the following criterion:
   i. Select the largest fractional variable with the earliest time period and minimum transportation cost.

4. Set the value of the selected fractional variable to nearest integer value. And, go to step 1.

This sequential procedure is repeated until all relaxed "$n$" values have been obtained integer values. The proposed heuristic does not guarantee feasible solutions to any set of data. Infeasibilities might occur, because the demand is not satisfied or certain capacity constraints are violated. To ensure feasibility, the smallest-sized vessel type is assigned to the route and time period with very small fractional values.

This procedure can be interpreted as a depth first search in a restricted branch and bound scheme for integer programming. A similar heuristic approach can be found in the papers by Barnhart et al. [1], Gunnarson et al. [7], Liu and Sherali [10], in which variable fixing heuristic is suggested for solving different industrial problems.

The proposed sequential fixing heuristic procedure is implemented with OPL Script [9] linked to the solver CPLEX [8] on a notebook with a Pentium IV 1.4 GHz processor having 1 GB of RAM. No additional parameter settings are used in CPLEX. The MIP problems were solved with a 1% optimality tolerance (the relative difference between the best known bound and the best known integer solution at which the search for better integer solutions stops).

## 5    Computational Results

In this section we present the results and analysis of the computational experiments for a real life shipment planning problem. The test cases we have used covers a period of three months. To validate the efficiency of the presented heuristic, a series of computational experiments were carried out. During each loop of the heuristic, integer restrictions on $n$ variables are enforced. Table 3 compares the heuristic with the mathematical model on different test problems with respect to objective value, computation time, and the number of iterations. The error rate of the heuristic to the optimal solution, that is computed by 100*(total cost obtained by CPLEX MIP solver- total cost obtained by the heuristic)/ total cost obtained by CPLEX MIP solver, remained relatively low, ranging 4~4.9 for the small-sized problems, 9.8~11.7 for the medium/large-sized problems.

This heuristic enables us to retain the quality of the solution obtained by direct solution of the problem to the extend possible and to handle larger problem instances that are intractable for the direct approach. The computational experiments described in this section were designed to evaluate the performance of the heuristic algorithm with respect to a series of test problems. Several aspects of the heuristic algorithm are worth discussing. The following facts may be summarized based on computational results. Computational results on small-sized problems indicate that the heuristic performed well in terms of both optimally approximation and computation time. For instance, for problem 5 the heuristic executes 16 loops and consumes 2454 simplex iterations and 7,13 CPU (s) to solve the problem, while the CPLEX-MIP package enumerates 591067 branch and bound nodes and performs 6967882 simplex iterations over 1281,45 CPU s.

**Table 3.** Comparison of solutions and computational effort for the OPL-CPLEX MIP package and the heuristic procedure, respectively

| Test Problems | CPLEX-MIP | | | | Heuristic Algorithm | | | | |
|---|---|---|---|---|---|---|---|---|---|
| | Total number of branch and bound nodes | Total number of simplex iterations | CPU time (s) | Objective Function Values | Total number of loops solved | Total number of simplex iterations | CPU time (s) | Objective Function Values | Error Rate |
| 1 | 2180 | 10878 | 1,49 | 11686733,04 | 28 | 1512 | 3,36 | 12161931,07 | 0,04 |
| 2 | 7323 | 55576 | 8,55 | 20994649,18 | 22 | 1980 | 3,96 | 22301567,64 | 0,05 |
| 3 | 365 | 1989 | 1,02 | 14306372,57 | 20 | 2150 | 3,8 | 14875136,22 | 0,04 |
| 4 | 58621 | 799550 | 125,56 | 20332456,46 | 38 | 3762 | 9,12 | 21244818,15 | 0,048 |
| 5 | 591067 | 6967882 | 1281,45 | 19901871,32 | 16 | 2454 | 7,13 | 20870631,99 | 0,049 |
| 6 | 154360 | 3783814 | 1288,13 | 18412671,32 | 28 | 4807 | 19,54 | 20207614,99 | 0,098 |
| 7 | 662780 | 11291917 | 2431,45 | 19770008,59 | 22 | 6810 | 22,28 | 21898936,66 | 0,108 |
| 8 | 624383 | 11258152 | 2730,93 | 23196357,83 | 47 | 11410 | 60,04 | 25764271,15 | 0,117 |
| 9 | 677544 | 11000400 | 3882,20 | 26148451,54 | 17 | 4848 | 14,61 | 28689682,36 | 0,098 |
| 10 | 1184959 | 12766068 | 8308,20 | out-of-memory | 81 | 19750 | 230,93 | 27157640,70 | - |

The error rate is 4.9% for this test problem. The error rates would be higher on larger problems. However, the magnitude may be acceptable when we want to obtain a good solution quickly for complex planning problems that can not be easily solved in an optimal way. Conclusions about heuristics performance are drawn by testing the algorithm on a wide collection of problem instances. Although the results are not very close to optimality for larger problems, our problem is a hard constrained real life problem.

While obtaining optimal solutions is desirable, deriving high quality solution quickly is essential for any practical application. As can be seen from the foregoing discussion, this section outlined the application of this heuristic procedure on a real life case along with presentation of the effectiveness of the heuristic algorithm.

## 6   Conclusion

In this paper, we developed a MILP model and a fixing variable based heuristic algorithm for the real shipment planning problem. An iterative variable fixing heuristic designed to quickly find an integer solution and to enhance the solvability of the proposed model. This heuristic operated by finding that integer value with the largest fractional value and then setting this to the nearest integer effectively fixing the number of vessels for some period. The problem was then resolved and the process repeated. This heuristic is similar in spirit to the constraint branching heuristic developed by Barnhart [1]. Computational results indicate that the heuristic procedure requires for far fewer total simplex iterations. The heuristic approach and CPLEX provide comparable solutions. A few words are now in order regarding recommendations for future work. Modification of the heuristic algorithm would be needed to provide better performance (reduce the computation time), improve bounds for the algorithm, and reduce further the error rate of the heuristic procedure to be able to obtain the optimal solution. It appears that more sophisticated techniques, e.g. decomposition techniques could be used to circumvent the memory problem encountered with the branch and bound algorithm and solve very large instances of the shipping problem.

# References

1. Barnhart C., Johnson E. L., Nemhauser G.L., Savelsbergh M.W.P., Vance, P.H.: Branch-and-price: Column generation for solving huge integer programs **46** (1998) 316–332
2. Bilgen, B., Ozkarahan, I: Strategic, tactical and operational production and distribution models: A review. International Journal of Technology Management (Special issue on Supply Chain Management: Integration Strategies based on New Technologies) **28** (2004) 151–171
3. Bilgen, B., Ozkarahan, I.: A mixed-integer programming model for bulk grain blending and shipping. International Journal of Production Economics, accepted for publication.
4. Bredström, D., Lundgren, J.T., Rönnqvist, M., Carlsson, D., Mason, A.: Supply chain optimization in the pulp mill industry-IP models, column generation and novel constraint branches. European Journal of Operational Research **156** (2004) 2–22.
5. Christiansen, M., Fagerholt, K., Ronen, D.: Ship routing and scheduling: Status and perspectives. Transportation Science **38** (2004) 1–18
6. Fleischmann, B.: Distribution and transport planning. In: H. Stadler, C. Kilger, (eds.): Supply Chain Management and Advanced Planning. Springer-Verlag, Berlin (2002) 167–181
7. Gunnarsson, H., Rönnqvist, M., Lundgren, J.T.: Supply chain modeling of forest fuel. European Journal of Operational Research **158** (2004) 103–123
8. ILOG CPLEX 8.0.: Users Manual. Gentilly, France: (2003) ILOG SA
9. ILOG OPL Studio 3.7.: Language Manual. Gentilly, France: (2003), ILOG SA
10. Liu, C.M., Sherali, H.D.: A coal shipping and blending problem for an electric utility company. Omega **28** (2000) 433–444
11. Mehrez, A., Hung, M.S., Ahn, B.H.: An industrial ocean-cargo shipping problem. Decision Sciences **26** (1995) 395–423
12. Persson Jan A., Göthe-Lundgren, M.: Shipment planning at oil refineries using column generation and valid inequalities. European Journal of Operational Research **163** (2005) 631–652

# Hybrid Heuristic Algorithm for GPS Surveying Problem

Stefka Fidanova

IPP – BAS, Acad. G. Bonchev str. bl.25A, 1113 Sofia, Bulgaria
stefka@parallel.bas.bg

**Abstract.** This paper introduces several approaches based on ant colony optimization for efficient scheduling the surveying activities of designing satellite surveying networks. These proposed approaches use a set of agents called ants that cooperate to iteratively construct potential observation schedules. Within the context of satellite surveying, a positioning network can be defined as a set of points which are coordinated by placing receivers on these point to determine sessions between them. The problem is to search for the best order in which these sessions can be observed to give the best possible schedule. The same problem arise in Mobile Phone Surveying networks. Several case studies have been used to experimentally assess the performance of the proposed approaches in terms of solution quality and computational effort.

## 1 Introduction

The continuing research on naturally occurring social systems offers the prospect of creating artificial systems that generate practical solutions to many Combinatorial Optimization Problems (COPs). Metaheuristic techniques have evolved rapidly in an attempt to find good solutions to these problems within a desired time frame [6]. They attempt to solve complex optimization problems by incorporating processes which are observed at work in real life [2,3]. For example, in the case of ants, using their simple individual interactions mediated by pheromone, they can collectively determine the shortest route from their nest to a food source without using visual cues [1]. When applied to satellite surveying, these techniques can assist surveyors in creating a better observation schedule for designing the whole positioning network.

The purpose of surveying is to determine the locations of points on the earth. Measuring tapes or chains require that the survey crew physically pass through all the intervening terrain to measure the distance between two points. Surveying methods have undergone a revolutionary change over that last few years with the deployment of the satellite navigation systems. The most widely known space systems are: the American Global Positioning System (GPS), the Russian GLObal Navigation Satellite System (GLONASS), and the forthcoming European Satellite Navigation System (GALILEO). In this paper, it is the use of GPS to establish surveying networks that is being investigated. GPS satellites continuously transmit electrical signals to the earth while orbiting the earth. A receiver,

T. Boyanov et al. (Eds.): NMA 2006, LNCS 4310, pp. 239–246, 2007.

with unknown position on the earth, has to detect and convert the signals transmitted from all of the satellites into useful measurements. These measurements would allow a user to compute a three-dimensional coordinates position for the location of the receiver. To maximize the benefit of using this technique, several procedures based on Ant Colony Optimization, have been developed and implemented to find an efficient schedule to improve the performance and explore the search space more effectively [8]. In this paper we implement an Ant Colony Optimization (ACO) algorithm applied to GPS surveying networks and the general case of the problem is addressed by presenting several case studies and the obtained numerical results. Our aim is to suggest various local search procedures, to combine them with ACO technique and to check which one is the best for this problem.

The rest of this paper is organized as follows. The general framework for GPS surveying network problem as a combinatorial optimization problem is described in Section 2. Then, the search strategy of the ACO is explained in Section 3. Various local search procedures are described in Section 4. The ACO algorithm coupled with local search procedures, applied to GPS surveying networks, is outlined and the general case of the problem is addressed by presenting several case studies and the obtained numerical results in Section 5. The paper ends with a summary of the conclusions and directions for future research.

## 2    Formulating the GPS Surveying Network Problem

A GPS network is distinctly different from a classical survey network in that no inter-visibility between stations is required. In GPS surveying, after defining the locations of the points for an area to be surveyed, GPS receivers will be used to map this area by creating a network of these coordinated points. These points, control stations within the context of surveying, are fixed on the ground and located by an experienced surveyor according to the nature of the land and the requirements of the survey [5]. At least two receivers are required to simultaneously observe GPS satellites, for a fixed period of time, where each receiver is mounted on each station. The immediate outcome of the observation is a session between these two stations. After completing the first stage of sessions observation and defining the locations of the surveyed stations, the receivers are moved to other stations for similar tasks till the whole network is completely observed according to an observation schedule. The total cost of carrying out the above survey, which is computed upon the criteria to be minimized, represents the cost of moving receivers between stations. The problem is to search for the best order, with respect to the time, in which these sessions can be observed to give the cheapest schedule V, i.e.:

$$Minimize: \ C(V) = \sum_{p \in R} C(S_p)$$

where: $C(V)$ is the total cost of a feasible schedule $V(N, R, U)$;

$S_p$ is the route of the receiver $p$ in a schedule;
$N$ is the set of stations $N = 1, \ldots, n$;
$R$ is the set of receivers $R = 1, \ldots, r$;
$U$ is the set of sessions $U = 1, \ldots, u$.

# 3   Ant Colony Optimization

Real ants foraging for food lay down quantities of pheromone (chemical cues) marking the path that they follow. An isolated ant moves essentially at random but an ant encountering a previously laid pheromone will detect it and decide to follow it with high probability and thereby reinforce it with a further quantity of pheromone. The repetition of the above mechanism represents the auto-catalytic behavior of real ant colony where the more the ants follow a trail, the more attractive that trail becomes.

The ACO algorithm uses a colony of artificial ants that behave as cooperative agents in a mathematics space were they are allowed to search and reinforce pathways (solutions) in order to find the optimal ones. The problem is represented by graph and the ants walk on the graph to construct solutions. After initialization of the pheromone trails, ants construct feasible solutions, starting from random nodes, then the pheromone trails are updated. At each step ants compute a set of feasible moves and select the best one (according to some probabilistic rules) to carry out the rest of the tour. The transition probability is based on the heuristic information and pheromone trail level of the move. The higher value of the pheromone and the heuristic information, the more profitable is to select this move and resume the search. In the beginning, the initial pheromone level is set to a small positive constant value $\tau_0$ and then ants update this value after completing the construction stage. ACO algorithms adopt different criteria to update the pheromone level. In our implementation we use Ant Colony System (ACS) [4] approach.

In ACS the pheromone updating consists of two stages: local update stage and global update stage.

## 3.1   Local Update Stage

While ants build their solution, at the same time they locally update the pheromone level of the visited paths by applying the local update rule as follows:

$$\tau_{ij} \leftarrow (1 - \rho)\tau_{ij} + \rho\tau_0 \tag{1}$$

Where $\tau_{ij}$ is an amount of the pheromone on the arc $(i, j)$ of the graph of the problem, $\rho$ is a persistence of the trail and term $(1 - \rho)$ can be interpreted as trail evaporation.

The aim of the local updating rule is to make better use of the pheromone information by dynamically changing the desirability of edges. Using this rule, ants will search in a wide neighborhood of the best previous solution. As shown in the formula, the pheromone level on the paths is highly related to the value of evaporation parameter $\rho$. The pheromone level will be reduced and this will reduce the chance that the other ants will select the same solution and consequently the search will be more diversified.

## 3.2   Global Updating Stage

When all ants have completed their solution, the pheromone level is updated by applying the global updating rule only on the paths that belong to the best solution since the beginning of the trial as follows:

$$\tau_{ij} \leftarrow (1 - \rho)\tau_{ij} + \Delta\tau_{ij} \tag{2}$$

$$\text{where } \Delta\tau_{ij} = \begin{cases} 1/C_{gb} \text{ if } (i,j) \in \text{best solution} \\ \\ 0 \qquad \text{otherwise} \end{cases},$$

$C_{gb}$ is the cost of the best solution from the beginning. This global updating rule is intended to provide a greater amount of pheromone on the paths of the best solution, thus intensify the search around this solution. The transition probability to select the next node is given as:

$$prob_{ij}^{k}(t) = \begin{cases} \dfrac{\tau_{ij}^{\alpha}\eta_{ij}^{\beta}}{\sum_{s \in allowed_k(t)} \tau_{is}^{\alpha}\eta_{is}^{\beta}} \text{ if } j \in allowed_k(t) \\ \\ 0 \qquad\qquad\qquad \text{otherwise} \end{cases} \tag{3}$$

where $\tau_{ij}$ is the intensity measure of the pheromone deposited by each ant on the path $(i,j)$, $\alpha$ the intensity control parameter, $\eta_{ij}$ is the visibility measure of the quality of the path $(i,j)$. This visibility is determined by $\eta_{ij} = 1/l_{ij}$, where $l_{ij}$ is the cost of move from session $i$ to session $j$. $\beta$ is the visibility parameter and $allowed_k(t)$ is the set of remaining feasible sessions. Thus the higher the value of $\tau_{ij}$ and $\eta_{ij}$, the more profitable it is to include item $j$ in the partial solution.

## 4   Local Search Strategy

The Local Search (LS) method (move-by-move method) perturbs a given solution to generate different neighborhoods using a move generation mechanism. LS attempts to improve an initial solution by a series of local improving changes. A move-generation is a transition from a solution $S$ to another one $S' \in V(S)$ in one step. These solutions are selected and accepted according to some pre-defined criteria. The returned solution $S'$ may not be optimal, but it is the best solution in its local neighborhood $V(S)$. A local optimal solution is a solution with the local minimal possible cost value. Knowledge of a solution space is the essential key to more efficient search strategies. These strategies are designed to use this prior knowledge and to overcome the complexity of an exhaustive search by organizing searches across the alternatives offered by a particular representation

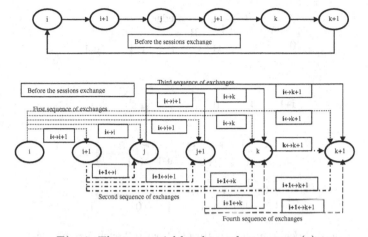

**Fig. 1.** The sequential local search structure (a)

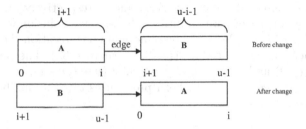

**Fig. 2.** Local search structure (b1) with 1-exchange of edges

of the solution. The main purpose of LS implementation is to speed up and improve the solutions constructed by the ACO.

In this paper, various local search structures that satisfy the requirements of the GPS network, have been developed and implemented to search the schedule space more effectively. The problem is represented by graph. The nodes correspond to the sessions and the edges represents the distance. A solution is a path in a graph (chain of nodes) which contains all nodes.

The first procedure [8,9], shown in a Figure 1, is based on the sequential session-interchange, the potential pair-swaps are examined in the order $\{1,2\}$, $\{1,3\},\ldots,\{1,u\},\{2,3\},\{2,4\},\{u-1,u\}$ (For comparison reasons this sequential local search procedure is called structure a).

In the second procedure as shown graphically in Figure 2, $A$ and $B$ are chains of nodes (solution slices), while $\{0,i\}$ and $\{i+1,u\}$ are the first and the last nodes of the chains $A$ and $B$ respectively. In this procedure, only one exchange has been performed of new edge $(u-1,0)$ each iteration and this can be done by selecting one edge $(i,i+1)$ to be removed (For comparison reasons this local search procedure is called structure b1).

In the third procedure as shown graphically in Figure 3, $A$, $B$ and $C$ are chains of nodes (part of the solution), while $\{0,i\}$, $\{i+1,j\}$, and $\{j+1,u-1\}$ are the

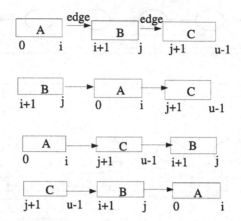

**Fig. 3.** Local search structure (b2) with 2-exchange of edges

first and the last nodes of the chains $A$, $B$ and $C$ respectively. In the case of a 2-exchange, there are several possibilities to build a new solution where the edges $(i, i+1)$ and $(j, j+1)$ are selected to be removed in one iteration (For comparison reasons this local search procedure is called structure b2). The possibilities for new edges are $(j, 0)$ and $(i, j+1)$, $(i, j+1)$ and $(u-1, i+1)$, $(u-1, i+1)$ and $(j, 0)$ where $0 \leq i < j \leq u - 1$. In both procedures $u$ is the number of sessions in the observed network.

## 5   Experimental Results

This section reports on the computational experience of the ACS coupled with various local search procedures using real GPS networks. The first network is a GPS network for Malta and consists of 38 sessions [8]. The initial schedule with a cost of 1405 minutes was composed. The second network is a GPS network for the Seychelles and consists of 71 sessions [9]. The initial schedule with a cost of 994 minutes was composed. The performance of the developed metaheuristic techniques were evaluated with respect to the schedule quality and computational effort using the following measure:

$$RRC = [(C_{INT} - C(V))/C_{INT}] * 100$$

Where:
   $RRC$ is the Relative Reduction of the Cost;
   $C_{INT}$ is the Cost of the Initial Schedule $V_{INT}$ obtained randomly.
   $C_a$ is the Cost of the metaheuristic schedule obtained by procedure $a$.
   $C_{b1}$ is the Cost of the metaheuristic schedule obtained by procedure $b1$.
   $C_{b2}$ is the Cost of the metaheuristic schedule obtained by procedure $b2$.
   $TS$ is the Cost of the metaheuristic schedule in [8,9] obtained by tabu search.
   $SA$ is the Cost of the metaheuristic schedule in [8,9] obtained by simulated annealing.

**Table 1.** Comparison of local search techniques applied to different types of GPS networks

| Data | Malta network | $RRC$ | Seychelles network | $RRC$ |
|---|---|---|---|---|
| $U$ | 38 | | 71 | |
| $C_{INT}$ | 1405 | 0 | 994 | 0 |
| $C_a$ | 880 | 37.37 | 859 | 13.58 |
| $C_{b1}$ | 685 | 51.24 | 807 | 18.81 |
| $C_{b2}$ | 850 | 39.50 | 871 | 12.37 |
| $TS$ | 1075 | 23.5 | 933 | 6.14 |
| $SA$ | 1375 | 2.14 | 969 | 2.52 |

The following parameter settings is used: $\rho = 0.4$, $\alpha = \beta = 1$. The initial amount of the pheromone was set to a fixed value $\tau_0 = 0.0005$ on all edges, while the number of iterations was set to 200. The reported results are average results over 20 runs. The developed technique has been coded in C++ and implemented on a Pentium 3 with 900MHz speed and 512 MB RAM.

The Table 1 shows the advantage of ACO techniques using various local search procedures. ACO technique using local search procedure (b1) produced the best results. With respect to the Malta network with size of 38 sessions, the RRC of the metaheuristic schedule produced by the ACO technique is 37.37% using local search procedure (a), compared to RRC of 51.24% using local search procedure (b1), RRC of 39.50% using local search procedure (b2). In the Seychelles network with size of 71 sessions, the RRC of the metaheuristic schedule produced by the ACO is 13.58% using local search procedure (a), compared to RRC of 18.81% using local search procedure (b1), RRC of 12.37% using local search procedure (b2). In this comparison, local search procedure (b1) gives better results than the other local search procedures for the both GPS networks. ACO is a constructive method and the selected set of neighbors has not so important role in achieving good results. Having better results using the procedure (b1) is due its features: more economic and less perturbs the pheromone. All ACO algorithm outperforms TS and SA.

## 6   Concluding Remarks and Future Work

In this paper, two local search procedures have been developed and compared with local search procedure from [8] and [9] (structure a). The comparison of the performance of the ACO with these procedures applied to different types and sizes of GPS networks is reported. The obtained results are encouraging and the ability of the developed techniques to generate rapidly high-quality solutions for observing GPS networks can be seen. For the future work, the developed local search procedures in this paper will be coupled with other metaheuristics to search for further improvement and provide a practical comparison, based on the success of the developed ACO algorithm for constructing optimal solutions for large GPS positioning networks which form the basic framework for geomatic information that support all the environmental activities in the countries.

**Acknowledgments.** The author is supported by European community grant BIS 21++.

# References

1. Beckers R., Deneubourg J.L., Gross S. (1992), *Trail and U-turns in the Selection of the Shortest Path by the Ants*, J. of Theoretical Biology 159,397–415.
2. Bonabeau E., Dorigo M., Theraulaz G. (1999),*Swarm Intelligence: From Natural to Artificial Systems*, Oxford University Press, New York.
3. Corne D., Dorigo M., Glover F. eds.(1999), *New Ideas in Optimization*, McCraw Hill, London.
4. Dorigo M., Gambardella L.M. (1997), *Ant Colony System: A Cooperative Learning Approach to the Traveling Salesman Problem*, IEEE Transactions on Evolutionary Computation 1, 53–66.
5. Leick, A. (1995): *GPS Satellite Surveying, 2nd. ed..* Wiley, Chichester, England.
6. Osman I.H., Kelly J.P. (1996): *Metaheuristics:An overview.* In Osman I.H., Kelly J.P. eds. Metaheuristics:Theory and Applications, Kluwer Acad. Publisher.
7. Reeves, C. R. (editor). (1993): *Modern Heuristic Techniques for Combinatorial Problems.* Blackwell Scientific Publications, Oxford, England.
8. Saleh H. A., and P. Dare,(2001): *Effective Heuristics for the GPS Survey Network of Malta: Simulated Annealing and Tabu Search Techniques.* Journal of Heuristics **N 7** (6), 533-549.
9. Saleh, H. A., and Dare, P. J. (2002): *Heuristic methods for designing Global Positioning System Surveying Network in the Republic of Seychelles.* The Arabian Journal for Science and Engineering **N 26**, 73-93.
10. Saleh, H. A. and Dare, P., (2000): *Local Search Strategy to produce Schedules for a GPS Surveying Network.* In: Eleventh Young Operational Research Conference (YOR11), University of Cambridge, Cambridge, UK, 28-30 March, (editor A. Tuson), 2000-ORS, UK, ISBN 0-903440-202, 87-103.

# A Hybrid Metaheuristic for a Real Life Vehicle Routing Problem

Panagiotis P. Repoussis*, Christos D. Tarantilis, and George Ioannou

Athens University of Economics & Business
prepousi@aueb.gr

**Abstract.** This paper presents a solution methodology to tackle a new realistic vehicle routing problem that incorporates heterogeneous fleet, multiple commodities and multiple vehicle compartments. The objective is to find minimum cost routes for a fleet of heterogeneous vehicles without violating capacity, loading and time window constraints. The solution methodology hybridizes in a reactive fashion systematic diversification mechanisms of Greedy Randomized Adaptive Search Procedures with Variable Neighborhood Search for intensification local search. Computational results reported justify the applicability of the methodology.

## 1  Introduction

The goal of this paper is to develop a solution method that addresses a realistic Vehicle Routing Problem (VRP) which, according to our knowledge, has not appeared in literature before. The problem can be stated as follows: a heterogeneous fixed fleet of depot returning vehicles with multiple storage compartments, transport different commodities to a set of customers of known demand within an interval during which delivery has to take place. Each customer is serviced only once by exactly one vehicle. The objective is to design a minimum cost fleet mix using the available fleet composition following routes of minimum distance, such that vehicle capacity, loading and time window constraints are satisfied.

The underlying mathematical problem can be accurately described as follows: Let $G = (V, A)$ be a complete graph, where $V = \{0, 1, ..., n+1\}$ is the node set, $A = \{(i,j) : 0 \leq i, j \leq n, i \neq j\}$ is the arc set and the depot is represented by nodes 0 and $n + 1$. All feasible vehicle routes correspond to paths in $G$ that start from 0 and end at $n + 1$. A set $K$ of available heterogeneous vehicles, equipped with $|M|$ compartments of known capacity $C_m^k$ ($m = 1, 2, ..., |M|$ and $k = 1, 2, ..., |K|$). Each customer $i$ is associated with a known demand, $d_i^q$, of a predefined commodity $q$ ($q = 1, 2, ..., |Q|$). Furthermore, each customer poses a time window $[a_i, b_i]$ that models the earliest and latest time that the service of $i$ can take place. The service of each customer must start within the associated time window, while the vehicle must stop at the customer's location for $s_i$. In case of early arrival at the location of $i$, the vehicle is allowed to wait until $a_i$.

---

* Correspondence: Evelpidon 47A, 11362, Athens, Greece.

T. Boyanov et al. (Eds.): NMA 2006, LNCS 4310, pp. 247–254, 2007.

There is a nonnegative travel cost $c_{ij}^k$, a travel time $t_{ij}^k$ and a distance $h_{ij}^k$ associated with each arc $(i, j)$ of set $A$, with respect to the vehicle $k \in K$. Furthermore, a cost $y_k$ is relevant to the activation of vehicle $k \in K$. The total number of customers is $n$, i.e., $n = |V| - 2$. Indices $i$, $j$ and $u$ refer to customers and take values between 1 and $n$. A time window is also associated with nodes 0 and $n+1$, i.e., $[a_0, b_0]=[a_{n+1}, b_{n+1}]=[E, L]$, where $E$ and $L$ represent the earliest possible departure from the depot and the latest possible return. Feasible solutions exist only if $a_0 = E \leq \min_{i \in V \setminus \{0\}} b_i - t_{0i}$ and $b_{n+1} = L \geq \min_{i \in V \setminus \{0\}} a_i + s_i + t_{i0}$. Additionally, let flow binary variables $x_{ij}^k$ model the sequence in which vehicles visit customers ($x_{ij}^k$ equals 1 if $i$ precedes $j$ visited by vehicle $k$; 0 otherwise), and $w_{ik}$ specify the arrival time at $i$ when serviced by vehicle $k$. Finally, binary variables $f_{mq}^k$ couple vehicle $k$'s compartment $m$ with a commodity $q$.

Each route must satisfy fixed fleet, loading and time window constraints. Fixed fleet constraint state that $\sum_{k \in K} \sum_{j \in N} x_{0j}^k \leq |K|$. The latter bounds the number of vehicles that can be deployed. Time window constraints state that $a_i \sum_{j \in \Delta^+(i)} x_{ij}^k \leq w_{ik} \leq b_i \sum_{j \in \Delta^+(i)} x_{ij}^k$ for all $k \in K$ and $i \in N$, where $\Delta^+(i)$ denotes the set of nodes $j$ such that arc $(i, j) \in A$. Loading poses the following condition: although all compartments can hold all commodities, during one route each compartment $m$ must hold a single commodity $q$. The latter implies that a route $r$ is feasible only if $\sum_{i \in r} d_i^q (\sum_{j \in r} x_{ij}^k) = \sum_{m \in M} f_{mq}^k C_m^k$ and $\sum_{q \in Q} f_{mq}^k \leq 1$ for all $q \in Q$.

The objective is to determine the fleet of vehicles such that fixed costs and travel costs are minimized, or similarly to determine the optimal fleet composition with minimum overall distribution costs within the fixed fleet restrictions. This combined objective reflects the trade off between fixed vehicle activation cost and variable distribution cost. Given the above-defined variables and parameters, the objective function of the problem can be formulated as follows:

$$\min \sum_{k \in K} \sum_{(i,j) \in A} c_{ij}^k x_{ij}^k + \sum_{k \in K} y_k \sum_{j \in N} x_{0j}^k. \tag{1}$$

In the literature, problems where vehicles are equipped with two or more compartments are known as Multiple Compartment VRP. The use of multiple compartments is relevant when vehicles transport several commodities which must remain separate. The latter complicates matters since the loading (packing) of commodities with different volume into a finite number of compartments (bins) of different capacity, using the least number of compartments or maximizing the total free available capacity of a vehicle, is not always straightforward [1]. Other routing problem related to the one studied in this research are the VRP with Time Windows (VRPTW) [2], the Fleet Size and Mix VRP [3] and the Heterogeneous Fixed Fleet VRP [4]. Finally, interested readers may refer to [5] for the latest application of metaheuristics on combinatorial optimization problems.

## 2  Solution Methodology

A hybrid metaheuristic which employs Greedy Randomized Adaptive Search Procedures (GRASP)[6] for diversification and Variable Neighborhood Search (VNS)[7] for intensification local search, is proposed in this work. GRASP is an iterative multi-start method which combines greedy heuristics, randomization and local search. The GRASP greedy randomization mechanism is characterized by a dynamic constructive heuristic and randomization. Initially, a solution is constructed iteratively by adding a new element to the partial incomplete solution. All elements are ordered in a list, called restricted candidate list ($RCL$) composed of the $\lambda$ highest quality elements, with respect to an adaptive greedy function. The probabilistic component is determined by randomly choosing one of the best element in the list, but not necessary the top of the list. Subsequently, the feasible solution found at each iteration is subject to local search. The overall procedure is repeated until some termination conditions are met. Given that GRASP constructions inject a degree of diversification to the search process, the improvement phase may consist of a VNS that is tuned for intensification local search. Recent developments propose such hybrids where construction is followed by sophisticated local search and post optimization procedures [5].

Common to most metaheuristic is that the balance between diversification and intensification determines their effectiveness. This implies quick identification of regions with high quality solutions. The repetitive sampling mechanism of GRASP, it is controlled by the size of $RCL$. In our implementation $RCL$ is cardinality based; thus, it comprises $\lambda$ elements with the best incremental costs. Obviously, cases where $\lambda = 1$ corresponds to pure greedy construction, while $\lambda \gg 1$ is equivalent to random construction. Let $\dot{s}$ and $\ddot{s}$ denote two solutions produced by two sequential construction phase iterations. A possible approach to measure quantitatively their diversity is to measure their dissimilarity $D_{\dot{s}}^{\ddot{s}}$. The latter can be defined as the number of different arcs between $\dot{s}$ and $\ddot{s}$, i.e.:

$$D_{\dot{s}}^{\ddot{s}} = \sum_{(i,j)\in A} \xi_{ij}, \qquad (2)$$

where the binary variable $\xi_{ij}$ is equal to 1 if $(i,j)$ is an arc of both $\dot{s}$ and $\ddot{s}$; 0 otherwise. The larger the $\lambda$ the larger the distance $D_{\dot{s}}^{\ddot{s}}$, and thus the better the sampling of the solution space. On the other hand, while $\lambda$ tends to increase from 1, the worse is the quality of the solutions produced, and thus, the more the computational effort needed by the local search phase to improve the incumbent solution's quality. Thus, appropriate choice of the $\lambda$ value is critical.

Given these definitions, a probabilistic learning mechanism is proposed, as an extension of the basic memoryless GRASP procedure, called Reactive GRASP [8], in which the size of $RCL$ is self tuned, capturing both trends discussed above. Let a non fixed size $\lambda$ take values, at each iteration, from a discrete set such that $\Lambda = \{\lambda_1, \lambda_2, \ldots, \lambda_\nu\}$. The probabilities associated with the choice of each value are initially equal to $B_\tau = 1/|\Lambda|$, where $\tau = 1, 2, \ldots, \nu$. Moreover, let $\dot{s}$ and $\ddot{s}$ be two sequential incumbent solutions, $A_\tau$ be the average objective function value and $D_\tau$ the average dissimilarity of all solutions found using

$\lambda = \lambda_\tau$. All probabilities are reevaluated once, by taking $B_\tau = \beta_\tau / \sum_{v=1}^{\nu} \beta_v$, where $\beta_\tau = A_\tau / \ddot{s} + D_{\ddot{s}}^{\ddot{s}} / D_\tau$ for $\tau = 1, 2, \ldots, \nu$. The first component, $A_\tau / \ddot{s}$, expresses the ratio between the overall average values found so far, $A_\tau$, and the value of the current solution $\ddot{s}$ using $\lambda_\tau$. Obviously, $A_\tau / \ddot{s}$ will increase when better (on average) solution values are found. Similarly, the larger the distance between $\dot{s}$ and $\ddot{s}$ against the average distance $D_\tau$, the larger their ratio. Therefore, the value of $\beta_\tau$ will be larger for values of $\lambda_\tau$ leading to the best valued and most diversified solutions (on average). Larger values of $\beta_\tau$ correspond to more suitable values of $\lambda$. Thus, the probabilities of these more appropriate values will then increase when they are reevaluated. The above reactive approach of reducing progressively the set $\Lambda$, improves significantly robustness and solution quality, due to greater diversification and less reliance on parameter tuning. Below, the proposed solution framework is provided as pseudocode. The termination condition bounds the allowed computational time to an upper limit.

**Reactive GRASP-VNS**
$\Lambda \leftarrow$InitializeSet($|N|/10$), $index \leftarrow 1$
**For** all elements $\lambda_i \in \Lambda$ **Do** Initialize $(D_i, A_i, B_i)$
**While** termination conditions not met do $\lambda \leftarrow \lambda_{index}$, $s \leftarrow \emptyset$
    **While** solution not complete **do**
        $RCL_{\lambda_{index}} \leftarrow$ Build Restricted Candidate List$(s)$
        $x \leftarrow$ Select Element At Random$(RCL_{\lambda_{index}})$
        $s \leftarrow s \cup \{x\}$, Update Greedy Function$(s)$
    **End while**
    VNS$(s)$, UpdateBestSolution$(s, elite)$, Reevaluate$(D_{index}, A_{index}, B_{index})$
    **If** $index = |\Lambda|$ AND $|\Lambda| > 1$ **Do**
        Remove $\lambda_i$ with the smallest $B_i$ from set $\Lambda$, $index \leftarrow 0$
    **Else** $index \leftarrow index + 1$
**Endwhile**

## 2.1 Construction Heuristic

The proposed construction phase adopts the generic parallel construction framework presented in [9] enhanced with additional customer selection criteria. Let $\pi_{ij,u}$ denote the insertion cost of an unassigned customer $u$ when inserted between $i$ and $j$ in a partial solution $\Omega$. For every feasible insertion position of $u$ into a route $\rho$, the minimum insertion cost $\pi_{\rho,u} = \min_{i,j \in \rho} \pi_{ij,u}$ is found. Similarly, the overall minimum insertion cost $\pi_{\rho^*,u}$ corresponds to the $\min_{\rho \in \Omega} \pi_{\rho,u}$ and denotes the best feasible insertion position at route $\rho^*$ of $u$. Subsequently, a penalty cost, $\Pi_u$ is calculated for every unassigned customer. This penalty can be viewed as a measure of the cost that would have to be paid later if the corresponding customer is not assigned to its current best position.

$$\Pi_u = \sum_{\rho \in \Omega} (\pi_{\rho,u} - \pi_{\rho^*,u}) . \tag{3}$$

Large $\Pi_u$ values indicate that $u$ should be considered first to avoid paying later a relatively high cost, while the insertion of customers with small penalty values can be delayed. Thus, for customers that cannot be feasibly inserted into a route, the insertion cost must be set to a large value $lv$, in order to force $\Pi_u$ to large values as well. In [9] $lv$ was set to infinity. Alternatively, we propose an intuitively intelligent approach that adaptively tunes the values of $lv$. In particular, $lv$ is set equal to the difference between the overall maximum and minimum insertion cost of all unassigned customers $u$ for the existing set of routes. Thereafter, whenever a customer $u$ cannot be feasibly inserted into a route $\rho$, the current penalty $\Pi_u$ is incremented by $max_{u \in V}\{max_{\rho \in \Omega}\{\pi_{\rho,u}\}\} - min_{u \in V}\{min_{\rho \in \Omega}\{\pi_{\rho,u}\}\}$. Finally, cost $\pi_{ij,u}$ is defined as a weighted combined result from several sub-metrics, i.e.:

$$\pi_{ij,u} = \vartheta_1 \pi_{ij,u}^1 + \vartheta_2 \pi_{ij,u}^2 + \vartheta_3 \pi_{ij,u}^3 + \vartheta_4 \pi_{ij,k}^4 , \qquad (4)$$

where $\vartheta_1, \vartheta_2, \vartheta_3$ and $\vartheta_4$ are nonnegative weights such that $\vartheta_1 + \vartheta_2 + \vartheta_3 + \vartheta_4 = 1$. Component $\pi_{ij,u}^1$ measures the distance increase caused by the insertion of $u$ [10]:

$$\pi_{ij,u}^1 = t_{iu} + t_{uj} - t_{ij} . \qquad (5)$$

Component $\pi_{ij,u}^2$ measures the time delay related to the vehicle arrival at $j$ [10]:

$$\pi_{ij,u}^2 = (w_{uk} + s_u + t_{uj}) - (w_{ik} + s_i + t_{ij}) . \qquad (6)$$

Component $\pi_{ij,u}^3$ maps large loads into small costs. Let $QC_{i,q}^k$ denote the available capacity of commodity $q$ which a vehicle $k$ can hold when it arrives at $i$. Then:

$$\pi_{ij,u}^3 = \min_{i \in \rho \cup \{u\}} QC_{i,q}^k - d_u^q . \qquad (7)$$

Finally, $\pi_{ij,u}^4$ accounts vehicle utilization. Let, $FC_u^k$ denote the total capacity of unutilized compartments. These unutilized compartments can accommodate all commodities, contrary to utilized compartments which can accommodate only the commodity already carried at the residual space left $RC_\rho^k$, if any. Thus:

$$\pi_{ij,u}^{4r} = RC_{\rho \cap \{u\}}^k - RC_{\rho \cup \{u\}}^k , \qquad (8)$$

where $RC_k^\rho$ is defined as $RC_k^\rho = \sum_{q \in Q} \left( min_{i \in \rho} QC_{i,q}^k - min_{i \in \rho} FC_i^k \right)$. Finally, in cases where infeasible vehicles are employed to service unassigned customers, the route elimination procedure proposed in [2] is applied to restore feasibility.

## 2.2   Multi Compartment-Commodity Loading Heuristic

Let $r$ denote a route performed by a vehicle $k$ of a partial solution $\Omega$, serving a set of customers $\Xi \subseteq V/\{0, n+1\}$. Let $D_q$ denote the aggregate quantity of a particular commodity $q$ of all customers $i \in \Xi$ such that $D_q = \sum_{i \in \Xi} d_i^q$. The proposed loading heuristic, in a semi parallel construction fashion, assigns quantities and couples compartments with particular commodities based on two criteria, for commodity and compartment selection. In particular, the objective at each iteration is to load first the commodity $q$ that maximizes $FC_r^k$, while the associate compartments assigned are those that minimize $RC_r^k$.

Initially, all compartments $m$ of vehicle $k$ are considered empty and $D_q$ is calculated for all $q$. Next, each unloaded $D_q$ is sequentially loaded temporarily into compartments and the associated $FC_r^k$ is determined in the following manner: First the empty unutilized $m'$ with maximum available capacity is identified. If $m'$ can accommodate only part of $D_q$ ($D_q \geq C_{m'}^k$), $m'$ is marked temporarily as utilized, the residual left is set to 0 and $D_q$ is reduced by $C_{m'}^k$ units. Otherwise, if $m'$ can hold all $D_q$ ($D_q < C_{m'}^k$), $m''$ that minimizes $RC_r^k$ is selected, it is marked temporarily as utilized and $D_q$ is set to 0. This procedure is repeated until $D_q$ equals 0. Consequently, the $FC_r^k$ obtained is stored, all temporarily marked compartments are re-initialized and the overall procedure is repeated for all unloaded $D_q$. Next, the $D_{\hat{q}}$ of $\hat{q}$ that maximizes $FC_r^k$ is loaded and the associated compartments are permanently marked as utilized. The latter procedure iterates until all commodities are loaded.

## 2.3   Variable Neighborhood Search

Variable Neighborhood Search (VNS) explicitly applies a strategy based on dynamically changing neighborhood structures [7]. Herein, the variant called Variable Neighborhood Descent (VND) is utilized. At the initialization step, a set of $\gamma$ neighborhood structures with increasing cardinality ($|N_1|<|N_2|\ldots|N_{\gamma_{max}}|$) is defined. Given an initial solution, the neighborhood index is initialized and the algorithm performs an exploration of the solution space for each respective neighborhood structure. At each iteration a best improvement local search is applied until a local minimum is found. If $f(s')<f(s)$, $s$ is replaced by $s'$ and the algorithm continues. Otherwise, $\gamma$ is incremented (moving phase). The process of changing neighborhoods with increasing cardinality, in case of no improvements, corresponds to diversification of the search close to the incumbent solution, since the neighborhood structure defines the topology of the search landscape.

**Variable Neighborhood Descent**
Select a set of neighborhood structures $N_\gamma$, $\gamma=1,2,\ldots,\gamma_{max}$
$s \leftarrow$ InitialSolution()
**While** no improvement can be obtained **do** $k \leftarrow 1$
   **While** $\gamma \leq \gamma_{max}$ **do**
      $s' \leftarrow$ FindBestNeighborhood($s' \in N_\gamma(s)$)
      **if** ($f(s')<f(s)$) **then** $s \leftarrow s'$, $\gamma \leftarrow 1$
      **else** $\gamma \leftarrow \gamma+1$
   **Endwhile**
   VehicleReassignment()
**Endwhile**

A vehicle reassignment mechanism is embedded within the VND scheme. The main objective is to improve the vehicle's capacity utilization through reciprocal exchange of the vehicles deployed between two routes. Moreover, such an approach allows the efficient exploration of the solution space since the sequence of customers served by vehicles are maintained intact. Finally, the definition of rich neighborhood structures increases the possibilities of finding higher quality local optimum solutions. In particular, the neighborhoods used are defined as a blend of well known local search moves, such as 2-Opt, 0-1 Relocate, Cross, 1-1 Interchange and Or-Opt.

# 3  Computational Results

For the evaluation of the proposed solution method, various experiments were conducted. All computational results reported herein are based on the benchmark data sets of [11]. These data sets proposed for the Fleet Size and Mix VRPTW. However, since we are not aware of benchmarks instances for the problem considered in this paper, we assumed the best fleet mix obtained in [11] to be the given fixed fleet. The latter approach was also followed in [12] and [13] for the Heterogeneous Fixed Fleet VRP. Moreover, based on the characteristics of the fleet mix of [11], new benchmark data sets generated. These data sets, along with the characteristics and specifications of the vehicle fleet (fleet composition, fixed costs, capacity per compartment etc.) are available online at (http://www.msl.aueb.gr/management_science/MCCHFFVRPTW/nbdc.html).

Initially, the single commodity case was examined, i.e., the relaxed, in terms of loading constraints, instance of the problem. The resulting problem is a Heterogeneous Fixed Fleet VRPTW and can be directly compared to the solutions produced by the heuristics with improvement (LS) proposed in [11]. Table 1 illustrates the performance of LS and the proposed reactive GRASP-VNS which produced substantially better distribution costs $(DC_1)$.

**Table 1.** Computational results for the multi commodity/compartment problem

| Instance | Fleet Mix | LS [11] | Reactive GRASP-VNS | | |
|---|---|---|---|---|---|
| | | $DC_1$ | $DC_1$ | $DC_2$ | $DC_3$ |
| R101a | $A^1 B^{11} C^{11} D^1$ | 5061,00 | 4492,20 | 4682,31 | - |
| R102a | $A^1 B^4 C^{14} D^2$ | 5013,00 | 4375,91 | 4553,42 | 4656,34 |
| R103a | $B^7 C^{15}$ | 4772,00 | 4213,08 | 4293,31 | 4464,57 |
| R104a | $B^9 C^{14}$ | 4455,00 | 4131,93 | 4311,33 | 4342,45 |
| RC101a | $A^7 B^7 C^7$ | 5687,00 | 5371,34 | 5583,45 | 5843,87 |
| RC102a | $A^5 B^6 C^8$ | 5649,00 | 5199,56 | 5534,75 | 5547,13 |
| RC103a | $A^{11} B^2 C^8$ | 5419,00 | 5111,23 | 5365,42 | 5389,94 |
| RC104a | $A^2 B^{13} C^3 D^1$ | 5189,00 | 5102,97 | 5493,61 | 5612,57 |

Furthermore, in order to consider loading constraints, two cases were examined with 2 $(DC_2)$ and 3 $(DC_3)$ commodities (see Table 1) using the same fleet mix as the given fixed fleet. In almost all cases the proposed method produced feasible solutions. However, contrary to the results obtained without loading constraints $(DC_1)$, the distribution costs increased accordingly, since the problem become more constrained. The latter intensifies for problem instances with 3 commodities. Indeed, for the problem instance R101, with tight time windows, no feasible solution could be obtained for the given fixed fleet.

# 4  Conclusions

This paper presented a robust and effective solution method to tackle a new realistic VRP which incorporated customers' time windows, heterogeneous fixed fleet of vehicles and multiple commodities and compartments. The proposed methodology hybridized the diversification mechanisms of GRASP with VNS for intensification local search, in a reactive fashion utilizing a probabilistic learning mechanism for the strategic sampling of the search space. Computational results justified the applicability of the methodology. In terms of further research, additional memory structures that exploit to a larger extent information gathered during the search, is a worth pursuing research direction.

**Acknowledgements.** This work is supported by the General Secretariat for Research and Technology of the Hellenic Ministry of Development under contract GSRT NM-67.

# References

1. Scholl, A., Klein, R., Jürgens, C.: BISON: a fast hybrid procedure for exactly solving the one-dimensional bin packing problem. Comp. & Opns. Res. **24**(1997) 627–645
2. Bräysy, O.: A Reactive Variable Neighborhood Search for the Vehicle Routing Problem with Time Windows. INFORMS J. Comp. **15**(2003) 347–368
3. Golden, B., Assad, A., Levy, L., Gheysens, F.: The fleet size and mix vehicle routing problem. Comp. & Opns. Res. **11**(1984) 49–66
4. Tarantilis, C.D., Kiranoudis, C.T., Vassiliadis, V.S.: A list based threshold accepting metaheuristic for the heterogeneous fixed fleet vehicle routing problem. J. Opl. Res. Soc. **54**(2003) 65–71
5. Gendreau, M., Potvin, J-Y.: Metaheuristics in Combinatorial Optimization, Anns. Opns. Res. **140**(2005) 189–213
6. Feo, T., and Resende, M.: Greedy randomized adaptive search procedures. J. Glb. Opt. **6**(1995) 109–154
7. Hansen, P., Mladenović, N.: Variable neighborhood search: Principles and applications, Eur. J. Opl. Res. **130**(2002) 449–467
8. Prais, M., Rideiro, C.C.: Parameter variation in GRASP procedures. Investigación Operativa **9**(2000) 1–20
9. Kontoravdis, G.,Bard, J.F.: A GRASP for the vehicle routing problem with time windows. ORSA J. Comp. **7**(1995) 10–23
10. Solomon, M.M.: Algorithms for the vehicle routing and scheduling problems with time window constraints. Opns. Res. **35**(1987) 254–265
11. Liu, F.H., Shen, S.Y.,: The Fleet Size and Mix Routing Problem with Time Windows. J. Opl Res. Soc. 50(1999) 721–732
12. Taillard, E.D.,: A heuristic column generation method for the heterogeneous fleet VRP. RAIRO **33**(1999) 1–14
13. Tarantilis, C.D., Kiranoudis, C.T., Vassiliadis, V.S.: A threshold accepting metaheuristic for the heterogeneous fixed fleet vehicle routing problem. Eur. J. Opl. Res. **152**(2004) 148–158.

# Multipopulation Genetic Algorithms: A Tool for Parameter Optimization of Cultivation Processes Models

Olympia Roeva

CBE "Prof. Ivan Daskalov" - BAS,
105 Acad. G. Bonchev Str., 1113 Sofia, Bulgaria
olympia@clbme.bas.bg

**Abstract.** This paper endeavors to show that genetic algorithms, namely Multipopulation genetic algorithms (MpGA), are of great utility in cases where complex cultivation process models have to be identified and, therefore, rational choices have to be made. A system of five ordinary differential equations is proposed to model biomass growth, glucose utilization and acetate formation. Parameter optimization is carried out using experimental data set from an *E. coli* cultivation. Several conventional algorithms for parameter identification (Gauss-Newton, Simplex Search and Steepest Descent) are compared to the MpGA. A general comment on this study is that traditional optimization methods are generally not universal and the most successful optimization algorithms on any particular domain, especially for the parameter optimization considered here. They have been fairly successful at solving problems of type which exhibit bad behavior like multimodal or nondifferentiable for more conventional based techniques.

## 1 Introduction

A major deficiency in computational approaches to design and optimization of bioprocess systems is the lack of applicable methods. Cultivation processes are complex highly nonlinear dynamic systems and their modeling and optimization is a complicated and rather time consuming task. The important part of model building is the choice of a certain optimization procedure for parameter estimation, so with a given set of experimental data to calibrate the model in order to reproduce the experimental results in the best possible way. This mathematical problem, so-called inverse problem, is a big challenge for the traditional optimization methods. Various meta-heuristics are used as an alternative to surmount the parameter estimation difficulties. Simulated annealing, tabu search, evolutionary algorithms like genetic algorithms (GAs) and evolution strategies, ant colony optimization, estimation of distribution algorithms, scatter search, path relinking, the greedy randomized adaptive search procedure, multi-start and iterated local search, guided local search, and variable neighborhood search are - among others - often listed as examples of classical meta-heuristics, and

T. Boyanov et al. (Eds.): NMA 2006, LNCS 4310, pp. 255–262, 2007.

they have individual historical backgrounds and follow different paradigms and philosophies [4,5,7,10].

The genetic algorithms are widespread optimization technique and they find applications in a large domain of problems. GAs are successfully applied in the areas of engineering, control, neural networks, signal processing and pattern recognition, molecular docking, parameter fitting, manufacturing, clustering, scheduling, robotics and machine learning [8,12,13]. A genetic algorithm requires only information concerning the quality of the solution produced by each parameter set (objective function value information). This characteristic differs from the optimization methods that require derivative information or, worse yet, complete knowledge of the problem structure and parameters. Since GAs do not require such problem-specific information, they are more flexible than the most search methods. A single population genetic algorithm is powerful and performs well on a broad class of problems. However, better results can be obtained by introducing many populations, so-called Multipopulation genetic algorithm (MpGA) [1,8,11]. The MpGA models the evolution of a species in a way more similar to the nature than the single population genetic algorithm.

In this paper a Multipopulation genetic algorithm for the identification of fed-batch cultivation of E. coli is proposed. The main purpose is to investigate if the MpGA can provide more efficient and reliable results for considered parameter identification problem. The input-output measurement data are used to estimate the model parameters such that a certain objective function is minimized. The results of three conventional search methods, namely Gauss-Newton (GN), Simplex Search (SS) and Steepest Descent (SD), as a "competing" set of algorithm candidates, are presented for comparison with the MpGA.

The paper is organized as follows: theoretical background of Multipopulation genetic algorithms and the algorithm parameters and operators used here are described in Section 2. Parameter identification problem concerning considered cultivation process is formulated in Section 3. The results and discussion are presented in Section 4. Conclusion remarks are done in Section 5.

## 2    Multipopulation Genetic Algorithm

Multipopulation genetic algorithm works on a coding of the parameter space. The coding is an essential part of the GAs design procedure and results in formation of strings composed of characters belonging to a finite alphabet. Having decided on the coding to be used (real-value encoding here), initial sets subpopulations (subpop = 5) are created at random. Each subpopulation contains a certain number of individuals (nind = 100). An objective function (usually referred to as fitness function in GAs terminology) serves as a measure of goodness of a string and is a functional of the function that will be optimized. In the next step individuals represented by their fitness function are ranked. Chromosomes from a population are selected according to their fitness: the better fitness, the bigger chance to be selected. Thus solutions from one population are taken and used to form a new population. This is motivated by the hope, that the new population will be better than the old one. A certain function performing selection (roulette-wheel selection here) concordant with the generation gap (ggap = 0.97) is used. Roulette-wheel selection is the simplest selection scheme, also called stochastic sampling with replacement. In the next step a set of operators is applied to the initial populations to generate successive generations. The selected individuals are then recombined (extended intermediate recombination) in order to form new offspring (children) crossing over the parents with a crossover probability (xovr = 0.7). Then, mutation (breeder mutation) takes place with mutation rate (mutr = 0.05). For the new individuals of the subpopulations the objective function values are calculated. Fitness-based reinsertion is used with given insertion rate (insr = 0.2). The new offspring is inserted in the population. The new generated population is used for a further run of the algorithm. The subpopulations evolve independently from each other for a certain number of generations - isolation time (miggen = 20). After the isolation time a number of individuals is distributed between the subpopulations (migration). The migration rate (migr = 0.2), the selection function of the individuals for migration and the scheme of migration determine how much genetic diversity can occur in the subpopulations and the exchange of information between subpopulations. The selection of the individuals for migration is fitness-based. The best fit of each subpopulation is selected for migration. Here, the individuals migrate based on neighborhood migration topology. Migration is made only between nearest neighbors. For each subpopulation, the possible immigrants are determined, according to the selection method, from adjacent subpopulations and a final selection made from this pool of individuals. This ensures that individuals will not migrate from a subpopulation to the same subpopulation. Fitness-based migration selects individuals according to their fitness level and replaces individuals in a subpopulation uniformly at random.

Natural evolution of the subpopulations continues until a predetermined number of generations (maxgen = 150) is reached.

Some adjustments of the genetic parameters, according to the regarded problem, have to be done to improve the optimization capability and the decision speed. The primary choice of genetic operators and parameters depends on the problem, as well as on the chosen encoding. An inappropriate choice of operators and parameters in the evolutionary process makes the GAs susceptible to premature convergence.

## 3 Model Parameter Identification of a Fed-Batch Cultivation of *Escherichia coli*

The cultivation condition of the fed-batch cultivation of *E. coli* and the experimental data have been published previously [6] as a result of teamwork of the *DFG Project* between *CBE "Prof. I. Daskalov"* - *BAS* and *University of Hannover, Germany*. The mathematical formulation of the nonlinear dynamic model of *E. coli* cultivation is described according to the mass balance as follows:

$$\frac{dX}{dt} = \mu_{max}\frac{S}{k_S + S}X - \frac{F}{V}X, \tag{1}$$

$$\frac{dS}{dt} = -\frac{1}{Y_{S/X}}\mu_{max}\frac{S}{k_S + S}X + \frac{F}{V}(S_{in} - S), \tag{2}$$

$$\frac{dA}{dt} = \frac{1}{Y_{A/X}}\mu_{max}\frac{S}{k_S + S}X - \frac{F}{V}A, \tag{3}$$

$$\frac{dV}{dt} = F, \tag{4}$$

where: $X$ is biomass concentration, [g/l]; $S$ - substrate concentration, [g/l]; $A$ - acetate concentration, [g/l]; $F$ - feeding rate, [l/h]; $V$ - bioreactor volume, [l]; $S_{in}$ - substrate concentration in the feeding solution, [g/l]; $\mu_{max}$ - maximum value of the specific growth rate, [1/h]; $k_S$ - saturation constant, [g/l]; $Y_{S/X}$ and $Y_{A/X}$ - yield coefficients.

As it shown, the model consisted of a set of four differential equations (Eqs. (1) - (4)) thus represented three dependent state variables, $\mathbf{x} = [X, \ S, \ A]$, and four parameters, $\mathbf{p} = [\mu_{max}, \ k_S, \ Y_{S/X}, \ Y_{A/X}]$. Parameter estimation problem of nonlinear dynamic system is stated as the minimization of a distance measure $J$ between experimental and model predicted values of the considered state variables:

$$J = J_X + J_S + J_A \to min, \tag{5}$$

where: $J_X = \sum_{i=1}^{N}[X_{exp}(i) - X_{mod}(i)]^2$; $J_S = \sum_{i=1}^{N}[S_{exp}(i) - S_{mod}(i)]^2$; $J_A = \sum_{i=1}^{N}[A_{exp}(i) - A_{mod}(i)]^2$; $N$ is the number of data for each state variable; $X_{exp}$, $S_{exp}$ and $A_{exp}$ represent the known experimental data; $X_{mod}$, $S_{mod}$ and $A_{mod}$ are model predictions with a given set of the 4 parameters.

## 4    Results and Discussion

Based on off-line measurements of biomass and acetate concentrations and on-line measurements of glucose concentration a parameter identification using MpGA, GN, SS and SD methods is performed. For fair comparison the *Matlab* implementation is considered for all methods. For the implementation of Simplex Search method, the *Matlab 5.3 Optimization Toolbox* function "*fmins*" is used. The function "*fmins*" uses the Nelder-Mead simplex (direct search) method. The Gauss-Newton optimization function used here is the "*leastsq*" function. For the Steepest Descent method the function "*fminu*" is used. Multipopulation genetic algorithm using *Genetic Algorithm Toolbox* [2] is applied. All the computations are performed using a PC/Pentium IV (3 GHz) platform running Windows XP.

The results from the identification are presented in Table 1. The best results are marked in bold. The estimates of the yield of glucose per biomass using considered search methods are in the interval $Y_{S/X} = [0.33 \ 0.67]$. Levisauskas et al. have reported similar results [9]. For yield of acetate per biomass SD method and MpGA are achieved the values of $Y_{A/X} = 0.021$ g/l and $Y_{A/X} = 0.014$ g/l.

**Table 1.** Search parameters utilized in the different algorithms

| Search parameter | SS | GN | SD | MpGA |
|:---:|:---:|:---:|:---:|:---:|
| $\mu_{max}$ | 0.52 | 0.47 | 0.39 | **0.53** |
| $k_S$ | 0.028 | 0.014 | 0.02 | **0.029** |
| $Y_{S/X}$ | 0.49 | 0.67 | 0.33 | **0.49** |
| $Y_{A/X}$ | 0.0048 | 0.0072 | 0.021 | **0.014** |

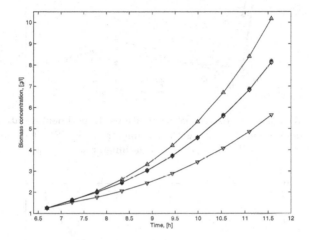

**Fig. 1.** Time profiles of the biomass concentrations. Experimental data are presented with ($*$) and models predicted data with: parameter set 1 (SS) - ($\diamond$), parameter set 2 (GN) - ($\triangle$), parameter set 3 (SD) - ($\triangledown$) and parameter set 4 (MpGA) - ($\circ$).

Analogous results are reported from Contiero et al. [3]. According to Zelic et al. [14] the values of the parameters $\mu_{max}$ and $k_S$ are in admissible boundaries too.

The model predictions of the state variables, based on four sets of search parameters are compared to the experimental data points of the *E. coli* cultivation in Figures 1, 2 and 3.

As it can be seen from the simulation results the MpGA and the SS method have achieved almost equal results about description of biomass and substrate variation (Figures 1 and 2). However, the SS could not find appropriate solution for acetate (Figure 3). The results also reveal an interesting fact. The SD gives a satisfactory result for the acetate fit. Unfortunately, this method failed in the modeling of biomass and substrate. The presented graphical results show that only MpGA copes with the present problem. It is worth to note the very good correlation between the experimental and predicted data.

In most cases, graphical comparisons clearly show the existence or absence of systematic deviations between model predictions and measurements. It is evident that a quantitative measure of the differences between calculated and measured values is an important criterion for the adequacy of a model. Detailed results for values of considered criteria (objective function value, number of iterations,

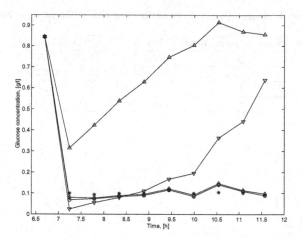

**Fig. 2.** Time profiles of the substrate concentrations. Experimental data are presented with (∗) and models predicted data with: parameter set 1 (SS) - (◇), parameter set 2 (GN) - (△), parameter set 3 (SD) - (▽) and parameter set 4 (MpGA) - (○).

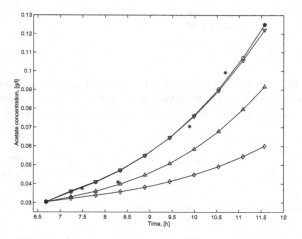

**Fig. 3.** Time profiles of the acetate concentrations. Experimental data are presented with (∗) and models predicted data with: parameter set 1 (SS) - (◇), parameter set 2 (GN) - (△), parameter set 3 (SD) - (▽) and parameter set 4 (MpGA) - (○).

number of floating point operations and CPU time) are given in Table 2 with the solution vectors shown in Table 1. The best result ($J = 5.2529$) is obtained using the MpGA after total computation time of 16.83 min. This shows that the proposed parameters of MpGA, especially population size, crossover and mutation rates, are correctly chosen. The second best search method is the SS method, which converged to a value of $J = 5.9167$. This result is 13% worst than the result obtained using MpGA. Concerning biomass and glucose prediction once again it could be seen that the SS has achieved very good results, close

Table 2. Results of the search methods

| Criterion | SS | GN | SD | MpGA |
|---|---|---|---|---|
| $J_X$ | 4.7100 | 1.5393e+003 | 3.2032e+003 | 4.0911 |
| $J_S$ | 1.2017 | 824.8283 | 86.2690 | 1.1616 |
| $J_A$ | 0.0051 | 0.0016 | 3.3479e-004 | 2.9073e-004 |
| $J$ | 5.9167 | 2.3633e+003 | 3.2895e+003 | 5.2529 |
| iterations | 248 | 766 | 523 | 150 |
| CPU time (sec) | 0.8091e+003 | 2.6196e+003 | 2.08871e+003 | 1.0098e+003 |
| floating point operations | 1.6848e+009 | 5.2321e+009 | 7.1736e+009 | 2.8813e+009 |

to MpGA results, but the difference in $J_A$ functions between the mentioned methods is very big (in an order) (see Table 2).

Table 2 shows that none of the other algorithms tested (GN and SD), could reach the vicinity of the above mentioned solutions. This is a clear sign of the very challenging nature of these problems. Due to the nonlinear and constrained nature of the system dynamics, the parameter estimation problems are very often multimodal (nonconvex). Therefore, if this inverse problem is solved via standard local methods, such as the standard Gauss-Newton method, it is very likely that the solution found will be of local nature.

## 5  Conclusion

The use of Multipopulation genetic algorithms in the parameter identification of nonlinear dynamical systems has been investigated in this paper. Since the correct solution of the inverse problems plays a key role in the development of dynamic models, the MpGA are used to estimate the unknown model parameters from *E. coli* fed-batch cultivation. The identification problem is formulated as an optimization problem. The mathematical model is presented by a system of five ordinary differential equations, describing biomass growth, glucose utilization and acetate formation. The estimates of Gauss-Newton, Simplex Search and Steepest Descent methods are presented too. The concurrent nature implies that GAs are much more likely to locate a global peak than the traditional techniques. The conventional search methods work extremely well provided a good starting point is known. In the problem considered here the proper initial values of parameters are unknown. The performance of the genetic algorithms is much less sensitive to the initial conditions. In fact, GAs make hundreds, or even thousands, of initial guesses. Simulation results reveal that accurate and consistent results can be obtained using Multipopulation genetic algorithms. The genetic algorithms property makes them suitable and more applicable for parameter estimation of cultivation models.

**Acknowledgements.** This work is partially supported from National Science Fund Project No. MI - 1505/2005.

# References

1. Cant'u-Paz, E.: Designing scalable multi-population parallel genetic algorithms. IllGAL Report 98009, The University of Illinois (1998)
2. Chipperfield, A. J., Fleming, P. J., Pohlheim, H., Fonseca, C. M.: Genetic algorithm toolbox for use with MATLAB. Department of Automatic Control and System Engineering, University of Sheffield, U. K (1994)
3. Contiero, J., Beatty, C ., Kumari, S., DeSanti, C. L., Strohl, W. L., Wolfe, A.: Effects of mutations in acetate metabolism on high-cell-density growth of *Escherichia coli*. Journal of Industrial Microbiology and Biotechnology **24** (2000) 421–430
4. Dorigo, M., Di Caro, G.: The ant colony optimization meta-heuristic. In: Corne, D, Dorigo, M., Glover, F. (eds).: New Idea in Optimization, McGrow-Hill (1999) 11–32
5. Fidanova, S.: Ant colony optimization: Additional reinforcement and convergence. Tech. report IRIDIA-2002-30, Free university of Bruxelles, Belgium, 12
6. Georgieva, O., Arndt, M., Hitzmann, B.: Modelling of *Escherichia coli* fed-batch fermentation, In "Bioprocess Systems'2001", Sofia, Bulgaria, October 1–3, (2001) I.61–I.64
7. Glover, F., Kochenberger, G. A.: Handbook of metaheuristics. Kluwer (2003)
8. Goldberg, D.: Genetic algorithms in search, optimization and machine learning. Addison-Weslcy Publishing Company, Massachusetts (1989)
9. Levisauskas, D., Galvanauskas, V., Henrich, S., Wilhelm, K., Volk, N., Lubbert, A.: Model-based optimization of viral capsid protein production in fed-batch culture of recombinant *Escherichia coli*, Bioprocess and Biosystems Engineering **25** (2003) 255–262
10. Raidl, G. R.: A unified view on hybrid metaheuristics. In: C. Blum et al., eds.: Proceedings of the Hybrid Metaheuristics Workshop, to appear in LNCS, Springer (2006)
11. Siarry, P., Petrowski, A., Bessaou, M.: A multipopulation genetic algorithm aimed at multimodal optimization. Advances in Engineering Software **33(4)** (2002) 207–213
12. Srinivas, M., Patnaik, L. M.: Genetic algorithms: A survey. In: IEEE Computer (1994) 17–26
13. Yi, W., Liu, Q., He, Y.: Dynamic distributed genetic algorithms. Congress on Evolutionary Computation (CEC'2000), San Diego, California, USA, July 16–19 (2000)
14. Zelic, B., Vasic-Racki, D., Wandrey, C., Takors, R.: Modeling of the pyruvate production with *Escherichia coli* in a fed-batch bioreactor. Bioprocess and Biosystems Engineering **26** (2004) 249–258

# Design of Equiripple 2-D Linear-Phase
# FIR Digital Filters Using Genetic Algorithm

Felicja Wysocka-Schillak

University of Technology and Agriculture,
Institute of Telecommunications,
al. Prof. S. Kaliskiego 7, 85-796 Bydgoszcz, Poland
felicja@mail.atr.bydgoszcz.pl

**Abstract.** The paper presents a method for designing 2-D linear-phase
FIR filters with an equiripple magnitude response. The filter design problem is transformed into an equivalent nonlinear optimization problem. In
order to improve the speed of convergence, a two-step solution procedure
of the considered problem is proposed. In the first step, a genetic algorithm is applied. The final point from the genetic algorithm is used as the
starting point for a local optimization method. The proposed technique
is applied to the design of 2-D FIR linear-phase filters with different
symmetries. Design examples are included.

## 1 Introduction

Two-dimensional (2-D) digital filtering is one of the most important processing
techniques in 2-D digital signal processing. An important class of 2-D digital
filters are finite impulse response (FIR) filters. FIR filters can be made to have
linear phase and they are free from stability problems. 2-D linear-phase FIR
filters have important applications in image processing where the phase of 2-D
signals usually needs to be preserved.

Several techniques have been developed for designing 2-D linear-phase FIR
filters that approximate desired magnitude specifications using different error
criteria [1]-[6]. This paper attempts to demonstrate that a genetic algorithm
(GA) [7] can be used as a tool for the design of 2-D linear-phase FIR filters according to the equiripple error criterion. GAs are probabilistic search techniques
based on the mechanics of natural genetics and natural selection. They have
strong robustness and general utility. Therefore, GAs are often used for solving
difficult nonlinear optimization problems and multi-objective optimizations.

In the paper, a new approach for the design of 2-D linear-phase FIR filters according to the equiripple error criterion is proposed. The filter coefficients vector
is defined and an objective function is introduced. Then, the 2-D filter design
problem is transformed into an equivalent nonlinear minimization problem. The
solution of the considered problem is achieved using a two-step procedure. In the
first step, the GA is applied. The final point from the GA is used as the starting
point for a local optimization method. Using a local optimization method, when

T. Boyanov et al. (Eds.): NMA 2006, LNCS 4310, pp. 263–270, 2007.
© Springer-Verlag Berlin Heidelberg 2007

the solution is close to the optiumum, results in improving the speed of convergence. The proposed approach is used for designing 2-D FIR filters that have centro-symmetric, circularly symmetric and octagonally symmetric magnitude responses. Numerical examples demonstrating the performance of the proposed method are given.

## 2    Formulation of the Design Problem

The frequency response of a 2-D FIR filter can be expressed as [1]

$$H(e^{j\omega_1}, e^{j\omega_2}) = \sum_{m=0}^{M-1} \sum_{n=0}^{N-1} h(m, n)e^{-j(n\omega_1 + m\omega_2)}, \tag{1}$$

where $h(m, n)$ is the rectangularly sampled impulse response of the filter, $M$ and $N$ represent the lengths of the filter, and $\omega_1$ and $\omega_2$ are the horizontal and vertical frequencies, respectively.

A 2-D FIR filter can be designed to possess linear-phase which makes it attractive in some applications such as e.g., image processing. It can be shown that if $h(m, n)$ is real and meets the constraint [3]

$$h(m, n) = h(M - 1 - m, N - 1 - n) \tag{2}$$

then the phase response of the filter is linear and the zero-phase frequency response $H(\omega_1, \omega_2)$ is centro-symmetric, i.e., $H(\omega_1, \omega_2) = H(-\omega_1, -\omega_2)$. In such a case, the filter is called centro-symmetric filter and its frequency response can be expressed in the form [3]

$$H(e^{j\omega_1}, e^{j\omega_2}) = H(\omega_1, \omega_2)e^{-j(\frac{M-1}{2}\omega_1 + \frac{N-1}{2}\omega_2)} \tag{3}$$

where the real function $H(\omega_1, \omega_2)$ is the zero-phase frequency response and it represents the magnitude response $A$ of the filter ($A = |H(\omega_1, \omega_2)|$). The complex exponential part denotes the linear-phase characteristic of the filter.

The 2-D FIR filter design problem is to find an impulse response $h(m, n)$, such that the zero-phase frequency response of the filter is the best approximation of the desired zero-phase frequency response $H_d(\omega_1, \omega_2)$ in a given sense.

The impulse response $h(m, n)$ of a centro-symmetric filter has twofold symmetry, so only approximately half of the points in it are independent. The region of support of $h(m, n)$ consists of three mutually exclusive regions: the origin $(0, 0)$, $R^+$ and $R^-$. The regions $R^+$ and $R^-$ are flipped with respect to the origin. As a result, the zero-phase frequency response can be expressed as [1]

$$H(\omega_1, \omega_2) = h(0, 0) + \sum_{(m,n) \in R^+} \sum 2h(m, n) \cos(m\omega_1 + n\omega_2). \tag{4}$$

Zero-phase frequency responses can posses also other types of symmetries. The presence of these symmetries can be used to reduce the number of independent filter coefficients that must be estimated in the design.

A 2-D zero-phase frequency response $H(\omega_1, \omega_2)$ possesses quadrantal symmetry if it satisfies the following condition [1]:

$$H(\omega_1, \omega_2) = H(-\omega_1, \omega_2) = H(\omega_1, -\omega_2) = H(-\omega_1, -\omega_2). \tag{5}$$

If, in addition, $H(\omega_1, \omega_2)$ satisfies

$$H(\omega_1, \omega_2) = H(\omega_2, \omega_1) \tag{6}$$

then it has octagonal symmetry.

Let the considered filter be a quadrantally symmetric filter. The zero-phase frequency response of this filter can be written in the form [4]

$$H(\omega_1, \omega_2) = \sum_{m=0}^{L_1} \sum_{n=0}^{L_2} a(m, n) \cos(m\omega_1) \cos(n\omega_2), \tag{7}$$

where $a(m, n)$ are free filter coefficients which can be expressed in terms of the impulse response $h(m, n)$ [1], and $L_1$ and $L_2$ are the numbers of free filter coefficients ($M = 2L_1 + 1, N = 2L_2 + 1$). Note that in this case the number of independent filter coefficients is reduced about four times in comparison with (1).

Let $\mathbf{Y}$ be a vector of filter coefficients. In case of a centro-symmetric filter, $\mathbf{Y} = [y_1, y_2, ..., y_{L+1}]^\top$, where

$$y_1 = h(0, 0), \tag{8}$$

$$y_i = h(m, n), \quad i = 1, 2, ..., (L + 1); m, n \in R^+. \tag{9}$$

In case of a quadrantally symmetric filter, $\mathbf{Y} = [y_1, y_2, ..., y_{(L_1+1)(L_2+1)}]^\top$, where

$$y_i = a(m, n), \quad i = 1, 2, ..., (L_1+1)(L_2+1); \quad m = 0, 1, ..., L_1; \quad n = 0, 1, ..., L_2. \tag{10}$$

Assume that the continuous $(\omega_1, \omega_2)$ - plane is discretized by using a $K_1 \times K_2$ rectangular grid $(\omega_{1k}, \omega_{2l})$, $k = 0, 1, ..., K_1 - 1, k = 0, 1, ..., K_2 - 1$. The desired zero-phase frequency response $H_d(\omega_{1k}, \omega_{2l})$ of the 2-D filter is:

$$H_d(\omega_{1k}, \omega_{2l}) = \begin{cases} 1 & \text{for } (\omega_{1k}, \omega_{2l}) \text{ in the passband } P, \\ 0 & \text{for } (\omega_{1k}, \omega_{2l}) \text{ in the stopband } S. \end{cases} \tag{11}$$

Let $H(\omega_1, \omega_2, \mathbf{Y})$ denote the zero-phase frequency response of the filter obtained by putting into equation (4) or (7), respectively, the coefficients given by vector $\mathbf{Y}$. In the passband $P$ and in the stopband $S$ the approximation is to be equiripple. The error function $E(\omega_{1k}, \omega_{2l}, \mathbf{Y})$ is:

$$E(\omega_{1k}, \omega_{2l}, \mathbf{Y}) = H(\omega_{1k}, \omega_{2l}, \mathbf{Y}) - H_d(\omega_{1k}, \omega_{2l}), \quad \omega_1, \omega_2 \in P \cup S. \tag{12}$$

The filter design problem can be formulated as follows: For a given zero-phase frequency response $H_d(\omega_{1k}, \omega_{2l})$ defined on a rectangular grid $K_1 \times K_2$, and prescribed values $N$ and $M$ find a vector $\mathbf{Y}$ for which the error function $E(\omega_{1k}, \omega_{2l}, \mathbf{Y})$ is equiripple.

As in case of 2-D FIR filters the equiripple solution may not be unique [1], the following condition on the maximum allowable approximation error $\delta > 0$ in the passband can be additionally imposed:

$$\forall \omega_{1k}, \omega_{2l} \in P \quad |H(\omega_{1k}, \omega_{2l}, \mathbf{Y}) - H_d(\omega_{1k}, \omega_{2l})| \leq \delta. \tag{13}$$

Adding the above condition results in obtaining the magnitude ripple equal or smaller than $\delta$.

The advantage of employing the symmetry constraints is the reduction of the number of independent filter coefficients to be calculated. Besides, in the case of a centro-symmetric filter, we only need to consider the error function in the half of the $(\omega_1, \omega_2)$- plane. In the case of a quadrantally symmetric filter, it is necessary to consider the error function only in the first quadrant of the $(\omega_1, \omega_2)$-plane. In the case of octagonally symmetric filter, the calculations need only to be done in a $45^\circ$ sector of the frequency plane.

## 3    Transformation of the Problem

In this section, we convert the considered filter design problem into an equivalent optimization problem. In order to do this, we introduce an objective function $X(\mathbf{Y})$ possessing the property that it has a minimum equal to zero when the error function $E(\omega_{1k}, \omega_{2l}, \mathbf{Y})$ is equiripple in the passband and in the stopband. The error function $E(\omega_{1k}, \omega_{2l}, \mathbf{Y})$ is equiripple when the absolute values $\Delta E_i(\mathbf{Y})$, $i = 1, 2, ..., J$ , of all the local extrema of the function $E(\omega_{1k}, \omega_{2l}, \mathbf{Y})$ in the passband and in the stopband, as well as the maximum values $\Delta E_{J+1}(\mathbf{Y})$ and $\Delta E_{J+2}(\mathbf{Y})$ of $E(\omega_{1k}, \omega_{2l}, \mathbf{Y})$ at the passband and stopband edges are equal, i.e.,

$$\Delta E_i(\mathbf{Y}) = \Delta E_k(\mathbf{Y}), \quad k, i = 1, 2, ..., J + 2. \tag{14}$$

We asume that the objective function $X(\Delta E_1, \Delta E_2, \ldots, \Delta E_{J+2})$ is defined as follows:

$$X_1(\Delta E_1, \Delta E_2, ..., \Delta E_{J+2}) = \sum_{i=1}^{J+2} (\Delta E_i - R)^2, \tag{15}$$

where

$$R = \frac{1}{J+2} \sum_{k=1}^{J+2} \Delta E_k. \tag{16}$$

is the arithmetic mean of all $\Delta E_k$, $k = 1, 2, ..., J + 2$.

Note that the function $X$ defined above is a non-negative function of $\Delta E_1$, $\Delta E_2, \ldots, \Delta E_{J+2}$, and it is equal to zero if and only if $\Delta E_1 = \ldots = \Delta E_{J+2}$. As $\Delta E_1, \Delta E_2, ..., \Delta E_{J+2}$ are functions of the vector $\mathbf{Y}$, the function $X$ can be used as an objective function in the considered optimization problem.

The equivalent optimization problem can be stated as follows: For given filter specifications, find a vector $\mathbf{Y}$ such that the function $X(\mathbf{Y})$ is minimized.

# 4    Two-Step Solution Procedure

In order to apply a local minimization method to solve the optimization problem formulated in the previous section, a starting point sufficiently close to the solution is necessary. In case of 2-D FIR filters, the equiripple solution may not be unique [1]. Global methods, such as GAs, are largely independent of the initial conditions. Besides, GAs are particularly effective when the goal is to find an approximate global minimum in case of high-dimensional, difficult optimization problems and objective functions that can have many local minima. That is why GAs are well suited for solving the considered optimization problem.

GAs are stochastic search and optimization techniques based on the mechanism of natural selection where stronger individuals would likely be the winners in a competing environment. The detailed description of a simple GA can be found in [7]. GAs operate on fixed length strings (chromosomes) representing possible solutions of a given optimisation problem. To start implementing a GA, an initial population is considered. Successive generations are produced by manipulating the solutions in the current populations. Each solutions has a fitness (an objective function) that measures its competence. New solutions are formed using crossover and mutation operations. According to the fitness value, a new generation is formed by selecting the better chromosomes from the parents and offspring, and rejecting other so as to keep the population size constant. The algorithm converges to the best chromosome, which represents the solution of the considered optimization problem.

In order to solve the optimization problem formulated in the previous section, a two-step procedure is proposed, i.e., a hybridization of the GA and a local optimization method. Such hybridization is described by Golberg [7]. It is useful in our case because GAs are slow in convergence, especially when the solution is close to the optiumum. In order to improve the speed of convergence, after a specified number of generations in the GA has been reached, a local optimization method, i.e., the Davidon, Fletcher, and Powell (DFP) method is applied to solve the considered problem. The final point from the GA is used as the starting point for the DFP method. The DFP method is a quasi-Newton method which approximates the inverse Hessian matrix [8].

Numerical calculations have shown that it is possible to achieve better convergence if, instead of the minimization problem formulated in the previous section, we apply the GA to the following least square approximation problem

$$E_2(\mathbf{Y}) = \sum_{(\omega_{1k},\omega_{2l}) \in P \cup S} \sum [H(\omega_{1k}, \omega_{2l}, \mathbf{Y}) - H_d(\omega_{1k}, \omega_{2l})]^2. \tag{17}$$

Then, the solution of this problem is used as a starting point for solving the problem of minimizing $X(\mathbf{Y})$ using the DFP method. The local extrema of the error function $E(\omega_{1k}, \omega_{2l}, \mathbf{Y})$ are determined by searching the grid.

As crossover and mutation operations are basic operations in the GA, the choice of the probability of crossover, the probability of mutation as well as the choice of the population size are very important. Their settings are dependent on the form of objective function. In the developed program, the population size is 30, the probability of crossover is 0.8, and the probability of mutation is 0.01.

## 5 Design Examples

In this section, design examples are presented to illustrate the performance of the proposed technique. In all examples, the desired magnitude response is 1 in the passband $P$, 0 in the stopband $S$ and varies linearly in the transition band $Tr$. A square grid of $101 \times 101$ points is used for discretizing the $(\omega_1, \omega_2)$ - plane.

As a first example, we consider the design of a quadrantally symmetric diamond shaped filter. The passband of the filter is situated between the points $(0, 0.7\pi)$, $(0.7\pi, 0)$, $(0, -0.7\pi)$ and $(-0.7\pi, 0)$ on the $(\omega_{1k}, \omega_{2l})$-plane. The filter is designed with $M = N = 11$. The width of the transition band is $0.20\pi$. The resulting magnitude response $A$ of the filter is shown in Fig. 1. The obtained ripple is $\delta = 0.099$ both in the passband and in the stopband.

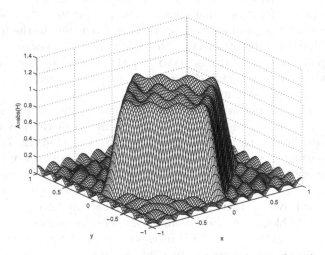

**Fig. 1.** Magnitude response of the diamond shaped filter designed in the first example $(x = \omega_1/\pi,\ y = \omega_2/\pi)$

In the second example, a centro-symmetric rotated elliptically symmetric filter is designed. The passband of the filter is an elliptic region with a rotation angle of $30^\circ$. The major and minor axes of the passband edge are $0.45\pi$ and $0.225\pi$, respectively. The major and minor axes of the stopband edge are $0.65\pi$ and $0.375\pi$. The filter is designed with $M = N = 19$. The resulting magnitude response $A$ of the filter is shown in Fig. 2. The obtained ripple is $\delta = 0.037$ both in the passband and in the stopband.

In order to compare the results with the results obtained in previous works, we design a circularly symmetric filter with the same specifications as in case of the design methods presented in [4] and in [2]. The passband of the filter is a

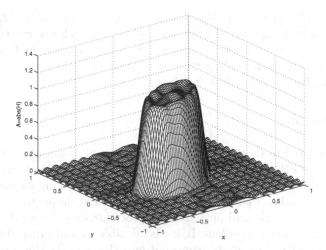

**Fig. 2.** Magnitude response of the rotated elliptically symmetric filter designed in the second example

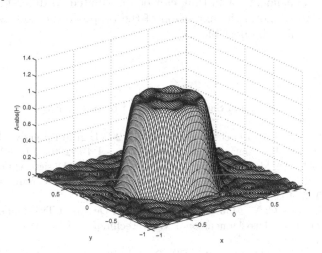

**Fig. 3.** Magnitude response of the circularly symmetric filter with $M = N = 15$

circular region centered at $(0, 0)$ with a radius $r_p$. The stopband corresponds to the region outside the circle with a radius $r_s$. As in [4], we design a lowpass filter with $r_p = 0.4\pi$ and $r_s = 0.6\pi$ for $M = N = 7$ and $M = N = 9$. The obtained ripples are equal 0.1193 and 0.1173, respectively. In case of the method presented in [4], the ripples are 0.1251 and 0.1165. As can be observed, the results obtained using both methods are approximately the same.

To compare the proposed method with the method described in [2], we design a lowpass filter with $r_p = 0.425\pi$ and $r_s = 0.575\pi$ for $M = N = 11$ and $M = N = 15$. In the case of the proposed method, the resulting ripples are 0.0720 and 0.0387, respectively. In [2], the maximum ripple $\delta_p$ in the passband is not the same as the maximum ripple $\delta_s$ in the stopband. For $M = N = $

11 - $\delta_p = 0.1247$ and $\delta_s = 0.1591$. For $M = N = 15$ - $\delta_p = 0.0822$ and $\delta_s = 0.1115$. Note that in case of the proposed method, in both designs the obtained ripples are considerable smaller. It may be the result of using the GA which is a global optimization method. As a representative design, the magnitude response of the filter with $M = N = 15$ is shown in Fig. 3.

# 6    Conclusions

We have attempted to show that the GA can be used as a tool in the design of 2-D linear-phase FIR filters according to the equiripple error criterion. A technique for the design of 2-D linear-phase FIR filters according to the equiripple error criterion has also been proposed. The technique is simple to implement because standard GA and local optimization procedures can be used to solve the considered minimization problem. It is also flexible as additional linear and/or nonlinear constraints can be incorporated into the optimization problem. The proposed technique can be parallelized because the original problem can be partitioned into independent parts that can be distributed to different processors for solution. Furthermore, the extension of the proposed method to the design of 2-D IIR filters may be possible.

# References

1. Lim, J. S.: Two-Dimensional Signal and Image Processing. Englewood Cliffs, Prentice Hall (1990)
2. Lu W.-S.: A Unified Approach for the Design of 2-D Digital Filters via Semidefinite Programming. IEEE Trans. on Circuits and Syst. - I **49** (2002) 814–826
3. Zhu, W.-P., Ahmad, M. O., Swamy, M. N.: A Least-Square Design Approach for 2-D FIR Filters with Arbitrary Frequency Response. IEEE Trans. Circuits and Syst.- II **46** (1999) 1027–1034
4. Hsieh, C.-H., Kuo, C.-M., Jou, Y.-D., Han, Y.-L.: Design of Two-Dimensional FIR Digital Filters by a Two-Dimensional WLS Technique. IEEE Trans. Circuits and Syst. - II **44** (1997) 248–356
5. Wysocka-Schillak, F.: Design of 2-D FIR Digital Filters Using a Parallel Machine. In: Lirkov I. et al. (eds.): Large-Scale Scientific Computing, Lecture Notes in Computer Science **2907**. Springer-Verlag, Berlin Heidelberg New York (2004) 412–418
6. Wysocka-Schillak, F.: Design of 2-D FIR Centro-Symmetric Filters With Equiripple Passband and Least-Squares Stopband. Proc. of the Europ. Conf. on Circuit Theory and Design, Cracow, Poland (2003) III-113–116
7. Goldberg D. E.: Genetic Algorithms in Search, Optimization, and Machine Learning. Addison Wesley, New York (1989)
8. Nocedal, J., Wright S. J.: Numerical Optimization. Springer-Verlag, Berlin Heidelberg New York (1999)

# A Simple and Efficient Algorithm for Eigenvalues Computation

René Alt

University Pierre et Marie Curie – Paris 6, Laboratoire LIP6, CNRS UMR 7606
4 Place Jussieu, Paris 75230, France
Rene.Alt@lip6.fr

**Abstract.** A simple algorithm for the computation of eigenvalues of real or complex square matrices is proposed. This algorithm is based on an additive decomposition of the matrix. A sufficient condition for convergence is proved. It is also shown that this method has many properties of the QR algorithm : it is invariant for the Hessenberg form, shifts are possible in the case of a null element on the diagonal. Some other interesting experimental properties are shown. Numerical experiments are given showing that most of the time the behavior of this method is not much different from that of the QR method, but sometimes it gives better results, particularly in the case of a bad conditioned real matrix having real eigenvalues.

**Keywords:** Linear algebra, eigenvalues computation, CESTAC method.

## 1  Introduction

In this paper the additive reduction (AR) algorithm for computation of the eigenvalues of a real or complex matrix is recalled. The main properties of this algorithm are shown, and numerical experiments allow us to compare it with the classical QR algorithm. The paper is organized as follows. First, the AR algorithm is described. Then it is shown that it is invariant for the Hessenberg form, and a sufficient condition for convergence in terms of this form is proved. Finally, some numerical experiments are given, showing that the method is very efficient and often even more than the QR method, particularly in the case of multiple eigenvalues or for computation of roots of polynomials using the companion matrix.

## 2  The Additive Reduction (AR) Algorithm

### 2.1  Definition of the Algorithm

Let us consider a real or complex square matrix $A$ with dimension $n$. One wishes to compute its eigenvalues. The algorithm, formerly proposed in [1] and called additive reduction algorithm (AR) is the following.

T. Boyanov et al. (Eds.): NMA 2006, LNCS 4310, pp. 271–278, 2007.
© Springer-Verlag Berlin Heidelberg 2007

1. Split the matrix $A$ into a sum of two triangular matrices $L$ and $U$ such that $L$ is the lower triangular part of $A$ including the diagonal and $U$ is the strict upper triangular part of $A$. So $A = L + U$.
2. Set $A_0 = A = L_0 + U_0$
3. Suppose that $L_{k-1}$ is invertible for $k > 0$, and compute

$$A_k = L_{k-1}^{-1} A_{k-1} L_{k-1} = L_{k-1} + L_{k-1}^{-1} U_{k-1} L_{k-1}. \tag{1}$$

4. Split $A_k = L_k + U_k$.
5. Perform iterations until reaching the tolerance prescribed or the maximum number of iterations specified.

It is clear that the eigenvalues are invariant in all the computed matrices. Thus, if the method converges to a lower triangular matrix, then the eigenvalues are found on the main diagonal of the limit matrix.

## 2.2    Invariance of the Algorithm for the Hessenberg Form

Let us recall that a matrix $A = (a_{i,j})$ has a lower Hessenberg form if and only if $a_{i,j} = 0$ for $j \geq i + 2$ . Any matrix is similar to a Hessenberg matrix which can be obtained in finite time by the algorithms of Givens or Householder. Both algorithm are close to each other and require $O(n^3)$ operations. Classically, the LR and QR algorithm, being invariant for the Hessenberg form, start with a transformation of the intial matrix into this form. Let us see now that this is also possible for the AR algorithm.

**Proposition 1.** *If at iteration $k$ the matrix $A_k$ has a Hessenberg form, then this property also prevails at iteration $k + 1$ of the AR algorithm.*

**Proof.** The proof is a simple straightforward calculation. Suppose that a matrix $A$ has a lower Hessenberg form, that is $a_{ij} = 0$ for $j \geq i + 2$, then from the definition of the AR algorithm, the next iterate is

$$B = L^{-1} A L = L + L^{-1} U L. \tag{2}$$

Set $V = UL$, then $v_{i,j} = \sum_{k=\max\{i+1,j\}}^{n} a_{i,k} \, a_{k,j}$. But, since $a_{i,k} = 0$ for $k \geq i+2$, we have $v_{i,j} = a_{i,i+1} a_{i+1,j}$, if $j \leq i+1$, and $v_{i,j} = 0$ for $j \geq i+2$. Hence $V = UL$ has a Hessenberg form. It is easy to check that the product of a lower diagonal matrix $L^{-1}$ with a lower Hessenberg matrix $V$ gives a lower Hessenberg matrix. Thus $Z = L^{-1}UL$ is a lower Hessenberg matrix, and so is $B = L + Z$. $\qquad\square$

## 3    Convergence of the Additive Reduction Method

Finding a necessary and sufficient condition for the convergence of the AR algorithm is still an open problem. However, a sufficient condition for convergence in the case of a lower Hessenberg matrix can be easily shown.

**Proposition 2.** *Let $A = (a_{i,j})$ be a lower complex Hessenberg matrix. Assume that the following properties are fulfilled:*

*1) All the eigenvalues $\lambda_1$, $\lambda_2$, ..., $\lambda_n$ of $A$ are distinct in moduli.*

*2) At each iteration $k$ the matrix $A_k$ is such that*

$$|a_{1,1}^{(k)}| > |a_{2,2}^{(k)}| > \cdots > |a_{n,n}^{(k)}|,$$

*i.e., the muduli of its diagonal elements form a strictly decreasing sequence.*

*3) The maximum ratio $r = \max_{i=1}^{n-1} |a_{i+1,i+1}^{(k)}|/|a_{i,i}^{(k)}|$ is bounded by a constant $C < 1$, independent of $k$.*

*Then the additive reduction method is convergent.*

Here the matrix A is supposed to be complex, the operations involve complex numbers and the notation $|x|$ denotes the modulus of the complex number $x$.

**Proof.** Proposition 2 follows from a theorem due to Bauer and Fike [4]. As a consequence of this theorem, if $A$ is diagonalisable, i.e., $A = P^{-1}\Lambda P$ with $\Lambda = diag(\{\lambda_i\}_{i=1}^n)$, then for each eigenvalue $\mu$ of $L = A - U$ there exists an eigenvalue $\lambda$ of $A$ such that:

$$|\lambda - \mu| < \|P\|.\|P^{-1}\|.\|U\|.$$

Thus, it suffices to prove that under the three precedent hypotheses the norm $\|U_k\|$ of the matrix $U_k$ tends to zero. This will prove that the eigenvalues of $L$ tend to the eigenvalues of $A$, as all they are of different muduli.

Using notation and the formulae from Proposition 1, it is clear that $U_k$ is the strictly triangular superior part of the matrix $Z$. Thus, for $i = 1, ..., n - 1$,

$$u_{i,i+1}^{(k)} = u_{i,i+1}^{(k-1)} \, a_{i+1,i+1}^{(k-1)}/a_{i,i}^{(k-1)}$$

$$u_{i,j}^{(k)} = 0 \quad \text{for} \quad j \geq i + 2.$$

Here, the superscript $(k)$ denotes the number of the iteration. From assumption 3) we obtain

$$|u_{i,i+1}^{(k)}| \leq C \, |u_{i,i+1}^{(k-1)}| \quad \text{with} \quad C < 1,$$

and thus

$$|u_{i,i+1}^{(k)}| \leq C^k \, |u_{i,i+1}^{(0)}| \quad \text{with} \quad C < 1.$$

Let us now choose the classical norm $\|U\| = \max_{1 \leq i \leq n} \sum_{j=1}^n |u_{i,j}|$. In the present case $\|A_k\|$ is a lower Hessenberg matrix, $U_k$ has only its first over diagonal different from zero, that is, $u_{i,j}^{(k)} = 0$ for $i < j - 1$ and for $i \geq j$. The same property is also true for $U_{k+1}$. We have

$$\|U_k\| = \max_{1 \leq i \leq n-1} |u_{i,i+1}^{(k)}| \leq C \max_{1 \leq i \leq n-1} |u_{i,i+1}^{(k-1)}| = \|U_{k-1}\|,$$

and consequently

$$\|U_k\| \leq \ C^k \, \|U_0\| \quad \text{with} \quad C < 1. \tag{3}$$

This proves Proposition 2. □

*Remark 1.* If $A$ is a real matrix with different real eigenvalues, then all the computations are done with real numbers and formula (3) shows that convergence still occurs as the main over diagonal of the lower Hessenberg iterates tends to zero. Thus this method cannot, in theory, lead to erroneous complex conjugate eigenvalues.

*Remark 2.* In the case of a matrix $A$ with eigenvalues having different moduli, all the numerical experiments that have been done have shown that in most cases, even if at the beginning the elements of the main diagonal of $A$ are not ordered in decreasing order of their moduli, this property becomes true at some iteration and stays true during all remaining iterations.

*Remark 3.* As was already mentioned, a necessary and sufficient condition for convergence is not known, although the numerical experiments have shown that the method may converge even in the case of multiple eigenvalues.

*Remark 4.* From the definition of the algorithm, it is clear that its complexity per iteration is $O(n^3)$, as each iteration requires solution of a triangular linear system. Thus, the complexity of AR is of the same order as the one of QR.

## 4    Case of a Null Element on the Main Diagonal

The detection of the fact that at some iteration $L_k$ is not invertible is here very easy, as $L_k$ is lower triangular.

In this case, in the same way that it is done in the classical $QR$ or $LU$ methods, the algorithm can be continued by adding an arbitrary constant to all the elements of the diagonal of $A_k$. This shift is then subtracted from the computed values at the end of the process.

## 5    Deflations of Columns and Termination of Iterations

As was said above, the eigenvalues are obtained on the main diagonal of a lower triangular matrix, which is the limit of the process. We already mentioned that, in most cases, on this diagonal the eigenvalues appear experimentally to be in decreasing order, i.e., the smallest one is in the last line and last column. Thus the iterations can be stopped using at least two criteria.

- Criterion 1: The elements of the diagonal of the matrix $A_k$ become stable to some extent, i.e.,

$$\max_{1 \le n} |a_{i,i}^{(k)} - a_{i,i}^{(k-1)}| \le \varepsilon. \tag{4}$$

- Criterion 2: The elements of the upper triangular part $U_k$ of $A_k$ are stochastic zeroes, i.e. only correspond to round-off errors.

Let us recall here that a stochastic zero is a real or a floating point number which contains an error greater than itself. In other words, it has no significant digit. In the theory of stochastic arithmetic ([8], [9], [2], [3], [7]), any stochastic number is represented as $(m, \sigma)$, where $m$ is the mean value and $\sigma$ is the standard deviation of a Gaussian distribution, i.e., $m$ represents the value and $\sigma$ is the error on $m$. A stochastic zero is such that $\sigma \geq m$. The corresponding notion in the theory of interval arithmetic is an interval containing zero. Remember also that the number of significant digits of the result of any floating point computation and hence detection of a stochastic zero can be obtained using the CESTAC method ([8], [9]).

The experiments show another feature of the method: in the case of a complex matrix $A$ and computations using complex numbers or of a real matrix having real eigenvalues, during the iterations the matrices $U_k$ and hence the matrices $A_k$ have their columns tending to zero column after column, starting from the last. This means that column $n$ tends to zero, then column $n - 1$ and so on to column 1.

In the case of a given real matrix $A$ and computations done using only real numbers, there may exist complex conjugate roots. But it also happens that the columns corresponding to real roots tend to zero in the same order from the last one as in the complex case and that the columns corresponding to two complex conjugate roots, say $\lambda_j$ and $\lambda_{j+1}$ tend to zero except the four values $a_{j,j}$, $a_{j,j+1}$, $a_{j+1,j}$, $a_{j+1,j+1}$. The characteristic polynomial of this $2 \times 2$ block sub-matrix provides the two complex eigenvalues. This property is exactly the same as in QR or LR algorithms.

This experimental property can be exploited to increase the speed of convergence of the algorithm in the following manner: once the last column (resp, the last two columns) of the matrix $U_k$ is (are) considered sufficiently small in norm, or that each of its (resp. their) components is equal to a stochastic zero, then the corresponding eigenvalue is obtained and it is possible to reduce the size of the current matrix $A_k$ by suppressing the last (resp. last two) line(s) and column(s). This process can be re-done each time when a real or two complex conjugate eigenvalues are obtained until the dimension of the final matrix becomes 1 or 2. The speed of convergence of the algorithm is thus highly increased. Moreover, this property also experimentally diminishes the propagation of round-off errors and increases the accuracy of the results.

# 6    Numerical Experiments

Several numerical experiments have been done for small size matrices ($n \leq 100$). Many of them are taken from ([6], [5], [10]). Three of them are reported here to show the efficiency of the AR method. The results are compared with those provided by the classical QR algorithm [11]. The QR program is taken from [12], it provides the maximum possible accuracy. Since the original program does not give the number of necessary iterations to reach the solution, a counter has been added in order to compare the speed of convergence of the two methods. Only a very naive programming has been done for the AR method, in particular, there is no preconditioning. The termination criterion is on the stability of the diagonal of the iterates, i.e., criterion 1 with formula (4).

**Example 1:** The Fibonacci matrix

$$A_1 = \begin{pmatrix} 3 & 5 & 8 & 13 & 21 \\ 5 & 8 & 13 & 21 & 34 \\ 8 & 13 & 21 & 34 & 55 \\ 13 & 21 & 34 & 55 & 89 \\ 21 & 34 & 55 & 89 & 144 \end{pmatrix}$$

The obtained results are given in Table 1. These results show that the two algorithms are equivalent concerning the number of iterations, but the AR algorithm produces only real eigenvalues, which is closer to the real situation, as the initial matrix is symmetric.

**Table 1.** Eigenvalues of the Fibonacci matrix

| index | $AR\ \varepsilon = 10^{-5}$ Nb.iter = 5 | $QR$ Nb.iter = 4 |
|-------|------------------------------------------|-------------------|
| 1 | $0.2311038\ 10^3$ | $0.2311038\ 10^3$ |
| 2 | $-0.1038494\ 10^0$ | $-0.1038494\ 10^0$ |
| 3 | $-0.2687906\ 10^{-14}$ | $-0.3158887\ 10^{-14}$ |
| 4 | $-0.5342609\ 10^{-16}$ | $-0.1528318\ 10^{-28} + i\ 0.1209090\ 10^{-21}$ |
| 5 | $-0.16677753\ 10^{-28}$ | $-0.1528318\ 10^{-28} - i\ 0.1209090\ 10^{-21}$ |

**Example 2.** ([10])

$$A_2 = \begin{pmatrix} 10 & 1 & 4 & 0 \\ 1 & 10 & 5 & -1 \\ 4 & 5 & 10 & 7 \\ 0 & -1 & 7 & 9 \end{pmatrix}$$

The computed eigenvalues are shown in Table 2.

**Table 2.** Eigenvalues of matrix $A_2$ computed with AR and QR methods

| index | $AR\ \varepsilon = 10^{-5}$, Nb. iter = 56 | QR Nb. iter = 3 |
|-------|---------------------------------------------|------------------|
| 1 | $0.1912248\ 10^2$ | $0.1912248\ 10^2$ |
| 2 | $0.1088245\ 10^2$ | $0.5342609\ 10^{-3}$ |
| 3 | $0.8994541\ 10^1$ | $0.1088282\ 10^2$ |
| 4 | $0.5342609\ 10^{-3}$ | $0.8994170\ 10^1$ |

One can see that the results obtained with the two methods AR and QR are very close to each other, except that the first one provides the eigenvalues in decreasing order. For the AR algorithm with termination criterion 1 and $\varepsilon = 10^{-5}$, the convergence speed is noticeably slower compared to that of QR.

**Example 3.** A matrix with double eigenvalues (Rutishauser [10])

$$A_3 = \begin{pmatrix} 6 & 4 & 4 & 1 \\ 1 & 6 & 4 & 4 \\ 4 & 1 & 6 & 4 \\ 1 & 4 & 4 & 4 \end{pmatrix}$$

The exact eigenvalues are 15, 5, 2, 2. Both methods give the exact values. Again, the eigenvalues provided by the QR algorithm are obtained in the order 5, 2, 15, 2, whereas with the AR algorithm they are obtained in decreasing order. For the AR algorithm the iterations are stopped with criterion 1 and $\varepsilon = 10^{-7}$. The number of iterations is 20. For QR the number of iterations is 4, so QR is faster again.

**Example 4.** A real matrix with real opposite eigenvalues.

$$A_4 = \begin{pmatrix} 1.5 & 1 & -2 & 1 \\ 1 & 0.5 & -3 & -2 \\ -2 & -3 & -0.5 & -1 \\ 1 & -2 & -1 & -1.5 \end{pmatrix}$$

The eigenvalues of $A_4$ are $-1.5$, $+1.5$, $-4.5$, $+4.5$. They are exactly computed by both AR and QR algorithms, with 19 iterations for AR and 4 iterations for QR. As in the preceding examples, AR performs slightly slower than QR. However, it must be reminded that the program for AR is a very simple and naive one, particularly concerning the termination criterion, while the program for QR is rather sophisticated.

# 7 Conclusion

In this paper we have shown that the eigenvalues of a real or complex square matrix can be computed with a method called additive reduction method, which is different from the classical methods. A sufficient condition for convergence has been proved, and it has been also proved that is has the same properties as the QR or LU algorithms. In particular, it is invariant for the Hessenberg form and has the same order of complexity per iteration as QR. Some other experimental properties is also given: the eigenvalues are obtained in decreasing order on the main diagonal of a lower triangular matrix, and consequently the algorithm provides them the last one (i.e. the smallest) first. Many numerical experiments have shown that the method is often as good as the classical general methods for eigenvalues, and in some cases even better, particularly for a real matrix with real eigenvalues. Some of these experiments are reported here. It is clear that, in many cases, this method is not really competitive with QR, but it shows that the eigenvalue problem can be also solved by a different approach from the classical one. Many questions are still open, for example, find a necessary and sufficient condition for convergence, explain why in most cases the eigenvalues are obtained the smallest first.

# References

1. Alt, R.: Un algorithme simple et efficace de calcul de valeurs propres. C. R. Acad Sc. Paris **306(1)** 1988 437–440
2. Alt, R., Markov, S.: On the Algebraic Properties of Stochastic Arithmetic. Comparison to Interval Arithmetic. In: W. Kraemer and J. Wolff von Gudenberg, (eds.): Scientific Computing, Validated Numerics, Interval Methods. Kluwer (2001) 331–341
3. Alt R., Lamotte J.-L., Markov, S.: Numerical Study of Algebraic Solutions to Linear Problems Involving Stochastic Parameters. Large Scale Scientific Computing conf. Lecture Notes in Computer Science Vol. 3743 (2006) 273–280
4. Bauer, F. L., Fike, C. T.: Norms and exclusion theorems. Numer. Math. **2** (1960) 137–141
5. Gregory, R., Kerney, D.: A collection of matrices for testing computational algorithms. Wiley, New-York (1969)
6. Higham, N.: Algorithm 694: A Collection of Test Matrices in MATLAB. ACM Transactions on Mathematical Software Vol. 17, no. 3 (1991) 289–305
7. Markov, S., Alt, R.: Stochastic arithmetic: Addition and multiplication by scalars. Applied Num. Math. **50** (2004) 475–488
8. Vignes, J., Alt, R.: An Efficient Stochastic Method for Round-Off Error Analysis. In: Accurate Scientific Computations. Lecture Notes in Computer Science Vol. 235. Springer (1985) 183–205
9. Vignes, J.: A Stochastic Arithmetic for Reliable Scientific Computation. Math. Comp. in Sim. **35** (1993) 233–261
10. Westlake, J. R.: A handbook of Numerical Matrix inversion and solution of linear equations. J. Wiley (1968) 136–157
11. Wilkinson, J.: Convergence of the LR, QR and related algorithms. Computer J. **4** (1965) 77–84
12. Wilkinson, J. H., Reinsch, C. (eds.): Linear Algebra. Handbook for Automatic Computation Vol. II. Springer Verlag, New York (1971)

# Numerical Computations with Hausdorff Continuous Functions

Roumen Anguelov[1] and Svetoslav Markov[2]

[1] University of Pretoria
Pretoria 0002, South Africa
roumen.anguelov@up.ac.za
[2] Institute of Mathematics and Informatics, BAS
"Acad. G. Bonchev" st., block 8, 1113 Sofia, Bulgaria
smarkov@bio.bas.bg

## 1 Introduction

Hausdorff continuous (H-continuous) functions appear naturally in many areas of mathematics such as Approximation Theory [11], Real Analysis [1], [8], Interval Analysis, [2], etc. From numerical point of view it is significant that the solutions of large classes of nonlinear partial differential equations can be assimilated through H-continuous functions [7]. In particular, discontinuous viscosity solutions are better represented through Hausdorff continuous functions [6]. Hence the need to develop numerical procedures for computations with H-continuous functions. It was shown recently, that the operations addition and multiplication by scalars of the usual continuous functions on $\Omega \subseteq \mathbb{R}^n$ can be extended to H-continuous functions in such a way that the set $\mathbb{H}(\Omega)$ of all Hausdorff continuous functions is a linear space [4]. In fact $\mathbb{H}(\Omega)$ is the largest linear space involving interval functions. Furthermore, multiplication can also be extended [5], so that $\mathbb{H}(\Omega)$ is a commutative algebra. Approximation of $\mathbb{H}(\Omega)$ by a subspace were discussed in [3]. In the present paper we consider numerical computations with H-continuous functions using ultra-arithmetical approach [9], namely, by constructing a functoid of H-continuous functions. For simplicity we consider $\Omega \subseteq \mathbb{R}$. In the next section we recall the definition of the algebraic operations on $\mathbb{H}(\Omega)$. The concept of functoid is defined in Section 3. In Section 4 we construct a functoid comprising a finite dimensional subspace of $\mathbb{H}(\Omega)$ with a Fourier base extended by a set of H-continuous functions. Application of the functoid to the numerical solution of the wave equation is discussed in Section 5.

## 2 The Algebra of H-Continuous Functions

The real line is denoted by $\mathbb{R}$ and the set of all finite real intervals by $\mathbb{IR} = \{[\underline{a}, \overline{a}] : \underline{a}, \overline{a} \in \mathbb{R}, \underline{a} \leq \overline{a}\}$. Given an interval $a = [\underline{a}, \overline{a}] \in \mathbb{IR}$, $w(a) = \overline{a} - \underline{a}$ is the width of $a$. An interval $a$ is called proper interval, if $w(a) > 0$ and point interval, if $w(a) = 0$. Identifying $a \in \mathbb{R}$ with the point interval $[a, a] \in \mathbb{IR}$, we consider $\mathbb{R}$ as a subset of $\mathbb{IR}$. Let $\Omega \subseteq \mathbb{R}$ be open. We recall [11] that an interval

T. Boyanov et al. (Eds.): NMA 2006, LNCS 4310, pp. 279–286, 2007.
© Springer-Verlag Berlin Heidelberg 2007

function $f : \Omega \to \mathbb{IR}$ is S-continuous if its graph is a closed subset of $\Omega \times \mathbb{R}$. An interval function $f : \Omega \to \mathbb{IR}$ is Hausdorff continuous (H-continuous) if it is an S-continuous function which is minimal with respect to inclusion, that is, if $\varphi : \Omega \to \mathbb{IR}$ is an S-continuous function and $\varphi \subseteq f$, then $\varphi = f$. Here inclusion is understood point-wise. We denote by $\mathbb{H}(\Omega)$ the set of H-continuous functions on $\Omega$. The following theorem states an essential property of the continuous functions which is preserved by the H-continuity [1].

**Theorem 1.** *Let $f, g \in \mathbb{H}(\Omega)$. If there exists a dense subset $D$ of $\Omega$ such that $f(x) = g(x)$, $x \in D$, then $f(x) = g(x)$, $x \in \Omega$.*

H-continuous functions are also similar to usual continuous real functions in that they assume point values on a residual subset of $\Omega$. More precisely, it is shown in [1] that for every $f \in \mathbb{H}(\Omega)$ the set $W_f = \{x \in \Omega : w(f(x)) > 0\}$ is of first Baire category and $f$ is continuous on $\Omega \setminus W_f$. Since a finite or countable union of sets of first Baire category is also a set of first Baire category we have:

**Theorem 2.** *Let $\mathcal{F}$ be a finite or countable set of H-continuous functions. Then the set $D_{\mathcal{F}} = \{x \in \Omega : w(f(x)) = 0, \ f \in \mathcal{F}\} = \Omega \setminus \bigcup_{f \in \mathcal{F}} W_f$ is dense in $\Omega$ and all functions $f \in \mathcal{F}$ are continuous on $D_{\mathcal{F}}$.*

For every S-continuous function $g$ we denote by $[g]$ the set of H-continuous functions contained in $g$, that is,

$$[g] = \{f \in \mathbb{H}(\Omega) : f \subseteq g\}.$$

Identifying $\{f\}$ with $f$ we have $[f] = f$ whenever $f$ is H-continuous. The S-continuous functions $g$ such that the set $[g]$ is a singleton, that is, it contains only one function, play an important role in the sequel. In analogy with the H-continuous functions, which are minimal S-continuous functions, we call these functions quasi-minimal. The following characterization of the quasi-minimal S-continuous functions is an easy consequence of Theorem 1.

**Theorem 3.** *If the function $f$ is S-continuous on $\Omega$ and assumes point values on a dense subset of $\Omega$, then $f$ is a quasi-minimal S-continuous function.*

The familiar operations of addition, multiplication by scalars and multiplication on the set of real intervals are defined for $[\underline{a}, \overline{a}], [\underline{b}, \overline{b}] \in \mathbb{IR}$ and $\alpha \in \mathbb{R}$ as follows:

$$[\underline{a}, \overline{a}] + [\underline{b}, \overline{b}] = \{a + b : a \in [\underline{a}, \overline{a}], b \in [\underline{b}, \overline{b}]\} = [\underline{a} + \underline{b}, \overline{a} + \overline{b}],$$
$$\alpha \cdot [\underline{a}, \overline{a}] = \{\alpha a : a \in [\underline{a}, \overline{a}]\} = [\min\{\alpha \underline{a}, \alpha \overline{a}\}, \max\{\alpha \underline{a}, \alpha \overline{a}\}],$$
$$[\underline{a}, \overline{a}] \times [\underline{b}, \overline{b}] = \{ab : a \in [\underline{a}, \overline{a}], b \in [\underline{b}, \overline{b}]\} = [\min\{\underline{a}\underline{b}, \underline{a}\overline{b}, \overline{a}\underline{b}, \overline{a}\overline{b}\}, \max\{\underline{a}\underline{b}, \underline{a}\overline{b}, \overline{a}\underline{b}, \overline{a}\overline{b}\}].$$

Point-wise operations for interval functions are defined in the usual way:

$$(f + g)(x) = f(x) + g(x), \ (\alpha \cdot f)(x) = \alpha \cdot f(x), \ (f \times g)(x) = f(x) \times g(x). \quad (1)$$

It is easy to see that the set of S-continuous functions is closed under the above point-wise operations while the set of H-continuous functions is not, see [2], [4]. Hence the significance of the following theorem.

**Theorem 4.** *For any $f, g \in \mathbb{H}(\Omega)$ and $\alpha \in \mathbb{R}$ the functions $f + g$, $\alpha \cdot f$ and $f \times g$ are quasi-minimal S-continuous functions.*

*Proof.* Denote by $D_{fg}$ the subset of $\Omega$ where both $f$ and $g$ assume point values. Then $f + g$ assumes point values on $D_{fg}$. According to Theorem 2 the set $D_{fg}$ is dense in $\Omega$, which in terms of Theorem 3 implies that $f + g$ is quasi-minimal. The quasi-minimality of $\alpha \cdot f$ and $f \times g$ is proved in a similar way.  □

We define the algebraic operations on $\mathbb{H}(\Omega)$ using Theorem 4. We denote these operations respectively by $\oplus$, $\odot$ and $\otimes$ so that distinction from the pointwise operations can be made.

**Definition 1.** *Let $f, g \in \mathbb{H}(\Omega)$ and $\alpha \in \mathbb{R}$. Then*

$$f \oplus g = [f + g], \quad \alpha \odot f = [\alpha \cdot f], \quad f \otimes g = [f \times g]. \tag{2}$$

**Theorem 5.** *The set $\mathbb{H}(\Omega)$ is a commutative algebra with respect to the operations $\oplus$, $\odot$ and $\otimes$ given in (2).*

The proof will be omitted; it involves standard techniques and is partially discussed in [5].

# 3    The Concept of Ultra-Arithmetical Functoid

Functoid is a structure resulting from the ultra-arithmetical approach to the solution of problems in functional spaces. The aim of ultra-arithmetic is the development of structures, data types and operations corresponding to functions for direct digital implementation. On a digital computer equipped with ultra-arithmetic, problems associated with functions are solvable, just as now we solve algebraic problems [9]. Ultra-arithmetic is developed in analogy with the development of computer arithmetic.

Let $\mathcal{M}$ be a space of functions and let $M$ be a finite dimensional subspace spanned by $\Phi_N = \{\varphi_k\}_{k=0}^N$. Every function $f \subset \mathcal{M}$ is approximated by $\tau_N(f) \in M$. The mapping $\tau_N$ is called rounding (in analogy with the rounding of numbers) and the space $M$ is called a screen of $\mathcal{M}$. Every rounding must satisfy the requirement (invariance of rounding on the screen): $\tau_N(f) = f$ for every $f \in M$. A function $f = \sum_{i=0}^N \alpha_i \varphi_i \in M$ can be represented by its coefficient vector $\nu(f) = (\alpha_0, \alpha_1, \ldots, \alpha_N)$. Therefore the approximation of the functions in $\mathcal{M}$ is realized through the mappings $\mathcal{M} \xrightarrow{\tau_N} M \xleftarrow{\nu} K^{N+1}$, where $K$ is the scalar field of $\mathcal{M}$ (i.e. $K = \mathbb{R}$ or $K = \mathbb{C}$). Since $\nu$ is a bijection we can identify $M$ and $K^{N+1}$ and consider only the rounding $\tau_N$.

In $\mathcal{M}$ we consider the operations addition ($+$), multiplication by scalars ($.$), multiplication of functions ($\times$) and integration ($\int$) defined in the conventional way. By the semimorphism principle $\tau_N$ induces corresponding operations in $M$:

$$f \boxed{\circ} g = \tau_N(f \circ g), \quad \circ \in \{+, ., \times\};$$

$$\boxed{\oint} f = \tau_N \left( \int f \right).$$

The structure $(M, \boxplus, \boxdot, \boxtimes, \boxed{\oint})$ is called an (ultra-arithmetical) functoid [10].

## 4    A Functoid in $\mathbb{H}(\Omega)$

To simplify matters we consider the space of all bounded H-continuous functions on $\Omega = (-1, 1)$. Furthermore, since we shall often use a shift of the argument, we assume that all functions are produced periodically (period 2) over $\mathbb{R}$ and denote the space under consideration by $\mathbb{H}_{per}(-1, 1)$. All algebraic operations on $\mathbb{H}_{per}(-1, 1)$ are considered in terms of Definition 1. For simplicity we denote them as the operations for reals. Namely, addition is "$+$" and a space is interpreted as multiplication, where the context shows whether this is a multiplication by scalars or product of functions. In particular, note that indicating the argument of a function in a formula does not mean point-wise operation. Denote by $s_1$ the H-continuous function given by

$$s_1(x) = \begin{cases} x, & \text{if } x \in (-1, 1), \\ [-1, 1], & \text{if } x = \pm 1; \end{cases}$$

and produced periodically over the real line. Since the integrals of $\underline{s}_1$ and $\overline{s}_1$ are equal over any interval the integral of $s_1$ is a usual real function. We construct iteratively the sequence of periodic splines $s_1, s_2, s_3, \dots$ using

$$s_{j+1} = \int s_j(x)dx + c,$$

$$\int_{-1}^{1} s_{j+1}(x)dx = s_{j+2}(1) - s_{j+2}(-1) = 0.$$

**Theorem 6.** *Let $f \in \mathbb{H}_{per}(-1, 1)$ be given. Assume that there exists a finite set $\Lambda = \{\lambda_1, \lambda_2, \dots, \lambda_m\} \subset (-1, 1]$ such that $f$ assumes real values and is $p$ times differentiable on $(-1, 1] \setminus \Lambda$ with the $p$-th derivative in $L^2(-1, 1)$. Then $f$ has a unique representation in the form*

$$f(x) = a_0 + \sum_{l=1}^{m} \sum_{j=1}^{p} a_{jl}s_j(x + 1 - \lambda_l) + \sum_{\substack{k=-\infty \\ k \neq 0}}^{\infty} b_k e^{ik\pi x}, \tag{3}$$

*where $\sum_{\substack{k=-\infty \\ k \neq 0}}^{\infty} b_k e^{ik\pi x}$ is $p$ times differentiable with its $p$-th derivative in $L^2(-1, 1)$.*

*Furthermore, the coefficients are given by:*

$$a_0 = \frac{1}{2} \int_{-1}^{1} f(x)dx,$$

$$a_{jl} = \frac{1}{2} \left( \frac{d^{j-1}f}{dx^{j-1}}(\lambda_l - 0) - \frac{d^{j-1}f}{dx^{j-1}}(\lambda_l + 0) \right), \quad j = 1, \dots, p, \; l = 1, \dots, m,$$

$$b_k = \frac{1}{2(ik\pi)^p} \int_{-1}^{1} \frac{d^p f(x)}{dx^p} e^{-ik\pi x} dx, \quad k = \pm 1, \pm 2, \dots$$

The proof uses standard techniques and will be omitted.

The function $f$ is approximated by

$$\rho_{Np}(f;x) = a_0 + \sum_{l=1}^{m}\sum_{j=1}^{p} a_{jl}s_j(x+1-\lambda_l) + \sum_{\substack{k=-N \\ k\neq 0}}^{N} b_k e^{ik\pi x}, \tag{4}$$

with a rounding error

$$|f(x) - \rho_{Np}(f;x)| = \left|\sum_{|k|>N} b_k e^{ik\pi x}\right| \le \sum_{|k|>N} |b_k|$$

$$\le \left(\sum_{|k|>N} (k\pi)^{2p}|b_k|^2\right)^{\frac{1}{2}} \left(\sum_{|k|>N} \frac{1}{(k\pi)^{2p}}\right)^{\frac{1}{2}} \tag{5}$$

$$\le \left(\frac{1}{2}\int_{-1}^{1}\left(\frac{d^p f(x)}{dx^p}\right)^2 dx - \left(\sum_{l=1}^{m} a_{pl}\right)^2 - \sum_{\substack{k=-N \\ k\neq 0}}^{N} (k\pi)^{2p}|b_k|^2\right)^{\frac{1}{2}} \left(\frac{2}{(2p-1)\pi^{2p}N^{2p-1}}\right)^{\frac{1}{2}}$$

$$= o\left(\frac{1}{N^{p-\frac{1}{2}}}\right).$$

Motivated by the above we consider a screen in $\mathbb{H}_{per}(-1,1)$ comprising the subspace $M$ spanned by the basis

$$\{s_0(x)\} \cup \{s_j(x+1-\lambda_l) : j = 0,1,..,p,\ l = 1,..m\} \cup \{e^{ik\pi x} : k = 0, \pm 1, ..., \pm N\},$$

where $p, m, N \in \mathbb{N}$ and $\{\lambda_1, \lambda_2, ..., \lambda_m\} \subset (-1, 1]$ are parameters with arbitrary but fixed values. Here $s_0$ is the function which is the constant 1 on $\mathbb{R}$. Defining a rounding from $\mathbb{H}_{per}(-1,1)$ to $M$ is still an open problem. However, for functions of the type described in Theorem 6 the rounding is defined through $\rho_{Np}$. Furthermore, to define a functoid we only need to know how to round the functions resulting from operations in $M$. For this purpose the rounding $\rho_{Np}$ is sufficient. Naturally, since $M$ is a subspace, it is closed under the operations addition and multiplication by scalars. Furthermore, to define multiplication of functions and integration we only need to define these operations on the elements of the basis. The products of the functions in the basis are given by

$$s_{q_1}(x+1-\lambda_{l_1})s_{q_2}(x+1-\lambda_{l_2})$$

$$= \sum_{j=q_1}^{q_1+q_2} \binom{j-1}{q_1-1} s_{q_1+q_2-j}(1+\lambda_{l_1}-\lambda_{l_2})s_j(x+1-\lambda_{l_1}) \tag{6}$$

$$+ \sum_{j=q_2}^{q_1+q_2} \binom{j-1}{q_2-1} s_{q_1+q_2-j}(1-\lambda_{l_1}+\lambda_{l_2})s_j(x+1-\lambda_{l_2}),$$

$$e^{ik_1\pi x}e^{ik_2\pi x} = e^{i(k_1+k_2)\pi x}, \tag{7}$$

$$s_q(x + 1 - \lambda_l)e^{in\pi x}$$

$$= \sum_{j=q}^{p} (-1)^n e^{i(\lambda_l - 1)\pi} \binom{j-1}{q-1} (in\pi)^{j-q} s_j(x + 1 - \lambda_l) + \sum_{\substack{k=-\infty \\ k \neq 0}}^{\infty} \beta_k e^{ik\pi x}, \quad (8)$$

where the coefficients $\beta_k$ in (8) are given by

$$\beta_k = \frac{(-1)^{k-n-1} n^{p-q}}{k^p (i\pi)^q} \sum_{r=0}^{q-1} \binom{p}{r} \left(\frac{n}{k-n}\right)^{q-r}, \quad \text{if } k \neq 0, n,$$

$$\beta_n = \binom{p}{q} (in\pi)^{-q}.$$

For the respective integrals we have

$$\int s_j(x)dx = s_{j+1}(x), \quad j = 1, ..., p, \quad (9)$$

$$\int e^{ik\pi x} dx = \frac{1}{ik\pi} e^{ik\pi x}, \quad k = 0, \pm 1, ..., \pm N. \quad (10)$$

Obviously, in formulas (6)–(9) we obtain splines $s_j$ with $j > p$ and exponents $e^{ik\pi x}$ with $|k| > N$, which need to be rounded. Using that

$$s_j(x + 1 - \lambda_l) = \sum_{\substack{k=-\infty \\ k \neq 0}}^{\infty} \frac{(-1)^{k-1} e^{i(1-\lambda_l)\pi}}{(ik\pi)^j} e^{ik\pi},$$

the rounding of integrals and products of functions in $\mathbb{H}_{per}(-1, 1)$ calculated via (6)–(9) is reduced to rounding a Fourier series which is done by truncation. Note that at any time we truncate a Fourier series of a function which is at least $p$ times differentiable with its $p$-th derivative in $L^2(-1, 1)$. Hence the uniform norm of the error is $o\left(\frac{1}{N^{p-\frac{1}{2}}}\right)$. This in particular implies that although the approximated functions are discontinuous at certain points undesirable effects such as the Gibbs phenomenon do not occur. Furthermore, the rate of approximation with respect to the uniform norm is the same as for $p$ times differentiable functions. It should be also noted that the integration of $s_0$, when it arises in practical problems, should be handled with special care as $\int s_0(x)dx = s_1(x)$ holds only on $(-1, 1)$.

## 5   Application to the Wave Equation

We consider the wave equation in the form

$$u_{tt}(x, t) - u_{xx}(x, t) = \rho(t)u(x, t) + \phi(x, t)$$
$$u(x, 0) = g_1(x), \quad u_t(x, 0) = g_2(x)$$

with periodic boundary conditions at $x = -1$ and $x = 1$, assuming that $g_1, g_2, \phi$ or some of their space derivatives may be discontinuous but the functions can be represented as a spline-Fourier series (3) of the space variable. An approximation to the solution is sought in the form

$$u(x,t) = a_0(t) + \sum_{l=1}^{m}\sum_{j=1}^{p}\sum_{\delta\in\{-1,0,1\}} a_{lj\delta}(t)s_j(x+\delta t+1-\alpha_l)$$

$$+ \sum_{\substack{k=-N\\k\neq 0}}^{N} b_k(t)e^{ik\pi x} \tag{11}$$

wherein $\alpha_l$, $l=1,\ldots,m$, are points in $(-1,1]$ where the data functions or some of their first $p-1$ derivatives may be discontinuous.

The following Newton-type iterative procedure is applied

$$u^{(r+1)} = (1-\lambda)u^{(r)} + \lambda\left(g + \frac{1}{2}\iint\limits_{G(x,t)} \rho u^{(r)}\right),$$

where $G(x,t)$ is the triangle with vertices $(x,t)$, $(x-l,0)$, $(x+t,0)$ and

$$g(x,t) = \frac{1}{2}\left(g_1(x+t) + g_1(x-t) + \int_{x-t}^{x+t} g_2(\theta)d\theta + \iint\limits_{G(x,t)} \phi(y,\theta)dyd\theta\right).$$

The essential part of each iteration is the evaluation of the integral. This can be done successfully using the arithmetic in the functoid discussed in the preceding section. We also have to choose some form of representation of the coefficients $a_{mj}(t)$, $b_k(t)$. Here we carry out the computations representing those coefficients as polynomials of $t$. The following formulas are used:

$$\iint\limits_{G(x,t)} \frac{\theta^q}{q!}s_j(y)dyd\theta = s_{j+q+2}(x+t) + (-1)^q s_{j+q+2}(x-t) - 2\sum_{\substack{l=0\\l-\text{even}}}^{q} \frac{t^{q-l}}{(q-l)!}s_{j+l+2}(x),$$

$$\iint\limits_{G(x,t)} \frac{\theta^q}{q!}s_j(y+\theta)dyd\theta = \sum_{l=0}^{q+1}\left(-\frac{1}{2}\right)^l \frac{t^{q+1-l}}{(q+1-l)!}s_{j+l+1}(x+t) - \left(-\frac{1}{2}\right)^{q+1}s_{j+q+2}(x-t),$$

$$\iint\limits_{G(x,t)} \frac{\theta^q}{q!}s_j(y-\theta)dyd\theta = \left(\frac{1}{2}\right)^{q+1}s_{j+q+2}(x+t) - \sum_{l=0}^{q+1}\left(\frac{1}{2}\right)^l \frac{t^{q+1-l}}{(q+1-l)!}s_{j+l+1}(x-t),$$

$$\iint\limits_{G(x,t)} \frac{\theta^q}{q!}e^{ik\pi y}dyd\theta = \frac{1}{(ik\pi)^{q+2}}\left(e^{ik\pi(x+t)} + (-1)^q e^{ik\pi(x-t)} - 2\sum_{\substack{l=0\\l-\text{even}}}^{q} \frac{t^{q-l}}{(ik\pi)^{q-l}(q-l)!}e^{ik\pi x}\right)$$

$$= 2\sum_{\substack{l=0\\l-\text{even}}}^{\infty} (ik\pi)^l \frac{t^{q+l+2}}{(q+l+2)!}e^{ik\pi x}.$$

The splines $s_j$ for $j > p$ as well as the infinite series in the last formula above are approximated by a partial sum of the resp. Fourier series using the rounding $\rho_{Np}$. As was shown in Section 4, the truncation error is $o\left(\frac{1}{N^{p-\frac{1}{2}}}\right)$. The main advantage of the method is that it produces highly accurate results for relatively small values of $p$ and $N$ for non-smooth data functions. Numerical experiments using $p = 5$ and $N = 5$ produce 4–5 correct decimal digits of the solution.

# 6    Conclusion

In this work we propose a new methodology for numerical computations with H-continuous functions. We propose a method based on the fact that H-continuous functions form a linear space when addition is defined in a suitable way. Our method makes use of the ultra-arithmetic approach for the construction of a relevant functoid. The method has been tested numerically for the solution of the wave equation for non-smooth boundary conditions. Highly accurate results have been achieved for rather small number of base functions, i. e. small dimensions of the underlying linear space. No Gibbs phenomenon occur.

# References

1. Anguelov, R.: Dedekind order completion of C(X) by Hausdorff continuous functions. Quaestiones Mathematicae **27** (2004) 153–170
2. Anguelov, R., Markov, S.: Extended Segment Analysis. Freiburger Intervall-Berichte **10**. Inst. Angew. Math., Univ. Freiburg i. Br. (1981) 1–63
3. Anguelov, R., Markov, S., Sendov, Bl.: On the Normed Linear Space of Hausdorff Continuous Functions. Lecture Notes in Computer Science **3743** (2005) 281–288
4. Anguelov, R., Markov, S., Sendov, Bl.: The Set of Hausdorff Continuous Functions—the Largest Linear Space of Interval Functions. Reliable Computing **12** (2006) 337–363
5. Anguelov, R., Markov, S., Sendov, Bl.: Algebraic Operations on the Space of Hausdorff Continuous Interval Functions, In: B. D. Bojanov, Ed.: Constructive Theory of Functions Varna 2005, Marin Drinov Acad. Publ. House (2006), Sofia, 35–44
6. Anguelov, R., Minani, F.: Interval Viscosity Solutions of Hamilton-Jacobi Equations. Technical Report UPWT 2005/3, University of Pretoria, 2005
7. Anguelov, R., Rosinger, E. E.: Hausdorff Continuous Solutions of Nonlinear PDEs through the Order Completion Method. Quaestiones Mathematicae **28**(3) (2005) 271–285
8. Anguelov, R., van der Walt, J. H.: Order Convergence Structure on C(X). Quaestiones Mathematicae **28**(4) (2005) 425–457
9. Epstein, C., Miranker, W. L., Rivlin, T. J.: Ultra Arithmetic, Part I: Function Data Types, Part 2: Intervals of Polynomials. Mathematics and Computers in Simulation **24** (1982) 1–18
10. Kaucher, E., Miranker, W.: Self-Validating Numerics for Function Space Problems. Academic Press, New York (1984)
11. Sendov, Bl.: Hausdorff Approximations. Kluwer (1990)

# Mixed Discretization-Optimization Methods for Nonlinear Elliptic Optimal Control Problems

Ion Chryssoverghi

Department of Mathematics, National Technical University of Athens
Zografou Campus, 15780 Athens, Greece
ichris@central.ntua.gr

**Abstract.** An optimal control problem is considered, for systems governed by a nonlinear elliptic partial differential equation, with control and state constraints. Since this problem may have no classical solutions, it is also formulated in the relaxed form. The classical problem is discretized by using a finite element method, where the controls are approximated by elementwise constant, linear, or multilinear, controls. Our first result is that strong accumulation points in $L^2$ of sequences of admissible and extremal discrete controls are admissible and weakly extremal classical for the continuous classical problem, and that relaxed accumulation points of sequences of admissible and extremal discrete controls are admissible and weakly extremal relaxed for the continuous relaxed problem. We then propose a penalized gradient projection method, applied to the discrete problem, and a corresponding discretization-optimization method, applied to the continuous classical problem, that progressively refines the discretization during the iterations, thus reducing computing time and memory. We prove that accumulation points of sequences generated by the first method are admissible and extremal for the discrete problem, and that strong classical (resp. relaxed) accumulation points of sequences of discrete controls generated by the second method are admissible and weakly extremal classical (resp. relaxed) for the continuous classical (resp. relaxed) problem. Finally, numerical examples are given.

## 1 The Continuous Optimal Control Problems

Let $\Omega$ be a bounded domain in $\mathbb{R}^d$, with Lipschitz boundary $\Gamma$. Consider the nonlinear elliptic state equation
$$Ay + f(x, y(x), w(x)) = 0 \text{ in } \Omega, \quad y(x) = 0 \text{ on } \Gamma,$$
where $A$ is the formal second order elliptic differential operator
$$Ay := -\sum_{j=1}^{d} \sum_{i=1}^{d} (\partial/\partial x_i)[a_{ij}(x)\partial y/\partial x_j].$$
The state equation will be interpreted in the following weak form
$$y \in V := H_0^1(\Omega) \text{ and } a(y, v) + \int_\Omega f(x, y(x), w(x))v(x)dx = 0, \forall v \in V,$$
where $a(\cdot, \cdot)$ is the usual bilinear form on $V \times V$ associated with $A$
$$a(y, v) := \sum_{i,j=1}^{d} \int_\Omega a_{ij}(x)(\partial y/\partial x_i)(\partial v/\partial x_j)dx.$$

T. Boyanov et al. (Eds.): NMA 2006, LNCS 4310, pp. 287–295, 2007.

Define the set of *classical controls*

$W := \{w : \Omega \to U \,|\, w \text{ measurable}\} \subset L^\infty(\Omega; \mathbb{R}^\nu) \subset L^2(\Omega; \mathbb{R}^\nu),$

where $U$ is a compact subset of $\mathbb{R}^\nu$, and the functionals

$G_m(w) := \int_\Omega g_m(x, y(x), w(x))dx, \quad m = 0, ..., q.$

The continuous classical optimal control problem $P$ is to minimize $G_0(w)$ subject to the constraints $w \in W$, $G_m(w) = 0$, $m = 1, ..., p$, $G_m(w) \leqslant 0$, $m = p + 1, ..., q$.

Next, define the set of *relaxed controls* (see [13,11])

$R := \{r : \Omega \to M_1(U) \,|\, r \text{ weakly measurable}\} \subset L_w^\infty(\Omega, M(U)) \equiv L^1(\Omega, C(U))^*,$

where $M(U)$ (resp. $M_1(U)$) is the set of Radon (resp. probability) measures on $U$. The set $R$ is endowed with the relative weak star topology, and $R$ is convex, metrizable and compact. If each classical control $w(\cdot)$ is identified with its associated Dirac relaxed control $r(\cdot) := \delta_{w(\cdot)}$, then $W$ may also be regarded as a subset of $R$, and $W$ is thus dense in $R$. For a given $\phi \in L^1(\Omega; C(U)) \equiv B(\bar\Omega, U; \mathbb{R})$, where $B(\bar\Omega, U; \mathbb{R})$ denotes the set of Caratheodory functions in the sense of Warga [13], and $r \in R$, we shall write for simplicity

$\phi(x, r(x)) := \int_U \phi(x, u)r(x)(du).$

The continuous relaxed optimal control Problem $\bar P$ is then defined by replacing $w$ by $r$ (with the above notation) and $W$ by $R$ in the continuous classical problem.

We suppose that the coefficients $a_{ij}$ satisfy the ellipticity condition

$$\sum_{i,j=1}^d a_{ij}(x)z_i z_j \geqslant \alpha_0 \sum_{i=1}^d z_i^2, \quad \forall z_i, z_j \in \mathbb{R}, \quad x \in \Omega,$$

with $\alpha_0 > 0$, $a_{ij} \in L^\infty(\Omega)$, and that the functions $f, f_y, f_u$ (resp $g_m, g_{my}, g_{mu}$) are defined on $\Omega \times \mathbb{R} \times U'$ (resp. on $\Omega \times \mathbb{R} \times U'$), with $U' \supset U$ open, measurable for fixed $y, u$, continuous for fixed $x$, and satisfy in $\Omega \times \mathbb{R} \times U$

$|f(x, y, u)| \leqslant c_1(1 + |y|^{\rho-1}), \quad 0 \leqslant f_y(x, y, u) \leqslant c_2(1 + |y|^{\rho-2}),$

$|f_u(x, y, u)| \leqslant c_3(1 + |y|^{\rho-1}),$

$|g_m(x, y, u)| \leqslant c_4(1 + |y|^\rho), \quad |g_{my}(x, y, u)| \leqslant c_5(1 + |y|^{\rho-1}),$

$|g_{mu}(x, y, u)| \leqslant c_6(1 + |y|^{\frac{\rho}{2}}),$

with $c_i \geqslant 0$, $2 \leqslant \rho < +\infty$ if $d = 1$ or 2, $2 \leqslant \rho < \frac{2d}{d-2}$ if $d \geqslant 3$.

For every $r \in R$, the state equation has a unique solution $y := y_r \in V$ (see [2]). The results of this section can be proved by using the techniques of [13,7]. The weak relaxed minimum principle in Theorem 2 is shown similarly to Theorem 2.2 in [9].

**Theorem 1.** *If the relaxed problem is feasible, then it has a solution.*

**Lemma 1.** *Dropping the index m in the functionals, the directional derivative of G defined on R (resp. W, with U convex), is given by*

$DG(r, \bar r - r) = \lim_{\alpha \to 0^+} \{[G(r + \alpha(\bar r - r)) - G(r)]/\alpha\}$

$= \int_\Omega H(x, y(x), z(x), r'(x) - r(x))dx, \text{ for } r, \bar r \in R,$

*(resp.* $DG(w, \bar w - w) = \lim_{\alpha \to 0^+} \{[G(w + \alpha(\bar w - w)) - G(w)]/\alpha\}$

$= \int_\Omega H_u(x, y(x), z(x), w(x))(\bar w(x) - w(x))dx, \text{ for } w, \bar w \in W$ *),*

*where the Hamiltonian is defined by*

$H(x, y, z, u) := -z\, f(x, y, u) + g(x, y, u),$

*and the adjoint state $z := z_r \in V$ (resp. $z := z_w$) satisfies the linear adjoint equation*

$$a(v, z) + (f_y(y, r)z, v) = (g_y(y, r), v),$$
$$(\text{resp. } a(v, z) + (f_y(y, w)z, v) = (g_y(y, w), v) ),$$
$$\forall v \in V, \text{with } y := y_r \text{ (resp. } y := y_w).$$

**Theorem 2.** *(Optimality Conditions) If $r \in R$ (resp. $w \in W$, with $U$ convex) is optimal for Problem $\bar{P}$ or $P$ (resp. Problem $P$), then $r$ (resp. $w$) is strongly extremal relaxed (resp. weakly extremal classical), i.e. there exist multipliers $\lambda_m \in \mathbb{R}$, $m = 0, ..., q$, with $\lambda_0 \geqslant 0$, $\lambda_m \geqslant 0$, $m = p+1, ..., q$, $\sum_{m=0}^{q} |\lambda_m| = 1$, such that*

(1)   $\sum_{m=0}^{q} \lambda_m DG_m(r, \bar{r} - r) \geqslant 0, \forall \bar{r} \in R,$

(2)   $\lambda_m G_m(r) = 0, \quad m = p+1, ..., q$      *(relaxed transversality conditions),*

*(resp.*

(3)   $\sum_{m=0}^{q} \lambda_m DG_m(w, \bar{w} - w) \geqslant 0, \forall \bar{w} \in W,$

(4)   $\lambda_m G_m(w) = 0, \quad m = p+1, ..., q$      *(classical transversality conditions)).*
*The condition (1) is equivalent to the strong relaxed pointwise minimum principle*

(5)   $H(x, y(x), z(x), r(x)) = \min_{u \in U} H(x, y(x), z(x), u), \text{ a.e. in } \Omega,$

*where the complete Hamiltonian $H$ and adjoint $z$ are defined with $g := \sum_{m=0}^{q} \lambda_m g_m$.*

*If $U$ is convex, then (5) implies the weak relaxed pointwise minimum principle*

(6)   $H_u(x, y, z, r(x))r(x) = \min_{\phi} H_u(x, y, z, r(x))\phi(x, r(x)), \text{ a.e. in } \Omega,$

*where the minimum is taken over the set $B(\bar{\Omega}, U; U)$ of Caratheodory functions in the sense of Warga [13], and (6) implies the global weak relaxed condition*

(7)   $\int_{\Omega} H_u(x, y, z, r(x))[\phi(x, r(x)) - r(x)]dx \geqslant 0, \forall \phi \in B(\bar{\Omega}, U; U).$
*A control $r$ satisfying (7) and (2) is called weakly extremal relaxed. The condition (3) is equivalent to the weak classical pointwise minimum principle*

(8)   $H_u(x, y(x), z(x), w(x))w(x) = \min_{u \in U} H_u(x, y(x), z(x), w(x))u, \text{ a.e. in } \Omega.$

# 2   Discretizations and Behavior in the Limit

We suppose in what follows that $\Omega$ is a polyhedron (for simplicity), and that $U$ is convex. For each integer $n \geqslant 0$, let $\{E_i^n\}_{i=1}^{N^n}$ be an admissible regular partition of $\bar{\Omega}$ into elements (e.g. $d$-simplices), with $h^n = \max_i[\text{diam}(E_i^n)] \to 0$ as $n \to \infty$. Let $V^n \subset V$ be the subspace of functions that are continuous on $\bar{\Omega}$ and linear (for $d$-simplices), or multilinear, on each element $E_i^n$. The set of discrete controls $W^n \subset W$ is defined as the subset of (not necessarily continuous) controls $w^n$ that are (optionally) constant, linear, or multilinear, on each element $E_i^n$, and (optionally) such that $\|\nabla w^n\|_\infty \leqslant L$. For $w^n \in W^n$, the corresponding discrete state $y^n := y_{w^n}^n \in V^n$ is the solution of the discrete state equation

$$a(y^n, v^n) + (f(y^n, w^n), v^n) = 0, \quad \forall v^n \in V^n.$$

For every $w^n \in W^n$, the discrete state equation (a nonlinear system) has a unique solution $y^n \in V^n$ (see [10]), and can be solved by iterative methods. The discrete functionals, defined on $W^n$, are given by
$G_m^n(w^n) = \int_\Omega g_m(x, y^n, w^n)dx$, $\quad m = 0, ..., q$.
The discrete control constraint is $w^n \in W^n$ and the discrete state constraints are
$G_m^n(w^n) = \varepsilon_m^n$, $\quad m = 1, ..., p$, $\quad G_m^n(w^n) \leqslant \varepsilon_m^n$, $\quad \varepsilon_m^n \geqslant 0$, $\quad m = p+1, ..., q$,
where the feasibility perturbations $\varepsilon_m^n$ are chosen numbers converging to zero, to be defined later. The discrete optimal control Problem $P^n$ is to minimize $G_0^n(w^n)$ subject to $w^n \in W^n$ and to the above state constraints.

**Theorem 3.** *The operator $w^n \mapsto y^n$, from $W^n$ to $V^n$, and the functionals $w^n \mapsto G_m^n(w^n)$, on $W^n$, are continuous. For every $n$, if Problem $P^n$ is feasible, then it has a solution.*

**Lemma 2.** *Dropping the index $m$, the directional derivative of $G^n$ is given by*
$DG^n(w^n, \bar{w}^n - w^n) = \int_\Omega H_u(x, y^n, z^n, w^n)(\bar{w}^n - w^n)dx$, *for $w^n, \bar{w}^n \in W^n$,*
*where the discrete adjoint state $z^n := z_{w^n}^n \in V^n$ satisfies the discrete adjoint equation*
$a(z^n, v^n) + (z^n f_y(y^n, w^n), v^n) = (g_y(y^n, w^n), v^n)$ $\quad \forall v^n \in V^n$, *with $y^n := y_{w^n}^n$.*
*Moreover, the operator $w^n \mapsto z^n$, from $W^n$ to $V^n$, and the functional $(w^n, \bar{w}^n) \mapsto DG^n(w^n, \bar{w}^n - w^n)$, on $W^n \times W^n$, are continuous.*

**Theorem 4.** *(Discrete Optimality Conditions) If $w^n \in W^n$ is optimal for Problem $P^n$, then $w^n$ is weakly discrete extremal classical (or discrete extremal), i.e. there exist $\lambda_m^n \in \mathbb{R}$, $m = 0, ..., q$, with $\lambda_0^n \geqslant 0$, $\lambda_m^n \geqslant 0$, $m = p+1, ..., q$, $\sum_{m=0}^{q} |\lambda_m^n| = 1$, such that*
(9) $\quad \sum_{m=0}^{q} \lambda_m^n DG_m^n(w^n, \bar{w}^n - w^n) = \int_\Omega H^n(y^n, z^n, \bar{w}^n - w^n)dx \geqslant 0$, $\forall \bar{w}^n \in W^n$,
(10) $\quad \lambda_m^n(G_m(w^n) - \varepsilon_m^n) = 0$, $\quad m = p+1, ..., q$,
*where $H^n$ and $z^n$ are defined with $g := \sum_{m=0}^{q} \lambda_m^n g_m$. The condition (9) is equivalent to the strong discrete classical elementwise minimum principle*
(11) $\quad \int_{E_i^n} H_u^n(y^n, z^n, w^n)w^n dx = \min_{u \in U} \int_{E_i^n} H_u^n(y^n, z^n, w^n)u dx$, $\quad i = 1, ..., N^n$.

**Proposition 1.** *(Control Approximation) For every $r \in R$ (resp. $w \in W$), there exists a sequence $(w^n \in W^n)$, regarded as a sequence in $R$ (resp. $W$), that converges to $r$ in $R$ (resp. $w$ in $L^2$ strongly).*

**Lemma 3.** *(Consistency) (i) If the sequence $(w^n \in W^n)$ converges to $r \in R$ in $R$ (resp. $w \in W$ in $L^2$ strongly), then $y^n \to y_r$ (resp. $y^n \to y_w$) in $V$ strongly, $G^n(w^n) \to G(r)$ (resp. $G^n(w^n) \to G(w)$), and $z^n \to z_r$ (resp. $z^n \to z_w$) in $L^p(\Omega)$ strongly and in $V$ strongly.*
*(ii) If the sequences $(w^n \in W^n)$ and $(\bar{w}^n \in W^n)$ converge to $w$ and $\bar{w}$, respectively, in $W$, then $DG^n(w^n, \bar{w}^n - w^n) \to DG(w, \bar{w} - w)$.*

In what follows, we suppose that the considered Problem $P$ or $\bar{P}$ is feasible. We now examine the behavior in the limit of extremal discrete controls. We shall construct sequences of perturbations $(\varepsilon_m^n)$ that converge to zero and such that the discrete problem $P^n$ is feasible for every $n$. Let $\bar{w}^n \in W^n$ be any solution of the following auxiliary problem without state constraints

$$c^n := \min_{w^n \in W^n} \{ \sum_{m=1}^{p} [G_m^n(w^n)]^2 + \sum_{m=p+1}^{q} [\max(0, G_m^n(w^n))]^2 \},$$

and set

$$\varepsilon_m^n := G_m^n(\bar{w}^n), \ m = 1, ..., p, \ \varepsilon_m^n := \max(0, G_m^n(\bar{w}^n)), \ m = p+1, ..., q.$$

It can be easily shown that $c^n \to 0$, hence $\varepsilon_m^n \to 0$ (see [9]). Then clearly $P^n$ is feasible for every $n$, for these $\varepsilon_m^n$. We suppose in what follows that the $\varepsilon_m^n$ are chosen as in the above minimum feasibility procedure. The following theorem can be proved by using convergence arguments similar to those of Theorem 6 in Section 3.

**Theorem 5.** *For each $n$, let $w^n$ be admissible and extremal for Problem $P^n$.*
*(i) In the definition of $W^n$, we suppose that $\|\nabla v^n\|_\infty \leqslant L, \ \forall v^n \in W^n$. Every accumulation point of $(w^n)$ in $R$ is admissible and weakly extremal relaxed for Problem $\bar{P}$.*
*(ii) Every strong accumulation point, if it exists, of the sequence $(w^n)$ in $L^2(\Omega)$ is admissible and weakly extremal classical for Problem $P$.*

## 3    Mixed Discretization-Optimization Methods

Let $(M_m^l)$, $m = 1, ..., q$, be positive increasing sequences such that $M_m^l \to \infty$ as $l \to \infty$, and define the penalized discrete functionals

$$G^{nl}(w^n) := G_0^n(w^n) + \tfrac{1}{2} \{ \sum_{m=1}^{p} M_m^l [G_m^n(w^n)]^2 + \sum_{m=p+1}^{q} M_m^l [\max(0, G_m^n(w^n))]^2 \}.$$

Let $\gamma \geqslant 0$, $b, c \in (0, 1)$, and let $(\beta^l)$, $(\zeta_k)$ be positive sequences, with $(\beta^l)$ decreasing and converging to zero, and $\zeta_k \leqslant 1$. The algorithm described below contains two options. In the case of the progressively refining version, we suppose that each element $E_{i'}^{n+1}$ is a subset of some $E_i^n$, in which case $W^n \subset W^{n+1}$.

**Algorithm** (Discrete Penalized Gradient Projection Methods)
*Step 1.* Set $k := 0$, $l := 1$, choose a value of $n$ and an initial control $w_0^{nl} \in W^n$.
*Step 2.* Find $v_k^{nl} \in W^n$ such that

$$e_k := DG^{nl}(w_k^{nl}, v_k^{nl} - w_k^{nl}) + (\gamma/2) \left\| v_k^{nl} - w_k^{nl} \right\|^2$$

$$= \min_{\bar{v}^n \in W^n} [DG^{nl}(w_k^{nl}, \bar{v}^n - w_k^{nl}) + (\gamma/2) \left\| \bar{v}^n - w_k^{nl} \right\|^2],$$

and set $d_k := DG^{nl}(w_k^{nl}, v_k^{nl} - w_k^{nl})$.
*Step 3.* If $|d_k| > \beta^l$, go to Step 4. Else, set $w^{nl} := w_k^{nl}$, $v^{nl} := v_k^{nl}$, $d^l := d_k$, $e^l := e_k$.
Version A: Set $w_k^{n,l+1} := w_k^{nl}$. Version B: Set $w_k^{n+1,l+1} := w_k^{nl}$, $n := n + 1$.
In both versions, set $l := l + 1$, and go to Step 2.
*Step 4.* (Modified Armijo Step Search) Find the lowest integer value $s \in \mathbb{Z}$, say $\bar{s}$, such that $\alpha(s) = c^s \zeta_k \in (0, 1]$ and $\alpha(s)$ satisfies

$$G^{nl}(w_k^{nl} + \alpha(s)(v_k^{nl} - w_k^{nl})) - G^{nl}(w_k^{nl}) \leqslant \alpha(s) b d_k,$$

and then set $\alpha_k := \alpha(\bar{s})$.

*Step 5.* Set $w_{k+1}^{nl} := w_k^{nl} + \alpha_k(v_k^{nl} - w_k^{nl})$, $k := k+1$, and go to Step 2.

This Algorithm contains two versions:

**Version A:** $n$ is a constant integer chosen in Step 1, i.e. a *fixed discretization* is chosen, and the $G_m^n, m = 1, ..., q$, are replaced by the perturbed ones $\tilde{G}_m^n = G_m^n - \varepsilon_m^n$.

**Version B:** This is a *progressively refining* method, i.e. $n \to \infty$, in which case we can take $n = 1$ in Step 1, hence $n = l$ in the Algorithm.

Version B has the advantage of reducing computing time and memory, and also of avoiding the computation of minimum feasibility perturbations $\varepsilon_m^n$ (see Section 2). It is justified by the fact that finer discretizations become progressively more essential as the iterate gets closer to an extremal control.

With $w^{nl}$ as defined in Step 3, define the sequences of multipliers

$$\lambda_m^{nl} := M_m^l G_m^n(w^{nl}), \quad m = 1, ..., p, \quad \lambda_m^{nl} := M_m^l \max(0, G_m^n(w^{nl})), \quad m = p+1, ..., q.$$

**Theorem 6.** *(i) In the definition of $W^n$, we suppose that $\|\nabla v^n\|_\infty \leqslant L$, for every $v^n \in W^n$. In Version B, let $(w^{nl})$ be a subsequence, regarded as a sequence in $R$, of the sequence generated by the Algorithm in Step 3 that converges to some $r$ in $R$, as $l \to \infty$ (hence $n \to \infty$). If the sequences $(\lambda_m^{nl})$ are bounded, then $r$ is admissible and weakly extremal relaxed for Problem $\bar{P}$.*

*(ii) In Version B, let $(w^{nl})$ be a subsequence of the sequence generated by the Algorithm in Step 3 that converges to some $w \in W$ in $L^2$ strongly, as $l \to \infty$ (hence $n \to \infty$). If the sequences $(\lambda_m^{nl})$ are bounded, then $w$ is admissible and weakly extremal classical for Problem $P$.*

*(iii) In Version A, let $(w^{nl})$ ($n$ fixed) be a subsequence of the sequence generated by the Algorithm in Step 3 that converges to some $w^n \in W^n$ as $l \to \infty$. If the sequences $(\lambda_m^{nl})$ are bounded, then $w^n$ is admissible and extremal for Problem $P^n$.*

**Proof.** It can first be shown by contradiction, similarly to Theorem 5.1 in [9], that $l \to \infty$ in the Algorithm, hence $d^l \to 0$, $e^l \to 0$, in Step 3, and $n \to \infty$ in Version B.

(i) Let $(w^{nl})$ be a subsequence (same notation) of the sequence generated in Step 3, that converges to some $r \in R$ as $l, n \to \infty$. Suppose that the sequences $(\lambda_m^{nl})$ are bounded and (up to subsequences) that $\lambda_m^{nl} \to \lambda_m$. By Lemma 3, we have

$$0 = \lim_{l \to \infty} \frac{\lambda_m^{nl}}{M_m^l} = \lim_{l \to \infty} G_m^n(w^{nl}) = G_m(r), \quad m = 1, ..., p,$$

$$0 = \lim_{l \to \infty} \frac{\lambda_m^{nl}}{M_m^l} = \lim_{l \to \infty} [\max(0, G_m^n(w^{nl}))] = \max(0, G_m(r)), \quad m = p+1, ..., q,$$

which show that $r$ is admissible. Now, by Steps 2 and 3 we have, for every $\tilde{v}^n \in W^n$

(12) $\int_\Omega H_u^{nl}(x, y^{nl}, z^{nl}, w^{nl})(\tilde{v}^n - w^{nl})dx + (\gamma/2) \int_\Omega |\tilde{v}^n - w^{nl}|^2 dx \geqslant d^l$,

where $H^{nl}, z^{nl}$ are defined with $g := \sum_{m=0}^{q} \lambda_m^{nl} g_m$. Define the elementwise constant vector functions $\tilde{w}^n(x) := w^n(\tilde{x}^n(x))$, where $\tilde{x}^n(x)$ is the barycenter of $E_i^n$, in each $\overset{o}{E_i^n}$. Clearly, $\tilde{x}^n \to x$ uniformly on $\Omega$, and by our assumption on $W^n$ and the mean value theorem, $\|\tilde{w}^n - w^n\|_\infty \leqslant Lh^n \to 0$. For every function

$\phi \in C(\bar{\Omega} \times U; U)$, we can then replace $\tilde{v}^n$ by $\phi(\tilde{x}^n, \tilde{w}^n)$ in (12). Using the above convergences, Lemma 3, and Proposition 2.1 in [5], we can pass to the limit in (12) and obtain

$$\int_\Omega H_u(x, y, z, r(x))[\phi(x, r(x)) - r(x)]dx + (\gamma/2)\int_\Omega [\phi(x, r(x)) - r(x)]^2 dx \geqslant 0,$$

for every $\phi \in B(\bar{\Omega}, U; U)$, where $H$ and $z$ are defined with $g := \sum_{m=0}^{q} \lambda_m g_m$, and this inequality holds also, by density, for every Caratheodory function $\phi \in B(\bar{\Omega} \times U; U)$. Replacing $\phi$ by $u + \mu(\phi - u)$, with $\mu \in (0, 1]$, dividing by $\mu$, and then taking the limit as $\mu \to 0^+$, we obtain the same inequality, but without the quadratic term. By the construction of the $\lambda_m^{nl}$, we clearly have in the limit

$\lambda_0 = 1, \lambda_m \geqslant 0, m = p+1, ..., q, \sum_{m=0}^{q} |\lambda_m| \neq 0$. On the other hand, if $G_m(w) < 0$, for some index $m \in \{p+1, ..., q\}$, then for large $l$ we have $G_m^{nl}(w^{nl}) < 0$ and $\lambda_m^l = 0$, hence $\lambda_m = 0$, i.e. the transversality conditions hold. Therefore, $r$ is also weakly extremal relaxed.

(ii) The proof follows the arguments of (i), with simpler involved convergences.

(iii) The admissibility of $w^n$ is proved as in (i). Passing here to the limit in the inequality resulting from Step 2, as $l \to \infty$, for $n$ fixed, and using Lemma 3, we get

$$\sum_{m=0}^{q} \lambda_m D\tilde{G}_m^n(w^n, v'^n - w^n) = \sum_{m=0}^{q} \lambda_m DG_m^n(w^n, v'^n - w^n) \geqslant 0, \quad \forall v'^n \in W^n,$$

and the discrete transversality conditions

$$\lambda_m^n \tilde{G}_m^n(w^n) = \lambda_m^n [G_m^n(w^n) - \varepsilon_m^n] = 0, \quad m = p+1, ..., q,$$

with multipliers $\lambda_m^n$ as in the discrete optimality conditions.

When applied to problems whose solutions are classical controls, the above classical (possibly penalized) gradient projection methods are often efficient (see examples below), if not as efficient as e.g., SQP methods (see [12]). They can be applied to a broad class of problems and often converge to admissible controls satisfying the classical discrete, or continuous, optimality conditions. Moreover, the progressively refining version increases the efficiency of the method. But when directly applied to nonconvex problems without classical solutions and such that the solutions of the relaxed form are non-classical relaxed controls, these methods may yield very slow convergence, due to highly oscillating involved controls. If the constraint set $U$ is convex, one can then reformulate the relaxed problem in the equivalent Gamkrelidze relaxed form, using convex combinations of Dirac controls involving a finite, usually small, number of classical controls. The above methods can then be applied to this extended classical control problem, with much better results (for details on this approach, see [9]). When $U$ is not convex, one can use methods generating relaxed controls for solving (in the relaxed form) such highly nonconvex problems (see [8]). For various approximation and optimization methods applied to distributed optimal control problems, see e.g. [1,3,4,6,7,8,9,12], and the references therein.

# 4   Numerical Examples

**Example 1.** Let $\Omega := (0,1)^2$. Define the reference controls and state
$$\bar{u}(x) := x_1 x_2, \quad \bar{v}(x) := 1 - x_1 x_2, \quad \bar{y}(x) := 8x_1 x_2(1 - x_1)(1 - x_2),$$
and consider the following optimal control problem, with state equation
$$-\Delta y + y\,|y|\,/2 + (1 + u - \bar{u}))y$$
$$-\bar{y}\,|\bar{y}|\,/2 - \bar{y} - 16[x_1(1 - x_1) + x_2(1 - x_2)] - (v - \bar{v}) = 0 \text{ in } \Omega, \; y(x) = 0 \text{ on } \Gamma,$$
control constraints $(u(x), v(x)) \in U := [0,1]^2$, $x \in \Omega$, and cost functional
$$G_0(u,v) := 0.5 \int_\Omega [(y - \bar{y})^2 + (u - \bar{u})^2 + (v - \bar{v})^2]dx.$$
Clearly, the optimal controls are $\bar{u}$, $\bar{v}$, the optimal state is $\bar{y}$, and the optimal cost is zero. The gradient projection method, without penalties, was applied to this problem using triangular elements (half squares of edge size $h = 1/80$) and trianglewise linear discrete controls, with $\gamma = 0.5$, and Armijo parameters $b = c = 0.5$. After 15 iterations, we obtained the results:
$$G_0^n(u_k^n, v_k^n) = 2.963 \cdot 10^{-4}, \quad d_k = -7.786 \cdot 10^{-12}, \quad \eta_k = 3.052 \cdot 10^{-5}, \quad \varepsilon_k = 4.878 \cdot 10^{-5},$$
where $\eta_k$ (resp. $\varepsilon_k$) is the discrete max-error for the controls (resp. states) at the vertices of the triangles (resp. midpoints of the triangle edges).

**Example 2.** With the same data and parameters as in Example 1, but with $U := [0, 0.7] \times [0.3, 1]$ and the additional state constraint
$$G_1(u,v) := \int_\Omega (y - 0.22)dx = 0,$$
the penalized gradient projection method yields after 63 iterations in $k$ the results:
$$G_0^n(u_k^{nl}, v_k^{nl}) = 2.468309525 \cdot 10^{-3}, \quad G_1^n(u_k^{nl}, v_k^{nl}) = 6.303 \cdot 10^{-6}, \quad d_k = -5.909 \cdot 10^{-6}.$$
Finally, the progressively refining method was also applied to the above problems, with successive step sizes $h = 1/20$, $1/40$, $1/80$, in three equal periods, and yielded results of similar accuracy, but required here less than half the computing time.

# References

1. Arnäutu, V., Neittaanmäki, P.: Discretization estimates for an elliptic control problem. Numer. Funct. Anal. Optim. **19** (1998) 431–464
2. Bonnans, J., Casas, E.: Un principe de Pontryagine pour le contrôle des systèmes semilineaires elliptiques. J. Diff. Equat. **90** (1991) 288–303
3. Casas, E., Troeltzsch, F.: Error estimates for the finite-element approximation of a semilinear elliptic control problem. Control Cybern. **90** (2002) 695–712
4. Casas, E., Mateos, M., Troeltzsch, F.: Error estimates for the numerical approximation of boundary semilinear elliptic control problems. Comp. Optim. Appl. **31** (2005) 193–220
5. Chryssoverghi, I.: Nonconvex optimal control of nonlinear monotone parabolic systems. Syst. Control Lett. **8** (1986) 55–62
6. Chryssoverghi, I., Bacopoulos, A.: Approximation of relaxed nonlinear parabolic optimal control problems. J. Optim. Theory Appl. **77** (1993) 31–50

7. Chryssoverghi, I., Kokkinis, B.: Discretization of nonlinear elliptic optimal control problems. Syst. Control Lett. **22** (1994) 227–234
8. Chryssoverghi, I., Bacopoulos, A., Kokkinis, B., Coletsos, J.: Mixed Frank-Wolfe penalty method with applications to nonconvex optimal control problems. J. Optim. Theory Appl. **94** (1997) 311–334
9. Chryssoverghi, I.: Discretization methods for semilinear parabolic optimal control problems. Int. J. Num. Anal. Modeling **3** (2006) 437–458
10. Lions, J.-L.: Quelques Méthodes de Résolution des Problèmes aux Limites non Linéaires. Dunod Paris (1969) (English translation by Le Van, Holt, Rinehart and Winston, New York (1972))
11. Roubíček, T.: Relaxation in Optimization Theory and Variational Calculus. Walter de Gruyter, Berlin (1997)
12. Troeltzsch, F.: An SQP method for optimal control of a nonlinear heat equation. Control Cybern. **23** (1994) 267–288
13. Warga, J.: Optimal control of Differential and Functional Equations. Academic Press, New York (1972)

# Stability and Bifurcation Analysis of a Nonlinear Model of Bioreactor

Neli Dimitrova[1] and Plamena Zlateva[2]

[1] Institute of Mathematics and Informatics, Bulgarian Academy of Sciences
Acad. G. Bonchev Str. Blok 8, 1113 Sofia, Bulgaria
nelid@bio.bas.bg
[2] Institute of Control and System Research, Bulgarian Academy of Sciences
Acad. G. Bonchev Str. Blok 2, P.O. Box 79, 1113 Sofia, Bulgaria
plamzlateva@icsr.bas.bg

**Abstract.** The stability characteristics of a nonlinear model of a continuously stirred tank bioreactor with cell recycle are studied. Assuming
that some practically important model parameters are uncertain, existence of bifurcations of equilibrium points is shown. The dynamic behaviour
of the system near bifurcation points is also demonstrated. Numerical
simulations in the computer algebra system *Maple* are presented.

## 1 Introduction

In dynamical systems the object of bifurcation theory is to study the changes
that vector fields undergo as parameters change. Bifurcations may occur at nonhyperbolic equilibrium points (steady states). An equilibrium point is called
hyperbolic, if none of the eigenvalues of the Jacobian at that point has zero
real part; otherwise the equilibrium point is called nonhyperbolic [10]. Local
stability/instability of a hyperbolic equilibrium point $(y_0, \lambda_0)$, depending on a
bifurcation parameter $\lambda$, is determined by examining the eigenvalues of the Jacobian $J(y_0, \lambda_0)$. Since hyperbolic steady states are structurally stable, varying
$\lambda$ slightly around $\lambda_0$ does not change the stability type of $(y_0, \lambda_0)$. For nonhyperbolic equilibrium points $(y_0, \lambda_0)$, small changes of $\lambda$ around $\lambda_0$ and $y$ around $y_0$
may lead to new phase portraits of the dynamic system [5], [10].

We shall apply the bifurcation theory to the equilibrium points of a nonlinear
model of a bioreactor with cell recycle. Thereby, we consider the simplest way
in which an equilibrium point can be nonhyperbolic, namely when the Jacobian
at that point has a single zero eigenvalue. In this case the orbit structure near
the equilibrium point is determined by one differential equation. In particular,
the steady states are solutions of one nonlinear algebraic equation. To find the
equilibrium points, we use a Newton type method with result verification [2],
[3], that produces rigorous bounds (intervals) for the exact solution, see also [8].

Using appropriate scaling of the model, we keep several practically important
parameters, one of which is chosen as a bifurcation parameter. Moreover, we
assume that two of the model parameters are unknown but bounded within

T. Boyanov et al. (Eds.): NMA 2006, LNCS 4310, pp. 296–303, 2007.

given intervals. Under these assumptions, we show that hysteresis, pitchfork and transcritical bifurcations at equilibrium points may occur.

In Section 2 we shortly present the model of the continuously stirred tank bioreactor with cell recycle. The bifurcation and stability characteristics of the steady states with respect to the model parameters are studied in Section 3. The dynamic behaviour of the system near bifurcation points is demonstrated numerically in Section 4. The numerical computations and graphical visualizations are carried out in *Maple*.

## 2    The Model of the Activated Sludge Process

The activated sludge wastewater treatment process is carried out in a system, which consists of an aeration tank and a secondary settler. It is assumed that the hydraulic characteristics of the aeration tank are those of a continuously stirred tank bioreactor with cell recycle. A large variety of mathematical models are known [4], that incorporate the basic biotransformation processes of the wastewater treatment plant [6]. There are also models that take account of the specific engineering design of the system [9]. Here we consider a simplified model of the activated sludge process, assuming that the purge fraction containing the biomass is operated from the settler, and not directly from the bioreactor [1]. The model is described by the following two nonlinear ordinary differential equations

$$\frac{dX}{dt} = \mu(S) \cdot X + rDX_r - (1+r)DX \tag{1}$$

$$\frac{dS}{dt} = -K\mu(S) \cdot X + D(S_{in} - S), \tag{2}$$

where the state variables $X = X(t)$ and $S = S(t)$ are biomass (activated sludge) and substrate (biological oxygen demand) concentration, $D$ is dilution rate, $S_{in}$ is influent substrate concentration, $X_r$ is recycle biomass concentration, $\mu(S)$ is the specific growth rate of the biomass, $r$ is sludge recycle ratio, $K$ is yield coefficient. Taking different model functions $\mu(S)$ leads to different models of the bioreactor. Here we consider the Haldane law $\mu(S) = \dfrac{\mu_m S}{K_s + S + \frac{S^2}{K_i}}$, where $\mu_m$ and $K_s$ are kinetic parameters, $K_i$ is inhibition constant.

The coefficients $K_s$ and $K_i$ have the meaning of substrate concentrations: $K_s$ exhibits the affinity of the biomass to the substrate, $K_i$ is related to the decrease of the specific growth rate because of excessive substrate concentration. Therefore, $K_i$ is always greater than the constant $K_s$ [7].

The scaling

$$s = \frac{S}{K_s}, \quad x = \frac{K}{K_s}X, \quad u = \frac{D}{\mu_m}, \quad s_{in} = \frac{S_{in}}{K_s},$$
$$\alpha = \frac{K_s}{K_i}, \quad x_r = \frac{K}{K_s}X_r, \quad \bar{t} = \mu_m t$$

transforms (1)–(2) into the dimensionless model

$$\frac{dx}{dt} \equiv \dot{x} = \bar{\mu}(s) \cdot x + rux_r - (1+r)ux \tag{3}$$

$$\frac{ds}{dt} \equiv \dot{s} = -\bar{\mu}(s) \cdot x + u(s_{in} - s), \tag{4}$$

with

$$\bar{\mu}(s) = \frac{s}{1 + s + \alpha s^2}.$$

In optimal operation of the wastewater treatment plant the recycle biomass concentration $x_r$ varies slightly between tight bounds. By this technological reason we have assumed $x_r$ to be a model parameter, which varies in given interval $[x_r] = [x_r^-, x_r^+] \geq 0$. The last inequality means that we do not exclude the possibility $x_r$ to become zero; this situation corresponds to feed interrupt of the recycle activated sludge into the bioreactor due to technological problems. We also assume $s_{in}$ to be uncertain, but located in a known interval $[s_{in}] = [s_{in}^-, s_{in}^+] > 0$. The following conditions are also satisfied due to physical evidence: $0 < s \leq s_{in}$, $x \geq 0$, $u > 0$, $0 < r < 1$, $0 < \alpha < 1$.

## 3   Stability and Bifurcations of the Equilibrium Points

In what follows we consider $u$ as a bifurcation parameter. Let $s_{in} \in [s_{in}]$ and $x_r \in [x_r]$ be arbitrary parameter values. The steady states $s = s(u)$, $x = x(u)$ are solutions of the system

$$\bar{\mu}(s) \cdot x + rux_r - (1+r)ux = 0 \tag{5}$$

$$-\bar{\mu}(s) \cdot x + u(s_{in} - s) = 0. \tag{6}$$

Adding (6) to (5), then expressing $x$ from the new equation, $x = \dfrac{s_{in} + rx_r - s}{1+r}$ and substituting in (6) yields the algebraic equation for $s$

$$G(s, u) \equiv -s^3 + As^2 - Bs + C = 0, \tag{7}$$

where

$$A = \frac{1}{(1+r)\alpha u} + s_{in} - \frac{1}{\alpha}, \quad B = \frac{s_{in} + rx_r}{(1+r)\alpha u} + \frac{1 - s_{in}}{\alpha}, \quad C = \frac{s_{in}}{\alpha}.$$

Suppose that $(s_0, u_0)$, $s_0 = s(u_0)$, satisfies

$$G(s_0, u_0) = 0 \quad \text{(equilibrium point condition)} \tag{8}$$

$$G'_s(s_0, u_0) = 0 \quad \text{(zero eigenvalue condition)}. \tag{9}$$

It is straightforward to see that the Jacobian of (3)–(4) has a single zero eigenvalue at $(s_0, x_0)$, $x_0 = (s_{in} + rx_r - s_0)/(1+r)$, with the other eigenvalue having a nonzero real part. Then the orbit structure near $(s_0, x_0, u_0)$ is determined by the associated center manifold equation $\dot{s} = G(s, u)$ [10]. The question we ask is what is the nature of the nonhyperbolic equilibrium point for $u$ close to $u_0$.

*Hysteresis bifurcations.* In order that the vector field $\dot{s} = G(s, u)$ undergoes a hysteresis bifurcation at $(s_0, u_0)$ it is necessary to have (8), (9) and $G''_{ss}(s_0, u_0) = 0$, $G'_u(s_0, u_0) \neq 0$, $G''_{su}(s_0, u_0) \neq 0$, $G'''_{sss}(s_0, u_0) \neq 0$; then in a sufficiently small neighborhood of $s_0$ there exists a smooth function $p$, such that $G(s, p(s)) = 0$ and $p'(s_0) = p''(s_0) = 0$, $p'''(s_0) \neq 0$ ([1], [5]).

The conditions (8), (9) and $G''_{ss}(s_0, u_0) = 0$ mean that $s_0$ should be a triple root of $G(s, u) = 0$. Using the elementary symmetric functions (Vieta formulae) we obtain the relations $3s_0 = A$, $3s_0^2 = B$, $s_0^3 = C$; the latter imply

$$s_0 = s_{in}{}^{1/3}\alpha^{-1/3}, \qquad u_0 = \frac{s_{in} + rx_r}{(1+r)(3s_{in}^{2/3}\alpha^{1/3} + s_{in} - 1)}, \qquad (10)$$

if and only if the parameters $(s_{in}, x_r, \alpha, r)$ satisfy

$$x_r = f_{\alpha,r}(s_{in}), \qquad f_{\alpha,r}(s_{in}) = \frac{1}{r} \cdot \frac{\left(\left(\alpha s_{in}^2\right)^{1/3} - 1\right)^3}{3\left(\alpha^2 s_{in}\right)^{1/3} - \alpha s_{in} + 1}, \qquad \alpha s_{in}^2 \geq 1. \quad (11)$$

Let (11) hold true. It is straightforward to see that

$$G'_u(s_0, u_0) \neq 0 \iff s_0 \neq s_{in} + rx_r \iff rx_r(1 - \alpha s_{in}) \neq \alpha s_{in}^2 - 1. \quad (12)$$

If $\alpha s_{in} = 1$, the last inequality in (12) is obviously fulfilled, since $\alpha < 1$. Let $\alpha s_{in} \neq 1$. Then

$$G'_u(s_0, u_0) \neq 0 \iff x_r \neq \frac{\alpha s_{in}^2 - 1}{r(1 - \alpha s_{in})} \iff \frac{\alpha s_{in}^2 - 1}{r(1 - \alpha s_{in})} \neq f_{\alpha,r}(s_{in}). \quad (13)$$

When $\alpha s_{in}^2 = 1$, then $x_r = 0$ follows from (11), but this is an obvious contradiction to (13). Therefore $G'_u(s_0, u_0) \neq 0$ if $(s_{in}, x_r, \alpha, r)$ satisfy (11) with $f_{\alpha,r}(s_{in}) > 0$. Further, $G''_{su}(s_0, u_0) \neq 0 \iff s_0 \neq \frac{1}{2}(s_{in} + rx_r)$ and similar considerations as above show that this inequality is also fulfilled when $(s_{in}, x_r, \alpha, r)$ satisfy (11) with $\alpha s_{in}^2 \neq 1$. Finally, $G'''_{sss}(s_0, u_0) \neq 0$ is obviously fulfilled.

Assume now that $\alpha$ and $r$ are fixed, but $x_r$ and $s_{in}$ vary in the intervals $[x_r] = [x_r^-, x_r^+]$ and $[s_{in}] = [s_{in}^-, s_{in}^+]$ respectively. Denote by $[f] = [f^-, f^+]$ the range of $f_{\alpha,r}(s_{in})$ on $[s_{in}]$, that is $[f] = \{f_{\alpha,r}(s_{in}) : s_{in} \in [s_{in}]\}$. Set $\Gamma = [x_r] \cap [f]$. The following cases are possible: (a) $\Gamma = [x_r]$, that is $[x_r] \subseteq [f]$; (b) $\Gamma = [f]$, that is $[f] \subseteq [x_r]$; (c) $\Gamma = [f^-, x_r^+]$, that is $x_r^- < f^- < x_r^+ < f^+$; (d) $\Gamma = [x_r^-, f^+]$, that is $f^- < x_r^- < f^+ < x_r^+$; (e) $\Gamma = \emptyset$, that is $x_r^+ < f^-$ or $f^+ < x_r^-$. We shall consider in detail the first case (a); the other possibilities are treated similarly.

Let $\Gamma = [x_r]$, that is, $f^- \leq x_r^- < x_r^+ \leq f^+$. Since $f_{\alpha,r}(s_{in})$ is continuous and monotone increasing, there exist points $\tilde{s}_{in}^-, \tilde{s}_{in}^+ \in [s_{in}]$ such that $x_r^- = f_{\alpha,r}(\tilde{s}_{in}^-)$, $x_r^+ = f_{\alpha,r}(\tilde{s}_{in}^+)$. Then for any $x_r \in [x_r]$ there exists $s_{in} \in [\tilde{s}_{in}^-, \tilde{s}_{in}^+] = [\tilde{s}_{in}]$ with $x_r = f_{\alpha,r}(s_{in})$, and vice versa. According to (11), hysteresis bifurcations occur only for these parameter values. Fig. 1(a) presents the graph of $f_{\alpha,r}(s_{in})$

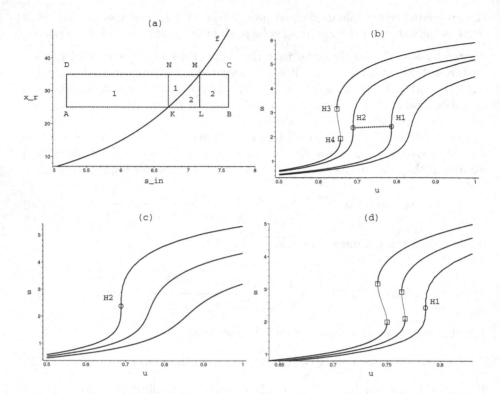

**Fig. 1.** (a) Parameter regions for hysteresis bifurcation; (b) Bifurcation diagrams on $KLMN$; (c) Bifurcation diagrams on $AKND$; (d) Bifurcation diagrams on $LBCM$ (thin lines – unstable curve branches; thick lines – stable curve branches)

(denoted by $f$); the rectangle $ABCD$ visualizes the interval vector $([s_{in}], [x_r])$, that is $A = (s_{in}^-, x_r^-)$, $B = (s_{in}^+, x_r^-)$, $C = (s_{in}^+, x_r^+)$, $D = (s_{in}^-, x_r^+)$; $KLMN$ corresponds to $([\tilde{s}_{in}], [x_r])$ with $K = (\tilde{s}_{in}^-, x_r^-)$, $L = (\tilde{s}_{in}^+, x_r^-)$, $M = (\tilde{s}_{in}^+, x_r^+)$, $N = (\tilde{s}_{in}^-, x_r^+)$; the curve $KM$ presents the set of parameter values, for which the hysteresis bifurcations occur (that is, a triple root of $G(s, u) = 0$ exists).

When $x_r \neq f_{\alpha,r}(s_{in})$, then $G''_{ss} = 0$ is no more valid. It is straightforward to see that for any pair of parameter values $(s_{in}, x_r) \in ([s_{in}], [x_r])$ with $x_r < f_{\alpha,r}(s_{in})$ there exist bifurcation values $u_3$ and $u_4$ such that $G(s_i, u_i) = 0$, $G'_s(s_i, u_i) = 0$, $i = 3, 4$, i. e. $s_3 = s(u_3)$ and $s_4 = s(u_4)$ are double roots of (7); on Fig. 1(a) these parameter regions are marked by '2'. When $(s_{in}, x_r) \in ([s_{in}], [x_r])$ are such that $x_r > f_{\alpha,r}(s_{in})$, equation (7) has a unique real solution $s(u)$ for any $u > 0$; on Fig. 1(a) these parameter regions are marked by '1'.

Fig. 1(b) presents bifurcation diagrams for $(s_{in}, x_r) \in ([\tilde{s}_{in}], [x_r])$, i. e. on the parameter domain $KLMN$. The curve, containing the point $H_1$ resp. $H_2$ corresponds to parameter values $(\tilde{s}_{in}^+, x_r^+) = M$ resp. $(\tilde{s}_{in}^-, x_r^-) = K$. The curve connecting $H_1, H_2$ presents the set of all hysteresis points when $x_r = f_{\alpha,r}(s_{in})$, $s_{in} \in [\tilde{s}_{in}^-, \tilde{s}_{in}^+]$, that is for points $(s_{in}, x_r)$ on the curve $KM$. The curve containing

$H_3$ and $H_4$ is computed for parameter values $(\tilde{s}_{in}^+, x_r^-) = L$; the points $H_3 = (u_3, s_3)$ and $H_4 = (u_4, s_4)$ visualize the double roots of (7), and they build the unstable curve branch $H_3 H_4$. The lowest curve corresponds to $(\tilde{s}_{in}^-, x_r^+) = N$ and presents a steady state solution $s(u)$ with no bifurcation points.

Fig. 1(c) visualizes solution curves for $s_{in} \in [s_{in}^-, \tilde{s}_{in}^-]$ and $x_r \in [x_r]$, i. e. on $AKND$; on this parameter region, the hysteresis bifurcation point $H_2$ disappears. On Fig. 1(d), bifurcation diagrams for $s_{in} \in [\tilde{s}_{in}^+, s_{in}^+]$, $x_r \in [x_r]$ (i. e. on $LBCM$) are shown; the hysteresis point $H_1$ splits up into two bifurcation points (double roots of (7)), denoted by boxes.

All bifurcation points $H_i$ ($i = 1, 2, 3, 4$) are computed for particular parameter values (see Section 4) using the verification Newton type algorithm [2].

If we return to the model of the bioreactor, the latter should be operated for parameter values $(s_{in}, x_r, \alpha, r)$ satisfying $x_r \geq f_{\alpha,r}(s_{in})$, i. e. on the parameter domain denoted by '1' on Fig. 1(a) and on the curve $KM$; there, for all (practically admissible) values of the controllable input $u$ the steady states $s(u)$ are stable and the system is predictable. Since $s_{in}$ and $x_r$ are not fixed, but vary in intervals, it may happen that they change and move to the region with $x_r < f_{\alpha,r}(s_{in})$ (denoted by '2' on Fig. 1(a)). This will cause changes in the stability characteristics of the bioreactor. An example for the dynamics of the model near the bifurcation point $H_4$ is given in the next section, see Fig. 2(b).

*Pitchfork and transcritical bifurcations.* For the vector field $\dot{s} = G(s, u)$ to undergo a pitchfork bifurcation at $(s_0, u_0)$, it is necessary to have (8), (9) and $G_{ss}''(s_0, u_0) = 0$, $G_u'(s_0, u_0) = 0$, $G_{su}''(s_0, u_0) \neq 0$, $G_{sss}'''(s_0, u_0) \neq 0$; then in a sufficiently small neighborhood of $s_0$ there exist exactly two smooth functions $p_1(s)$ and $p_2(s)$ such that $G(s, p_i(s)) = 0$, $p_i(s_0) = u_0$, $p_i'(s_0) = 0$ and $p_i''(s_0) \neq 0$ ($i = 1, 2$). Under the above conditions, if $G_{ss}''(s_0, u_0) \neq 0$, then the vector field $\dot{s} = G(s, u)$ is said to undergo a transcritical bifurcation at $(s_0, u_0)$; the functions $p_i(s)$ satisfy $G(s, p_i(s)) = 0$, $p_i(s_0) = u_0$, $p_i'(s_0) \neq 0$, $i = 1, 2$ [10].

From the previous considerations it follows that pitchfork bifurcation occurs if (13) holds with equality, that is if $\alpha s_{in}^2 = 1$ and therefore $x_r = 0$ is valid.

For $x_r = 0$ the equation (7) takes the form

$$(s_{in} - s)\, g(s, u) = 0, \qquad g(s, u) = s^2 + \left( \frac{1}{\alpha} - \frac{1}{(1+r)\alpha u} \right) s + \frac{1}{\alpha}. \qquad (14)$$

The equation $g(s, u) = 0$ has a double root for $u = u' = ((1+r)(2\sqrt{\alpha} + 1))^{-1}$. When $\alpha s_{in}^2 = 1$, this double root is equal to $s_{in}$ at $u_0 = \dfrac{s_{in}}{(1+r)(2+s_{in})} = u'$; therefore $(s_{in}, u_0)$ is the pitchfork bifurcation point. If $\alpha s_{in}^2 \neq 1$, then $G_{ss}''' = 0$ is no more valid. In this case $g(s, u) = 0$ has a root $s(u'') = s_{in}$ for $u'' = \dfrac{\bar{\mu}(s_{in})}{1+r}$. Hence, $(s_{in}, u'')$ is a transcritical bifurcation point.

Assume now that $s_{in}$ is enclosed by an interval $[s_{in}] > 0$ such that $\alpha [s_{in}]^2 = \alpha[(s_{in}^-)^2, (s_{in}^+)^2] \ni 1$. There exists a unique $\hat{s}_{in} \in [s_{in}]$ with $\alpha \hat{s}_{in}^2 = 1$. Then $(\hat{s}_{in}, \hat{u}_0)$ with $\hat{u}_0 = \dfrac{\hat{s}_{in}}{(1+r)(2+\hat{s}_{in})}$ is the unique pitchfork bifurcation. For

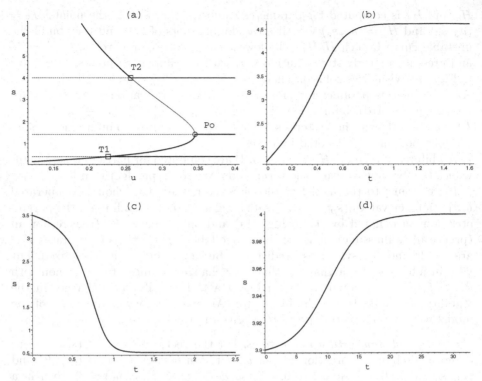

**Fig. 2.** (a) Pitchfork and transcritical bifurcation diagram (thin lines – unstable curve branches; thick lines – stable curve branches); (b) Dynamic behaviour near the bifurcation point $H_4$; (c), (d) Dynamic behaviour near the transcritical point $T_2$

all parameter values $s_{in} \in [s_{in}] \setminus \{\hat{s}_{in}\}$ the points $(s_{in}, u'')$ are transcritical bifurcations. In the case when $\alpha[s_{in}]^2 \not\ni 1$, no pitchfork but only transcritical bifurcations are present. Fig. 2(a) presents the bifurcation diagram of the equilibrium solutions $s$ versus $u$. The point $P_0 = (\hat{u}_0, \hat{s}_{in})$ is the pitchfork bifurcation; $T_1 = (u_1, s_{in}^-)$, $u_1 = \overline{\mu}(s_{in}^-)/(1+r)$ and $T_2 = (u_2, s_{in}^+)$, $u_2 = \overline{\mu}(s_{in}^+)/(1+r)$ are (only two from the set of all) transcritical bifurcations.

## 4   Dynamic Behaviour Near Bifurcation Points

We shall demonstrate numerically the changes in the orbit structure as the bifurcation parameter passes through the bifurcation values. Let $\alpha = 0.5$ and $r = 0.2$.

First we take $s_{in} \in [5.2, 7.6]$, $x_r \in [25, 35]$ and obtain $[f] = [7.88, 48.29]$. This is the case $(a)$ of the hysteresis bifurcation, with $\tilde{s}_{in}^- = 6.7$, $\tilde{s}_{in}^+ = 7.16$. For the bifurcation point $H_4 = (u_4, s_4)$ from Fig. 1(b) we have $u_4 \in [0.655, 0.656]$, $s_4 \in [1.93, 1.94]$. In $\dot{s} = G(s, u)$ let us substitute $u$ by the value 0.7, which is slightly larger than $u_4$; with $s(0) = 1.7$ close to $s_4$, Fig. 2(b) presents the solution

$s(t)$. As one can see, the phase curve $s(t)$ "jumps" and stabilizes to a steady state on the upper stable curve branch, emanating from the bifurcation point $H_3$.

To demonstrate the dynamics near the pitchfork and the transcritical bifurcations (see Fig. 2(a)), we take $x_r = 0$ and $s_{in} \in [0.4, 4]$. For $P_0 = (\hat{u}_0, \hat{s}_{in})$ it is obvious that if we start with initial value $s(0)$ close to $\hat{s}_{in}$ and take $u$ slightly larger than $\hat{u}_0$, then $s(t)$ will tend to $\hat{s}_{in}$. This result is not visualized due to place limitations. For the point $T_2 = (u_2, s_{in}^+)$ we have $u_2 \in [0.256, 0.257]$. Fig. 2(c),(d) present two different portraits of the solution $s(t)$ for $u = 0.26$ slightly larger than $u_2$, first with a starting value $s(0) = 3.5$ and second with $s(0) = 3.9$. In the first case $s(t)$ stabilizes to a steady state on the stable curve branch $T_1 P_0$, in the second case $s(t)$ approaches $s_{in}^+$. The reason is that the direction field of $\dot{s} = (s_{in} - s(t))g(s(t), u)$ is determined by the sign of $g(s, u)$. More precisely, $g(s, u) < 0$ if $(s, u)$ lies in the region to the left of the curve $T_1 P_0 T_2$ and thus $s(t)$ decreases (Fig. 2(c)); if $(s, u)$ lies in the region to the right of $T_1 P_0 T_2$ then $g(s, u) > 0$ and hence $s(t)$ increases (Fig. 2(d)). Another situation that may occur is due to the fact that $s_{in}$ is not fixed, but varies in $[s_{in}]$. It may happen that $s_{in}$ suddenly changes and becomes equal to $s(t_0)$ at some time $t_0 > 0$, leading in this way to wash-out state $s(t) = s_{in}$ for all $t > t_0$. No wash-out states occur if $u < u_1 = \overline{\mu}(s_{in}^-)/(1 + r)$, $s_{in} = s_{in}^-$ and $s(0) < s_{in}^-$; then the phase curve $s(t)$ approaches the stable curve branch emanating to the left of the point $T_1$.

**Acknowledgements.** The authors are grateful to the anonymous referees for the invaluable advices and comments.

# References

1. Ajbar, A., Gamal, I.: Stability and Bifurcation of an Unstructured Model of a Bioreactor with Cell Recycle. Math. Comput. Modelling **25(2)** (1997) 31–48
2. Dimitrova N. S.: On a Parallel Method for Enclosing Real Roots of Nonlinear Equatins. In: I. T. Dimov, Bl. Sendov, P. S. Vassilevski (eds.), Numerical Methods and Applications, 78–84. World Scientific (1994)
3. Dimitrova N. S., Markov, S. M.: On Validated Newton Type Method for Nonlinear Equations. Interval Computations **2** (1994) 27–51
4. Dunn, I., Heinzle, E., Ingham, J., Prenosil, J.: Biological Reaction Engineering. VCH Weiheim, Germany (1992)
5. Golubitsky, M., Schaeffer, D.: Singularities and Groups in Bifurcation Theory vol. 1. Springer, New York (1985)
6. Henze, M., Gujer, W., Mino, T., Matsuo, T., Wentzel, M., Marais, G.: Activated sludge model no. 2. IWA Scientific and Technical Report No 3, London (1995)
7. Leslie Grady Jr., C. P.: Lecture Course on Biological Principles of Environmental Engineering. Clemson University http://www.ces.clemson.edu/grady/851-notes/
8. Neumaier, A.: Interval Methods for Systems of Equations Cambridge University Press (1990)
9. Puteh, M., Minekawa, K., Hashimoto, N., Kawase, Y.: Modeling of Activated Sludge Wastewater Treatment Process. Bioprocess Engineering **21** (1999) 249–254
10. Wiggins, S.: Introduction to Applied Nonlinear Dynamical Systems and Chaos. Texts in Applied Math. **2**. Springer (1990)

# Discrete Approximations of Singularly Perturbed Systems

Tzanko Donchev[1] and Vasile Lupulescu[2]

[1] Department of Mathematics, University of Architecture and Civil Engineering
1 "Hr. Smirnenski" str., 1046 Sofia, Bulgaria
[2] Department of Mathematics, University Constantin Brâncuşi
1 "Republicii" str., 210152 Târgu Jiu, Romania
tdd51us@yahoo.com, vasile@utgjiu.ro

**Abstract.** In the paper we study discrete approximations of singularly perturbed system in a finite dimensional space. When the right-hand side is almost upper semicontinuous with convex compact values and one-sided Lipschitz we show that the distance between the solution set of the original and the solution set of the discrete system is $O\left(h^{\frac{1}{2}}\right)$.

## 1 Introduction

We study Euler discrete approximations of two time scale system having the form:

$$\begin{pmatrix} \dot{x} \\ \varepsilon \dot{y} \end{pmatrix} \in F(t, x(t), y(t)), \quad \begin{matrix} x(0) = x^0 \\ y(0) = y^0, \end{matrix} \tag{1}$$

where $F : I \times \mathbb{R}^n \times \mathbb{R}^m \rightrightarrows \mathbb{R}^n \times \mathbb{R}^m$ is bounded on the bounded sets with nonempty convex and compact values.

There is a large number of papers devoted to the existence of the limit of the solution set $Z(\varepsilon)$ of (1) when $\varepsilon \to 0^+$. We mention only [3,6,9,10,13] and the references therein.

The multifunction $F(t, \cdot, \cdot)$ is said to be One-Sided Lipschitz (OSL) if there exist positive constants $A$, $B$, $C$, $\mu$ such that for every $x_1$, $x_2 \in \mathbb{R}^n$, every $y_1$, $y_2 \in \mathbb{R}^m$ and every $(f_1, g_1) \in F(t, x_1, y_1)$ there exists $(f_2, g_2) \in F(t, x_2, y_2)$ such that

$$\langle x_1 - x_2, f_1 - f_2 \rangle \le A|x_1 - x_2|^2 + B|y_1 - y_2|^2,$$
$$\langle y_1 - y_2, g_1 - g_2 \rangle \le C|x_1 - x_2|^2 - \mu|y_1 - y_2|^2.$$

This condition is introduced in [2]. It is used in the approximation of differential inclusions and in singular perturbation theory in [3,6,11,12,14,16].

We refer to [4,8,12,14] and the references therein for discretization of differential inclusions.

Denote by $\mathbb{B}_n$ the unit ball in $\mathbb{R}^n$. We refer to [1] for all the concepts used here, but not explicitly presented.

T. Boyanov et al. (Eds.): NMA 2006, LNCS 4310, pp. 304–311, 2007.

The multifunction $F(\cdot,\cdot)$ defined on $I \times K$ is said to be almost USC, when for every $\varepsilon > 0$ there exists a compact $I_\varepsilon \subset I$ with $\operatorname{meas}(I_\varepsilon) > 1 - \varepsilon$ (here meas is Lebesgue measure), such that $F$ is USC on $I_\varepsilon \times K$.

Further we assume that $F$ is:

**A1.** Almost USC with nonempty convex compact values (of course if $F$ is autonomous it is USC).

**A2.** $F$ is bounded on the bounded sets.

**A3.** $F$ is OSL.

We show that under the above assumptions **A1, A2, A3**, one has accuracy $O\left(\sqrt{\dfrac{h}{\varepsilon}}\right)$ of the Euler Approximation Schema *EAS*.

If we assume also that $F(\cdot,\cdot)$ is locally Lipschitz (and of course does not depend on $t$), then the accuracy of *EAS* is $O\left(\dfrac{h}{\varepsilon}\right)$.

We note the recent paper of Grammel [11], where the system (1) is studied, when its "slow component" is single valued, i.e. $F(x,y) = \{f(x,y) \times G(x,y)\}$. He assumes that $f(\cdot,\cdot)$ and $G(\cdot,\cdot)$ are Lipschitz and, moreover, there exist constants $0 < \alpha < \beta$ for every $x, y_1, y_2$ and every $v_1 \in G(x,y_1)$ there exists $v_2 \in G(x,y_2)$ such that:

$$\langle v_1 - v_2, y_2 - y_1 \rangle \leq -\alpha |y_2 - y_1|^2, \quad |v_1 - v_2| \leq \beta |y_2 - y_1|.$$

Grammel obtains ([11, Theorem 4]) $O\left(\dfrac{h}{\varepsilon}\right)$ of (*EAS*).

## 2  Discrete Approximations

In this section we study discrete approximations of the initial value problem (1) in case of one and two time scale discretization.

**Lemma 1.** *Under* **A1, A2, A3** *there exist constants $M$ and $N$ such that $|x(t)| + |y(t)| \leq M$ and $|F(t, x(t) + \mathbb{B}_n, y(t) + \mathbb{B}_m)| \leq N$ for every $\varepsilon > 0$ and for every solution $(x(\cdot), y(\cdot))$ of*

$$\begin{pmatrix} \dot{x} \\ \varepsilon\dot{y} \end{pmatrix} \in \overline{co}\, F(t, x(t) + \mathbb{B}_n, y(t) + \mathbb{B}_m), \quad \begin{matrix} x(0) = x' \in x^0 + \mathbb{B}_n \\ y(0) = y' \in y^0 + \mathbb{B}_m. \end{matrix}$$

The proof can be found in [5] and it is omitted.

We are going to prove the following lemma of Filippov–Pliss type (compare with Lemma 5 of [3], where particular case is considered).

**Lemma 2.** *(Lemma of Plis) Under* **A1, A2, A3** *there exists a constant $P$ such that:*

*For every solution $(u, v)$ of the system:*

$$\begin{pmatrix} \dot{u} \\ \varepsilon\dot{v} \end{pmatrix} \in F(t, u(t) + \delta\mathbb{B}_n, v(t) + \delta\mathbb{B}_m), \quad \begin{matrix} u(0) = u^0 \\ v(0) = v^0, \end{matrix} \tag{2}$$

there exists a solution $(x, y)$ of (1) such that denoting: $|x(t) - u(t)|^2 = r(t)$ and $|y(t) - v(t)|^2 = s(t)$, one has:

$$\dot{r}(t) \leq 2Ar + 2Bs + P\delta, \quad r(0) = |x^0 - u^0|^2$$
$$\varepsilon\dot{s}(t) \leq 2Cr - 2\mu s + P\delta, \quad s(0) = |y^0 - v^0|^2.$$

*Proof.* It is easy to see that there exist measurable functions $|\alpha(t)| \leq \delta$ and $|\beta(t)| \leq \delta$ such that:

$$\begin{pmatrix} \dot{u} \\ \varepsilon\dot{v} \end{pmatrix} \in F(t, u(t) + \alpha(t), v(t) + \beta(t)), \quad \begin{matrix} u(0) = u^0 \\ v(0) = v^0. \end{matrix}$$

We define the multifunction:

$$G(t, x, y) = \{(f, g) \in F(t, x, y):$$
$$\langle u(t) + \alpha(t) - x, \dot{u}(t) - f \rangle \leq A|u(t) + \alpha(t) - x|^2 + B|v(t) + \beta(t) - y|^2,$$
$$\varepsilon\langle v(t) + \beta(t) - y, \dot{v}(t) - g \rangle \leq C|u(t) + \alpha(t) - x|^2 - \mu|v(t) + \beta(t) - y|^2\}.$$

It is easy to see that $G(\cdot, \cdot, \cdot)$ is almost USC with nonempty convex, compact values.

Let $(x(\cdot), y(\cdot))$ be a solution of:

$$\begin{pmatrix} \dot{x} \\ \varepsilon\dot{y} \end{pmatrix} \in G(t, x(t), y(t)), \quad \begin{matrix} x(0) = x^0 \\ y(0) = y^0. \end{matrix}$$

Using standard calculations one can show that:

$$\frac{d}{dt}|x(t) - u(t)|^2 \leq 2A|x(t) - u(t)|^2 + 2B|y(t) - v(t)|^2 + P\delta,$$
$$\varepsilon\frac{d}{dt}|y(t) - v(t)|^2 \leq 2C\frac{d}{dt}|x(t) - u(t)|^2 - 2\mu|y(t) - v(t)|^2 + P\delta.$$

The proof is therefore complete. □

Given a subdivision $\Delta^K \overset{def}{=} \{0 = t_0 < t_1 < \ldots < t_K = 1\}$, where $t_i = ih$ and $h = \dfrac{1}{K}$. Consider the Euler scheme:

$$(f(t), g(t)) \in F(t, x^i, y^i), x^i = x(t_i), \; y^i = y(t_i),$$
$$x(t) = x^i + \int_{t_i}^t f(s)\, ds, \; t \in [t_i, t_{i+1}],$$
$$\varepsilon y(t) = \varepsilon y^i + \int_{t_i}^t g(s)\, ds. \tag{3}$$

Denote by $Sol(DI)$ the solution set of (1) and by $Sol(AI)$ the solution set of (3). We prove the following theorem:

**Theorem 1.** *Under **A1, A2, A3** the following estimation holds true:*

$$D_H\left(Sol(DI), Sol(AI)\right) \leq O\left(\sqrt{\frac{h}{\varepsilon}}\right), \tag{4}$$

*where $D_H(\cdot, \cdot)$ is the Hausdorff distance.*

*Proof.* Let $(x(\cdot), y(\cdot)) \in Sol(AI)$. Due to Lemma 1, $x(\cdot)$ is $N$–Lipschitz and $y(\cdot)$ is $\dfrac{N}{\varepsilon}$–Lipschitz. Therefore, $|x_i - x(t)| \leq Nh$ and $|y_i - y(t)| \leq \dfrac{Nh}{\varepsilon}$. Hence $(x(\cdot), y(\cdot))$ is a solution of (2) with $\delta$ replaced by $\dfrac{Nh}{\varepsilon}$. It follows from Lemma 2 that $dist\left((x(\cdot), y(\cdot)), Sol(DI)\right) \leq O\left(\sqrt{\dfrac{h}{\varepsilon}}\right)$.

Let $(\tilde{x}(\cdot), \tilde{y}(\cdot)) \in Sol(DI)$. We will find the corresponding solution $(x, y)$ successively on the intervals $[t_i, t_{i+1}]$ for $i = 1, 2, \ldots, K$. Suppose that we have found a solution $(x, y)$ of (3) such that $|x(t) - \tilde{x}(t)| + |y(t) - \tilde{y}(t)| \leq O\left(\sqrt{\dfrac{h}{\varepsilon}}\right)$ on $[0, t_i]$. On $[t_i, t_{i+1}]$ we take selections

$$\begin{pmatrix} f(t) \\ \varepsilon g(t) \end{pmatrix} \in F(t, x^i, y^i), \quad \begin{matrix} x^i = x(t_i) \\ y^i = y(t_i), \end{matrix}$$

such that

$$\langle \tilde{x}(t) - x^i, \dot{\tilde{x}}(t) - f(t) \rangle \leq A|\tilde{x}(t) - x^i|^2 + B|\tilde{y}(t) - y^i|^2 \tag{5}$$
$$\varepsilon\langle \tilde{y}(t) - y^i, \dot{\tilde{y}}(t) - g(t) \rangle \leq C|\tilde{x}(t) - x^i|^2 - \mu|\tilde{y}(t) - y^i|^2.$$

We define $x(\cdot)$ and $y(\cdot)$ on $[t_i, t_{i+1}]$ as in (3). Consequently $\dot{x}(t) = f(t)$, $\varepsilon\dot{y}(t) = g(t)$.

After some technical calculations we derive:

$$\langle \tilde{x}(t) - x(t), \dot{\tilde{x}}(t) - \dot{x}(t) \rangle \leq A|\tilde{x}(t) - x(t)|^2 + B|\tilde{y}(t) - y(t)|^2 + \frac{Ph}{\varepsilon}$$
$$\varepsilon\langle \tilde{y}(t) - y(t), \dot{\tilde{y}}(t) - \dot{y}(t) \rangle \leq C|\tilde{x}(t) - x(t)|^2 - \mu|\tilde{y}(t) - y(t)|^2 + \frac{Ph}{\varepsilon}, \tag{6}$$

where $P > 0$ is a constant. We let $|x(t) - \tilde{x}(t)|^2 = r(t)$ and $|y(t) - \tilde{y}(t)|^2 = s(t)$. Consequently,

$$\dot{r}(t) \leq 2Ar + 2Bs + \frac{Ph}{\varepsilon}, \quad r(0) = 0$$
$$\varepsilon\dot{s}(t) \leq 2Cr - 2\mu s + \frac{Ph}{\varepsilon}, \quad s(0) = 0.$$

Due to the second inequality, we have that either $\dot{s}(t) < 0$ (that is, $s(\cdot)$ decreases, since $r(\cdot)$ and $s(\cdot)$ are continuous), or $s(t) \leq \dfrac{Cr}{\mu} + \dfrac{Ph}{\varepsilon\mu}$. Hence,

$$\dot{r}(t) \leq \left(2A + \frac{2BC}{\mu}\right)r + \frac{Ph}{\varepsilon}\left(1 + \frac{2B}{\mu}\right).$$

Now it is standard to prove that $r(t) \leq O\left(\dfrac{h}{\varepsilon}\right)$ and $s(t) \leq O\left(\dfrac{h}{\varepsilon}\right)$. One can continue in the same way to extend $x(\cdot)$ and $y(\cdot)$ on the whole interval $I$ such that $|x(t) - \tilde{x}(t)| \leq O\left(\sqrt{\dfrac{h}{\varepsilon}}\right)$ and $|y(t) - \tilde{y}(t)| \leq O\left(\sqrt{\dfrac{h}{\varepsilon}}\right)$.    □

Now we consider the problem (1), when $F$ does not depend on $t$, i.e. the autonomous case:

$$\binom{\dot{x}}{\varepsilon \dot{y}} \in F(x(t), y(t)), \quad \begin{matrix} x(0) = x_0 \\ y(0) = y_0. \end{matrix} \tag{7}$$

The discrete approximation scheme will be changed to:

$$(\alpha, \beta) \in F(x^i, y^i), x^i = x(t_i), \ y^i = y(t_i),$$
$$x(t) = x^i + (t - t_i)\alpha, \ \varepsilon y(t) = \varepsilon y^i + (t - t_i)\beta, \ t \in [t_i, t_{i+1}]. \tag{8}$$

Notice that in (3) $(f(\cdot), g(\cdot))$ may be non-constant on $[t_i, t_{i+1}]$. The solution set of (8) will be denoted by $Sol(Au)$.

**Corollary 1.** *If $F$ is USC with convex compact values and if $F$ is OSL, then*
$$D_H(Sol(DI), Sol(Au)) \leq O\left(\sqrt{\dfrac{h}{\varepsilon}}\right).$$

*Proof.* We know from Theorem 1 that $D_H(Sol(DI), Sol(AI)) \leq O\left(\sqrt{\dfrac{h}{\varepsilon}}\right)$. Let $(x, y)$ be a solution of (3). It is easy to see that there exists a solution $(\hat{x}, \hat{y})$ of (8) such that $x(t_i) = \hat{x}(t_i)$ and $y(t_i) = \hat{y}(t_i)$. This implies that $|x(t) - \hat{x}(t)| \leq Nh$ and $|y(t) - \hat{y}(t)| \leq \dfrac{Nh}{\varepsilon}$. The latter completes the proof.    □

Now we study discrete approximation when, however, there are two subdivisions:

For slow variables $x - \Delta^K \overset{def}{=} \{0 = t_0 < t_1 < \ldots < t_K = 1\}$, where $t_i = ih$ and $h = \dfrac{1}{K}$.

For fast variables $y - \Delta^{K_\varepsilon} \overset{def}{=} \{0 = t_0 < t_{0,1} < \ldots < t_{0,k_\varepsilon} = t_1 < t_{1,1} < \ldots <$
$t_{1,k_\varepsilon} = t_2 < \ldots < t_{K,k_\varepsilon} = t_K = 1\}$, where $k_\varepsilon = -\left[-\dfrac{1}{\varepsilon}\right]$.

The discretization scheme becomes:

$$(f^j, g^j) \in F(x^i, y^{i,j}), x^{i,j} = x(t_{i,j}), \ y^{i,j} = y(t_{i,j}),$$
$$x(t) = x^{i,j} + (t - t_{i,j})f^j, \ \varepsilon y(t) = \varepsilon y^{i,j} + (t - t_{i,j})g^j, \ t \in [t_{i,j}, t_{i,j+1}], \tag{9}$$

with a solution set $Sol(Dis)$. Notice that (9) is different from (3) with step $h = \dfrac{1}{k_\varepsilon K}$ only that in the first row of (9) we take one value of $x-x^i$ on $[t_i, t_{i+1}]$.

The following theorem is then valid:

**Theorem 2.** *Under the assumptions of Theorem 1 the next inequality holds:*

$$D_H(Sol(DI), Sol(Dis)) \leq O\left(\sqrt{h}\right). \tag{10}$$

*Proof.* The proof is very similar of the proof of Theorem 1 and it will be only sketched.

First we find approximate solutions $(u(\cdot), v(\cdot))$ using the following scheme: On $[t_{i,j}, t_{i,j+1}]$ we take measurable:

$$(f^j(t), g^j(t)) \in F(u^i, v^{i,j}), \; u^{i,j} = u(t_{i,j}), \; v^{i,j} = v(t_{i,j}),$$

$$u(t) = u^i + \int_{t_i}^t f(s) \, ds, \; t \in [t_{i,j}, t_{i,j+1}],$$

$$\varepsilon v(t) = \varepsilon v^{i,j} + \int_{t_{i,j}}^t g(s) \, ds. \tag{11}$$

Notice that for the step is not greater than $h\varepsilon$. Hence the discrete solution $(u, v)$ satisfies:

$$\begin{pmatrix} \dot{u} \\ \varepsilon \dot{v} \end{pmatrix} \in F(t, u(t) + Nh\mathbb{B}_n, v(t) + Nh\mathbb{B}_m), \; \begin{matrix} u(0) = u^0 \\ v(0) = v^0, \end{matrix}$$

and Lemma 2 applies.

If $(x, y)$ is a solution of (1), we choose $f(t), g(t)$ such that

$$\langle x(t) - u^i, \dot{x}(t) - f(t) \rangle \leq A|x(t) - u^i|^2 + B|y(t) - v^{i,j}|^2$$
$$\varepsilon \langle y(t) - v^{i,j}, \dot{y}(t) - g(t) \rangle \leq C|x(t) - u^i|^2 - \mu|y(t) - v^{i,j}|^2.$$

Dealing as in the proof of Theorem 1 we derive $|x(t) - u(t)| \leq O(\sqrt{h})$ and $|y(t) - v(t)| \leq O(\sqrt{h})$. Hence we have proved that the Hausdorff distance between the solution set $Sol(11)$ of (11) and $Sol(DI)$ is not greater than $O\left(\sqrt{h}\right)$.

Now we use the same proof as the proof of Corollary 1 to see that $D_H\left(Sol(dtsc), Sol(11)\right) \leq O(h)$. □

## 3 Concluding Remarks

It is easy to see that there are many interesting questions, which were not studied in the paper. A reason (not the only one) is the length restriction. Below we present some notions for problems not discussed here, some of them will be subjects of other our papers.

**Note 1.** Assume that $F(\cdot, \cdot)$ is locally Lipschitz and replace the OSL condition by the stronger one:

*There exist positive constants $A, B, C, \mu$ such that:*

For every $(x, y), (x_1, y_1) \in \mathbb{R}^n \times \mathbb{R}^m$ and every $(f, g) \in F(x, y)$ there exists $(f_1, g_1) \in F(x_1, y_1)$ which satisfy:

$$\langle x_1 - x, f_1 - f \rangle \leq A|x_1 - x|^2 + B|y_1 - y||x - x_1|, \tag{12}$$
$$\varepsilon \langle y_1 - y, g_1 - g \rangle \leq C|x_1 - x||y_1 - y| - \mu|y_1 - y|^2.$$

**Theorem 3.** *Under the assumptions of Theorem 1 (with OSL condition replaced by (12)) the next inequality holds:*

$$D_H(Sol(DI), Sol(dtsc)) \leq O(h). \tag{13}$$

The proof is omitted. It is very similar to the proof of Theorem 1, however, one should follow the method presented in [4] and the variant of Filippov–Pliss lemma given there.

**Note 2.** As it is shown in [7] when $F$ is autonomous under our conditions there exists a constant $L > 0$ such that for every $\varepsilon > 0$ there exists a solution $(x^\varepsilon, y^\varepsilon)$ such that $x^\varepsilon$ is $L$–Lipschitz on $I$ and $y^\varepsilon$ is $L$–Lipschitz on $[\sqrt{\varepsilon}, 1]$. Therefore it would be interesting to present a scheme with Euler approximation of such a solution.

**Note 3.** When $F(x, y) = G(x, y, u) \times H(x, y, u)$, it is proved in [3,6] that the solution set $Z(\varepsilon)$ has a limit $Z(0)$. The fast solution set converges to some set of Young measures, which are limit occupational measures of appropriate differential inclusion. It is interesting to investigate approximation of such measures.

**Note 4.** In [15] Mordukhovich obtains necessary optimality conditions of Euler–Lagrange form using discrete approximations. It would be interesting to prove similar result in the case of singularly perturbed systems.

# References

1. Deimling, K.: Multivalued Differential Equations. De Gruyter, Berlin (1992)
2. Donchev, T.: Functional differential inclusions with monotone right-hand side. Nonlinear analysis **16** (1991) 543–552
3. Donchev, T., Dontchev, A.: Singular perturbations in infinite-dimensional control systems. SIAM J. Control Optim. **42** (2003) 1795 – 1812
4. Donchev T., Farkhi E.: Stability and Euler approximations of one sided Lipschitz convex differential inclusions. SIAM J. Control Optim. **36** (1998) 780–796
5. Donchev T., Slavov I.: Singularly perturbed functional differential inclusions. Set-Valued Analysis **3** (1995) 113–128
6. Donchev, T., Slavov, I.: Averaging method for one sided Lipschitz differential inclusions with generalized solutions. SIAM J. Control Optim. **37** (1999) 1600–1613
7. Dontchev, A., Donchev, T., Slavov, I.: A Tikhonov-type theorem for singularly perturbed differential inclusions. Nonlinear Analysis **26** (1996) 1547–1554
8. Dontchev, A., Farkhi, E.: Error estimates for discretized differential inclusions. Computing **41** (1989) 349–358
9. Filatov, O., Hapaev, M.: Averaging of Systems of Differential Inclusions. Moskow University Press, Moskow (1998) [in Russian]

10. Gaitsgory, V.: Suboptimization of singularly perturbed control systems. SIAM J. Control Optim. **30** (1992) 1228–1249
11. Grammel, G.: On the Time-Discretization of Singularly Perturbed Uncertain Systems. In: I. Lirkov, S. Margenov, J.Wasniewski (eds.): LNCS **3743** Springer (2006) 297–304
12. Grammel, G.: Towards fully discretized differential inclusions. Set Valued Analysis **11** (2003) 1–8
13. Grammel, G.: Singularly perturbed differential inclusions: an averaging approach. Set-Valued Analysis **4** (1996) 361–374
14. Lempio F., Veliov, V.: Discrete approximations of differential inclusions. Bayreuter Mathematische Schiften, Heft **54** (1998) 149–232
15. Mordukhovich B.: Discrete approximations and refined Euler-Lagrange conditions for differential inclusions. SIAM J. Control. Optim. **33** (1995) 882–915
16. Veliov, V.: Differential inclusions with stable subinclusions. Nonlinear Analysis TMA **23** (1994) 1027–1038

# A Generalized Interval LU Decomposition for the Solution of Interval Linear Systems

Alexandre Goldsztejn[1] and Gilles Chabert[2]

[1] University of Central Arkansas, Conway 72035 Arkansas, USA
AGoldsztejn@uca.edu
[2] Projet Coprin, INRIA, 2004 route des Lucioles, 06902 Sophia Antipolis, France
Gilles.Chabert@sophia.inria.fr

**Abstract.** Generalized intervals (intervals whose bounds are not constrained to be increasingly ordered) extend classical intervals providing better algebraic properties. In particular, the generalized interval arithmetic is a group for addition and for multiplication of zero free intervals. These properties allow one constructing a LU decomposition of a generalized interval matrix $\mathbf{A}$: the two computed generalized interval matrices $\mathbf{L}$ and $\mathbf{U}$ satisfy $\mathbf{A} = \mathbf{LU}$ with equality instead of the weaker inclusion obtained in the context of classical intervals. Some potential applications of this generalized interval LU decomposition are investigated.

## 1 Introduction

The LU decomposition of a square real matrix $A$ consists of computing two matrices $L$ and $U$ being respectively lower and upper triangular and satisfying $A = LU$. Such a decomposition eases the solution of many problems such as solving linear equations, inverting matrices, etc..

The Gauss elimination algorithm allows constructing LU decompositions. It has been generalized to interval matrices (see e.g. [1]). The interval Gauss elimination is widely used in order to construct an outer approximation of united solution sets, i.e. $\{x \in \mathbb{R}^n \mid (\exists A \in \mathbf{A})(\exists b \in \mathbf{b})(Ax = b)\}$, where boldface letters mean interval matrices and vectors. In the study of the interval Gauss elimination, an interval LU decomposition has been introduced (cf. [1]). This interval LU decomposition does not satisfy $\mathbf{A} = \mathbf{LU}$ but only the weaker relation $\mathbf{A} \subseteq \mathbf{LU}$, which is well suited for outer approximation of united solution sets.

Generalized intervals are intervals whose bounds are not constrained to be increasingly ordered (e.g. $[-1, 1]$ is a proper interval while $[1, -1]$ is an improper interval). They have been introduced in [2,3] in order to improve the mathematical structure of intervals. The generalized interval arithmetic (also called Kaucher arithmetic) is a group for addition and for multiplication of zero free generalized intervals. Thanks to the framework of generalized intervals, a new LU decomposition of a generalized interval matrix $\mathbf{A}$ is introduced. It satisfies $\mathbf{A} = \mathbf{LU}$.

T. Boyanov et al. (Eds.): NMA 2006, LNCS 4310, pp. 312–319, 2007.

The paper is organized as follows: Section 2 gives an overview of generalized intervals and their arithmetic. Section 3 presents the generalized interval LU decomposition, and Section 4 shows how it can be used to compute approximations of linear AE-solution sets.

## 2 Generalized Intervals

Generalized intervals are intervals whose bounds are not constrained to be ordered, for example $[-1, 1]$ and $[1, -1]$ are generalized intervals. They have been introduced in [2,3] so as to improve the algebraic structure of intervals, while maintaining the inclusion monotonicity. The set of generalized intervals is denoted by $\mathbb{KR}$ and is divided into three subsets:

- The set of *proper intervals* with bounds ordered increasingly. These proper intervals are identified with classical intervals. The set of proper intervals is denoted $\mathbb{IR} := \{[a, b] \mid a \leq b\}$. *Strictly* proper intervals satisfy $a < b$.
- The set of *improper intervals* with bounds ordered decreasingly. It is denoted by $\overline{\mathbb{IR}} := \{[a, b] \mid a \geq b\}$. *Strictly* proper intervals satisfy $a > b$.
- The set of *degenerated intervals* $\{[a, b] \mid a = b\} = \mathbb{IR} \cap \overline{\mathbb{IR}}$. Degenerated intervals are identified to reals.

Therefore, from a set of reals $\{x \in \mathbb{R} \mid a \leq x \leq b\}$, one can build the two generalized intervals $[a, b]$ and $[b, a]$. It will be convenient to switch from one to the other keeping the underlying set of reals unchanged. To this purpose, the following three operations are introduced: the dual operation is defined by dual $[a, b] = [b, a]$; the proper projection is defined by pro $[a, b] = [\min\{a, b\}, \max\{a, b\}]$; the improper projection is defined by imp $[a, b] = [\max\{a, b\}, \min\{a, b\}]$.

The generalized intervals are partially ordered by an inclusion which extends the inclusion of classical intervals. Given two generalized intervals $\mathbf{x} = [\underline{\mathbf{x}}, \overline{\mathbf{x}}]$ and $\mathbf{y} = [\underline{\mathbf{y}}, \overline{\mathbf{y}}]$, the inclusion is defined by $\mathbf{x} \subseteq \mathbf{y} \iff \underline{\mathbf{y}} \leq \underline{\mathbf{x}} \wedge \overline{\mathbf{x}} \leq \overline{\mathbf{y}}$. For example, $[-1, 1] \subseteq [-1.1, 1.1]$ (this matches the set inclusion), $[1.1, -1.1] \subseteq [1, -1]$ (the inclusion between the underlying sets of real is reversed for improper intervals) and $[2, 0.9] \subseteq [-1, 1]$. As degenerated intervals are identified to reals, if $\mathbf{x}$ is proper then $x \in \mathbf{x} \iff x \subseteq \mathbf{x}$. On the other hand, if $\mathbf{x}$ is strictly improper then for all $x \in \mathbb{R}$ the inclusion $x \subseteq \mathbf{x}$ is false.

The generalized interval arithmetic (also called Kaucher arithmetic) extends the classical interval arithmetic. Its definition can be found in [4,5]. When only proper intervals are involved, this arithmetic coincides with the interval arithmetic: $\mathbf{x} \circ \mathbf{y} = \{x \circ y \in \mathbb{R} \mid x \in \mathbf{x}, y \in \mathbf{y}\}$. When proper and improper intervals are involved, some new expressions are used. For example, $[a, b] + [c, d] = [a+c, b+d]$ and if $a, b, c, d \geq 0$ then $[a, b] \times [c, d] = [a \times c, b \times d]$. The following useful property provides some bounds on the proper projection of the results of the generalized interval arithmetic. Let us consider $\mathbf{x}, \mathbf{y} \in \mathbb{KR}$ and $\circ \in \{+, -, \times, /\}$. If (pro $\mathbf{x}$) $\circ$ (pro $\mathbf{y}$) is defined then $\mathbf{x} \circ \mathbf{y}$ is defined and it satisfies

$$\text{pro}\,(\mathbf{x} \circ \mathbf{y}) \subseteq (\text{pro }\mathbf{x}) \circ (\text{pro }\mathbf{y}). \tag{1}$$

Generalized interval arithmetic has better algebraic properties than the classical interval arithmetic: the addition in $\mathbb{KR}$ is a group. The opposite of an interval $\mathbf{x}$ is $-\operatorname{dual} \mathbf{x}$, i.e.,

$$\mathbf{x} + (-\operatorname{dual} \mathbf{x}) = \mathbf{x} - \operatorname{dual} \mathbf{x} = [0, 0]. \tag{2}$$

The multiplication in $\mathbb{KR}$ restricted to generalized intervals whose proper projection does not contain 0 is also a group. The inverse of such a generalized interval $\mathbf{x}$ is $1/(\operatorname{dual} \mathbf{x})$, i.e.,

$$\mathbf{x} \times (1/\operatorname{dual} \mathbf{x}) = \mathbf{x}/(\operatorname{dual} \mathbf{x}) = [1, 1]. \tag{3}$$

Although addition and multiplication in $\mathbb{KR}$ are associative, they are not distributive. The addition and multiplication in $\mathbb{KR}$ are linked by the following distributivity laws (see [6,7,5]). Whatever are $\mathbf{x}, \mathbf{y}, \mathbf{z} \in \mathbb{KR}$,

- conditional distributivity:

$$\mathbf{x} \times \mathbf{y} + (\operatorname{imp} \mathbf{x}) \times \mathbf{z} \subseteq \mathbf{x} \times (\mathbf{y} + \mathbf{z}) \subseteq \mathbf{x} \times \mathbf{y} + (\operatorname{pro} \mathbf{x}) \times \mathbf{z}. \tag{4}$$

The three following particular cases will be of practical interest in this paper.

- subdistributivity: if $\mathbf{x} \in \mathbb{IR}$ then $\mathbf{x} \times (\mathbf{y} + \mathbf{z}) \subseteq \mathbf{x} \times \mathbf{y} + \mathbf{x} \times \mathbf{z}$;
- superdistributivity: if $\mathbf{x} \in \overline{\mathbb{IR}}$ then $\mathbf{x} \times (\mathbf{y} + \mathbf{z}) \supseteq \mathbf{x} \times \mathbf{y} + \mathbf{x} \times \mathbf{z}$;
- distributivity: if $x \in \mathbb{R}$ then $x \times (\mathbf{y} + \mathbf{z}) = x \times \mathbf{y} + x \times \mathbf{z}$.

Another useful property of the Kaucher arithmetic is its monotonicity with respect to the inclusion: whatever are $\circ \in \{+, \times, -, \div\}$ and $\mathbf{x}, \mathbf{y}, \mathbf{x}', \mathbf{y}' \in \mathbb{KR}$,

$$\mathbf{x} \subseteq \mathbf{x}' \wedge \mathbf{y} \subseteq \mathbf{y}' \implies (\mathbf{x} \circ \mathbf{y}) \subseteq (\mathbf{x}' \circ \mathbf{y}'). \tag{5}$$

The next example illustrates the way these properties will be used in the sequel.

*Example 1.* Consider the expression $\mathbf{x} + \mathbf{uv} \subseteq \mathbf{y}$. Subtracting $\operatorname{dual}(\mathbf{uv}) = (\operatorname{dual} \mathbf{u})(\operatorname{dual} \mathbf{v})$ to each side preserves the inclusion: $\mathbf{x} + \mathbf{uv} - \operatorname{dual}(\mathbf{uv}) \subseteq \mathbf{y} - (\operatorname{dual} \mathbf{u})(\operatorname{dual} \mathbf{v})$. As $-\operatorname{dual}(\mathbf{uv})$ is the opposite of $\mathbf{uv}$, the following inclusion is eventually proved to hold: $\mathbf{x} \subseteq \mathbf{y} - (\operatorname{dual} \mathbf{u})(\operatorname{dual} \mathbf{v})$.

Finally, generalized interval vectors $\mathbf{x} \in \mathbb{KR}^n$ and generalized interval matrices $\mathbf{A} \in \mathbb{KR}^{n \times n}$ together with their additions and multiplications are defined similarly to their real and classical interval counterparts.

## 3    Generalized Interval LU Decomposition

A Gauss elimination algorithm for generalized interval matrices is first presented. It will be used to construct the generalized interval LU decomposition. Consider a generalized interval matrix $\mathbf{A} = (\mathbf{A}_{ij})$. In order to obtain a generalized interval LU decomposition, the Gauss elimination algorithm is applied in the following way: if $0 \notin \text{pro } \mathbf{A}_{11}$, for $i \in [2..n]$ multiply the first row by $\mathbf{A}_{i1}/(\text{dual } \mathbf{A}_{11})$. As $\mathbf{A}_{11}/(\text{dual } \mathbf{A}_{11}) = 1$ the following row is obtained:

$$\left( \mathbf{A}_{i1} \, , \, \frac{\mathbf{A}_{12}\mathbf{A}_{i1}}{\text{dual } \mathbf{A}_{11}} \, , \cdots , \, \frac{\mathbf{A}_{1n}\mathbf{A}_{i1}}{\text{dual } \mathbf{A}_{11}} \right). \tag{6}$$

The second step consists in subtracting the dual of the previously computed row to the $i^{th}$ row of the matrix $\mathbf{A}$. As $\mathbf{A}_{i1} - \text{dual } \mathbf{A}_{i1} = 0$ one obtains

$$\left( 0 \, , \, \mathbf{A}_{i2} - \frac{(\text{dual } \mathbf{A}_{12})(\text{dual } \mathbf{A}_{i1})}{\mathbf{A}_{11}} \, , \cdots , \, \mathbf{A}_{in} - \frac{(\text{dual } \mathbf{A}_{1n})(\text{dual } \mathbf{A}_{i1})}{\mathbf{A}_{11}} \right). \tag{7}$$

Once this transformation is applied for each $i \in [2..n]$, the following interval matrix $\mathbf{A}$ is obtained:

$$\mathbf{A} := \begin{pmatrix} \mathbf{A}_{11} & \mathbf{A}_{12} & \cdots & \mathbf{A}_{1n} \\ 0 & \mathbf{A}'_{22} & \cdots & \mathbf{A}'_{2n} \\ \vdots & \vdots & \ddots & \vdots \\ 0 & \mathbf{A}'_{n2} & \cdots & \mathbf{A}'_{nn} \end{pmatrix}, \quad \text{where } \mathbf{A}'_{ij} := \mathbf{A}_{ij} - \frac{(\text{dual } \mathbf{A}_{1j})(\text{dual } \mathbf{A}_{i1})}{\mathbf{A}_{11}}. \tag{8}$$

As in the context of real numbers and classical intervals, this leads to a LU decomposition of the generalized interval matrix $\mathbf{A}$. This LU decomposition can be formulated in the following way:

$$\mathbf{L}_{ii} = 1 \quad \text{and} \quad \mathbf{L}_{ij} = 0 \quad \text{for} \quad i < j, \tag{9}$$

$$\mathbf{L}_{ij} = \left( \mathbf{A}_{ij} - \sum_{k<j} \text{dual } (\mathbf{L}_{ik}\mathbf{U}_{kj}) \right) / (\text{dual } \mathbf{U}_{ii}) \quad \text{for } j < i, \tag{10}$$

$$\mathbf{U}_{ij} = 0 \quad \text{for} \quad i > j, \tag{11}$$

$$\mathbf{U}_{ij} = \mathbf{A}_{ij} - \sum_{k<i} \text{dual } (\mathbf{L}_{ik}\mathbf{U}_{kj}) \quad \text{for } i \leq j. \tag{12}$$

The previous expressions allow constructing a generalized interval LU decomposition of $\mathbf{A}$: the recursive construction is started by the first row of $\mathbf{U}$ which is trivially computed using (12). It is equal to the first row of $\mathbf{A}$. Then, provided that $0 \notin \text{pro } \mathbf{U}_{ii}$, the $i^{th}$ column of $\mathbf{L}$ using (10) and the $i^{th}$ row of $\mathbf{U}$ is constructed using (12). This process is recursively repeated for $i + 1$.

**Proposition 1.** *Let $\mathbf{A} \in \mathbb{KR}^{n \times n}$ be a generalized interval matrix. Provided that the generalized interval matrices $\mathbf{L}$ and $\mathbf{U}$ defined by (9-12) can be constructed, they satisfy $\mathbf{A} = \mathbf{LU}$.*

*Proof.* Consider $i, j \in [1..n]$ such that $i \leq j$. Then by equation (12)

$$\mathbf{U}_{ij} = \mathbf{A}_{ij} - \sum_{k<i} \text{dual } (\mathbf{L}_{ik}\mathbf{U}_{kj}). \tag{13}$$

Adding $\sum_{k<i} \mathbf{L}_{ik}\mathbf{U}_{kj}$ to each side of the equality, $\mathbf{U}_{ij} + \sum_{k<i} \mathbf{L}_{ik}\mathbf{U}_{kj} = \mathbf{A}_{ij}$. As $\mathbf{L}_{jj} = 1$ and $\mathbf{L}_{ik} = 0$ for $i < k$, $\sum_{k\in[1..n]} \mathbf{L}_{ik}\mathbf{U}_{kj} = \mathbf{A}_{ij}$. Considering the equation (10), we argue similarly in the case $i, j \in [1..n]$, with $i > j$. ☐

Two examples of LU decomposition are now presented.

*Example 2.* Consider the interval matrix

$$\mathbf{A} = \begin{pmatrix} [9,11] & [-1,1] & [-1,1] \\ [-11,11] & [8,12] & [-2,2] \\ [-11,11] & [-12,12] & [7,13] \end{pmatrix}. \tag{14}$$

The computations of the generalized interval LU decomposition are detailed: the first row of $\mathbf{U}$ is the first row of $\mathbf{A}$. Now, using (10) and (12),

$$\mathbf{L}_{21} = \mathbf{A}_{21}/(\text{dual } \mathbf{U}_{11}) = [-11,11]/[11,9] = [-1,1], \tag{15}$$
$$\mathbf{L}_{31} = \mathbf{A}_{31}/(\text{dual } \mathbf{U}_{11}) = [-11,11]/[11,9] = [-1,1], \tag{16}$$
$$\mathbf{U}_{22} = \mathbf{A}_{22} - \text{dual} (\mathbf{L}_{21}\mathbf{U}_{12}) = [9,11], \tag{17}$$
$$\mathbf{U}_{23} = \mathbf{A}_{23} - \text{dual} (\mathbf{L}_{21}\mathbf{U}_{13}) = [-1,1], \tag{18}$$
$$\mathbf{L}_{32} = (\mathbf{A}_{32} - \text{dual} (\mathbf{L}_{31}\mathbf{U}_{12}))/(\text{dual } \mathbf{U}_{22}) = [-1,1], \tag{19}$$
$$\mathbf{U}_{33} = \mathbf{A}_{33} - \text{dual} (\mathbf{L}_{31}\mathbf{U}_{13}) - \text{dual} (\mathbf{L}_{32}\mathbf{U}_{23}) = [9,11]. \tag{20}$$

Therefore, we obtain following interval matrices:

$$\mathbf{L} = \begin{pmatrix} 1 & 0 & 0 \\ [-1,1] & 1 & 0 \\ [-1,1] & [-1,1] & 1 \end{pmatrix} \quad \text{and} \quad \mathbf{U} = \begin{pmatrix} [9,11] & [-1,1] & [-1,1] \\ 0 & [9,11] & [-1,1] \\ 0 & 0 & [9,11] \end{pmatrix}. \tag{21}$$

These interval matrices satisfy $\mathbf{A} = \mathbf{LU}$.

*Example 3.* Consider the interval matrix

$$\mathbf{A} = \begin{pmatrix} [2,3] & 1 & 0 \\ 1 & [2,3] & 1 \\ 0 & 1 & [2,3] \end{pmatrix}. \tag{22}$$

The following interval matrices correspond to the generalized interval LU decomposition of $\mathbf{A}$:

$$\mathbf{L} = \begin{pmatrix} 1 & 0 & 0 \\ [0.5, \frac{1}{3}] & 1 & 0 \\ 0 & [\frac{2}{3}, 0.375] & 1 \end{pmatrix} \quad \text{and} \quad \mathbf{U} = \begin{pmatrix} [2,3] & 1 & 0 \\ 0 & [1.5, \frac{8}{3}] & 1 \\ 0 & 0 & [\frac{4}{3}, 2.625] \end{pmatrix}. \tag{23}$$

These interval matrices satisfy $\mathbf{A} = \mathbf{LU}$. Notice that $\mathbf{L}$ is not proper anymore.

As a direct consequence of the property (1), given a generalized interval matrix $\mathbf{A} \in \mathbb{KR}^{n\times n}$, if (pro $\mathbf{A}$) has a classical interval LU decomposition (as defined in [1]) then $\mathbf{A}$ has a generalized interval LU decomposition.

## 4  Approximation of Linear AE-Solution Set

We define linear AE-solution sets using a different convention than the one proposed in [5]. The justification of this new convention is out of the scope of this
paper (cf. [8]). Let $\mathbf{A} \in \mathbb{KR}^{n \times n}$ be a generalized interval matrix, and $\mathbf{b} \in \mathbb{KR}^n$
be a generalized interval vector. Define $\mathbf{A}^\forall, \mathbf{A}^\exists \in \mathbb{IR}^{n \times n}$ and $\mathbf{b}^\forall, \mathbf{b}^\exists \in \mathbb{IR}^n$ by

$$\mathbf{A}^\exists_{ij} := \begin{cases} \mathbf{A}_{ij} & \text{if } \mathbf{A}_{ij} \in \mathbb{IR}, \\ 0 & \text{otherwise.} \end{cases} \qquad \mathbf{A}^\forall_{ij} := \begin{cases} \text{pro } \mathbf{A}_{ij} & \text{if } \mathbf{A}_{ij} \in \overline{\mathbb{IR}}, \\ 0 & \text{otherwise.} \end{cases}$$

$$\mathbf{b}^\exists_i := \begin{cases} \mathbf{b}_i & \text{if } \mathbf{b}_i \in \mathbb{IR}, \\ 0 & \text{otherwise.} \end{cases} \qquad \mathbf{b}^\forall_i := \begin{cases} \text{pro } \mathbf{b}_i & \text{if } \mathbf{b}_i \in \overline{\mathbb{IR}}, \\ 0 & \text{otherwise.} \end{cases}$$

So $\mathbf{A} = \mathbf{A}^\exists + (\text{dual } \mathbf{A}^\forall)$ and $\mathbf{b} = \mathbf{b}^\exists + (\text{dual } \mathbf{b}^\forall)$. Then, the linear AE-solution
set $\Sigma(\mathbf{\Lambda}, \mathbf{b})$ is the following subset of $\mathbb{R}^n$:

$$(\forall A^\forall \in \mathbf{A}^\forall)(\forall b^\forall \in \mathbf{b}^\forall)(\exists A^\exists \in \mathbf{A}^\exists)(\exists b^\exists \in \mathbf{b}^\exists)\ (A^\forall + A^\exists)x = (b^\forall + b^\exists). \quad (24)$$

We can exhibit the following two special cases: if $\mathbf{A}$ and $\mathbf{b}$ are proper, then
$\Sigma(\mathbf{A}, \mathbf{b}) = \{x \in \mathbb{R}^n | \exists A \in \mathbf{A}, \exists b \in \mathbf{b}, Ax = b\}$, which is called a *united solution
set*. While, if $\mathbf{A}$ is improper and $\mathbf{b}$ is proper, then $\Sigma(\mathbf{A}, \mathbf{b}) = \{x \in \mathbb{R}^n | \forall A \in$
pro $\mathbf{A}, \exists b \in \mathbf{b}, Ax = b\}$, which is called a *tolerable solution set*. Shary has discovered a very useful characterization of linear AE-solution sets (cf. [5, Theorem 5.1,
p. 375]). Let us reproduce his proof with our new conventions. (24) can be writ-
ten $(\forall A^\forall \in \mathbf{A}^\forall)(\forall b^\forall \in \mathbf{b}^\forall)(\exists A^\exists \in \mathbf{A}^\exists)(\exists b^\exists \in \mathbf{b}^\exists)\ A^\forall x - b^\forall = b^\exists - A^\exists x$. Now as
every quantified parameter has only one occurrence in the expression, interval
arithmetic leads to the exact range and we obtain the following equivalent condi-
tion: $\mathbf{A}^\forall x - \mathbf{b}^\forall \subseteq \mathbf{b}^\exists - \mathbf{A}^\exists x$. Now, *using the group property of the Kaucher arithmetic*, the previous condition is equivalent to $\mathbf{A}^\forall x + \text{dual}(\mathbf{A}^\exists x) \subseteq \mathbf{b}^\exists + \text{dual } \mathbf{b}^\forall$.
Finally, notice that dual $(\mathbf{A}^\exists x) = (\text{dual } \mathbf{A}^\exists)x$ and use the distributivity w.r.t. $x$
to obtain

$$x \subset \Sigma(\mathbf{A}, \mathbf{b}) \iff (\text{dual } \mathbf{A})x \subseteq \mathbf{b}. \quad (25)$$

This characterization only differs from [5, Theorem 5.1, p. 375] by the new conventions we use. It can be used in order to obtain generalized interval operators
that help to approximate linear AE-solution sets. In particular, outer approximations can be done using the generalized interval Gauss-Seidel iteration introduced in [9,5]. The generalized interval LU decomposition is now used to build
inner and outer approximations of linear AE-solution sets $\Sigma(\mathbf{A}, \mathbf{b})$.

**Theorem 1.** *Let $\mathbf{A} \in \mathbb{KR}^{n \times n}$, $\mathbf{b} \in \mathbb{KR}^n$. Let $\mathbf{L}$ and $\mathbf{U}$ be the generalized interval
LU decomposition of $\mathbf{A}$. Then define the generalized interval vectors $\mathbf{x}, \mathbf{x}', \mathbf{y}, \mathbf{y}' \in
\mathbb{KR}^n$ such that for all $i \in [1..n]$*

$$\mathbf{y}_i = \mathbf{b}_i - \sum_{j<i} \mathbf{L}_{ij}(\text{dual } \mathbf{y}_j) \quad \text{and} \quad \mathbf{x}_i = \left(\mathbf{y}_i - \sum_{j>i} \mathbf{U}_{ij}(\text{dual } \mathbf{x}_j)\right)/\mathbf{U}_{ii}.$$

$$\mathbf{y}'_i = \mathbf{b}_i - \sum_{j<i} \mathbf{L}_{ij}\mathbf{y}'_j \quad \text{and} \quad \mathbf{x}'_i = \left(\mathbf{y}'_i - \sum_{j>i} \mathbf{U}_{ij}\mathbf{x}'_j\right)/\mathbf{U}_{ii},$$

*Both following properties hold:*

(i) *If* **L** *is proper and* **x** *is proper then* $\mathbf{x} \subseteq \Sigma(\mathbf{A}, \mathbf{b})$.

(ii) *Suppose that* **U** *and* **L** *are improper. If* $\mathbf{x}'$ *is proper then* $\Sigma(\mathbf{A}, \mathbf{b}) \subseteq \mathbf{x}'$. *Otherwise,* $\Sigma(\mathbf{A}, \mathbf{b}) = \emptyset$.

*Proof.* As a direct consequence of the construction of the LU decomposition, if $\mathbf{A} = \mathbf{L}\mathbf{U}$ then $(\text{dual } \mathbf{A}) = (\text{dual } \mathbf{L})(\text{dual } \mathbf{U})$.

(i) By definition of **y** and **x** we have $(\text{dual } \mathbf{L})\mathbf{y} = \mathbf{b}$ and $(\text{dual } \mathbf{U})\mathbf{x} = \mathbf{y}$. Therefore we obtain $(\text{dual } \mathbf{L})((\text{dual } \mathbf{U})\mathbf{x}) = \mathbf{b}$. As **L** is supposed to be proper, $(\text{dual } \mathbf{L})$ is improper. Consider any $x \in \mathbf{x}$. A componentwise application of the super-distributivity shows $((\text{dual } \mathbf{L})(\text{dual } \mathbf{U}))x \subseteq (\text{dual } \mathbf{L})((\text{dual } \mathbf{U})x)$. That is $(\text{dual } \mathbf{A})x \subseteq \mathbf{b}$ and finally $x \in \Sigma(\mathbf{A}, \mathbf{b})$ thanks to (25).

(ii) Assume $x \in \Sigma(\mathbf{A}, \mathbf{b})$. Using (25) we obtain $((\text{dual } \mathbf{L})(\text{dual } \mathbf{U}))x \subseteq \mathbf{b}$. As both $(\text{dual } \mathbf{L})$ and $(\text{dual } \mathbf{U})$ are proper, **b** has to be proper. Also, the subdistributivity law proves $(\text{dual } \mathbf{L})((\text{dual } \mathbf{U})x) \subseteq ((\text{dual } \mathbf{L})(\text{dual } \mathbf{U}))x$, so $(\text{dual } \mathbf{L})((\text{dual } \mathbf{U})x) \subseteq \mathbf{b}$. As $(\text{dual } \mathbf{U})$ is proper, $\mathbf{u} := (\text{dual } \mathbf{U})x$ is proper too. Claim: $\mathbf{u} \subseteq \mathbf{y}'$. We shall prove it by induction. The first line of $(\text{dual } \mathbf{L})\mathbf{u} \subseteq \mathbf{b}$ yields $\mathbf{u}_1 \subseteq \mathbf{b}_1$ while $\mathbf{b}_1 = \mathbf{y}'_1$ by definition of $\mathbf{y}'$. Now fix $i \in \{2, \ldots, n-1\}$ and suppose $j < i \Rightarrow \mathbf{u}_j \subseteq \mathbf{y}'_j$. The $i^{th}$ line of $(\text{dual } \mathbf{L})\mathbf{u} \subseteq \mathbf{b}$ yields $\mathbf{u}_i + \sum_{j<i}(\text{dual } \mathbf{L}_{ij})\mathbf{u}_j \subseteq \mathbf{b}_i$. Therefore $\mathbf{u}_i \subseteq \mathbf{b}_i - \sum_{j<i} \mathbf{L}_{ij}(\text{dual } \mathbf{u}_j) \subseteq \mathbf{b}_i - \sum_{j<i} \mathbf{L}_{ij}\mathbf{u}_j$, the second inclusion being a consequence of inclusion monotonicity $(\text{dual } \mathbf{u}_j \subseteq \mathbf{u}_j$ because $\mathbf{u}_j$ is proper). Finally, using the induction hypothesis and inclusion monotonicity, we obtain $\mathbf{u}_i \subseteq \mathbf{b}_i - \sum_{j<i} \mathbf{L}_{ij}\mathbf{y}'_j$ which is equal to $\mathbf{y}'_i$ by definition. So far, we have proved $(\text{dual } \mathbf{U})x \subseteq \mathbf{y}'$. Using induction in a similar way as previously, on can prove $x \subseteq \mathbf{x}'$. As a consequence, if $\mathbf{x}'$ is proper then $\Sigma(\mathbf{A}, \mathbf{b}) \subseteq \mathbf{x}'$. Otherwise $\Sigma(\mathbf{A}, \mathbf{b}) = \emptyset$. $\qquad\square$

Due to the conditions required for the LU decomposition in Theorem 1, outer approximation is likely to succeed when **A** is improper, i.e. when a tolerable solution set is under consideration. While, inner approximation is likely to succeed when **A** is proper, i.e. when a united solution set is under consideration. Under these conditions, experimentations have shown that the generalized interval LU decomposition applied to diagonally dominant matrices centered around the identity satisfies the conditions of Theorem 1. However, the generalized interval Gauss-Seidel operator and the direct resolution of the equation $(\text{dual } \mathbf{A})\mathbf{x} = \mathbf{b}$ seem to provide close but better results for this class of matrices. It still remains that in presence of several solution sets sharing the same interval matrix, the generalized interval LU decomposition needs to be computed only once, thereby sparing computation time.

In some situations, the generalized interval LU decomposition can provide a much sharper outer approximation than the generalized interval Gauss-Seidel (GIGS) operator. This is illustrated in the next example.

*Example 4.* Consider the tridiagonal matrix **A** and vector **b** defined by

$$\mathbf{A} = \begin{pmatrix} \mathbf{a} & \mathbf{a} & 0 \\ \mathbf{a} & \mathbf{a} & \mathbf{a} \\ 0 & \mathbf{a} & \mathbf{a} \end{pmatrix} \quad \text{and} \quad \mathbf{b} = \begin{pmatrix} [0, 10] \\ [-10, 0] \\ [-5, 5] \end{pmatrix}$$

with $\mathbf{a} = [1.1, 0.9]$ and the tolerable solution set $\Sigma(\mathbf{A}, \mathbf{b})$. Both the GIGS and the generalized interval LU decomposition have a bad behavior on this example. However, exchanging raws 2 and 3 leads to better behavior. Starting from the initial box $(\pm 1000, \pm 1000, \pm 1000)^\top$ the GIGS operator gives raise to $([-18.4091, 27.5], [-22.5, 22.5], [-27.5, 18.4091])^\top$. The generalized interval LU decomposition raises the following outer approximation:

$$([-12.8099, 11.157], [-9.12847, 19.5718], [-13.6364, 4.54545])^\top,$$

which is much sharper than the one computed by the GIGS operator.

It must be noted that changing slightly the uncertainties in the interval matrix of the previous example leads to a matrix $\mathbf{L}$ which is not proper any more. While this good result is promising, some additional developments will be necessary to deal with more general situations.

## 5 Conclusion

An interval LU decomposition has been proposed in the framework of generalized intervals. Thanks to the group structure of the generalized interval arithmetic, this new generalized interval LU decomposition satisfies $\mathbf{A} = \mathbf{LU}$. In addition to the theoretical interest of such a generalized interval LU decomposition, we proved that it can be used to construct some approximations of linear AE-solution sets. On some examples, these approximations may be much more accurate than the one computed by the generalized interval Gauss-Seidel operator. However, some stringent conditions about the proper/improper qualities of $\mathbf{L}$ and $\mathbf{U}$ restrict the application of the decomposition. Some work still have to be conducted to generalize these particular application cases.

## References

1. Neumaier, A.: Interval Methods for Systems of Equations. Cambridge Univ. Press, Cambridge (1990)
2. Ortolf, H.J.: Eine Verallgemeinerung der Intervallarithmetik. Geselschaft fuer Mathematik und Datenverarbeitung, Bonn **11** (1969) 1–71
3. Kaucher, E.: Uber metrische und algebraische Eigenschaften einiger beim numerischen Rechnen auftretender Raume. PhD thesis, Karlsruhe (1973)
4. Kaucher, E.: Interval Analysis in the Extended Interval Space $\mathbb{IR}$. Computing, Suppl. **2** (1980) 33–49
5. Shary, S.: A new technique in systems analysis under interval uncertainty and ambiguity. Reliable computing **8** (2002) 321–418
6. Popova, E.: Multiplication distributivity of proper and improper intervals. Reliable computing **7**(2) (2001) 129–140
7. SIGLA/X group: Modal intervals. Reliab. Comp. **7** (2001) 77–111
8. Goldsztejn, A., Chabert, G.: On the approximation of linear AE-solution sets. In: 12th GAMM IMACS International Symposion on Scientific Computing, Computer Arithmetic and Validated Numerics, Duisburg, Germany (2006)
9. Shary, S.: Interval Gauss-Seidel Method for Generalized Solution Sets to Interval Linear Systems. Reliable computing **7** (2001) 141–155

# On the Relationship Between the Sum of Roots with Positive Real Parts and Polynomial Spectral Factorization

Masaaki Kanno[1], Hirokazu Anai[2], and Kazuhiro Yokoyama[3]

[1] CREST, Japan Science and Technology Agency
4-1-8, Honcho, Kawaguchi-shi, Saitama, 332-0012, Japan
M.Kanno.99@cantab.net
[2] Fujitsu Laboratories Ltd / CREST JST
4-1-1 Kamikodanaka, Nakahara-ku, Kawasaki, 211-8588, Japan
anai@jp.fujitsu.com
[3] Rikkyo University / CREST JST
3-34-1 Nishi Ikebukuro, Toshima-ku, Tokyo, 171-8501, Japan
yokoyama@rkmath.rikkyo.ac.jp

**Abstract.** This paper is concerned with the relationship between the sum of roots with positive real parts (SORPRP) of an even polynomial and the polynomial spectral factor of the even polynomial. The SORPRP and its relationship to Gröbner bases are firstly reviewed. Then it is shown that the system of equations satisfied by the coefficients of the polynomial spectral factor is directly related to a Gröbner basis. It is then demonstrated by means of an $\mathcal{H}_2$ optimal control problem that the above fact can be used to facilitate guaranteed accuracy computation.

## 1 Introduction

In various fields of science and engineering, there is an increasing interest in the use of symbolic computation [1] and guaranteed accuracy computation [2] that have been developed in the computer science field. The systems and control area is no exception [3,4,5], and a number of approaches which make use of, e.g., interval methods and computer algebra systems have been proposed. Regarding computer algebra systems, not only the capability of symbolic manipulation but also the recent development of algebraic methods are employed [6,7,8].

The aim of this paper is to show another example of the use of an algebraic method for a control problem, which facilitates the execution of guaranteed accuracy computation. An intriguing connection is revealed between the sum of roots with positive real parts (SORPRP) of an even polynomial and polynomial spectral factorization, an important mathematical tool in control and signal processing. That is, the coefficients of the spectral factor can be expressed as polynomials in the SORPRP of a certain polynomial and, furthermore, another polynomial that defines the SORPRP can be obtained. It is then demonstrated that the proven fact can be utilized to solve a control problem and to express

T. Boyanov et al. (Eds.): NMA 2006, LNCS 4310, pp. 320–328, 2007.
© Springer-Verlag Berlin Heidelberg 2007

quantities to be sought as polynomials in the SORPRP, thus enabling guaranteed accuracy computation. The paper is organized as follows. Section 2 reviews the SORPRP of an even polynomial and defines polynomials having the SORPRP as one of their roots. In Section 3, the formulation of the polynomial spectral factorization problem is stated, and the connection between the SORPRP and the coefficients of the spectral factor is shown by means of the theory of Gröbner bases. Section 4 then applies the result to a particular $\mathcal{H}_2$ optimal control problem. Some concluding remarks are made in Section 5.

## 2    Sum of Roots with Positive Real Parts of an Even Polynomial

We consider an even polynomial $f(x)$ of degree $2m$ in $\mathbb{Q}[x]$ with non-zero constant term. For any root $\alpha$ of $f(x)$, $-\alpha$ is also a root of $f(x)$. Assume that there are no roots on the imaginary axis. Then, there are $m$ roots, say $\alpha_1, \ldots, \alpha_m$, which have positive real parts, and also $m$ roots, say $\alpha_{m+1}, \ldots, \alpha_{2m}$, which have negative real parts and $\alpha_{m+i} = -\alpha_i$ for $1 \le i \le m$. Thus, we can write

$$f(x) = a_{2m}x^{2m} + a_{2m-2}x^{2m-2} + \cdots + a_2x^2 + a_0 \tag{1}$$
$$= a_{2m}\prod_{i=1}^{2m}(x - \alpha_i) = a_{2m}\prod_{i=1}^{m}(x^2 - \alpha_i^2) \ ,$$

where $a_{2k} \in \mathbb{Q}$ for $0 \le k \le m$, $a_{2m} \ne 0$ and $a_0 \ne 0$.

Our first target is to compute $\sigma := \alpha_1 + \ldots + \alpha_m$, which is the sum of all roots with positive real parts, without computing individual $\alpha_i$'s. For simplicity, we call $\sigma$ the *SORPRP* of $f$. It is apparent that the real part of $\sigma$ is positive. Moreover, since, for each non-real root of $f(x)$, its complex conjugate has the same real part, we have the following.

**Lemma 1 ([9]).** *The quantity $\sigma$ is real and positive.*

Now we define polynomials which have $\sigma$ as one of their roots.

**Definition 1.** *Let $\mathcal{P} = \{(\epsilon_1, \ldots, \epsilon_m) \mid \epsilon_i \in \{1, -1\}\}$ and $\mathcal{C}$ be the set consisting of all distinct linear sums $\epsilon_1\alpha_1 + \cdots + \epsilon_m\alpha_m$, $(\epsilon_1, \ldots, \epsilon_m) \in \mathcal{P}$. Then, the characteristic polynomial $S_f(z)$ of the SORPRP is defined as*

$$S_f(z) := \prod_{(\epsilon_1, \ldots, \epsilon_m) \in \mathcal{P}}(z - (\epsilon_1\alpha_1 + \cdots + \epsilon_m\alpha_m)) \ .$$

*Also, the minimal polynomial $\bar{S}_f$ of the SORPRR is defined as the square-free part of $S_f$:*

$$\bar{S}_f(z) := \prod_{C \in \mathcal{C}}(z - C) \ .$$

Some comments on the properties of $S_f(z)$ and $\bar{S}_f(z)$ are in order. Considering the action of the Galois group of $\mathbb{Q}(\alpha_1, \ldots, \alpha_{2m})$ on $\{\alpha_1, \ldots, \alpha_{2m}\}$, we can see that $S_f(z)$ and $\bar{S}_f(z)$ belong to $\mathbb{Q}[z]$. The degree of $S_f(z)$ is always $2^m$, which is understood from the cardinality of $\mathcal{P}$. As there are cases where $\epsilon_1\alpha_1+\cdots+\epsilon_m\alpha_m$ and $\epsilon'_1\alpha_1 + \cdots + \epsilon'_m\alpha_m$ coincide for distinct $(\epsilon_1,\cdots,\epsilon_m)$ and $(\epsilon'_1,\cdots,\epsilon'_m)$, the square-free part $\bar{S}_f(z)$ can be smaller than $S_f(z)$. Finally, it is obvious that the SORPRP $\sigma$ of $f(x)$ coincides with the maximal real root of $S_f(z)$ (and of $\bar{S}_f(z)$).

## 3   Polynomial Spectral Factorization and SORPRP

We first review *polynomial spectral factorization* of an even polynomial used in control and signal processing.

**Definition 2.** *Given $f(x)$ as in (1), we call the following decomposition the spectral factorization of $f$:*

$$a_{2m}f(x) = g(x)\bar{g}(x) \tag{2}$$

*where*

$$g(x) = b_m x^m + b_{m-1}x^{m-1} + \cdots + b_1 x + b_0 = a_{2m}\prod_{i=1}^{m}(x+\alpha_i) \ ,$$

$$\bar{g}(x) = b_m x^m - b_{m-1}x^{m-1} + \cdots + (-1)^m b_0 = a_{2m}\prod_{i=1}^{m}(x-\alpha_i) \ ,$$

$$\operatorname{Re}\alpha_i > 0 \ , \ b_m = a_{2m} \ .$$

In this section, we investigate the relationship between the coefficients of the spectral factor $g(x)$ and the SORPRP of $f(x)$ based on Gröbner basis theory [10] and then show how to compute $S_f(z)$. Comparing the coefficients in (2), we get a system of equations, which has a useful property.

**Lemma 2.** *Consider each $b_i$, $0 \le i < m$, as a variable. Then, polynomial spectral factorization gives the following system of equations:*

$$(-1)b_{m-1}^2+2b_m b_{m-2} - a_{2m}a_{2(m-1)} = 0 \ ,$$

$$(-1)^2 b_{m-2}^2-2b_{m-1}b_{m-3} - 2a_{2m}b_{m-4} - a_{2m}a_{2(m-1)} = 0 \ ,$$

$$\vdots$$

$$(-1)^k b_{m-k}^2+2\sum_{1\le i\le 2k-1, i\ne k}(-1)^i b_{m-i}b_{m-2k+i}$$
$$+ 2a_{2m}b_{m-2k} - a_{2m}a_{2(m-k)} = 0 \ \ for \ \ 2k \le m \ ,$$

$$\vdots$$

$$(-1)^k b_{m-k}^2+2\sum_{2k-m\le i\le m, i\ne k}(-1)^i b_{m-i}b_{m-2k+i}$$
$$- a_{2m}a_{2(m-k)} = 0 \ \ for \ \ 2k > m \ ,$$

$$\vdots$$

$$(-1)^{m-1}b_1^2+(-1)^{m-2}2b_2 b_0 - a_{2m}a_2 = 0 \ ,$$

$$(-1)^m b_0^2-a_{2m}a_0 = 0 \ .$$

*With respect to the graded reverse lexicographic order $b_{m-1} \succ \cdots \succ b_0$, the set $\mathcal{G}$ of polynomials on the left hand sides in the above system forms the reduced Gröbner basis of the ideal generated by itself.*

For the definition of the graded reverse lexicographic order, readers are referred to [10, Chapter 2, §2].

We call the ideal $\langle \mathcal{G} \rangle$ the *ideal of spectral factorization*. Then, as $\sigma$ is associated with the variable $b_{m-1}$ (i.e., $b_{m-1} = a_{2m}\sigma$), we have the following.

**Lemma 3.** *The ideal of spectral factorization is 0 dimensional and the number of its zeros with multiplicities counted is $2^m$. Also, the minimal polynomial $N_f$ of $b_{m-1}$ modulo $\langle \mathcal{G} \rangle$ is a factor of $S_f(b_{m-1}/a_{2m})$ and has $\bar{S}_f(b_{m-1}/a_{2m})$ as its factor.*

Thus, we can compute $N_f$ by using the Gröbner basis $\mathcal{G}$. If the ideal $\langle \mathcal{G} \rangle$ is a radical ideal [10, Chapter 4, §2], then $N_f$ coincides with $\bar{S}_f(b_{m-1}/a_{2m})$.

In particular, if the cardinality of $\mathcal{C}$ is $2^m$, that is, when we have distinct $\epsilon_1 \alpha_1 + \cdots + \epsilon_m \alpha_m$ for distinct $(\epsilon_1, \ldots, \epsilon_m) \in \mathcal{P}$, the ideal of spectral factorization has another type of basis with respect to the lexicographic ordering, the so called *shape basis* in the following form:

$$\left\{ \hat{S}_f(b_{m-1}), \; b_{m-2} - \hat{h}_{m-2}(b_{m-1}), \; \ldots, \; b_0 - \hat{h}_0(b_{m-1}) \right\} \;, \tag{3}$$

where $\hat{h}_i$ is a polynomial of degree strictly less than $2^m$. Hence it turns out that all coefficients $b_i$ can be described as polynomials in $b_{m-1}$. Since $b_{m-1} = a_{2m}\sigma$, (3) means that each coefficient of the spectral factor is described as a polynomial in $\sigma$:

$$S_f(\sigma) = 0 \;, \quad b_{m-1} = a_{2m}\sigma \;, \quad b_{m-2} = h_{m-2}(\sigma) \;, \quad \ldots \;, \quad b_0 = h_0(\sigma) \;,$$

where $S_f(\sigma) := \hat{S}_f(a_{2m}\sigma)$ and $h_i(\sigma) := \hat{h}_i(a_{2m}\sigma)$. In general, we can efficiently compute a shape basis from the set $\mathcal{G}$ of polynomials by the basis-conversion (change-of-order) technique [10, Appendix D, §2].

*Remark 1 (Parametric polynomial case).* We can deal with a parametric polynomial $f(x, \mathbf{p}) \in \mathbb{Q}[x, \mathbf{p}]$ in a similar manner, under the assumption that, for any admissible $\mathbf{p}$, the leading coefficient of $f$ does not vanish and, further, $f$ does not have roots on the imaginary axis. Readers are referred to [11] for an exposition of an algorithm dealing with the parametric case.

## 4   Application to $\mathcal{H}_2$ Optimal Control Problem

In this section, the developed result is demonstrated on a control problem in order to show that the result can facilitate guaranteed accuracy computation. The normalized $\mathcal{H}_2$ optimal control problem is considered here and it is shown that the quantities to be computed can be expressed as polynomials in the SORPRP of an even polynomial derived from the input data. The problem is

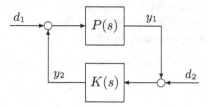

**Fig. 1.** Standard feedback configuration

formulated as follows. In the feedback configuration depicted in Fig. 1, given the $n$-th order, strictly proper transfer function $P(s)$ of a single-input-single-output linear, time-invariant plant, the task is to design a controller (denoted by its transfer function $K(s)$), which stabilizes the closed-loop system and minimizes the $\mathcal{H}_2$-norm of the transfer function matrix $T_{wz}(s)$ from $w = (d_1\ d_2)^\top$ to $z = (y_1\ y_2)^\top$, i.e.,

$$T_{wz}(s) = \frac{1}{1 - PK}\begin{pmatrix} P & PK \\ PK & K \end{pmatrix}.$$

That is, the value of the optimal performance level

$$J^* := \min_{K\ \text{stabilizing}} \left\| T_{wz}(P, K) \right\|_2^2$$

and the controller $K_{\text{opt}}(s)$ that achieves $J^*$ are sought. Remember that the $\mathcal{H}_2$-norm of a system $G(s)$ is defined as

$$\|G(s)\|_2 := \left( \frac{1}{2\pi} \int_{-\infty}^{\infty} \text{tr}\left\{ G^*(j\omega)G(j\omega) \right\} d\omega \right)^{\frac{1}{2}}$$

and that the square of the $\mathcal{H}_2$-norm of a system is equal to the energy of the system output when an impulse signal is applied to the input. So the $\mathcal{H}_2$-norm indicates how promptly the system attenuates disturbance.

Now it is shown with the aid of a numerical example that, when $P(s)$ is given in $\mathbb{Q}(s)$, $J^*$ and the coefficients of $K_{\text{opt}}$ can be expressed as polynomials in the SORPRP by making use of the result developed in the preceding sections. As a numerical example, the following system is used:

$$P(s) = \frac{s - 7}{s(s - 1)} =: \frac{P_N(s)}{P_D(s)},$$

where $P_N$, $P_D$ are coprime polynomials and $P_D$ is monic. If the plant is $n$-th order, the degree of $P_D$ is $n$ and that of $P_N$ is strictly smaller than $n$; in this particular numerical example, the plant is second order, i.e., $n = 2$. Firstly, the spectral factor $M_D(s)$ of the even polynomial

$$P_N(-s)P_N(s) + P_D(-s)P_D(s) \tag{4}$$

is found. In particular, the coefficients $m_i$ of $M_D(s) := s^2 + m_1 s + m_0$ are expressed as polynomials in the SORPRP $\sigma$ of (4). For this numerical example, the even polynomial to be factored is

$$s^4 - 2s^2 + 49 .$$

Equating the coefficients of the above polynomial and $M_D(s)M_D(-s)$, a set of polynomial equations in $m_i$ is obtained, which immediately yields a Gröbner basis. Computing the shape basis,

$$S_f(\sigma) = \sigma^4 - 4\sigma^2 - 192 = 0 , \tag{5}$$

$$m_1 = \sigma , \quad m_0 = \frac{\sigma^2}{2} - 1$$

are derived.

Next, in order to find the optimal controller, polynomials $V_N(s)$, $U_N(s)$ of degrees $n$ and at most $n-1$, respectively, are found that satisfy

$$P_D(s)V_N(s) - P_N(s)U_N(s) = \{M_D(s)\}^2 .$$

From the coprimeness of $P_N$ and $P_D$, there is only one pair of $V_N$ and $U_N$ satisfying the requirements. The coefficients of such $V_N$, $U_N$ can be obtained by solving a set of linear equations, and these coefficients can be expressed as polynomials in $\sigma$. Then the optimal controller is given as $K_{opt}(s) = U_N(s)/V_N(s)$. In the case of the numerical example, the approach yields

$$K_{opt}(s) = \frac{\left(\frac{1}{6}\sigma^3 + \frac{1}{3}\sigma^2 + 1\right)s + 7}{s^2 + (2\sigma + 1)s + \frac{1}{6}\sigma^3 + \frac{7}{3}\sigma^2 + 2\sigma} .$$

It is emphasized here that all the coefficients of $K_{opt}$ are found as polynomials in $\sigma$.

Now, $J^\star$ is to be computed. It can be derived that

$$T_{wz}(P, K_{opt}) = -\frac{1}{M_D^2}\begin{pmatrix} P_N V_N & P_N U_N \\ P_N U_N & P_D U_N \end{pmatrix} =: \begin{pmatrix} T_{11} & T_{12} \\ T_{12} & T_{22} \end{pmatrix} .$$

Based on the fact that $\|T_{wz}(P, K_{opt})\|_2^2 = \|T_{11}\|_2^2 + 2\|T_{12}\|_2^2 + \|T_{22}\|_2^2$, $\|T_{11}\|_2^2$ etc. are computed separately. Write

$$T_{11}(s) = -\frac{P_N(s)V_N(s)}{M_D(s)^2} = \frac{b_{2n-1}s^{2n-1} + b_{2n-2}s^{2n-2} + \cdots + b_0}{s^{2n} + a_{2n-1}s^{2n-1} + a_{2n-2}s^{2n-2} + \cdots + a_0} ,$$

where $a_i$ and $b_j$ are determined from the coefficients of $M_D^2$ and $-P_N V_N$, respectively (which further implies that $a_i$, $b_j$ can be written as polynomials in $\sigma$). Then, the $\mathcal{H}_2$-norm of $T_{11}$ can be calculated as follows. Write $T_{11}$ in state-space description and in, e.g., the observer canonical form:

$$
T_{11}(s) = \left[\begin{array}{ccccc|c}
-a_{2n-1} & 1 & 0 & \cdots & 0 & b_{2n-1} \\
-a_{2n-2} & 0 & 1 & \cdots & 0 & b_{2n-2} \\
\vdots & \vdots & \vdots & \ddots & \vdots & \vdots \\
-a_1 & 0 & 0 & \cdots & 1 & b_1 \\
-a_0 & 0 & 0 & \cdots & 0 & b_0 \\
\hline
1 & 0 & 0 & \cdots & 0 & 0
\end{array}\right] =: \left[\begin{array}{c|c} A & B \\ \hline C & 0 \end{array}\right] .
$$

By using the solution $L_o$ of the Lyapunov equation

$$
A^* L_o + L_o A + C^* C = 0 , \tag{6}
$$

$\|T_{11}\|_2^2$ can be computed as $\|T_{11}\|_2^2 = \operatorname{tr}\{B^* L_o B\}$. Since the elements of $A$ and $C$ are polynomials in $\sigma$ and the Lyapunov equation is equivalent to a set of linear equations, the elements of $L_o$ can be expressed as rational functions in $\sigma$. By way of (5), the inverses of the denominators can be found in $\mathbb{Q}[\sigma]$ and thus the rational functions can be converted to polynomials in $\sigma$. Therefore, it is seen that $\|T_{11}\|_2^2$ can be expressed as a polynomial in $\sigma$ (whose degree is strictly smaller than that of $S_f$). For the numerical example under consideration, the solution $L_o$ of (6) can be expressed as

$$
L_o = \begin{bmatrix}
\ell_1 & 0 & -\ell_2 & 0 \\
0 & \ell_2 & 0 & -\ell_3 \\
-\ell_2 & 0 & \ell_3 & 0 \\
0 & -\ell_3 & 0 & \ell_4
\end{bmatrix} ,
$$

where $\ell_i$ are the solution of

$$
\begin{bmatrix}
-2\sigma & \sigma^3 - 2\sigma & 0 & 0 \\
1 & 2 - 2\sigma^2 & 49 & 0 \\
0 & 2\sigma & -\sigma^3 + 2\sigma & 0 \\
0 & -1 & 2\sigma^2 - 2 & -49
\end{bmatrix}
\begin{bmatrix} \ell_1 \\ \ell_2 \\ \ell_3 \\ \ell_4 \end{bmatrix}
=
\begin{bmatrix} -\frac{1}{2} \\ 0 \\ 0 \\ 0 \end{bmatrix} . \tag{7}
$$

As has been stated, $\ell_i$ can be obtained as polynomials in $\sigma$:

$$
\begin{bmatrix} \ell_1 \\ \ell_2 \\ \ell_3 \\ \ell_4 \end{bmatrix}
=
\begin{bmatrix}
\frac{73}{36864}\sigma^3 - \frac{85}{9216}\sigma \\
-\frac{1}{36864}\sigma^3 + \frac{13}{9216}\sigma \\
\frac{25}{1806336}\sigma^3 - \frac{37}{451584}\sigma \\
-\frac{97}{88510464}\sigma^3 + \frac{1837}{22127616}\sigma
\end{bmatrix} . \tag{8}
$$

(In the case of high order systems, solving the Lyapunov equation (6) symbolically can be problematic in terms of computation time. However, the approach taken here is preferable in that it solves a set of linear equations in fewer variables and the elements of the matrix to be inverted is polynomials in only one variable, see (7).) The result (8) yields $\left\|T_{11}\right\|_2^2 = \frac{65}{768}\sigma^3 + \frac{5}{8}\sigma^2 + \frac{859}{192}\sigma + 8$. Repeating the same calculation for $\left\|T_{12}\right\|_2^2$ and $\left\|T_{22}\right\|_2^2$, the optimal performance $J^\star$ is obtained:

$$J^\star := \left\|T_{wz}(P, K_{\mathrm{opt}})\right\|_2^2 = \frac{2}{9}\sigma^3 + \frac{10}{9}\sigma^2 + \frac{26}{3}\sigma + \frac{64}{3} . \tag{9}$$

In this particular numerical example, it can be confirmed that $\sigma = 4$ is the true (exact) solution and, using this value, the true $K_{\mathrm{opt}}$ and $J^\star$ can be recovered. However, it is in general impossible to find the exact root of $S_f$ and thus impossible to compute $K_{\mathrm{opt}}$ and $J^\star$ exactly. Nevertheless, $\sigma$ is a real root of a polynomial and, therefore, can be computed with guaranteed accuracy [12]. Since the quantities to be computed in the $\mathcal{H}_2$ optimal control problem are expressed as polynomials in $\sigma$, they can be found with guaranteed accuracy by means of interval arithmetic [4].

## 5   Conclusion

This paper has seen an interesting connection between the SORPRP of an even polynomial and the coefficients of the spectral factor of the even polynomial. A Gröbner basis is directly obtained from a system of equations that the coefficients of the spectral factor must satisfy. It is demonstrated on the normalized $\mathcal{H}_2$ optimal control problem that this fact can be employed to express the optimal performance level and the controller that achieves it as polynomials in the SORPRP. It is planned to investigate the feasibility of this approach to the case of a plant with real parameters, and it is expected that the approach can be extended so that it can carry out the sensitivity analysis of the optimal cost and the optimal controller to the parameters.

## References

1. Cohen, A.M., ed.: Computer Algebra in Industry: Problem Solving in Practice. John Wiley & Sons, Chichester (1993)
2. Adams, E., Kulisch, U., eds.: Scientific Computing with Automatic Result Verification. Volume 189 of Mathematics in Science and Engineering, London, Academic Press (1993)
3. Munro, N., ed.: Symbolic Methods in Control System Analysis and Design. Volume 56 of IEE Control Engineering Series. The Institution of Electrical Engineers, Stevenage (1999)
4. Jaulin, L., Kieffer, M., Didrit, O., Walter, E.: Applied Interval Analysis. Springer-Verlag, London (2001)
5. Kanno, M., Smith, M.C.: Validated numerical computation of the $\mathcal{L}_\infty$-norm for linear dynamical systems. Journal of Symbolic Computation **41**(6) (2006) 697–707
6. Hanzon, B., Maciejowski, J.M.: Constructive algebra methods for the $L_2$-problem for stable linear systems. Automatica **32**(12) (1996) 1645–1657

7. Zettler, M., Garloff, J.: Robustness analysis of polynomials with polynomial parameter dependency using Bernstein expansion. IEEE Transactions on Automatic Control **43**(3) (1998) 425–431
8. Anai, H., Yanami, H., Sakabe, K., Hara, S.: Fixed-structure robust controller synthesis based on symbolic-numeric computation: Design algorithms with a CACSD toolbox (invited paper). In: Proceedings of CCA/ISIC/CACSD 2004 (Taipei, Taiwan). (2004) 1540–1545
9. Anai, H., Hara, S., Yokoyama, K.: Sum of roots with positive real parts. In: Proceedings of the ACM SIGSAM International Symposium on Symbolic and Algebraic Computation, ISSAC2005 (2005) 21–28
10. Cox, D., Little, J., O'Shea, D.: Ideals, Varieties, and Algorithms. 2nd edn. Springer-Verlag, New York, NY (1996) 2nd printing(1998).
11. Yokoyama, K.: Stability of parametric decomposition. In Takayama, N., Iglesias, A., eds.: Proceedings of Second International Congress on Mathematical Software ICMS 2006. Volume 4151 of Lecture Notes in Computer Science, Springer-Verlag (2006) to appear 391–402
12. Collins, G.E.: Infallible calculation of polynomial zeros to specified precision. In Rice, J.R., ed.: Mathematical Software III, New York, NY, Academic Press (1977) 35–68

# Lie Brackets and Stabilizing Feedback Controls

Mikhail I. Krastanov

Institute of Mathematics and Informatics, Bulgarian Academy of Sciences,
1113 Sofia, Bulgaria
krast@bas.bg
http://www.math.bas.bg/~krast/

**Abstract.** The relation between a class of high-order control variations and the asymptotic stabilizability of a smooth control system is briefly discussed. Assuming that there exist high-order control variations "pointing" to a closed set at every point of some its neighborhood, an approach for constructing stabilizing feedback controls is proposed. Two illustrative examples are also presented.

## 1 High-Order Variations and Asymptotic Stabilizability

The traditional approach towards proving sufficient controllability conditions has been to construct "control variations". If one can construct control variations in all possible directions, then the reachable set ought to be a full neighborhood of the starting point. For example, let us consider the following control system:

$$\dot{x}(t) = f_0(x(t)) + \sum_{i=1}^{m} u_i(t) f_i(x(t)), \ u_i(t) \in [-1, 1], \tag{1}$$

where $f_i$, $i = 0, \ldots, m$, are smooth vector fields defined on a neigbourhood of the point $x_0 \in \mathbb{R}^n$ with $f_0(x_0) = 0$. Let $u(\cdot) = (u_1(\cdot), \ldots, u_m(\cdot))$ be an integrable function defined on the interval $[0, T]$, whose components take values from $[-1, 1]$. An absolutely continuous function $x(\cdot)$ with $x(0) = x_0$ and satisfying (1) for almost every $t$ from $[0, T]$ is called admissible trajectory of (1) defined on $[0, T]$, starting from the point $x_0$ and corresponding to the control $u(\cdot)$. By $\mathcal{R}(x, T)$ we denote all points of $\mathbb{R}^n$ reachable from the point $x$ by means of admissible trajectories of (1) defined on $[0, T]$ and starting from the point $x$.

To introduce the set $E^+(x_0)$ of *high-order control variations* to the reachable set of the control system (1) at the point $x_0$, we need the following notation: $\mathrm{Exp}(Z_t)x_0$ denotes the value of the solution of the equation

$$\dot{x}(\tau) = Z_t(x(\tau)), \ x(0) = x_0,$$

at time $\tau = 1$, where $\{Z_t : t \in R_+\}$ is a given family of smooth vector fields, defined on $\mathbb{R}^n$ and depending continuously on $t \geq 0$.

**Definition 1.** *It is said that the smooth vector field $g$ is a high-order control variation to the reachable set of the control system (1) at the point $x_0$, if there* *exist a positive real $T$, a neighbourhood $\Omega$ of $x_0$, two families of smooth vector fields $a_t$ and $b_t$ parameterized by $t \geq 0$, and a continuous function $p : \mathbb{R}_+ \to \mathbb{R}_+$ such that for each $x \in \Omega$ and each $t \in [0, T]$*

$$Exp\,(tg + a_t + b_t)\,(x) \in \mathcal{R}(x, p(t)),$$

*where*

$$\|a_t(x)\| \leq M\,t^\theta\,\|x - x_0\|, \quad \|b_t(x)\| \leq N\,t^\sigma, \quad p(t) < P\,t^\lambda,$$

*for some positive constants $M$, $N$, $P$, $\theta$, $\sigma > 1$ and $\lambda$. We denote by $E^+(x_0)$ the set of all high-order control variations to the reachable set of the control system (1) at the point $x_0$.*

*Remark 1.* Different definitions of high-order control variations can be found in [4], [6], [7], [9], [11], [12] etc. All these definitions use the notion of a Lie bracket. Let $f$ and $g$ be smooth vector fields defined on $R^n$. Then the Lie bracket $[f, g]$ is defined as

$$[f, g](x) := \frac{\partial g}{\partial x}(x)f(x) - \frac{\partial f}{\partial x}(x)g(x).$$

It should be mentioned that the notion of Lie brackets is extended to the non-smooth case in [4].

The following lemmas provide constructions of elements of the set $E^+(x_0)$:

**Lemma 1.** *The set $E^+(x_0)$ is a convex cone.*

**Lemma 2.** *The vector fields $f_i$, $\pm[f_i, f_j]$, $[f_i, [f_i, f_0]]$, $i, j = 1, \ldots, m$, are elements of the set $E^+(x_0)$.*

**Lemma 3.** *Let $\pm g \in E^+(x_0)$. Then $\pm[g, f_0] \in E^+(x_0)$.*

The following lemma shows how the set $E^+(x_0)$ is related to the small-time local controllability (STLC):

**Lemma 4.** *Let $g^1, \ldots, g^k \in E^+(x_0)$ and*

$$0 \in \text{int co}\,\{g^i(x_0) :\; i = 1, \ldots, k\},$$

*where "int" and "co" mean the interior and the convex hull, respectively. Then the control system (1) is STLC.*

*Remark 2.* The proofs of slightly different versions of these lemmas can be found in [6], [7], [9], [11]. These proofs are based mainly on the so-called Campbell-Baker-Hausdorff formula. On the other hand, the proofs of the sufficient STLC condition in [1] and [10] show that under suitable assumptions all elements of the Lie algebra generated by the vector fields $f_0, f_1, \ldots, f_m$ belong to the set $E^+(x_0)$. For different applications of the control theory, it is very important to realize them explicitly by using admissible controls.

It has been shown by Brockett (cf. [2]) that, contrary to the case of linear systems, many small-time locally controllable nonlinear system can not be stabilized by means of stationary continuous feedback law. To get around this problem, we use some ideas of Hermes presented in [5]: Let us assume that there exist a positive number $\rho_0 > 0$ and elements $g^1, \ldots, g^k \in E^+(x_0)$ such that

$$0 \in \text{int co} \left\{ g^i(x) : \ i = 1, \ldots, k \right\} \text{ for each point } x \in B_{\rho_0}(x_0), \tag{2}$$

where $B_{\rho_0}(x_0)$ denotes the closed ball centered at $x_0$ with radius $\rho_0$.

Without loss of generality, we may think that there exist a positive real $T$, some families of smooth vector fields $a^i_t$ and $b^i_t$ parameterized by $t \geq 0$, and continuous functions $p_i : R_+ \to R_+$, $i = 1, \ldots, k$, such that for each $x \in B_{\rho_0}(x_0)$ and each $t \in [0, T]$

$$\text{Exp} \left( tg^i + a^i_t + b^i_t \right)(x) \ \in \ \mathcal{R}(x, p_i(t)),$$

where

$$\|a_t(x)\| \ \leq \ M \, t^\theta \, \|x - x_0\|, \ \|b_t(x)\| \ \leq \ N \, t^\sigma, \ p(t) < P \, t^\lambda,$$

for some positive constants $M$, $N$, $P$, $\theta$, $\sigma > 1$ and $\lambda$. Moreover, we assume that $T > 0$ and $\rho_0 > 0$ are so small that for each $t_i \in [0, T]$, $i = 1, \ldots, k$, and for each $x \in B_{\rho_0}(x_0)$ the following inclusion holds true

$$\text{Exp} \left( t_1 g^1 + a^1_{t_1} + b^1_{t_1} \right) \cdots \text{Exp} \left( t_k g^k + a^k_{t_k} + b^k_{t_k} \right)(x) \in \mathcal{R} \left( x, \sum_{i=1}^{k} p_i(t_i) \right).$$

Let us fix $\rho_1 \in (0, \rho_0)$ so small that $\dfrac{\rho_1}{\rho_0} \leq T$. Let $x$ be an arbitrary point of the ball $B_{\rho_1}(x_0)$ with $x \neq x_0$. Then the vector $\rho_0 \dfrac{x_0 - x}{\|x_0 - x\|}$ belongs to $B_{\rho_0}(x_0)$. According to (2), this vector can be presented as a convex combination of $g_i(x)$, $i = 1, \ldots, k$. Let

$$\rho_0 \frac{x_0 - x}{\|x_0 - x\|} = \sum_{i=1}^{k} \alpha_i(x) g^i(x) \text{ with } \alpha_i(x) \geq 0 \ \text{ and } \ \sum_{i=1}^{k} \alpha_i(x) = 1.$$

Then

$$x_0 - x = \sum_{i=1}^{k} \beta_i(x) g^i(x) \text{ with } \beta_i(x) \geq 0 \ \text{ and } \ \sum_{i=1}^{k} \beta_i(x) = \frac{\|x_0 - x\|}{\rho_0}. \tag{3}$$

We set $\beta(x) := \max \left\{ \beta_i(x) : i = 1, \ldots, k \right\}$. Our choice of $\rho_1$ and (3) imply that $\beta(x) \leq T$ for each $x \in B_{\rho_1}(x_0)$. We set

$$\Xi(x) := \text{Exp} \left( \beta_1(x) g^1 + a^1_{\beta_1(x)} + b^1_{\beta_1(x)} \right) \cdots \text{Exp} \left( \beta_k(x) g^k + a^k_{\beta_k(x)} + b^k_{\beta_k(x)} \right)(x),$$

and obtain that

$$\Xi(x) \in \mathcal{R}(x, T_x), \text{ with } T_x = \sum_{i=1}^{k} p_i \left(\beta_i(x)\right).$$

Applying the Campbell-Baker-Hausdorff formula, we can present $\Xi(x)$ as follows:

$$\Xi(x) = x + \sum_{i=1}^{k} \beta_i(x)g_i(x) + A(x) + B(x)$$

with $\|A(x)\| \leq A \left(\beta(x)\right)^\alpha \|x - x_0\|$ and $\|B(x)\| \leq B \left(\beta(x)\right)^{1+\beta}$ for some positive constants $A$, $B$, $\alpha$ and $\beta$. Let us choose $\rho \in (0, \rho_1)$ so small that

$$A \left(\frac{\rho}{\rho_0}\right)^\alpha + B \left(\frac{\rho^\beta}{\rho_0^{1+\beta}}\right) < \frac{1}{2}.$$

From the last inequality and from (3), we obtain that $\Xi(x) - x_0 = A(x) + B(x)$ for each point $x \in B_\rho(x_0)$ and

$$\|\Xi(x) - x_0\| \leq A \left(\beta(x)\right)^\alpha \|x - x_0\| + B \left(\beta(x)\right)^{1+\beta}$$

$$\leq \left(A \left(\frac{\rho}{\rho_0}\right)^\alpha + B \frac{\rho^\beta}{\rho_0^{1+\beta}}\right) \|x - x_0\| < \frac{1}{2}\|x - x_0\|. \tag{4}$$

The last inequality motivates a computational procedure: Staring from arbitrary point $x_1 \in B_\rho(x_0)$, we set $x_2 := \Xi(x_1)$. The estimate (4) implies that $\|x_2 - x_0\| \leq \frac{1}{2}\|x_1 - x_0\| \leq \rho$. So, we can set $x_3 := \Xi(x_1)$ to obtain that

$$\|x_3 - x_0\| \leq \frac{1}{2}\|x_2 - x_0\| \leq \frac{1}{2^2}\|x_1 - x_0\| \leq \rho.$$

Continuing in the same manner, we obtain a sequence $\{x_i\}_{i=1}^{\infty}$ whose elements belong to $B_\rho(x_0)$ and $\|x_i - x_0\| \leq \frac{1}{2^{i-1}}\|x_1 - x_0\|$ for $i = 1, 2, 3, \ldots$ So, this sequence is convergent to $x_0$. In this sense, this procedure can be applied to generate a stabilizing feedback control. In the next section we present a similar procedure to construct stabilizing controls with respect to a closed set.

## 2    Stabilizing Feedback Controls with Respect to a Set

Let $S$ be an arbitrary closed subset of $\mathbb{R}^n$. Let $\delta_0 > 0$ and $S_{\delta_0}$ be the closed neighbourhood of the set $S$ consisting of all points $x$ such that $\text{dist}_S(x) \leq \delta_0$ (here $\text{dist}_S(x)$ denotes the distance between the point $x$ and the set $S$). If $x$ is an arbitrary point of $\mathbb{R}^n$, we set

$$\pi_S(x) := \{y \in S : \|x - y\| = \text{dist}_S(x)\}.$$

Let $y$ belong to the boundary $\partial S$ of the set $S$. A vector $\xi \in \mathbb{R}^n$ is called a proximal normal to $S$ at $y$ provided that there exists $r > 0$ so that the point $y + r\xi$ has closest point $y$ in $S$. The set of all proximal normals at a point $y$ is a cone denoted by $N_S^p(y)$ (for more details cf. [3]).

We consider the following control system in $S_{\delta_0}$:

$$\dot{x}(t) = f(x(t), u(t)), \tag{5}$$

where $x(t)$ is the state and $u(t) \in U \subset \mathbb{R}^m$ is the control. By $\mathcal{R}(x,t)$ we denote all points of $\mathbb{R}^n$ reachable from the point $x$ by means of admissible trajectories of (5) defined on $[0,T]$ and starting from the point $x$. First, we define a notion of high-order control variations for the control system (5) with respect to the closed set $S$:

**Definition 2.** *We denote by $E^+(S)$ the set of all Lipschitz continuous functions $A : S_{\delta_0} \to \mathbb{R}^n$ such that for each of them there exist a positive real $T$, two continuous functions $a : S_{\delta_0} \times [0,T] \to \mathbb{R}^n$ and $b : S_{\delta_0} \times [0,T] \to \mathbb{R}^n$, and a non-decreasing continuous function $p : [0,T] \to \mathbb{R}$ with $p(0) = 0$ such that for each $x \in S_{\delta_0}$ and each $t \in [0,T]$*

$$x + tA(x) + a(t,x) + b(t,x) \in \mathcal{R}(x, p(t))$$

*with*

$$\|a(t,x)\| \leq M\, t^\theta\, dist_S(x), \ \|b(t,x)\| \leq N\, t^\sigma, \ p(t) < P\, t^\lambda,$$

*for some positive constants $M$, $N$, $P$, $\theta$, $\sigma > 1$ and $\lambda$.*

*Remark 3.* A generalization of Lemmas $1 \div 4$ as well some other assertions that are useful for constructing high-order control variations for the control system (5) with respect to the closed set $S$ are proved in [7].

Let $\mathcal{V}$ be a subset of elements of $E^+(S)$. The set $\mathcal{V}$ is said to be regular when: 1) all elements of $\mathcal{V}$ are Lipschitz continuous functions defined on a neighbourhood of the set $S$ with the same Lipschitz constant; 2) all functions related to the elements of $E^+(S)$ (according to Definition 2) are uniformly bounded on this neighbourhood. In our further consideration we shall use only regular subsets of the set $E^+(S)$. This assumption is technical and guarantees the existence of suitable trajectories of the considered control system (5) defined on some fixed interval $[0,T]$. This is especially important for the case of unbounded set $S$.

The proof of the main result of [8] implies the following

**Theorem 1.** *Suppose that $S$ is a nonempty closed subset of $\mathbb{R}^n$, $\mathcal{V}$ is a regular subset of $E^+(S)$ and $\mu > 0$. Let us assume that whenever $x \in S_{\delta_0} \setminus S$, $y \in \pi(x)$ and $\xi \in N_S^p(y)$ there exists $A \in \mathcal{V}$ for which*

$$< \xi, A(y) > \ \leq \ -\mu.\|\xi\|. \tag{6}$$

*Then there exist real numbers $q \in (0,1)$ and $\delta > 0$ such that for each point $x \in S_\delta$ there exist $t_x > 0$ and an admissible control function $u_x : [0, t_x] \to U$*

such that the solution $z(\cdot, x, u_x)$ of (5) starting from $x$ and corresponding to the control $u_x$ is well defined on $[0, t_x]$ and satisfies the inequality

$$dist_S(z(t_x, x, u_x)) \leq q.dist_S(x).$$

The proof of Theorem 1 is constructive. It allows us to calculate explicitly a sequence $x_1, x_2, \ldots$, of points belongsing to $S_\delta$ that converges to a point from the set $S$. In this sense, Theorem 1 may be applied to generate a stabilizing feedback control with respect to $S$.

## 3    Illustrative Examples

In this section we present two illustrative examples. All computations are performed using the computer algebra system Maple V.

*Example 1.*

Let us consider the following three-dimensional control system:

$$\begin{aligned}
\dot{x}_1 &= u \cos(x_3), \quad x_1(0) = 0, \quad u \in [-1, 1], \\
\dot{x}_2 &= \sin(x_1), \quad x_2(0) = 0, \\
\dot{x}_3 &= \sin(x_2), \quad x_3(0) = 0,
\end{aligned}$$

We set $x = (x_1, x_2, x_3)$, $X(x) = (0, \sin(x_1), \sin(x_2))$, $Y = (\cos(x_3), 0, 0)$. The vectors $Y(x)$, $[Y, X](x)$ and $[X, [Y, X]](x)$ are linearly independent at every point $x \in B_{\rho_0}(0)$ whenever $\rho_0 > 0$ is sufficiently small. Moreover, Lemma 2 and Lemma 3 imply that $\pm Y$ and $\pm[Y, X]$ and $\pm[X, [Y, X]]$ belong to the set $E^+(0)$. This allows us to stabilize this control system in a neighborhood of the origin. For simulations, at each step, starting from the point $x = (x_1, x_2, x_3)$ we move time $t_x$ using a suitable control $u_x$ depending on $x$. The calculated end-point is staring point for the next calculation. For the sample run, the computations end after 26 steps.

*Example 2.*

Let $S = \{(x, y) \in R^2 | \ 0 \leq x \leq 1, \ -\dfrac{1}{2} \leq y \leq 0\}$ and let us consider the following control system

$$\begin{aligned}
\dot{x} &= u, \quad\quad\quad\quad u \in [-1, 1], \\
\dot{y} &= y + 1 - x + v, \ v \in [-1, 1].
\end{aligned}$$

The origin is a boundary point of $S$ and the vector $n = (0, 1)$ is a proximal normal to $S$ at the origin. It can be checked that the scalar products of the vector $n$ and all admissible velocities at points of the form $z = (0, y)$, with $y \geq 0$, are not negative. So, in order to move towards the set $S$, we need to use high-order control variations. For example, let $T \in [0, 1]$, $t \in (0, T/2)$ and $z = (0, y)^\top$ with $y > 0$. We set $v_t(s) = -1$ for every $s \in [0, t]$ and

$$u_t(s) = \begin{cases} 1, & \text{if } s \in [0, t] \\ -1, & \text{if } s \in [t, 2t]. \end{cases}$$

**Table 1.** Simulation results for Example 1. At each step, starting from the point $x = (x_1, x_2, x_3)$ we move along a suitable chosen trajectory time $t_x$. The calculated end-point is a starting point for the next iteration.

| step | $x_1$ | $x_2$ | $x_3$ | $t_x$ |
|------|-------|-------|-------|-------|
| 1 | 0.8 | -0.4 | 0.2 | 1.9488 |
| 2 | $7.7881.10^{-2}$ | $3.2591.10^{-1}$ | $8.5059.10^{-1}$ | 2.6059 |
| ... | ... | ... | ... | ... |
| 25 | $2.2867.10^{-5}$ | $4.0206.10^{-4}$ | $4.7193.10^{-2}$ | 0.8471 |
| 26 | $1.3531.10^{-5}$ | $2.3338.10^{-4}$ | $3.2613.10^{-2}$ | 0.7662 |

**Table 2.** Simulation results for Example 2 with four different starting points $z_0 = (x_0, y_0)$. Each end-point $z_1 = (x_1, y_1)$ is reached from the corresponding starting point $z_0$ in time $t_{z_0}$.

| $x_0$ | $y_0$ | $t_{z_0}$ | $x_1$ | $y_1$ |
|-------|-------|-----------|-------|-------|
| 1.5 | 0.5 | 1.284211638 | 1.0 | $-2.949259209.10^{-5}$ |
| -0.2 | 0.5 | 5.371304140 | $3.0087043967.10^{-14}$ | $-1.891439399.10^{-4}$ |
| 0.8 | 0.5 | 7.166944604 | 0.800000000000000154 | $-2.454711917.10^{-5}$ |
| 0.1 | 0.5 | 2.057154154 | 0.099999999999999976 | $-3.737032781.10^{-5}$ |

It can be directly checked that the trajectory $z_t(\cdot) = (x_t(\cdot), y_t(\cdot))$ starting from $z$ and corresponding to the controls $u_t(\cdot)$ and $v_t(\cdot)$ is defined on $[0, 2t]$ and

$$z_t(2t) = z + a(t, z) - t^2 A(z) + O(t^3, z), \qquad (7)$$

where $A(z) = (0, -1)$, $a(t, z) = (0, (e^{2t} - 1)y)$ and $O(t^3, z) = (0, 2e^t - e^{2t} - 1 + t^2)$. Taking into account that $d_{S_1}(z) = y$, we obtain that

$$|a(t, z)| \leq M.t.d_{S_1}(z), \quad |O(t^3, z)| \leq N.t^3 \tag{8}$$

for suitably chosen positive numbers $M$ and $N$. These estimates and (7) imply that $A$ belongs to the set $E^+(S)$. Since the origin is the unique metric projection of the point $z$ on $S$ and $\langle n, A(0) \rangle = -1 = -\|n\|$, the condition (6) also holds true. Similarly, one can check that all the assumptions of Theorem 1 hold true.

## 4   Conclusion

Starting from states close to a set $S$ we want to steer $S$ and to stay always close to $S$. Unfortunately, open-loop controls are very sensitive to disturbances and can lead to very bad practical results. To avoid errors due to different disturbances, one can try to find a stabilizing feedback control law. To get around the problem of impossibility to stabilize nonlinear systems by a continuous autonomous feedback, we use some ideas of Hermes presented in [5] to define a suitable class of high-order control variations and to use them for constructing stabilizing feedback controls.

## References

1. Agrachev, A., Gamkrelidze, R.: Local controllability and semigroups of diffeomorphisms. Acta Applicandae Mathematicae **32** (1993) 1–57
2. Brockett, R.: Asymptotic stability and feedback stabilization. In: R. Brockett, R. Millmann, H. Sussmann, eds.: Differential Geometric Control Theory. Progr. Math. **27** (1983). Birkhäuser, Basel-Boston, 181–191
3. Clarke, F., Ledyaev, Yu., Stern R., Wolenski, P.: Nonsmooth analysis and control theory. Graduate Text in Mathematics **178** Springer-Verlag, New York (1998)
4. Frankowska, H: Local controllability of control systems with feedback. J. Optimiz. Theory Appl. **60** (1989) 277–296
5. Hermes, H.,: On the synthesis of a stabilizing feedback control via Lie algebraic methods. SIAM J. Control Optimiz. **16** (1978) 715–727
6. Hermes, H.: Lie algebras of vector fields and local approximation of attainable sets. SIAM J. Control Optimiz. **18** (1980) 352–361
7. Krastanov, M., Quincampoix, M.: Local small-time controllability and attainability of a set for nonlinear control systems. ESAIM: Control. Optim. Calc. Var. **6** (2001) 499–516
8. Krastanov, M. I.: A sufficient condition for small-time local attainability of a set, Control and Cybernetics **31**(3) (2002) 739-750
9. Krastanov, M. I., Veliov, V. M.: On the controllability of switching linear systems. Automatica **41** (2005) 663–668
10. Sussmann, H. J.: A general theorem on local controllability. SIAM J. Control Optimiz. **25** (1987) 158–194
11. Veliov, V., M. Krastanov, M. I.: Controllability of piecewise linear systems. Systems & Control Letters **7** (1986) 335–341
12. Veliov, V.: On the controllability of control constrained linear systems. Math. Balk., New Ser. **2** (1988) 147–155

# Interval Based Morphological Colour Image Processing

Antony T. Popov

St. Kliment Ohridski University of Sofia, Faculty of Mathematics and Informatics,
5 James Bourchier Blvd., BG-1164 Sofia, Bulgaria
atpopov@fmi.uni-sofia.bg

**Abstract.** In image analysis and pattern recognition fuzzy sets play the role of a good model for segmentation and classifications tasks when the regions and the classes cannot be strictly defined. Fuzzy morphology has been shown to be a very eficient tool in processing and segmentation of grey-scale images. In this work we show that using interval modelling we can apply efficiently fuzzy morphological operations to colour images. In this case intervals help us to avoid the problem of lack of total ordering in multidimensional Euclidean spaces, in particular in the three dimensional RGB colour space.

## 1 Mathematical Morphology – Preliminaries

Generally speaking, the word morphology refers to the study of forms and structures. In image processing, morphology is the name of a specific methodology for analysing the geometric structure. It provides a systematic approach to analyse the geometric characteristics of signals and images, and has been applied widely to many applications such as edge detection, object segmentation, noise suppression, esp. in the case of salt-and-pepper noise [11]. Beginning in the mid-60's and especially in the mid-70's, it became much more closely affiliated with the work of Georges Matheron and Jean Serra from Ecole Normale Superieure des Mines de Paris who studied the properties of the porous media with respect to their geometrical structure [11]. At first, the morphological operations have been applied to binary images, which can be naturally interpreted as sets. So, let us assume that our objects belong to a linear space $M$. Considering two subsets of $M$, namely $A$ and $B$ the following operations are introduced:

- Minkowski addition of $A$ and $B$, $A \oplus B$ is defined as $A \oplus B = \bigcup_{b \in B} (A + b) = \{ a + b \mid a \in A, b \in B \}$, and Minkowski difference of $A$ by $B$, $A \ominus B$ is defined as $A \ominus B = \bigcap_{b \in B} (A - b) = \{ x \in M \mid B + x \subseteq A \}$. For the properties of these operations see [4].
- Opening of $A$ by $B$ is defined as $A \circ B = (A \ominus B) \oplus B$, while closing of $A$ by $B$ is defined as $A \bullet B = (A \oplus B) \ominus B$.

The operation $\delta_A(X) = A \oplus X = X \oplus A$ is referred to as *dilation of the set $X$ by the structuring element $A$* and $\varepsilon_A(X) = X \ominus A$ - *erosion of the set $X$ by the*

T. Boyanov et al. (Eds.): NMA 2006, LNCS 4310, pp. 337–344, 2007.

*structuring element* $A$. Openings and closings are generally used as filters for denoising of binary images.

Later, morphological operations have been defined for grey-scale images. Any grey-scale image can be represented mathematically as a function on a given domain, or as a topographic surface over the same 2D domain such that the height of the point is equal to the pixel value. Thus grey-scale morphological operations can be defined naturally by binary operations on the closed subgraph of the functions which lead to the following formulas [11,4]:

$$(\delta_g(f))(x) = \sup_{x \in M}(f(x-h)+g(h)), \quad (\varepsilon_g(f))(x) = \inf_{x \in M}(f(x+h)-g(h)), \quad (1)$$

where $f$ and $g$ are functions mapping the linear space $M$ to the compactified real line $\bar{\mathbf{R}} = \mathbf{R} \cup \{\infty, -\infty\}$, supposing that $s+t = -\infty$ if $s = -\infty$ or $t = -\infty$, and $s - t = \infty$ if $s = \infty$ or $t = -\infty$.

Later Serra [12] and Heijmans [4] have shown that the operations of mathematical morphology can be formulated on any complete lattice, so there exists a well developed algebra that can be employed for representing abstract morphological operations by analog with the binary and grey-scale ones.

A set $\mathcal{L}$ with a partial ordering "$\leq$" is called a *complete lattice* if every subset $\mathcal{H} \subseteq \mathcal{L}$ has a supremum $\bigvee \mathcal{H} \in \mathcal{L}$ (least upper bound) and infimum (greatest lower bound) $\bigwedge \mathcal{H} \in \mathcal{L}$. An operator $\varphi : \mathcal{L} \mapsto \mathcal{M}$, where $\mathcal{L}$ and $\mathcal{M}$ are complete lattices, is called dilation if it distributes over arbitrary suprema: $\varphi(\bigvee_{i \in I} X_i) = \bigvee_{i \in I} \varphi(X_i)$, and erosion if it distributes over arbitrary infima. Erosions and dilations are increasing operations [4]. An operator $\psi \in \mathcal{L}^*$ is called a closing if it is increasing, idempotent ($\psi^2 = \psi$) and extensive ($\psi(X) \geq X$). An operator $\psi \in \mathcal{L}^*$ is called an opening if it is increasing, idempotent and anti-extensive ($\psi(X) \leq X$) [4]. A pair of operators $(\varepsilon, \delta)$, is called an adjunction, if for every two elements $X, Y \in \mathcal{L}$ it follows that $\delta(X) \leq Y \iff X \leq \varepsilon(Y)$. In [4] it is proved that if $(\varepsilon, \delta)$ is an adjunction, then $\varepsilon$ is erosion and $\delta$ is dilation. If $(\varepsilon, \delta)$ is an adjunction, then the composition $\varepsilon\delta$ is a closing, and $\delta\varepsilon$ is an opening. In the case of binary operations the lattice $\mathcal{L}$ under consideration contains as elements the subsets of a linear space $M$. The couple of binary operations with the same structuring element $(\varepsilon_A, \delta_A)$, as well as the grey-scale operations couple $(\varepsilon_g, \delta_g)$, are examples of adjunctions.

## 2    Fuzzy Morphological Operations

Fuzzy mathematical morphology has been developed to soften the classical binary morphology so as to make the operators less sensitive to image imprecision. It can also be viewed simply as an alternative grey-scale morphological theory with a very significant advantage: In practical applications the result of the operations cannot jump over the predefined pixel range, for example [0..255], as it can happen using the operations defined by equation (1).

Consider now the set $E$ called the universal set. A fuzzy subset $A$ of the universal set $E$ can be considered as a function $\mu_A : E \mapsto [0,1]$, called the membership function of $A$. $\mu_A(x)$ is referred to as the degree of membership of the point $x$ to the set $A$. The ordinary subsets of $E$, sometimes called 'crisp sets', can be considered as a particular case of a fuzzy set with membership function taking only the values 0 and 1. This definition leads to two possible interpretations:

– in image representation the value of the membership function $\mu_A(x)$ at a point $x$ may be interpreted as the grey level value associated with that point of the image plane,

– in pattern recognition, the value $0 \le \mu_A(x) \le 1$ indicates the probability that the point $x$ is in the foreground of an image.

The usual set-theoretical operations can be defined naturally on fuzzy sets: Union and intersection of a collection of fuzzy sets is defined as supremum, resp. infimum of their membership functions. Also, we say that $A \subseteq B$ if $\mu_A(x) \le \mu_B(x)$ for all $x \in E$. The complement of $A$ is the set $A^c$ with membership function $\mu_{A^c}(x) = 1 - \mu_A(x)$ for all $x \in E$. Further, for simplicity we will write $A(x)$ instead of $\mu_A(x)$. If the universal set $E$ is linear, like the $n$-dimensional Euclidean vector space $\mathbf{R}^n$ or the space of integer vectors with length $n$, then any geometrical transformation (like scaling, translation, rotation) of a fuzzy set can be defined simply by transforming its $\alpha$–cuts [7].

We say that the function $c(x,y) : [0,1] \times [0,1] \mapsto [0,1]$ is conjunctor if $c$ is increasing in the both arguments, $c(0,1) = c(1,0) = 0$, and $c(1,1) = 1$.

We say that the function $i(x,y) : [0,1] \times [0,1] \mapsto [0,1]$ is implicator if $i$ is increasing in $y$ and decreasing in $x$, $i(0,0) = i(1,1) = 1$, and $i(1,0) = 0$.

We say that the conjucnctor - implication pair is adjoint if $c(b,y) \le x$ is true if and only if $y \le i(b,x)$. The notion of conjucnctor - implication adjunction has its origin in fuzzy logic in solving *if-then-else* and *modus ponens* clauses. For fixed $b$ function $f(x) = i(b,x)$ is an erosion, and its adjoint dilation is $g(y) = c(b,y)$.

Then having an adjoint conjunctor - implicator pair, as proposed in [5], we can define an adjoint pair of fuzzy erosion and dilation:

$$\delta_B(A)(x) = \sup_y c(B(x-y), A(y)),$$

$$\varepsilon_B(A)(x) = \inf_y i(B(y-x), A(y)).$$

Heijmans [5] has proposed a number of following conjunctor - implicator pairs to construct morphological operations. Here we give examples of two of them:

$$c(b,y) = \min(b,y), \quad i(b,x) = \begin{cases} x, & x < b, \\ 1, & x \ge b. \end{cases}$$

These operations are known as operations of Gödel-Brouwer.

$$c(b,y) = \max(0, b+y-1), \quad i(b,x) = \min(1, x-b+1).$$

These operations are suggested by Lukasiewicz.

Most often the first conjunctor - implicator pair is used. The respective dilation has the form

$(\delta_B(A))(x) = \sup_b \min(A(b), B(x-b))$. In this case we can denote $\delta_A(B) = \delta_B(A) = A \oplus B$, because this operation can be obtained also directly from the binary Minkowski addition using the *extension principle*.

Note that in these cases the conjunctor is symmetric, i.e. it is a t-norm [7], and therefore we have $\delta_A(B) = \delta_B(A)$ as in the binary morphology.

## 3    Morphological Operations for Colour Images

We saw that the basic morphological operations are expressed as products of the suprema and the infima of the lattice under study. When we deal with colour images, we work in fact in a multidimensional space (usually $\mathbf{R}^3$ or $\mathbf{Z}^3$), where a natural ordering of the elements cannot be achieved. Therefore we try to introduce some heuristics and to compromise with the accuracy at acceptable level to guarantee the lattice properties and therefore to ensure idempotent opening and closing filtering.

An useful implementation of a basic subjective colour model is the HSV (hue, saturation, value) cone [10] and its slight modification HLS [2]. It is based on such intuitive colour characteristics as tint, shade and tone (or family (hue), purity (saturation) and intensity (value/luminance)). The coordinate system is cylindrical, and the colours are defined inside a cone for the case of HSV and a double cone in the case of HLS. The hue value H runs from 0 to $2\pi$ counterclockwise from red. The saturation S is the degree of strength or purity and varies from 0 to 1. The saturation shows how much white is added to the colour, so S=1 makes the purest colour (no white). Brightness (value) V also ranges from 0 to 1, where 0 is the black. If S=0 the hue is undefined, and therefore the colour is achromatic, namely a shade of grey. However, when we use the HSV or HLS model, the main obstacle is the fact that the hue is measured as an angle, and it is not defined for the levels of grey. In [2] a circular 'ordering' of the hue modulo $2\pi$ is defined, however this ordering does not lead to a complete lattice. Therefore in this case the obtained operators provide good results for some segmentation tasks, but the usage of openings and closings for denoising and filtering is risky. In general, there are no clear mathematical reasons for hue ordering. However, from psychophysiological point of view one may order the colours in the following way - red, magenta, blue, yellow, cyan, green, based on the way humans perceive the hue of the colour. Red is considered to be the smallest, since it stimulates the eye less than the other colours. Contrary, green mostly stimulates the eye [6]. In the last approach the 'ordered' colour hues can be represented approximately by 6 triangular fuzzy numbers representing the major colours in the sequence mentioned above.

In [3] an interesting approach for creating colour morphology is presented based on L*a*b* (CIELAB) colour space representation (the two models are related by the fact that $H = \arctan\left(\frac{b^*}{a^*}\right)$ ). There the authors divide this space into equipotential surfaces. Unfortunately, the best order for the colour vectors in the same equipotential surface is not obvious. In order to obtain a complete ordering of the colour vectors, they make use of the lexicographical order. Therefore in this case we have idempotent closing and opening filters. However, the transformation from RGB to L*a*b* is non-linear and time consuming. Moreover, there exists no unique inverse transform. The inverse transform depends on the way we characterize the white point. If one knows the illumination conditions used when acquiring the image, then the specification of the white point is simple. However, if the illumination conditions are unknown, a heuristic hypothesis should be made.

An alternative approach is to use the YCrCb colour model [10]. To obtain the parameters Y, Cr and Cb we use a simple linear combination of R,G, B values. Note that Y represents the *lightness* , and should not be mistaken with the yellow colour in the RGB model notation. The parameter Cr encodes the red-cyan sensation, with value $\approx 0$ for the cyan colour and $\approx 1$ for the red. The parameter Cb encodes the yellow-blue sensation with $\approx 0$ indicating yellow and $\approx 1$ indicating blue. Without lack of generality we can assume that R,G and B values are represented as points in an unit cube, namely $0 \leq R, G, B \leq 1$. The YCrCb colour space is also a unit cube with transformation formulas:

$$Y = 0.299\,R + 0.577\,G + 0.114\,B, \tag{2}$$

$$Cr = 0.5\,R - 0.411\,G - 0.081\,B + 0.5, \tag{3}$$

$$Cb = 0.5\,B - 0.169\,R - 0.325\,G + 0.5 . \tag{4}$$

Henceforth it is clear that the transformation between RGB and YCrCb models is linear and easy to compute.

When we use the HSV model usually we give priority to the value (V), since if anyone looks at the V-map of a colour image, he can usually distinguish the different objects on the image as looking on a black -and- white TV. Much better grey-scale representation of a colour image is produced by its Y - map. Consider the Shannon entropy for a fuzzy set represented grey-scale image $X$ with size $M \times N$ pixels:

$$E(X) = -\frac{1}{MN} \sum_{i=1}^{M} \sum_{j=1}^{N} X(i,j) \ln X(i,j).$$

This function is known as a measure of information content and is widely used in coding, statistics as well as in fuzzy set theory and image processing (finding optimal filters). Although it is not strictly mathematically proven, the experiments taken over 20 different colour images show that the entropy value for the Y-map is greater than the entropy value for the V-map. For instance, for the image on Figure 1 the respective values are 0.2920 for HSV and 0.3092 for YCrCb. So the higher information content is one of the reasons to prefer the YCrCb model.

In general, in colour image processing less priority is usually given to the chrominance maps - V,S in HSV model, or Cr, Cb in YCrCb. When working with HSV model, the next priority is given to the hue, because it contains mostly the colour information. The least priority is given to the saturation because it is correlated with the other two components and its role as a parameter in image processing tasks is sometimes criticized – one can refer for instance to the work

[2]. This is another reason to prefer the YCrCb model, where the components Cr and Cb have equal weights.

Let us divide the interval $[0, 1]$ into $N$ equal pieces $I_i = \left(\frac{i-1}{N}, \frac{i}{N}\right]$. $(0 \in I_1)$. Let us also suppose that for a pixel $x$ $Cb(x) \in I_j$ and $1 - Cr(x) \in I_i$. Note that we use the negation of $Cr$ to obtain an ordering closer to the one presented in [6]. In Table 1 one can see a $N \times N$ table with a zigzag path going through its cells.

**Table 1.** Zigzag tracing of the CrCb parameter space

| 1 | 4 | 5 | ... |
|---|---|---|-----|
| 2 | 3 | 6 | ... |
| 9 | 8 | 7 | ... |
| 10 | ... | ... | ... |
| ... | ... | ... | ... |

Thus we code approximately with accuracy $1/N$ the $Cr$ and $Cb$ values by the number of the step at which we visit the respective cell. Let us consider the cell $(i, j)$ and let $n = \max(i, j)$ and $m = \min(i, j)$. Then for the number of the step we can easily prove by induction that

$$T(i, j) = \begin{cases} n^2 - m + 1, & n- \text{ even and } n = i, \text{ or } n- \text{ odd and } n = j \\ n^2 - 2n + m + 1, & \text{otherwise.} \end{cases}$$

Further on, for simplicity, for any pixel $x$ we will denote the respective integer number from the table by $T(x)$. Then if given a colour image $X$, we define the transformation

$$(\chi(X))(x) = \frac{N^2[(N^2 - 1)Y(x)] + T(x) - 1}{N^4 - 1},$$

which is a real number between 0 and 1. Then it is clear that having $\chi(X)$, we can find $Y(x)$ with accuracy $1/N^2$ and $Cr(X)$ and $Cb(x)$ with $1/N$ simply by taking the quotient and the reminder of the division of $[\chi(X)(N^4 - 1)]$ by $N^2$. Here $[t]$ means the integer part of $t$. The last transformation we denote by $\chi^{-1}$. Then it is obvious that $\chi^{-1}(\chi(X))$ gives an approximation of the original colour image $X$, while for any grey-scale image $\chi(\chi^{-1}(Y)) = Y$.

Then we could order the colour images, namely say that $A \prec B$ if $\chi(A) \leq \chi(B)$. In this case, if $A \prec B$ and $B \prec A$ doesn't mean that $A = B$, but means that they are close and lie in the same equivalence class. Thus we present a more precise approximate order than the one described in [6]. Also, now we can give correct definition of colour fuzzy morphology, i.e. when given an adjoint conjunctor - implicator pair we can define dilation-erosion adjunction as:

$$\delta_B(X)(x) = \chi^{-1}\left[\bigvee_y c(B(x - y), (\chi(X))(y))\right], \tag{5}$$

$$\varepsilon_B(X)(x) = \chi^{-1}\left[\bigwedge_y i(B(y - x), (\chi(X))(y))\right]. \tag{6}$$

Here $B$ can be any fuzzy structuring element, i.e. an image which pixel values are real numbers between 0 and 1.

To compute easily $\chi^{-1}$, we represent the function $T(i,j)$ by a $N^2 \times 2$ table in which the number of the row $s$ means the current step of the zigzag line, while its columns hold the numbers of the intervals $i$ and $j$ such that $T(i,j) = s$. In the example shown in the next section we use value $N = 16$. Thus combining the upper erosions and dilations we are able to construct idempotent morphological filtering operations (opening and closing) which do not produce colours in the output image that are not present in the input image.

## 4    Experiments and Further Research

In this work a new approach for construction of morphological filters for colour images is presented. It is based on interval approximation of the colour space. Thus we meet the requirements for image quality and we control the colour accuracy. The same approximation can be applied also in fuzzy algorithms for image enhancement and binarization.

In figure 1 one can see a colour picture of a lizard statue in Barcelona (size $540 \times 360$ pixels). Next its dilation, erosion, closing and opening by $5 \times 5$ flat square structuring element are presented. Note that a structuring element is called flat if it takes values only 0 and 1 on its domain. Analogical experiment has been made with the same image and the same structuring element in [3] and anyone can compare visually the results. The down right subfigure shows the result of closing operation based on the CIELAB model using equipotential surfaces. Note that for flat structuring elements the choice of the    t-norm   - implicator pair is not essential, which follows easily from the properties of fuzzy t-norms and the uniqueness of the adjoint erosion [9]. The second applcation of the opening (closing) operation on the opened (closed) image docs not affect it due to the idempotence of thus generated opening and closing filters. Thus in our case we check experimentally our theoretical result for the opening and closing idempotence.

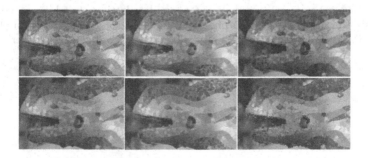

**Fig. 1.** From top to bottom and left to right: original, dilation, erosion; opening, closing and closing through L*a*b*

Operations with non-flat structuring elements are rarely used in image filtering, because it is not clear a priori how an incremental operation will affect the colours. Further experiments with non-flat operators will be made to show the efficiency of our approach in *texture analysis* by fractal dimension estimation or granulometry [13]. Our method can be applied also for contour extraction by morphological gradient and top-hat transform [4] or to study approximate connectivity [9] and convexity [8] on colour images.

# References

1. Bloch, I, Maître, H.: Fuzzy mathematical morphologies: A comparative study. Pattern Recognition **28(9)** (1995) 1341–1387
2. Hanbury, A., Serra, J.: Mathematical morphology in the HLS colour space. In: T. Cootes, C. Taylor (eds): Proceedings of the 12th British Machine Vision Conference - September 2001, Manchester, UK. Springer (2001) 451–460
3. Hanbury, A., Serra, J.: Mathematical morphology in the CIELAB Space. Journal of Image Analysis and Stereology **21** (2002) 201–206
4. Heijmans, H. J. A. M.: Morphological image operators. Academic Press, Boston (1994)
5. Deng, Ting-Quan, Heijmans, H. J. A. M.: Grey-scale morphology based on fuzzy logic. CWI Report PNA-R0012, Amsterdam, October 2000
6. Louverdis, G., Andreadis, I., Tsalides, Ph.: New fuzzy model for morphological colour image processing. IEE Proceedings – Vision, Image and Signal Processing **149(3)** (2002) 129–139
7. Nguyen, H. T., Walker, E. A. : A first course in fuzzy logic (2nd edn). CRC Press, Boca Raton FL (2000)
8. Popov, A. T.: Convexity indicators based on fuzzy morphology. Pattern Recognition Letters **18 (3)** (1997) 259 – 267.
9. Popov, A. T.: Aproximate connectivity and mathematical morphology. In: J. Goutsias, L. Vincent, D. S. Bloomberg (eds.): Mathematical Morphology and its Applications to Image and Signal Processing. Kluwer (2000) 149–158
10. Rogers, D. F.: Procedural elements for computer graphics (2nd edn). WCB McGraw - Hill (1998)
11. Serra, J.: Image analysis and mathematical morphology. Academic Press, London (1982)
12. Serra, J.: Mathematical morphology for complete lattices. In : J. Serra, ed.: Image analysis and mathematical morphology vol. 2. Academic Press, London (1988)
13. Soille, P.: Morphological image analysis (2nd edn). Springer-Verlag, Berlin (2002)

# Solving Linear Systems Whose Input Data Are Rational Functions of Interval Parameters*

Evgenija D. Popova

Institute of Mathematics & Informatics, Bulgarian Academy of Sciences
Acad. G. Bonchev str., block 8, BG-1113 Sofia, Bulgaria
epopova@bio.bas.bg

**Abstract.** The paper proposes an approach for self-verified solving of linear systems involving rational dependencies between interval parameters. A general inclusion method is combined with an interval arithmetic technique providing inner and outer bounds for the range of monotone rational functions. The arithmetic on proper and improper intervals is used as an intermediate computational tool for eliminating the dependency problem in range computation and for obtaining inner estimations by outwardly rounded interval arithmetic. Supporting software tools with result verification, developed in the environment of CAS Mathematica, are reported.

## 1 Introduction

Consider a linear algebraic system

$$A(p) \cdot x = b(p), \tag{1a}$$

where the elements of the $n \times n$ matrix $A(p)$ and the vector $b(p)$ are, in general, nonlinear functions of $k$ parameters varying within given intervals

$$a_{ij}(p) = a_{ij}(p_1, \ldots, p_k), \quad b_i(p) = b_i(p_1, \ldots, p_k), \quad i, j = 1, \ldots, n \tag{1b}$$

$$p \in [p] = ([p_1], \ldots, [p_k])^\top. \tag{1c}$$

The set of solutions to (1a–1c), called parametric solution set, is

$$\Sigma^p = \Sigma (A(p), b(p), [p]) := \{x \in \mathbb{R}^n \mid A(p) \cdot x = b(p) \text{ for some } p \in [p]\}. \tag{2}$$

$\Sigma^p$ is bounded if $A(p)$ is nonsingular for every $p \in [p]$. For a nonempty bounded set $\Sigma \subseteq \mathbb{R}^n$, define interval hull by $\square \Sigma := [\inf \Sigma, \sup \Sigma]$. Since it is quite expensive to obtain $\Sigma^p$ or $\square \Sigma^p$, we seek an interval vector $[y]$ for which it is guaranteed that $[y] \supseteq \square \Sigma^p \supseteq \Sigma^p$.

In this paper we combine the inclusion theory, developed by S. Rump in [8], with methods for sharp range estimation of continuous and monotone rational

---

* This work was supported by the Bulgarian National Science Fund under grant No. MM-1301/03.

T. Boyanov et al. (Eds.): NMA 2006, LNCS 4310, pp. 345–352, 2007.

functions, in order to compute inner and outer bounds for $\Box \, \Sigma^p(A(p), b(p), [p])$ whenever the elements of $A(p)$, $b(p)$ are rational functions of the parameters $p$. By now there are no other general-purpose self-validating methods for solving linear systems involving nonlinearly dependent interval parameters. For some other approaches see the literature cited in [5]. Section 2 contains some basic properties of the arithmetic on proper and improper intervals which is used in this work as an intermediate computational tool for eliminating the dependency problem in range computation and for obtaining inner estimations by outwardly rounded interval arithmetic. Section 3 presents the inclusion theory in real and floating-point arithmetic as well as the implementation algorithm. Section 4 demonstrates the proposed approach on a numerical example and reports some implemented software tools. The behavior of the presented general-purpose parametric fixed-point iteration method applied to linear systems involving nonlinear dependencies is explored in [5], [6]. Therein the methodology and software tools discussed in this paper are applied to practical problems from structural engineering and the results obtained by various methods are compared.

**Notations.** $\mathbb{R}^n, \mathbb{R}^{n \times m}$ denote the set of real vectors with $n$ components and the set of real $n \times m$ matrices, respectively. By normal (*proper*) interval we mean a real compact interval $[a] = [a^-, a^+] := \{a \in \mathbb{R} \mid a^- \leq a \leq a^+\}$. By $\mathbb{IR}^n, \mathbb{IR}^{n \times m}$ we denote interval $n$-vectors and interval $n \times m$ matrices. The end-point functionals $(\cdot)^-, (\cdot)^+$ and the function $\mathrm{mid}([a^-, a^+]) := (a^- + a^+)/2$ are applied to interval vectors and matrices componentwise. $I$ denotes the identity matrix. For interval quantities, the operations between them are always interval operations. We assume the reader is familiar with conventional interval arithmetic [3].

## 2  The Arithmetic on Proper and Improper Intervals

The set of proper intervals $\mathbb{IR}$ is extended in [2] by the set $\{[a^-, a^+] \mid a^-, a^+ \in \mathbb{R}, a^- \geq a^+\}$ of *improper* intervals obtaining thus the set $\mathbb{I}^*\mathbb{R} = \{[a^-, a^+] \mid a^-, a^+ \in \mathbb{R}\}$ of all ordered couples of real numbers called here generalized intervals. The conventional arithmetic and lattice operations, order relations and other functions are isomorphically extended onto the whole set $\mathbb{I}^*\mathbb{R}$ [2]. *Modal interval analysis* [1] imposes a logical-semantic background on generalized intervals (considered there as modal intervals) and allows giving a logical meaning to the interval results. The conventional interval arithmetic can be obtained as a projection of the generalized interval arithmetic on $\mathbb{IR}$.

An element-to-element symmetry between proper and improper intervals is expressed by the "Dual" operator $\mathtt{Dual}([a]) := [a^+, a^-]$ for $[a] = [a^-, a^+] \in \mathbb{I}^*\mathbb{R}$. Dual is applied componentwise to vectors and matrices. For $[a], [b] \in \mathbb{I}^*\mathbb{R}$

$$\mathtt{Dual}(\mathtt{Dual}([a])) = [a], \quad \mathtt{Dual}([a] \circ [b]) = \mathtt{Dual}([a]) \circ \mathtt{Dual}([b]), \quad \circ \in \{+, -, \times, /\}.$$

The generalized interval arithmetic structure possesses group properties with respect to addition and multiplication operations.

Let $\mathbb{F} \subset \mathbb{R}$ be the set of floating-point numbers and $\mathbb{IF}, \mathbb{I}^*\mathbb{F}$ be the corresponding sets of floating-point intervals. Denote by $\bigtriangledown, \bigtriangleup : \mathbb{R} \longrightarrow \mathbb{F}$ the directed roundings toward $-\infty$, resp. $+\infty$ specified by the IEEE floating-point standard. Outward ($\Diamond$) and inward ($\bigcirc$) roundings $\Diamond, \bigcirc : \mathbb{I}^*\mathbb{R} \longrightarrow \mathbb{I}^*\mathbb{F}$ are defined as

$$\Diamond([a]) := [\nabla(a^-), \Delta(a^+)] \supseteq [a], \qquad \bigcirc[a] := [\Delta(a^-), \nabla(a^+)] \subseteq [a]. \quad (3)$$

If $\circ \in \{+, -, \times, /\}$ is an arithmetic operation in $\mathbb{I}^*\mathbb{R}$ and $[a], [b] \in \mathbb{I}^*\mathbb{F}$, the corresponding computer operations $\Diamond\!\!\!\!\Diamond\,, \odot : \mathbb{I}^*\mathbb{F} \times \mathbb{I}^*\mathbb{F} \longrightarrow \mathbb{I}^*\mathbb{F}$ are defined by

$$[a] \odot [b] := \bigcirc([a] \circ [b]) \subseteq [a] \circ [b] \subseteq \Diamond([a] \circ [b]) =: [a] \Diamond\!\!\!\!\Diamond\ [b]. \quad (4)$$

The following additional properties (cf. [1]) show that inner numerical approximations can be obtained at no additional cost only by outward directed rounding and the `Dual` operator in $\mathbb{I}^*\mathbb{F}$. For $[a], [b] \in \mathbb{I}^*\mathbb{F}$, $\circ \in \{+, -, \times, /\}$

$$\bigcirc([a]) = \mathtt{Dual}(\Diamond(\mathtt{Dual}[a])); \qquad [a] \odot [b] = \mathtt{Dual}(\mathtt{Dual}[a] \Diamond\!\!\!\!\Diamond\ \mathtt{Dual}[b]). \quad (5)$$

For more details on the theory, implementation and applications of generalized interval arithmetic consult [1], [2], [4].

Let $f : D_f \subseteq \mathbb{R}^n \longrightarrow \mathbb{R}$ be a real-valued rational function continuous in a domain $[x] \in \mathbb{IR}^n$, $[x] \subseteq D_f$. Denote the range of $f$ over $[x]$ by $r_f([x]) = \{f(x) \mid x \in [x]\}$. Since $f$ is continuous, $r_f([x])$ is interval. In [1] the estimations of functional ranges are connected to an enhanced interpretation of quantified propositions which has many promising applications but is out of the scope of this work. Here, generalized interval arithmetic is used for range computation over a domain of proper intervals. Let the interval function $R_f([x]) : \mathbb{I}^*\mathbb{R}^n \longrightarrow \mathbb{I}^*\mathbb{R}$ be defined by the expression of $f$ where real variables are replaced by generalized intervals and real operations are replaced by operations between generalized intervals. For $[x] \in \mathbb{IR}^n$, $R_f([x])$ is the classical interval extension of $f$.

A real function $f(x, y) : \mathbb{R}^{1+m} \longrightarrow \mathbb{R}$ is $x$-uniformly monotonic for $x$ on a domain $([x], [y])$ if it is monotonic for $x$ on $[x]$, and it keeps the same monotonicity for all $y \in [y]$. A real function $f$ is $x$-totally monotonic for a multi-incident variable $x \in \mathbb{R}$ if $f$ is uniformly monotonic for this variable and for each one of its incidences (considering each incidence as an independent variable).

**Theorem 1.** (special case of [1, Thm 5.2]) *Let $f : \mathbb{R}^n \longrightarrow \mathbb{R}$ be a real-valued rational function continuous in a given interval vector $[x] \in \mathbb{IR}^n$, and multi-incident on its variables. Let $R_f([x])$ be defined on $[x]$ and let there exist splitting $x = (xn, xt)$ such that $f$ be totally monotonic for the components of $xt$. Let $[xt^*]$ be the enlarged vector of $[xt]$, such that each incidence of every component of $xt$ is included in $[xt^*]$ as independent component, but transformed into its dual if the corresponding incidence-point has a monotonicity type opposite to the global one of the corresponding $xt$-component. Then*

$$r_f([x]) \subseteq R_f([xn], [xt^*])) \subseteq R_f([x]).$$

In case a function $f$ is totally monotonic in all its variables, we have a sharp range estimation, specified in more details by the following theorem. For a set of indices $\mathcal{I} = \{i_1, \ldots, i_n\}$, the vector $(x_{i_1}, \ldots, x_{i_n})^\top$ will be denoted by $x_\mathcal{I}$.

**Theorem 2.** *Let $f : \mathbb{R}^n \longrightarrow \mathbb{R}$ be a rational function continuous in a given interval vector $[x] \in \mathbb{IR}^n$, multi-incident on its variables and totally monotonic on all variables. Define two sets of indices $\mathcal{P} = \{i_1, \ldots, i_q\}$, $\mathcal{N} = \{i_1, \ldots, i_r\}$ such that $\mathcal{P} \cap \mathcal{N} = \emptyset$, $\mathcal{P} \cup \mathcal{N} = \{1, \ldots, n\}$, and $f$ be $\leq$-isotone for $x_i : i \in \mathcal{P}$, $f$ be $\leq$-antitone for $x_i : i \in \mathcal{N}$. Let for each variable $x_i$, $1 \leq i \leq n$, there exist splitting $x'_i = (x'_{i1}, \ldots, x'_{ik})$, $x''_i = (x''_{i1}, \ldots, x''_{im})$ of the incidences of $x_i$. Let $f^*(x'_{\mathcal{P}}, x''_{\mathcal{P}}, x'_{\mathcal{N}}, x''_{\mathcal{N}})$ correspond to the expression of $f$ with explicit reference to the incidences of every variable, $f^*$ be continuous on $[x'_{\mathcal{P}}] \times [x''_{\mathcal{P}}] \times [x'_{\mathcal{N}}] \times [x''_{\mathcal{N}}]$, and $R_{f^*}([x'_{\mathcal{P}}], [x''_{\mathcal{P}}], [x'_{\mathcal{N}}], [x''_{\mathcal{N}}])$ be defined. If $f^*(x'_{\mathcal{P}}, x''_{\mathcal{P}}, x'_{\mathcal{N}}, x''_{\mathcal{N}})$ is $\leq$-isotone for the components of $x'_{\mathcal{P}}, x'_{\mathcal{N}}$ and $\leq$-antitone for the components of $x''_{\mathcal{P}}, x''_{\mathcal{N}}$, then*

$$r_f([x]) = R_{f^*}([x'_{\mathcal{P}}], \text{Dual}([x''_{\mathcal{P}}]), \text{Dual}([x'_{\mathcal{N}}]), [x''_{\mathcal{N}}]) \subseteq R_f([x]). \qquad (6)$$

From now on, referring to Theorems 1, 2 we shall use the following notations. For a function $f : D_f \subseteq \mathbb{R}^n \longrightarrow \mathbb{R}$, which is specified by an expression $f(x)$, the corresponding expression $f^*(x_{\mathcal{I}}, \text{Dual}(x_{\mathcal{J}}))$ is called *dual-transformed expression* where $\mathcal{I}$, $\mathcal{J}$ are index sets involving the indexes of the incidences for all the variables $x$ such that $\mathcal{J}$ contains the indexes of those variable instances which are dual-transformed according to the application of Theorems 1, 2, and $\mathcal{I}$ contains the indexes of not dual-transformed variable instances.

The inclusion properties (3), (4) allow a rigorous implementation of Theorems 1, 2 on the computer providing inner and outer inclusions of the true range

$$\bigcirc R_{f^*}(\bigcirc[x_{\mathcal{I}}], \text{Dual}(\bigcirc[x_{\mathcal{J}}])) \subseteq r_f([x]) \subseteq \Diamond R_{f^*}(\Diamond[x_{\mathcal{I}}], \text{Dual}(\Diamond[x_{\mathcal{J}}])). \quad (7)$$

Hereafter an arithmetic expression preceded by a rounding symbol $(\bigcirc, \Diamond)$ implies that all operations are performed in floating-point in the specified rounding mode. By applying (5) it is possible to obtain inner inclusion only by outwardly rounded operations. Thus, the left inclusion relation in (7) becomes

$$\bigcirc R_{f^*}(\bigcirc[x_{\mathcal{I}}], \text{Dual}(\bigcirc[x_{\mathcal{J}}])) = \text{Dual}(\Diamond R_{f^*}(\Diamond(\text{Dual}[x_{\mathcal{I}}]), \qquad (8)$$
$$\text{Dual}(\Diamond(\text{Dual}[x_{\mathcal{J}}])))) \subseteq r_f([x]).$$

## 3   Inclusion Theorems

The inclusion theorems for the solution set of a parametric linear system present a direct consequence from the theory for nonparametric problems developed by S. Rump and discussed in many works (cf. [8] and the literature cited therein). The next theorem is a general formulation of the enclosure method for parametric linear systems.

**Theorem 3.** *Consider the parametric linear system defined by (1a–1c). Let $R \in \mathbb{R}^{n \times n}$, $[y] \in \mathbb{IR}^n$, $\tilde{x} \in \mathbb{R}^n$ be given and define $[z] \in \mathbb{IR}^n$, $[C] \in \mathbb{IR}^{n \times n}$ by*

$$[z] := \square\{R(b(p) - A(p)\tilde{x}) \mid p \in [p]\}, \qquad [C] := \square\{I - R \cdot A(p) \mid p \in [p]\}.$$

*Define $[v] \in \mathbb{R}^n$ by means of the following Einzelschrittverfahren*

$$1 \leq i \leq n \;:\; [v_i] := \{[z] + [C] \cdot [u]\}_i, \qquad u := (v_1, ..., v_{i-1}, y_i, ..., y_n)^\top.$$

*If $[v] \subsetneqq [y]$, then $R$ and every matrix $A(p)$ with $p \in [p]$ are regular, and for every $p \in [p]$ the unique solution $\hat{x} = A^{-1}(p)b(p)$ of (1a–1c) satisfies $\hat{x} \in \tilde{x} + [v]$. With $[d] := [C] \cdot [v] \in \mathbb{R}^n$ the following inner estimation of $\Box \Sigma^p$ holds true*

$$[\tilde{x} + [z]^- + [d]^+, \; \tilde{x} + [z]^+ + [d]^-] \subseteq [\inf(\Sigma^p), \; \sup(\Sigma^p)].$$

The affine-linear dependencies between the parameters in $A(p), b(p)$ allow an explicit representation of the ranges of $z(p), C(p)$ by interval expressions, as done in [8] for $z(p)$. In case of arbitrary nonlinear dependencies, computing $[z]$ and $[C]$ in Theorem 3 requires sharp range enclosure for nonlinear functions. This is a key problem and there exists a huge amount of methods and techniques devoted to this problem, no one being universal. Here we restrict ourselves to linear systems where the elements of $A(p)$ and $b(p)$ are rational functions of the uncertain parameters (in this case the elements of $z(p)$ and $C(p)$ are also rational functions of $p$) and apply Theorems 2, 1 for enclosing the ranges of $z_i(p)$, $C_{ij}(p)$. It may seem quite restrictive to require that the elements of $z(p), C(p)$ be monotone functions of the parameters on some interval domains. However, [5], [6] demonstrate that there are realistic practical problems which can be solved successfully by this approach.

The second part of Theorem 3, establishes how to compute a componentwise inner estimation of the parametric solution set. An interval vector $[x] \in \mathbb{R}^n$ is called componentwise inner approximation for some set $\Sigma \in \mathbb{R}^n$ if

$$\inf_{\sigma \in \Sigma} \sigma_i \leq x_i^- \quad \text{and} \quad x_i^+ \leq \sup_{\sigma \in \Sigma} \sigma_i, \qquad \text{for every } 1 \leq i \leq n.$$

An inner inclusion $[x] \subset \Box \Sigma$ is important for estimating the quality of the computed outer enclosure, that is how much such an enclosure overestimates the exact hull of the solution set. In order to have a guaranteed inner inclusion all the computations should be done in computer arithmetic with directed roundings.

Basic goals of self-validating methods are to deliver rigorous results by computations in finite precision arithmetic, including the proof of existence (and possibly uniqueness) of a solution. To achieve this goal the inclusion theorems should be verifiable on computers. With the definitions of rounded floating-point interval arithmetic and due to its inclusion properties the following holds true.

**Theorem 4.** *Consider the parametric linear system defined by (1a–1c) with $p \in [p] \in \mathbb{IF}^k$. Let $R \in \mathbb{F}^{n \times n}$, $[y] \in \mathbb{IF}^n$, $\tilde{x} \in \mathbb{F}^n$ be given. Define*

$$z(p) := R\,(b(p) - A(p)\tilde{x}), \qquad C(p) := I - R \cdot A(p).$$

*and suppose that the elements of $z(p)$, $C(p)$ are real-valued rational functions. With the assumptions of Theorems 1, 2 and the notations thereafter, define $[z] \in \mathbb{IF}^n$, $[C] \in \mathbb{IF}^{n \times n}$ by*

$$[z_i] = \Diamond z_i^*([p_{\mathcal{I}_i}], \mathrm{Dual}[p_{\mathcal{J}_i}])$$
$$[C_{ij}] = \Diamond C_{ij}^*([p_{\mathcal{I}_{ij}}], \mathrm{Dual}[p_{\mathcal{J}_{ij}}]), \qquad i, j = 1, \ldots, n.$$

*Define* $[v] \in \mathbb{IF}^n$ *by means of the following Einzelschrittverfahren*

$$1 \le i \le n \ : \ [v_i] := \{\Diamond([z] + [C] \cdot [u])\}_i, \qquad u := (v_1, ..., v_{i-1}, y_i, ..., y_n)^\top.$$

*If* $[v] \subsetneqq [y]$*, then $R$ and every matrix $A(p)$ with $p \in [p]$ are regular, and for every $p \in [p]$ the unique solution $\widehat{x} = A^{-1}(p)b(p)$ of (1a–1c) satisfies $\widehat{x} \in \Diamond(\tilde{x} + [v])$. With $[d] := \Diamond([C] \cdot [v]) \in \mathbb{IF}^n$ the following inner estimation of $\square \, \Sigma^p$ holds true*

$$\left[\triangle\left(\tilde{x} + (\bigcirc[z])^- + [d]^+\right), \ \triangledown(\tilde{x} + (\bigcirc[z])^+ + [d]^-)\right] \subseteq [\inf(\Sigma^p), \ \sup(\Sigma^p)].$$

In the implementation of the above Theorem we choose $R \approx A^{-1}(p_m)$ and $\tilde{x} \approx A^{-1}(p_m) \cdot b(p_m)$, where $p_m = \mathrm{mid}([p])$.

The inclusions of the residual vector $[z]$ and the iteration matrix $[C]$ should be sharp. Suppose that `rangeExpr(f(p),[p])` is a function which verifies the conditions of Theorem 1 and yields the corresponding dual-transformed expression $f^*(p_\mathcal{I}, \mathtt{Dual}p_\mathcal{J})$ for a rational function $f(p)$ continuous on $[p]$. The evaluation of this expression in rounded generalized interval arithmetic gives a corresponding inner/outer inclusion of the true range of $f(p)$, as presented in (7), (8).

When aiming to compute a self-verified enclosure of the solution to a linear system by the above inclusion method, an iteration scheme, usually called fixed-point iteration, is proven to be very useful [7]. To force $[v] \subsetneqq [y]$, the concept of $\varepsilon$*-inflation* is performed by the function `blow([a],ε)`.

**Algorithm.** Guaranteed Inner and Outer Inclusions of the Solution Set Hull for Linear System Whose Input Data are Rational Functions of Interval Parameters.

1. *Initialization.* $\check{p} := \mathrm{mid}\,([p]); \quad \check{A} := A(\check{p}); \quad \check{b} := b(\check{p});$
   Compute $R \approx \check{A}^{-1}; \quad \tilde{x} = R \cdot \check{b}.$
2. *Enclosures.*
   2.1 Compute the analytic expressions
   $$z(p) := R\left(b(p) - A(p) \cdot \tilde{x}\right); \qquad C(p) := I - R \cdot A(p);$$
   2.2 Apply algebraic simplification to $z(p)$ and $C(p)$ in order to reduce the number of incidences of the variables.
   2.3 Obtain the corresponding dual-transformed expressions. For $i, j = 1..n$
   $$z_i^*(p_{\mathcal{I}_i}, \mathtt{Dual}(p_{\mathcal{J}_i})) = \mathtt{rangeExpr}(z_i(p), [p])$$
   $$C_{ij}^*(p_{\mathcal{I}_{ij}}, \mathtt{Dual}(p_{\mathcal{J}_{ij}})) = \mathtt{rangeExpr}(C_{ij}(p), [p]);$$
   2.4 For $i, j = 1, \ldots, n$
   $$[z]_i = \Diamond z_i^*([p_{\mathcal{I}_i}], \mathtt{Dual}[p_{\mathcal{J}_i}]); \qquad [C]_{ij} = \Diamond C_{ij}^*([p_{\mathcal{I}_{ij}}], \mathtt{Dual}[p_{\mathcal{J}_{ij}}]);$$
3. *Verification.*
   $[x] := [z];$
   **repeat**
   $\qquad [y] := [x] := \mathtt{blow}([x], \varepsilon)$
   $\qquad$ **for** $i = 1$ to $n$ **do** $\ [x_i] := \Diamond([z_i] + [C_i] \cdot [x])$
   **until** $[x] \subsetneqq [y]$ or max iteration exceeded
   **If** $[x] \subsetneqq [y]$ **then** $\square\, \Sigma^p \subseteq \Diamond(\tilde{x} + [x]);$ **else** the algorithm fails.

4. *Inner Estimation of the Outer Enclosure.* (If $[x] \subsetneq [y]$)

$[y] = \text{Dual} (\Diamond (\tilde{x} + \Diamond z^*(\text{Dual}[p_{\mathcal{I}}], [p_{\mathcal{J}}]) + [C] \cdot [x]))$;
for $i = 1$ to $n$, If $[y_i] \notin \mathbb{IR}$ then $[y_i] = \emptyset$;
$[y] \subseteq \square \Sigma^p \subseteq \Diamond(\tilde{x} + [x])$.

## 4 Mathematica Software, Numerical Example

The above algorithm and the function $\texttt{rangeExpr(f(p),[p])}$ are implemented in the environment of *Mathematica* (cf. [5]). In order to provide a broad access to the available parametric solvers, a web interface is developed and can be found at $\texttt{http://cose.math.bas.bg/webComputing/}$.

Consider a linear system with

$$A(p) = \begin{pmatrix} -(p_1 + p_2)/p_4 & p_5 \\ p_2 p_4 & p_3/p_5 \end{pmatrix}, \quad b(p) = \begin{pmatrix} 1 \\ 1 \end{pmatrix},$$

where $p_1, p_3 \in [0.96, 1.04]$, $p_2 \in [1.92, 2.08]$, $p_4, p_5 \in [0.48, 0.52]$. In the initialization step we have

$$\check{p} = (1,2,1,0.5,0.5)^\top; \quad R = \begin{pmatrix} -0.16 & 0.04 \\ 0.08 & 0.48 \end{pmatrix}; \quad \tilde{x} = (-0.12, 0.56)^\top.$$

After an algebraic simplification, the residual vector is

$$z(p) = \begin{pmatrix} -0.12 + 0.0192(p_1 + p_2)/p_4 + 0.0048p_2p_4 - 0.0224p_3/p_5 + 0.0896p_5 \\ 0.56 - 0.0096(p_1 + p_2)/p_4 + 0.0576p_2p_4 - 0.2688p_3/p_5 - 0.0448p_5 \end{pmatrix}.$$

The function $\texttt{rangeExpr}$ proves the total monotonicity of both components with respect to all the parameters. For the first component function we have one incidence of $p_3$ and the same monotonicity for all the incidences of $p_2$ and $p_5$, so that there will be no dual transformation for these parameters. The first component function is globally $\leq$-antitone with respect to $p_4$ while $\leq$-isotone for the first $p_4$ incidence and $\leq$-antitone w.r.t. the second $p_4$ incidence. Analogously, the second component function is globally $\leq$-isotone with respect to $p_2$, $p_5$ and has different monotonicity w.r.t. their incidences. Thus, the $\texttt{rangeExpr}$ function yields the following dual-transformed expressions for $z^*(p, \text{Dual})$

$$\begin{pmatrix} -0.12 + 0.0192(p_1 + p_2)/p_4 + 0.0048p_2\text{Dual}(p_4) - 0.0224p_3/p_5 + 0.0896p_5 \\ 0.56 - 0.0096(p_1 + \text{Dual}(p_2))/p_4 + 0.0576p_2p_4 - 0.2688p_3/p_5 - 0.0448\text{Dual}(p_5) \end{pmatrix}.$$

The evaluation of the above $z^*(p, \text{Dual})$ in generalized interval arithmetic gives the following enclosure with outwardly rounded arithmetic (for simplicity the results are presented with 6 digits accuracy)

$$[z] = ([-0.0143946, 0.0148305], [-0.0500198, 0.0466349])^\top.$$

The interval evaluation of $z(p)$ overestimates the evaluation of $z^*(p, \text{Dual})$ by 2.5%, resp. 9.1% for the vector components. The evaluation of $C^*(p, \text{Dual})$ in generalized interval arithmetic gives the following enclosure

$$[C] = \begin{pmatrix} [-0.079936, 0.0739102] & [-0.00986667, 0.00935385] \\ [-0.0514757, 0.0509653] & [-0.0784, 0.0722462] \end{pmatrix}.$$

The verification iteration converges in one iteration, yielding

$$([-0.132555, -0.107016], [0.515136, 0.601717])^\top \subseteq \Box \varSigma^p \subseteq$$
$$([-0.136242, -0.103329], [0.505062, 0.611791])^\top.$$

The overestimation of the exact solution set hull is 11.19%, 9.51% for the components.

## 5    Conclusion

Here we proposed an approach for self-verified solving of linear systems involving rational dependencies between interval parameters. The applied technique for sharp range enclosure, based on generalized interval arithmetic, is by no means obligatory. Combining the iteration method with more sophisticated tools for range enclosure will certainly expand its scope of application to systems involving more complicated dependencies. The implemented software tools are the only, by now, that provide a broad access to the presented methodology. A more detailed presentation, exploration of the behavior of the presented approach, as well as its application to some practical examples can be found in [5], [6].

## References

1. Gardenes, E., Sainz, M. A., Jorba, L., Calm, R., Estela, R., Mielgo, H., Trepat, A.: Modal intervals. Reliable Computing **7**(2) 2001 77–111
2. Kaucher, E.: Interval Analysis in the Extended Interval Space $IR$. Computing Suppl. **2** (1980) 33–49
3. Moore, R.: Methods and Applications of Interval Analysis. SIAM (1979)
4. http://www.math.bas.bg/~epopova/directed.html
5. Popova, E. D.: Solving Linear Systems whose Input Data are Rational Functions of Interval Parameters. Preprint 3/2005, Institute of Mathematics and Informatics, BAS, Sofia, 2005. (http://www.math.bas.bg/~epopova/papers/05PreprintEP.pdf)
6. Popova, E., Iankov, R., Bonev, Z.: Bounding the Response of Mechanical Structures with Uncertainties in All the Parameters. In: R.Muhannah, R.Mullen (eds): Proc. NSF Workshop on Reliable Engineering Computing, Svannah, 2006, 245–265
7. Rump, S.: New Results on Verified Inclusions. In: Miranker, W. L., Toupin, R. (eds.): Accurate Scientific Computations. Springer LNCS **235** (1986) 31–69
8. Rump, S. M.: Verification Methods for Dense and Sparse Systems of Equations. In: J. Herzberger, ed.: Topics in Validated Computations. Elsevier Science B. V. (1994) 63–135

# Differential Games and Optimal Tax Policy

Rossen Rozenov[1] and Mikhail I. Krastanov[2]

[1] Department of Mathematics and Informatics, University of Sofia
James Bourchier Boul. 5, 1126 Sofia, Bulgaria
rossen_rozenov@yahoo.com
[2] Institute of Mathematics and Informatics, Bulgarian Academy of Sciences
Acad. G. Bonchev Str. Bl. 8, 1113 Sofia, Bulgaria
krast@math.bas.bg

**Abstract.** In this paper we consider the problem of finding the solution of a differential game in relation with choosing an optimal tax policy rule. The paper extends the existing literature in two directions. First, instead of treating the tax base as given, in our formulation it is a control variable for the government. Secondly, we impose a phase constraint of mixed type for the considered problem of taxation. We present new conditions under which the solution of the differential game is found explicitly. The obtained optimal tax policy is time-consistent.

## 1  Introduction

The issue of time-consistency of optimal policy is central in modern economic theory. Since the influential work by Kydland and Prescott [4] economists have attempted different approaches to resolve the inconsistency problem. One possible strategy is to consider the interaction between the policy-maker and the agent in a dynamic game setup. Cohen and Michel [1], for example, by inspecting different types of equilibria found that a time-consistent outcome corresponds to a feedback Nash equilibrium, while the open-loop Stackelberg equilibrium corresponds to a time-inconsistent policy. It turns out, however, that in some cases the open-loop Stackelberg equilibrium may produce time-consistent results. Using a simple model of an income tax, Xie [6] demonstrates that one of the boundary conditions that had been previously assumed in the economic literature is not necessary for optimality. Karp and Ho Lee [3] generalized Xie's findings by considering an arbitrary function of capital as a tax base.

This paper extends the results in [3] and [6] in two directions. First, instead of solving the optimization problem of the government with respect to the tax rate by treating the tax base as given, in our formulation the tax rule is a control variable for the government. In other words, we seek a function of capital which is optimal for the government and at the same time it agrees with the problem of the consumer. We obtain new conditions under which the solution is found explicitly and the optimal tax policy is time-consistent. Secondly, since in many applications it may be reasonable to constrain the state variable to be non-decreasing, we examine how the solution changes when constraints of mixed type are present. In the particular example of taxation, the justification of the constraint could be that for both the government and the consumer it may be undesirable to allow capital to be eventually eaten up.

In the next section we prove a proposition which gives the general solution of the differential game and after that we apply this proposition to specific examples.

## 2    General Solution

There are two players – the government and the representative consumer[1]. These two players solve two dynamic optimization problems which are interrelated.

The government decides about the rule by which to tax the consumer. This rule must be consistent with the behaviour of the representative agent. In other words, the government will seek a function which agrees with the problem of the consumer and at the same time it maximizes government's utility. The utility of the government in our model depends both on the amount of collected taxes and on the consumption of the representative agent.

The representative agent chooses the optimal path of his or her consumption by taking into account the tax set by the government. Formally, the consumer maximizes

$$\int_0^\infty e^{-\rho t} U(c(t)) dt \tag{1}$$

with respect to $c(t)$ and subject to

$$\dot{k}(t) = f(k(t)) - b^*(t) - c(t), \tag{2}$$

$$\dot{k}(t) \geq 0, k(0) = k_0,$$

where $b^*(t)$ denotes the solution of the government's problem. The government solves

$$\int_0^\infty e^{-\rho t} [U(c^*(t)) + V(b(t))] dt \rightarrow \max \tag{3}$$

with respect to $b(t)$ and subject to

$$\dot{k}(t) = f(k(t)) - b(t) - c^*(t), \tag{4}$$

$$\dot{k}(t) \geq 0, k(0) = k_0,$$

where $c^*(t)$ is the optimal consumption of the consumer. The solution of the game is a pair of functions $(c^*, b^*)$ which maximize the utilities of the consumer

---

[1] Think of an economy populated by a large number of identical individuals. Although the influence of each consumer is negligible, the fact that consumers know that they are identical allows them to form their decisions as if they had strategic power.

and the government, respectively, and when substituted in (2) and (4) lead to the same capital accumulation dynamics[2].

Here the continuously differentiable function $f(k)$ is the production function with the standard properties that $f(k) > 0$, $f'(k) > 0$ and $f''(k) < 0$ for $k > 0$. The utility functions $U(c)$ and $V(b)$ are concave and continuously differentiable, and the time preference parameter $\rho$ is strictly positive. We also require that $c(t) \geq 0$ and $k(t) \geq 0$ in order for the problem to be meaningful.

The problems of the consumer and the government as stated above are examples of optimal control problems. In Proposition 1 below we make use of the sufficiency results obtained by Seierstad and Sydsaeter in [5]. More precisely, we use a combination of Theorem 6 and Theorem 10 and replace their transversality condition with $\lim_{t \to \infty} e^{-\rho t} \pi(t) k(t) = 0$ since both the state variable $k$ and the co-state variable $\pi$ take non-negative values.

**Proposition 1.** *Assume that the following condition holds true for some constant $\alpha$:*

$$(f(k) - \bar{b}(k))U'(\bar{c}(k)) = \alpha, \tag{5}$$

*where $\bar{b}(k)$ is the solution of the following differential equation*

$$U'\left(\frac{\rho(f(k) - \bar{b}(k))}{f'(k) - \bar{b}'(k)}\right) = \frac{\alpha}{f(k) - \bar{b}(k)} \tag{6}$$

*with the initial condition $\bar{b}(0) = 0$ and*

$$\bar{c}(k) = \rho \frac{f(k) - \bar{b}(k)}{f'(k) - \bar{b}'(k)}. \tag{7}$$

*Let $f(k) - \bar{b}(k)$ and $f(k) - \bar{c}(k)$ be concave functions of $k$. For $f(k) - \bar{b}(k)$ we also require that it is positive for positive $k$ and that its first derivative is positive. Further, let $F(k) = f(k) - \bar{b}(k) - \bar{c}(k)$ be Lipschitz continuous.*
*(A) If $f(k_0) - \bar{b}(k_0) - \bar{c}(k_0) \geq 0$ and*

$$V''(\bar{b}(k)) = \frac{V'(\bar{b}(k))[\rho - f'(k) + \bar{c}'(k)] - U'(\bar{c}(k))\bar{c}'(k)}{\bar{b}'(k)(f(k) - \bar{b}(k) - \bar{c}(k))}, \tag{8}$$

*for each $k \geq k_0$, then the solution of the differential game is given by $(c^*(k), b^*(k))$ with*

$$b^*(k) = \bar{b}(k),$$

$$c^*(k) = \bar{c}(k),$$

*whenever the corresponding transversality condition for the government's problem holds true.*

---

[2] The solution corresponds to a stationary Markovian Nash equilibrium [2].

**(B)** If $f(k_0) - \bar{b}(k_0) - \bar{c}(k_0) < 0$, then the solution of the differential game is given by the pair of constants $(c^*, b^*)$, where the admissible $c^*$ and $b^*$ are determined from the following relations:

$$c^* = f(k_0) - b^*, \tag{9}$$

$$U'(f(k_0) - b^*) = V'(b^*). \tag{10}$$

*Remark 1.* The transversality condition for the problem of the government requires that $\lim_{t \to \infty} e^{-\rho t} \lambda(t) k(t) = 0$. There are two possibilities: (i) $F(k) > 0$ for each $k$ and (ii) there exists some $\bar{k} \geq k_0$ such that $F(\bar{k}) = 0$. In the second case the transversality condition clearly holds since capital is bounded and because of (19) the costate variable is also bounded. Regarding the first case, where capital is a strictly increasing function, let us assume that there exist some constants $P$ and $Q$ such that

$$kU'(c^*(k))c^{*'}(k) \geq P, \tag{11}$$

$$\frac{f(k) - b^*(k) - c^*(k)}{k} + \rho - f'(k) + c^{*'}(k) \leq Q \tag{12}$$

for all sufficiently large $k$. Define $z(t) = k(t)\lambda(t)$. Then $\dot{z}(t) \leq Qz(t) - P$, and so

$$z(t) \leq \frac{P}{Q} + e^{Qt}\left(k_0 V'(b^*(k_0)) - \frac{P}{Q}\right).$$

Hence, the transversality condition will be satisfied whenever $Q < \rho$.

*Proof.* **A)** Let $k(t)$ be the solution of (2). Since the differential equation (2) has an unique solution, if $F(k_0) > 0$, then $F(k(t)) > 0$ for each $t \geq 0$. Otherwise, there necessarily must be a point, say $k' \neq k_0$, at which the right-hand side becomes equal to zero, meaning that $k'$ is a stationary point. However, this contradicts the uniqueness of the solution. Similarly, $F(k_0) = 0$ implies that $F(k(t)) = 0$ for each $t \geq 0$.

Define the current value Hamiltonian for the consumer's problem by

$$H^F(k, c, \pi) = U(c) + \pi(f(k) - b^*(k) - c),$$

and the current value Lagrangian by

$$L^F(k, c, \pi, q) = H(k, c, \pi) + q(f(k) - b^*(k) - c).$$

The assumption $F(k_0) \geq 0$ ensures that the constraint does not become active and therefore, the complementary slackness condition holds with $q(t) = 0$. Hence, the Lagrangian reduces to the Hamiltonian and the sufficient optimality conditions for the consumer's problem are

$$\pi(t) = U'(c^*(k(t))). \tag{13}$$

$$\dot{\pi}(t) = \rho\pi(t) - \pi(t)f'(k(t)) + \pi(t)b^{*\prime}(k(t)) \tag{14}$$

and the transversality condition $\lim_{t\to\infty} e^{-\rho t}\pi(t)k(t) = 0$.

Let the function $\pi$ be determined by (13). Then we can write (5) as

$$(f(k(t)) - b^{*}(k(t)))\pi(t) = \alpha. \tag{15}$$

By taking the time derivative of both sides in expression (15) we obtain

$$\dot{\pi}(t)(f(k(t)) - b^{*}(k(t))) + \pi(t)(f'(k(t)) - b^{*\prime}(k(t)))\dot{k}(t) = 0 \tag{16}$$

Substituting $\dot{k}(t)$ with the right-hand side of differential equation (2) with $c$ as in (7) and solving for $\dot{\pi}$ gives

$$\dot{\pi}(t) = \pi(t)\rho - \pi(t)(f'(k(t)) - b^{*\prime}(k(t))), \tag{17}$$

which is exactly (14).

The properties of the production and utility functions and the assumption about $f(k) - b^{*}(k)$ ensure the concavity of the Hamiltonian and the constraint. To finish the proof, it remains to verify the transversality condition. By letting $g(k) = f(k) - b^{*}(k)$ and $y(t) = \dfrac{k(t)}{g(k(t))}$, it follows that

$$\dot{y}(t) = \frac{y(k(t))\dot{k}(t) - g'(k(t))k(t)\dot{k}(t)}{g(k(t))^2} = \left(1 - \frac{c^{*}(k(t))}{g(k(t))}\right)\left(1 - \frac{g'(k(t))k(t)}{g(k(t))}\right).$$

Since $g(k) - c^{*}(k) \geq 0$, $g(k) > 0$ and $c^{*}(k) \geq 0$, we obtain that $0 \leq 1 - \dfrac{c^{*}(k)}{g(k)} \leq 1$.

Because of $g(0) = 0$ and the concavity of $g(k)$, it follows that $1 - \dfrac{g'(k)k}{g(k)} \geq 0$. Since $g(\cdot)$ and $g'(\cdot)$ take positive values for positive $k$, we have that $1 - \dfrac{g'(k)k}{g(k)} < 1$.

All these inequalities imply that $0 \leq \dot{y}(t) \leq 1$, and hence $0 < y(t) \leq y_0 + t$. The last two inequalities and (15) imply that

$$\lim_{t\to\infty} e^{-\rho t}\pi(t)k(t) = \lim_{t\to\infty} e^{-\rho t}\alpha\frac{k(t)}{g(k(t))} = \lim_{t\to\infty} \alpha y(t)e^{-\rho t} = 0.$$

Thus the transversality condition also holds true.

This completes the proof for the consumer's problem. Next we need to establish that whenever relationship (8) is satisfied, the tax rule $b^{*}$ is optimal for the government.

The current value of Hamiltonian for the government's problem is

$$H^{L}(k, b, \lambda) = U(c^{*}(k)) + V(b) + \lambda(f(k) - b - c^{*}(k)). \tag{18}$$

In order for $b^{*}(k)$ to be the optimal control for the government, the following must be satisfied (assuming the transversality condition holds true):

$$\lambda(t) = V'(b^*(k(t))), \tag{19}$$

and

$$\dot{\lambda}(t) = \rho\lambda(t) - U'(c^*(k(t)))c^{*\prime}(k(t)) - \lambda(t)f'(k(t)) + \lambda(t)c^{*\prime}(k(t)). \tag{20}$$

Differentiating the expression in (19) with respect to $t$ gives

$$\dot{\lambda}(t) = V''(b^*(k(t)))b^{*\prime}(k(t))\dot{k}(t).$$

By substituting $V''(b^*(k))$ from (8) above and by replacing $\dot{k}$ with the right-hand side of the differential equation (4) we obtain exactly (20).

The same arguments as in the consumer's problem establish the concavity of the Hamiltonian and the constraint functions.

**B)** If $F(k_0) < 0$, then the previous solution is no longer feasible. Again we begin with the consumer's problem and we show that for

$$\pi(t) = \frac{g'(k_0)U'(g(k_0))}{\rho} = \bar{\pi} \tag{21}$$

and

$$q(t) = \frac{U'(g(k_0))[\rho - g'(k_0)]}{\rho} = \bar{q}, \tag{22}$$

where $\bar{\pi}$ and $\bar{q}$ are constants, the optimal consumption given by $c^* = f(k_0) - b^* = g(k_0)$ solves the problem of the consumer as stated in the proposition.

The Lagrangian of the consumer is:

$$L^F(k, c, \pi, q) = U(c) + \pi(f(k) - b^* - c) + q(f(k) - b^* - c).$$

Note that for $c = c^*$ the differential equation (2) becomes $\dot{k}(t) = 0$, hence $k(t) = k_0$. One can now directly check that with the above choice of $c^*$, $\bar{\pi}$ and $\bar{q}$ all the sufficiency conditions stated in [5] are satisfied: The condition $\dfrac{\partial L(k, c^*, \pi, q)}{\partial c} = 0$ follows from the equalities

$$U'(c^*) = U'(g(k_0)) = \frac{g'(k_0)U'(g(k_0))}{\rho} + \frac{U'(g(k_0))[\rho - g'(k_0)]}{\rho} = \bar{\pi} + \bar{q}.$$

Since

$$\dot{\pi}(t) = 0 = \rho\frac{g'(k_0)U'(g(k_0))}{\rho} - U'(g(k_0))g'(k_0) = \rho\bar{\pi} - (\bar{\pi} + \bar{q})g'(k_0),$$

the so defined $\pi$ satisfies the costate equation. The Hamiltonian and the constraint are concave and since both $\pi$ and $k$ are constant at the optimum, the transversality condition also holds true. Finally, consider the problem of the government when $c^* = g(k_0)$. Like in the case of the consumer it is straightforward to show that for

$$\lambda(t) = \frac{U'(c^*)f'(k_0)}{\rho} = \bar{\lambda}, \tag{23}$$

and

$$\phi(t) = V'(f(k_0) - c^*) - \frac{U'(c^*)f'(k_0)}{\rho} = \bar{\phi}, \qquad (24)$$

with $\lambda$ and $\phi$ being the co-state variable and the Lagrangian multiplier, respectively, the optimal tax rule $b^*$ solves the problem of the government. The line of reasoning is the same as before and we skip the details of the proof.

Note, however, that at $b = b^*$ the Lagrangian of the government reduces to

$$L^L(k, b^*, \lambda, \phi) = U(f(k_0) - b^*) + V(b^*).$$

Since in this case $b^*$ is determined by the equality

$$U'(f(k_0) - b^*) = V'(b^*), \qquad (25)$$

we obtain that the derivative of the Lagrangian with respect to the control variable at $b^*$ should be equal to zero. This completes the proof.     $\square$

## 3     Examples

**Example 1** (cf. [6]). Let $f(k) = Ak$, $U(c) = \ln(c)$ and $V(b) = \ln(b)$.

(A) Assume that the parameters of the model are such that $A \geq \frac{3\rho}{2}$. We will show that the solution to the differential game is $(c^*(k) = \rho k, b^*(k) = \frac{\rho k}{2})$ for $\alpha = \frac{A}{\rho} - \frac{1}{2}$.

Clearly the optimal consumption $c^*$ satisfies equation (7) and the tax rule $b^*$ solves differential equation (6), which for this example takes the form

$$b^{*'}(k) = A - \alpha\rho, \quad \text{with initial condition } b^*(0) = 0.$$

Equation (8) holds and the transversality condition is satisfied since the left-hand side of (12) is equal to $\frac{\rho}{2} < \rho$. Substituting the optimal controls in the differential equation for capital and solving it we obtain $k(t) = k_0 e^{(A - \frac{3\rho}{2})t}$. Since by assumption $A \geq \frac{3\rho}{2}$, capital is non-decreasing.

(B) Next, we proceed with the case when $A < \frac{3\rho}{2}$. Clearly, the previous solution is no longer feasible since it implies that capital will eventually go to zero. Thus, we need to apply part (B) of Proposition 1, which establishes that optimal consumption should be $c^* = f(k_0) - b^*$.

With logarithmic utility functions condition (10) leads to $b^* = \frac{Ak_0}{2}$, and for optimal consumption we get $c^* = Ak_0 - b^* = \frac{Ak_0}{2}$.

**Example 2.** Now consider an example where the production function is non-linear. Let $U(c) = 2\sqrt{c}$, $V(b) = 2\sqrt{2b}$ and $f(k) = k + \alpha\sqrt{2\rho k}$. To simplify the calculations, we have chosen this value for $\rho$, though it seems unrealistic. Similar to Example 1, one may check that the pair $(c^*(k) = 2k, b^*(k) = k)$ solves the game. Equation (6) in this case becomes

$$b^{*\prime}(k) = 1 + \frac{\alpha\sqrt{2}}{2\sqrt{k}} - \frac{\alpha}{k + \sqrt{2k} - b^*(k)},$$

and $b^* = k$ is a solution of the above differential equation. Also, it is easy to see that $c^* = 2k$ satisfies (7) and that (8) also holds. Since capital is bounded, the transversality condition is guaranteed. Note that in the non-linear example the behaviour of the state variable is different from the linear case. The differential equation for capital is

$$\dot{k}(t) = \alpha\sqrt{2k} - 2k.$$

This equation has two equilibrium points: 1) $k_z = 0$ and 2) $k_e = \alpha^2/2$. If $k_0 < \alpha^2/2$, capital grows initially and converges to the equilibrium value $k_e = \alpha^2/2$. For example, for $\alpha = 10$, capital converges to $k_e = 50$. If we change the initial condition so that $k_0 > 50$ and keep the previous value of $\alpha$, then capital will be decreasing and this means that part B) of the proposition has to be applied. The solution will change as follows:

$$b^*(k) = \frac{2k_0 + 10\sqrt{2k_0}}{3},$$

$$c^*(k) = \frac{k_0 + 20\sqrt{2k_0}}{3}.$$

# References

1. Cohen, D., Michel, P.: How Should Control Theory Be Used to Calculate a Time-Consistent Government Policy? Review of Economic Studies **55**(2) (1988) 263–274
2. Dockner, E., Jorgensen, S., Van Long, N., Sorger, G.: Differential Games in Economics and Management Science. Cambridge University Press (2000)
3. Karp, L., Ho Lee,In : Time-Consistent Policies. Journal of Economic Theory **112**(2) (2003) 353–364
4. Kydland, F., Prescott, E.: Rules Rather than Discretion: The inconsistency of Optimal Plans. Journal of Political Economy **85**(3) (1977) 473–492
5. Seierstad, A., Sydsaeter, K.: Sufficient Conditions in Optimal Control Theory. International Economic Review **18**(2) (1977) 367–391
6. Xie, D.: On Time Consistency: A Technical Issue in Stackelberg Differential Games. Journal of Economic Theory **76**(2) (1997) 412–430

# Evolutionary Optimization Method for Approximating the Solution Set Hull of Parametric Linear Systems

Iwona Skalna

Department of Applied Computer Science,
University of Mining and Metallurgy AGH
ul. Gramatyka 10, 30-067 Cracow, Poland

**Abstract.** Systems of parametric interval equations are encountered in many practical applications. Several methods for solving such systems have been developed during last years. Most of them produce both outer and inner interval solutions, but the amount of overestimation, resp. underestimation is not exactly known. If a solution of a parametric system is monotonic and continuous on each interval parameter, then the method of combination of endpoints of parameter intervals computes its interval hull. Recently, a few polynomial methods computing the interval hull were developed. They can be applied if some monotonicity and continouity conditions are fulfilled. To get the most accurate inner approximation of the solution set hull for problems with any bounded solution set, an evolutionary optimization method is applied.

## 1    Introduction

The behavior of the loaded truss structure or analog linear circuit can be described by a system of parameter-dependent linear equations [2,5,6,7,16,17]. Assuming some model parameters are unknown and lie in given intervals will lead to a parametric system of linear interval equations. Several methods for solving such systems have been developed during last years [5,6,10,12,15,16]. Most of them compute both outer and inner interval approximations of the solution set hull, but the amount of overestimation, resp. underestimation is not exactly known. It can be estimated by a comparison of inner and outer approximations of the solution set hull [12,15].

If a solution of the parametric system is monotonic and continuous on each parameter interval, then the method of combination of endpoints of parameter intervals (CEPI in brief) [2,14] computes interval hull of the parametric solution set, that is the tightest interval vector containing this set. Unfortunately, because the complexity of the CEPI method – $2^k$ real systems have to be solved, where $k$ is the number of interval parameters – increases at an exponential rate as a factor of the number of parameters, it can be applied to problems with small number of parameters. Recently, a few polynomial methods computing interval hull were developed [14]. They can be applied if the solution of the parametric system is continous and monotonic on each interval parameter. In [13] Popova

T. Boyanov et al. (Eds.): NMA 2006, LNCS 4310, pp. 361–368, 2007.

shows, that when some sufficient conditions are fulfilled, the interval hull (or some bounds) of the parametric solution set can be easily computed.

To get the best inner approximation of the solution set hull for problems with bounded (non-monotonic and non-continous) solution set, and a lot of interval parameters, an evolutionary optimization method (EOM in brief) is applied in Section 3. In Section 4 some numerical examples are presented and used to evaluate the results of the EOM method. They include two examples of parametric systems with non-monotonic solution set, and two test examples of truss structures. The computations performed show that the EOM method produces a high-quality approximation of the interval hull of a parametric solution set.

## 2   Preliminaries

Let $\mathbb{IR}$, $\mathbb{IR}^n$, $\mathbb{IR}^{n \times n}$ denote the set of real compact intervals, interval vectors, and interval square matrices, respectively [9]. Italic faces will denote real quantities, while bold italic faces will denote their interval counterparts.

Consider linear algebraic system

$$A(p)x = b(p) \tag{1}$$

with coefficients being functions that are linear in parameters

$$a_{ij}(p) = \omega_{ij0} + \sum_{\nu=1}^{k} \omega_{ij\nu} \cdot p_\nu, \quad b_j(p) = \omega_{0j0} + \sum_{\nu=1}^{k} \omega_{0j\nu} \cdot p_\nu, \tag{2}$$

$(i, j = 1, \ldots, n)$; where $p = \{p_1, \ldots, p_k\}^\top \in \mathbb{R}^k$ is a vector of parameters, $\omega \in (\mathbb{R}^{k+1})^{((n+1) \times n)}$ is a matrix of real $(k+1)$-dimensional vectors.

Assuming some model parameters are unknown and lie in given intervals $\boldsymbol{p}_i \ni p_i$ $(i = 1, \ldots, k)$ would give a family of systems (1) which is usually written in a symbolic compact form

$$A(\boldsymbol{p})x = b(\boldsymbol{p}), \tag{3}$$

and is called *parametric interval linear system*.

Parametric solution set of the system (3) is defined [4,15] as

$$S(\boldsymbol{p}) = \{ x \mid A(p)x = b(p), \, p \in \boldsymbol{p} \}. \tag{4}$$

If the solution set is bounded, then its interval hull exists and is defined as

$$\square S(\boldsymbol{p}) = [\inf S(\boldsymbol{p}), \sup S(\boldsymbol{p})] = \bigcap \{ y \in \mathbb{IR}^n \mid S(\boldsymbol{p}) \subseteq y \}. \tag{5}$$

In order to guarantee that the solution set is bounded, the matrix $A(\boldsymbol{p})$ must be regular, i.e. $A(p)$ must be regular for all $p \in \boldsymbol{p}$.

A vector $\boldsymbol{x} = [\underline{x}, \overline{x}] \in \mathbb{IR}^n$ is called inner approximation of $S \subseteq \mathbb{R}^n$ if

$$\inf_{s \in S} s_i \leqslant \underline{x}_i \quad \text{and} \quad \sup_{s \in S} s_i \geqslant \overline{x}_i, \quad i = 1, \ldots, n,$$

resp. outer approximation of $S \subseteq \mathbb{R}^n$ if

$$\inf_{s \in S} s_i \geqslant \underline{x}_i \quad \text{and} \quad \sup_{s \in S} s_i \leqslant \overline{x}_i, \quad i = 1, \ldots, n.$$

# 3 Evolutionary Optimization

The problem of computing optimal inner approximation of the solution set hull of the parametric linear system (3) can be written as a problem of solving $2n$ constrained optimization problems: for $i = 1, \ldots, n$,

$$\min \left\{ f(p) = \left( A(p)^{-1} b(p) \right)_i \mid p \in \boldsymbol{p} \right\} \tag{6}$$

and

$$\max \left\{ f(p) = \left( A(p)^{-1} b(p) \right)_i \mid p \in \boldsymbol{p} \right\}, \tag{7}$$

where $\boldsymbol{p} \in \mathbb{IR}^k$ is a vector of interval parameters.

**Theorem 1.** *Let $A(\boldsymbol{p})$ be regular, $p \in \mathbb{IR}^k$, and $x_{\min}^i$, $x_{\max}^i$ denote the global solutions of the $i$-th minimization (6), resp. maximization (7) problems. Then the interval vector $\boldsymbol{x} = [x_{\min}, x_{\max}] = \left( [x_{\min}^i, x_{\max}^i] \right)_{i=1}^n = \square S(\boldsymbol{p})$.*

*Proof.* $\subseteq$: $x_{\min}^i$, $x_{\max}^i \in S(\boldsymbol{p})_i$, hence $[x_{\min}^i, x_{\max}^i] \subseteq \square S(\boldsymbol{p})_i$. $\supseteq$: take any $x \in S(\boldsymbol{p})$, then $x = A(p)^{-1} b(p)$ for some $p \in \boldsymbol{p}$. Since for each $p \in \boldsymbol{p}$ $x_{\min}^i \leqslant (A(p)^{-1} b(p))_i \leqslant x_{\max}^i$, then $S(\boldsymbol{p})_i \subseteq [x_{\min}^i, x_{\max}^i]$ and hence
$$\square S(\boldsymbol{p})_i \subseteq [x_{\min}^i, x_{\max}^i]. \qquad \square$$

The optimization problems (6) and (7) will be solved using an evolutionary approach [3,8,11]. As a result of the minimization (maximization) problem one will obtain a value greater of equal (less or equal) to the actual minimum (maximum). The final result will be the inner approximation of the solution set hull.

## 3.1 Evolutionary Algorithm Description

Optimization is performed using the evolutionary algorithm shown in Fig. 1. Each evolutionary algorithm requires some input parameters. These are: population size ($pop_{size}$), crossover rate ($c_r$), mutation rate ($m_r$), number of generations ($n_g$). All of them have great influence on the result of the optimization, but the choice of the best values is still a matter of trial. Suggestions for parameter values can be found in the literature [1,8,11].

For $t = 0, \ldots n_g$, the population $P(t) = \left\{ p_1^t, \ldots, p_{n_g}^t \right\}$ consists of individuals characterized by $k$-dimensional vectors of real numbers $p_i^t = \{p_{i1}^t, \ldots, p_{ik}^t\}^{\mathrm{T}}$ with $p_{ij}^t \in \boldsymbol{p}_j$, $i = 1, \ldots, pop_{size}, j = 1, \ldots, k$. The elements $p_{ik}^0$ of the initial population $P(0)$ are generated randomly based on the uniform distribution.

The two following genetic operators are employed [8]:

- *non-uniform mutation* - this one-argument operator vouch for the system adaptation ability. If the element $p_j$ of the individual $p$ is chosen for mutation, then $p' = \{p_1, \ldots, p_j', \ldots, p_k\}^{\mathrm{T}}$ with

$$p_j' = \begin{cases} p_j + \left( \overline{p}_j - p_j \right) r \left( 1 - t/n_g \right)^b, & \text{if } q < 0.5 \\ p_j + \left( p_j - \underline{p}_j \right) r \left( 1 - t/n_g \right)^b, & \text{if } q \geqslant 0.5, \end{cases}$$

where $r$, $q$ are random numbers from $[0, 1]$, $t$ is a number of generation, $n_g$ is a number of generations, and $b$ is a parameter of the system describing the level of heterogeneity; the probability that mutation factor is close to zero increases as $t$ increases from 0 to $n_g$;

- *arithmetic crossover* - this two-argument operator is defined as linear combination of two vectors. If the parents $p^1$ and $p^2$ are chosen for crossover, then the offsprings are

$$p^{1\prime} = rp^1 + (1 - r)p^2, \qquad p^{2\prime} = (1 - r)p^1 + rp^2,$$

where $r$ is a random number from $[0, 1]$; arithmetic crossover guarantees that the elements of the vectors $p^{1\prime}$ and $p^{2\prime}$ lie in parameter intervals.

Let $P'(t)$ denote the population after the crossover process, and $P''(t)$ - the population after the mutation process. The best (wrt fitness function), (7)) $pop_{size}$ individuals, from the combined population $P(t) \cup P''(t)$, form a new population $P(t+1)$.

```
Initialize parameters
t := 0
Initialize population P(t) of pop_size size
while t < n_g do
    P'(t) ⟵ Crossover_with_c_r_rate(P(t))
    P''(t) ⟵ Mutate_with_m_r_rate(P'(t))
    Evaluate the fitness f(P''(t))
    P(t + 1) ⟵ Select_pop_size(P(t) ∪ P''(t))
    t:=t+1
end while
```

**Fig. 1.** Pseudo-code of an evolutionary algorithm

## 4    Numerical Examples

In this section the results of the EOM are presented. Two small parametric linear systems with non-monotonic solution set and two exemplary truss structures are included to evaluate the performance of the EOM method. Several variants of the input parameter values have been examined. The best results have been obtained for the following values: population size $pop_{size} = 10$ (exs. 1, 2), $pop_{size} = 16$ (exs. 3, 4), crossover rate $c_r = 0.1$, mutation rate $m_r = 0.9$, $b = 96$, number of generations $n_g = 80$ (exs. 1, 2), $n_g = 100$ (exs. 3, 4).

**Example 1**
Assume the two-dimensional parametric linear system is of the form:

$$\begin{bmatrix} p_1 & 1 + p_2 \\ -2 & 3p_1 - 1 \end{bmatrix} \cdot \begin{bmatrix} x_1 \\ x_2 \end{bmatrix} = \begin{bmatrix} 2p_1 \\ 0 \end{bmatrix}, \quad \boldsymbol{p}_1 = [0, 1], \boldsymbol{p}_2 = [0, 1]. \tag{8}$$

The numerical results are presented in Table 1. Column 2 contains the mid-point solution, columns 3 - the result of the EOM method, column 4 - the interval hull (IH). In case of two-dimensional system the interval hull can be easily computed. From (8) one gets two equations $p_1x_1 + (1 + p_2)x_2 = 2p_1$ and $-2x_1 + (3p_1 - 1)x_2 = 0$. Eliminating $p_1$ from these equations gives the equation $2x^2 + 3(1 + p_2)y^2 + xy - 4x - 2y = 0$ which, with $p_2 \in [0, 1]$, define the pencil of ellipses. The intersection of the pencil with the united solution set gives the parametric solution set.

Table 1. Numerical results for Example 1

| $x_0$ | | EOM | IH |
|---|---|---|---|
| $x_1$ | 0.1538 | [−0.087, 1] | [−0.087, 1] |
| $x_2$ | 0.6154 | [0, 1.026] | [0, 1.026] |

## Example 2
The three-dimensional parametric linear system is of the form:

$$\begin{bmatrix} p_1 & p_2 + 1 & -p_3 \\ p_2 + 1 & -3 & p_1 \\ 2 - p_3 & 4p_2 + 1 & 1 \end{bmatrix} \cdot \begin{bmatrix} x_1 \\ x_2 \\ x_3 \end{bmatrix} = \begin{bmatrix} 2p_1 \\ p_3 - 1 \\ -1 \end{bmatrix}, \tag{9}$$

where $p_1 \in \mathbf{p}_1 = [0, 1]$, $p_2 \in \mathbf{p}_2 = [0, 1]$, $p_3 \in \mathbf{p}_3 = [0, 1]$.

The numerical results of the EOM method, the direct method (DM) [17], and the Monte Carlo method (MC) (100000 samples), are presented in Table 2.

Table 2. Numerical results for Example 2

| | $x_0$ | MC | EOM | DM |
|---|---|---|---|---|
| $x_1$ | 0.286 | [−0.866, 2.641] | [−1, 2.667] | [−2.184, 4.685] |
| $x_2$ | 0.048 | [−0.65, 0.328] | [−0.667, 0.333] | [−0.84, 1.337] |
| $x_3$ | −1.571 | [−5.592, 0.679] | [−5.667, 1] | [−11.163, 2.663] |

## Example 3. (25-bars plane truss structure)
Consider the truss structure shown in Fig. 2. Young's modulus E= $2.1 \times 10^{11}$ [Pa], cross-section area C= 0.004 [m²]. Assume the stiffness of all bars is uncertain by ±5%. This gives 25 interval parameters. The vector of displacements $d$ is a solution of the parametric system $K(p)d = q(p)$, where $K(p)$ is parameter-dependent stiffness matrix, $q(p)$ is parameter-dependent vector of forces. Numerical results are presented in table 3. Column 2 contains the midpoint solution ($d_0$), columns 3, 4 - the result, resp. the relative error ($rerr = (\bar{d} - \underline{d})/(2 \cdot d_0)$) of the Monte Carlo method (100000 samples), columns 5, 6 - the result, resp. the relative error of the EOM method, columns 7, 8 - the result, resp. the relative error of the DM method.

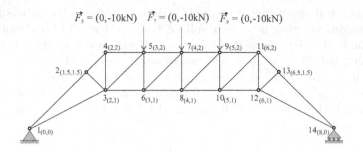

$\vec{F_5} = (0,-10\text{kN})$    $\vec{F_7} = (0,-10\text{kN})$    $\vec{F_9} = (0,-10\text{kN})$

**Fig. 2.** 25-bar plane truss structure

**Table 3.** Numerical results for example 3

| | $d_0$ [$\times 10^{-5}$] | MC [$\times 10^{-5}$] | $rerr$[%] | EOM [$\times 10^{-5}$] | $rerr$[%] | DM [$\times 10^{-5}$] | $rerr$[%] |
|---|---|---|---|---|---|---|---|
| $d_2^x$ | 62.2 | [60.26, 64.34] | 3.27 | [59.12, 65.57] | 5.19 | [56.47, 67.9] | 9.2 |
| $d_2^y$ | −74.32 | [−76.95, −71.95] | 3.37 | [−78.33, −70.66] | 5.16 | [−79.99, −68.63] | 7.64 |
| $d_3^x$ | 50.83 | [49.16, 52.69] | 3.46 | [48.33, 53.57] | 5.16 | [47.27, 54.37] | 6.98 |
| $d_3^y$ | −85.68 | [−88.6, −83.05] | 3.24 | [−90.33, −81.45] | 5.18 | [−92.86, −78.49] | 8.39 |
| $d_4^x$ | 66.66 | [64.66, 68.96] | 3.22 | [63.34, 70.31] | 5.23 | [61.01, 72.3] | 8.47 |
| $d_4^y$ | −82.83 | [−85.6, −80.31] | 3.19 | [−87.33, −78.73] | 5.19 | [−89.45, −76.18] | 8.01 |
| $d_5^x$ | 63.81 | [61.85, 65.98] | 3.23 | [60.62, 67.31] | 5.24 | [58.56, 69.04] | 8.21 |
| $d_5^y$ | −102.7 | [−106.03, −99.81] | 3.03 | [−108.17, −97.74] | 5.08 | [−111.82, −93.56] | 8.89 |
| $d_6^x$ | 55.12 | [53.28, 57.15] | 3.51 | [52.41, 58.08] | 5.15 | [51.41, 58.8] | 6.7 |
| $d_6^y$ | −103.18 | [−106.5, −100.29] | 3.01 | [−108.65, −98.21] | 5.06 | [−112.53, −93.81] | 9.07 |
| $d_7^x$ | 59.52 | [57.59, 61.56] | 3.33 | [56.53, 62.79] | 5.26 | [54.57, 64.46] | 8.31 |
| $d_7^y$ | −108.93 | [−112.32, −105.9] | 2.95 | [−114.66, −103.74] | 5.01 | [−119.04, −98.8] | 9.29 |
| $d_8^x$ | 59.88 | [57.88, 62.1] | 3.52 | [56.95, 63.1] | 5.14 | [56.03, 63.71] | 6.41 |
| $d_8^y$ | −108.45 | [−111.85, −105.41] | 2.97 | [−114.18, −103.26] | 5.03 | [−118.57, −98.32] | 9.33 |
| $d_9^x$ | 54.76 | [53.01, 56.89] | 3.54 | [51.99, 57.78] | 5.28 | [50.05, 59.45] | 8.58 |
| $d_9^y$ | −101.98 | [−105.03, −99.10] | 2.90 | [−107.42, −97.06] | 5.08 | [−111.56, −92.41] | 9.39 |
| $d_{10}^x$ | 64.16 | [62.11, 66.53] | 3.44 | [61.03, 67.61] | 5.13 | [60.14, 68.17] | 6.28 |
| $d_{10}^y$ | −100.56 | [−103.63, −97.69] | 2.95 | [−105.96, −95.64] | 5.13 | [−109.91, −91.19] | 9.31 |
| $d_{11}^x$ | 50.47 | [48.64, 52.45] | 3.77 | [47.81, 53.38] | 5.52 | [45.89, 55.04] | 9.06 |
| $d_{11}^y$ | −82.83 | [−85.62, −80.19] | 3.27 | [−87.36, −78.69] | 5.23 | [−90.46, −75.21] | 9.21 |
| $d_{12}^x$ | 67.02 | [64.9, 69.4] | 3.35 | [63.75, 70.62] | 5.12 | [62.73, 71.3] | 6.39 |
| $d_{12}^y$ | −84.25 | [−87.06, −81.69] | 3.18 | [−88.87, −80.05] | 5.23 | [−92.33, −76.18] | 9.58 |
| $d_{13}^x$ | 56.01 | [54, 58.28] | 3.82 | [53.13, 59.16] | 5.38 | [51.43, 60.58] | 8.17 |
| $d_{13}^y$ | −73.25 | [−75.8, −70.86] | 3.37 | [−77.23, −69.62] | 5.20 | [−80.18, −66.3] | 9.47 |
| $d_{14}^x$ | 117.13 | [113.57, 121.14] | 3.23 | [111.55, 123.29] | 5.01 | [110.18, 124.07] | 5.93 |

**Example 4.** (*Baltimore bridge 1820*)

Every bar of the bridge (Fig. 3) has Young's modulus E= $2.1 \times 10^{11}$ [Pa], cross-section area C= 0.004 [m²]. Assuming the stiffness of all bars is uncertain by ±5% will give 45 interval parameters. The computational time of the EOM method increased about 7 times. Most coordinates of the vector solution, produced by the

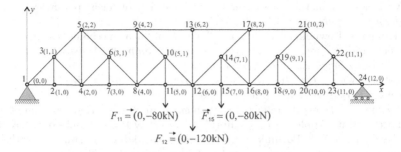

**Fig. 3.** Baltimore bridge 1820

EOM method, were exactly equal to the interval hull. The remaining coordinates differed only slightly from the exact hull solution. The result of the EOM method is not included because of the limit for the number of pages.

## 5  Conclusions

The problem of solving parametric linear systems has been considered in Section 2. In Section 3 the evolutionary optimization method EOM for approximating from below the solution set hull of parametric linear systems has been described. Computations performed in Section 4 show that the EOM is a powerful tool for solving such systems. The EOM method produced a very accurate approximation of the interval hull of all parametric solution sets considered. It is simple and quite efficient. The main advantage of the EOM method is that it can be applied to any parametric linear system with bounded solution set. The EOM method can be used to solve problems with a lot of interval parameters. However, since the accuracy of the EOM method is not exactly known it should be used in conjunctions with methods that compute inner and outer approximations. The comparison of the EOM method with existing methods solving parametric linear systems will be a subject of future work.

## References

1. T. Bäck, D.B. Fogel, and Z. Michalewicz. *Handbook of Evolutionary Computations.* Oxford University Press, 1997.
2. A. Dreyer. Combination of symbolic and interval-numeric methods for analysis of analog circuits. *Proc. 8 International Workshop on Symbolic Methods and Applications in Circuit Design (SMACD 2004)*, september 2004. Wroclaw, Poland.
3. D.E. Goldberg. *Genetic Algorithms in Search, Optimization and Machine Learning.* Addison-Wesley, Reading, Massachusetts, 1989.
4. C. Jansson. Interval linear systems with symmetric matrices, skew-symmetric matrices and dependencies in the right hand side. *Computing*, 46(3):265–274, 1991.
5. L.V. Kolev. Outer solution of linear systems whose elements are affine functions of interval parameters. *Reliable Computing*, 6:493–501, 2002.

6. L.V. Kolev. A method for outer interval solution of linear parametric systems. *Reliable Computing*, 10:227–239, 2004.

7. Z. Kulpa, A. Pownuk, and I. Skalna. Analysis of linear mechanical structures with uncertainties by means of interval methods. *Computer Assisted Mechanics and Engineering Sciences*, 5(4):443–477, 1998.

8. Z. Michalewicz. *Genetic Algorithms + Data Structures = Evolution Programs*. Springer-Verlag, Berlin, Germany, 1996.

9. A. Neumaier. *Interval Methods for Systems of Equations*. Encyclopedia of Mathematics and its Applications. Cambridge University Press, Cambridge, UK, 1990.

10. A. Neumaier and A. Pownuk. Linear systems with large uncertainties, with applications to truss structures. *(submitted for publication)*.

11. A. Osyczka. *Evolutionary Algorithm for Single and Multicriteria Design Optimization*. Studies in Fuzziness and Soft Computing Physica-Verlag, Heidelberg New York, 2002.

12. E.D. Popova. On the solution of parametrised linear systems. *Scientific Computing, Validated Numerics, Interval Methods*, pages 127–138, 2001.

13. E.D. Popova. Quality of the solution of parameter-dependent interval linear systems. *ZAMM*, 82(10):723–727, 2002.

14. A. Pownuk. Efficient method of solution of large scale engineering problems with interval parameters. *NSF workshop on Reliable Engineering Computing*, September 15-17, 2004, Savannah, Georgia, USA.

15. S.M. Rump. Verification methods for dense and sparse systems of equations. In Jürgen Herzberger, editor, *Topics in validated computations: proceedings of IMACS-GAMM International Workshop on Validated Computation, Oldenburg, Germany, 30 August–3 September 1993*, volume 5 of *Studies in Computational Mathematics*, pages 63–135, Amsterdam, The Netherlands, 1994. Elsevier.

16. I. Skalna. Methods for solving systems of linear equations of structure mechanics with interval parameters. *Computer Assisted Mechanics and Engineering Sciences*, 10(3):281–293, 2003.

17. I. Skalna. A method for outer interval solution of parametrized systems of linear interval equations. *Reliable Computing*, 12(2):107–120, 2006.

# Formulae for Calculation of Normal Probability

Vesselin Gushev and Geno Nikolov

Department of Mathematics, University of Sofia,
5 James Bourchier Boulevard, 1164 Sofia, Bulgaria
{geno,v_gushev}@fmi.uni-sofia.bg

**Abstract.** Various approximate formulae for calculation of the normal probability integral

$$P(x) = \frac{1}{\sqrt{2\pi}} \int_0^x e^{-t^2/2} dt, \quad x \geq 0,$$

are given. Our formulae provide a good approximation to $P(x)$ over the entire range $0 < x < \infty$, hence they can be used in practice instead of the usual numerical tables. Error bounds based on the Peano kernel technique are proved.

**Keywords:** Normal probability, error function, quadrature formulae, Peano kernels.

## 1   Introduction

For a random variable $X$ with standard normal distribution, the probability that $0 < X < x$ is given by

$$P(x) = \frac{1}{\sqrt{2\pi}} \int_0^x e^{-t^2/2} dt.$$

The integral in $P(x)$ is probably the best-known example of an integral that cannot be evaluated through the Leibnitz-Newton rule. An alien to $P(x)$ is the error function

$$\operatorname{erf}(x) = \frac{2}{\sqrt{\pi}} \int_0^x e^{-t^2} dt = 2P(\sqrt{2}x). \tag{1}$$

The standard statistics textbooks incorporate four- or five-place tables with computed values of $P(x)$. Most mathematical handbooks (see, e.g., [1,4,5]) provide a detailed information about $P(x)$ and $\operatorname{erf}(x)$. We quote two series representations of $\operatorname{erf}(x)$ (formulae (7.1.5) and (7.1.6) in [1]), which, in view of (1), have their obvious counterparts for $P(x)$:

$$\operatorname{erf}(x) = \frac{2}{\sqrt{\pi}} \sum_{n=0}^{\infty} \frac{(-1)^n x^{2n+1}}{n!(2n+1)}, \tag{2}$$

T. Boyanov et al. (Eds.): NMA 2006, LNCS 4310, pp. 369–377, 2007.

$$\operatorname{erf}(x) = \frac{2}{\sqrt{\pi}} e^{-x^2} \sum_{n=0}^{\infty} \frac{2^n x^{2n+1}}{(2n+1)!!}. \tag{3}$$

The series in (2) and (3) are fast converging, and can be used for effective calculation of $\operatorname{erf}(x)$. It is desirable, however, to have a closed-type approximate formula for $\operatorname{erf}(x)$ (or for $P(x)$), supplied with a good uniform (i.e., independent of $x$) error bound. Any formula of this kind may be viewed as a coding of the information about the function $\operatorname{erf}(x)$ (or $P(x)$), and can be used for a quick recovery of the function value at any point $x$, $0 < x < \infty$. Some rational approximate formulae for $\operatorname{erf}(x)$, with error bound ranging between $5 \times 10^{-4}$ and $1.5 \times 10^{-7}$, are given in [1].

In this paper, we present some new approximate formulae for $P(x)$. The key step to their derivation is the representation

$$P^2(x) = \frac{1}{2\pi} \int_0^x \int_0^x e^{-(t^2+u^2)/2} du \, dt = \frac{1}{\pi} \int_0^x \int_0^t e^{-(t^2+u^2)/2} du \, dt.$$

Passage to polar coordinates yields

$$P^2(x) = \frac{1}{\pi} \int_0^{\pi/4} \int_0^{x \sec \varphi} e^{-r^2/2} r \, dr \, d\varphi = \frac{1}{4} - \frac{1}{\pi} \int_0^{\pi/4} e^{-(x \sec \varphi)^2/2} d\varphi, \tag{4}$$

hence the evaluation of $P(x)$ is reduced to the calculation of the last integral. The integrand

$$g(\varphi) := e^{-\frac{x^2}{2\cos^2 \varphi}}, \tag{5}$$

possesses the property that

$$g^{(k)}(\varphi) = g(\varphi) p_k(t), \tag{6}$$

where $p_k$ is an algebraic polynomial in variable $t = \tan \varphi$, defined recurrently by

$$p_0(t) \equiv 1, \quad p_{k+1}(t) = (t^2 + 1)[p_k'(t) - a^2 t p_k(t)], \quad k = 0, 1, \ldots \tag{7}$$

This representation reveals that quadrature formulae involving one or two interior nodes and multiple nodes at the endpoints would be an appropriate tool for the evaluation of $\int_0^{\pi/4} g(\varphi) d\varphi$. Such quadrature formulae, along with some error bounds based on the Peano kernel technique, are given in Section 3. The resulting approximate formulae for $P(x)$ are presented in the next section. In Section 4 we establish some estimates for the sup-norm of $g^{(4)}$ and $g^{(6)}$. These estimates are used in Section 5 for the derivation of theoretical bounds for the error of our formulae. Although the obtained error bounds well overestimate the real error magnitude, they are of some value as they apply to the whole interval $(0, \infty)$ and do not rely on comparison with any tabulated values of $P(x)$.

## 2    Approximate Formulae for $P(x)$

Our approximate formulae are denoted by $F_i(x)$, $i = 1, \ldots, 13$, and are listed below.

$$F_1(x) = \frac{1}{2}\left[1 - \frac{1}{2}\left(e^{-x^2/2} + \left(1 + \frac{\pi x^2}{12}\right)e^{-x^2}\right)\right]^{1/2},$$

$$F_2(x) = \frac{1}{2}\left[1 - \frac{1}{4}\left(e^{-x^2/2} + 2e^{-(2-\sqrt{2})x^2} + \left(1 + \frac{\pi x^2}{24}\right)e^{-x^2}\right)\right]^{1/2},$$

$$F_3(x) = \frac{1}{2}\left[1 - \frac{1}{6}\left(e^{-x^2/2} + 2e^{-2(2-\sqrt{3})x^2} + 2e^{-2x^2/3} + \left(1 + \frac{\pi x^2}{36}\right)e^{-x^2}\right)\right]^{1/2},$$

$$F_4(x) = \frac{1}{2}\left[1 - \frac{1}{6}\left(e^{-x^2/2} + 4e^{-(2-\sqrt{2})x^2} + e^{-x^2}\right)\right]^{1/2},$$

$$F_5(x) = \frac{1}{2}\left[1 - \frac{1}{30}\left(7e^{-x^2/2} + 16e^{-(2-\sqrt{2})x^2} + \left(7 + \frac{\pi x^2}{4}\right)e^{-x^2}\right)\right]^{1/2},$$

$$F_6(x) = \frac{1}{2}\left[1 - \frac{1}{2}\left(e^{-x^2/2} + \left(1 + \frac{\pi x^2}{12} - \frac{\pi^3(x^6 - 6x^4 + 5x^2)}{2880}\right)e^{-x^2}\right)\right]^{1/2},$$

$$F_7(x) = \frac{1}{2}\left[1 - \frac{1}{4}\left(e^{-x^2/2} + 2e^{-(2-\sqrt{2})x^2}\right.\right.$$
$$\left.\left. + \left(1 + \frac{\pi x^2}{24} - \frac{\pi^3(x^6 - 6x^4 + 5x^2)}{23040}\right)e^{-x^2}\right)\right]^{1/2},$$

$$F_8(x) = \frac{1}{2}\left[1 - \frac{1}{6}\left(e^{-x^2/2} + 2e^{-2(2-\sqrt{3})x^2} + 2e^{-2x^2/3}\right.\right.$$
$$\left.\left. + \left(1 + \frac{\pi x^2}{36} - \frac{\pi^3(x^6 - 6x^4 + 5x^2)}{77760}\right)e^{-x^2}\right)\right]^{1/2},$$

$$F_9(x) = \frac{1}{2}\left[1 - \frac{1}{70}\left(\left(19 - \frac{\pi^2 x^2}{192}\right)e^{-x^2/2} + 32e^{-(2-\sqrt{2})x^2}\right.\right.$$
$$\left.\left. + \left(19 + \pi x^2 + \frac{\pi^2(x^4 - 2x^2)}{48}\right)e^{-x^2}\right)\right]^{1/2},$$

$$F_{10}(x) = \frac{1}{2}\left[1 - \frac{1}{630}\left(\left(187 - \frac{3\pi^2 x^2}{32}\right)e^{-x^2/2} + 256e^{-(2-\sqrt{2})x^2}\right.\right.$$
$$\left.\left. + \left(187 + \frac{47\pi x^2}{4} + \frac{3\pi^2(x^4 - 2x^2)}{8} + \frac{\pi^3(x^6 - 6x^4 + 5x^2)}{192}\right)e^{-x^2}\right)\right]^{1/2},$$

$$F_{11}(x) = \frac{1}{2}\left[1 - \frac{1}{8}\left(e^{-x^2/2} + 3e^{-2(2-\sqrt{3})x^2} + 3e^{-2x^2/3} + e^{-x^2}\right)\right]^{1/2},$$

$$F_{12}(x) = \frac{1}{2}\left[1 - \frac{1}{80}\left(13e^{-x^2/2} + 27e^{-2(2-\sqrt{3})x^2} + 27e^{-2x^2/3}\right.\right.$$
$$\left.\left. + \left(13 + \frac{\pi x^2}{12}\right)e^{-x^2}\right)\right]^{1/2},$$

$$F_{13}(x) = \frac{1}{2}\left[1 - \frac{1}{2240}\left(\left(391 - \frac{\pi^2 x^2}{24}\right)e^{-x^2/2} + 729e^{-2(2-\sqrt{3})x^2} + 729e^{-2x^2/3}\right.\right.$$
$$\left.\left. + \left(391 + 13\pi x^2 + \frac{\pi^2(x^4 - 2x^2)}{6}\right)e^{-x^2}\right)\right]^{1/2}.$$

Many other approximations to $P(x)$ can be obtained through the scheme described in the next section and use of appropriate quadrature formulae.

## 3    Quadrature Formulae

We start this section with recalling some basic facts about quadratures (for more details, see, e.g., [3]). A quadrature formula $Q$ for approximate calculation of

$$I[f] = \int_a^b f(x)\,dx$$

is called any linear functional $Q$ of the form

$$Q[f] = \sum_{i=1}^n \sum_{j=0}^{\nu_i-1} a_{ij} f^{(j)}(x_i),$$

where $\{x_i\}_{i=1}^n$ are the nodes of $Q$, the natural numbers $\{\nu_i\}_{i=1}^n$ define their multiplicities, and $\{a_{ij}\}$ are the coefficients of $Q$. Usually it is assumed (and we assume here, too) that $a \le x_1 < x_2 < \cdots < x_n \le b$.

A quadrature formula $Q$ is said to have *algebraic degree of precision* $m$ (in short, $ADP(Q) = m$), if its remainder $R[Q; f]$,

$$R[Q; f] := I[f] - Q[f]$$

vanishes whenever $f$ is an algebraic polynomial of degree not exceeding $m$, and $R[Q; f] \ne 0$ if $deg(f) = m + 1$.

If $ADP(Q) = m \ge 0$ and $\max_i\{\nu_i\} = \nu$, then for every $r \in \mathbb{N}$, $\nu \le r \le m+1$, the *r-th Peano kernel of $Q$* is defined by

$$K_r[Q; t] := R\Big[Q; \frac{(\cdot - t)_+^{r-1}}{(r-1)!}\Big], \quad a \le t \le b,$$

where $x_+ = \max\{x, 0\}$. If $f^{(r)}$ is integrable on $[a, b]$, then the remainder $R[Q; f]$ admits the representation

$$R[Q; f] = \int_a^b K_r[Q; t] f^{(r)}(t)\,dt.$$

In particular, if $f \in C^r[a, b]$ and $\max_{x \in [a,b]} |f^{(r)}(x)| =: M_r[f]$, then

$$|R[Q; f]| \le M_r[f] \int_a^b |K_r[Q; t]|\,dt. \tag{8}$$

Let $Q_{(n)}$ be the compound quadrature formula, obtained by applying $Q$ to the $n$ subintervals $[a_i, a_{i+1}]$ $(i = 0, \ldots, n-1)$, where $a_i = a + i(b-a)/n$, and summing the results. Then we have

$$|R[Q_{(n)}; f]| \le \frac{M_r[f]}{n^r} \int_a^b |K_r[Q; t]|\,dt. \tag{9}$$

We list below the quadrature formulae we used for the derivation of the approximate formulae $\{F_i(x)\}$, along with the error bounds obtained through (8) and (9), which will be needed. Two classical Newton-Cotes quadrature formulae stand at the beginning of our list: the Simpson formula and the "3/8"-rule.

$$Q_1[f] = \frac{b-a}{6}\left[f(a) + 4f\left(\frac{a+b}{2}\right) + f(b)\right], \quad |R[Q_1; f]| \leq \frac{(b-a)^5}{2880}M_4[f],$$

$$Q_2[f] = \frac{b-a}{8}\left[f(a) + 3f\left(\frac{2a+b}{3}\right) + 3f\left(\frac{a+2b}{3}\right) + f(b)\right],$$
$$|R[Q_2; f]| \leq \frac{(b-a)^5}{6480}M_4[f].$$

The next two quadrature formulae are partial sums in the Euler-MacLaurin summation formula.

$$Q_3[f] = \frac{b-a}{2}[f(a) + f(b)] + \frac{(b-a)^2}{12}[f'(a) - f'(b)],$$
$$|R[Q_3; f]| \leq \frac{(b-a)^5}{720}M_4[f],$$

$$Q_4[f] = \frac{b-a}{2}[f(a)+f(b)] + \frac{(b-a)^2}{12}[f'(a) - f'(b)] - \frac{(b-a)^4}{720}[f'''(a)-f'''(b)],$$
$$|R[Q_4; f]| \leq 8.16 \times 10^{-4}(b-a)^5 M_4[f].$$

The remaining quadrature formulae in our list are of interpolatory type with equidistant abscissae and multiple end-point nodes. We mention that quadrature formula $Q_5$ has been already used by Bagby [2] to obtain formula $F_5(x)$.

$$Q_5[f] = \frac{b-a}{30}\left[7f(a) + 16f\left(\frac{a+h}{2}\right) + 7f(b)\right] + \frac{(b-a)^2}{60}[f'(a) - f'(b)],$$
$$|R[Q_5; f]| \leq \frac{(b-a)^5}{14580}M_4[f],$$

$$Q_6[f] = \frac{b-a}{70}\left[19f(a) + 32f\left(\frac{a+b}{2}\right) + 19f(b)\right] + \frac{(b-a)^2}{35}[f'(a) - f'(b)]$$
$$+ \frac{(b-a)^3}{840}[f''(a) + f''(b)],$$
$$|R[Q_6 ; f]| \leq 2.38 \times 10^{-7}(b-a)^7 M_6[f],$$

$$Q_7[f] = \frac{b-a}{630}\left[187f(a) + 256f\left(\frac{a+b}{2}\right) + 187f(b)\right]$$
$$+ \frac{47(b-a)^2}{1260}[f'(a) - f'(b)] + \frac{(b-a)^3}{420}[f''(a) + f''(b)]$$
$$+ \frac{(b-a)^4}{15120}[f'''(a) - f'''(b)],$$
$$|R[Q_7 ; f]| \leq 8.55 \times 10^{-8}(b-a)^7 M_6[f],$$

$$Q_8[f] \;=\; \frac{b-a}{80}\left[13f(a) + 27f\left(\frac{2a+b}{3}\right) + 27f\left(\frac{a+2b}{3}\right) + 13f(b)\right]$$
$$+\frac{(b-a)^2}{120}[f'(a) - f'(b)],$$

$$|R[Q_8 ; f]| = \frac{(b-a)^5}{73728}M_4[f],$$

$$Q_9[f] \;=\; \frac{b-a}{2240}\left[391f(a) + 729f\left(\frac{2a+b}{3}\right) + 729f\left(\frac{a+2b}{3}\right) + 391f(b)\right]$$
$$+\frac{13(b-a)^2}{1120}[f'(a) - f'(b)] + \frac{(b-a)^3}{3360}[f''(a) + f''(b)],$$

$$|R[Q_9 ; f]| \le 2.77 \times 10^{-8}(b-a)^7 M_6[f].$$

We point out that the error estimates for quadrature formulae $Q_4$, $Q_6$, $Q_7$ and $Q_9$ are obtained numerically, as there is no explicit expression for the zeros of the corresponding Peano kernels. Although some of quadratures above possess Peano kernels of higher order, our error estimates are obtained only through their fourth or sixth Peano kernels. The reason is that for $k > 6$ the expression for $g^{(k)}$ is too cumbersome, and it is difficult to find good estimates for $M_k[g]$.

Our approximate formulae $\{F_i(x)\}_{i=1}^{13}$ for the normal probability integral $P(x)$ are obtained from (4), where the integral

$$\int_0^{\pi/4} g(\varphi)\, d\varphi, \quad g(\varphi) = e^{-\frac{x^2}{2\cos^2\varphi}},$$

is replaced by some of the quadrature formulae above (with $a = 0$ and $b = \pi/4$). The correspondence between $\{F_i\}$ and $\{Q_j\}$ is shown in Table 1.

**Table 1.** The correspondence between $\{F_i\}$ and $\{Q_j\}$

| $F_1$ | $F_2$ | $F_3$ | $F_4$ | $F_5$ | $F_6$ | $F_7$ | $F_8$ | $F_9$ | $F_{10}$ | $F_{11}$ | $F_{12}$ | $F_{13}$ |
|---|---|---|---|---|---|---|---|---|---|---|---|---|
| $Q_3$ | $Q_{3(2)}$ | $Q_{3(3)}$ | $Q_1$ | $Q_5$ | $Q_4$ | $Q_{4(2)}$ | $Q_{4(3)}$ | $Q_6$ | $Q_7$ | $Q_2$ | $Q_8$ | $Q_9$ |

## 4   Estimates for $M_k[g]$

As we are going to apply the error estimates for the quadrature formulae in the previous section, we need estimates for the magnitude of the derivatives of $g$. To this end, we substitute

$$c := x^2 \ (c \ge 0), \quad u := t^2 + 1 = \frac{1}{\cos^2\varphi} \ (1 \le u \le 2).$$

By using formulae (5), (6) and (7), we obtain for $k = 2, 4, 6$ the representation

$$g^{(k)}(\varphi) = e^{-cu/2}h_k(c, u), \quad \text{where } h_k(c, u) = \sum_{i=1}^{k}(-1)^k d_{ik}(u)(cu)^k.$$

The coefficients $\{d_{ik}(u)\}$ are polynomials in $u$, and each of them is monotonically increasing for $u \in [1, 2]$. We describe below the way we estimate $g^{(k)}$.

## 4.1   The Case $k = 4$

The coefficients $\{d_{i4}(u)\}$ are:     $d_{44}(u) = (u - 1)^2$, $d_{34}(u) = 6(3u^2 - 5u + 2)$, $d_{24}(u) = 75u^2 - 100u + 28$ and $d_{14}(u) = 60u^2 - 60u + 8$. We divide the interval $[1, 2]$ into 40 subintervals $[u_j, u_{j+1}]$, with $u_j = 1 + j/40$ ($j = 0, \ldots, 39$). Making use of the monotonicity of coefficients $\{d_{i4}(u)\}$, we obtain the inequality

$$M_4[g]$$
$$\leq \max_{0 \leq j \leq 39} \Big\{ \max_{\tau \geq 0}\{e^{-\tau/2}[d_{44}(u_{j+1})\tau^4 - d_{34}(u_j)\tau^3 + d_{24}(u_{j+1})\tau^2 - d_{14}(u_j)\tau]\},$$
$$- \min_{\tau \geq 0}\{e^{-\tau/2}[d_{44}(u_j)\tau^4 - d_{34}(u_{j+1})\tau^3 + d_{24}(u_j)\tau^2 - d_{14}(u_{j+1})\tau]\} \Big\}.$$

From $\tau = x^2 u \leq 2x^2$ we also obtain

$$M_4[g] \leq 2x^2$$
$$\times \max_{0 \leq j \leq 39} \Big\{ \max_{\tau \geq 0}\{e^{-\tau/2}[d_{44}(u_{j+1})\tau^3 - d_{34}(u_j)\tau^2 + d_{24}(u_{j+1})\tau - d_{14}(u_j)]\},$$
$$- \min_{\tau \geq 0}\{e^{-\tau/2}[d_{44}(u_j)\tau^3 - d_{34}(u_{j+1})\tau^2 + d_{24}(u_j)\tau - d_{14}(u_{j+1})]\} \Big\}.$$

The examination of the univariate functions in the right-hand sides of the above inequalities yields the estimates

$$M_4[g] \leq 54.3 \quad \text{and} \quad M_4[g] \leq 256x^2.$$

## 4.2   The Case $k = 6$

The coefficients $\{d_{i6}(u)\}_{i=1}^6$ in this case are

$$d_{66}(u) = (u - 1)^3, \quad d_{56}(u) = 15(u - 1)^2(3u - 2),$$

$$d_{46} = 5(u - 1)(129u^2 - 172u + 52), \quad d_{36}(u) = 15(231u^3 - 462u^2 + 280u - 48),$$

$$d_{26} = 4(1575u^3 - 2625u^2 + 1204u - 124), \quad d_{16}(u) = 8(315u^3 - 420u^2 + 126u - 4).$$

We divide the interval $[1, 2]$ into two hundred subintervals of equal length and apply the same idea as in the previous case to reduce the estimation of $M_6[g]$ to examination of univariate functions. The resulting estimates are

$$M_6[g] \leq 4183 \quad \text{and} \quad M_6[g] \leq 17408x^2.$$

Although we do not use estimation with the second order Peano kernel, for the sake of completeness we quote the estimates for $M_2[g]$, obtained in the same way as above (by dividing $[1, 2]$ into ten subintervals of equal length):

$$M_2[g] \leq 1.93 \quad \text{and} \quad M_2[g] \leq 8x^2.$$

## 5  Error Estimates

To obtain error bounds independent on $x$, we exploit the two kind of estimates for $M_k[g]$ from the previous section. A general observation we shall use is that $P(x)/x$ and $F_i(x)/x$, $i = 1, \ldots, 13$, are decreasing functions of $x$ for $x \geq 0$. The verification is straightforward and therefore is omitted. Below we demonstrate the way we prove our error estimates on two of our approximate formulae .

Consider formula $F_1(x)$. We have

$$|P(x) - F_1(x)| = \frac{|R[Q_3; g]|}{\pi(P(x) + F_1(x))} \leq \frac{\pi^4 \times M_4[g]}{4^5 \times 720 \times (P(x) + F_1(x))}.$$

We distinguish between two cases. If $0 < x < 0.461$, then we apply the estimate $M_4[g] \leq 256x^2$ to obtain

$$|P(x) - F_1(x)| \leq \frac{\pi^4 \times 256 \times 0.461^2}{4^5 \times 720 \times (P(0.461) + F_1(0.461))} < 2.04 \times 10^{-2}.$$

If $x \geq 0.461$, then we make use of the estimate $M_4[g] \leq 54.3$ to obtain

$$|P(x) - F_1(x)| \leq \frac{\pi^4 \times 54.3}{4^5 \times 720 \times (P(0.461) + F_1(0.461))} < 2.03 \times 10^{-2}.$$

Thus, we obtained the estimate $|P(x) - F_1(x)| < 2.04 \times 10^{-2}$, valid for every $x \geq 0$. The observed maximum of $|P(x) - F_1(x)|$ is approximately $1.82 \times 10^{-3}$, i.e., we have overestimated the sharp error bound about 11 times.

Our second error estimate is for formula $F_{13}(x)$. We have

$$|P(x) - F_{13}(x)| = \frac{|R[Q_9; g]|}{\pi(P(x) + F_{13}(x))} \leq \frac{\pi^6 \times 2.77 \times 10^{-8} M_6[g]}{4^7 \times (P(x) + F_{13}(x))}.$$

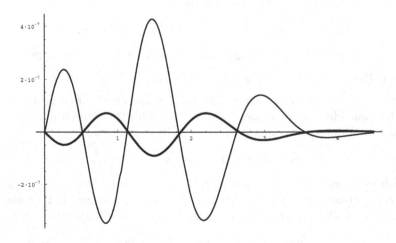

**Fig. 1.** Graphs of $P(x) - F_{10}(x)$ (thin line) and $P(x) - F_{13}(x)$ (thick line)

**Table 2.** Error estimates and observed maximal error of $\{F_i\}$

| Formula | $F_1$ | $F_2$ | $F_3$ | $F_4$ | $F_5$ |
|---|---|---|---|---|---|
| Error bound | $2.04 \times 10^{-2}$ | $1.27 \times 10^{-3}$ | $2.5 \times 10^{-4}$ | $5.06 \times 10^{-3}$ | $1.0 \times 10^{-3}$ |
| $\max\|P(x) - F_i(x)\|$ | $1.82 \times 10^{-3}$ | $1.4 \times 10^{-4}$ | $2.9 \times 10^{-5}$ | $4.2 \times 10^{-4}$ | $3.04 \times 10^{-5}$ |
| Formula | $F_6$ | $F_7$ | $F_8$ | $F_9$ | $F_{10}$ |
| Error bound | $1.2 \times 10^{-2}$ | $7.44 \times 10^{-4}$ | $1.47 \times 10^{-4}$ | $1.56 \times 10^{-4}$ | $5.59 \times 10^{-5}$ |
| $\max\|P(x) - F_i(x)\|$ | $7.11 \times 10^{-4}$ | $1.63 \times 10^{-5}$ | $1.59 \times 10^{-6}$ | $3.24 \times 10^{-6}$ | $4.26 \times 10^{-7}$ |
| Formula | $F_{11}$ | $F_{12}$ | $F_{13}$ | $(F_{10} + 5F_{13})/6$ | — |
| Error bound | $2.25 \times 10^{-3}$ | $1.98 \times 10^{-4}$ | $1.81 \times 10^{-5}$ | — | — |
| $\max\|P(x) - F_i(x)\|$ | $1.95 \times 10^{-4}$ | $7.33 \times 10^{-6}$ | $8.97 \times 10^{-8}$ | $9.1 \times 10^{-9}$ | — |

If $0 < x < 0.49$, then

$$|P(x) - F_{13}(x)| \leq \frac{\pi^6 \times 2.77 \times 10^{-8} \times 17408 \times 0.49^2}{4^7 \times (P(0.49) + F_{13}(0.49))} < 1.81 \times 10^{-5}.$$

The same estimate holds true in the case $x \geq 0.49$, since

$$|P(x) - F_{13}(x)| \leq \frac{\pi^6 \times 2.77 \times 10^{-8} \times 4183}{4^7 \times (P(0.49) + F_{13}(0.49))} < 1.81 \times 10^{-5}.$$

The observed maximum of $|P(x) - F_{13}(x)|$ does not exceed $8.97 \times 10^{-8}$, therefore the overestimation factor is about 200.

The error bounds for the remaining approximate formulae are obtained in the same way. These bounds, as well as the observed maximal errors are given in Table 2. Notice that all the formulae $\{F_i(x)\}_{i=1}^{13}$ tend to the right value of $P(x)$ as $x \to 0$ or $x \to \infty$. Figure 1 depicts the actual error of formulae $F_{10}$ and $F_{13}$. It is seen that for most values of $x$ the two errors are of opposite sign. This implies the empirical formula $F_{14}(x) = (F_{10}(x) + 5F_{13}(x))/6$, with observed maximal error $9.1 \times 10^{-9}$.

**Acknowledgments.** The second named author was supported in part by the Swiss National Science Foundation (SCOPES Joint Research Project no. IB7320–111079 "New Methods for Quadrature"), and by the Sofia University Research Foundation through Contract 86/2006.

# References

1. Abramowitz, M., Stegun, I.: Handbook of Mathematical Functions, Dover New York 1972
2. Bagby, R. J.: Calculating Normal Probability. Amer. Math. Monthly, **102(1)** (1995) 46–49
3. Brass, H.: Quadraturverfahren. Vandenhoeck & Ruprecht, Göttingen (1975)
4. Korn, G. A., Korn, T. M.: Mathhematical Handbook for Scientist and Engineers, MGraw–Hill Book Company, New York Toronto London (1961)
5. Gradsteyn, I. S., Ryznik, I. M.: Tables of Integrals, Sums, Series and Products, 4th edn. Academic Press, New York (1980)

# Iterative-Collocation Method for Integral Equations of Heat Conduction Problems

Lechosław Hącia

Institute of Mathematics, Faculty of Electrical Engineering,
Poznan University of Technology, Piotrowo 3a, 60-965 Poznan, Poland
lhacia@math.put.poznan.pl

**Abstract.** The integral equations studied here play very important role in the theory of parabolic initial-boundary value problems (heat conduction problems) and in various physical, technological and biological problems (epidemiology problems). This paper is concerned with the iterative-collocation method for solving these equations. We propose an iterative method with corrections based on the interpolation polynomial of spatial variable of the Lagrange type with given collocation points. The coefficients of these corrections can be determined by a system of Volterra integral equations. The convergence of the presented algorithm is proved and an error estimate is established. The presented theory is illustrated by numerical examples and a comparison is made with other methods.

**Keywords:** integral equations in space-time, iterative-collocation method, interpolating polynomial, correction function.

**AMS Subject Classification:** 65R20.

A mathematical model in the heat-conduction theory is reduced to the following integral equations in space-time [4]

$$u(x,t) = f(x,t) + \int_0^t \int_M k(x,t,y,s)u(y,s)dyds \,. \tag{1}$$

We shall consider that integral equation in the space-time domain, where $f$ is a given function in domain $D = M \times [0,T]$ ($M$ - a compact subset of m-dimensional Euclidean space) and $u$ is unknown function in $D$. The given kernel $k$ is defined in domain $\Omega = \{(x,t,y,s) : x,y \in M, 0 \le s \le t \le T\}$. Numerical methods for these equations were presented in papers [1-5]. In this paper we propose a new numerical method for equation

$$u = f + Ku \,, \tag{2}$$

where

$$(Ku)(x,t) = \int_0^t \int_a^b k(x,t,y,s)u(y,s)dyds \,. \tag{3}$$

T. Boyanov et al. (Eds.): NMA 2006, LNCS 4310, pp. 378–385, 2007.

# 1    Iterative-Collocation Method with a System of Volterra Equations

From a numerical point of view, the method of successive approximations is of limited use. We can introduce some correction c to obtain the following method

$$u_k = f + K(u_{k-1} + c),$$

where $c$ can be defined in various ways.

The method presented here is restricted to the following integral equation

$$u(x,t) = f(x,t) + \int_0^t \int_a^b k(x,t,y,s)u(y,s)dyds, \qquad (4)$$

where the given functions $f$ and $k$ are defined in domains $D = [a,b] \times [0,T]$ and $\Omega = \{(x,t,y,s) : a \le x, y \le b, 0 \le s \le t \le T\}$, respectively.

We seek for approximate solution of equation (4) of the form

$$u_k^n(x,t)f(x,t) + \int_0^t \int_a^b k(x,t,y,s)[u_{k-1}^n(y,s) + p_k^n(y,s)]dyds, \ k = 1,2,\ldots \ (5)$$

where

$$p_k^n(y,t) = \sum_{j=0}^n a_{jk}(t)\varphi_j(x), \qquad (6)$$

$u_0^n$ is any function defined on $[a,b] \times [0,T]$, and $\varphi_j$ are some basis functions. The unknown coefficients $a_{jk}$ must satisfy the following conditions:

$$p_k^n(x_j,t) - \Delta^n u_k(x_j,t), \ k = 1,2,\ldots \qquad (7)$$

and

$$\Delta^n u_k(x_j,t) = u_k^n(x_j,t) - u_{k-1}^n(x_j,t), \ j = 0,1,\ldots,n,$$

where $x_j$ are collocation points such that

$$\varphi_i(x_j) = \begin{cases} 1 ,i = j \\ 0 ,i \neq j. \end{cases}$$

The basis functions $\varphi_i$ can be defined as Lagrange fundamental polynomials

$$l_i(x)\frac{(x-x_0)\ldots(x-x_{i-1})(x-x_{i+1})\ldots(x-x_n)}{(x_i-x_0)\ldots(x_i-x_{i-1})(x_i-x_{i+1})\ldots(x_i-x_n)}, \qquad i = 0,1,\ldots,n.$$

From (6) and (7) we get

$$p_k^n(x_j,t) = a_{jk}(t). \qquad (8)$$

Then

$$p_k^n(x,t) = \sum_{j=0}^{n} p_k^n(x_j, t)\varphi_j(x)$$

is a collocation polynomial of the Lagrange type with respect to variable $x$ for almost every $t \in [0, T]$. From (6), (7) and (4) we get

$$a_{jk}(t)g_{jk}(t) + \int_0^t \int_a^b k(x_j, t, y, s)p_k^n(y, s)dyds, \tag{9}$$

where

$$g_{jk}(t) \int_0^t \int_a^b k(x_j, t, y, s)[u_{k-1}^n(y, s) - u_{k-2}^n(y, s) - p_{k-1}^n(y, s)]dyds.$$

With the notation

$$c_{ij}(t, s) = \int_a^b k(x_j, t, y, s)\varphi_i(y)dy \tag{10}$$

we obtain the system of linear integral equations of Volterra type of the second kind

$$a_{jk}(t)g_{jk}(t) + \sum_{i=0}^{n} \int_0^t c_{ij}(t, s)a_{ij}(s)ds \tag{11}$$

with unknown functions $a_{0k}, a_{1k}, \ldots, a_{nk}$ and given $g_{0k}, g_{1k}, \ldots, g_{nk}$ and $c_{ij}$ $(i, j = 0, 1, \ldots, n)$. We introduce also notations

$$V_k(y, s) = u(y, s) - u_{k-1}^n(y, s) - p_k^n(y, s), \quad k = 1, 2, \ldots,$$

$$k_n(x, t, y, s) = \sum_{j=0}^{n} k(x_j, t, y, s)\varphi_j(x),$$

where the points $x_j$ $(j = 0, 1, \ldots, n)$ are the roots of the $(n+1)$-th orthogonal polynomial. Let $R$ be the space of the Riemann integrable functions on $D$ with a norm

$$\|u\|_R = \sup_{(x,t)\in D} \{|u(x,t)|\}$$

and let $L_w^2$ be the $L_2$-space associated with a bounded positive weight function $w(x)$, equipped with norms

$$\|u\|_{L_w^2} = \left(\int_0^T \int_a^b |u(x,t)|^2 w(x)dxdt\right)^{1/2} \quad \text{and} \quad \|u\|_{L^2} = \left(\int_0^T \int_a^b |u(x,t)|^2 dxdt\right)^{1/2}.$$

**Remark** [4]. The integral equation (4) has a unique solution for $g \in R(D)$ and $k \in C(\Omega)$. It is of the form

$$u(x,t) = g(x,t) + \int_0^t \int_a^b r(x,t,y,s)g(y,s)dyds,$$

where $r(x,t,y,s) = \sum_{n=1}^{\infty} k^{(n)}(x,t,y,s)$, and the iterated kernels are defined by

$$k^{(1)}(x,t,y,s) = k(x,t,y,s),$$
$$k^{(n)}(x,t,y,s) = \int_0^t \int_a^b k(x,t,z,w)k^{(n-1)}(z,w,y,s)dzdw \text{ for } n = 2,3,\ldots$$

**Lemma 1 ([6]).** *If* $u(x) \in R[a,b]$, *then*

$$\|p^n u - u\| = \left[ \int_a^b |p^n u(x) - u(x)|^2 w(x)dx \right]^{\frac{1}{2}} \to 0 \quad as \quad n \to \infty,$$

*where*

$$p^n u(x) = \sum_{i=0}^{n} u(x_j)\varphi_i(x)$$

*is the Lagrange interpolating polynomial with knots* $x_i$ $(i = 0,1,2,\ldots,n)$ *being the zeros of some orthogonal polynomials with respect to the weight-function* $w(x)$ *in* $[a,b]$, *and* $\varphi_i(x)$ *are Lagrange's basis functions.*

**Theorem 1.** *Let the following assumptions be satisfied:*

*a) The points* $x_i$ $(i = 0,1,2,\ldots,n)$ *are zeros of an orthogonal polynomial on* $[a,b]$ *with respect to the weight-function* $w(x)$;
*b)* $\Delta u_k$ *is a Riemann integrable function in* $D$.

*Then*

$$\|p_k^n - \Delta u_k\|_{L_w^2} \to 0 \quad as \quad n \to \infty,$$

*where*

$$p_k^n(x,t) = \sum_{i=0}^{n} \Delta u_k(x_i,t)l_i(x)$$

*is the Lagrange interpolating polynomial with respect to variable* $x$ *built on the knots* $x_i$ *and the basis functions* $l_i(x)$ $(i = 0,1,2,\ldots,n)$.

**Proof.** By Lemma 1 we obtain

$$\|p_k^n(\cdot,t) - \Delta u_k(\cdot,t)\| \to 0 \quad as \quad n \to \infty \quad \text{for almost every } t \in [0,T].$$

Introducing the notation

$$\|p_k^n(\cdot,t) - \Delta u_k(\cdot,t)\|^2 = W_{k,n}^2(t),$$

we get

$$W_{k,n}^2(t) = \int_a^b |p_k^n(x,t) - \Delta u_k(x,t)|^2 w(x)dx \to 0 \quad \text{as} \quad n \to \infty.$$

Then (see [7])

$$\int_0^T W_{k,n}^2(t)dt = \int_0^T \int_a^b |p_k^n(x,t) - \Delta u_k(x,t)|^2 w(x)dxdt \to 0,$$

and

$$\|W_{k,n}\|_{L^2}^2 = \|p_k^n - \Delta u_k\|_{L_w^2}^2 \to 0 \quad \text{as} \quad n \to \infty.$$

Hence

$$\|p_k^n - \Delta u_k\|_{L_w^2}^2 \to 0 \quad \text{as} \quad n \to \infty. \qquad \square$$

**Theorem 2.** *If $u \in R(D)$ and*

$$C = \sup_{(x,t)\in D} \int_0^t \int_a^b \frac{|k(x,t,y,s)|^2}{w(y)} dy ds < \infty,$$

*then*

$$q_n = \frac{K_n M}{1 - M K_n} \longrightarrow 0 \quad as \quad n \to \infty,$$

*where $M = 1 + N$,*

$$N = \int_0^T \int_a^b \int_0^t \int_a^b \frac{|r(x,t,y,s)|^2}{w(y)} w(x)dydsdxdt,$$

$$K_n = \int_0^T \int_a^b \int_0^t \int_a^b \frac{|k_n(x,t,y,s) - k(x,t,y,s)|^2}{w(y)} w(x)dydsdxdt,$$

*and the sequence $\{u_k^n\}$ defined by formula (5) tends to the unique solution of equation (4). Moreover, the following estimates hold true:*
*a)*

$$\|u_k^n - u\|_R \le C q_n^{k-1} \|V_1\|_{L_w^2},$$

*b)*

$$\|u_k^n - u\|_{L_w^2} \le q_n^{k-1} L \|V_1\|_{L_w^2},$$

*where*

$$L = \int_0^T \int_a^b \int_0^t \int_a^b \frac{|k(x,t,y,s)|^2}{w(y)} w(x)dydsdxdt.$$

**Proof.** From the presented method we get

$$p_k^n(x,t) = \int_0^t \int_a^b k_n(x,t,y,s) \left[ u_{k-1}^n(y,s) - u_{k-2}^n(y,s) \right.$$
$$\left. + p_k^n(y,s) - p_{k-1}^n(y,s) \right] dyds \,.$$

After some calculations we find that $V_k$ satisfies the Volterra-Fredholm integral equation

$$V_k(x,t) = g_k(x,t) + \int_0^t \int_a^b k(x,t,y,s) \cdot V_k(y,s) dyds \,,$$

where $g_k$ is defined by the formula:

$$g_k(x,t) = \int_0^t \int_a^b \left[ k_n(x,t,y,s) - k(x,t,y,s) \right] \left[ V_k(y,s) - V_{k-1}(y,s) \right] dyds \,.$$

According to the remark above, the solution of the last equation we can written in the form

$$V_k(x,t) = g_k(x,t) + \int_0^t \int_a^b r(x,t,y,s) \cdot g_k(y,s) dyds \,,$$

where $r$ is a resolvent kernel of $k$.
Then

$$\|V_k\|_{L_w^2} \le \|g_k\|_{L_w^2} (1 + N) \,.$$

Hence

$$\|g_k\|_{L_w^2} \le (\|V_k\|_{L_w^2} + \|V_{k-1}\|_{L_w^2}) K_n \,,$$

and

$$\|V_k\|_{L_w^2} \le q_n \|V_{k-1}\|_{L_w^2} \,.$$

By induction we obtain

$$\|V_k\|_{L_w^2} \le q_n^{k-1} \|V_1\|_{L_w^2}$$

and hence, the error estimates. The convergence of the method follows from Theorem 1, since $K_n \to 0$ and $q_n \to 0$ as $n \to \infty$.    $\square$

## 2    Numerical Experiments

The corrections $p_k^n$ can be determined from the system of Volterra integral equations (11), solvable by quadrature formulas. The tables below contain the relative errors $\delta$ at points $t_j$ and $x_i$:

$$\delta = \left| \frac{u_k^n(x_i,t_j) - u(x_i,t_j)}{u(x_i,t_j)} \right| \,.$$

Here, $n + 1$ means the number of basis functions and $k$ is the a number of iterations.

**Example 1**

$$u(x,t)\mathrm{e}^{-t}x^2 - \frac{2}{3}t^3x^2 + \int_0^t \int_{-1}^1 x^2 t^2 \mathrm{e}^s u(y,s)\,dy\,ds$$

$n = 5, k = 3$

| $t$ | $x$ | | |
|---|---|---|---|
| | $\pm 1$ | $\pm 0.4$ | $\pm 0.2$ |
| 0.1 | $-0.14751 \cdot 10^{-12}$ | $-0.14751 \cdot 10^{-12}$ | $-0.14751 \cdot 10^{-12}$ |
| 0.2 | $-0.14581 \cdot 10^{-9}$ | $-0.14581 \cdot 10^{-9}$ | $-0.14581 \cdot 10^{-9}$ |
| 0.3 | $-0.12935 \cdot 10^{-7}$ | $-0.129357 \cdot 10^{-7}$ | $-0.129357 \cdot 10^{-7}$ |
| 0.4 | $-0.401118 \cdot 10^{-6}$ | $-0.401118 \cdot 10^{-6}$ | $-0.401118 \cdot 10^{-6}$ |
| 0.5 | $-0.677362 \cdot 10^{-5}$ | $-0.67736 \cdot 10^{-5}$ | $-0.677362 \cdot 10^{-5}$ |
| 0.6 | $-0.000076935$ | $-0.000076935$ | $-0.000076935$ |
| 0.7 | $-0.00066408$ | $-0.00066408$ | $-0.00066408$ |
| 0.8 | $-0.004724257$ | $-0.004724257$ | $-0.004724257$ |
| 0.9 | $-0.029458851$ | $-0.029458851$ | $-0.029458851$ |
| 1 | $-0.1700621$ | $-0.1700621$ | $-0.1700621$ |

$n = 5, k = 5$

| $t$ | $x$ | | |
|---|---|---|---|
| | $\pm 1$ | $\pm 0.4$ | $\pm 0.2$ |
| 0.1 | $-0.40061 \cdot 10^{-19}$ | $-0.40061 \cdot 10^{-19}$ | $-0.40061 \cdot 10^{-19}$ |
| 0.2 | $-0.77735 \cdot 10^{-15}$ | $-0.77735 \cdot 10^{-15}$ | $-0.77735 \cdot 10^{-15}$ |
| 0.3 | $-0.4406 \cdot 10^{-12}$ | $-0.4406 \cdot 10^{-12}$ | $-0.4406 \cdot 10^{-12}$ |
| 0.4 | $-0.56572 \cdot 10^{-10}$ | $-0.56572 \cdot 10^{-10}$ | $-0.56572 \cdot 10^{-10}$ |
| 0.5 | $-0.31376 \cdot 10^{8}$ | $-0.31376 \cdot 10^{-8}$ | $-0.31376 \cdot 10^{-8}$ |
| 0.6 | $-0.10109 \cdot 10^{-6}$ | $-0.101099 \cdot 10^{-6}$ | $-0.101099 \cdot 10^{-6}$ |
| 0.7 | $-0.22403 \cdot 10^{-5}$ | $-0.22403 \cdot 10^{-5}$ | $-0.22403 \cdot 10^{-5}$ |
| 0.8 | $-0.00003824$ | $-0.00003824$ | $-0.00003824$ |
| 0.9 | $-0.00055116$ | $-0.00055116$ | $-0.00055116$ |
| 1 | $-0.00731353$ | $-0.00731353$ | $-0.00731353$ |

$n = 5, k = 8$

| $t$ | $x$ | | |
|---|---|---|---|
| | $\pm 1$ | $\pm 0.4$ | $\pm 0.2$ |
| 0.1 | $-0.44206 \cdot 10^{-29}$ | $-0.30699 \cdot 10^{-29}$ | $-0.27629 \cdot 10^{-29}$ |
| 0.2 | $-0.66183 \cdot 10^{-23}$ | $-0.66183 \cdot 10^{-23}$ | $-0.66183 \cdot 10^{-23}$ |
| 0.3 | $-0.56654 \cdot 10^{-19}$ | $-0.56654 \cdot 10^{-19}$ | $-0.56654 \cdot 10^{-19}$ |
| 0.4 | $-0.54062 \cdot 10^{-16}$ | $-0.54062 \cdot 10^{-16}$ | $-0.54062 \cdot 10^{-16}$ |
| 0.5 | $-0.14921 \cdot 10^{-13}$ | $-0.14921 \cdot 10^{-13}$ | $-0.14921 \cdot 10^{-13}$ |
| 0.6 | $-0.17824 \cdot 10^{-11}$ | $-0.17824 \cdot 10^{-11}$ | $-0.17824 \cdot 10^{-11}$ |
| 0.7 | $-0.107139 \cdot 10^{-9}$ | $-0.107139 \cdot 10^{-9}$ | $-0.107139 \cdot 10^{-9}$ |
| 0.8 | $-0.24428 \cdot 10^{-8}$ | $-0.24428 \cdot 10^{-8}$ | $-0.24428 \cdot 10^{-8}$ |
| 0.9 | $0.15146 \cdot 10^{-6}$ | $0.15146 \cdot 10^{-6}$ | $0.15146 \cdot 10^{-6}$ |
| 1 | $0.0000212474$ | $0.0000212474$ | $0.0000212474$ |

In this example the presented method gives higher accuracy than Galerkin method and Galerkin-Fourier method (see [2, 3]).

**Example 2**

$$u(x,t)x\sin(t) + \frac{1}{3}x^2t^2e^2\cos(t) - \frac{1}{3}x^2t^2e^2\sin(t) - \frac{1}{3}x^2t^2 + \int\limits_{-1}^{1}\int\limits_{0}^{t} x^2yt^2e^su(y,s)dyds$$

$n = 5, k = 3$

| t | x | | |
|---|---|---|---|
| | ±1 | ±0.4 | ±0.2 |
| 0.1 | −0.00001168 | −0.00007011 | −0.00002337 |
| 0.2 | −0.00004918 | −0.00002951 | −0.00009837 |
| 0.3 | −0.00011649 | −0.00002329 | −0.00002329 |
| 0.4 | −0.00021800 | −0.00013082 | −0.00004360 |
| 0.5 | −0.00035848 | −0.00021509 | −0.00007169 |
| 0.6 | −0.00054295 | −0.00032577 | −0.00010859 |
| 0.7 | −0.00077656 | −0.00046593 | −0.00015531 |
| 0.8 | −0.00106430 | −0.00063858 | −0.00021286 |
| 0.9 | −0.00141010 | −0.00084606 | −0.00028202 |
| 1 | −0.00180881 | −0.00108529 | −0.00036176 |

Increasing $n$ and $k$ gives better results.

## Acknowledgments

The author is thankful to the reviewer for the useful remarks which contributed for improving this paper.

## References

1. Brunner, H.: On the numerical solution of nonlinear solution of Volterra equations by collocation methods. SIAM J. Numer. Anal. **27** (1990) 987–1000
2. Hącia, L.: Projection methods for integral equations in epidemic. Mathematical Modelling and Analysis **7(2)** (2002) 229–240
3. Hącia, L.: Computational methods for Volterra-Fredholm integral equations. Computational Methods in Science and Technology **8(2)** (2002) 13-26
4. Hącia, L.: Integral equations in some problems of electrotechnics. Plenary Lecture - 4-th International Conference APLIMAT 2005 153-164, Bratislava 2005
5. Kauthen, J. P.: Continuous time collocation method for Volterra-Fredholm integral equations. Numer. Math. **56** (1989) 409-424
6. Lucev, E.M.: Collocation-iterative method of solving linear integral equations of the second kind. Investigation methods for differential and integral equations. Kiev (1989) 132-138 (in Russian)
7. Piskorek, A.: Integral equations - theory and applications. PWN, Warsaw (1997) [in Polish]

# Numerical Computation of the Markov Factors for the Systems of Polynomials with the Hermite and Laguerre Weights*

Lozko Milev

Department of Mathematics, University of Sofia,
Blvd. James Bourchier 5, 1164 Sofia, Bulgaria
milev@fmi.uni-sofia.bg

**Abstract.** Denote by $\pi_n$ the set of all real algebraic polynomials of degree at most $n$, and let $U_n := \{e^{-x^2}p(x) : p \in \pi_n\}$, $V_n := \{e^{-x}p(x) : p \in \pi_n\}$. It was proved in [9] that $M_k(U_n) := \sup\{\|u^{(k)}\|_{\mathbb{R}}/\|u\|_{\mathbb{R}} : u \in U_n, u \not\equiv 0\} = \|u_{*,n}^{(k)}\|_{\mathbb{R}}$, $\forall n, k \in \mathbb{N}$, and $M_k(V_n) := \sup\{\|v^{(k)}\|_{\mathbb{R}_+}/\|v\|_{\mathbb{R}_+} : v \in V_n, v \not\equiv 0\} = \|v_{*,n}^{(k)}\|_{\mathbb{R}_+}$, where $\|\cdot\|_{\mathbb{R}}$ ($\|\cdot\|_{\mathbb{R}_+}$) is the supremum norm on $\mathbb{R}$ ($\mathbb{R}_+ := [0, \infty)$) and $u_{*,n}$ ($v_{*,n}$) is the Chebyshev polynomial from $U_n$ ($V_n$). We prove here the convergence of an algorithm for the numerical construction of the oscillating weighted polynomial from $U_n$ ($V_n$), which takes preassigned values at its extremal points. As an application, we obtain numerical values for the Markov factors $M_k(U_n)$ and $M_k(V_n)$ for $1 \le n \le 10$ and $1 \le k \le 5$.

## 1 Introduction

Denote by $\pi_n$ the set of all real algebraic polynomials of degree not exceeding $n$, and by $\|\cdot\|_I$ the supremum norm in a given interval $I \subseteq \mathbb{R}$, $\|f\|_I := \sup_{x \in I} |f(x)|$.

Markov-type inequalities on $\mathbb{R}$ have the form

$$\|(\mu p)'\| \le C_n(\mu)\|\mu p\|, \quad \forall p \in \pi_n,$$

where $\mu$ is a weight function and $\|\cdot\|$ is a norm. In connection with the research in the field of the weighted approximation by polynomials, Markov-type inequalities have been proved for various weights and norms. See [2], [3], [6], [10], and the references therein.

In the case of supremum norm, exact Markov-type inequalities for the weight functions $\mu(x) = e^{-x^2}$ on $\mathbb{R}$ and $\mu(x) = e^{-x}$ on $\mathbb{R}_+ := [0, \infty)$, were proved in [7] and [4] respectively.

Recently, the above-mentioned results were extended in [9] to derivatives of arbitrary order. Let $U_n := \{e^{-x^2}p(x) : p \in \pi_n\}$ and $V_n := \{e^{-x}p(x) : p \in \pi_n\}$. We

---

* The research was supported by the Bulgarian Ministry of Education and Science under Contract MM-1402/2004.

T. Boyanov et al. (Eds.): NMA 2006, LNCS 4310, pp. 386–393, 2007.

use the notation $u_{*,n}$ $(v_{*,n})$ for the Chebyshev polynomial from $U_n$ $(V_n)$. It is known that $u_{*,n}$ is the unique polynomial from $U_n$, which has norm equal to one and there exist $n+1$ points $t_0^* < \cdots < t_n^*$ such that $u_{*,n}(t_k^*) = (-1)^{n-k}$. It was proved in [9] that

$$\|u^{(k)}\|_{\mathbb{R}} \leq \|u_{*,n}^{(k)}\|_{\mathbb{R}} \|u\|_{\mathbb{R}}, \quad \forall n, k \in \mathbb{N}, \ \forall u \in U_n, \tag{1}$$

and

$$\|v^{(k)}\|_{\mathbb{R}_+} \leq \|v_{*,n}^{(k)}\|_{\mathbb{R}_+} \|v\|_{\mathbb{R}_+}, \quad \forall n, k \in \mathbb{N}, \ \forall v \in V_n. \tag{2}$$

Moreover, the equality in (1) ((2)) is attained if and only if $u = c u_{*,n}$ $(v = c v_{*,n})$. The inequalities (1) and (2) can be considered as characterizations of the Markov factors

$$M_k(U_n) := \sup\left\{ \frac{\|u^{(k)}\|_{\mathbb{R}}}{\|u\|_{\mathbb{R}}} : u \in U_n, u \not\equiv 0 \right\},$$

and

$$M_k(V_n) := \sup\left\{ \frac{\|v^{(k)}\|_{\mathbb{R}_+}}{\|v\|_{\mathbb{R}_+}} : v \in V_n, v \not\equiv 0 \right\},$$

in terms of the corresponding weighted Chebyshev polynomial, namely

$$M_k(U_n) = \|u_{*,n}^{(k)}\|_{\mathbb{R}}, \quad M_k(V_n) = \|v_{*,n}^{(k)}\|_{\mathbb{R}_+}. \tag{3}$$

In contrast with some other situations (e.g., the classical Markov inequality), the explicit expressions for $u_{*,n}$ and $v_{*,n}$ as well as the exact values of $M_k(U_n)$ and $M_k(V_n)$ are not known. Therefore, it is of interest to have a numerical method for finding the above extremal polynomials and quantities.

We consider a more general problem. It was proved in [8] (cf. [11], [5], [1]) that given a vector $\mathbf{h} = (h_0, \ldots, h_n)$ with positive components, there exist a unique $u(x) = u(\mathbf{h}; x) \in U_n$ and a unique set of points $\mathbf{t}(\mathbf{h}) = \{t_k(\mathbf{h})\}_{k=0}^n$, $t_0(\mathbf{h}) < \cdots < t_n(\mathbf{h})$, such that

$$u(t_k(\mathbf{h})) = (-1)^{n-k} h_k, \ k = 0, \ldots, n,$$

$$\tag{4}$$

$$u'(t_k(\mathbf{h})) = 0, \ k = 0, \ldots, n.$$

Clearly, $u_{*,n}(x) = u(\mathbb{1}; x)$, where $\mathbb{1} = (1, 1, \ldots, 1) \in \mathbb{R}^{n+1}$.

For a numerical solution of the nonlinear system (4), we use an algorithm, proposed by Prof. Ranko Bojanic (for the case of algebraic polynomials). We prove in Sect. 2 a theorem, which shows that the method is globally convergent and the rate of convergence is quadratic. In the proof we use some ideas from [12], where the above algorithm was studied in the case of algebraic polynomials and for the smoothest interpolation problem. We apply the algorithm to find numerically $u_{*,n}$ and $v_{*,n}$ for $1 \leq n \leq 10$. Then, using (3) we compute $M_k(U_n)$ and $M_k(V_n)$ for $1 \leq n \leq 10$ and $1 \leq k \leq 5$. The results are given in Sect. 3.

## 2    An Algorithm for Interpolation at Extremal Points

We first introduce some additional notations. Let

$$T_n := \{\mathbf{t} = (t_0, \ldots, t_n) \in \mathbb{R}^{n+1} : \ -\infty < t_0 < \cdots < t_n < \infty\}.$$

Given $\mathbf{t} \in T_n$ and $\mathbf{y} = (y_0, \ldots, y_n) \in \mathbb{R}^{n+1}$, we denote by $u(\mathbf{t}, \mathbf{y}; x)$ the unique polynomial from $U_n$, which satisfies the interpolation conditions

$$u(\mathbf{t}, \mathbf{y}; t_k) = (-1)^{n-k} y_k, \ \ k = 0, \ldots, n. \tag{5}$$

We denote by $\|\mathbf{f}\| := \max_{0 \leq k \leq n} |f_k|$ the max norm of a vector $\mathbf{f} = (f_0, \ldots, f_n) \in \mathbb{R}^{n+1}$.

The following algorithm can be used for the numerical construction of the polynomial $u(\mathbf{h}; x)$ from (4).

**Algorithm 1**

1. Choose an arbitrary $\mathbf{t}^{(0)} \in T_n$ and set $u_0(x) := u(\mathbf{t}^{(0)}, \mathbf{h}; x)$.
2. For $m = 0, 1, \ldots$, do:
   (a) Find the zeros $s_0^{(m)} < \cdots < s_n^{(m)}$ of $u_m'(x)$ and set $\mathbf{s}^{(m)} := (s_0^{(m)}, \ldots, s_n^{(m)})$;
   (b) Set $\mathbf{t}^{(m+1)} := \mathbf{s}^{(m)}$ and $u_{m+1}(x) := u(\mathbf{t}^{(m+1)}, \mathbf{h}; x)$.

In the next theorem we study the properties of the sequence $\{\mathbf{t}^{(m)}\}$ generated by the algorithm.

**Theorem 1.** *The sequence $\{\mathbf{t}^{(m)}\}$ converges to the set $\mathbf{t}(\mathbf{h})$ of the extremal points of $u(\mathbf{h}; x)$ for every $\mathbf{t}^{(0)} \in T_n$. There exists a constant $C$ such that*

$$\|\mathbf{t}^{(m+1)} - \mathbf{t}(\mathbf{h})\| \leq C \|\mathbf{t}^{(m)} - \mathbf{t}(\mathbf{h})\|^2, \ m = 0, 1, \ldots,$$

*for any sufficiently good initial approximation $\mathbf{t}^{(0)}$.*

*Proof.* Let $u_m(x) = e^{-x^2} p_m(x), m = 0, 1, \ldots$, where

$$p_m(x) = p_m(\mathbf{t}^{(m)}, \mathbf{h}; x) = A_m x^n + \cdots \in \pi_n.$$

It follows from the definition of $u_m$ that $A_m > 0$ for every $m \geq 0$. We claim that

$$A_m \geq A_{m+1} \text{ for every } m \geq 0. \tag{6}$$

To prove this, we first remind the usual notation $f[x_0, \ldots, x_n]$ for the divided difference of the function $f$ at the points $x_0, \ldots, x_n$. Let us set $\omega(x) := (x -$

$s_0^{(m)}) \cdots (x - s_n^{(m)})$. Using the fact that $(-1)^{n-k} u_m(s_k^{(m)}) \geq h_k$, $k = 0, \ldots, n$, we get

$$A_m = p_m[s_0^{(m)}, \ldots, s_n^{(m)}] = \sum_{k=0}^{n} \frac{p_m(s_k^{(m)})}{\omega'(s_k^{(m)})} = \sum_{k=0}^{n} \frac{(-1)^{n-k} p_m(s_k^{(m)})}{|\omega'(s_k^{(m)})|}$$

$$\geq \sum_{k=0}^{n} \frac{(-1)^{n-k} p_{m+1}(s_k^{(m)})}{|\omega'(s_k^{(m)})|} = p_{m+1}[s_0^{(m)}, \ldots, s_n^{(m)}] = A_{m+1},$$

which finishes the proof of (6).

The inequalities (6) imply that $\{A_m\}$ is a bounded (in fact, a convergent) sequence. Our next goal is to prove that $\{t^{(m)}\}$ is a bounded sequence, too. Assume the contrary. Then $\tau_m := \max\{|t_0^{(m)}|, |t_n^{(m)}|\}$ is an unbounded sequence. Without loss of generality we can suppose that $\tau_m \to \infty$ as $m \to \infty$. We have

$$A_m = p_m[t_0^{(m)}, \ldots, t_n^{(m)}] = \sum_{k=0}^{n} \frac{h_k e^{(t_k^{(m)})^2}}{|\omega_m'(t_k^{(m)})|},$$

where $\omega_m(x) := (x - t_0^{(m)}) \cdots (x - t_n^{(m)})$.

Let $h_s := \min_{0 \leq k \leq n} h_k$. Since

$$|\omega_m'(t_k^{(m)})| \leq (t_n^{(m)} - t_0^{(m)})^n \leq (2\tau_m)^n,$$

we obtain

$$A_m \geq \frac{h_s}{(2\tau_m)^n} \sum_{k=0}^{n} e^{(t_k^{(m)})^2} \geq \frac{h_s}{2^n} \frac{e^{\tau_m^2}}{\tau_m^n} \to \infty \text{ as } m \to \infty,$$

which is a contradiction with the boundedness of $\{A_m\}$.

Next we consider an arbitrary convergent subsequence of $\{t^{(m)}\}$. For the sake of simplicity, we denote it again by $\{t^{(m)}\}$. Let $t^* := \lim_{m \to \infty} t^{(m)}$. Note that $t^* \in T_n$, i.e. $|t_i^* - t_{i-1}^*| > 0$, $i = 1, \ldots, n$. Indeed, it follows by induction from the interpolation conditions (5) that for any fixed $k < n$ the divided differences $\gamma_{i,k}^{(m)} := p_m[t_i^{(m)}, \ldots, t_{i+k}^{(m)}]$, $i = 0, \ldots, n-k$, change sign alternatively. There exists a constant $d = d(t^{(0)}, h)$, such that $\|t^{(m)}\| \leq d$, for every $m = 0, 1, \ldots$ Then we have

$$|\gamma_{i,k}^{(m)}| = \frac{|\gamma_{i+1,k-1}^{(m)} - \gamma_{i,k-1}^{(m)}|}{t_{i+k}^{(m)} - t_i^{(m)}} \geq \frac{1}{2d}(|\gamma_{i+1,k-1}^{(m)}| + |\gamma_{i,k-1}^{(m)}|)$$

$$\geq \frac{1}{2d} \max\{|\gamma_{i+1,k-1}^{(m)}|, |\gamma_{i,k-1}^{(m)}|\}.$$

Hence,

$$A_m \geq \frac{1}{(2d)^{n-1}} \max_{0 \leq i \leq n-1} |\gamma_{i,1}^{(m)}|.$$

Thus, if we suppose that $|t_{j+1}^{(m)} - t_j^{(m)}| \to 0$ for some $j$, then

$$A_m \geq \frac{1}{(2d)^{n-1}} \frac{h_{j+1}e^{(t_{j+1}^{(m)})^2} + h_j e^{(t_j^{(m)})^2}}{t_{j+1}^{(m)} - t_j^{(m)}} \geq \frac{1}{(2d)^{n-1}} \frac{h_{j+1} + h_j}{t_{j+1}^{(m)} - t_j^{(m)}} \to \infty,$$

a contradiction.

Let $u_\star(x) := u(\mathbf{t}^\star, \mathbf{h}; x)$. By Cramer's rule, the coefficients of the interpolating polynomial $u(\mathbf{t}, \mathbf{h}; x)$ are continuous functions of $\mathbf{t}$ in a neighbourhood of $\mathbf{t}^\star$. This implies that for each $j \geq 0$ we have uniformly on $\mathbb{R}$

$$u^{(j)}(\mathbf{t}, \mathbf{h}; x) \to u_\star^{(j)}(x) \text{ as } \mathbf{t} \to \mathbf{t}^\star. \tag{7}$$

In particular, $u_m'(x) \to u_\star'(x)$ and since the zeros $\{s_i^{(m)}\}_{i=0}^n$ of $u_m'$ and $\{s_i^\star\}_{i=0}^n$ of $u_\star'$ are simple, we conclude that $\mathbf{s}^{(m)} \to \mathbf{s}^\star := (s_0^\star, \ldots, s_n^\star)$, i.e. $\mathbf{t}^{(m+1)} \to \mathbf{s}^\star$. The uniqueness of the limit gives $\mathbf{t}^\star = \mathbf{s}^\star$. So, we have

$$u_\star(t_k^\star) = (-1)^{n-k} h_k, \quad k = 0, \ldots, n,$$

and

$$u_\star'(t_k^\star) = 0, \quad k = 0, \ldots, n.$$

The uniqueness of the solution of the problem (4) implies $\mathbf{t}^\star = \mathbf{t}(\mathbf{h})$ and $u_\star = u(\mathbf{h}; \cdot)$. The convergence of the algorithm is proved.

Next we shall prove that the method converges quadratically. It follows from (4),(5) and (7) that there exist an $\epsilon_0 = \epsilon_0(\mathbf{h}) > 0$ and constants $C_i = C_i(\mathbf{h}, \epsilon_0) > 0$, $i = 1, 2, 3$, with the following properties:

(i) For every $\mathbf{t} \in B(\mathbf{t}(\mathbf{h}), \epsilon_0) := \{\mathbf{t} \in \mathbb{R}^{n+1} : \|\mathbf{t} - \mathbf{t}(\mathbf{h})\| \leq \epsilon_0\}$ we have $\operatorname{sign} u(\mathbf{h}; t_k) = (-1)^{n-k}$, $k = 0, \ldots, n$.

(ii) $|u(\mathbf{h}; t_k) - u(\mathbf{h}; t_k(\mathbf{h}))| \leq C_1 \|\mathbf{t} - \mathbf{t}(\mathbf{h})\|^2$, $k = 0, \ldots, n$, provided $\mathbf{t} \in B(\mathbf{t}(\mathbf{h}), \epsilon_0)$.

(iii) If $\mathbf{t} \in B(\mathbf{t}(\mathbf{h}), \epsilon_0)$ then

$$|s_k - t_k(\mathbf{h})| \leq C_2 \|u'(\mathbf{t}, \mathbf{h}; \cdot) - u'(\mathbf{h}; \cdot)\|_{\mathbb{R}}, \quad k = 0, \ldots, n,$$

where $s_0 < \cdots < s_n$ are the zeros of $u'(\mathbf{t}, \mathbf{h}; x)$.

(iv) The estimates

$$\|u^{(j)}(\mathbf{t}, \mathbf{y_1}; \cdot) - u^{(j)}(\mathbf{t}, \mathbf{y_2}; \cdot)\|_{\mathbb{R}} \leq C_3 \|\mathbf{y_1} - \mathbf{y_2}\|, \quad j = 0, 1,$$

hold true, provided $\mathbf{t} \in B(\mathbf{t}(\mathbf{h}), \epsilon_0)$ and $\mathbf{y_1}, \mathbf{y_2} \in \mathbb{R}^{n+1}$.

Now, we are ready to determine the rate of convergence of the sequence $\{\mathbf{t}^{(m)}\}$, generated by the algorithm. Let us suppose that $\mathbf{t} \in B(\mathbf{t}(\mathbf{h}), \epsilon_0)$. It follows from (i) that $u(\mathbf{h}; x)$ can be represented in the form

$$u(\mathbf{h}; x) = u(\mathbf{t}, \mathbf{y}; x),$$

where $\mathbf{y} = (y_0, \ldots, y_n)$, with $y_k := |u(\mathbf{h}; t_k)|$, $k = 0, \ldots, n$. Using (iv) and (ii), we get

$$\|u^{(j)}(\mathbf{t}, \mathbf{h}; \cdot) - u^{(j)}(\mathbf{h}; \cdot)\|_{\mathbb{R}} \leq C_3 \|\mathbf{h} - \mathbf{y}\| \leq C_1 C_3 \|\mathbf{t} - \mathbf{t}(\mathbf{h})\|^2, \quad j = 0, 1.$$

Then (iii) implies $|s_k - t_k(\mathbf{h})| \leq C\|\mathbf{t} - \mathbf{t}(\mathbf{h})\|^2$, $k = 0, \ldots, n$, with $C := C_1 C_2 C_3$. The last inequality shows that if $\|\mathbf{t}^{(m)} - \mathbf{t}(\mathbf{h})\| = \epsilon \leq \epsilon_0$, then $\|\mathbf{t}^{(m+1)} - \mathbf{t}(\mathbf{h})\| \leq C\epsilon^2$. The theorem is proved. $\qquad\square$

Next we discuss briefly the case of the weight $\mu(x) = e^{-x}$ on $\mathbb{R}_+$. According to [8], if the vector $\mathbf{h} = (h_0, h_1, \ldots, h_n)$ satisfies $h_k > 0, k = 0, \ldots, n$, then there exists a unique $v(x) = v(\mathbf{h}; x) \in V_n$ along with $n + 1$ points $0 =: t_0(\mathbf{h}) < \cdots < t_n(\mathbf{h})$ such that

$$v(t_k(\mathbf{h})) = (-1)^{n-k} h_k, \quad k = 0, \ldots, n,$$

and

$$v'(t_k(\mathbf{h})) = 0, \quad k = 1, \ldots, n.$$

The above algorithm can easily be modified to find $v(\mathbf{h}; x)$. The global convergence and the quadratic convergence rate of the method remain valid and can be proved in a similar way.

**Remark 1.** Let us denote by $\mathcal{M}$ the class of all weight functions $\mu$ such that: $\mu \in C^1(\mathbb{R})$, $\mu(x) > 0$ for all $x \in \mathbb{R}$, $\mu(x)x^n \to 0$ as $|x| \to \infty$, $n = 0, 1, 2, \ldots$, and $\mu'/\mu$ is decreasing on $\mathbb{R}$. Given a weight $\mu \in \mathcal{M}$, we set

$$U_n(\mu) := \{u : u = \mu p, \ p \in \pi_n\}.$$

According to [7, Lemma 2], if a polynomial $u \in U_n(\mu)$ has $n$ simple real zeros, then there are exactly $n + 1$ distinct real numbers where $u'$ vanishes. By Rolle's theorem, the $n + 1$ zeros of $u'$ are the points of local extrema of $u$. Using the method of Fitzgerald and Schumaker [5], it can be proved that if $\mu \in \mathcal{M}$ then the interpolation problem (4) has a unique solution $u(\mu, \mathbf{h}; x)$ in $U_n(\mu)$.

Furthermore, we define the class

$$\mathcal{M}_1 := \{\mu \in \mathcal{M} : \ \mu \in C^2(\mathbb{R}), \ \mu'(x)x^n \to 0 \text{ as } |x| \to \infty, \ n = 0, 1, 2, \ldots\}.$$

Algorithm 1 can be used to find $u(\mu, \mathbf{h}; x)$, provided $\mu \in \mathcal{M}_1$. Theorem 1 is valid also in this more general situation.

Similar results can be obtained for a class of weights on $\mathbb{R}_+$, namely $\mathcal{M}_2$, which consists of all functions $\mu$ such that: $\mu \in C^2(\mathbb{R}_+)$, $\mu(x) > 0$ for all $x \in \mathbb{R}_+$, $\mu^{(k)}(x)x^n \to 0$ as $x \to \infty$ for $k = 0, 1$ and $n = 0, 1, 2, \ldots$, $\mu'(0) \leq 0$, and $\mu'/\mu$ is decreasing on $\mathbb{R}_+$.

## 3  Numerical Computation of the Markov Factors

In order to compute $M_k(U_n)$, we proceed as follows:

1. We compute $u_{*,n} = u(\mathbb{1}; \cdot)$ by Algorithm 1. As a termination criterion we use $\|\mathbf{t}^{(m+1)} - \mathbf{t}^{(m)}\| \leq 10^{-10}$.
2. We compute $u_{*,n}^{(k)}, u_{*,n}^{(k+1)}$, and the zeros $x_1^{(k+1)} < \cdots < x_{n+k+1}^{(k+1)}$ of $u_{*,n}^{(k+1)}$.
3. We have

$$M_k(U_n) = \|u_{*,n}^{(k)}\|_{\mathbb{R}} = \max_{1 \leq i \leq n+k+1} \{|u_{*,n}^{(k)}(x_i^{(k+1)})|\}.$$

The quantities $M_k(V_n)$ are computed similarly.

The following two tables contain the numerical values of $M_k(U_n)$ and $M_k(V_n)$ for $1 \le n \le 10$ and $1 \le k \le 5$.

**Table 1.** Markov factors $M_k(U_n)$

|  | $k = 1$ | $k = 2$ | $k = 3$ | $k = 4$ | $k = 5$ |
|---|---|---|---|---|---|
| $n = 1$ | 2,33164 | 4,55086 | 13,9899 | 38,1386 | 139,899 |
| $n = 2$ | 2,87650 | 9,18224 | 26,3720 | 98,1869 | 335,455 |
| $n = 3$ | 3,61041 | 12,4016 | 48,0742 | 173,370 | 744,859 |
| $n = 4$ | 4,04882 | 16,9372 | 67,7545 | 301,889 | 1273,81 |
| $n = 5$ | 4,56796 | 20,4303 | 96,1517 | 435,455 | 2162,81 |
| $n = 6$ | 4,94261 | 24,8119 | 121,721 | 631,048 | 3164,86 |
| $n = 7$ | 5,36362 | 28,4415 | 155,027 | 826,118 | 4646,97 |
| $n = 8$ | 5,69551 | 32,7326 | 185,544 | 1087,03 | 6233,64 |
| $n = 9$ | 6,05825 | 36,4421 | 222,999 | 1344,78 | 8397,91 |
| $n = 10$ | 6,35913 | 40,6765 | 257,832 | 1670,29 | 10662,5 |

**Table 2.** Markov factors $M_k(V_n)$

|  | $k = 1$ | $k = 2$ | $k = 3$ | $k = 4$ | $k = 5$ |
|---|---|---|---|---|---|
| $n = 1$ | 4,59112 | 8,18224 | 11,7734 | 15,3645 | 18,9556 |
| $n = 2$ | 8,46860 | 25,1574 | 51,0663 | 86,1955 | 130,545 |
| $n = 3$ | 12,4060 | 52,5873 | 143,204 | 305,916 | 562,383 |
| $n = 4$ | 16,3663 | 90,5857 | 313,511 | 823,798 | 1808,90 |
| $n = 5$ | 20,3383 | 139,191 | 587,455 | 1847,27 | 4760,30 |
| $n = 6$ | 24,3170 | 198,420 | 990,544 | 3642,05 | 10832,0 |
| $n = 7$ | 28,3001 | 268,283 | 1548,31 | 6532,24 | 22094,6 |
| $n = 8$ | 32,2863 | 348,788 | 2286,31 | 10900,3 | 41402,9 |
| $n = 9$ | 36,2746 | 439,938 | 3230,08 | 17187,2 | 72526,8 |
| $n = 10$ | 40,2646 | 541,737 | 4405,20 | 25892,2 | 120280 |

# References

1. Bojanov, B.: A generalization of Chebyshev polynomials. J. Approx. Theory **26** (1979) 293–300
2. Bojanov, B.: Markov-type inequalities for polynomials and splines. In: C. K. Chui, L.L. Schumaker, J. Stökler (Eds.), Approximation Theory X: Abstract and Classical Analysis. Vanderbilt University Press, Nashville TN (2002)
3. Borwein P., Erdélyi, T.: Polynomials and Polynomial Inequalities. Graduate Texts in Mathematics vol. 161. Springer-Verlag, New York Berlin Heidelberg (1995)
4. Carley, H., Li, X., Mohapatra, R. N.: A sharp inequality of Markov type for polynomials associated with Laguerre weight. J. Approx. Theory **113** (2001) 221–228
5. Fitzgerald, C. H., Schumaker, L. L.: A differential equation approach to interpolation at extremal points. J. Analyse Math. **22** (1969) 117–134

6. Levin, E., Lubinsky, D. S.: Orthogonal Polynomials for Exponential Weights. CMS Books in Mathematics vol. 4. Springer-Verlag, New York Berlin Heidelberg (2001)
7. Li, X., Mohapatra, R. N., Rodriguez, R. S.: On Markov's inequality on IR for the Hermite weight. J. Approx. Theory **75** (1993) 267–273
8. Milev, L.: Weighted polynomial inequalities on infinite intervals. East J. Approx. **5** (1999) 449–465
9. Milev, L., Naidenov, N.: Exact Markov inequalities for the Hermite and Laguerre weights. J. Approx. Theory **138** (2006) 87–96
10. Milovanović, G. M., Mitrinović, D. S., Rassias, Th. M.: Topics in Polynomials: Extremal Problems, Inequalities, Zeros. World Scientific, Singapore (1994)
11. Mycielski, J., Paszkowski, S.: A generalization of Chebyshev polynomials. Bull. Acad. Polonaise Sci. Série Math. Astr. et Phys. **8** (1960) 433–438
12. Naidenov, N.: Algorithm for the construction of the smoothest interpolant. East J. Approx. **1** (1995) 83–97

# Connection of Semi-integer Trigonometric Orthogonal Polynomials with Szegő Polynomials*

Gradimir V. Milovanović, Aleksandar S. Cvetković, and
Zvezdan M. Marjanović

Department of Mathematics, Faculty of Electronic Engineering,
University of Niš, P.O. Box 73, 18000 Niš, Serbia

**Abstract.** In this paper we investigate connection between semi-integer orthogonal polynomials and Szegő's class of polynomials, orthogonal on the unit circle. We find a representation of the semi-integer orthogonal polynomials in terms of Szegő's polynomials orthogonal on the unit circle for certain class of weight functions.

## 1 Introduction

Let us denote by $\mathcal{T}_{n+1/2}$ the linear span of trigonometric functions

$$\cos x/2, \sin x/2, \cos(1+1/2)x, \sin(1+1/2)x, \ldots, \cos(n+1/2)x, \sin(n+1/2)x.$$

The elements of $\mathcal{T}_{n+1/2}$ are called trigonometric polynomials of semi-integer degree. For the convenience sake we define $\mathcal{T}_{-1/2} = \{0\}$. Clearly, $\mathcal{T}_{n+1/2}$ is a linear space of dimension $2n+2$. We introduce inner product by

$$(f, g) = \int_{-\pi}^{\pi} f(x)g(x)w(x)dx, \quad f, g \in \mathcal{T}_{n+1/2}, \tag{1}$$

where $w$ is a non-negative weight function on $(-\pi, \pi]$, which equals zero only on a set of Lebesgue measure zero.

Next, we define the following set

$$\mathcal{T}_{n+1/2}^{a,b} = a\cos(n+1/2)x + b\sin(n+1/2)x + \mathcal{T}_{n-1/2},$$

where $a, b \in \mathbb{R}$ are fixed with the property $|a| + |b| > 0$. The case $a = b = 0$ is not interesting since in that case $\mathcal{T}_{n+1/2}^{0,0} = \mathcal{T}_{n-1/2}$. Given the inner product (1), we can pose a question of finding $A_n \in \mathcal{T}_{n+1/2}^{a,b}$, such that

$$\int_{-\pi}^{\pi} A_n(x)t(x)w(x)dt = 0, \quad t \in \mathcal{T}_{n-1/2}.$$

It turns out that this problem has a unique solution.

* The authors were supported in part by the Serbian Ministry of Science and Environmental Protection (Project: Orthogonal Systems and Applications, grant number #144004) and the Swiss National Science Foundation (SCOPES Joint Research Project No. IB7320–111079 "New Methods for Quadrature").

T. Boyanov et al. (Eds.): NMA 2006, LNCS 4310, pp. 394–401, 2007.
© Springer-Verlag Berlin Heidelberg 2007

**Lemma 1.** *There exist a unique* $A_n \in \mathcal{T}_{n+1/2}^{a,b}$, *such that*

$$\int_{-\pi}^{\pi} A_n(x)t(x)w(x)dx = 0, \quad t \in \mathcal{T}_{n-1/2}.$$

**Proof.** Any polynomial $t \in \mathcal{T}_{n+1/2}^{a,b}$ can be represented as

$$t(x) = a\cos(n+1/2)x + b\sin(n+1/2)x + \sum_{k=0}^{n-1}[a_k\cos(k+1/2)x + b_k\sin(k+1/2)x].$$

In order to be orthogonal to $\mathcal{T}_{n-1/2}$, its coefficients have to satisfy the following system of linear equations

$$\sum_{k=0}^{n-1}\big[a_k(\cos(k+1/2)x, \cos(\ell+1/2)x) + b_k(\sin(k+1/2)x, \cos(\ell+1/2)x)\big]$$

$$= -(a\cos(n+1/2)x + b\sin(n+1/2)x, \cos(\ell+1/2)x),$$

$$\sum_{k=0}^{n-1}\big[a_k(\cos(k+1/2)x, \sin(\ell+1/2)x) + b_k(\sin(k+1/2)x, \sin(\ell+1/2)x)\big]$$

$$= -(a\cos(n+1/2)x + b\sin(n+1/2)x, \sin(\ell+1/2)x),$$

for $\ell = 0, \ldots, n-1$. The solution of this system is unique since its matrix is a Gram matrix formed by linearly independent vectors (see [2, p. 224]).     □

From now on, we denote by $A_n \in \mathcal{T}_{n+1/2}^{a,b}$ the trigonometric polynomial orthogonal to $\mathcal{T}_{n-1/2}$. When we want to emphasize the dependence on $a$ and $b$, we write $A_n^{a,b}$ for $A_n \in \mathcal{T}_{n+1/2}^{a,b}$. From the proof of Lemma 1 it is clear that we have

$$A_n^{a,b}(x) = aA_n^{1,0} + bA_n^{0,1}. \tag{2}$$

In [1], we used notation $A_n^C = A_n^{1,0}$ and $A_n^S = A_n^{0,1}$ for obvious reasons.

Consider the following quadrature rule

$$\int_{-\pi}^{\pi} w(x)t(x)dx = \sum_{k=0}^{2n} w_k p(x_k), \quad t \in \mathcal{T}_{2n},$$

where $\mathcal{T}_n$ denotes linear span of trigonometric functions $1, \cos x, \sin x, \ldots, \cos nx$, $\sin nx$. In [4] and [1] it is proved that such quadrature rule exists, it has positive weights $w_k$, $k = 0, \ldots, 2n$, and its nodes $x_k$, $k = 0, \ldots, 2n$, are zeros of the polynomial $A_n \in \mathcal{T}_{n+1/2}^{a,b}$, orthogonal to $\mathcal{T}_{n-1/2}$, where $a, b \in \mathbb{R}$, $|a| + |b| > 0$, are arbitrary.

We note that a basis of $\mathcal{T}_{n+1/2}^{a,b}$ is the set

$$\{A_k^{a,b}, A_k^{-b,a} \mid k = 0, \ldots, n-1, |a| + |b| > 0, a, b \in \mathbb{R}\} \cup \{A_n^{a,b}\}.$$

This is obvious since

$$\cos(k+1/2)x + t(x) = \frac{a}{a^2+b^2}A_k^{a,b} - \frac{b}{a^2+b^2}A_k^{-b,a},$$

$$\sin(k+1/2)x + t(x) = \frac{b}{a^2+b^2}A_k^{a,b} + \frac{a}{a^2+b^2}A_k^{-b,a},$$

where $t \in \mathcal{T}_{k-1/2}$, $k = 0, \ldots, n$.

In [1], it is proved that the sequences of polynomials $A_n^C$ and $A_n^S$ satisfy the following five-term recurrence relations

$$A_n^C = (2\cos x - \alpha_n^1)A_{n-1}^C - \alpha_n^2 A_{n-2}^C - \beta_n^1 A_{n-1}^S - \beta_n^2 A_{n-2}^S, \tag{3}$$

$$A_n^S = (2\cos x - \delta_n^1)A_{n-1}^S - \delta_n^2 A_{n-2}^S - \gamma_n^1 A_{n-1}^C - \gamma_n^2 A_{n-2}^C.$$

Using relations (2), we can prove easily the following Lemma.

**Lemma 2.** *Polynomials $A_n^{a,b}$ and $A_n^{-b,a}$ satisfy the following five term recurrence relations*

$$A_n^{a,b} = (2\cos x - \tilde{\alpha}_n^1)A_{n-1}^{a,b} - \tilde{\alpha}_n^2 A_{n-2}^{a,b} - \tilde{\beta}_n^1 A_{n-1}^{-b,a} - \tilde{\beta}_n^2 A_{n-2}^{-b,a},$$

$$A_n^{-b,a} = (2\cos x - \tilde{\delta}_n^1)A_{n-1}^{-b,a} - \tilde{\delta}_n^2 A_{n-2}^{-b,a} - \tilde{\gamma}_n^1 A_{n-1}^{a,b} - \tilde{\gamma}_n^2 A_{n-2}^{a,b},$$

*where*

$$\tilde{\alpha}_n^1 = \frac{a^2\alpha_n^1 + ab(\beta_n^1 + \gamma_n^1) + b^2\delta_n^1}{a^2 + b^2}, \quad \tilde{\alpha}_n^2 = \frac{a^2\alpha_n^2 + ab(\beta_n^2 + \gamma_n^2) + b^2\delta_n^2}{a^2 + b^2}$$

$$\tilde{\beta}_n^1 = \frac{a^2\beta_n^1 - ab(\alpha_n^1 - \delta_n^1) - b^2\gamma_n^1}{a^2 + b^2}, \quad \tilde{\beta}_n^2 = \frac{a^2\beta_n^2 - ab(\alpha_n^2 - \delta_n^2) - b^2\gamma_n^2}{a^2 + b^2}$$

$$\tilde{\delta}_n^1 = \frac{b^2\alpha_n^1 - ab(\beta_n^1 + \gamma_n^1) + a^2\delta_n^1}{a^2 + b^2}, \quad \tilde{\delta}_n^2 = \frac{b^2\alpha_n^2 - ab(\beta_n^2 + \gamma_n^2) + a^2\delta_n^2}{a^2 + b^2},$$

$$\tilde{\gamma}_n^1 = \frac{a^2\gamma_n^1 - ab(\alpha_n^1 - \delta_n^1) - b^2\beta_n^1}{a^2 + b^2}, \quad \tilde{\gamma}_n^2 = \frac{a^2\gamma_n^2 - ab(\alpha_n^2 - \delta_n^2) - b^2\beta_n^2}{a^2 + b^2}$$

**Proof.** Just use connection

$$A_n^C = \frac{aA_n^{a,b} - bA_n^{-b,a}}{a^2 + b^2}, \quad A_n^S = \frac{bA_n^{a,b} + aA_n^{-b,a}}{a^2 + b^2},$$

in (3), and solve linear system for $A_n^{a,b}$ and $A_n^{-b,a}$. □

## 2   Connection to the Szegő Polynomials

In the rest of this paper we shall need the following lemma which gives a factorization of the positive trigonometric polynomials. Recall that $\mathcal{T}_n$ is linear space spanned by the set of trigonometric functions $\cos kx, \sin kx$, $k = 0, \ldots, n$. By $\mathcal{P}_n$, $n \in \mathbb{N}_0$, we denote the set of all algebraic polynomials of degree at most $n$. The next lemma (in a slightly different formulation) can be found in [3, p. 4].

**Lemma 3.** *Let $t_n \in \mathcal{T}_n$, $n \in \mathbb{N}$, be a trigonometric polynomial of exact degree $n$, which is strictly positive on the interval $(-\pi, \pi]$. Then there exist a unique (up to a multiplicative constant of modulus one) algebraic polynomial $H_n \in \mathcal{P}_n$ of exact degree $n$, such that*

$$t_n(x) = e^{-inx} H_n(e^{ix}) H_n^*(e^{ix}), \quad where \quad H_n^*(z) = z^n \overline{H_n(1/\overline{z})}, \qquad (4)$$

*and all the zeros of $H_n$ are of modulus smaller then one.*

**Proof.** Let

$$t_n(x) = \sum_{k=0}^{n} \left( a_k \cos kx + b_k \sin kx \right), \quad |a_n| + |b_n| > 0,$$

then we can expand

$$t_n(x) = e^{-inx} T_n(e^{ix}) = e^{-inx} \sum_{k=0}^{n} \left( \frac{a_k - ib_k}{2} e^{i(n+k)x} + \frac{a_k + ib_k}{2} e^{i(n-k)x} \right),$$

where $T_n$ is an algebraic polynomial of exact degree $n$.

For the polynomial $T_n$ we have

$$T_n^*(z) = z^{2n} \overline{T_n(1/\overline{z})} = z^{2n} \sum_{k=0}^{n} \left( \frac{a_k + ib_k}{2} \frac{1}{z^{n+k}} + \frac{a_k - ib_k}{2} \frac{1}{z^{n-k}} \right) = T_n(z).$$

As a consequence we have that if $T_n(z) = 0$, then also $T_n(1/\overline{z}) = 0$. Note that $T_n(0) = (a_n + ib_n)/2 \neq 0$. Hence, $T_n$ has $n$ zeros of modulus smaller then one and $n$ zeros of modulus bigger then one. Notice that $T_n$ can not have a zero $e^{ix}$, $x \in \mathbb{R}$, since in that case $t_n(x) = e^{-inx} T_n(e^{ix}) = 0$, which is impossible according to the assumptions of the theorem.

Denote by $z_k$, $k = 1, \ldots, n$, the zeros of $T_n$ of modulus smaller then one, then we have

$$T_n(z) = \frac{a_n - ib_n}{2} \prod_{k=0}^{n} (z - z_k)(z - 1/\overline{z}_k).$$

To ensure that $H_n$ is of exact degree $n$ with all its zeros lying inside the unit circle, we set

$$H_n(z) = A \prod_{k=1}^{n} (z - z_k),$$

with some $A \in \mathbb{C}$. Then

$$H_n^*(z) = z^n \overline{A} \prod_{k=1}^{n} (1/z - \overline{z}_k) = \overline{A} \prod_{k=1}^{n} (-\overline{z}_k) \prod_{k=1}^{n} (z - 1/\overline{z}_k),$$

and obviously

$$H_n(z) H_n^*(z) = |A|^2 \prod_{k=1}^{n} (-\overline{z}_k) \prod_{k=1}^{n} (z - z_k)(z - 1/\overline{z}_k).$$

In order to get the desired representation, there must hold

$$|A|^2 = \frac{a_n - \mathrm{i}b_n}{2\prod_{k=1}^n(-\overline{z}_k)}. \tag{5}$$

Hence, it remains to prove that the quantity on the right is positive. We have

$$t_n(x) = e^{-\mathrm{i}nx}T_n(e^{\mathrm{i}x}) = \frac{a_n - \mathrm{i}b_n}{2}\prod_{k=1}^n(e^{\mathrm{i}x} - z_k)(1 - e^{-\mathrm{i}x}/\overline{z}_k)$$

$$= \frac{a_n - \mathrm{i}b_n}{2\prod_{k=1}^n(-\overline{z}_k)}\prod_{k=1}^n|e^{\mathrm{i}x} - z_k|^2,$$

hence it is positive. Since we imposed only condition on the modulus of $A$, we are free to choose its argument.    □

Now, we are ready to prove the following theorem.

**Theorem 1.** *Let $t_\ell \in \mathcal{T}_\ell$, $\ell \in \mathbb{N}_0$, be a trigonometric polynomial of exact degree $\ell$, strictly positive on $(-\pi, \pi]$. Then, for $n \geq \ell$, the polynomial $A_n^{a,b} \in \mathcal{T}_{n+1/2}^{a,b}$, orthogonal with respect to the weight function $w(x) = 1/t_\ell(x)$, is given by*

$$A_n^{a,b}(x) = \frac{a - \mathrm{i}b}{2}e^{\mathrm{i}(n-\ell+1/2)x}h_\ell(e^{\mathrm{i}x}) + \frac{a + \mathrm{i}b}{2}e^{-\mathrm{i}(n-\ell+1/2)x}\overline{h_\ell(e^{\mathrm{i}x})},$$

*where $h_\ell$ is the monic version of the polynomial $H_\ell$ from Lemma 3. The coefficients in five term recurrence terms are given by*

$$\widetilde{\alpha}_n^1 = \widetilde{\beta}_n^1 = \widetilde{\gamma}_n^1 = \widetilde{\delta}_n^1 = 0, \quad n \geq \ell + 1,$$
$$\widetilde{\beta}_n^2 = \widetilde{\gamma}_n^2 = 0, \quad \widetilde{\alpha}_n^2 = \widetilde{\delta}_n^2 = 1, \quad n \geq \ell + 2.$$

**Proof.** To prove orthogonality of $A_n^{a,b}$ to $\mathcal{T}_{n-1/2}$, it suffices to establish its orthogonality to the set $e^{\mathrm{i}(k+1/2)x}$, $k = n - 1, \ldots, 0, \ldots, -n$. With $c = (a - \mathrm{i}b)/2$, we have

$$\int_{-\pi}^{\pi} \frac{e^{\mathrm{i}(k+1/2)x}A_n^{a,b}(x)}{t_\ell(x)}dx = -\mathrm{i}\oint_C \frac{z^{k-1}(cz^{n-\ell+1}h_\ell(z) + \overline{c}z^{-n+\ell}\overline{h}_\ell(z^{-1}))}{H_\ell(z)\overline{H}_\ell(z^{-1})}dz$$

$$= -\frac{\mathrm{i}c}{A}\oint_C \frac{z^{n+k}}{H_\ell^*(z)}dz - \frac{\mathrm{i}\overline{c}}{A}\oint_C \frac{z^{-n+k+\ell-1}}{H_\ell(z)}dz = 0, \quad z = e^{\mathrm{i}x},$$

where $C$ denotes the unit circle, and $A$ is the leading coefficient in $H_\ell$. The first integral equals zero, due to the fact that, for $k = n-1, \ldots, 0, \ldots, -n$ the integrand does not have singularities inside the unit circle. The second integral equals zero since for $k = n-1, \ldots, 0, \ldots, -n$ the integrand does not have singularities outside the unit circle and it is of order at least $z^{-2}$ as $z$ tends to infinity.

Using similar methods as in [1], it can be proved that the coefficients in the five term recurrence relations are given uniquely as the solutions of the following systems of linear equations

$$J_{n-1,n-j}^{a,b} = \tilde{\alpha}_n^j I_{n-j}^{a,b} + \tilde{\beta}_n^j I_{n-j}, \quad J_{n-1,n-j} = \tilde{\alpha}_n^j I_{n-j} + \tilde{\beta}_n^j I_{n-j}^{-b,a},$$

$$J_{n-j,n-1} = \tilde{\gamma}_n^j I_{n-j}^{a,b} + \tilde{\delta}_n^j I_{n-j}, \quad J_{n-1,n-j}^{-b,a} = \tilde{\gamma}_n^j I_{n-j} + \tilde{\delta}_n^j I_{n-j}^{-b,a},$$

for $j = 1, 2$, and $n > 1$, where

$$J_{n,n}^{a,b} = (2\cos x A_n^{a,b}, A_n^{a,b}), \quad J_{n,n-1}^{a,b} = I_n^{a,b}, \quad I_{n-j}^{a,b} = (A_{n-j}^{a,b}, A_{n-j}^{a,b}) = \|A_{n-j}^{a,b}\|^2,$$

$$I_{n-j} = (A_{n-j}^{a,b}, A_{n-j}^{-b,a}), \quad J_{n,n} = (2\cos x A_n^{a,b}, A_n^{-b,a}), \quad J_{n,n-1} = J_{n-1,n} = I_n.$$

Next we calculate the norm of the polynomial $A_k^{a,b}$. We have

$$\|A_n^{a,b}\|^2 = I_n^{a,b} = \int_{-\pi}^{\pi} \frac{(A_n^{a,b}(x))^2}{t_n(x)} dx = -\frac{ic^2}{A} \oint_C \frac{z^{2n-\ell}h_\ell(z)}{H_\ell^*(z)} dz - 2\frac{i|c|^2}{|A|^2} \oint_C \frac{dz}{z}$$

$$-\frac{i\overline{c}^2}{\overline{A}} \oint_C \frac{z^{-2(n-\ell)-2}\overline{h_\ell}(z^{-1})}{H_\ell(z)} dz = \frac{4\pi|c|^2}{|A|^2} = \pi\frac{a^2+b^2}{|A|^2},$$

where the first and the third integrals are equal to zero since their integrands are analytic inside and outside the unit circle, respectively, with integrand in the third integral being of order at least $z^{-2}$ at infinity.

For the integral $J_{n,n}^{a,b}$, we have

$$J_{n,n}^{a,b} = \int_{-\pi}^{\pi} \frac{2\cos x (A_n^{a,b}(x))^2}{t_\ell(x)} dx = -\frac{ic^2}{A} \oint_C \frac{(z+z^{-1})z^{2n-\ell}h_\ell(z)}{H_\ell^*(z)} - \frac{i|c|^2}{|A|^2} \oint \frac{z^2+1}{z^2} dz$$

$$-\frac{i\overline{c}^2}{\overline{A}} \oint \frac{(z+z^{-1})z^{-2(n-\ell)-2}\overline{h_\ell}(z^{-1})}{H_\ell(z)} dz = 0,$$

using the same argumentation as in calculating the previous integral. Next, denoting $d - (b + ia)/2$, we have

$$I_n = \int_{-\pi}^{\pi} \frac{A_n^{a,b}(x)A_n^{-b,a}(x)}{t_\ell(x)} dx = -\frac{icd}{A} \oint_C \frac{z^{2n-\ell}h_\ell(z)}{H_\ell^*(z)} dz - \frac{i(c\overline{d}+\overline{c}d)}{|A|^2} \oint_C \frac{dz}{z}$$

$$-\frac{i\overline{cd}}{\overline{A}} \oint_C \frac{z^{-2(n-\ell)-2}\overline{h_\ell}(z^{-1})}{H_\ell(z)} dz = 0,$$

where the first and the third integrals are zero by the same arguments as above, and for the second one we have

$$c\overline{d} + \overline{c}d == -\frac{ab - ab - i(a^2+b^2) + ab - ab + i(a^2+b^2)}{4} = 0.$$

Finally, we have

$$J_{n,n} = \int_{-\pi}^{\pi} \frac{2\cos x A_n^{a,b}(x)A_n^{-b,a}(x)}{t_\ell(x)} dx = -\frac{icd}{A} \oint_C \frac{(z+z^{-1})z^{2n-\ell}h_\ell(z)}{H_\ell^*(z)} dz$$

$$-\frac{i(c\overline{d}+\overline{c}d)}{|A|^2} \oint \frac{z^2+1}{z^2} dz - \frac{i\overline{cd}}{\overline{A}} \oint_C \frac{z^{-2(n-\ell)-2}\overline{h_\ell}(z^{-1})}{H_\ell(z)} dz = 0.$$

Using the systems of linear equations for the five term recurrence coefficients, we get

$$J^{a,b}_{n-1,n-1} = 0 = \tilde{\alpha}^1_n I^{a,b}_{n-1} + \tilde{\beta}^1_n I_{n-1} = \tilde{\alpha}^1_n \|A^{a,b}_{n-1}\|^2,$$

$$J_{n-1,n-1} = 0 = \tilde{\alpha}^1_n I_{n-1} + \tilde{\beta}^1_n I^{-b,a}_{n-1,n-1} = \tilde{\beta}^1_n \|A^{-b,a}_{n-1}\|^2,$$

$$J^{a,b}_{n-1,n-2} = I^{a,b}_{n-1} = \|A^{a,b}_{n-1}\|^2 = \tilde{\alpha}^2_n I^{a,b}_{n-2} + \tilde{\beta}^2_n I_{n-2} = \tilde{\alpha}^2_n \|A^{a,b}_{n-2}\|^2,$$

$$J_{n-1,n-2} = I_{n-1} = 0 = \tilde{\alpha}^2_n I_{n-2} + \tilde{\beta}^2_n I^{-b,a}_{n-2} = \tilde{\beta}^2_n \|A^{-b,a}_{n-2}\|^2,$$

$$J_{n-1,n-1} = 0 = \tilde{\gamma}^1_n I^{a,b}_{n-1} + \tilde{\delta}^1_n I_{n-1} = \tilde{\gamma}^1_n \|A^{a,b}_{n-1}\|^2,$$

$$J^{-b,a}_{n-1,n-1} = 0 = \tilde{\gamma}^1_n I_{n-1} + \tilde{\delta}^1_n I^{-b,a}_{n-1} = \tilde{\delta}^1_n \|A^{-b,a}_n\|^2,$$

$$J_{n-2,n-1} = I_{n-1} = 0 = \tilde{\gamma}^2_n I^{a,b}_{n-2} + \tilde{\delta}^2_n I_{n-2} = \tilde{\gamma}^2_n \|A^{a,b}_{n-2}\|^2,$$

$$J^{-b,a}_{n-2,n-1} = I^{-b,a}_{n-1} = \|A^{-b,a}_{n-1}\|^2 = \tilde{\gamma}^2_n I_{n-2} + \tilde{\delta}^2_n I^{-b,a}_{n-2} = \tilde{\delta}^2_n \|A^{-b,a}_{n-2}\|^2.$$

Since the norms of the polynomials $A^{a,b}_n$ and $A^{-b,a}_n$ are the same, and different from zero, we get what is stated. □

As we can see, we established a connection with Szegő's polynomials. Recall that Szegő's polynomials (see [3, p. 287]) are defined to be orthogonal on the unit circle with respect to the inner product

$$(p,q) = \frac{1}{2\pi} \int_{-\pi}^{\pi} p(e^{ix})\overline{q(e^{ix})} w(x)dx, \quad p,q \in \mathcal{P}.$$

In [3, p. 289], it can be found that for the special type of weights $w(x) = 1/t_\ell(x)$, $t_\ell \in \mathcal{T}_\ell$, where $t_\ell$ is strictly positive on $(-\pi, \pi]$, Szegő monic polynomials can be expressed as

$$\phi_n(e^{ix}) = e^{i(n-\ell)x} h_\ell(e^{ix}), \quad n \geq \ell.$$

Hence, we have established the following Theorem.

**Theorem 2.** *The trigonometric polynomial* $A^{a,b}_n \in \mathcal{T}^{a,b}_{n+1/2}$, $n \geq \ell$, *orthogonal with respect to the strictly positive weight function* $w(x) = 1/t_\ell(x)$, $t_\ell \in \mathcal{T}_n$, *can be represented as*

$$A^{a,b}_n(x) = \frac{a-ib}{2} e^{ix/2} \phi_n(e^{ix}) + \frac{a+ib}{2} e^{-ix/2} \overline{\phi_n(e^{ix})}, \quad n \geq \ell,$$

*where* $\phi_n$ *is the respective Szegő polynomial orthogonal on the unit circle.*

*Moreover, the norm of the polynomial* $A^{a,b}_n$, $n \geq \ell$, *is given by*

$$\|A^{a,b}_n\|^2 = \pi(a^2+b^2) \exp\left(-\frac{1}{2\pi} \int_{-\pi}^{\pi} \log t_\ell(x)dx\right).$$

**Proof.** We need to prove only the statement about the norm. In [3, p. 300-304], it is proven that the norm of the monic Szegő polynomial is given by

$$\|\phi_n\|^2 = \exp\left(-\frac{1}{2\pi} \int_{-\pi}^{\pi} \log t_\ell(x)dx\right).$$

According to the proof of Theorem 1, we have

$$\|A^{a,b}_n\|^2 = \pi(a^2+b^2)\|\phi_n\|^2.$$ □

# References

1. Milovanović, G. V., Cvetković, A. S., Stanić, M. P.: Trigonometric quadrature formulae and orthogonal systems (submitted)
2. Milovanović, G. V., Djordjević, R. Ž.: Linear Algebra. Faculty of Electronical Engineering, Niš (2004) [In Serbian]
3. Szegő, G.: Orthogonal Polynomials. Amer. Math. Soc. Colloq. Publ. **23**, 4th edn.. Amer. Math. Soc., Providence R. I. (1975)
4. Turetzkii, A. H.: On quadrature formulae that are exact for trigonometric polynomials. East J. Approx. **11** (2005) 337–359 (English translation from Uchenye Zapiski, Vypusk 1(149), Seria math. Theory of Functions, Collection of papers, Izdatel'stvo Belgosuniversiteta imeni V.I. Lenina, Minsk (1959) 31–54)

# Trigonometric Orthogonal Systems and Quadrature Formulae with Maximal Trigonometric Degree of Exactness*

Gradimir V. Milovanović[1], Aleksandar S. Cvetković[1], and Marija P. Stanić[2]

[1] Department of Mathematics, Faculty of Electronic Engineering,
University of Niš, P.O. Box 73, 18000 Niš, Serbia
[2] Department of Mathematics and Informatics, Faculty of Science,
University of Kragujevac, P.O. Box 60, 34000 Kragujevac, Serbia

**Abstract.** Turetzkii [Uchenye Zapiski, Vypusk **1** (149) (1959), 31–55, (English translation in East J. Approx. **11** (2005) 337–359)] considered quadrature rules of interpolatory type with simple nodes, with maximal trigonometric degree of exactness. For that purpose Turetzkii made use of orthogonal trigonometric polynomials of semi–integer degree.

Ghizzeti and Ossicini [Quadrature Formulae, Academie-Verlag, Berlin, 1970], and Dryanov [Numer. Math. **67** (1994), 441–464], considered quadrature rules of interpolatory type with multiple nodes with maximal trigonometric degree of exactness. Inspired by their results, we study here $s$–orthogonal trigonometric polynomials of semi–integer degree. In particular, we consider the case of an even weight function.

## 1 Introduction

In [8], Turetzkii considered quadrature rules of interpolatory type of the form

$$\int_0^{2\pi} f(x)w(x)dx \approx \sum_{\nu=0}^{2n} w_\nu f(x_\nu),$$

where $w$ is a weight function, integrable and non-negative on the interval $[0, 2\pi)$, vanishing there only on a set of a measure zero. The maximal trigonometric degree of exactness of such quadrature formulae is $2n$, in which case they are called *Gaussian quadratures*. In [6], a simple generalization for these rules dealing with the translations of the interval $(0, 2\pi)$ was given. In the sequel we shall work with the interval $(-\pi, \pi)$.

Let the weight function $w$ be integrable and non-negative on the interval $(-\pi, \pi)$, vanishing there only on a set of a measure zero. The nodes of the

---

* The authors were supported in part by the Serbian Ministry of Science and Environmental Protection (Project: Orthogonal Systems and Applications, grant number #144004) and the Swiss National Science Foundation (SCOPES Joint Research Project No. IB7320–111079 "New Methods for Quadrature").

T. Boyanov et al. (Eds.): NMA 2006, LNCS 4310, pp. 402–409, 2007.
© Springer-Verlag Berlin Heidelberg 2007

Gaussian quadrature formulae are the zeros from $[-\pi, \pi)$ of the trigonometric polynomial of semi–integer degree $n + 1/2$, which is of the form

$$\sum_{\nu=0}^{n} \left( c_\nu \cos \left( \nu + \frac{1}{2} \right) x + d_\nu \sin \left( \nu + \frac{1}{2} \right) x \right), \quad c_\nu, d_\nu \in \mathbb{R}, \ |c_n| + |d_n| \neq 0, \quad (1)$$

being orthogonal on $(-\pi, \pi)$ with respect to the weight function $w$ to every trigonometric polynomial of semi–integer degree less than or equal to $n - 1/2$. Such a trigonometric polynomial with given leading coefficients $c_n$ and $d_n$, is uniquely determined (see [8, §3]) and it has exactly $2n + 1$ distinct simple zeros in $[-\pi, \pi)$ (see [8] and [6]).

We denote by $\mathcal{T}_n^{1/2}$ the set of all trigonometric polynomials of semi integer degree at most $n+1/2$, i.e., the linear span of the set $\cos(\nu+1/2)x$, $\sin(\nu+1/2)x$, $\nu = 0, 1, \ldots, n$, and by $\mathcal{T}_n$ the set of all trigonometric polynomials of degree at most $n$.

In [6] two choices of leading coefficients for orthogonal trigonometric polynomials of semi–integer degree were considered: $c_n = 1$, $d_n = 0$ and $c_n = 0$, $d_n = 1$, and five–term recurrence relations for such orthogonal trigonometric polynomials were found. Also, a numerical method for constructing such quadratures based on the five–term recurrence relations was presented.

Obviously,

$$A_{n+1/2}(x) = A \prod_{k=0}^{2n} \sin \frac{x - x_k}{2}, \quad A = \text{const} \neq 0, \quad (2)$$

is a trigonometric polynomial of semi–integer degree $n + 1/2$. Conversely, every trigonometric polynomial of semi–integer degree $n + 1/2$ of the form (1) can be represented in the form (2) (see [8, Lemma 1]), with

$$A = (-1)^n 2^{2n} i (c_n - i d_n) e^{i/2 \sum_{\nu=0}^{2n} x_\nu}, \quad (3)$$

where $x_0, x_1, \ldots, x_{2n}$ are zeros of the trigonometric polynomial (1) that lie in the strip $0 \leq \operatorname{Re} x < 2\pi$.

Quadrature formulae of interpolatory type of the form

$$\int_{-\pi}^{\pi} f(x)dx = \sum_{\nu=0}^{2n} \sum_{j=0}^{2s} A_{j,\nu} f^{(j)}(x_\nu) + R(f), \quad (4)$$

where $s$ is a non-negative integer, were considered in [4]. Dryanov [2] generalized quadrature rules (4) in such a way that the nodes have different multiplicities. The trigonometric degree of exactness of a quadrature formula of the form (4) is $(2n + 1)(s + 1) - 1$, i.e., $R(f) = 0$ for all $f \in \mathcal{T}_{(2n+1)(s+1)-1}$, if the nodes $x_0, x_1, \ldots, x_{2n}$ are chosen such that (see [4,2])

$$\int_{-\pi}^{\pi} \left( \prod_{\nu=0}^{2n} \sin \frac{x - x_\nu}{2} \right)^{2s+1} \cos \left( \ell + \frac{1}{2} \right) x \, dx = 0, \quad \ell = 0, 1, \ldots, n - 1,$$

$$\int_{-\pi}^{\pi} \left( \prod_{\nu=0}^{2n} \sin \frac{x - x_\nu}{2} \right)^{2s+1} \sin \left( \ell + \frac{1}{2} \right) x \, dx = 0, \quad \ell = 0, 1, \ldots, n - 1.$$

Obviuosly, for $s = 0$, the maximal trigonometric degree of exactness is $2n$.

The product $\prod_{\nu=0}^{2n} \sin(x - x_\nu)/2$ is a trigonometric polynomial of semi–integer degree $n + 1/2$, so that the nodes $x_0, x_1, \ldots, x_{2n}$ of the quadrature rule (4) must coincide with zeros of a trigonometric polynomial of semi–integer degree $n + 1/2$ such that its $(2s+1)$-st power is orthogonal on $(-\pi, \pi)$ with respect to the weight function $w(x) = 1$ to every trigonometric polynomial from $\mathcal{T}_{n-1}^{1/2}$. Ghizzeti and Ossicini [4] proved that such a polynomial is of type

$$c \cos(n + 1/2)x + d \sin(n + 1/2)x.$$

In the sequel we consider trigonometric polynomials of semi–integer degree satisfying $s$−orthogonality conditions on $(-\pi, \pi)$ with respect to an admissible weight function $w$. A special attention is directed to the case of even weight function $w$. Also, we give some remarks when $w(x) = 1$.

## 2    $S$−Orthogonal Trigonometric Polynomials

Let $w$ be a given weight function on $(-\pi, \pi)$, $s \in \mathbb{N}_0$, and let $\{c_n\}$ and $\{d_n\}$, $n = 0, 1, \ldots$, be two given sequences of numbers such that $(c_n, d_n) \neq (0, 0)$ for all $n \in \mathbb{N}_0$.

We want to construct the sequence $\{A_{s,n+1/2}(x)\}_{n \in \mathbb{N}_0}$, where $A_{s,n+1/2}(x)$ is a trigonometric polynomial of semi–integer degree $n + 1/2$ with leading coefficients $c_n$ and $d_n$, i.e., $A_{s,n+1/2}(x) = c_n \cos(n + 1/2)x + d_n \sin(n + 1/2)x + \cdots$, such that

$$\int_{-\pi}^{\pi} (A_{s,n+1/2}(x))^{2s+1} A_{s,m+1/2}(x) \, w(x) \, dx = 0, \quad \text{if } m < n,$$

or, equivalently,

$$\int_{-\pi}^{\pi} (A_{s,n+1/2}(x))^{2s+1} \Pi_{n-1/2}(x) \, w(x) \, dx = 0, \tag{5}$$

for arbitrary $\Pi_{n-1/2}(x) \in \mathcal{T}_{n-1}^{1/2}$.

We call such trigonometric polynomials as $s$−*orthogonal trigonometric polynomials* with respect to the weight function $w$ on $(-\pi, \pi)$.

**Theorem 1.** *There exists a unique sequence* $\{A_{s,n+1/2}(x)\}_{n \in \mathbb{N}_0}$ *of trigonometric polynomials of semi–integer degree with given leading coefficients, $s$−orthogonal on $(-\pi, \pi)$ with respect to the weight function $w$.*

*Proof.* In order to prove the existence and uniqueness of $A_{s,n+1/2}(x)$, we need some well–known facts about the best approximation (see [1, p. 58–60]).

Let $X$ be a Banach space with real or complex scalars and $Y$ be a closed linear subspace of $X$. For each $f \in X$, the error of approximation $E(f)$ of $f$ by elements from $Y$ is $E(f) = \inf_{g \in Y} \|f - g\|$. If this infimum is attained for some $g = g_0$, then $g_0$ is called a *best approximation* to $f$ from $Y$. For each finite dimensional subspace $X_n$ of $X$ and each $f \in X$, there is a best approximation to $f$ from $X_n$. If $X$ is a strictly convex space (characterized by the property $f_1 \neq f_2$, $\|f_1\| = \|f_2\| = 1$, $\alpha_1, \alpha_2 > 0$, $\alpha_1 + \alpha_2 = 1 \Rightarrow \|\alpha_1 f_1 + \alpha_2 f_2\| < 1$), then each $f \in X$ has at most one element of best approximation in each closed linear subspace $Y \subset X$.

We set $X = L^{2s+2}[-\pi, \pi]$,

$$u = w(x)^{1/(2s+2)}(c_n \cos(n + 1/2)x + d_n \sin(n + 1/2)x) \in L^{2s+2}[-\pi, \pi],$$

and fix the following $2n$ linearly independent elements in $L^{2s+2}[-\pi, \pi]$,

$$u_j = w(x)^{1/(2s+2)} \cos(j + 1/2)x, \quad v_j = w(x)^{1/(2s+2)} \sin(j + 1/2)x,$$

where $j = 0, 1, \ldots, n - 1$. Let $Y = \text{span}\{u_0, v_0, u_1, v_1, \ldots, u_{n-1}, v_{n-1}\}$. Here, $Y$ is a finite dimensional subspace of $X$, so for each vector of $X$ there exists a best approximation from $Y$, i.e., there exist $2n$ constants $\alpha_j, \beta_j$, $j = 0, 1, \ldots, n - 1$, such that the error

$$\left\| u - \sum_{j=0}^{n-1}(\alpha_j u_j + \beta_j v_j) \right\| = \left( \int_{-\pi}^{\pi} \left( c_n \cos(n + 1/2)x + d_n \sin(n + 1/2)x \right. \right.$$

$$\left. \left. - \sum_{j=0}^{n-1} \left( \alpha_j \cos(j + 1/2)x + \beta_j \sin(j + 1/2)x \right) \right)^{2s+2} w(x)\, dx \right)^{1/(2s+2)},$$

is minimal, i.e., for every $n$ and for every choice of $(c_n, d_n) \neq (0, 0)$, there exists a trigonometric polynomial of semi–integer degree $n + 1/2$

$$A_{s,n+1/2}(x) = c_n \cos(n + 1/2)x + d_n \sin(n + 1/2)x$$

$$- \sum_{j=0}^{n-1} \left( \alpha_j \cos(j + 1/2)x + \beta_j \sin(j + 1/2)x \right),$$

such that

$$\int_{-\pi}^{\pi} (A_{s,n+1/2}(x))^{2s+2} w(x)\, dx$$

is minimal. Since the space $L^{2s+2}[-\pi, \pi]$ is strictly convex, it follows that the problem of best approximation has the unique solution and the polynomial is unique.

There follows that for each of the following $2n$ functions

$$F_j^C(\lambda) = \int_{-\pi}^{\pi} \left( A_{s,n+1/2}(x) + \lambda \cos(j + 1/2)x \right)^{2s+2} w(x)\, dx, \quad j = 0, 1, \ldots, n - 1,$$

$$F_j^S(\lambda) = \int_{-\pi}^{\pi} \left( A_{s,n+1/2}(x) + \lambda \sin(j + 1/2)x \right)^{2s+2} w(x)\, dx, \quad j = 0, 1, \ldots, n - 1,$$

its derivative must be equal zero for $\lambda = 0$. Thus, we get

$$\int_{-\pi}^{\pi} (A_{s,n+1/2}(x))^{2s+1} \sin(j+1/2)x \, w(x) \, dx = 0, \quad j = 0, 1, \ldots, n-1, \quad (6)$$

$$\int_{-\pi}^{\pi} (A_{s,n+1/2}(x))^{2s+1} \cos(j+1/2)x \, w(x) \, dx = 0, \quad j = 0, 1, \ldots, n-1, \quad (7)$$

which means that the polynomial $A_{s,n+1/2}(x)$ satisfies (5).     □

**Theorem 2.** *The trigonometric polynomial $A_{s,n+1/2}(x)$, which is $s$-orthogonal on $(-\pi, \pi)$ with respect to the weight function $w$, has exactly $2n+1$ simple zeros in $[-\pi, \pi)$.*

*Proof.* The trigonometric polynomial $A_{s,n+1/2}(x)$ has on $[-\pi, \pi)$ at least one zero of odd multiplicity. If we assume the contrary, for $n \geq 1$ we obtain the following contradiction to (5)

$$\int_{-\pi}^{\pi} (A_{s,n+1/2}(x))^{2s+1} \cos \frac{x}{2} \, w(x) \, dx \neq 0,$$

since $A_{s,n+1/2}(x) \cos x/2$ does not change its sign on $[-\pi, \pi)$.

Let us suppose that the number of zeros of $A_{s,n+1/2}(x)$ in $[-\pi, \pi)$ of odd multiplicities is $2m-1$, for $m \leq n$. Denote those zeros by $y_1, \ldots, y_{2m-1}$ and set

$$\Pi(x) = \prod_{k=1}^{2m-1} \sin \frac{x - y_k}{2}.$$

Since $\Pi(x) \in \mathcal{T}_{n-1}^{1/2}$, there should hold $\int_{-\pi}^{\pi} (A_{s,n+1/2}(x))^{2s+1} \Pi(x) w(x) \, dx = 0$, but this is impossible, since the integrand does not change its sign on $[-\pi, \pi)$.

If we assume that the number of zeros of $A_{s,n+1/2}(x)$ in $[-\pi, \pi)$ of odd multiplicities is $2m$, for $m \leq n-1$, denoting those zeros by $y_1, \ldots, y_{2m}$ and setting

$$\Pi(x) = \cos \frac{x}{2} \prod_{k=1}^{2m} \sin \frac{x - y_k}{2},$$

we obtain again a contradiction. The number of zeros of $A_{s,n+1/2}(x)$ in $[-\pi, \pi)$ of odd multiplicities cannot be equal to $2n$, because $A_{s,n+1/2}(x)$ has $2n+1$ zeros.

Therefore, the trigonometric polynomial $A_{s,n+1/2}$ must have $2n+1$ zeros of odd multiplicities, i.e., it has $2n+1$ simple zeros on $[-\pi, \pi)$.     □

Similarly as in proof of Theorem 1 one can prove that for any positive integer $n$ there exists a unique trigonometric polynomial of semi–integer degree $n+1/2$ of the form $A_{s,n+1/2}^C(x) = \sum_{\nu=0}^{n} c_\nu^{(n)} \cos(\nu+1/2)x$, involving only cos functions, with a given leading coefficient $c_n^{(n)}$, such that its $(2s+1)$-st power is orthogonal to $\cos(k+1/2)x$, for $k = 0, 1, \ldots, n-1$, with respect to weight $w$ on $(-\pi, \pi)$. Also, for every positive integer $n$ there exists a unique trigonometric polynomial of semi–integer degree $n + 1/2$, involving only sin functions, i.e., of the form $A_{s,n+1/2}^S(x) = \sum_{\nu=0}^{n} d_\nu^{(n)} \sin(\nu+1/2)x$, with a given leading coefficient $d_n^{(n)}$, such that its $(2s+1)$-st power is orthogonal to $\sin(k+1/2)x$, $k = 0, 1, \ldots, n-1$ with respect to $w$ on $(-\pi, \pi)$.

## 2.1   Even Weight Functions

Now, we consider the case of an even weight function $w$, i.e., when $w(-x) = w(x)$, $x \in (-\pi, \pi)$. Such weight functions are interesting, because the problem of the symmetric weights can be reduced to algebraic polynomials.

We start with the following simple lemma.

**Lemma 1.** *If the weight function $w$ is even, then for all non-negative integers $n$, the s-orthogonal trigonometric polynomial $A^{1,0}_{s,n+1/2}$ with leading coefficients $c_n = 1$ and $d_n = 0$ coincides with $A^{C}_{s,n+1/2}$ with leading coefficient $c^{(n)}_n = 1$. Also, the s-orthogonal trigonometric polynomial $A^{0,1}_{s,n+1/2}$ with leading coefficients $c_n = 0$ and $d_n = 1$ coincides with $A^{S}_{s,n+1/2}$ with leading coefficient $d^{(n)}_n = 1$.*

*Proof.* Since $A^{C}_{s,n+1/2}$ is an even function, as well as $w$, we easily obtain that $A^{C}_{s,n+1/2}$ is orthogonal to $\sin(k+1/2)x$ for $k = 0, 1, \ldots, n-1$ with respect to $w$ on the symmetric interval $(-\pi, \pi)$. Thus, $A^{C}_{s,n+1/2}$ satisfies $s$-orthogonality condition (5). In addition, $A^{C}_{s,n+1/2}$ is a trigonometric polynomial of semi–integer degree $n + 1/2$ with leading coefficients $c_n = 1$ and $d_n = 0$. According to the uniqueness of such trigonometric polynomials we conclude that $A^{1,0}_{s,n+1/2}$ coincides with $A^{C}_{s,n+1/2}$.

The assertion for $A^{S}_{s,n+1/2}$ can be obtained similarly.    □

**Theorem 3.** *The zeros of the trigonometric polynomial $A_{s,n+1/2}$ with leading coefficients $c_n = 1$ and $d_n = 0$, s-orthogonal on $(-\pi, \pi)$ with respect to an even weight function $w$, are given by*

$$x_0 = -\pi, \quad x_{2n+1-\nu} = -x_\nu = \arccos \tau_\nu, \quad \nu = 1, \ldots, n,$$

*where $\tau_\nu$, $\nu = 1, \ldots, n$, are the zeros of the algebraic polynomial $C_n$, s-orthogonal on $(-1, 1)$ with respect to the weight function $\sqrt{(1+x)^{2s+1}/(1-x)}\, w(\arccos x)$.*

*Proof.* According to Lemma 1, the trigonometric polynomial $A_{s,n+1/2}$, $n = 0, 1, \ldots$, is an even function of the following form

$$A_{s,n+1/2}(x) = \sum_{\nu=0}^{n} c^{(n)}_\nu \cos(\nu + 1/2)x, \quad c^{(n)}_n = 1.$$

Obviously $A_{s,n+1/2}(-\pi) = 0$, i.e., $x_0 = -\pi$. Since $A_{s,n+1/2}(-x) = A_{s,n+1/2}(x)$, the remaining zeros are located symmetrically in $(-\pi, \pi)$. From the $s$-orthogonality conditions for $A_{s,n+1/2}$, we have

$$\int_0^\pi (A_{s,n+1/2}(x))^{2s+1} A_{s,k+1/2}(x) w(x) dx = 0, \quad n \in \mathbb{N}, \ k = 0, 1, \ldots, n-1. \quad (8)$$

Substituting $x := \arccos x$, after some elementary transformations (see [6]), we obtain

$$A_{s,k+1/2}(\arccos x) = \sqrt{\frac{1+x}{2}} C_k(x), \quad C_k(x) = \sum_{\nu=0}^{k} c^{(k)}_\nu (T_\nu(x) - (1-x)U_{\nu-1}(x)),$$

where $T_\nu$ and $U_\nu$, $\nu \in \mathbb{N}_0$ are the Chebyshev polynomials of the first and the second kind, respectively. From (8) we obtain

$$\int\limits_{-1}^{1} (C_n(x))^{2s+1} C_k(x) \sqrt{\frac{(1+x)^{2s+1}}{1-x}} \, w(\arccos x) dx = 0, \quad n \in \mathbb{N}, \ k = 0, 1, \ldots, n-1.$$

According to the well-known fact that the $s$-orthogonal algebraic polynomial $C_n$ has $n$ simple zeros in $(-1,1)$ (see [5] and [3]), we get what is stated.  $\square$

Analogously the following result can be proved.

**Theorem 4.** *The zeros of the trigonometric polynomial $A_{s,n+1/2}$ with leading coefficients $c_n = 0$ and $d_n = 1$, $s$-orthogonal on $(-\pi, \pi)$ with respect to an even weight function $w$, are given by*

$$x_0 = 0, \quad x_{2n+1-\nu} = -x_\nu = \arccos \tau_\nu, \quad \nu = 1, \ldots, n,$$

*where $\tau_\nu$, $\nu = 1, \ldots, n$, are the zeros of the algebraic polynomial $S_n$, $s$-orthogonal on $(-1,1)$ with respect to the weight function $\sqrt{(1-x)^{2s+1}/(1+x)} \, w(\arccos x)$.*

**Weight function $w(x) = 1$.** For this special case we give some remarks. Let $A_{s,n+1/2}^{1,0}$ and $A_{s,n+1/2}^{0,1}$ be the $s$-orthogonal trigonometric polynomials with leading coefficients $c_n = 1$, $d_n = 0$ and $c_n = 0$, $d_n = 1$, respectively. From Theorems 3 and 4, we have

$$A_{s,n+1/2}^{1,0}(\arccos x) = \sqrt{\frac{1+x}{2}} \, C_n(x), \quad A_{s,n+1/2}^{0,1}(\arccos x) = \sqrt{\frac{1-x}{2}} \, S_n(x),$$

where $C_n(x)$ and $S_n(x)$ are the $s$-orthogonal algebraic polynomials on $(-1,1)$ with respect to the weight functions

$$w_3(t) = (1+t)^{1/2+s}(1-t)^{-1/2} \quad \text{and} \quad w_4(t) = (1-t)^{1/2+s}(1+t)^{-1/2},$$

respectively. It is shown (see [7]) that the Chebyshev polynomials of the third kind $V_n$, and of the fourth kind $W_n$, defined by

$$V_n(x) = \frac{\cos(n+\frac{1}{2})\theta}{\cos\frac{\theta}{2}}, \quad W_n(x) = \frac{\sin(n+\frac{1}{2})\theta}{\sin\frac{\theta}{2}}, \quad x = \cos\theta,$$

are the $s$-orthogonal polynomials with respect to the weight functions $w_3$ and $w_4$, respectively. Hence, we get explicit expressions for the trigonometric polynomials $A_{s,n+1/2}^{1,0}$ and $A_{s,n+1/2}^{0,1}$ as follows

$$A_{s,n+1/2}^{1,0}(x) = \cos\left(n+\frac{1}{2}\right)x, \quad A_{s,n+1/2}^{0,1}(x) = \sin\left(n+\frac{1}{2}\right)x.$$

Now, it is easy to see that the zeros of $A_{s,n+1/2}^{1,0}$ from $[-\pi, \pi)$ are given by

$$x_0 = -\pi, \quad x_{2n+1-\nu} = -x_\nu = \frac{2\nu+1}{2n+1}\pi, \quad \nu = 0, 1, \ldots, n-1,$$

and of $A_{s,n+1/2}^{0,1}$ by

$$x_0 = 0, \quad x_{2n+1-\nu} = -x_\nu = \frac{2\nu}{2n+1}\pi, \quad \nu = 1, \ldots, n.$$

# References

1. DeVore, R. A., Lorentz, G.G.: Constructive Approximation. Springer–Verlag, Berlin Heildeberg (1993)
2. Dryanov, D. P.: Quadrature formulae with free nodes for periodic functions. Numer. Math. **67** (1994) 441–464
3. Gautschi, W.: Orthogonal Polynomials, Computation and Approximation. Oxford University Press (2004)
4. Ghizzeti, A., Ossicini, A.: Quadrature Formulae. Academie-Verlag, Berlin (1970)
5. Milovanović, G. V.: Quadratures with multiple nodes, power orthogonality, and moment-preserving spline approximation. Numerical analysis 2000, Vol. V, Quadrature and orthogonal polynomials (W. Gautschi, F. Marcellan, and L. Reichel, eds.). J. Comput. Appl. Math. **127** (2001) 267–286
6. Milovanović, G. V., Cvetković, A. S., Stanić, M. P.: Trigonometric quadrature formulae and orthogonal systems (submitted)
7. Ossicini, A., Rosati, F.: Funzioni caratteristiche nelle formule di quadratura gaussiane con nodi multipli. Boll. Un. Math. Ital. **11(4)** (1975) 224–237
8. Turetzkii, A. H.: On quadrature formulae that are exact for trigonometric polynomials. East J. Approx. **11** (2005) 337–359 (English translation from Uchenye Zapiski, Vypusk 1 (149), Seria math. Theory of Functions, Collection of papers, Izdatel'stvo Belgosuniversiteta imeni V.I. Lenina, Minsk (1959) 31–54)

# On the Calculation of the Bernstein-Szegő Factor for Multivariate Polynomials

Nikola Naidenov

University of Sofia, Department of Mathematics,
Blvd. James Bourchier 5, 1164 Sofia, Bulgaria
nikola@fmi.uni-sofia.bg

**Abstract.** Let $\mathbb{R}^d$ be the Euclidean space with the usual norm $|.|_2$, $\mathcal{P}_n^d$ be the set of all polynomials over $\mathbb{R}^d$ of degree $n$, and $K \subset \mathbb{R}^d$ be a convex body. An algorithm for calculation of the Bernstein-Szegő factor:

$$BS(K) := \sup_{\substack{\mathbf{x}\in\text{int}(K) \\ P\in\mathcal{P}_n^d, n\in\mathbb{N}}} \left\{ \frac{|\text{grad}P(\mathbf{x})|_2 w(K)\sqrt{1 - \alpha^2(K,\mathbf{x})}}{n\sqrt{||P||_{C(K)}^2 - P^2(\mathbf{x})}} \right\}$$

is considered, where $w(K)$ is the width of $K$ and $\alpha(K,\mathbf{x})$ is the generalized Minkowsky functional. It is known that $BS(K) \in [2, 2\sqrt{2}]$. On the basis of computer experiments, we show that the existing in the literature hypothesis, that $BS(K) = 2$ for any convex body $K \subset \mathbb{R}^d$, fails to hold.

## 1 Introduction

In 1991, Y. Sarantopoulos ([4]) obtained a fine multivariate extension of the classical Bernstein -Szegő inequality for algebraic polynomials of degree $n$:

$$|P_n'(x)| \leq \frac{n\sqrt{||P_n||_{C[-1,1]}^2 - P_n^2(x)}}{\sqrt{1 - x^2}}, \quad x \in (-1, 1).$$

To formulate his result, let $X$ be a normed vector space with norm $|.|_X$ and $\mathcal{P}_n(X)$ – the set of all polynomials of degree $n$ over $X$ (see [3] for a precise definition). Let $K \subset X$ be a centrally symmetric convex body (bounded closed convex set with nonempty interior). Without loss of generality we may assume that 0 is the center of $K$. Then the Minkowski functional with respect to $K$, $|\mathbf{x}|_K := \inf\{\lambda > 0 : \mathbf{x} \in \lambda K\}$, is also a norm over $X$. Next, for any $\mathbf{v} \in X$ let $\tau(K, \mathbf{v}) := \sup\{\lambda : \exists \mathbf{y}, \mathbf{z} \in K \text{ such that } \mathbf{z} = \mathbf{y} + \lambda\mathbf{v}\}$ be the maximal chord of $K$ in direction $\mathbf{v}$ and let $w(K) := \inf\{\tau(K, \mathbf{v}) : |\mathbf{v}|_X = 1\}$ be the width of K. Then, for every $\mathbf{x} \in \text{int}(K)$ and $P_n \in \mathcal{P}_n(X)$, we have

(1)
$$|\text{grad}\, P_n(\mathbf{x})|_{X^*} \leq \frac{2n\sqrt{||P_n||_{C(K)}^2 - P_n^2(\mathbf{x})}}{w(K)\sqrt{1 - |\mathbf{x}|_K^2}},$$

where $\text{grad}P_n(\mathbf{x})$ is the linear operator from $X^*$ for which $\langle\text{grad}P_n(\mathbf{x}), \mathbf{y}\rangle$, $\mathbf{y} \in X$,

T. Boyanov et al. (Eds.): NMA 2006, LNCS 4310, pp. 410–418, 2007.
© Springer-Verlag Berlin Heidelberg 2007

is the directional derivative of $P_n$ at $\mathbf{x}$ in direction $\mathbf{y}$. Moreover, the constant 2 in the inequality (1) is the best possible.

Next, for any convex body $K \subset X$ and $\mathbf{x} \in \text{int}(K)$ the generalized Minkowski functional $\alpha(K, \mathbf{x})$ can be defined by $\alpha(K, \mathbf{x}) := \sqrt{1 - \gamma^2(K, \mathbf{x})}$, where $\gamma(K, \mathbf{x}) :=$ $\inf\left\{ 2\frac{\sqrt{|\mathbf{x}-\mathbf{a}|_x\,|\mathbf{x}-\mathbf{b}|_x}}{|\mathbf{b}-\mathbf{a}|_x} : \mathbf{a}, \mathbf{b} \in \partial K \text{ and } \mathbf{x} \in [\mathbf{a}, \mathbf{b}] \right\}$ measures the distance between $\mathbf{x}$ and $\partial K$. (See [3] for more details about the different measures of a convex body.) With these notations we have the following result of Kroó and Révész [1]:

$$(2) \qquad |\text{grad} P_n(\mathbf{x})|_{X^*} \le \frac{Cn\sqrt{\|P_n\|^2_{C(K)} - P_n^2(\mathbf{x})}}{w(K)\sqrt{1 - \alpha^2(K, \mathbf{x})}},$$

where $C = 2\sqrt{2}$, $\mathbf{x} \in \text{int}(K)$, $K$ is an arbitrary convex body and $P_n \in \mathcal{P}_n(X)$. We see that the constant in (2) differs from that in (1) by a factor $\sqrt{2}$. Révész and Sarantopoulos raised the following

**Conjecture:** *The best possible constant in (2) is $C = 2$ (as in the case of centrally symmetric bodies).*

The initial motivation of the present study was to check out this conjecture numerically. It turns out that the conjecture fails to hold. Although the question was answered, the used algorithm is of interest itself. It can be useful for other extremal problems for polynomials under uniform restrictions.

## 2    The Algorithm and Its Realization

Let $C^*$ be the best possible (the smallest) constant in (2). For every specific $X, K$ and $\mathbf{x}$ the best constant in (2) is different and we denote it by $C(X, K, \mathbf{x})$. Clearly $C^* = \sup_{X, K, \mathbf{x}} C(X, K, \mathbf{x})$. Of course, we are not able to find numerically $C(X, K, \mathbf{x})$ for any $X$ and $K$. We restrict ourselves to the case $X = \mathbb{R}^2$ normed by $|\mathbf{x}| = \sqrt{x^2 + y^2}$, $\mathbf{x} = (x, y)$, and $K = \Delta$ – the standard triangle $\Delta OAB$ with vertices $O = (0, 0)$, $A = (1, 0)$, $B = (0, 1)$. (In a certain sense, see [3], the simplex is the least centrally symmetric convex body.) We have, e.g. [2], that $w(\Delta) = \frac{1}{\sqrt{2}}$ and $1 - \alpha^2(\Delta, \mathbf{x}) = 4\min\{x(1-x), y(1-y), (x+y)(1-x-y)\}$. Which expression in the min is active depends on that in which triangle: $\Delta OMB$, $\Delta OMA$ or $\Delta AMB$, is located $\mathbf{x}$, where $M = (\frac{1}{3}, \frac{1}{3})$ is the centroid of $\Delta$ (see Picture 1). Next, if $\|\cdot\| := \|\cdot\|_{C(\Delta)}$ and $\mathcal{P}_n^2 := \mathcal{P}(\mathbb{R}^2)$, from [2] we know that the constant

$$C^\Delta := \sup\left\{ \frac{|\text{grad}\, P_n(\mathbf{x})| w(\Delta)\sqrt{1 - \alpha^2(\Delta, \mathbf{x})}}{n\sqrt{\|P_n\|^2_{C(\Delta)} - P_n^2(\mathbf{x})}} : \mathbf{x} \in \text{int}(\Delta),\ P_n \in \mathcal{P}_n^2,\ n \in \mathbb{N} \right\}$$

belongs to the interval $[2, \sqrt{3 + \sqrt{5}}]$, $\sqrt{3 + \sqrt{5}} = 2.2882\ldots$ Actually, in our algorithm we fix $n$ and $x$ and vary only $P_n \in \mathcal{P}_n^2$. In view of the estimations in [2], the best choice for the point $\mathbf{x}$ is on the medians MA or MB. Without loss of generality we set $\|P_n\| = 1$. Then, for any fixed $\mathbf{x} \in \text{int}(\Delta)$ and $n \in \mathbb{N}$ we denote

$$G_n(P) := \frac{\sqrt{(P'_x)^2 + (P'_y)^2}\sqrt{4\min\{x(1-x), y(1-y), (x+y)(1-x-y)\}}}{n\sqrt{2}\sqrt{1 - P^2(\mathbf{x})}}$$

and $C_n^{\Delta}(\mathbf{x}) := \max\left\{G_n(P) : P \in \mathcal{P}_n^2, \ ||P|| = 1\right\}$. Clearly, $C_n^{\Delta}(\mathbf{x}) \le C^{\Delta} \le C^*$.

Let $f° := f/||f||$. The algorithm for maximization of $G_n(P°)$ is as follows. First we take some initial $P$ with $||P|| = 1$. Then in a ball with center $P$ and radius $r$ we look for a better polynomial $(P + dP)°$, $||dP|| = r$. The direction of the variation $dP$ we take arbitrarily but in a certain subspace of $\mathcal{P}_n^2$, described below. Actually we try with the both polynomials $P \pm dP$.

Let $N = \binom{n+2}{2}$ be the dimension of $\mathcal{P}_n^2$ and $dP^*$ be the direction of maximal increase of $G_n(P)$. One can estimate that arbitrary choice of $dP$ decreases the speed of the algorithm $O(\sqrt{N})$ times (in mean), comparing it with the best choice $dP^*$. However, the computation of $dP^*$ requires N times more calculations.

If we maximize a smooth function (say $G_n(P)$) and $\langle dP, dP^* \rangle \ne 0$, then for a sufficiently small $r$, in one of the directions $\pm dP$ we will gain an increase of $G_n$. Then, there is no need of special treatment and we can use any standard maximizing procedure. But we consider $G(P/||P||)$ and $||P+t.dP||_{C(\Delta)}$ is only one-sidedly differentiable function. So, the uniform restriction requires a special approach. We shall restrict $dP$ to vary in the following set of admissible directions:

$$W := \left\{w \in \mathcal{P}_n^2 : \ w \text{ vanishes at the critical points of } P\right\},$$

where the set of critical points of $P$ is $Ep := \{\mathbf{t} \in \Delta : \ |P(\mathbf{t})| = ||P|| = 1\}$.

Recall the univariate case: If $P, w \in \mathcal{P}_n^1$ are such that $w(t_i) = 0$, $i = 1, ..., d$, where $\{t_i\}_1^d$ are the critical points of $P$ on the interval $[a, b]$, then $||P + \epsilon w||_{C[a,b]}$ increases by $o(\epsilon)$ instead of $O(\epsilon)$ for arbitrary $w$.

Thus we smoothed our problem, but also we restricted the possible variations $dP$. Let us see what kind of restriction is this. Approximately, for a sufficiently small $||dP||$, the number of critical points can only increase and the points move slightly. In view of the univariate analog we expect that the extremal polynomial has "many" alternation points. So, restricting ourselves to the set of "highly alternating" polynomials, we hope that we will not miss the extremal polynomial. (But, in the algorithm there is a possibility to escape from an unappropriate restriction, excluding a critical point.) The conditions $dP(t_i) = 0$ for all critical points $\{t_i\}$ of $P$ form a linear system of equations for the coefficients of $dP$. If $d$ is the number of the critical points of $P$ and the system is regular, then $\dim(W) = N - d$.

The algorithm is realized on the system MATHEMATICA. The program works in interactive regime as the operator can perform a series of commands (calculations) from a given list.

- The main procedure 'A' generates an admissible $dP \in W$ with $||dP|| = r$ and tries to find a better polynomial $(P \pm dP)°$. If it succeeds, then $P := (P \pm dP)°$. If not, it tries again with other $dP$, up to 30 times. This is one step. Next we can start procedure A again as meanwhile we can change the parameters $r$ and $d$.

- The procedure 'Ar' tries to improve $P$ using the previous direction $dP$.
- The main "ruling parameters" are $r$ and $d$. The procedure 'D' doubles $r$ while 'H' halves it. 'D1' - increases $d$ by 1, while 'Dm1' decreases it. Note that
    - if two or more repetitions of Ar improve $P$, then it is better to increase $r$;
    - the parameter $d$ can be controlled by the graph of $P(x, y)$.
- When procedure A does not lead to improvement of $P$ anymore, we can start a justifying procedure 'Just'. The purpose of this procedure is the following. Since all calculations are approximate, then at the critical points $Ep = \{t_i\}_1^d$ we have $|P(t_i)| \approx 1$, not exactly (with one exception). That is why we try to improve $P$, excluding consecutively the points $t_i$ from $Ep$, and using a procedure similar to A, but with $r$ adapted according to $1 - |P(t_i)|$, see Picture 2 (which shows the univariate analog). However, it is possible $|P(t_i)|$ to decrease. This means that the point $t_i$ must be excluded from $Ep$.

Picture 1.

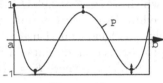

Picture 2.

Picture 3.

## 3    Counterexamples to the Conjecture $C^* = 2$

**Example 1:** (see Picture 4a)
$P = 1.00000 - 7.9866x + 7.97362x^2 - 4.78417y + 13.0868xy + 2.86128y^2$,
$n = 2$, $\mathbf{x} = (0.92, 0.04)$, $G_2(P) = 2.0053$.

**Example 2:** (see Picture 4b)
$P = 0.99999 - 17.8567x + 47.2389x^2 - 31.2420x^3 - 3.78921y + 67.7998xy - 83.45657x^2y - 3.77587y^2 - 38.3024xy^2 + 7.56249y^3$,
$n = 3$, $\mathbf{x} = (0.88, 0.06)$, $G_3(P) = 2.0150$.

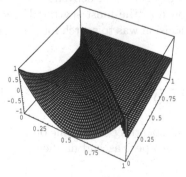

Picture 4a.

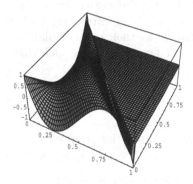

Picture 4b.

Note that the estimate in [2] is maximal when $\mathbf{x} \to A$ or $B$. However, in our case the maximum is attained at an inner point of $MA$ and $C_2^A(\mathbf{x}) \to 2$ for $\mathbf{x} \to A$.

# 4    Problems That Arise at the Time of Programming

1) *Calculation of the norm of a polynomial* $Pt(x,y)$, $Pt = P \pm dP$

It is important to calculate the norm of $Pt$ highly accurate, having in mind that the denominator in $G_n(P)$ can be arbitrarily close to zero. On the other hand this calculation is one of the most time-consuming parts of the programm. A formal approach to this problem is: a) Solve numerically the system of equations $P'_x = P'_y = 0$ for the interior extrema. b) Investigate $|Pt|$ on the three line segments OA, AB, BO. However I preferred a more direct approach: A numerical optimization of $|Pt|$. The corresponding procedure 'NormP' is built on two levels - global and local. We divide $\Delta$ by a square net of points $\{\mathbf{x}_k\}$, see Picture 3, and for those of them for which $|Pt(\mathbf{x}_k)| > 0.95M$, where $M = \max_{j<k} |Pt(\mathbf{x}_j)|$, we start a justifying procedure 'Norm1[x,y]' which makes one or several steps by the steepest ascend method. Meanwhile I use also the critical points of the polynomial $P$. A related problem is to find limits for the ratio of the discrete and continuous norms $||P||_{C(\hat{K})}/||P||_{C(K)}$, where $\hat{K} = \{(ih, jh) \in K : i, j \in \mathbb{Z}\}$.

2) *Determination of the set of critical points*

By definition it is $Ep = \{\mathbf{t} \in \Delta : |P(\mathbf{t})| = ||P|| = 1\}$. But, our polynomial $P$ is only an approximation to the extremal one, so that we have $|P(\mathbf{t})| \approx 1$, $\mathbf{t} \in Ep$. Moreover, in the bivariate case the equality $|P(\mathbf{t})| = ||P||$ is possible to hold over a whole line from which we need only finite number of points to construct $dP : dP(\mathbf{t}_k) = 0$, $k = 1, ..., d$. So, we come to the question which points to accept for critical ones. One of the first ideas was to take these $d$ points of a square net over $\Delta$ for which the difference $\delta_k = 1 - |P(\mathbf{t}_k)|$ is the smallest. However, it can happen that too many points concentrate around that one for which $\delta_k$ is minimal. Hence, we come to the following problem: For a given discrete set of points, say $\hat{K}_1 := \{\mathbf{t} \in \hat{K} : |P(\mathbf{t})| > 0.95\}$, how to choose a subset of $d$ elements which are "the most scattered". Such points can be for example the Fekete points or those connected with the Chebyshev constants for the set. It is interesting to find a criterion for "the most scattered" points which allows their fast calculation. My final decision was to search for the points $Ep = \{\mathbf{t}_i(x_i, y_i)\}_1^d \subset \hat{K}_1$ such that

$$\sum_{k=1}^{d}(1 + \epsilon - |P(\mathbf{t}_k)|) \left( \sum_{j \neq k}^{d} \frac{1}{(x_j - x_k)^2 + (y_j - y_k)^2} + 0.1 \right) \to \min, \quad \epsilon \to 0,$$

as the minimization is approximate one without spending much time on it. Next, if necessary, the points can be justified by the steepest ascend method.

3) *Verification of the result*

Some examples for $n = 2$, $\mathbf{x} = (0.96, 0.02)$ show that the function $G_n(P^\circ)$ may have several (essentially different) local maxima. So, how to check if a maximum we obtain is the global one? Also, when do we have to stop searching for better polynomials $P$, and how far is the result from a local maximum?

To these questions I can only point out some heuristical arguments: We shall consider $|\text{grad}\,P(\mathbf{x})|$ instead of the normalized gradient $G_n(P)$, since both quantities are asymptotically equivalent as $n \to \infty$, up to multiplication by a factor independent of $P$, $(w(K)\sqrt{1 - \alpha^2(K, \mathbf{x})}/n)$. At least, so is the case in the Bernstein-Szegő inequality. Then, when $P$ runs over the ball $\|P\|_{C(\Delta)} \leq 1$, the linear operator $\text{grad}\,P(\mathbf{x})$ runs over some convex and centrally symmetric set $S$ in $\mathbb{R}^2$, see Picture 5. If $\text{grad}\,P^0(\mathbf{x}) = ON$ is a local maximum of $|\text{grad}\,P(\mathbf{x})|$, then $N$ gives a local maximum of the distance $|OM| : M \in \partial S$. (This follows from the convexity.) So, the boundary $\partial S$ is divided into parts around every local maximum, such that if $\text{grad}\,P(\mathbf{x})$ belongs to this part, then the algorithm "directs" $P$ to the corresponding local maximum. We will call this part *the support* of the local maximum. It is clear that a very small local maximum can not have a big support. Let us estimate the summary support of the "big" local maxima. Precisely, with $r = d(S)/2$ and $\epsilon \in (0,1)$ we will estimate the quantity

$$Q := \frac{1}{|\partial S|} \sum_{\alpha} \left\{ |\text{supp}(P^\alpha)| : P^\alpha \text{ is a local max. with } |\text{grad}P(\mathbf{x})| \geq r(1 - \epsilon) \right\}.$$

Let $P^*$ be a global maximum of $\text{grad}\,P(\mathbf{x})$, see Picture 6, and let $P^0$ be the first local maximum, clockwise from $P^*$ with $|\text{grad}P^0(\mathbf{x})| < r(1 - \epsilon)$ (assuming it exists). Let $M$ be the initial point of the $\text{supp}(P^0)$. Clearly, $|OM| < r(1 - \epsilon)$ and there is a supporting line $\beta$ to $S$ at $M$ such that $\beta \perp OM$. Consider now the point $T$ such that $VT$ is the tangent from $V$ toward the inner circumference. $\Delta OTV$ is a rectangular triangle. Then necessarily M is outside $\Delta OTV$, because otherwise the supporting line $\beta$ would separate the points $O$ and $V$ from $S$. Next, it is easily seen that $|\widehat{VM}| \geq |VM| \geq |VT|$. Also, $|\partial S| \leq 2\pi r$. Then, in view of $-V \in S$ and the analogous construction symmetrical to $OV$ we can write

$$Q \geq \frac{4|VT|}{2\pi r} = \frac{2}{\pi}\sqrt{2\epsilon - \epsilon^2} \sim \frac{2\sqrt{2}}{\pi}\sqrt{\epsilon} \text{ as } \epsilon \to 0.$$

Assume that probability to hit a given local maximum is proportional to its support. Then, in order to get a value of $|\text{grad}\,P(\mathbf{x})|$, which is "$\epsilon$ - close" to the global maximum, it suffices to make $O(\frac{1}{\sqrt{\epsilon}})$ independent starts of the algorithm.

S: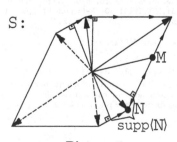

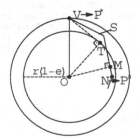

Picture 5.          Picture 6.

About the second question: Since the quantity $G(P/||P||)$ is one-sidedly differentiable with respect to $P$ in any direction, then some estimations of $|G'_n(P)|$ and $|G''_n(P)|$ can help to estimate how close is it to a local extremum. (Also, I expect a geometric convergence rate, as in the steepest ascend method.)

## 5   The Programm

(* The programm maximizes the normalized gradient of the polynomials of 2 variables in a point $(x_0, y_0)$ of the triangle T with vertices $\{(0,0),(0,1),(1,0)\}$ under the restriction $||P||_{C(T)} \leq 1$. *)

```
(n=3;Mon=Sum[x^i*y^j,{i,0,n},{j,0,n-i}];nm=Length[Mon];Ep={};O1=0.1;h=0.000001;)
```
(*n-degree; Mon-array of the monomials; O1,h-little numbers.*)

```
(Norm1[x0_,y0_]:=(x1=x0;y1=y0;Do[(M1=Abs[Pt/.{x->x1,y->y1}];gx=(Abs[Pt/.{x->x1+h,y->y1}]-M1)
   /h;gy=(Abs[Pt/.{x->x1,y->y1+h}]-M1)/h;ngrad=Sqrt[gx^2+gy^2];gxr=ro*gx/ngrad;gyr=ro*gy/
   ngrad;x2=x1+gxr;y2=y1+gyr;dder=ro*ngrad;boun=False;If[y2<0,(y2=0;x2=x1+Sign[gx]*ro;x2=
   Min[Max[x2,0],1];boun=True)];If[x2<0,(x2=0;y2=y1+Sign[gy]*ro;y2=Min[Max[y2,0],1];boun=
   True)];If[x2+y2>1,(y2=(1-x1+y1)/2+Sign[gy-gx]*ro*0.71;y2=Min[Max[y2,0],1];x2=1-y2;boun=
   True)];If[boun,(gxr=x2-x1;gyr=y2-y1;dder=gx*gxr+gy*gyr)];M2=Abs[Pt/.{x->x2,y->y2}];a=M2
   -M1-dder;If[M2-M1<-a,(tp=-dder/(2a);x2=x1+tp*gxr;y2=y1+tp*gyr;M2=M1+dder*tp/2)];x1=x2;
   y1=y2;),{stn1}];M2);
NormP:=(M=Max[(ro=1/N[Sqrt[2]*maxn*stn1];Norm1[x0,y0]),Table[Norm1[Ep[[k,1]],Ep[[k,2]]],
   {k,d}]];For[i=0,i<=maxn,i++,For[j=0,j<=maxn-i,j++,If[Abs[Pt/.{x->i/maxn,y->j/maxn}]>
   lev1*M,M=Max[M,Norm1[i/maxn,j/maxn]]] ]]; );    )
```
(* NormP - calculates $M =$ the norm of the polynomial $Pt$ over T. $maxn$ - the size of the net of points in T. Norm1 - the norm of $Pt$ in a neighborhood of $(x0, y0)$ in T with radius $ro$, also returns $(x1, y1)$ - the point where the norm attains. $stn1$ - the number of steps for Norm1. *)

```
G[P_]:=(Sqrt[4Min[x0(1-x0),y0(1-y0),(x0+y0)(1-x0-y0)]]/Sqrt[2]*Sqrt[(D[P,x]/.{x->x0,y->y0}
   )^2+(D[P,y]/.{x->x0,y->y0})^2]/(n*Sqrt[1-(P/.{x->x0,y->y0})^2]))
```
(* The normalized gradient of $P$ at $(x0, y0)$. It is supposed that $||P|| = 1$. *)

```
(Ep0:=(Pt=P;Do[(x2=Ep[[k,1]];y2=Ep[[k,2]];If[(Ep[[k,4]]<maxn^2),Norm1[x2,y2]];Ep[[k]]={x2,y2,
   Abs[P/.{x->x2,y->y2}]},0},{k,d}];For[k=1,k<=d,k++,(z=(Ep[[k,1]]
   -Ep[[k1,1]])^2+(Ep[[k,2]]-Ep[[k1,2]])^2;If[z<h,(Ep[[k,3]]=0;1/h),1/z)],{k1,d}] ]);
Epc:=(ro=0.5/(maxn*stn1);Ep0;For[i=0,i<=maxn,i++,For[j=0,j<=maxn-i,j++,(bul=True;x2=i/maxn;
   y2=j/maxn;p=Abs[P/.{x->x2,y->y2}];If[(p>lev1)&&(i(maxn-i)+j(maxn-j)>0),Pt=P;p=Norm1[x2,y2]];
   Sk0=O1+Sum[(z=(Ep[[k,1]]-x2)^2+(Ep[[k,2]]-y2)^2;If[z<h,(z=h;bul=False)],1/z),{k,d}];If[bul,
   (For[k=1,k<=d,k++,(l=Ep[[k]];R=(lev4-p)*(Sk0-1/((l[[1]]-x2)^2+(l[[2]]-y2)^2))-(lev4-l[[3]])
   *l[[4]];If[R<0,(Ep[[k]]={x2,y2,p,0};Ep0;Break[])])]    )]    )]]    )    )
```
(* Ep(Extremal points) -$\{\{xk, yk, \delta k, .\}, ...\}$, $d$ of number approximate critical points with coordinates $(xk, yk)$. $\delta k = 1 - |P(xk, yk)|$ and the fourth numbers in Ep depend on the location of $(x_k, y_k)$ with respect to other points. Epc(Extremal points correction): It is supposed that $||P|| = 1$ and $d \geq 1$. First in Ep0 are recalculated the thirds and fourths data of Ep, because the polynomial P, or the number $d$ can be new. Next, the point $(x2, y2)$ traverses the net and we look for a replacement $(x2, y2) \rightarrow (xk, yk)$, which improves the configuration of Ep in the sense to minimize the expression:

$$\textstyle\sum_k (lev4 - |P(xk, yk)|)\Big(O1 + \sum_{i \neq k} 1/((xk - xi)^2 + (yk - yi)^2)\Big),$$

(e.g. $lev4 = 1.001$). O1 is added for the case $d = 1$. *)

```
(A:=(If[d>0,(Epc;CPc=NullSpace[Table[(Mon[[i]]/.{x->Ep[[k1,1]],y->Ep[[k1,2]]}),{k1,d},{i,nm}]]
   ;)];Do[(If[d>0,Cpc=Sum[(2Random[]-1)*CPc[[j]],{j,Length[CPc]}]/nm;,Cpc=Table[(2Random[]-1),
   {i,nm}]];pc=Sum[Cpc[[i]]*Mon[[i]],{i,nm}];Pt=P+r*pc;NormP;Pt1=Pt/M;G1=N[G[Pt1]];Pt=P-r*pc;
   NormP;Pt2=Pt/M;G2=N[G[Pt2]];sgn=0;rep=G1+G2-3G0+Max[G1,G2];If[G1>G0,(P=Expand[Pt1];G0=G1;
   sgn=1)];If[G2>G0,(P=Expand[Pt2];G0=G2;sgn=-1)];If[sgn!=0,(Print[G0," for ",k," it.",
   If[rep>0," repeat?",""]];Break[])]),{k,30}]);
Ar:=(Pt=P+sgn*r*pc;NormP;Pt=Pt/M;G1=N[G[Pt]];If[G1>G0,(P=Expand[Pt];G0=G1;
   Print[G0," repeated"])]);
B:=(r=2*r);H:=(r=r/2);
D1:=(Ep=Append[Ep,{Random[],0,0,0}];d=d+1);Dm1:=(Ep=Delete[Ep,-1];d=d-1);)
```
(* A - searches for a better polynomial than $P$ in a ball with a radius $r$. Cpc(Coefficients of correcting polynomial $pc$). For $d > 0$, CPc is a table with rows the coefficients of the admissible correcting polynomials, i.e. which vanish at the points Ep. Ar - searches for a better polynomial in the previous direction. B, H - change $r$. D1, Dm1 - change $d$ *)

```
Just:=(ro=1./(3maxn*stn1);Ep0;Print[Table[Ep[[k,3]],{k,d}]];Tb=Table[(Mon[[i]]/.{x->Ep[[k,1]],
   y->Ep[[k,2]]}),{k,d},{i,nm}];For[k=1,k<=d,k++,(CPc=NullSpace[Delete[Tb, k]];Do[(Cpc=Sum[
   (2Random[]-1)*CPc[[j]],{j,Length[CPc]}]/nm;pc=Sum[Cpc[[j]]*Mon[[j]],{j,nm}];r1=((1-Abs[P])/
   pc)/.{x->Ep[[k,1]],y->Ep[[k, 2]]};Pt=P+r1*pc;NormP;Pt1=Pt/M;G1=N[G[Pt1]];Pt=P-r1*pc;NormP;
   Pt2=Pt/M;G2=N[G[Pt2]];If[G1>G0,(P=Expand[Pt1];G0=G1;Break[])];If[G2>G0,(P=Expand[Pt2];G0=G2;
   Break[])]),{i,3}])];Ep0;Print[Table[Abs[P/.{x->Ep[[k,1]],y->Ep[[k,2]]}],{k, d}]];   )
```
(* Just - Justify the extremal points *)

```
(x0=0.8;y0=0.1;d=0;r=0.5;maxn=10;stn1=1;lev1=0.95;lev4=1.001;Pt=Sum[(2Random[]-1)*Mon[[j]],
   {j,nm}];NormP;P=(Pt/M);G0=(G[P]//N))
```
(* General parameters $(x0, y0), maxn, \ldots$ and initial values of $d, r$. Also, the first approximation. *)

To start using the programm we must active the above definitions. As we see they concerns the case $n = 3$ and $(x_0, y_0) = (0.8, 0.1)$. So, in order to use other parameters, we have to fill them in the corresponding places. Except $n$, the other parameters can be changed later also, but then, we have to recalculate $G0$, which contains the current maximal gradient (see the change of $maxn$ below).

An example of using the program:

| Input | Output | Comment |
|---|---|---|
| In[8]:=A | 0.419273 for 1 it. repeat? | the value of $G(P)$ |
| In[9]:=Ar | – | no result |
| In[10]:=A | 0.481999 for 1 it. repeat? | |
| In[11]:=Ar | – | no result |
| In[12]:=A | 0.510531 for 8 it. | |
| In[13]:=D1 | 1 | the value of d |
| In[14]:=A | 0.517901 for 1 it. repeat? | |
| ... | ... | |
| In[18]:=Ar | 0.537307 repeated | |
| In[19]:=B | 1 | the value of r |
| In[20]:=A | 0.555567 for 1 it. repeat? | |
| In[21]:=Ar | – | no result |
| Plot3D[If[x+y<=1,P,0],{x,0,1},{y,0,1}, | graph | |
| PlotRange->{-1,1},PlotPoints->50] | | |
| In[23]:=D1 | 2 | the value of d |
| In[24]:=A | 0.56953 for 1 it. repeat? | |
| In[30]:=Ar | 0.647994 repeated | |
| In[31]:=B | 4 | |
| In[32]:=A | 0.684358 for 1 it. repeat? | |
| In[83]:=A | 1.54598 for 8 it. | |
| In[84]:=Plot3D[...] | graph | |
| In[85]:=D1;A | 1.54701 for 16 it. | |
| In[86]:=Just | {0.99.., 0.96.., 1.0001, ., .} | The values of $|P|$ at the points |
| | {0.99.., 0.99.., 0.9998, ., .} | of Ep, before and after Just. |
| In[90]:=maxn=15;Pt=P;NormP; | | 1.0001 shows an error in $\|P\|$. |
| P=Pt/M;G0=G[P];M | 1. | From In[90] to In[115] we had |
| In[96]:=A | – | almost no increase of G0. We |
| In[100]:=Dm1 | 5 | succeed to continue after an |
| In[108]:=H | 16 | increasing of the precision of |
| In[109]:=A | 1.58188 for 7 it. repeat? | calculation of $\|P\|$, decreasing |
| etc. | etc. | $d$ and double halving $r$. |

# Acknowledgments

The author was supported by the Sofia University Research Foundation under Contract no. UF–83/2006.

# References

1. Kroó, A., Révész, Sz. Gy.: On Bernstein and Markov-type inequalities for multivariate polynomials on convex bodies. J. Approx. Theory **99** (1999) 134–152
2. Milev, L. B., Révész, Sz. Gy.: Bernstein's Inequality for multivariate polynomials on the standard simplex. J. Ineq. Appl. **2005:2** (2005) 145–163
3. Révész, Sz. Gy., Sarantopoulos, Y.: A generalized Minkowski functional with applications in approximation theory. J. Convex Analysis **11 (2)** (2004) 303–334
4. Sarantopoulos, Y.: Bounds on the derivatives of polynomials on Banach spaces. Math. Proc. Cambr. Phil. Soc. **110** (1991) 307–312

# Quadrature Formula Based on Interpolating Polynomials: Algorithmic and Computational Aspects

Dana Simian[1] and Corina Simian[2]

[1] University "Lucian Blaga" of Sibiu, Faculty of Sciences
5-7 dr. I. Raţiu str, 550012 Sibiu, România
[2] University Babeş - Bolyai of Cluj-Napoca, Faculty of Mathematics and Informatics,
1 M. Kogălniceanu str., 400084 Cluj-Napoca, România

**Abstract.** The aim of this article is to obtain a quadrature formula for functions in several variables and to analyze the algorithmic and computational aspects of this formula. The known information about the integrand is $\{\lambda_i(f)\}_{i=1}^n$, where $\lambda_i$ are linearly independent linear functionals. We find a form of the coefficients of the quadrature formula which can be easy used in numerical calculations. The main algorithm we use in order to obtain the coefficients and the remainder of the quadrature formula is based on the Gauss elimination by segments method. We obtain an expression for the exactness degree of the quadrature formula. Finally, we analyze some computational aspects of the algorithm in the particular case of the Lagrange conditions.

## 1 Introduction

Let $\mathcal{A}_0$ be the set of analytic functions at the origin and $\mathcal{F} = \{f \mid f : D \subset R^d \to R\} \subset \mathcal{A}_0$. Let $\Lambda = \{\lambda_i : \mathcal{F} \to R \mid i = 1, \ldots n\}$ be a set of linearly independent linear functionals, and $f \in \mathcal{F}$ an arbitrary function. If we consider an interpolation formula: $f = L_\Lambda f + R_\Lambda f$, where $L_\Lambda$ is an interpolation operator, $R_\Lambda$ is the corresponding remainder operator and $\Lambda$ are the interpolation conditions, than we can obtain a quadrature formula by integrating this formula on the domain $D$. We are interested in finding a general quadrature formula of the following form:

$$I(f) = \int_D f(x)dx = \sum_{i=1}^n A_i \cdot \lambda_i(f) + R_n(f), \tag{1}$$

where $A_i \in R$ are the quadrature formula's coefficients, $\{\lambda_i\}_{i=1}^n$ are the used functionals and $R_n(f)$ is the quadrature formula's remainder. In order to obtain this formula, we start from a multivariate polynomial interpolation scheme, introduced by C. de Boor and A. Ron in [1] and called "least interpolation". The least interpolation scheme is given by a pair $(\Lambda, H_\Lambda \downarrow)$. The interpolation polynomial space, $H_\Lambda \downarrow$ is defined as $H_\Lambda \downarrow = \text{span}\{g \downarrow \mid g \in H_\Lambda\}$, where $H_\Lambda = \text{span}\{\lambda^\vee; \lambda \in \Lambda\}$, $\lambda^\vee$ is the generating function of the functional $\lambda$, i.e.,

T. Boyanov et al. (Eds.): NMA 2006, LNCS 4310, pp. 419–426, 2007.
© Springer-Verlag Berlin Heidelberg 2007

$$\lambda^\nu(x) = \sum_{\alpha \in N^d} \frac{D^\alpha \lambda^\nu(0)}{\alpha!} x^\alpha = \sum_{\alpha \in N^d} \frac{\lambda(m_\alpha)}{\alpha!} x^\alpha, \tag{2}$$

and $\alpha = (\alpha_1, \ldots, \alpha_d) \in N^d$, $\alpha! = \alpha_1! \cdot \ldots \cdot \alpha_d!$ and $m_\alpha(x) = x^\alpha$.

The notations and elements we will operate with, will be presented next. For an analytic function $g$ we denote by $g\downarrow$ the least term of $g$, that is, the homogeneous polynomial of minimal degree in the power series of $g$. $L_\Lambda(f)$ is the unique polynomial from $H_\Lambda\downarrow$, which matches $f$ on $\Lambda$, that is

$$\lambda(f) = \lambda(L_\Lambda(f)) \tag{3}$$

In order to define least-interpolation, in [1] it is defined the pair between an analytic function and a polynomial,

$$\langle f, p \rangle = (p(D)f)(0) = \sum_{|\alpha| \le \deg(p)} \frac{D^\alpha p(0) D^\alpha f(0)}{\alpha!}, \tag{4}$$

with $p(D)$ being the differential operator with constant coefficients associated to $p$. If $p = \sum_{|\alpha| \le \deg(p)} c_\alpha (\cdot)^\alpha$, then $p(D) = \sum_{|\alpha| \le \deg(p)} c_\alpha D^\alpha$. The pair (4) is a veritable inner product on polynomial spaces.

The action of a functional $\lambda$ on a function $f$ is $\langle \lambda^\nu, f \rangle = \lambda(f)$.

If $H$ is a space of functions analytic at the origin, we can find (see [1]) a basis $(g_i)_{i=1,\ldots,\dim(H)}$ of $H$ such that $\langle g_i, g_j\downarrow \rangle = 0$, $\forall i \ne j$. The set $(g_i\downarrow)_{i=1,\ldots,\dim(H)}$, is a basis of $H\downarrow$. If $(p_i)_{i=1,\ldots,\dim(H)}$ is a known basis of $H$, we start with $g_1 = p_1$ and repeat the following two steps for $i = 2, \ldots, \dim(H)$:

Step 1: $g_j \leftarrow p_j - \sum_{k=1}^{j-1} g_k \frac{\langle p_j, g_k\downarrow \rangle}{\langle g_k, g_k\downarrow \rangle}$.

Step 2: For each $i < j$ having the property that $\deg g_j\downarrow > \deg g_i\downarrow$, we recalculate

$$g_i \leftarrow g_i - g_j \frac{\langle g_i, g_j\downarrow \rangle}{\langle g_j, g_j\downarrow \rangle}.$$

If the functionals $\lambda_i$, $i = 1, \ldots, n$ are linearly independent, then $\dim(H_\Lambda) = n$ and the set $(\lambda_i^\nu)_{i=1}^n$ forms a basis for $H_\Lambda$. We express the functions $g_j$ from the previous algorithm in this basis:

$$g_j(x) = \sum_{i=1}^n c_{j,i} \lambda_i^\nu(x) \tag{5}$$

**Theorem 1.** *With $c_{j,i}$ as given in (5), the fundamental polynomials in the least interpolation scheme are*

$$\varphi_i = \sum_{j=1}^n \frac{c_{j,i} g_j\downarrow}{\langle g_j, g_j\downarrow \rangle}. \tag{6}$$

**Proof.** Let us consider the line matrices

$$G = [g_i]_{i=1}^n, \quad \Lambda^\nu = [\lambda_i^\nu]_{i=1}^n, \quad \Phi = [\varphi_i]_{i=1}^n, \quad M = \left[\frac{g_j\downarrow}{\langle g_j, g_j\downarrow\rangle}\right]_{j=1}^n,$$

and let $C = [c_{i,j}]_{i,j=1}^n$, $C^{-1} = [\tilde{c}_{i,j}]_{i,j=1}^n$. Let us denote $G \cdot (C^{-1})^\top = (u_i)_{i=1}^n$ and $M \cdot C = (v_j)_{j=1}^n$. A formal calculus gives $G = \Lambda^\nu \cdot C^\top$, $\Phi = M \cdot C$, and

$$\langle \Lambda^\nu, \Phi \rangle = \langle G(C^\top)^{-1}, M \cdot C \rangle = \langle G(C^{-1})^\top, M \cdot C \rangle = \langle u_i, v_j \rangle$$

$$= \left\langle \sum_{k=1}^n g_k \tilde{c}_{i,k}, \sum_{l=1}^n \frac{g_l\downarrow}{\langle g_l, g_l\downarrow\rangle} c_{l,j} \right\rangle = \sum_{k=1}^n \tilde{c}_{i,k} \cdot c_{k,j} = \delta_{i,j}.$$

Hence, $\lambda_i(\varphi_j) = \delta_{i,j}$ for every $i,j \in \{1, \ldots, n\}$. Moreover, $\varphi_i \in H_\Lambda\downarrow$.     □

**Theorem 2.** *The coefficients $A_i$ of quadrature formula (1) are given by*

$$A_i = \int_D \varphi_i(x)dx = \sum_{j=1}^n \frac{c_{j,i}}{\langle g_j, g_j\downarrow\rangle} \cdot \int_D g_j\downarrow (x)dx \qquad (7)$$

**Proof.** We use the fundamental polynomials and express

$$L_\Lambda(f) = \sum_{i=1}^n \varphi_i \cdot \lambda_i(f)$$

Integrating this formula on $D$ and using (6), we obtain (7).     □

Formula (7) is only of theoretical importance. We look for an algorithm which allows us to calculate numerically these coefficients.

## 2   Algorithmic Aspects. Main Results

From Theorem 2 we observe that, in order to calculate the coefficients $A_i$, we need to find the fundamentals polynomials $\varphi_i$ or the basis $(g_i\downarrow)_{i=1}^n$, together with the coefficients $c_{i,j}$ and the products $\langle g_j, g_j\downarrow\rangle$. We shall use the Gauss elimination by segments algorithm, presented in [4] and [5], in order to obtain the fundamental polynomials required in formula (7). This algorithm is used in [4] for obtaining the coefficients of the interpolating polynomial $L_\Lambda$, for $\Lambda = \{\delta_{\theta_i} \mid i = 1, \ldots, n, \; \theta_i \in \Theta \subset R^2, \; \theta_i \neq \theta_j\}$. The main advantage of this method is that it is not necessary to know apriori the degree of the interpolating polynomial. We present briefly this method, for the case of an arbitrary set of interpolation conditions $\Lambda$.

The interpolation problem (3) can be reformulated in algebraic setting, by the system: $V \cdot C = F$, with $V = [\lambda_i(x^\alpha)]$; $F = [\lambda_i(f)]$; $C = [c_\alpha]$ the coefficients of the interpolation polynomial $L_\Lambda$, $i \in \{1, \ldots, n\}$; $\alpha \in N^d$. We make a segmentation of the matrix $V$ in the form $V = [V_0, V_1, \ldots]$. The segment $V_j$ has $r_j$ columns, with $r_j$ being the dimension of the subspace of homogeneous polynomials of degree $j$ in $d$ variables, $r_j = \dim(\Pi_j^0)$. The columns in the segments of the matrix $V$ are indexed using a multiindex $\alpha \in N^d$. The segment $V_j$ corresponds to the multiindex $\alpha$ with $|\alpha| = \alpha_1 + \ldots + \alpha_d = j$. In every segment, we take the inverse lexicographical order for this multiindex. Inside any segment it is defined a proper inner product. We denote by $\langle \cdot, \cdot \rangle_j$ the inner product associated to the segment $j$. Any element of the segment $V_j$ is a vector with $r_j$ components. Using elementary operations on the segment $V_j$ we make a factorization of this segment in the form $V_j = [\mathcal{U}_{jj}\ \mathcal{R}_{jj}\ 0]^\top$, $\mathcal{R}_{jj}$ is a diagonal block whose lines are orthogonal vectors with respect the inner product $\langle \cdot, \cdot \rangle_j$. Except for the trivial case, there are many factorizations of the segmented matrix, but the blocks $\mathcal{R}_{jj}$ depend only of the matrix $V$ and of the chosen segmentation. In our case, the inner product used for the factorization of the matrix $V_j$ is $\langle a, b \rangle_j = \sum_{|\alpha|=j} \dfrac{a_\alpha \cdot b_\alpha}{\alpha!}$;

$a = (a_\alpha)$, $b = (b_\alpha)$, $\alpha \in N^d, a, b \in R^{r_j}$. We obtain the following factorization: $V = L \cdot U \cdot G = L \cdot R$, where $L$ is an invertible matrix, $U$ is a upper triangular and invertible matrix and $G$ is a matrix segmented in the same way as $V$.

We suppose that at the step $j$, we are in the segment $k_j$. We denote by $R_{i,k_j}$ the vector with the elements of R situated in the segment $k_j$ on the line $i$. The following operations are carried out:

1. We look for a pivot line. The pivot line maximizes the inner product $\langle R_{i,k_j}, R_{j,k_j} \rangle_{k_j}$, for $i > j - 1$. If this scalar product is zero for every $i > j$, there are no pivot lines in the segment $k_j$. If necessary, the pivot line is brought on the position $j$.

2. We calculate $U_{i,j} = \dfrac{\langle R_{i,k_j}, R_{j,k_j} \rangle_{k_j}}{\langle R_{j,k_j}, R_{j,k_j} \rangle_{k_j}}$, $i \le j$.

3. We carry out orthogonalization procedure for vectors $R_{i,k_j}$ and $R_{j,k_j}$ for $i > j$, i.e., calculate $L_{i,j} = \dfrac{\langle R_{i,k_j}, R_{j,k_j} \rangle_{k_j}}{\langle R_{j,k_j}, R_{j,k_j} \rangle_{k_j}}$, $R_{i,k_j} = R_{i,k_j} - L_{i,j} \cdot R_{j,k_j}$, $i > j$. This step represents the "elimination" in Gauss elimination by segments method.

4. We go to step $j + 1$, by searching for a new pivot line in the segment $R_{k_j}$. When a pivot line does not exist anymore in the segment $R_{k_j}$, and $j \le n$, we pass to the next segment.

The number $k_m = \max\limits_{j \in \{1,\ldots,n\}} (k_j)$ represents the number of the segments ran through in the elimination process and in the same time it represents thee maximal degree of the polynomials from the interpolation space $H_\Lambda\!\downarrow$.

In [4] it is proved that

$$g_j = \sum_{\alpha \in N^d} \frac{(\cdot)^\alpha}{\alpha!} G_{j,\alpha}; \ j \in \{1, \ldots, n\}, \tag{8}$$

$$g_j\!\downarrow = \sum_{|\alpha|=k_j} \frac{(\cdot)^\alpha}{\alpha!} G_{j,\alpha}; \ j \in \{1, \ldots, n\}, \tag{9}$$

$$\langle g_i, g_j\downarrow\rangle = \frac{\delta_{i,j}}{U_{j,j}}. \tag{10}$$

Using the previous notation we prove one of the main theorems of this section.

**Theorem 3.** *Let*

$$B = \mathrm{diag}(U) \cdot (L \cdot U)^{-1}. \tag{11}$$

*Then, with $g_j\downarrow$ as in (9), the fundamental polynomials $\varphi_k$, $k \in \{1,\dots,n\}$ are given by*

$$\varphi_k = \sum_{j=1}^{n} g_j\downarrow \cdot B_{j,k}. \tag{12}$$

**Proof.** We shall prove that the polynomials $q_k = \sum_{j=1}^{n} g_j\downarrow \cdot B_{j,k}$ satisfy the equality $\lambda_i(q_k) = \delta_{i,k}$, i.e., $q_k = \varphi_k$. Using (2) and the definition of the matrix $V$, we obtain $V = [D^\alpha \lambda_i^\nu(0)]$. Hence $L \cdot U \cdot G = [D^\alpha \lambda_i^\nu(0)]$. Taking into account (8), we see that $\lambda_i^\nu = \sum_{j=1}^{n}(L \cdot U)_{i,j} \cdot g_j$. Therefore,

$$\lambda_i(q_k) = \langle \lambda_i^\nu, q_k \rangle = \sum_{j=1}^{n}(L \cdot U)_{i,j} \cdot B_{j,k}\langle g_j, g_j\downarrow\rangle$$

$$= \sum_{j=1}^{n}(L \cdot U)_{i,j} \cdot (L \cdot U)_{j,k}^{-1} = \delta_{i,k}. \qquad \square$$

**Theorem 4.** *With the notations from Theorem 3, and from Gauss elimination by segments method, the coefficients of the quadrature formula (1) are given by*

$$A_k = \sum_{j=1}^{n} B_{j,k} \sum_{|\alpha|=k_j} \frac{G_{j,\alpha}}{\alpha!} \cdot I_\alpha, \tag{13}$$

*with*

$$I_\alpha = \int_D x^\alpha \, dx. \tag{14}$$

In order to evaluate the remainder in the quadrature formula (1) we need the following proposition, obtained by a generalization of the results from [2].

**Proposition 1.** *The remainder of the least interpolation formula, based on the functionals $\Lambda$, is given by*

$$(R(f))(x) = \left\langle e_x(t) - \sum_{j=1}^{n} g_j(t) \frac{\langle e_x, g_j\downarrow\rangle}{\langle g_j, g_j\downarrow\rangle}, f(t) \right\rangle, \tag{15}$$

*with $e_x(t) = e^{x \cdot t}$, $x, t \in \mathbb{R}^d$.*

**Proof.** The proof uses the fact that the operator

$$L_\Lambda^*(f) = \sum_{j=1}^{n} g_j \frac{\langle f, g_j \downarrow \rangle}{\langle g_j, g_j \downarrow \rangle} \tag{16}$$

is the dual of the operator $L_\Lambda$ with respect to the inner product (4), that is, $\langle L_\Lambda^*(f), p \rangle = \langle f, L_\Lambda(p) \rangle$, $\forall p \in H_\Lambda \downarrow$.

**Theorem 5.** *For the remainder of the least interpolation formula, the following estimate holds true:*

$$\|R_\Lambda(f)\| \leq \sup_{x \in D}(|f(x)|) + \sum_{j=1}^{n} M_j \cdot \frac{\sup_{x \in D} |g_j \downarrow (x)|}{\langle g_j, g_j \downarrow \rangle}, \tag{17}$$

*where* $M_j = \sum_{i=1}^{n} |c_{j,i}| \cdot |\lambda_i(f)|$ *and* $c_{j,i}$ *are given in (5).*

**Proof.** From (15), taking into account that $\langle e_x, f \rangle = f(x)$ we obtain

$$(R_\Lambda(f))(x) = f(x) - \sum_{j=1}^{n} \langle g_j, f \rangle \cdot \frac{g_j \downarrow}{\langle g_j, g_j \downarrow \rangle},$$

$$\|(R_\Lambda(f))\| \leq \sup_{x \in D} |f(x)| + \sum_{j=1}^{n} \frac{\sup_{x \in D} |g_j \downarrow (x)|}{\langle g_j, g_j \downarrow \rangle} \cdot |\langle g_j, f \rangle|.$$

Finally, from (5) we have $\langle g_j, f \rangle = \left\langle \sum_{i=1}^{n} c_{j,i} \cdot \lambda_i^\nu, f \right\rangle = \sum_{i=1}^{n} c_{j,i} \cdot \lambda_i(f)$. $\square$

We are interested in analysing formula (17) from a computational point of view. The Gauss elimination by segments algorithm gives us both the basis $(g_j \downarrow)_{i=1}^{n}$ and the inner products $\langle g_j, g_j \downarrow \rangle$. We study the possibility to obtain the coefficients $c_{j,i}$ from (5) using the outputs of this algorithm. Let $C = [c_{i,j}]_{i,j=1}^{n}$ and $C^{-1} = [\tilde{c}_{i,j}]_{i,j=1}^{n}$. The generating functions are $\lambda_i^\nu = \sum_{j=1}^{n} \tilde{c}_{i,j} \cdot g_j$. Let $m_\alpha = x^\alpha$, $\alpha \in N^d$, be the first $n$ monomials taken in inverse lexicographical order. Then $\lambda_i(m_\alpha) = \langle \lambda_i^\nu, m_\alpha \rangle = \sum_{j=1}^{n} \tilde{c}_{i,j} \cdot D^\alpha g_j(0)$. From (8) we can write $D^\alpha g_j(0) = G_{j,\alpha}$.

Therefore, the coefficients $\tilde{c}_{i,j}$ are given by the system $\sum_{j=1}^{n} \tilde{c}_{i,j} \cdot G_{j,\alpha} = \lambda_i(m_\alpha)$ which can be solved using the classical Gauss elimination method. Thus, by inversion of the matrix $C^{-1}$ we obtain the coefficients $c_{i,j}$.

**Proposition 2.** *Using the Gauss elimination by segments method, the coefficients* $c_{j,i}$ *from (5) are*

$$c_{j,i} = B_{j,i} \cdot U_{j,j}. \tag{18}$$

**Proof.** We use (9) and (10) in (7), and then compare the result with (13). $\square$

**Theorem 6.** *Using the notation from this section, an estimate of the remainder of quadrature formula (1) is*

$$\|R_n(f)\| \leq S(D) \cdot \sup_{x \in D} |f(x)| + \sum_{j=1}^{n} \frac{M_j}{\langle g_j, g_j \downarrow \rangle} \cdot \sum_{|\alpha|=k_j} \frac{G_{j,\alpha}}{\alpha!} \cdot I_\alpha,$$

*where $I_\alpha$ are the moments from (14) and $S(D) = \int_D dx$.*

**Proof.** We integrate inequality (17) over $D$, and take into account (9). Let us mention that $k_j$ is in fact the degree of the polynomial $g_j \downarrow$. □

## 3    The Exactness Degree of the Quadrature Formula

**Definition 1.** *The quadrature formula (1) is said to have exactness degree $r \in \mathbb{N}$, if $R(p) = 0$ for every $p \in \Pi_k$, $k \leq r$, and there exists a polynomial $p$ of degree $r + 1$, such that $R(p) \neq 0$.*

Let $\lambda \notin \Lambda$ be an arbitrary functional. We define

$$\varepsilon_{\Lambda,\lambda}^\nu = (1 - L_\Lambda^*)(\lambda^\nu), \tag{19}$$

with $L_\Lambda^*$ given in (16) and let $r = \deg(\varepsilon_{\Lambda,\lambda}^\nu \downarrow)$.

In [6] we proved that if $H_\Lambda \downarrow \neq \Pi_m^d$, then for every polynomial $p \in \Pi_k$, $k < r$, $\lambda(p - L_\Lambda(p)) = 0$, and $r$ is the largest integer with this property. Here we prove even more, namely, that

$$\deg(\varepsilon_{\Lambda,\lambda}^\nu \downarrow) = \min\{\deg(p)|p \in \Pi^d; \ \lambda(p) \neq 0; \ p \in \ker(\Lambda)\}. \tag{20}$$

**Theorem 7.** *The degree of exactness of the formula (1) is equal to $m$, if there is an integer $m \in N$ such that $H_\Lambda \downarrow = \Pi_m^d$, and is equal to $r - 1$ in the other cases, with $r$ defined by*

$$r = \deg(\varepsilon_{\Lambda,\delta_x}^\nu \downarrow),$$

*or, equivalently,*

$$r = \min\{ \deg(p) \,|\, p(x) \neq 0, \ \forall p \in \ker(\Lambda); \ \delta_x \notin \Lambda\}$$

**Proof.** We use (19) and (20) with $\lambda = \delta_x$, $x \in D$, $\delta_x \notin \Lambda$. □

We observe that a possibility for obtaining exactness degree equal to $m$ in the general quadrature formula (1) is to use sets of functionals for which $H_\Lambda \downarrow = \Pi_m^d$. Such a set of functionals is given in [7] and supplies us with the following result.

**Proposition 3.** *The quadrature formula (1) has exactness degree $m$, if the conditions $\Lambda = \Lambda_\Theta$ are given by*

$$\Lambda_\Theta = \left\{ \lambda_{j,\theta_k} | \lambda_{j,\theta_k}(f) = f^{[j]}(\theta_k), \ \theta_k \in \Theta \subset R^2 \right\}, \ j \in \{0,\ldots,m\}, \ k \in \{0,\ldots j\},$$

*where $f^{[k]} = \sum_{|\alpha|=k} \frac{D^\alpha f(0)(\cdot)^\alpha}{\alpha!}.$*

## 4  Lagrange Case

In this section we analyze the particular quadrature formula obtained from (1) using a set of Lagrange functionals

$$\Lambda = \{\delta_{\theta_i} | \theta_i \neq \theta_j, \forall i \neq j, i = 1, \ldots, n, \ \theta_i \in \mathbb{R}^d\}.$$

We are looking for computational details in Gauss elimination by segments algorithm. The starting point in this algorithm is the matrix $V = [\lambda_i(x^\alpha)]$. We are looking for possibilities to construct recursively this matrix, segment after segment, starting with the segment number 0, which is a column vector with all elements equal to one. In the Lagrange case we can obtain the segment $V_{k+1}$ of the segmented matrix $V$, from the segment $V_k$. The segment $V_k$ will have $r_k = \dim(\Pi_k^0)$ elements. Let $V_{k,i} = (a_{k,\alpha_1}, \ldots, a_{k,\alpha_{r_k}})$ be the line $i$ of the segments $V_k$, with $\alpha_i \in N^d$, $|\alpha_i| = k$, ordered in inverse lexicographical order. The elements of the line $i$ of the next segment $V_{k+1,i}$ are obtained in the following way.

The first element $a_{k,(k,0,\ldots,0)}$ generates $d$ elements:

$a_{k+1,(k+1,0,\ldots,0)} = a_{k,(k,0,\ldots,0)} \cdot \theta_{i,1}$
$a_{k+1,(k,1,0,\ldots,0)} = a_{k,(k,0,\ldots,0)} \cdot \theta_{i,2}$
. . .

$a_{k+1,(k,0,\ldots,0,1)} = a_{k,(k,0,\ldots,0)} \cdot \theta_{i,d}$
with $\theta_i = (\theta_{i,1}, \ldots, \theta_{i,d})$.
The element $a_{k,(k-1,1,0,\ldots,0)}$, will generate $d - 1$ elements:
$a_{k+1,(k-1,2,0,\ldots,0)} = a_{k,(k-1,1,0,\ldots,0)} \cdot \theta_{i,2}$
$a_{k+1,(k-1,1,1,0,\ldots,0)} = a_{k,(k-1,1,0,\ldots,0)} \cdot \theta_{i,3}$
. . .

$a_{k+1,(k-1,1,0,\ldots,0,1)} = a_{k,(k-1,1,0,\ldots,0)} \cdot \theta_{i,d}$,
and so on.

## References

1. de Boor C., Ron A.: On multivariate polynomial interpolation. Constr. Approx. **6** (1990) 287–302.
2. de Boor C.: On the error in multivariate polynomial interpolation. Math. Z. **220** (1992) 221–230.
3. de Boor C., Ron A.: The least solution for the polynomial interpolation problem. Math. Z. **220** (1992) 347–378.
4. de Boor C., Ron A.: Computational aspects of polynomial interpolation in several variables. Math. Comp. **58** (1992) 705–727.
5. de Boor C.: Gauss elimination by segments and multivariate polynomial interpolation. In: Approximation and Computation: A Festschrift in Honor of Walter Gautschi, Birkhäuser Verlag (1994) 87–96.
6. Simian D.: The λ- Error Order in Multivariate Interpolation. Lectures Notes in Computer Science, Springer Berlin Heildelberg New York (2004) 478–486.
7. Simian D., Simian C.: Some Results in Multivariate Interpolation. Acta Universitatis Apulensis **11** (2006) 47–57. Alba Iulia, Romania.

# Stability of Semi-implicit Atmospheric Models with Respect to the Reference Temperature Profile

Andrei Bourchtein[1], Ludmila Bourchtein[1], and Maxim Naumov[2]

[1] Mathematics Department, Pelotas State University, Brazil
burstein@terra.com.br
[2] Department of Computer Science, Purdue University, USA
naumov@purdue.edu

**Abstract.** The dependence of the linear stability of two-time-level semi-implicit schemes on choice of the reference temperature profile is studied. Analysis is made for large time steps, keeping general form of such model parameters as the number of vertical levels, their distribution, and the values of the viscosity coefficients. The obtained results reveal more restrictive conditions on the reference temperature profile than those for three-time-level schemes. Nevertheless, general conclusions are consistent with the previous analysis: instability generated by inappropriate choice of the temperature profile is absolute and the scheme stability can be recovered by setting the reference temperature to be warmer than the actual one.

## 1 Introduction

The complexity and nonlinearity of processes of atmospheric dynamics have direct effect on the choice of the numerical methods used for computation of approximated solutions of the respective mathematical models. Explicit schemes are rarely employed because of excessive restriction on time step reflecting presence of the fast acoustic and gravity waves. On the other hand, fully implicit schemes are not used due to complexity of nonlinear systems arising at each time step. Therefore, the most popular numerical approach is semi-implicit method, which allows large time steps and reduces the implicit part of the scheme to solution of linear systems [7, 11, 14].

Since early applications of the semi-implicit method in the multi-level atmospheric models the phenomenon of absolute instability discovered by Burridge [4] attracted attention of the researches [5, 12, 13]. The essence of the problem consists of appearance of instability in the part of equations responsible for fast gravity waves, which are approximated implicitly. It was shown that this behavior is caused by explicit treatment of the deviations from the reference vertical temperature profile. It is essential to keep the explicit approximation of these deviations to maintain rather simple structure of the implicit equations at each time step. Various numerical experiments and theoretical analysis were

T. Boyanov et al. (Eds.): NMA 2006, LNCS 4310, pp. 427–434, 2007.
© Springer-Verlag Berlin Heidelberg 2007

performed to clarify how to avoid this instability. It was discovered numerically in [4] that instability does not appear if one choose the temperature of the reference profile warmer than the actual one. This result was later confirmed in some particular cases of analytical models [2, 5, 12].

The main attention in these studies was paid to three-time level models because they were more popular in atmospheric modeling until the 90's. In the late 80's it was shown that two-time-level schemes can support rather simple design of the three-time-level models and still assure more accurate solutions as applied to shallow water equations [9, 15]. Therefore, since the 90's different atmospheric centers start to adopt two-time-level baroclinic schemes and some tests were employed to reveal that a similar condition of a warmer reference temperature guarantees absolute stability of the gravity waves [6, 8, 10, 16]. However, the problem has not been solved analytically neither for three-time level nor for two-time-level schemes. In this study we derive analytical conditions of the absence and presence of absolute instability for the case related to large time steps. We apply the techniques presented in [2] for study of two-time-level models and the obtained results reveal slightly more restrictive conditions on the reference temperature profile than those obtained in [2].

## 2 Semi-implicit Two-Time-Level Scheme and Characteristic Equation

Using the pressure vertical coordinate $p$, Cartesian horizontal coordinates $x$ and $y$, and the time coordinate $t$, the momentum, continuity and thermodynamic equations of the hydrostatic atmosphere linearized about a state of rest can be written as follows [7]:

$$\left(\partial_t - \alpha\nabla^2\right) D = -\nabla^2\Phi, \quad \omega = -\int_{p_{up}}^{p} D\,dp, \tag{1}$$

$$\left(\partial_t - \beta\nabla^2\right)\Phi = R\int_{p}^{p_{lw}} \hat{\gamma}\omega d\left(lnp\right) + R\hat{T}\left(p_{lw}\right)\omega\left(p_{lw}\right). \tag{2}$$

Here $D = u_x + v_y$ is the horizontal divergence, $u, v$ are the horizontal components and $\omega = dp/dt$ is the vertical component of velocity, $\Phi = gz$ is geopotential, $g$ is the gravitational acceleration, $z$ is the height of the pressure surface, $T$ is the temperature, $\Gamma_d = g/c_p$ is the adiabatic lapse rate, $\hat{\gamma} = R\hat{T}(\Gamma_d - \hat{\Gamma})/gp$ and $\hat{\Gamma} = (gp/R\hat{T})\hat{T}_p = -\hat{T}_z$ is the vertical lapse rate of the reference temperature profile $\hat{T}$, $R$ is the gas constant, $c_p$ is the specific heat at constant pressure, $\alpha$ and $\beta$ are the viscosity coefficients (simulating turbulence effects or numerical dissipation), $\nabla^2 = \partial_{xx} + \partial_{yy}$ is the horizontal Laplace operator, and $p_{up}, p_{lw}$ are upper and lower pressure boundaries, respectively.

Introducing the most popular Lorenz staggered vertical grid [1], which divides the considered atmosphere in $K$ vertical layers with boundaries $p_{k+1/2}$

$$p_{up} = p_{1/2} < p_{3/2} < ... < p_{k-1/2} < p_{k+1/2} < ... < p_{K-1/2} < p_{K+1/2} = p_{lw}$$

and with inner levels $p_k$, satisfying the natural inequalities

$$p_{k-1/2} < p_k < p_{k+1/2}, \quad k = 1, ..., K,$$

we can discretize equations (1), (2) as follows:

$$(\partial_t - \alpha\nabla^2)D_k = -\nabla^2\Phi_k, \quad \omega_{k+1/2} = -\sum_{i=1}^{k} D_i(p_{i+1/2} - p_{i-1/2}), \qquad (3)$$

$$(\partial_t - \beta\nabla^2)\Phi_k = R\left[\sum_{i=k+1}^{K}(\hat{\gamma}\omega)_{i-1/2}\ln\frac{p_i}{p_{i-1}} + (\hat{\gamma}\omega)_{K+1/2}\ln\frac{p_{K+1/2}}{p_K}\right] + R(\hat{T}\omega)_{K+1/2},$$
$$(4)$$

where $k = 1, ..., K$ and summation is defined to be zero if the lower limit of the summation index exceeds the upper limit. Using the vector functions

$$\mathbf{D} = (D_1, ..., D_K)^T, \quad \mathbf{\Phi} = (\Phi_1, ..., \Phi_K)^T, \quad \omega = (\omega_{3/2}, ..., \omega_{K+1/2})^T,$$

we can rewrite (3), (4) in the form

$$\left(\partial_t - \alpha\nabla^2\right)\mathbf{D} = -\nabla^2\mathbf{\Phi}, \quad \omega = -\mathbf{BD}, \quad (\partial_t - \beta\nabla^2)\mathbf{\Phi} = \hat{\mathbf{A}}\omega. \qquad (5)$$

Here $\hat{\mathbf{A}}$ and $\mathbf{B}$ are the $K{\times}K$ upper and lower triangular matrices with the entries

$$a_{j,k} = \hat{a}_k, j \leq k; \; a_{j,k} = 0, j > k;$$

$$\hat{a}_k = R\hat{\gamma}_{k+1/2}\ln\frac{p_{k+1}}{p_k}, k = 1, ..., K{-}1; \; \hat{a}_K = R\hat{\gamma}_{K+1/2}\ln\frac{p_{K+1/2}}{p_K} + R\hat{T}_{K+1/2}; \quad (6)$$

$$b_{j,k} = b_k, j \geq k; \; b_{j,k} = 0, j < k; \; b_k = p_{k+1/2} - p_{k-1/2}, k = 1, ..., K. \quad (7)$$

Note that $\hat{a}_k$ depend on reference temperature profile and therefore matrix $\hat{\mathbf{A}}$ can be considered as the value of matrix function $\mathbf{A}(T)$ at $T = \hat{T} : \hat{\mathbf{A}} = \mathbf{A}(\hat{T})$.

Substituting $\mathbf{D}$ for $\omega$, we reduce the system (5) to a simpler form

$$(\partial_t - \alpha\nabla^2)\mathbf{D} = -\nabla^2\mathbf{\Phi}, (\partial_t - \beta\nabla^2)\mathbf{\Phi} = -\hat{\mathbf{C}}\mathbf{D}, \qquad (8)$$

where $\hat{\mathbf{C}} = \hat{\mathbf{A}}\mathbf{B}$ is the vertical structure matrix, which depends on reference temperature profile and vertical discretization. It can be shown that $\hat{\mathbf{C}}$ is oscillatory matrix and, therefore, all its eigenvalues are real and positive [3]. This is essential property of the matrix $\hat{\mathbf{C}}$ used in the following analysis of linear stability.

To keep the essence of semi-implicit time-differencing in atmospheric models, we represent the actual reference temperature profile $\hat{T}(p)$ in the form $\hat{T}=\bar{T}+\tilde{T}$, where $\bar{T}$ is basic profile and $\tilde{T}$ is its deviation. The only term of the system (8) that depends on $\hat{T}$ is the matrix $\hat{\mathbf{C}}$ on the right hand side of the second equation. Therefore, we represent $\hat{\mathbf{C}}=\bar{\mathbf{C}}+\tilde{\mathbf{C}}$, where $\bar{\mathbf{C}}=\mathbf{C}(\bar{T})=\mathbf{A}(\bar{T})\mathbf{B}=\bar{\mathbf{A}}\mathbf{B}$ is the basic matrix and $\tilde{\mathbf{C}}$ is the deviation matrix. Matrix $\hat{\mathbf{C}}$ will be called full or actual matrix. In this paper we will consider only the cases when actual and basic temperature reference profiles are statically stable, that is, $\hat{\Gamma} < \Gamma_d$ and $\bar{\Gamma} < \Gamma_d$. Otherwise, the primitive differential problem is not well posed. These conditions imply nonnegativity of matrices $\hat{\mathbf{A}}$ and $\bar{\mathbf{A}}$, and, consequently, $\hat{\mathbf{C}}$ and $\bar{\mathbf{C}}$.

According to semi-implicit time discretization the basic matrix term is approximated implicitly and the deviation one is extrapolated explicitly:

$$\frac{\mathbf{D}^{n+1} - \mathbf{D}^n}{\tau} - \alpha\nabla^2 \frac{\mathbf{D}^{n+1} + \mathbf{D}^n}{2} = -\nabla^2 \frac{\mathbf{\Phi}^{n+1} + \mathbf{\Phi}^n}{2}, \tag{9}$$

$$\frac{\mathbf{\Phi}^{n+1} - \mathbf{\Phi}^n}{\tau} - \beta\nabla^2 \frac{\mathbf{\Phi}^{n+1} + \mathbf{\Phi}^n}{2} = -\bar{\mathbf{C}}\frac{\mathbf{D}^{n+1} + \mathbf{D}^n}{2} - (\hat{\mathbf{C}} - \bar{\mathbf{C}})\frac{3\mathbf{D}^n - \mathbf{D}^{n-1}}{2}. \tag{10}$$

Here, $\tau$ is the time step and superscripts $n-1, n$, and $n+1$ denote the values at the "old" $(n-1)\tau$, "current" $n\tau$ and "new" $(n+1)\tau$ time levels, respectively.

Applying the von Neumann method of stability analysis we consider particular solution in the wave form

$$\begin{pmatrix} \mathbf{D} \\ \mathbf{\Phi} \end{pmatrix}^n (x, y) = \begin{pmatrix} \mathbf{W} \\ \mathbf{H} \end{pmatrix} \mu^n \exp(im_x x + im_y y),$$

where $K$-order vectors $\mathbf{W}, \mathbf{H}$ describe the vertical structure of the amplitudes of the individual wave with the wave numbers $(m_x, m_y)$ and $\mu$ is the amplification factor describing the behavior of the amplitudes with respect to time. For stability of the numerical scheme the amplification factors should lie in the unit disk for any pair of the wave numbers. Substituting this representation in (9), (10), we obtain the linear algebraic system for the vectors $\mathbf{W}, \mathbf{H}$:

$$\frac{\mu - 1}{\tau}\mathbf{W} + \alpha m^2 \frac{\mu + 1}{2}\mathbf{W} = m^2 \frac{\mu + 1}{2}\mathbf{H}, \tag{11}$$

$$\frac{\mu^2 - \mu}{\tau}\mathbf{H} + \beta m^2 \frac{\mu^2 + \mu}{2}\mathbf{H} = -\bar{\mathbf{C}}\frac{\mu^2 + \mu}{2}\mathbf{W} - (\hat{\mathbf{C}} - \bar{\mathbf{C}})\frac{3\mu - 1}{2}\mathbf{W}, \tag{12}$$

where $m^2 = m_x^2 + m_y^2 \geq 0$. The system (11), (12) has non-trivial solution iff its determinant is equal to zero:

$$\det\{\left[(\mu^2 + \mu)\bar{\mathbf{C}} + (3\mu - 1)(\hat{\mathbf{C}} - \bar{\mathbf{C}})\right]\tau^2 m^2 (\mu + 1)$$

$$+ \left[2(\mu - 1) + \tau\alpha m^2(\mu + 1)\right]\left[2(\mu - 1) + \tau\beta m^2(\mu + 1)\right]\mu\mathbf{I}\} = 0. \tag{13}$$

Note that the same kind of the characteristic equation will be obtained if any commonly used space discretization is applied to (9), (10). Therefore, semi-discrete system (9), (10) keeps all properties of the completely discrete approximations.

## 3   Some Analytical Solutions and Numerical Tests

If there are no deviations from basic temperature, that is, $\hat{\mathbf{C}} = \bar{\mathbf{C}}$, then equation (13) has $K$-fold zero root. Since that root satisfies stability condition, we can suppose that $\mu \neq 0$ to rewrite (13) in the form

$$\det\{\bar{\mathbf{C}} - \lambda\mathbf{I}\} = 0,$$

where

$$\lambda = -\frac{[2(\mu-1)+\tau\alpha m^2(\mu+1)][2(\mu-1)+\tau\beta m^2(\mu+1)]}{\tau^2 m^2(\mu+1)^2} \quad (14)$$

are the positive eigenvalues of the matrix $\bar{\mathbf{C}}$ ranging from 0 to $10^5$ under usual temperature conditions [3]. The last relation can be considered as the second order equation for $\mu$ including positive parameter $\lambda$. In this case, it is simple to show that all amplification factors $\mu$ lie in the unit disk. Therefore, we obtain well-known result that Crank-Nicholson approximation of the system (8) is absolutely stable.

Let us consider the important case of large values of the time step $\tau$. Evaluating the limit in (13) as $\tau$ approaches infinity, we obtain

$$\det\{[(\mu^2+\mu)\bar{\mathbf{C}}+(3\mu-1)(\dot{\mathbf{C}}-\bar{\mathbf{C}})]m^2(\mu+1)+\alpha\beta m^4(\mu+1)(\mu^2+\mu)\mathbf{I}\} = 0. \quad (15)$$

First, we note that $\mu = -1$ is $K$-fold root, which does not violate the stability of the scheme. Therefore (15) can be simplified to the form:

$$\det\{[(\mu^2+\mu)\bar{\mathbf{C}}+(3\mu-1)(\hat{\mathbf{C}}-\bar{\mathbf{C}})]+\alpha\beta m^2(\mu^2+\mu)\mathbf{I}\} = 0. \quad (16)$$

Since (16) is still too hard for exact analysis, we consider the case $\alpha\beta = 0$. Then (16) assumes the form

$$\det[(\mu^2+\mu)\bar{\mathbf{C}}+(3\mu-1)(\hat{\mathbf{C}}-\bar{\mathbf{C}})] = 0.$$

Due to definition of the matrices $\bar{\mathbf{C}}$ and $\hat{\mathbf{C}}$ we have

$$\det[(\mu^2+\mu)\bar{\mathbf{A}}+(3\mu-1)(\hat{\mathbf{A}}-\bar{\mathbf{A}})]\det\mathbf{B} = 0.$$

Since $\det\mathbf{B} \neq 0$ and the matrices $\bar{\mathbf{A}}$ and $\hat{\mathbf{A}}$ are upper triangular, the last equation transforms to

$$\prod_{k=1}^{K}[(\mu^2+\mu)\bar{a}_k+(3\mu-1)(\hat{a}_k-\bar{a}_k)] = 0. \quad (17)$$

The solutions of (17) are

$$\mu_{k\pm} = \frac{1}{2}\left[2-3d_k \pm \sqrt{9d_k^2-8d_k}\right], \quad d_k = \frac{\hat{a}_k}{\bar{a}_k} > 0, \quad k = 1, ..., K.$$

If $9d_k^2-8d_k \leq 0$, that is, $d_k \leq 8/9$, then

$$|\mu_{k\pm}|^2 = 1 - d_k < 1.$$

If $d_k > 8/9$, then $|\mu_{k-}| > |\mu_{k+}|$ and solution of inequality $|\mu_{k-}| \leq 1$ shows that the last is true iff $d_k \leq 1$. Joining two considered evaluations, we obtain that the scheme is stable for the large time steps iff

$$0 < \hat{a}_k \leq \bar{a}_k, \quad k = 1, ..., K. \quad (18)$$

Due to (6), the last inequality can be rewritten as follows

$$\hat{T}_{k+1/2}(\Gamma_d - \hat{\Gamma}_{k+1/2}) \leq \bar{T}_{k+1/2}(\Gamma_d - \bar{\Gamma}_{k+1/2}) , \quad k = 1, ..., K - 1; \qquad (19)$$

$$[\hat{T}(\Gamma_d - \hat{\Gamma} + \xi)]_{K+1/2} \leq [\bar{T}(\Gamma_d - \bar{\Gamma} + \xi)]_{K+1/2}, \ \xi_{K+1/2} = \frac{g}{R}\left(\frac{1}{p_{K+1/2}} \ln \frac{p_{K+1/2}}{p_K}\right)^{-1}.$$

$$(20)$$

If actual and basic temperature profiles have constant lapse rates $\hat{\Gamma}$ and $\bar{\Gamma}$, then

$$\hat{T}_{k+1/2} = \hat{T}_{K+1/2}\left(\frac{p_{k+1/2}}{p_{K+1/2}}\right)^{R\hat{\Gamma}/g} , \bar{T}_{k+1/2} = \bar{T}_{K+1/2}\left(\frac{p_{k+1/2}}{p_{K+1/2}}\right)^{R\bar{\Gamma}/g} , k = 1, ..., K-1$$

and conditions (19) take the form

$$\left(\frac{p_{k+1/2}}{p_{K+1/2}}\right)^{R(\hat{\Gamma} - \bar{\Gamma})/g} \leq \frac{\bar{T}_{K+1/2}(\Gamma_d - \bar{\Gamma})}{\hat{T}_{K+1/2}(\Gamma_d - \hat{\Gamma})} , \quad k = 1, ..., K - 1. \qquad (21)$$

If we assume that (21) should be satisfied for arbitrary vertical grid, then it is equivalent to condition

$$\bar{\Gamma} \leq \hat{\Gamma}. \qquad (22)$$

In fact, if (22) is broken, then the function in the left hand side of (21) has the negative exponent and, therefore, there exist such small values of $p_{k+1/2}$ that the exponential function will be greater than the constant in the right hand side of (21). Note that (22) is not necessary condition for fixed vertical discretization with $p_{up} > 0$, because for the fixed smallest pressure value $p_{1/2} = p_{up}$ one can find such lapse rates $\bar{\Gamma}$ and $\hat{\Gamma}$ that (22) is not satisfied, but (21) holds.

It worth to note that obtained stability conditions (18) are two times more restrictive for basic temperature profile than those for three-time-level scheme [2]. However, the resulting condition (22), which assures stability for arbitrary distribution of the vertical levels under constant lapse rates is the same. The difference between conditions for two- and three-time-level models can be shown numerically for specific vertical grids when condition (22) is broken. One of these cases is presented in Fig.1 and 2. The modulus of amplification factor is drawn as function of time step for three- and two-time-level models (Fig.1 and 2, respectively). The same vertical discretization characteristics were used for both experiments, namely: vertical grid composed of two levels, basic temperature profile defined by surface temperature $\bar{T}_{5/2} = 273K$ and constant lapse rate $\bar{\Gamma} = 0.006K/m$, and actual temperature profile defined by surface temperature $\hat{T}_{5/2} = 273K$ and constant lapse rate $\hat{\Gamma} = 0.005K/m$. Four different curves in each graph correspond to different values of the viscosity coefficients: $\alpha = \beta = 0$ - solid curve, $\alpha = \beta = 1$ - dashed curve, $\alpha = \beta = 10$ - dot-dashed curve, and $\alpha = \beta = 100$ - dotted curve. Evidently, condition (22) is violated, however three-time level scheme is absolutely stability, while two-time-level scheme becomes unstable. The used implicit viscosity can recover stability of the two-time-level scheme at least for small time steps if sufficiently great viscosity coefficients are applied ($\alpha = \beta \geq 10$).

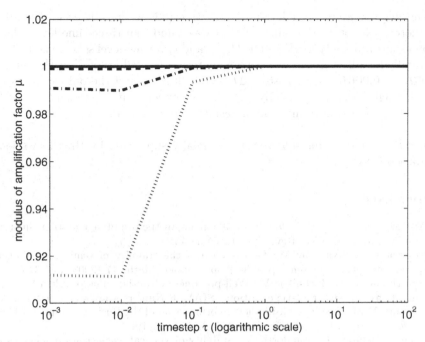

**Fig. 1.** Amplification factor as function of time step for three-time-level scheme

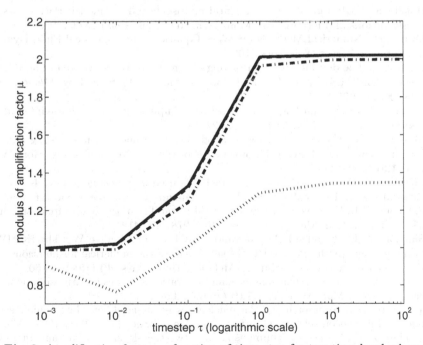

**Fig. 2.** Amplification factor as function of time step for two-time-level scheme

Studying the areas of violation of condition (22) we have fixed all the above parameters, except for the values of $\hat{\Gamma}$. It was found that three-time-level scheme becomes unstable only for $\hat{\Gamma} \leq 0.003K/m$ and two-time-level scheme is unstable practically for all values $\hat{\Gamma} < \bar{\Gamma}$ (we have tested the values of $\hat{\Gamma}$ in the interval $[0.005K/m, 0.006K/m]$ with step $\Delta\hat{\Gamma} = 0.0005K/m$ and the only value of absolute stability was $\hat{\Gamma} = 0.006K/m$). Similar experiments were carried out for thirty-level vertical grid and similar results were obtained.

*Acknowledgements.* This research was partially supported by Brazilian science foundation CNPq.

# References

1. Arakawa A., Suarez M.J.: Vertical differencing of the primitive equations in sigma coordinates. Mon. Wea. Rev. **111** (1983) 34-45.
2. Bourchtein A., Naumov M.: Dependence of the stability of semi-implicit atmospheric models on reference profile. Appl. Numer. Math. **47** (2003) 325-343.
3. Bourchtein A., Kadychnikov V.: Well-posedness of initial value problem for vertically discretized hydrostatic equations. SIAM J. Numer. Anal. **41** (2003) 195-207.
4. Burridge D.M.: A split semi-implicit reformulation of the Bushby-Timpson 10 level model. Quart. J. Roy. Meteor. Soc. **101** (1975) 777-792.
5. Cote J., Beland M., Staniforth A.: Stability of vertical discretization schemes for semi-implicit primitive equations models. Mon. Wea Rev. **111** (1983) 1189-1207.
6. Cote J., Gravel S., Methot A., Patoine A., Roch M., Staniforth A.: The operational CMC-MRB global environmental multiscale (GEM) model. Part I: Design considerations and formulation, Mon. Wea. Rev. **126** (1998) 1373-1395.
7. Durran D.: Numerical Methods for Wave Equations in Geophysical Fluid Dynamics, Springer-Verlag, New York (1999).
8. Hortal M.: The development and testing of a new two-time-level semi-Lagrangian scheme (SETTLS) in the ECMWF forecast model. Quart. J. Roy. Meteor. Soc. **128** (2002) 1671-1687.
9. Mcdonald A.: A semi-Lagrangian and semi-implicit two time level integration scheme. Mon. Wea. Rev. **114** (1986) 824-830.
10. McDonald A., Haugen J.: A two-time-level, three-dimensional semi-Lagrangian, semi-implicit, limited-area gridpoint model of the primitive equations. Mon. Wea. Rev. **120** (1992) 2603-2621.
11. Mesinger F., Arakawa A.: Numerical methods used in atmospheric models. GARP Publications Series, WMO/ICSU Joint Organizing Committee, Geneva (1976).
12. Simmons A.J., Hoskins B.J., Burridge D.M.: Stability of the semi-implicit method of time integration. Mon. Wea. Rev. **106** (1978) 405-412.
13. Simmons A.J., Burridge D.M., Jarraud M., Girard C., Wergen W.: The ECMWF medium-range prediction models: Development of the numerical formulations and the impact of increased resolution. Meteor. Atmos. Phys. **40** (1989) 28-60.
14. Staniforth A., Cote J.: Semi-Lagrangian integration schemes for atmospheric models - A review. Mon. Wea. Rev. **119** (1991) 2206-2223.
15. Temperton C., Staniforth A.: An efficient two-time-level semi-Lagrangian semi-implicit integration scheme. Quart. J. Roy. Meteor.Soc. **113** (1987) 1025-1039.
16. Temperton C., Hortal M., Simmons A.J.: A two-time-level semi-Lagrangian global spectral model. Quart. J. Roy. Meteor. Soc. **127** (2001) 111-126.

# Using Singular Value Decomposition
# in Conjunction with
# Data Assimilation Procedures

Gabriel Dimitriu

University of Medicine and Pharmacy, Faculty of Pharmacy,
Department of Mathematics and Informatics,
700115 Iasi, Romania
dimitriu@umfiasi.ro

**Abstract.** In this study we apply the singular value decomposition
(SVD) technique of the so-called 'observability' matrix to analyse the information content of observations in 4D-Var assimilation procedures. Using a simple one-dimensional transport equation, the relationship
between the optimal state estimate and the right singular vectors of
the observability matrix is examined. It is shown the importance of the
value of the variance ratio, between the variances of the background and
the observational errors, in maximizing the information that can be extracted from the observations by using Tikhonov regularization theory.
Numerical results are presented.

## 1 Introduction

One of the popular methods used to examine the information content in dynamical systems is the singular value decomposition (SVD). In this paper we describe
a technique which uses the SVD of the so-called 'observability' matrix in 4D-Var
assimilation procedures ([7], [8]), and then apply it to a simple one-dimensional
transport model. We examine the critical features in this assimilation process,
assuming for simplicity the linearity of the model and of the observation operator,
and the prior (background) error and observational errors are considered to be
uncorrelated with fixed variance.

The 4D-Var data assimilation scheme is used here for a system governed by
the following discrete linear equations ([1], [2], [9], [10]):

$$\mathbf{c}_{k+1} = M(t_{k+1}, t_k)\mathbf{c}_k, \qquad \mathbf{c}_k^{obs} = H_k\mathbf{c}_k + \delta_k, \qquad (1)$$

for $k = 0, \ldots, N - 1$. Here $M(t_{k+1}, t_k)$ denotes the system matrix describing the
evolution of the states from $t_k$ to time $t_{k+1}$, $\mathbf{c}_k^{obs} \in I\!\!R^{p_k}$ represents a vector of
$p_k$ observations at time $t_k$, and $H_k \in I\!\!R^{n \times p_k}$ is the observation operator that
generally includes transformations and grid interpolations. The observational
errors $\delta_k \in I\!\!R^{p_k}$ are assumed to be unbiased, serially uncorrelated, Gaussian
random vectors with covariance matrices $R_k \in I\!\!R^{p_k \times p_k}$.

T. Boyanov et al. (Eds.): NMA 2006, LNCS 4310, pp. 435–442, 2007.
© Springer-Verlag Berlin Heidelberg 2007

A prior estimate, or 'background estimate', $c_0^b$ of the initial state $c_0$ is assumed to be known and the initial random errors $(c_0 - c_0^b)$ are assumed to be Gaussian with covariance matrix $B_0 \in \mathbb{R}^{n \times n}$.

The aim of the data asimilation is to minimize the square error between the model predictions and the observed system states, weighted by the inverse of the covariance matrices, over the assimilation interval. The initial state $c_0$ is treated as the required control variable in the optimization process. Thus, the objective function associated with the data assimilation for (1) is expressed by

$$ \mathcal{J} = \frac{1}{2}(c_0 - c_0^b)^T B_0^{-1}(c_0 - c_0^b) + \frac{1}{2}(\mathbf{H}c_0 - c^{obs})^T \mathbf{R}^{-1}(\mathbf{H}c_0 - c^{obs}), \quad (2) $$

where
$$ \mathbf{H} = [H_0^T, (H_1 M(t_1, t_0))^T, \dots, (H_{N-1} M(t_{N-1}, t_0))^T]^T, \quad (3) $$
$$ (c^{obs})^T = [(c_0^{obs})^T, (c_1^{obs})^T, \dots, (c_{N-1}^{obs})^T], \quad (4) $$

and $\mathbf{R}$ is a block diagonal matrix with diagonal blocks equal to $R_j$ corresponding to the observation operators $H_j$. The matrix $\mathbf{H}$ is known as the *observability matrix* ([7]). The solution to the optimization problem is then given explicitly by

$$ c_0 = c_0^b + \left(B_0^{-1} + \mathbf{H}^T \mathbf{R}^{-1} \mathbf{H}\right)^{-1} \mathbf{H}^T \mathbf{R}^{-1} \mathbf{d}, \quad \text{where} \quad \mathbf{d} = c^{obs} - \mathbf{H}c_0^b. \quad (5) $$

## 2   Definition of the Model

Our model under investigation is a pure one-dimensional transport equation defined by the following partial differential equation ([10]):

$$ \frac{\partial \mathbf{c}}{\partial t} = -\mathcal{V} \frac{\partial \mathbf{c}}{\partial x}, \quad x \in [0, 2\pi], \quad t \in [0, T], \quad \mathbf{c}(x, 0) = f(x). \quad (6) $$

The initial condition $f$ and function $\mathcal{V}$ are chosen to be $f(x) = \sin(x)$ and $\mathcal{V}(x) = 6x(2\pi - x)/(4\pi^2)$. Then the exact (analytical) solution of (6) is given by $c_{exact}(x, t) = \sin(2\pi x/(x + (2\pi - x)\exp(3t/\pi)))$. Details about the numerical aspects and implementation of an data assimilation algorithm for this model are presented in [10].

## 3   Singular Value Decomposition, Filter Factors and Regularization Methods

Assume that the correlation matrices in (2) are diagonal matrices given by $B_0 = \sigma_b^2 I$ and $\mathbf{R} = \sigma_o^2 I$, with $I$ denoting the identity matrix of appropriate order. Here $\sigma_b^2$ and $\sigma_o^2$ represent the variances for the background and observational errors, respectively. We now define the singular value decomposition (SVD) of the observability matrix $\mathbf{H}$ given by (3) to be

$$ \mathbf{H} = \mathbf{U}\mathbf{\Lambda}\mathbf{V}^T, \quad (7) $$

where $\mathbf{\Lambda} = \mathrm{diag}\{\lambda_j\}$. The scalars $\lambda_j$ are the singular values of $\mathbf{H}$, and the left and right singular vectors $\mathbf{v}_j$ and $\mathbf{u}_j$ are given by the columns of $\mathbf{V}$ and $\mathbf{U}$, respectively. Applying the SDV in (5) enables us to obtain the optimal analysis as

$$\mathbf{c}_0 = \mathbf{c}_0^b + \sum_j \frac{\lambda_j^2}{\mu^2 + \lambda_j^2} \frac{\mathbf{u}_j^T \mathbf{d}}{\lambda_j} \mathbf{v}_j \,, \tag{8}$$

where $\mu^2 = \sigma_o^2/\sigma_b^2$. The increments made to the prior estimate $\mathbf{c}_0^b$ by the assimilation procedure are thus expressed by a linear combination of the right singular vectors of $\mathbf{H}$, weighted by the two factors

$$f_j = \frac{\lambda_j^2}{\mu^2 + \lambda_j^2} \,, \qquad c_j = \frac{\mathbf{u}_j^T \mathbf{d}}{\lambda_j} \,. \tag{9}$$

If there is no background (prior) estimate constraining the solution, so that $\mu^2 = 0$, then the 'filter factor' $f_j = 1$, for all $j$ (see Figure 1), and the weight $c_j$ given to each singular vector in the increment is proportional to the angle between the 'innovation' vector $\mathbf{d}$ and the corresponding left singular vector $\mathbf{u}_j$. Large values of $c_j$ indicate that the corresponding singular vector has a significant contribution to the correct reconstruction of the system state. If a background (prior) estimate is given, so that $\mu^2 > 0$, then the weighting on each singular vector is reduced by the corresponding filter factor $f_j$. For noisy observations

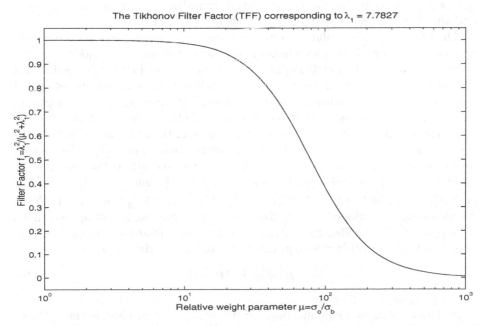

**Fig. 1.** Tikhonov Filter Factor (TFF) corresponding to the largest singular value of the observability matrix $\mathbf{H}$ versus the relative weight parameter $\mu$

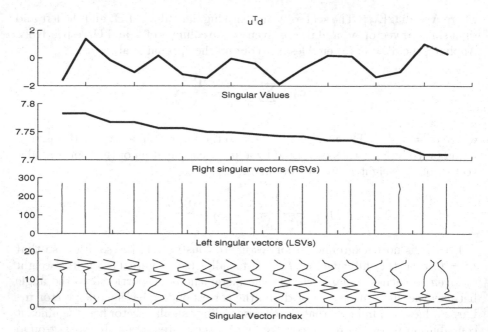

**Fig. 2.** Results of the SVD applied to the observability matrix **H**, corresponding to each singular vector index

$(\mu^2 \gg \lambda_j^2)$, the contribution to the analysis from the singular vector is damped and the observational information is strongly filtered (Figure 1).

The left and right singular vectors corresponding to our numerical example described in the next section are shown in the plots of Figure 2. We notice that for noisy observations, the weights $c_j$ are seen to grow as the singular values decay. If the corresponding singular vectors are not sufficiently filtered, then the estimated states may be very inaccurate due to the noise. This filtering is a vital aspect of the assimilation process, as both background and observations contain errors.

Several techniques for determining $\mu^2$ are given in the literature. Good choices for $\mu^2$ can be made by using Tikhonov regularization theory ([5]). We first reformulate the objective function (2) for the variational asimilation problem by making a change of variables. We let $C_B$ and $\mathbf{C}_R$ be such that $B_0 = \sigma_b^2 C_B$, $\mathbf{R} = \sigma_o^2 \mathbf{C}_R$, and define $\xi = C_B^{-1/2}(\mathbf{c}_0 - \mathbf{c}_0^b)$, $\tilde{\mathbf{H}} = \mathbf{C}_R^{-1/2}\mathbf{H}C_B^{1/2}$ and $\tilde{\mathbf{d}} = \mathbf{C}_R^{-1/2}\mathbf{d}$. As we already mentioned before, for simplicity in our numerical approach, we set $C_B$ and $\mathbf{C}_R$ to be identity matrices. For the linear model (1), minimizing the objective function (2) is then equivalent to minimizing the function

$$\tilde{\mathcal{J}} = \mu^2\|\xi\|_2^2 + \|\tilde{\mathbf{H}}\xi - \tilde{\mathbf{d}}\|_2^2,\tag{10}$$

where $\|\cdot\|_2$ denotes 2-norm.

A simple (although computationally still expensive) method is the L-Curve technique illustrated in Figure 3. For more precise computation, the Generalized Cross Validation (GCV) technique provides an algorithm for determining the

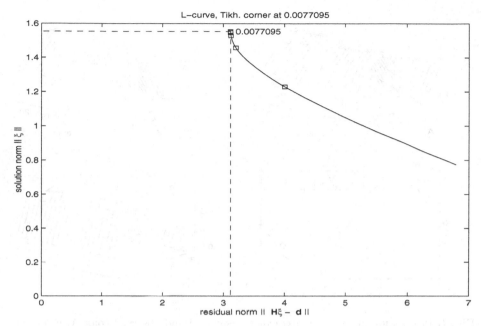

**Fig. 3.** L-curve based on Tikhonov theory, with the detection of the corner at the value 0.0077095

point of maximum curvature ([3]). Generalized cross-validation offers a way to estimate appropriate values of parameters such as the truncation parameter $k$ in truncated singular value decomposition.

## 4  Numerical Results and Interpretation

This section presents numerical results in data assimilation applying SVD theory and gives some interpretations. Figure 4 contains the plots of the exact and computed states of the model (6), together with the initial condition. By means of a perturbed initial condition (plot C) we generated the observed state, $\mathbf{c}^{obs}$. We set in our numerical approach $T = 1$, and used 9 discretization points in space and 24 points in time interval $[0, 1]$.

The choice of the specified value of the variance ratio $\mu^2$ is crucial to enable the extraction of the maximum information available in the observations. Figure 5 indicates that the assimilation accuracy of the initial condition $\mathbf{c}_0$ strongly depends on the appropriate choice of $\mu^2$. This plot presents numerical results by using a finer mesh: 19 discretization points in space and 60 time points.

For the Tikhonov regularization, we seek to obtain a curve that would have a flat portion where a good value of the regularization parameter $\mu^2$ can be located. Figure 3 contains the results of using L-curve method to obtain the value of the regularization parameter $\mu^2$ ([6]). This value was detected at $\mu^2 = 7.7095 \cdot 10^{-3}$.

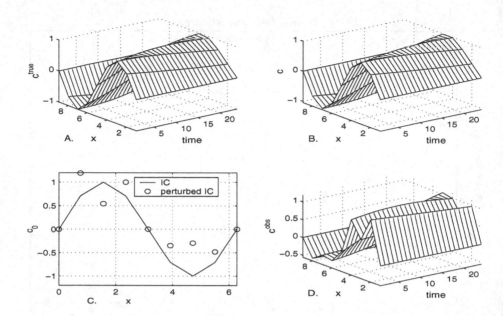

**Fig. 4.** Profiles for the model state: exact, computed and observed (plot A, B and D, respectively). Plot C contains the initial condition for the exact and perturbed solution.

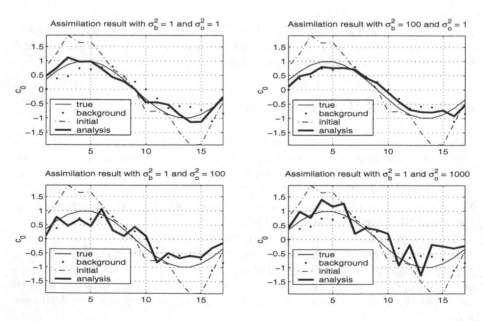

**Fig. 5.** Assimilation results of the initial condition $c_0$ for different values of the variance ratio $\mu^2$

**Table 1.** Regularization data for $\tilde{\mathbf{H}} = \mathbf{C}_R^{-1/2} \mathbf{H} C_B^{1/2}$ matrix

| Point $i$ | Variance ratio: $\mu_i^2$ | Solution Norm: $\|\xi_{\mu_i^2}\|$ | Relative Error: $\|\xi_{\mu_i^2} - \xi^{exact}\|/\|\xi^{exact}\|$ | Difference $\|\xi_{\mu_i^2}\| - \|\xi_{\mu_{i-1}^2}\|$ |
|---|---|---|---|---|
| 1 | 7.78265e+0 | 7.76665e-1 | 5.02071e-1 | – |
| 25 | 4.46679e+0 | 1.17087e+0 | 2.49336e-1 | 1.36635e-2 |
| 50 | 2.50505e+0 | 1.41224e+0 | 9.45874e-2 | 6.29749e-3 |
| 75 | 1.40487e+0 | 1.51016e+0 | 3.18119e-2 | 2.27156e-3 |
| 100 | 7.87876e-1 | 1.54382e+0 | 1.02283e-2 | 7.47413e-4 |
| 125 | 4.41854e-1 | 1.55472e+0 | 3.23970e-3 | 2.38483e-4 |
| 150 | 2.47799e-1 | 1.55818e+0 | 1.02120e-3 | 7.53488e-5 |
| 175 | 1.38970e-1 | 1.55928e+0 | 3.21408e-4 | 2.37323e-5 |
| 200 | 7.79365c-2 | 1.55962e+0 | 1.01110c-4 | 7.46754e-6 |
| 225 | 4.37081e-2 | 1.55973e+0 | 3.18028e-5 | 2.34899e-6 |
| 250 | 2.45122e-2 | 1.55976e+0 | 1.00027e-5 | 7.38825e-7 |
| 275 | 1.37468e-2 | 1.55977e+0 | 3.14602e-6 | 2.32375e-7 |
| 300 | 7.70946e-3 | 1.55978e+0 | 9.89474e-7 | 7.30858e-8 |

The data contained in Table 1 were performed using a mesh grid of 300 points. The mesh grid is constructed using the method employed in the Matlab code *lcurve.m* of Hansen ([4]). The method can be described as follows. Let $\mu_1 = \lambda_{max}(\tilde{\mathbf{H}})$ and $\mu_l$ be positive but smaller than $\max\{\lambda_{min}(\tilde{\mathbf{H}}), \lambda_{max}(\tilde{\mathbf{H}}) \cdot \varepsilon\}$, where $\varepsilon$ is the machine roundoff unit, $\lambda_{max}(\tilde{\mathbf{H}})$ and $\lambda_{min}(\tilde{\mathbf{H}})$ are the largest and smallest singular values of $A$. We want to fill $l-2$ numbers $\mu_2, \ldots, \mu_{l-1}$ betweeen $\lambda_1$ and $\lambda_l$. Let $\nu_i = \ln(1/\mu_i)$, $i = 1, 2, \ldots, l$. Since $\ln x$ is an increasing function of $x$, we have $\nu_1 < \nu_2 < \cdots < \nu_l$. Let $h = (\nu_l - \nu_1)/(l-1) = \ln(\mu_1/\mu_l)^{l-1}$. Put $\nu_i = \nu_1 + (i-1)h$, $i = 1, \ldots, l$. Then $\nu_i$'s form a uniform mesh grid. Converting $\nu_i$ back to $\mu_i$, we obtain a mesh grid for $\mu$: $\mu_1 > \mu_2 > \cdots > \mu_l$, with $\mu_i = \mu_1 (\mu_l/\mu_1)^{(i-1)/(l-1)}$, $i = 1, 2, \ldots, l$.

Table 1 gives some of the 300 data points along with some other quantities. Here, the fourth column lists, $\|\xi_{\mu_i^2} - \xi^{exact}\|/\|\xi^{exact}\|$, and the fifth column lists $\|\xi_{\mu_i^2}\| - \|\xi_{\mu_{i-1}^2}\|$. Notice that $(\|\xi_{\mu_i^2}\| - \|\xi_{\mu_{i-1}^2}\|)/h$ gives an approximate slope of the curve in Figure 3 at $\nu_i$. It can be observed that the smallest difference and therefore the minimum slope over the mesh grid occures at index $i = 300$, at which $\mu_{300}^2 \approx 7.70946 \times 10^{-3}$. It can be seen that the relative error reaches its minimum value $9.89474 \times 10^{-7}$. For this problem, we may regard $\mu_{300}^2$ as the optimal value of the regularization parameter for this choice of the $\mu^2$ grid.

## 5 Conclusions

An analysis of the 4D-Var assimilation procedure using a singular value decomposition (SVD) technique was presented, and the results of the numerical experiments were interpreted in terms of the singular values and singular vectors of the 'observability' matrix of the studied system. We showed that the filtering effect of the background state is controlled by a 'regularization' parameter defined as the variance ratio between the observation and background error variances. We also established that a good choice of the regularization parameter can significantly improve the reconstruction of the states in unobserved regions. Applicable techniques and results for selecting a good choice of this parameter, based on Tikhonov regularization theory and generalized cross-validation ([7], [8]) are also described.

# References

1. G. Dimitriu, R. Cuciureanu. Mathematical aspects of data assimilation for atmospheric chemistry models, In: *Advances in Air Pollution Modeling for Environmental Security*, (I. Faragó, À. Havasi, K. Georgiev, eds.), NATO Science Series, 54, Springer, Berlin, 93–103, 2005.
2. M. Ghil, P. Malanotte-Rizzoli. In *Data Assimilation in Meteorology and Oceanography. Advances in Geophysics*, R. Dmowska, B. Saltzman (eds.), 33, 141–266, 1991.
3. G. H. Golub, M. Heath, G. Wahba. Generalized cross-validation as a method for choosing a good ridge parameter. *Technometrics*, 21, 215–223, 1979.
4. P. C. Hansen. *Regularization Tools, a Matlab Package for Analysis and Solution of Discrete Ill-posed Problems*, Version 3.0 for Matlab 5.2, Technical Report UNIC-92-03, Danish Computing Center for Research and Education, Technical University of Denmark, June 1992 (Revised June 1998).
5. P. C. Hansen. *Rank-Deficient and Discrete Ill-Posed Problems: Numerical Aspects of Linear Inversion*, SIAM, Philadelphia, 1998.
6. P. C. Hansen, D. P. O'Leary. The use of the L-curve in the regularization of discrete ill-posed problems. *SIAM J. Sci. Computing*, 14, 1487–1503, 1993.
7. C. Johnson. *Information content of observations in variational data assimilation*, PhD Thesis, The University of Reading, 2003.
8. C. Johnson, N. K. Nichols, B. J. Hoskins. *A Singular Vector Perspective of 4D-VAR: Filtering and Interpolating*, Numerical Analysis Report NA 1/04, Department of Mathematics, The University of Reading, 2004.
9. J. N. Thépaut, R. N. Hoffman, P. Courtier. Interactions of dynamics and observations in a 4D variational assimilation, *Mon. Weather Rev.*, 121, 3393–3414, 1993.
10. Z. Zlatev, J. Brandth, A. Havasi. Implementation issues related to variational data assimilation: some preliminary results and conclusions, *Working Group on Matrix Computations and Statistics*, Copenhagen, Denmark, April 1-3, 2005.

# New Operator Splitting Methods and Their Analysis

István Faragó

Eötvös Loránd University, Pázmány P. s. 1/c, 1117 Budapest, Hungary

**Abstract.** In this paper we give a short overview of some traditional operator splitting methods. Then we introduce two new methods, namely the additive splitting and the iterated splitting. We analyze these methods and compare them to the traditional ones.

## 1  Introduction

Operator splitting is a powerful method for the numerical investigation of complex (physical) time-dependent models, where the stationary (elliptic) part consists of a sum of several simpler operators (processes). As an example, we can recall the phenomenon of the air-pollution process [5] and the Maxwell equations [4]. These tasks are very complicated because it is described by a system of nonlinear partial differential equations. Therefore, the analytical solution is impossible to find, moreover, the numerical modelling with direct discretization is also hopeless from a practical point of view.

Operator splitting is a widely used technique for solving such complex problems. The basic idea is to split the original problem into a sequence of smaller ("simpler") problems. The general scheme of this approach can be formulated as follows:

1. We select a small positive time step ($\tau$), and divide the whole time interval into subintervals of length $\tau$;
2. On each subinterval we consecutively solve the time-dependent problems, each of which involves only one physical process;
3. We pass to the next time sub-interval.

We mention that the different problems are connected via the initial conditions.

In our investigation we will assume that there are only two operators, i.e. we will demonstrate our methods on the Cauchy problem of the form

$$\left.\begin{aligned}\frac{dw(t)}{dt} &= (A+B)w(t), \quad t \in (0,T] \\[2mm] w(0) &= w_0.\end{aligned}\right\} \tag{1}$$

We assume that the operators are bounded linear operators, and hence the exact solution is $w(t) = \exp(t(A+B))w(0)$.

T. Boyanov et al. (Eds.): NMA 2006, LNCS 4310, pp. 443–450, 2007.

## 2    Traditional Operator Splitting Methods

In this section we describe the different known operator splitting methods which have been widely used in different applications. For more details, see [1], [3].

### 2.1    Sequential Splitting

The *sequential splitting*, which is historically the first one, has a simple algorithm: first we solve the Cauchy problem with the first operator and the original initial condition, and next the Cauchy problem corresponding to the second operator, using the solution of the first problem at the end-point of the time-subinterval as initial condition. This means that the algorithm is the following:

$$\frac{dw_1^n}{dt}(t) = Aw_1^n(t), \qquad (n-1)\tau < t \leq n\tau,$$

$$w_1^n((n-1)\tau) = w_{sp}^N((n-1)\tau), \tag{2}$$

and

$$\frac{dw_2^n}{dt}(t) = Bw_2^n(t), \qquad (n-1)\tau < t \leq n\tau,$$

$$w_2^n((n-1)\tau) = w_1^n(n\tau), \tag{3}$$

for $n = 1, 2, \ldots N$. Then the split solution at the mesh-points $t = n\tau$ is defined as

$$w_{sp}^N(n\tau) = w_2^n(n\tau).$$

Here $w_{sp}^N(0) = w_0$ is given from the initial condition.

### 2.2    Strang–Marchuk Splitting

The next operator splitting, which is called *Strang–Marchuk splitting*, has a more complicated algorithm. The basic idea is that we divide the split time-subinterval into two parts. Then, as in the sequential splitting, successively we solve the problems:

1. on the first half with the operator $A$;
2. on the whole interval with the operator $B$;
3. on the second half again with the operator $A$.

The different tasks are again connected with the initial conditions. The method has the following computational algorithm:

$$\frac{dw_1^n}{dt}(t) = Aw_1^n(t), \ (n-1)\tau < t \leq (n-0.5)\tau,$$

$$w_1^n((n-1)\tau) = w_{sp}^N((n-1)\tau), \tag{4}$$

$$\frac{dw_2^n}{dt}(t) = Bw_2^n(t), \qquad (n-1)\tau < t \le n\tau,$$

$$w_2^n((n-1)\tau) = w_1^n((n-0.5)\tau),$$

(5)

$$\frac{dw_3^n}{dt}(t) = Aw_3^n(t), \ (n-0.5)\tau < t \le n\tau,$$

$$w_3^n((n-0.5)\tau) = w_2^n(n\tau).$$

(6)

Then the split solution at the mesh-points is defined as

$$w_{sp}^N(n\tau) = w_3^n(n\tau).$$

Here $n = 1, 2, \ldots N$, and $w_{sp}^N(0) = w_0$.

## 2.3    Symmetrically Weighted Sequential Splitting

Let us notice that the sequential splitting is not symmetric w.r.t. the ordering of the operators. The third operator splitting, called *symmetrically weighted sequential splitting*, is a combination of two sequential splittings with different ordering. This makes the splitting symmetric and, as we will see later, more accurate. The method has the following computational algorithm:

$$\frac{du_1^n(t)}{dt} = Au_1^n(t), \qquad (n-1) < t \le n\tau,$$

$$u_1^n((n-1)\tau) = w_{sp}^N((n-1)\tau);$$

(7)

$$\frac{du_2^n(t)}{dt} = Bu_2^n(t), \qquad (n-1) < t \le n\tau,$$

$$u_2^n((n-1)\tau) = u_1^n(n\tau);$$

(8)

and

$$\frac{dv_1^n(t)}{dt} = Bv_1^n(t), \qquad (n-1) < t \le n\tau,$$

$$v_1^n((n-1)\tau) = w_{sp}^N((n-1)\tau);$$

(9)

$$\frac{dv_2^n(t)}{dt} = Av_2^n(t), \qquad (n-1) < t \le n\tau,$$

$$v_2^n((n-1)\tau) = v_1^n(n\tau).$$

(10)

The split solution at the mesh-points is defined as

$$w_{sp}^N(n\tau) := \frac{u_2^n(n\tau) + v_2^n(n\tau)}{2}$$

(11)

for $n = 1, 2, \ldots N$, where $w_{sp}^N(0) = w_0$.

## 2.4   Some Remarks on the Operator Splitting Methods

The operator splitting methods have several specific properties. Some of them are listed below.

a. Operator splitting methods can be viewed as time-discretization methods, which define exponential approximations to the exact solution at the mesh-points $t = n\tau$. Hence, as for any arbitrary time-discretization method, we can introduce the notion of the discretization error, called *local splitting error*. It is defined as

$$Err_{sp}(\tau) = w(\tau) - w_{sp}^N(\tau).$$

b. When $\tau$ tends to zero, the local splitting error should tend to zero, too. Its order defines the so-called *order of the splitting*, namely, when $Err_{sp}(\tau) = \mathcal{O}(\tau^{p+1})$ then the splitting is called a $p$-th order splitting. One can prove that for the traditional splittings we have:
  1. the sequential splitting is of first order;
  2. the Strang–Marchuk splitting is of second order;
  3. the symmetrically weighted sequential splitting is of second order.

Hence, the highest accuracy which can be achieved by these splittings is equal to two.

c. Between two mesh-points the above defined traditional splitting methods are not consistent, i.e., we obtain approximations to the exact solution only at the mesh-points. In order to show this, let us analyze the sequential splitting. (The proof for the other splittings is similar.) Using the formulas (14) and (15) on the time interval $[0, \tau]$ we obtain

$$w_2^n(t) = \exp(tB)\exp(\tau A)w_0. \tag{12}$$

Hence, for the local splitting error at any time $t \in [0, \tau]$ we get

$$w(t) - w_2^n(t) = (t - \tau)Aw_0 + \mathcal{O}(t^2). \tag{13}$$

This shows that we have consistency only for $t = \tau$.

## 3   New Operator Splittings

The property c. of the traditional splittings, mentioned in Section 2.4, implies the following disadvantage. When we apply some numerical method to the sub-problems with time discretization step-size $\Delta t$, which is much less than $\tau$, we cannot use the intermediate numerical results as an approximation to the original solution, because these methods are not consistent on the split time-subinterval. According to property b., a further problem of the traditional splitting methods is that (in lack of commutativity of the operators) we cannot achieve high-order (i.e., of order more than two) accuracy. This order barrier may cause serious restriction during the applications.

These problems gave the motivation to create some new operator splitting methods with improved properties. In what follows we define two new splitting methods.

## 3.1 Additive Splitting

This method is based on a simple idea: we solve the different sub-problems by using the same initial function. We obtain the split solution by the use of these results and the initial condition. The algorithm is the following:

$$\frac{dw_1^n}{dt}(t) = Aw_1^n(t), \qquad (n-1)\tau < t \leq n\tau,$$

$$w_1^n((n-1)\tau) = w_{sp}^N((n-1)\tau),$$

(14)

and

$$\frac{dw_2^n}{dt}(t) = Bw_2^n(t), \qquad (n-1)\tau < t \leq n\tau,$$

$$w_2^n((n-1)\tau) = w_{sp}^N((n-1)\tau).$$

(15)

Then the split solution at the mesh-points is defined as

$$w_{sp}^N(n\tau) = w_1^n(n\tau) + w_2^n(n\tau) - w_{sp}^N((n-1)\tau).$$

Here $n = 1, 2, \ldots N$, where $w_{sp}^N(0) = w_0$.

One can see the main advantage of this method at first sight: it can be parallelized on the operator level in a natural way (like the symmetrically weighted sequential splitting).

The local splitting error (for bounded operators) can be investigated directly.

**Theorem 1.** *The additive splitting is a first order accurate splitting method.*

*Proof.* The solution of the additive splitting at $t = \tau$ is defined as

$$w_{sp}^N(\tau) = [\exp(A\tau) + \exp(B\tau) - I]w_0,$$

(16)

where $I$ denotes the identity operator. Hence, we get

$$w_{sp}^N(\tau) = \left(I + A\tau + \frac{1}{2}A^2\tau^2 + I + B\tau + \frac{1}{2}B^2\tau^2 - I\right)w_0 + \mathcal{O}(\tau^3) =$$

$$= \left(I + (A+B)\tau + \frac{1}{2}(A^2 + B^2)\tau^2\right)w_0 + \mathcal{O}(\tau^3).$$

(17)

Comparing this expression with the similar Taylor expansion of the exact solution, we get the local splitting error

$$Err_{sp}(\tau) = \frac{1}{2}\left((AB + BA)\tau^2\right)w_0 + \mathcal{O}(\tau^3),$$

(18)

which proves the statement.

The following statement shows that the additive splitting approximates the exact solution not only at the mesh-points.

**Corollary 1.** *The additive splitting approximates the exact solution on the whole split time interval* $[0, \tau]$.

*Proof.* The exact solution of the additive splitting at $t \in [0, \tau]$ is

$$w_{\text{sp}}^N(t) = [\exp(At) + \exp(Bt) - I]w_0. \tag{19}$$

Hence, at any point of the time interval we have

$$Err_{\text{sp}}(t) = \frac{1}{2}\big((AB + BA)t^2\big)w_0 + \mathcal{O}(t^3), \tag{20}$$

which proves the statement.

## 3.2  Iterated Splitting

The method was introduced in [2]. This splitting method can be implemented as the continuous variant of the well-known ADI method. In this sense, it is the iterative improvement of the split result on some fixed split time interval $[n\tau, (n+1)\tau]$, and its algorithm reads as follows:

$$\frac{dw_i^n(t)}{dt} = Aw_i^n(t) + Bw_{i-1}^n(t), \quad \text{with } w_i^n(t^n) = w_{\text{sp}}^N(n\tau) \tag{21}$$

$$\frac{dw_{i+1}^n}{dt} = Aw_i^n(t) + Bw_{i+1}^n(t), \quad \text{with } w_{i+1}^n(t^n) = w_{\text{sp}}^N(n\tau) \tag{22}$$

for $i = 1, 3, 5, \ldots, 2m - 1$, where $w_0^n$ is a fixed starting function for the iteration. (The index $i$ denotes the number of the iteration on the fixed $n$-th time subinterval.) Then the split solution at the mesh-points is defined as

$$w_{\text{sp}}^N((n+1)\tau) = w_{2m}^n((n+1)\tau).$$

This method can be considered as an operator splitting method because we decompose the original problem into a sequence of two simpler sub-problems, in which the first sub-problem should be solved for the first operator, while the second sub-problem for the second operator. This splitting is formally similar to the sequential splitting, but here each split sub-problem contains the other operator as well with some previously defined approximate solution.

**Theorem 2.** *Assume that on the time interval* $[0, \tau]$ *the starting function* $w_0^0(t)$ *for the iterated splitting satisfies the condition*

$$w_0^0(0) = w_0. \tag{23}$$

*Then the iterated splitting is consistent.*

*Proof.* For $i = 1$ the exact solution of (21) is

$$w_1^0(t) = \exp(tA)w_0 + \int_0^t \exp((t-s)A)Bw_0^0(s)ds, \quad t \in [0, \tau]. \tag{24}$$

Using the obvious relation

$$w_0^0(s) = w_0^0(0) + \mathcal{O}(s) = w_0 + \mathcal{O}(s),\tag{25}$$

for the second term on the right side of (24) we have

$$\int_0^t \exp((t-s)A)Bw_0^0(s)ds = \int_0^t \left[\sum_{n=0}^{\infty} \frac{1}{n!}(t-s)^n A^n\right]B(w_0 + \mathcal{O}(s))ds =\tag{26}$$

$$= \int_0^t Bw_0 ds + \mathcal{O}(t^2) = tBw_0 + \mathcal{O}(t^2).$$

On the other hand,

$$\exp(tA)w_0 = (I + tA)w_0 + \mathcal{O}(t^2).\tag{27}$$

Hence, putting the relations (26) and (27) into (24), we get

$$w_1^0(t) = (I + tA + tB)w_0 + \mathcal{O}(t^2),\tag{28}$$

which proves the consistency.

**Corollary 2.** *Under the condition (23) the iterated splitting is consistent already after the first iteration. Moreover, it is consistent on the whole split time interval $[0, \tau]$. We can also observe that condition (23) is not only a sufficient, but also a necessary condition of the consistency in the first iterated step. Clearly, the simplest and most practical choice is*

$$w_0^0(t) = w_0, \quad t \in [0, \tau].\tag{29}$$

Next we examine the order of the local splitting error when we apply the second step of the method, i.e., we solve (22) for $i = 1$, using $w_0^0(t)$ from (21).

**Theorem 3.** *On the time interval $[0, \tau]$ one complete step (21)–(22), under the condition (24) results in a second order accuracy of the iterated splitting method.*

*Proof.* Clearly, the exact solution of the problem (22) can be written as

$$w_2^0(t) = \exp(tB)w_0 + \int_0^t \exp((t-s)B)Aw_1^0(s)ds.\tag{30}$$

Now, using the expression (28) for $w_1^0(s)$, we get

$$\exp((t-s)B)Aw_1^0(s) = [I - (t-s)B](A + A^2 s + ABs)w_0 + \mathcal{O}(s^2)) =\tag{31}$$

$$= (A + (t-s)BA + sA^2 + sAB)w_0 + \mathcal{O}(s^2)).$$

Hence, integrating on $[0, t]$, we get

$$\int_0^t \exp((t-s)B)Aw_1^0(s)ds = \left[At + \frac{1}{2}t^2(BA + A^2 + AB)\right]w_0 + \mathcal{O}(t^3).\tag{32}$$

Using the Taylor series for $\exp tB$ and (32), we obtain the statement:

$$w_2^0(t) = (I + tB + \frac{1}{2}t^2 B^2)w_0 + (At + \frac{1}{2}t^2 BA + \frac{1}{2}t^2 A^2 + \frac{1}{2}t^2 AB)w_0 + \mathcal{O}(t^3)) =$$

$$\left[ I + t(A + B) + \frac{1}{2}t^2(A + B)^2 \right] w_0 + \mathcal{O}(t^3).$$

$$(33)$$

By mathematical induction, continuing the process, the Reader can prove

**Theorem 4.** *Assume that on the time interval $[0, \tau]$ we execute the iteration (21)–(22), where the initial guess is chosen according to the condition (24). Then each iterative step increases the order of the local splitting error by one.*

# 4   Conclusions and Closing Remarks

We have introduced new consistent splittings, namely, the additive splitting and the iterated splitting. The main benefit of the first splitting is its algorithmic simplicity, more precisely its natural parallelization on the operator level. The main benefit of the iterated splitting is its high accuracy. However, these methods have some drawbacks, too. We list some of them.

1. Stability of the additive splitting is a crucial problem and to prove it seems to be very difficult. In practice, we have obtained a stable (and hence convergent) solution only for small time steps.
2. The additive splitting has always first order accuracy, even for commuting operators. Hence, it is not accurate enough for many practical applications.
3. The iterated splitting cannot be parallelized on the operator level.
4. The iterated splitting requires numerical integration in each iterative step, which can spoil the theoretical order of the method. Therefore, the number of the iteration should coincide with the order of the applied numerical integration method. (However, we emphasize that the values at the intermediate integration points give consistent approximations.)

# References

1. Csomós, P., Faragó, I., Havasi , Á..: Weighted sequential splittings and their analysis, Comput. Math. Appl., 50 (2005) 1017-1031.
2. Faragó, I., Geiser, J. Iterative operator- splitting methods for linear problems, Weierstass Institute für Angewandte Analysis, 1043 (2005) 1-18. International Journal of Computational Science and Engineering (to appear)
3. Hundsdorfer, W., Verwer, J. G.: Numerical solution of time-dependent advection-diffusion-reaction equations. Springer, Berlin, 2003.
4. Horváth, R.: Operator splittings for the numerical solution of the Maxwell's equation, Lecture Notes in Computer Science, 3743 (2006) 363-371.
5. Zlatev, Z.: Computer treatment of large air pollution models. Kluwer Academic Publishers, Dordrecht-Boston-London, 1995.

# On an Implementation of the TM5 Global Model on a Cluster of Workstations

Krassimir Georgiev

Institute for Parallel Processing,
Bulgarian Academy of Sciences
Acad. G. Bonchev, Bl. 25-A, 1113 Sofia, Bulgaria
georgiev@parallel.bas.bg

**Abstract.** TM5 is a global chemistry Transport Model. It allows two-way nested zooming which leads to possibility to run the model on relatively very fine space grid ($1° \times 1°$) over selected regions (Europe is most often used in up to now experiments but North America, Africa and Asia can be treated separately or in combinations). The boundary conditions for the zoomed subdomains are provided consistently from the global model. The TM5 model is a good tool for studying some effects due to the grid refinement on global atmospheric chemistry issues (intercontinental transport of air pollutants, etc.).

The huge increase in the number of the multi-processor platforms and their differences leads to a need of different approaches in order to meet the requirements for the optimality of the computer runs. The paper is devoted to an implementation of a parallel version of the TM5 model on a cluster of SUN Workstations and to the developing of a new parallel algorithm. It is based on the decomposition, in some sense, of the computational domain supposing that the zoomed regions are more than one. If it is assumed that the number of zoomed regions is $N$ and the number of the processors available is $p$. The processors arc divided in $N/p$ groups and each group is responsible for the whole computational domain and one of the zoomed regions. Some communications are needed in order to impose the inner boundary conditions. The new algorithm has better parallel feathers than the old one which is used in the inner level.

Some results concerning the CPU time, speed up and efficiency can be found.

**Keywords:** air pollution modelling, global models, parallel computing.

**Subject classifications:** 65Y05.

## 1 Introduction

TM5 is a three dimensional global chemistry Transport Model. It allows *two-way nested zooming* ([2]) which leads to possibility to run the model on relatively very fine space grid ($1° \times 1°$) (longitude $\times$ latitude) over selected regions. Up to now Europe is most often used in the experiments done (see Fig. 1), but North America, South America, Africa and Asia can be treated separately or in combinations

T. Boyanov et al. (Eds.): NMA 2006, LNCS 4310, pp. 451–457, 2007.

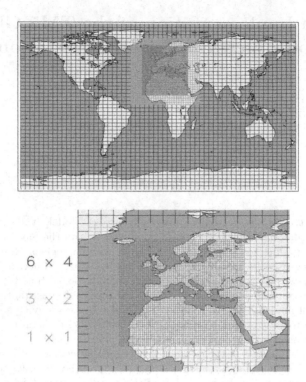

**Fig. 1.** Global computational domain and zooming over Europe

with the same resolution. The coarsest space resolution is ($6° \times 4°$). In order to allow the smooth transition between both discretizations ($3° \times 2°$) grid has been added (see Fig. 1). The boundary conditions for the zoomed subdomains are provided consistently from the global model. The model consists of 25 vertical layers (see Fig. 2): five layers in the boundary layer, ten layers in the free troposphere and ten layers in the stratosphere. There is no zooming in the vertical direction and all regions use the same vertical layer structure (hybrid sigma-pressure coordinate system). TM5 is an offline model which uses preprocessed meteorological fields from ECMWF (European Center for Medium-Range Weather Forecast), Reading, UK. The model is a good tool for studying some effects due to the grid refinement on global atmospheric chemistry issues (intercontinental transport of air pollutants, interhemisphere exchange, effects of the grid refinement on the budgets of the chemically active compounds, etc. ([2]).

The first of this row TM models was developed in 1988 (see [1]). Like in its predecessor, the TM3 model, the symmetrical operating splitting is used in TM5 model. The following subproblems are solved separately:

- transport;
- emissions;
- depositions and chemistry.

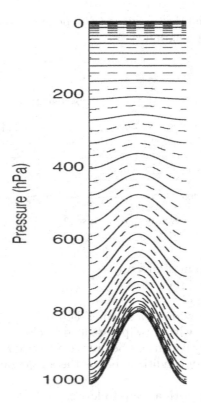

**Fig. 2.** The vertical structure of TM5 model

The splitting combining with *tailored implicit schemes* ([6]) leads to avoiding the small time steps due to the stiff problems in chemistry and the vertical mixing operators. Unfortunately, the symmetrical splitting can not always be presented in a zooming algorithm. It can be seen that in the subdomains with mesh refinement the algorithm is only partly symmetrical (see e.g. [2], p. 421, Tabl. 1). The advection numerical scheme is based on the *slope scheme* ([4]) and is mass-conserving. The convection parameterization is based on the algorithm developed by Tiedke which can be found in [5].

TM5 model is coded in Fortran 90. The model has implemented and tested on different high-performance computer platforms (IBM p690+, SGI Origin 3800, MAC OSX, etc.). The Message Passing Interface (MPI) is used as a communication tool. The parallel version of the model is run parallel over tracers as well as over vertical layers. The first one leads to good speedup in transport while the second leads to good speedup in chemistry ([3]). During a single run, each processor is responsible for one or more tracers during the transport phase, and the same processor is responsible for one or more vertical layers during the chemical part of the model. Between these two main parts of the numerical algorithm big amount of data are swapped between processors. The last one is the bottleneck of the parallel version of TM5 model.

Each time step for the three regions (European focused model) proceeds according to the following diagram ([3]):

$$t \longrightarrow \longrightarrow \longrightarrow \longrightarrow \longrightarrow \longrightarrow \longrightarrow \longrightarrow \longrightarrow t + \Delta T/2$$

$6 \times 4 \qquad xyz - - - - - - - - - - - - - - - - - - - - - - - - - - \hat{v}s\hat{c}$

$3 \times 2 \qquad \downarrow\downarrow\downarrow xyz - - - - - - - - - vs\hat{c} \, c\hat{s}v \, zyx - - - - - - - - - \Uparrow$

$1 \times 1 \qquad \downarrow\downarrow\downarrow xyz \, vs\hat{c} \, c\hat{s}v \, zyx \uparrow \qquad\qquad \downarrow\downarrow\downarrow \hat{c}sv \, zyx \, xyz \, vs\hat{c} \Uparrow$

$$t + \Delta T/2 \longrightarrow \longrightarrow \longrightarrow \longrightarrow \longrightarrow \longrightarrow \longrightarrow \longrightarrow t + \Delta T$$

$6 \times 4 \qquad c\hat{s}v \, xyz - - - - - - - - - - - - - - - - - - - - - - - -$

$3 \times 2 \qquad \Downarrow xyz \, vs\hat{c} - - - - - - - - - c\hat{s}v \, zyx - - - - - - - - - - \uparrow$

$1 \times 1 \qquad \Downarrow c\hat{s}\hat{v} \, zyx \, xyz \, vs\hat{c} \Uparrow \qquad\qquad \Downarrow xyz \, vs\hat{c} \, c\hat{s}v \, zyx \quad \uparrow$

where

- $x$   - advection step in $x$ direction;
- $y$   - advection step in $y$ direction;
- $z$   - vertical exchange;
- $\uparrow$   - update parent without a swap of data;
- $\Uparrow$   - update parent with a swap parent to levels;
- $\downarrow\downarrow\downarrow$ - write boundary conditions before the advection step (no swap needed);
- $\Downarrow$   - write boundary conditions before the advection step (a swap child to tracer needed);
- $\hat{c}$   - chemistry step with a swap to levels;
- $\hat{s}$   - sources treatment with a swap to tracer.

The huge increase in the number of the multi-processor platforms and their differences leads to a need of different approaches in order to meet the requirements for the optimality of the specific computer runs.

## 2    An Implementation of a Parallel Version of the TM5 Model on a Cluster of SUN Workstations

As was mentioned above different parallel version of the TM5 model were implemented on a number of parallel computers but not on a SUN clusters. The SUN cluster which was used for the implementation of one parallel version of the TM5 model is located at Joint Research Center (JRS), Ispra, Italy. It consists of three nodes and each node consists of four processors. So, the total number of the processors available for the specific run is 12. SUN HPC ClusterTools software which requires Solaris operating environment is used. For the communications between processors Sun MPI, which is a highly optimized version of the Message Passing Interface (MPI) communication library, is preferred. Among the highlights of Sun MPI is the full F77, C and C++ support, and only basis F90 support. The Sun Parallel File System provides high-performance file input/output multiprocessor applications running in a cluster-based, distributed memory environment. It is optimized for large files and complex data structures and therefore it is applicable in solving large and huge scientific problems.

Many successive runs were done in order to obtain more relevant output results for the parallel performance of the model on the SUN platform. The averaged results are summarized in the next table where

- *the computing times* $(T_p)$ which are measured in seconds,
- *the speedups* which are defined as the computing time on $p$ processors divided by the computing time on one processor $(S_p = \dfrac{T_p}{T_2})$ and
- *the parallel efficiency* $(E_p = \dfrac{S_p}{p})$

are presented.

**Table 1.** TM5 on SUN cluster - some performance results

| No. of processors | CPU time (in sec) | Speedup | Efficiency |
|:---:|:---:|:---:|:---:|
| 1 | 25 200 | – | – |
| 2 | 13 500 | 1.87 | 0.93 |
| 4 | 8 820 | 2.86 | 0.71 |
| 9 | 7 380 | 3.41 | 0.38 |

It is well seen from the output results that effective runs from the point of view of the parallel computations with TM5 model can be done up to four processors, i. e. inside one node of the SUN cluster. According to us there are two main reasons for that. First reason is that the communications between the processors are so much that there is no good balance between computations and communications when the number of the processors grows. The second one, but may be more important, is that in fact, the existing SUN cluster is a hybrid computer architecture. The cluster is distributed memory computer architecture (communications between nodes) while each node is a shared memory. Moreover, communications inside the nodes are much faster that these between the nodes. These explain the relatively good speedups and parallel efficiency, when the number of the processors used is two or four, and the jump to the very bad parallel characteristics, when the number of the used processor is nine.

## 3 Some Preliminary Notes on the New Parallel Version of the TM5 Algorithm

The new parallel version of the TM5 model is based on the existing TM5 parallel code. In some sense it can be considered as a domain decomposition approach to the existing already algorithm. The motivation for that more and more often the TM5 model is used with more than one zooming area. The aim is not to change the original TM5 source code too much. As was mentioned in the beginning of this report the parallel version of the TM5 model is run parallel over the tracers as well as over the vertical layers when multiprocessors computer systems are used. So, the current status of the model is determined by so called active communicator (see [3]). Let us mentioned, for seek of completeness, that the communicator is a MPI concept that denotes a group of processors which are sharing information. In the existing TM5 parallel version in addition to the main MPI communicator (**MPI_COMM_WORLD** – contains all available for the single run processors and is always active) are added two more communicators. The first one, **COM_TRAC**, is the communicator which contains all computer processors which are assigned to handle tracers (advection - diffusion part, etc.). The second one, **COM_LEV**, is the communicator which contains all computer processors which are assigned to handle levels (chemical part of the algorithm). Now, we will add one more communicator - **COMM_AREA**. This new communicator contains all computer processors which are assigned to handle the corresponding zooming area. The simple explanation of this idea (when the number of zooming areas is two) is given bellow but it is important to mention that when the is only one zooming area the new parallel algorithm is working like the existing one and **COMM_AREA** contains all available computer processors for the corresponding run.

Let us first assume that the number of the zooming areas is $N_{zoom}$ and the number of the available processors (PE's) is p. Then the PE's are divided into $N_{zoom}$ groups, and each group is supposed to be responsible for the whole computational domain and one zooming area based in some sense on the TM5 parallel code.

For seek of or simplicity from now on we will assume that the zooming areas in the algorithm are two, and they are Europe (*eur*) and North America (*nam*). Moreover, we will suppose that a four processors parallel computer will be used for making computations. Using **MPI_COMM_RANK** the processors will get ranks (*myid*): 0, 1, 2 and 3. The processors are divided into two groups. The first group consists of processors with ranks 0 and 1, and it is responsible for the global domain (*glb*) with a zoom over Europe. The rest of processors with ranks 2 and 3 belong to the second group, which is responsible for the global domain with a zoom over North America. On the base of this separation of the PE's new communicators (**COMM_AREA**) are defined. Each processor has in fact two ranks (*myid*; *myid_area*) – first one is corresponding to **MPI_COMM_WORLD** (*myid*) and the second one is corresponding to the particular group communicator **COMM_AREA** (*myid_area*). In our case PE1(0; 0), PE2(1; 1), PE3(2; 0), PE4(3; 1). Let us mention that we will use the following note *PE*(myid; myid_area)*. As there are two kinds of communications - global and local (into the groups) there are two root processors for each group in the new parallel algorithm. The global root is as a rule PE1 (*myid*=0;). The group roots are the first processors in the line. So, for the first group *root_area*= (0; 0) and for the second group *root_area*=(2; 0). The two existing in TM5 code communicators - **COM_TRAC** and **COM_LEV** now are based **NOT** on **MPI_COMM_WORLD** communicator but on the specific for the group **COMM_AREA** communicator and they have their own root processors – *root_t* and *root_k*.

## 4   Conclusions

A parallel version of the global chemistry Transport model TM5 was installed and checked on a SUN cluster. The test experiments show that when the "classical" (with one zoom region) version is performed the runs are successful. The speedup and the parallel efficiency are very good when up to four processors are used for a single run. The test experiments with the new parallel algorithm and its computer implementation are not finished yet. The first computer tests show that the new communicator work well and erectly distributes the input/output data and the corresponding computations between the available processors. Some problems concerning the reading of the input data are found. These problems not concerns the algorithm but the spacial storage of the input data.

## Acknowledgment

The research reported in this paper was supported mainly by the DG Joint Research Center of the European Commission and partly by the Bulgarian IST Center of Competence in 21st century – BIS-21++funded by the European Commission in FP6 INCO via Grant 016639/2005.

## References

1. M. Heinmann, P. Monfray, G. Polian (1988), *Long–range transport of* $^{222}Rn$ – *a test for 3D tracer models*, Chem. Geol., **70** , pp. 88 – 98
2. M. Krol, S. Houweling, B. Bergman, M. van den Broek, A. Segers, P. van Velthoven, W. Peters, F. Dentener, P. Bergamaschi (2005), *The two–way nested global chemistry's–transport model TM5: algorithm and applications*, Atmos. Chem. Phys., **5** , pp. 417–432
3. W. Peters (2003), *TM5 manual for parallel version*, http://www.phys.uu.nl/~tm5/ TM5_PW/tm5_manual.pdf
4. G. Russel, J. Lerner (1981), *A new finite–difference scheme for the tracer transport equation*, J. Appl. Meteorol., **20** , pp. 1483–1498
5. M. Tiedtke (1989), *A comprehensive mass flux scheme for cumulus parameterization in large scale models*, Mon. Wea. Rev., **177** , pp. 1779–1800
6. J. Verver, E. Spee, J. Blom, W. Hundsdorfer (1999), *A second order Rosenbrock method applied to photochemical dispersion problems*, SIAM J. Sci. Comp., **20** , pp. 1456–1480

# On the Sign-Stability of the Finite Difference Solutions of One-Dimensional Parabolic Problems

Róbert Horváth[*]

University of West-Hungary,
Erzsébet u. 9, 9400 Sopron, Hungary
rhorvath@ktk.nyme.hu

**Abstract.** In the numerical solutions of partial differential equations, the preservation of the qualitative properties of the original problem is a more and more important requirement. For 1D parabolic equations, one of this properties is the so-called sign-stability: the number of sign-changes of the solution cannot increase in time. This property is investigated for the finite difference solutions, and a sufficient condition is given to guarantee the numerical sign-stability. We prove sufficient conditions for the sign-stability and sign-unstability of tridiagonal matrices.

**Keywords:** Parabolic Problems, Numerical Solution, Qualitative Properties, Sign-Stability.

**Subject Classification:** 65N06, 65N22, 65N30.

## 1 Introduction

Let us consider the initial boundary value problem

$$\frac{\partial v}{\partial t} - \frac{\partial}{\partial x}\left(\kappa \frac{\partial v}{\partial x}\right) + \gamma v = 0, \quad (x,t) \in (0,1) \times (0,\infty), \tag{1}$$

$$v(x,0) = v_0(x), \quad x \in (0,1), \tag{2}$$

$$v(0,t) = v(1,t) = 0, \quad t \geq 0, \tag{3}$$

where the continuous function $\gamma : [0,1] \to \mathbb{R}$ possesses the property $0 < \gamma_{min} \leq \gamma(x) \leq \gamma_{max}$; the function $\kappa : [0,1] \to \mathbb{R}$ fulfills the property $0 < \kappa_{min} \leq \kappa(x) \leq \kappa_{max}$ and has continuous first derivatives. A function $v : [0,1] \times \mathbb{R}_0^+ \to \mathbb{R}$ is called the solution of problem (1)-(3) if it is sufficiently smooth and satisfies equalities (1)-(3).

Equation (1) is a heat conduction equation with linear source function. The variables $x$ and $t$ play the role of the spatial and the time coordinates, respectively. Operator splitting technique is generally applied for the solution of air-pollution models. The original problem is split into subproblems, then these

---

[*] The author of the paper was supported by the National Scientific Research Found (OTKA) N. T043765, K61800.

T. Boyanov et al. (Eds.): NMA 2006, LNCS 4310, pp. 458–465, 2007.

subproblems are solved cyclically, where the solution of one of the subproblems will be the input of another one ([12,13]). Equation (1) can be considered also as one of the subproblems of air-pollution models.

The heat conduction process and the heat conduction equation have a number of characteristic qualitative properties, such as the nonnegativity preservation, maximum-minimum principle, maximum norm contractivity, etc. It is a natural requirement that the numerical equivalents of these properties must be valid for the numerical solution of problem (1)-(3) too (e.g. [2,3,6]).

In the thirties, Sturm ([10]) and Pólya ([8]) showed for the case $\gamma = 0$, $\kappa = 1$ that the number of the sign-changes of the functions $x \mapsto v(x,t)$ ($x \in [0,1]$) does not grow in $t$. This property is called sign-stability. It was shown in [5] that the finite difference method with the uniform spatial step-size $h$ and with the $\theta$ time-discretization method with the time-step $\tau$ is sign-stable for problems with $\gamma = 0$, $\kappa = 1$ if the relation

$$\frac{\tau}{h^2} \le \frac{1}{4(1-\theta)} \tag{4}$$

is satisfied. If $\theta = 1$, then there is no upper bound for the quotient $\tau/h^2$. In [7], we showed that condition (4) is the necessary and sufficient condition of the uniform (independent of $h$) sign-stability. We extended the result also for finite element solutions. In [11], a sufficient condition of the sign-stability is given for the explicit finite difference solution of a semilinear parabolic problem. The proof is based on a six-page-long linear algebraic consideration about the sign-stability of positive tridiagonal matrices. In this paper, we shorten this proof essentially. We formulate the linear algebraic and numerical equivalents of the sign-stability and give sufficient conditions that guarantee the sign-stability for the finite difference numerical solutions of problem (1)-(3). We do not suppose the uniformity of the spatial mesh.

In the next section, the sign-stability is defined for matrices, and its sufficient condition is given for special tridiagonal matrices. These results are applied for the finite difference solution of (1)-(3) in Section 3.

## 2   Sign-Stability of Tridiagonal Matrices

Let $n$ be a fixed natural number. In order to simplify the notations, we introduce the sets $N = \{1, \ldots, n\}$, $J_i = \{i-1, i, i+1\}$ ($i = 2, \ldots, n-1$), $J_1 = \{1, 2\}$ and $J_n = \{n-1, n\}$. The elements of a matrix $\mathbf{A} \in \mathbb{R}^{n \times n}$ are denoted by $a_{i,j}$ or $(\mathbf{A})_{i,j}$. Moreover, we extend this notation: if one of the indices does not belong to $N$, then $a_{i,j}$ is defined to be zero.

For a given vector $\mathbf{x} = [x_1, \ldots, x_n]^\top \in \mathbb{R}^n$, let us denote the number of sign-changes in the sequence $x_1, x_2, \ldots, x_n$, where we leave out the occurrent zero values, by $\mathcal{S}(\mathbf{x})$. If $\mathbf{x}$ is the zero vector, then we set $\mathcal{S}(\mathbf{x}) = -1$. Naturally, the trivial relations $-1 \le \mathcal{S}(\mathbf{x}) \le n-1$ and $\mathcal{S}(\mathbf{x}) = \mathcal{S}(-\mathbf{x})$ hold.

**Definition 1.** *A matrix* $\mathbf{A} \in \mathbb{R}^{n \times n}$ *is said to be sign-stable (resp. sign-unstable) if the condition* $\mathcal{S}(\mathbf{Ax}) \le \mathcal{S}(\mathbf{x})$ *(resp.* $\mathcal{S}(\mathbf{Ax}) \ge \mathcal{S}(\mathbf{x})$*) is fulfilled for all vectors* $\mathbf{x} \in \mathbb{R}^n$.

Clearly, the inverse of a regular sign-stable matrix is sign-unstable, and vice versa. Moreover, the product of sign-stable matrices are sign-stable and the same is true also for sign-unstable matrices. It is known that totally nonnegative matrices (all minors are nonnegative) or minordefinite matrices (no two minors with same order that have different sign) are sign-stable ([4,9]).

**Theorem 1.** *Let* $\mathbf{A} \in \mathbb{R}^{n \times n}$ *be a tridiagonal matrix with the properties*
   *(P1)* $a_{i,j} > 0$ *if* $i \in N$ *and* $j \in J_i$,
   *(P2)* $a_{i,i} \geq a_{i,i-1} + a_{i,i+1}$ *if* $i \in N$ *(weak row-diagonal dominance).*
*Then the matrix* $\mathbf{A}$ *is sign-stable.*

*Proof.* The proof is based on the LU-decomposition of $\mathbf{A}$. It will be shown that both $\mathbf{L}$ and $\mathbf{U}$ are sign-stable. This trivially implies the statement of the theorem.

It can be seen easily that, under the assumptions $(P1)$ and $(P2)$, the LU-decomposition of $\mathbf{A}$ can be performed with the Gauss-elimination process without pivoting. The matrix $\mathbf{U}$ has the properties:
   (U0) $\mathbf{U}$ is an upper triangular matrix,
   (U1) the main and the upper diagonal of $\mathbf{U}$ are positive,
   (U2) all the other elements are zero.
We notice that the relation $u_{i,i} \geq u_{i,i+1}$ $(i \in N)$ is also valid but this fact will not be used in the remainder of the proof. For the matrix $\mathbf{L}$, the following properties are valid:
   (L0) $\mathbf{L}$ is a lower triangular matrix,
   (L1) the main and the lower diagonal of $\mathbf{L}$ are positive,
   (L2) all the other elements are zero.
We prove that the matrices $\mathbf{U}$ having the structure $(U0) - (U2)$ are sign-stable. The proof is based on the principle of induction. One can verify easily that $\mathbf{U}$ is sign-stable for $n = 2$. Let us assume that the statement is true for $k \times k$ matrices, and let us consider a $(k + 1) \times (k + 1)$ matrix $\mathbf{U}$. We have to show that $\mathcal{S}(\mathbf{y}) \leq \mathcal{S}(\mathbf{x})$ with $\mathbf{y} = \mathbf{U}\mathbf{x}$ for all vectors $\mathbf{x} \in \mathbb{R}^{k+1}$. Applying the standard MATLAB notations we introduce the matrix $\tilde{\mathbf{U}} = \mathbf{U}(1:k, 1:k)$ and the vectors $\tilde{\mathbf{x}} = \mathbf{x}(1:k)$, $\tilde{\mathbf{y}} = \mathbf{y}(1:k)$. Thus, let $\mathbf{x} \in \mathbb{R}^{k+1}$ be an arbitrary fixed vector. We consider three different cases keeping in mind the induction hypothesis.

Case A. If $x_{k+1} = 0$, then we have $\mathcal{S}(\mathbf{y}) = \mathcal{S}(\tilde{\mathbf{y}}) = \mathcal{S}(\tilde{\mathbf{U}}\tilde{\mathbf{x}}) \leq \mathcal{S}(\tilde{\mathbf{x}}) = \mathcal{S}(\mathbf{x})$.

Case B. If $x_{k+1} > 0$, then we distinguish three different cases again.

Case B1. If $x_k = 0$, then either $x_{k-1} = x_{k-2} = \ldots = x_1 = 0$ (in this case $\mathcal{S}(\mathbf{y}) \leq \mathcal{S}(\mathbf{x})$ is trivial) or there exists an index $1 \leq i \leq k - 1$ such that $x_i \neq 0$ and $x_{i+1} = \ldots = x_k = 0$. If $x_i > 0$, then we obtain the relations $\mathcal{S}(\mathbf{y}) = \mathcal{S}(\tilde{\mathbf{y}}) = \mathcal{S}(\tilde{\mathbf{U}}\tilde{\mathbf{x}}) \leq \mathcal{S}(\tilde{\mathbf{x}}) = \mathcal{S}(\mathbf{x})$. If $x_i < 0$, then we have $\mathcal{S}(\mathbf{y}) = \mathcal{S}(\tilde{\mathbf{y}}) = \mathcal{S}(\tilde{\mathbf{U}}\tilde{\mathbf{x}}) + 1 \leq \mathcal{S}(\tilde{\mathbf{x}}) + 1 = \mathcal{S}(\mathbf{x})$.

Case B2. If $x_k > 0$, then the following estimations are valid. $\mathcal{S}(\mathbf{y}) = \mathcal{S}(\tilde{\mathbf{y}}) = \mathcal{S}(\tilde{\mathbf{U}}\tilde{\mathbf{x}}) \leq \mathcal{S}(\tilde{\mathbf{x}}) = \mathcal{S}(\mathbf{x})$.

Case B3. If $x_k < 0$, then we have to distinguish three different cases.

Case B3a. If $y_k > 0$, then the relation $\mathcal{S}(\mathbf{y}) = \mathcal{S}(\tilde{\mathbf{y}}) \leq \mathcal{S}(\tilde{\mathbf{U}}\tilde{\mathbf{x}}) + 1 \leq \mathcal{S}(\tilde{\mathbf{x}}) + 1 = \mathcal{S}(\mathbf{x})$ is valid.

Case B3b. If $y_k < 0$, then we have $\mathcal{S}(\mathbf{y}) = \mathcal{S}(\tilde{\mathbf{y}}) + 1 = \mathcal{S}(\tilde{\mathbf{U}}\tilde{\mathbf{x}}) + 1 \leq \mathcal{S}(\tilde{\mathbf{x}}) + 1 = \mathcal{S}(\mathbf{x})$.

Case B3c. If $y_k = 0$, then we have $\mathcal{S}(\mathbf{y}) \leq \mathcal{S}(\tilde{\mathbf{y}}) + 1 \leq \mathcal{S}(\tilde{\mathbf{U}}\tilde{\mathbf{x}}) + 1 \leq \mathcal{S}(\tilde{\mathbf{x}}) + 1 = \mathcal{S}(\mathbf{x})$.

Case C. If $x_{k+1} < 0$, then multiplying the vector $\mathbf{x}$ by -1 we obtain Case B.

Thus, based on the induction principle, the sign-stability is valid for matrices $\mathbf{U}$ with the properties $(U0)-(U2)$. The sign-stability of $\mathbf{L}$ can be proved similarly considering the cases $x_1 = 0$, $x_1 > 0$ and $x_1 < 0$. This completes the proof.

*Remark 1.* In the above theorem, we proved the sign-stability directly. An alternative could be that we show the total nonnegativity of $\mathbf{A}$ based on the fact that a tridiagonal matrix $\mathbf{A} \in \mathbb{R}^{n \times n}$ is totally nonnegative if and only if $a_{i,i+1}, a_{i+1,i} \geq 0$ $(i = 1, \ldots, n-1)$ and each main-minor of the matrix is nonnegative ([4]).

**Theorem 2.** *Let $\mathbf{A} \in \mathbb{R}^{n \times n}$ be a tridiagonal matrix with the properties*
*(Q1) $a_{i,i} > 0$*
*(Q2) $a_{i,j} < 0$ if $i \in N$ and $j \in J_i$, $j \neq i$,*
*(Q3) $a_{i,i} \geq |a_{i,i-1}| + |a_{i,i+1}|$ if $i \in N$ (weak row-diagonal dominance).*
*Then the matrix $\mathbf{A}$ is sign-unstable.*

*Proof.* The proof is based on the LU-decomposition of $\mathbf{A}$. It will be shown that both $\mathbf{L}$ and $\mathbf{U}$ are sign-unstable. This trivially implies the statement of the theorem.

It can be seen easily that, under the assumptions $(Q1) - (Q3)$, the LU-decomposition of $\mathbf{A}$ can be performed with the Gauss-elimination process without pivoting. The matrix $\mathbf{U}$ has the following properties:

$(U0^\star)$ $\mathbf{U}$ is an upper triangular matrix,

$(U1^\star)$ the main diagonal is positive, the upper diagonal is negative,

$(U2^\star)$ all the other elements are zero.

We notice that the relation $u_{i,i} \geq |u_{i,i+1}|$ $(i \in N)$ is also valid but this fact will not be used in the remainder of the proof. For the matrix $\mathbf{L}$, the following properties are true

$(L0^\star)$ $\mathbf{L}$ is a lower triangular matrix,

$(L1^\star)$ the main diagonal is positive, the lower diagonal is negative,

$(L2^\star)$ all the other elements are zero.

We prove that the matrices $\mathbf{U}$ having the structure $(U0^\star) - (U2^\star)$ are sign-unstable. The proof is based on the principle of induction. One can verify easily that $\mathbf{U}$ is sign-unstable for $n = 2$. Let us assume that the statement is true for $k \times k$ matrices, and let us consider a $(k+1) \times (k+1)$ matrix $\mathbf{U}$. We have to show that $\mathcal{S}(\mathbf{y}) \geq \mathcal{S}(\mathbf{x})$ with $\mathbf{y} = \mathbf{U}\mathbf{x}$ for all vectors $\mathbf{x} \in \mathbb{R}^{k+1}$. We introduce the matrix $\tilde{\mathbf{U}} = \mathbf{U}(1:k, 1:k)$ and the vectors $\tilde{\mathbf{x}} = \mathbf{x}(1:k)$, $\tilde{\mathbf{y}} = \mathbf{y}(1:k)$. Thus, let $\mathbf{x} \in \mathbb{R}^{k+1}$ be an arbitrary fixed vector. We consider three different cases keeping in mind the induction hypothesis.

Case A. If $x_{k+1} = 0$, then we have $\mathcal{S}(\mathbf{y}) = \mathcal{S}(\tilde{\mathbf{y}}) = \mathcal{S}(\tilde{\mathbf{U}}\tilde{\mathbf{x}}) \geq \mathcal{S}(\tilde{\mathbf{x}}) = \mathcal{S}(\mathbf{x})$.

Case B. If $x_{k+1} > 0$, then we distinguish three different cases again.

Case B1. If $x_k = 0$, then either $x_{k-1} = x_{k-2} = \ldots = x_1 = 0$ (in this case $\mathcal{S}(\mathbf{y}) \geq \mathcal{S}(\mathbf{x})$ is trivial) or there exists an index $1 \leq i \leq k-1$ such that $x_i \neq 0$ and $x_{i+1} = \ldots = x_k = 0$. If $x_i > 0$, then we obtain the relations

$\mathcal{S}(\mathbf{y}) = \mathcal{S}(\tilde{\mathbf{y}}) + 1 = \mathcal{S}(\tilde{\mathbf{U}}\tilde{\mathbf{x}}) + 2 \geq \mathcal{S}(\tilde{\mathbf{x}}) + 2 = \mathcal{S}(\mathbf{x}) + 2 \geq \mathcal{S}(\mathbf{x})$. If $x_i < 0$, then we have $\mathcal{S}(\mathbf{y}) = \mathcal{S}(\tilde{\mathbf{y}}) + 1 = \mathcal{S}(\tilde{\mathbf{U}}\tilde{\mathbf{x}}) + 1 \geq \mathcal{S}(\tilde{\mathbf{x}}) + 1 = \mathcal{S}(\mathbf{x})$.

Case B2. If $x_k < 0$, then the following estimations are valid. $\mathcal{S}(\mathbf{y}) = \mathcal{S}(\tilde{\mathbf{y}}) + 1 = \mathcal{S}(\tilde{\mathbf{U}}\tilde{\mathbf{x}}) + 1 \geq \mathcal{S}(\tilde{\mathbf{x}}) + 1 = \mathcal{S}(\mathbf{x})$.

Case B3. If $x_k > 0$, then we have to distinguish three different cases.

Case B3a. If $y_k > 0$, then the relation $\mathcal{S}(\mathbf{y}) = \mathcal{S}(\tilde{\mathbf{y}}) = \mathcal{S}(\tilde{\mathbf{U}}\tilde{\mathbf{x}}) \geq \mathcal{S}(\tilde{\mathbf{x}}) = \mathcal{S}(\mathbf{x})$ is valid.

Case B3b. If $y_k < 0$, then we have $\mathcal{S}(\mathbf{y}) = \mathcal{S}(\tilde{\mathbf{y}}) + 1 \geq \mathcal{S}(\tilde{\mathbf{U}}\tilde{\mathbf{x}}) \geq \mathcal{S}(\tilde{\mathbf{x}}) = \mathcal{S}(\mathbf{x})$.

Case B3c. If $y_k = 0$, then we have $\mathcal{S}(\mathbf{y}) \geq \mathcal{S}(\tilde{\mathbf{U}}\tilde{\mathbf{x}}) \geq \mathcal{S}(\tilde{\mathbf{x}}) = \mathcal{S}(\mathbf{x})$.

Case C. If $x_{k+1} < 0$, then multiplying the vector $\mathbf{x}$ by -1 we obtain Case B.

Thus, based on the induction principle, the sign-unstability is valid for all matrices $\mathbf{U}$ having the properties $(U0^\star)$-$(U2^\star)$. The sign-unstability of $\mathbf{L}$ can be proved similarly considering the cases $x_1 = 0$, $x_1 > 0$ and $x_1 < 0$. This completes the proof.

## 3    Sign-Stability of the Finite Difference Methods

In order to obtain a finite difference spatial approximation, we define a spatial mesh $\Omega_h$ with the grid points $0 = x_0 < x_1 < \ldots < x_n < x_{n+1} = 1$, dividing the interval $[0, 1]$ into $n + 1$ subintervals. Denoting the semi-discretization of the solution $v$ of (1)-(3) at a grid point $x_i$ by $v_i(t)$, we approximate (1) at the point $x_i$ $(i \in N)$ as

$$\frac{dv_i(t)}{dt} - v_{ixx}(t) + \gamma_i v_i(t) = 0, \tag{5}$$

where

$$v_{ixx}(t) = \frac{2}{h_{i+1} + h_{i-1}} \left( \kappa_{i+1/2} \frac{v_{i+1}(t) - v_i(t)}{h_{i+1}} - \kappa_{i-1/2} \frac{v_i(t) - v_{i-1}(t)}{h_{i-1}} \right). \tag{6}$$

The distance between the points $x_{i+1}$ and $x_i$ is denoted by $h_{i+1}$. The value $\kappa_{i+1/2}$ denotes the approximate value of $\kappa$ on the segment $[x_i, x_{i+1}]$ (typically the midpoint value), $\gamma_i$ denotes the approximate value of $\gamma$ at the point $x_i$ (typically $\gamma_i = \gamma(x_i)$). Applying equation (5) for $i \in N$, we arrive at a Cauchy problem for the system of ordinary differential equations

$$\frac{d\mathbf{v}(t)}{dt} + \mathbf{K}\mathbf{v}(t) = \mathbf{0}, \quad \mathbf{v}(0) = [v_0(x_1), \ldots, v_0(x_n)]^\top, \tag{7}$$

where $\mathbf{v}(t) = [v_1(t), \ldots, v_n(t)]^\top$ and $\mathbf{K}$ is a sparse $n \times n$ matrix with the elements

$$k_{i,i-1} = \frac{-2\kappa_{i-1/2}}{h_{i-1}(h_{i-1} + h_{i+1})}, \quad k_{i,i+1} = \frac{-2\kappa_{i+1/2}}{h_{i+1}(h_{i-1} + h_{i+1})}, \tag{8}$$

$$k_{i,i} = \gamma_i + \frac{2\kappa_{i+1/2}}{h_{i+1}(h_{i-1} + h_{i+1})} + \frac{2\kappa_{i-1/2}}{h_{i-1}(h_{i-1} + h_{i+1})}, \quad i \in N.$$

In order to get a fully discrete numerical scheme, we choose a time-step $\tau$ and denote the approximation to $\mathbf{v}(j\tau)$ by $\mathbf{v}^j$ $(j = 0, 1, \ldots)$. To discretize (7) in time, we apply the $\theta$-method ($\theta \in [0, 1]$ is a given parameter) and obtain the system of linear algebraic equations

$$\frac{\mathbf{v}^{j+1} - \mathbf{v}^j}{\tau} + \theta\mathbf{K}\mathbf{v}^{j+1} + (1 - \theta)\mathbf{K}\mathbf{v}^j = 0 \qquad (9)$$

($j = 0, 1, \ldots$). The choice $\theta = 0$ results in an explicit method (explicit Euler method), the case $\theta = 1$ gives the implicit Euler method and the case $\theta = 1/2$ is the Crank-Nicolson method. Iteration (9) can be rewritten with the notations $\mathbf{A}_1 = \mathbf{I} + \theta\tau\mathbf{K}$, $\mathbf{A}_2 = \mathbf{I} - (1 - \theta)\tau\mathbf{K}$ and $\mathbf{A} = \mathbf{A}_1^{-1}\mathbf{A}_2$ ($\mathbf{I} \in \mathbb{R}^{n\times n}$ is the unit matrix) as

$$\mathbf{v}^{j+1} = \mathbf{A}\mathbf{v}^j. \qquad (10)$$

The matrix $\mathbf{A}_1$ is regular, namely it is a nonsingular $M$-matrix ([1]), which has nonnegative inverse.

Naturally presents itself the following definition.

**Definition 2.** *We say that the finite difference method for problem (1)-(3) is sign-stable on a fixed mesh $\Omega_{h,\tau}$ if the matrix $\mathbf{A} \in \mathbb{R}^{n\times n}$ in (10) is a sign-stable matrix.*

**Theorem 3.** *If the condition*

$$\tau \leq \frac{1}{(1 - \theta)(\gamma_{\max} + 4\kappa_{\max}/h_{\min}^2)} \qquad (11)$$

*is satisfied, then the finite difference method for problem (1)-(3) is sign-stable on a fixed mesh $\Omega_{h,\tau}$. If $\theta = 1$, then there is no upper bound for the time-step. Here $h_{\min}$ denotes the minimal step-size in the spatial discretization.*

*Proof.* It is enough to show that the matrix $\mathbf{A}_1$ is sign-unstable and $\mathbf{A}_2$ is sign-stable. The sign-unstability of $\mathbf{A}_1$ follows from the form (8) of the elements of the matrix $\mathbf{K}$ and Theorem 2. The matrix $\mathbf{A}_2$ is a tridiagonal matrix. If $\theta = 1$, then $\mathbf{A}_2 = \mathbf{I}$ and the unit matrix is trivially sign-stable. Thus the iteration matrix $\mathbf{A}$ is sign-stable without any restriction. If $\theta \neq 1$, then it is sufficient to show that the conditions $(P1)$-$(P2)$ are valid under the condition (11). The upper and lower diagonals of $\mathbf{A}_2$ are positive. We show the weak row-diagonal dominance, which clearly yields that the diagonal is also positive. This implies the sign-stability of $\mathbf{A}_2$ based on Theorem 1. Thus, we have

$$(\mathbf{A}_2)_{i,i} - (\mathbf{A}_2)_{i,i-1} - (\mathbf{A}_2)_{i-1,i} = 1 - (1 - \theta)\tau(\gamma_i - 2(k_{i-1,i} + k_{i,i+1})) \geq \quad (12)$$

$$\geq 1 - \frac{\gamma_i - 2(k_{i-1,i} + k_{i,i+1})}{\gamma_{\max} + 4\kappa_{\max}/h_{\min}^2} \geq 0 \quad (i \in N),$$

where we make use of the relations $|k_{i-1,i}| \leq \kappa_{\max}/h_{\min}^2$, $|k_{i,i+1}| \leq \kappa_{\max}/h_{\min}^2$. This completes the proof.

*Remark 2.* For uniform spatial meshes and for problems with $\gamma = 0$ and $\kappa = 1$, condition (11) gives condition (4). For fixed number of spatial grid points, the uniform mesh produces the largest possible time-step.

*Remark 3.* We stress that Theorem 3 gives only a sufficient condition for the sign-stability. For a fixed spatial mesh, we can choose greater value for $\tau$ than the value given by equation (11). For instance, let us consider the explicit Euler method on the uniform spatial mesh with $h = 1/4$ for a problem with $\gamma = 0$ and $\kappa = 1$. In this case the iteration matrix $\mathbf{A}$ has the form

$$\mathbf{A} = \text{tridiag}\,[16\tau, 1 - 32\tau, 16\tau].$$

It can be checked easily that this matrix is sign-stable if and only if $\tau \leq 1/48$, which is a less restrictive condition than $\tau \leq 1/64$ obtained from condition (11).

*Remark 4.* The simple example discussed in the previous remark shows also that the sign-stability is a stricter condition than the preservation of the nonnegativity of the initial function. Let us choose the time-step to be $\tau = 5/192$. Then $\mathbf{A}$ is a nonnegative matrix, namely $\mathbf{A} = \text{tridiag}\,[5/12, 1/6, 5/12]$, but it is not sign-stable: for the vector $\mathbf{x} = [1/2, 1, -1]^\top$, for which $\mathcal{S}(\mathbf{x}) = 1$, $\mathbf{Ax} = [1/12, -1/24, 1/4]^\top$ with $\mathcal{S}(\mathbf{Ax}) = 2$.

*Remark 5.* The two important linear algebraic results of Theorem 1 and Theorem 2 can be applied in the numerical solutions of semilinear parabolic problems and for problems with time-dependent coefficients too.

*Remark 6.* Guaranteeing the sign-stability for the first spatial derivative of the solution we can give conditions under which the number of the peaks of the temperature function will decrease. Under these conditions the Crank-Nicolson method does not produce spurious oscillations.

# References

1. A. BERMAN, R. J. PLEMMONS, *Nonnegative Matrices in the Mathematical Sciences*, Academic Press, New York 1979.
2. I. Faragó, *Nonnegativity of the Difference Schemes*, Pure Math. Appl. 6 (1996), 38–56.
3. I. Faragó, R. Horváth, S. Korotov, *Discrete Maximum Principle for Linear Parabolic Problems Solved on Hybrid Meshes*, Appl. Num. Math., Volume 53, Issues 2-4, May 2005, 249–264.
4. F.R. GANTMACHER, M.G. KREIN, *Oszillationsmatrizen, Oszillationskerne und kleine Schwingungen mechanischer Systeme*, Berlin, Akademie-Verlag, 1960.
5. K. GLASHOFF, H. KRETH, *Vorzeichenstabile Differenzenverfaren für parabolische Anfangsrandwertaufgaben*, Numerische Mathematik 35 (1980), 343-354.
6. R. Horváth, *Maximum Norm Contractivity in the Numerical Solution of the One-Dimensional Heat Equation*, Appl. Num. Math. 31 (1999), 451–462.
7. R. Horváth, *On the Sign-Stability of the Numerical Solution of the Heat Equation*, Pure Math. Appl. 11 (2000), 281–291.
8. G. PÓLYA, *Qualitatives über Wärmeausgleich*, Ztschr. f. angew. Math. und Mech., 13 (1933), 125-128.
9. I.J. SCHOENBERG, *Über variationsvermindernde lineare Transformationen*, Math. Z. 32, 1930.

10. CH. STURM, *Journal de Math. pures et appliquées*, 1 (1836), 373-444.
11. M. TABATA, *A Finite Difference Approach to the Number of Peaks of Solutions for Semilinear Parabolic Problems*, J. Math. Soc. Japan 32 (1980), 171-192.
12. J.G. Verwer, W. Hundsdorfer, J.G. Blom, *Numerical Time Integration for Air Pollution Models*, Report of CWI, MAS-R9825 (1982).
13. Z. Zlatev, *Computer Treatment of Large Air Pollution Models*, Kluwer Academic Publishers, Dordrecht-Boston-London 1995.

# Comprehensive Modelling of PM$_{10}$ and PM$_{2.5}$ Scenarios for Belgium and Europe in 2010

C. Mensink, F. Deutsch, L. Janssen, R. Torfs, and J. Vankerkom

VITO - Flemish Institute for Technological Research,
Boeretang 200, B-2400 Mol, Belgium
clemens.mensink@vito.be

**Abstract.** The extended EUROS model has been used to calculate concentrations of PM$_{10}$ and PM$_{2.5}$ for Europe for the years 2002 and 2010 using a recent emission scenario. The obtained results for Belgium show decreases in PM-concentrations between 5 and 26% in this period, depending on the location. The contribution of anthropogenic sources in Flanders to annual averaged PM$_{10}$ concentrations amounts to 17% in 2002 and 15% in 2010 on average. The most important contribution to PM$_{10}$ concentrations originates from agricultural activities in Flanders, whereas the sector "traffic" is the dominant source for anthropogenic PM$_{2.5}$ in Flanders.

**Keywords:** Air quality modelling, aerosols, fine particulate matter, PM$_{2.5}$, PM$_{10}$, emission scenarios, abatement strategies.

## 1  Introduction

Within the framework of the CAFE (Clean Air For Europe) initiative, the air quality standards for 2010 and beyond are discussed, with the aim to reduce exposure of people to high concentrations of inhalable particles in the air. These particles are associated with strong adverse health effects ([6,2]). In order to support related air quality policy on PM$_{10}$ and PM$_{2.5}$ in Belgium, the integrated air quality modelling system EUROS has been extended to model fine particulate matter (PM). Currently, modelling of mass and chemical composition of aerosols in two size fractions (PM$_{2.5}$ and PM$_{10-2.5}$) is possible. The chemical composition is expressed in terms of 7 components: *ammonium, nitrate, sulphate, elementary carbon, primary inorganic compounds, primary organic compounds* and *secondary organic compounds (SOA)*. A validation of the model for PM$_{10}$ was performed for various episodes in 2002 and 2003 ([1]). EUROS was originally developed at RIVM in the Netherlands and is now, coupled with a state-of-the-art user interface, an operational tool for policy support at the Interregional Cell for the Environment (IRCEL) in Brussels.

The extended version of the EUROS model was applied for 2002 and 2010 in order to assess current and future changes in aerosol concentrations and compositions over Belgium and Europe. The resulting impacts were evaluated for the three regions in Belgium, namely Flanders, the Walloon region and Brussels.

T. Boyanov et al. (Eds.): NMA 2006, LNCS 4310, pp. 466–474, 2007.

The individual contributions of the various emission sectors were evaluated and compared for 2002 and 2010.

Section 2 describes the EUROS model in more detail. It is explained how the model was extended with an advanced module dealing with the complex formation of aerosols. The reference sector emissions for 2002 as used in the model are presented and the selection of the emission scenarios for 2010 is discussed. The results are shown and discussed in section 3, focusing on the differences in $PM_{10}$ and $PM_{2.5}$ concentrations between 2002 and 2010 and on the impact of the individual sector contributions. Conclusions with respect to policy support in Belgium are presented in section 4.

## 2 Methodology

### 2.1 The EUROS Model

The base grid of the Eulerian grid model EUROS covers nearly whole Europe with a resolution of $60 \times 60km$. Since EUROS allows local grid refinement, a subgrid of approximately $700 \times 500km$ covering Belgium and the surrounding regions with a resolution of $15 \times 15km$ was used to perform the simulations in more detail. As far as the meteorology is concerned, the model uses three-dimensional input datasets derived from the ECMWF meteorological reanalysed datasets. The vertical structure of the atmosphere is represented in EUROS by four layers: ground layer, mixing layer, reservoir layer and top layer.

A detailed emission module describes the emission of six pollutant categories $(NO_x, NMVOC, SO_2, NH_3$ and two size fractions of particles $(< 2.5\mu m; 2.5 - 10\mu m))$ for 7 different emission sectors (traffic, residential heating, refineries, solvent use, combustion, industry and agriculture). Both point sources and area sources are included. Emission data are obtained from EMEP/CORINAIR ([8]) for the base grid and additional data from the emission inventory for Flanders ([9]). Table 1 shows the Belgian emissions for 2002.

Table 1. Belgian emissions (in Mg) for the year 2002 according to EMEP, Expert emissions W-05emis02-V5 (2005-03-10)

| EUROS-sector | $NH_3$ | $NMVOC$ | $NO_x$ | $SO_2$ | $PM_{2.5}$ | $PM_{10-2.5}$ |
|---|---|---|---|---|---|---|
| 1 combustion | 95 | 2035 | 47428 | 52131 | 1381 | 2570 |
| 2 residential | 360 | 5134 | 22175 | 22947 | 4770 | 1884 |
| 3 refineries | 0 | 14944 | 0 | 0 | 108 | 1308 |
| 4 industry | 3439 | 37846 | 66085 | 74406 | 15154 | 15630 |
| 5 solvent use | 0 | 72146 | 0 | 0 | 0 | 0 |
| 6 traffic | 1370 | 96814 | 164286 | 8491 | 11054 | 2219 |
| 7 agriculture | 73737 | 1082 | 25 | 26 | 1533 | 6389 |
| Total | 79000 | 230000 | 300000 | 158000 | 34000 | 30000 |

## 2.2  Modelling $PM_{10}$ and $PM_{2.5}$

The aerosol modelling is based on the Caltech Atmospheric Chemistry Mechanism (CACM, [3]) and the Model of Aerosol Dynamics, Reaction, Ionization and Dissolution (MADRID 2, [10]). CACM is the first gas phase chemical mechanism describing the formation of precursors of secondary organic aerosols (SOA) in the atmosphere by a mechanistic approach. Hence it offers possibilities e.g. to distinguish between different contributions to total SOA, such as SOA originating from anthropogenic and from biogenic VOC-emissions. The actual aerosol module, MADRID 2, contains various algorithms originating from other state-of-the-art aerosol models showing good performance. Additionally, MADRID 2 contains also some newly developed modules such as the equilibrium module for secondary organic aerosols (AEC-SOA-module, [7]) which treats hydrophilic and hydrophobic organic precursor components separately. Dynamic processes (e.g. mass transfer and nucleation) are included in MADRID 2 as well.

A first step in the extension of EUROS towards aerosol modelling was the implementation of CACM. It comprises 361 reactions among 122 components. Apart from the ozone chemistry it also contains the reactions of various generations of organic compounds. These reactions can generate semi-volatile reaction products which can equilibrate into the solid phase. 42 of these condensable products are treated in CACM. 31 products originate from anthropogenic NMVOC-emissions and 11 products originate from biogenic NMVOC-emissions, e.g. monoterpenes. Various routines of EUROS (e.g. NMVOC-split, background concentrations) were adjusted to CACM.

In MADRID 2 thermodynamic equilibrium calculations are carried out via ISORROPIA ([5]) for inorganic compounds and via the AEC-SOA-module for organic compounds. Mass transfer between gas phase and solid phase can be taken into account via different approaches. In this work, the CIT hybrid approach (Meng et al., 1998) was used in which a full equilibrium is assumed but the condensing mass is distributed according to a growth law depending on particle size. As only two size fractions were simulated ($PM_{2.5}$ and $PM_{10-2.5}$), both coagulation and condensational growth of particles were omitted because they lead only to little exchange between the two size fractions in comparison to the gas/particle mass transfer. Nucleation of new particles was treated by calculating the relative rates of new particle formation and condensation onto existing particles. Deposition of particles was calculated following a resistance approach.

## 2.3  Emission Scenarios

For 2010, the Current Legislation emission scenario or CAFE-baseline scenario (CAFE_2010_CLE) was implemented. This scenario was provided by the EMEP Centre for Integrated Assessment Modelling (CIAM) at the International Institute for Applied Systems Analysis (IIASA) in Laxenburg, Austria. A spatially distributed version was supplied by EMEP on a 50 × 50 km grid through their website (www.emep.int). Table 2 shows the Belgian emissions for 2010 associated with the CAFE_2010_CLE-scenario.

**Table 2.** Belgian emissions (in Mg) for the year 2010 as provided by EMEP/CAFE (CAFE_2010_CLE)

| EUROS-sector | $NH_3$ | $NMVOC$ | $NO_x$ | $SO_2$ | $PM_{2.5}$ | $PM_{10-2.5}$ |
|---|---|---|---|---|---|---|
| 1 combustion | 157 | 1370 | 23834 | 18279 | 432 | 165 |
| 2 residential | 290 | 4072 | 23649 | 10141 | 2690 | 367 |
| 3 refineries | 0 | 9720 | 0 | 0 | 4 | 30 |
| 4 industry | 3504 | 37771 | 66987 | 66113 | 16284 | 11356 |
| 5 solvent use | 0 | 60781 | 0 | 0 | 0 | 0 |
| 6 traffic | 1064 | 35540 | 117696 | 3949 | 6864 | 2171 |
| 7 agriculture | 74372 | 1137 | 28 | 31 | 1953 | 5802 |
| Total | 79387 | 150390 | 232194 | 98513 | 28227 | 19891 |

# 3 Results and Discussion

## 3.1 Changes in PM$_{10}$ and PM$_{2.5}$ Concentrations Between 2002 and 2010

Figure 1 shows the PM$_{10}$ concentrations for 2002 and 2010 as modeled by EU-ROS. Table 3 shows the relative changes or reductions obtained in the three Belgian regions and in Belgium itself. One can observe that in the southern part of Belgium (Wallonia) stronger reductions are obtained than in the northern part of Belgium (Flanders), especially for PM$_{10}$. This difference is mainly caused by the primary PM emission reductions in the neighboring countries: the current legislation (CLE) scenario predicts much stronger reductions for 2010 in France than in the Netherlands. This can also be observed in Figure 2, where the relative change in concentrations between 2002 and 2010 are shown for PM$_{10}$ and PM$_{2.5}$. The PM$_{10}$-concentrations in the Netherlands show a smaller reduction percentage (light blue color) than its surrounding countries (dark blue and pink color). It is clear that this situation has an immediate impact on the distribution of PM concentrations in Belgium.

Computation results show that the contributions from anthropogenic sources in Flanders are responsible for 17,1% of the annual averaged PM$_{10}$ concentrations in Flanders in 2002. In 2010 this contribution drops to 15,2%. For PM$_{2.5}$ these contributions are 13,9% in 2002 and 11,9% in 2010 respectively. Table 4 shows the

**Table 3.** Relative difference of PM$_{10}$- and PM$_{2.5}$-concentrations between 2002 and 2010 in the Belgian regions

| | max. change [%] | | min. change [%] | | avg. change [%] | |
|---|---|---|---|---|---|---|
| Location | $PM_{10}$ | $PM_{2.5}$ | $PM_{10}$ | $PM_{2.5}$ | $PM_{10}$ | $PM_{2.5}$ |
| Flanders | **-21.9** | **-20.7** | -4.8 | -8.0 | **-11.6** | **-14.8** |
| Walonia | -26.4 | -24.0 | -9.2 | -12.5 | -12.8 | -14.9 |
| Brussels | -12.8 | -16.5 | -11.9 | -15.1 | -12.3 | -15.8 |
| Belgium | -26.4 | -24.0 | -4.8 | -8.0 | -12.3 | -14.9 |

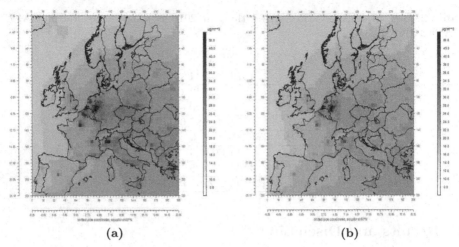

(a)                                    (b)

**Fig. 1.** Annual averaged $PM_{10}$ concentrations in 2002 in (a) and 2010 in (b)

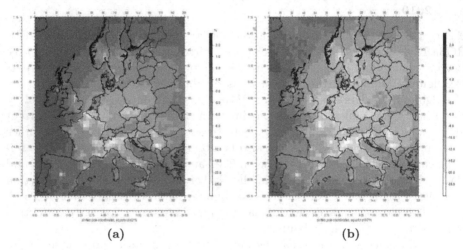

(a)                                    (b)

**Fig. 2.** Relative change (%)between 2002 and 2010 for the annual averaged $PM_{10}$-concentrationsin (a) and the annual averaged $PM_{2.5}$-concentrations in (b)

**Table 4.** Reductions [%] in $PM_{10}$- and $PM_{2.5}$-concentrations obtained by setting all anthropogenic emissions in Flanders to zero in 2010

| Location | max. change [%] | | min. change [%] | | avg. change [%] | |
|---|---|---|---|---|---|---|
| | $PM_{10}$ | $PM_{2.5}$ | $PM_{10}$ | $PM_{2.5}$ | $PM_{10}$ | $PM_{2.5}$ |
| Flanders | **-22.3** | **-25.3** | -2.4 | -1.5 | **-15.2** | **-11.9** |
| Walonia | -18.0 | -15.8 | -1.0 | -1.1 | -2.6 | -14.9 |
| Brussels | -19.8 | -20.6 | -11.0 | -12.1 | -15.4 | -15.8 |
| Belgium | -22.3 | -25.3 | -1.0 | -1.1 | -8.2 | -14.9 |

minimal, maximum and average reductions obtained in 2010 in all three regions and Belgium when all anthropogenic emissions in Flanders are set to zero.

## 3.2   Changes in Sector Contributions Between 2002 and 2010

Table 5 shows the results of the calculation of the individual sector contributions for 2002 and 2010. The sectors industry, refineries and solvent use were treated as one sector. By subsequently setting all emissions to zero in each of the sectors, their impact on the $PM_{10}$ and $PM_{2.5}$ concentrations was computed. Figure 3 again clearly shows that more than 80% of the concentrations observed in Flanders are due to non-anthropogenic sources in Flanders or sources outside Flanders, i.e. sources that can not be controlled by regional policy measures. With respect to the anthropogenic sources in Flanders (contributing less than 20%), "agriculture" is the dominant source for $PM_{10}$, whereas "traffic" is the dominant source for $PM_{2.5}$.

**Table 5.** Relative contributions [%] of the anthropogenic sources per sector to $PM_{10}$- and $PM_{2.5}$-concentrations obtained by setting all anthropogenic emissions in one sector in Flanders to zero

| Anthropogenic contributions from sectors in Flanders | $PM_{10}(\%)$ 2002 | $PM_{2.5}(\%)$ 2002 | $PM_{10}(\%)$ 2010 | $PM_{2.5}(\%)$ 2010 |
|---|---|---|---|---|
| Combustion | 0.7 | 1.1 | 0.3 | 0.5 |
| Residential | 2.4 | 3.7 | 1.4 | 2.4 |
| Refineries & industry & solvent | 0.9 | 1.2 | 0.8 | 1.2 |
| Trafic | 4.0 | 5.9 | 3.7 | 5.6 |
| Agriculture | 7.5 | 0.5 | 7.6 | 1.0 |
| Total | 15.5 | 12.4 | 13.8 | 10.7 |

Note also that the contribution of the sector "agriculture" to the $PM_{10}$ concentrations is rather high. This is possibly an overestimation due to the high emission factors that are currently used in Flanders. On the other hand, the contribution of the sector "industry" is possibly underestimated. Recent findings have demonstrated the importance of industrial PM emissions of diffusive nature. So far these sources of PM have not been included in the emission inventory, thus their impact is not taken into account.

## 3.3   Non-linear Sector Contributions

Sector contributions that consist only of primary emissions of particulate matter show a linear reduction in concentrations when diminishing the emissions. In that case the particulate matter acts as a tracer: the dust particles are transported and dispersed in ambient air, but do not take part in any chemical reaction. However, this is different in the case of formation of *secondary* aerosols. Because of the non-linear processes that take place when secondary aerosols are formed (nitrate, sulphate, ammonium and SOA) a small reduction in a gaseous compound (e.g.

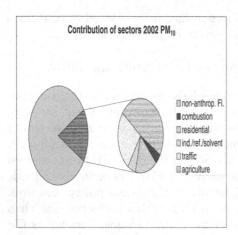
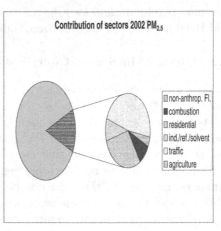

**Fig. 3.** Contribution of anthropogenic emissions in Flanders and "other" emissions to PM$_{10}$-concentrations and PM$_{2.5}$-concentrations in Flanders in 2002

$SO_2$) does not necessarily lead to the same amount of reduction of the secondary compound (e.g. sulfate).

Another "non-linear" aspect is the formation of aerosols by contributions from two compounds that are delivered by two individual sectors. This means that emissions from both sectors are effectively necessary for the formation of the secondary compound. An example is the formation of ammonium-nitrate through emission contributions from the sectors traffic and agriculture. On one hand, a lot of $NO_x$ is emitted by traffic. On the other hand, ammonia is mostly emitted through agricultural activities (see also Table 1). When the two sector contributions are considered separately, hardly any formation of secondary aerosols is observed. When both emission sources are combined, the $NO_x$ and $NH_3$ will react (after transformation in the atmosphere) and ammoniumnitrate is formed, being one of the most common reaction products in the secondary formed aerosol. Based on this effect, it may be assumed that in the computed sector contributions, the amount of secondary formed aerosol may be underestimated for some of the sectors.

In order to estimate this possible underestimation and the effects of non-linear phenomena, the contributions of all 5 sectors were added up and compared with the case where all emissions in Flanders have been removed at once. From Table 5 one can see that in 2010 the sum of the individual sector contributions adds up to 13,8% for PM$_{10}$ and 10,7% for PM$_{2.5}$. From Table 4 we learn that the average reduction obtained in 2010 by setting all sector emission in Flanders to zero is 15,2% and 11,9% respectively. Thus the effect of "non-linear" contributions is estimated to be an additional 1,4% for PM$_{10}$ and 1,2% for PM$_{2.5}$, representing an increase of 10% - 11%.

A further possible contribution to non-linear effects is related to the spatial distribution of the emissions and concentrations. From Tables 3 and 4 it can be observed that the maximum impact of emission reductions can be much larger than the average impact. Especially the sectors "combustion" and "refineries, industry and solvent use" show relatively small average contributions (as seen in Table 5), although locally these contributions can rise to 10% or more. This might locally lead to enhanced formation of secondary aerosols.

Finally we want to make some remarks regarding the assumptions made and caveats observed in this study. First of all the emission factor used to estimate the emissions for the sector "agriculture" is very uncertain. At the moment this emission factor is in revision. A new (and possibly lower) estimate is expected to have a serious impact on the results, as can be deducted from Figure 3 and Table 5. Secondly, as reported before, diffusive emission sources (e.g. fugitive emissions stemming from handling and storing activities) are not taken into account, although recently they are gaining more importance in abatement strategies with the aim to comply with the limit values for particulate matter (EU directive 1999/30/EU). A final remark concerns the emission scenario that was selected to evaluate the situation in 2010. When carrying out this study, the CLE scenario was still in discussion in the context of CAFE and can therefore not be considered as the definite choice.

## 4   Conclusions

We applied the extended version of the EUROS model to evaluate the impact of emission reductions on the concentrations of PM$_{10}$ and PM$_{2.5}$ in Flanders and Belgium for 2010. Individual sector contributions were assessed and the current and future changes in aerosol concentrations and compositions over Belgium and Europe were addressed. Contributions from anthropogenic sources in Flanders are found to be responsible for 17,1% of the annual averaged PM$_{10}$ concentrations in Flanders in 2002. In 2010 this contribution drops to 15,2%. For PM$_{2.5}$ these contributions are 13,9% in 2002 and 11,9% in 2010 respectively. Non-linear effects can not be neglected and were found to be in the order of 10%. The results demonstrate the severe limitations with respect to impact of national policy measures for relatively small countries such as Belgium. It emphasizes the need for better compliance with international agreements on transboundary air pollution.

## References

1. Deutsch F, Janssen L, Vankerkom J, Lefebre F, Mensink C, Fierens F, Dumont G, Roekens E (2006),*Modelling changes of aerosol compositions over Belgium and Europe*, Int. J. Env. Poll., in press.
2. Dockery, DW, Pope, CA III, Xiping, X, Spengler, JD, Ware, JH, Fay, MA, Ferries, BG Jr., Speizer, FE (1993), *An association between air pollution and mortality in six US cities.* N. Engl J Med, **329(24)**, pp. 1753-1759.
3. Griffin R.J., Dabdub D. and Seinfeld J.H. (2002), *Secondary organic aerosol 1. Atmospheric chemical mechanism for production of molecular constituents*, J. Geophys. Res. **107(D17)**, 4332, doi:10.1029/2001JD000541.
4. Meng Z., Dabdub D. and Seinfeld J.H. (1998), *Size-resolved and chemically resolved model of atmospheric aerosol dynamics*, J. Geophys. Res. **103**, pp. 3419-3435.

5. Nenes A., Pandis S.N. and Pilinis C. (1998), *ISORROPIA: a new thermodynamic equilibrium model for multiphase multicomponent inorganic aerosols*, Aquatic Geochemistry **4**, pp. 123-152.
6. Pope CA 3rd, Burnett RT, Thun MJ et al. (2002), *Lung cancer, cardiopulmonary mortality and long-term exposure to fine particulate air pollution.* JAMA, 2002, **287**, pp. 1132-1141.
7. Pun B.K., Griffin R.J., Seigneur C. and Seinfeld J.H. (2002), *Secondary organic aerosol 2. Thermodynamic model for gas/particle partitioning of molecular constituents*, J. Geophys. Res. **107 (D17)**, 4333, AAC 4-1 - 4-15.
8. Vestreng, V. et al. (2005), *Inventory Review 2005. Emission data reported to LR-TAP and NEC Directive, initial review of HMs and POPs*, EMEP/EEA Joint Review Report, EMEP/MSC-W Note 1, July 2005.
9. VMM (2004), *Lozingen in de lucht 1990-2003*, Vlaamse Milieumaatschappij, Aalst, Belgium, 185 pp. + appendix (in Dutch).
10. Zhang Y., Pun B., Vijayaraghavan K., Wu S.-Y., Seigneur C., Pandis S.N., Jacobson M.Z., Nenes A. and Seinfeld J.H. (2004), *Development and application of the Model of Aerosol Dynamics, Reaction, Ionization, and Dissolution (MADRID)*, J. Geophys. Res. **109**, D01202, doi:10.1029/2003JD003501.

# Parallel and GRID Implementation of a Large Scale Air Pollution Model

Tzvetan Ostromsky[1] and Zahari Zlatev[2]

[1] Institute for Parallel Processing, Bulgarian Academy of Sciences,
Acad. G. Bonchev str., bl. 25-A, 1113 Sofia, Bulgaria
ceco@parallel.bas.bg
http://parallel.bas.bg/ceco/
[2] National Environmental Research Institute, Department of Atmospheric
Environment, Frederiksborgvej 399 P. O. Box 358, DK-4000 Roskilde, Denmark
zz@dmu.dk
http://www.dmu.dk/AtmosphericEnvironment

**Abstract.** Large-scale environmental models are powerful tools, designed to meet the increasing demand in various environmental studies.
The atmosphere is the most dynamic component of the environment,
where the pollutants and other chemical species actively interact with
each other, and can quickly be moved in a very long distance. Therefore
the advanced modeling is usually done in a large computational domain.
Moreover, all relevant physical, chemical and photochemical processes
should be taken into account, which heavily depend on the meteorological conditions. All this makes the air pollution modeling a huge and
rather difficult computational task, requiring a large amount of computational power. The most powerful supercomputers have been used for
the development and test runs of such a model, the Danish Eulerin Model
(DEM). Distributed parallel computing via MPI is one of the most efficient techniques in achieving good performance and getting results in
real time. The quickly advancing GRID computing technology is another
powerful tool that can be used to reach higher level of performance of
such a huge model. Both techniques and their inherent problems are discussed in this paper. Results of numerical experiments are presented and
analysed and some conclusions are drown, based on the experiments.

## 1 Introduction

The problem for air pollution modelling has been studied for years [8,9,15]. An
air pollution model is generally described by a system of partial differential
equations for calculating the concentrations of a number of chemical species
(pollutants and other components of the air that interact with the pollutants)
in a large 3-D domain (part of the atmosphere above the studied geographical
region). The main physical and chemical processes (horizontal and vertical wind,
diffusion, chemical reactions, emissions and deposition) should be adequately
represented in the system.

T. Boyanov et al. (Eds.): NMA 2006, LNCS 4310, pp. 475–482, 2007.
© Springer-Verlag Berlin Heidelberg 2007

The Danish Eulerian Model (DEM) [1,10,14,15] is mathematically represented by the following system of partial differential equations:

$$\frac{\partial c_s}{\partial t} = -\frac{\partial(uc_s)}{\partial x} - \frac{\partial(vc_s)}{\partial y} - \frac{\partial(wc_s)}{\partial z} +$$

$$+ \frac{\partial}{\partial x}\left(K_x \frac{\partial c_s}{\partial x}\right) + \frac{\partial}{\partial y}\left(K_y \frac{\partial c_s}{\partial y}\right) + \frac{\partial}{\partial z}\left(K_z \frac{\partial c_s}{\partial z}\right) + \qquad (1)$$

$$+ E_s + Q_s(c_1, c_2, \ldots c_q) - (k_{1s} + k_{2s})c_s, \quad s = 1, 2, \ldots q .$$

where

- $c_s$ – the concentrations of the chemical species;
- $u$, $v$, $w$ – the wind components along the coordinate axes;
- $K_x$, $K_y$, $K_z$ – diffusion coefficients;
- $E_s$ – the emissions;
- $k_{1s}$, $k_{2s}$ – dry / wet deposition coefficients;
- $Q_s(c_1, c_2, \ldots c_q)$ – non-linear functions describing the chemical reactions between species under consideration [4] .

## 2  Splitting into Submodels

The above rather complex system (1) is split into three subsystems (submodels), according to the major physical and chemical processes as well as the numerical methods applied in their solution. These are the **horizontal advection and diffusion (2)**; **chemistry, emissions and deposition (3)**; and the **vertical exchange (4)** submodels (see below).

$$\frac{\partial c_s^{(1)}}{\partial t} = -\frac{\partial(uc_s^{(1)})}{\partial x} - \frac{\partial(vc_s^{(1)})}{\partial y} + \frac{\partial}{\partial x}\left(K_x \frac{\partial c_s^{(1)}}{\partial x}\right) + \frac{\partial}{\partial y}\left(K_y \frac{\partial c_s^{(1)}}{\partial y}\right) \qquad (2)$$

$$\frac{\partial c_s^{(2)}}{\partial t} = E_s + Q_s(c_1^{(2)}, c_2^{(2)}, \ldots c_q^{(2)}) - (k_{1s} + k_{2s})c_s^{(4)} \qquad (3)$$

$$\frac{\partial c_s^{(3)}}{\partial t} = -\frac{\partial(wc_s^{(3)})}{\partial z} + \frac{\partial}{\partial z}\left(K_z \frac{\partial c_s^{(3)}}{\partial z}\right) \qquad (4)$$

The discretization of the spatial derivatives in the right-hand-sides of the sub-models (2) – (4) results in forming three systems of ordinary differential equations. More details about the numerical methods, used in the submodels, can be found in [1,6,7,15].

## 3  Parallelization Strategy

The MPI standard library routines are used to parallelize this model. The MPI (Message Passing Interface, [5]) was initially developed as a standard communication library for distributed memory computers. Later, proving to be efficient, portable and easy to use, it became one of the most popular parallelization tools for application programming. Now it can be used on much wider class of parallel systems, including shared-memory computers and clustered systems (each node of the cluster being a separate shared-memory machine). Thus it provides high level of portability of the code.

Our MPI parallelization is based on the space domain partitioning [12,13]. The space domain is divided into several sub-domains (the number of the sub-domains being equal to the number of MPI tasks). Each MPI task works on its own sub-domain. On each time step there is no data dependency between the MPI tasks on both the chemistry and the vertical exchange stages. This is not so with the advection-diffusion stage. Spatial grid partitioning between the MPI tasks requires overlapping of the inner boundaries and exchange of certain boundary values on the neighboring subgrids for proper treatment of the boundary conditions. The subdomains are usually too large to fit into the fast cache memory of the target processor. In order to achieve good data locality, the smaller (low-level tasks are grouped in chunks where appropriate for more efficient cache utilization. A parameter CHUNKSIZE is provided in the code, which should be tuned with respect to the cache size of the target machine. More detailed description of the main computational stages and the parallelization strategy can be found in [1,10,12,13,15]

## 4   Performance and Scalability of the Parallel Code

Results of parallel execution of the 2D MPI version of DEM for one month on the SUN HPC system at DTU are presented in Table 1. The target system Sun-Fire E25K consists of 72 UltraSPARC-IV dual-core CPU-s (1350 MHz), i.e. 144 CPU in total. This is the largest SMP server available for Scientific Computing in Denmark. The MPI parallel code scales very well as can be seen from the

**Table 1.** Results of parallel execution of the 2-D version of DEM for one month on a cluster of SunFire E25k computers at DTU. The time for waiting on the queue is given in the second column. The total user time and the times of the main computational stages in seconds, as well as the corresponding **(speed-up)** (given next in brackets), are shown in the last 3 columns.

| The 2-D DEM on a SunFire E25k machine,   CHUNKSIZE=32 | | | | | |
|---|---|---|---|---|---|
| PE's | Wait time [sec.] | Run time [sec.] | User time time [sec.] | Advection **(speed-up)** | Chemistry |
| 1  | 13    | 1800 | 1798 |         | 307 |         | 1374 |         |
| 2  | 3     | 904  | 902  | (1.99)  | 155 | (1.98)  | 702 | (1.96)  |
| 4  | 158   | 456  | 454  | (3.96)  | 78  | (3.94)  | 346 | (3.97)  |
| 8  | 9740  | 249  | 247  | (7.28)  | 41  | (7.49)  | 178 | (7.72)  |
| 12 | 9634  | 182  | 181  | (9.93)  | 31  | (9.90)  | 120 | (11.45) |
| 16 | 9451  | 161  | 152  | (11.82) | 24  | (12.79) | 91  | (15.10) |
| 24 | 9184  | 116  | 107  | (16.80) | 17  | (18.06) | 60  | (22.90) |
| 32 | 10943 | 93   | 87   | (20.67) | 14  | (21.93) | 46  | (29.87) |

speed-ups in Table 1. The system, although not heavily loaded, has a lot of users and request of more processors can cause queueing of the job for several hours. This time is given in the second column of the table.

## 5  Running DEM on Various GRID Cites

As one can see from the above results, DEM can be run efficiently on various supercomputers. Moreover, the MPI code scales rather well on various parallel machines with up to 32 PE-s. If the job is not run on a dedicated queue, however, one should take into account the time, which the job spends waiting in a queue. For relatively short jobs (as, for example, 2-D DEM for one month period) this time can be much longer than the execution time of the job. Moreover, the parallel queues requireing more than 8 PE-s, the waiting time increases quickly with increasing the number of required processors, so the time saved due to the speed-up is entirely "eaten" by the time for waiting in a queue.

On the other hand, there is a vast amount of low-cost computer resources, distriuted all over the world, with free computational power. Some of them are even faster compared to a single processor of the SunFire supercomputer from the previous section, as seen from Table 2. Thanks to the novel GRID technology (based on the power of Internet), this scattered free resources can be used as a powerful computing system. If a similar job is submitted to various GRID cites, it has a good chance to be executed with almost no delay in the queue. The execution time, however, will vary with respect to the speed of the particular machine. The results of such experiment are given in 2.

The sequential 2-D version of DEM has been submitted to all cites open to the Earth Science Research group (ESR) of the EGEE grid project through an appropriate queue (resource). Almost the half of them (14 out of 31) started running within 5 min. from the submition, 4 – within 1 hour from the submition, and 13 – more than an hour from the submition; 6 of the cites aborted or failed to execute the job, the rest 25 runs were successful. Comparing the results in both tables, one can see that most of the GRID cites finished the job in shorter time than the parallel supercomputer on 8 or more PE's if the queueing time is also taken into account, in spite of the much shorter running times on the supercomputer. This happens, because the parallel supercomputers are often bisy and the job has to wait on a long queue, especially if a large number of processors is required.

The possibility of some of the GRID cites to run parallel jobs (MPI jobs in particular) has not been used yet. This is a task for the near future. By using this possibility we can expect to decrease significantly the run time of the job (in dependence with the degree of parallelism). The waiting time, however, will probably increase due to the larger resourse requirements, which in general require more time to be satisfied. This is a common rule, valid for any multiprocessor system.

**Table 2.** Results from running the 2-D version of DEM with 96 × 96 grid for 1 month on the available GRID cites. The address of the cites is given in the first column, the waiting time (while the job is in state "scheduled") is given in the second column, the time for execution (wall clock) – in the third column. The same test problem, with the same parameters, as in the previous section, is used in this experiment. The performance of most cites (20 out of 31) seems to be similar or better than those of a single PE of the SunFire E25k supercomputer, used in the previous section experiment.

| GRID cite [web address] | Wait time [sec.] | Run time [sec.] | User time [sec.] | Advection [sec.] | Chemistry [sec.] |
|---|---|---|---|---|---|
| atlasce01.na.infn.it | 41652 | 937 | 935 | 186 | 718 |
| cclcgceli01.in2p3.fr | 725 | 5706 | 2815 | 916 | 1750 |
| ce.epcc.ed.ac.uk | 78939 | aborted | | | |
| ce.grid.tuke.sk | 60 | 1763 | 1751 | 279 | 1408 |
| ce.hep.ntua.gr | 136 | 1741 | 1736 | 256 | 1426 |
| ce.phy.bg.ac.yu | 44569 | 2169 | 2163 | 515 | 1530 |
| ce.ui.savba.sk | 51 | 1509 | 1506 | 221 | 1238 |
| ce001.grid.bas.bg | 37111 | 1701 | 1697 | 257 | 1386 |
| ce01.ariagni.hellasgrid.gr | 68 | 1461 | 1460 | 226 | 1185 |
| ce01.isabella.grnet.gr | 152 | 2161 | 2155 | 421 | 1629 |
| ce01.kallisto.hellasgrid.gr | 21520 | 1466 | 1465 | 226 | 1189 |
| ce01.marie.hellasgrid.gr | 59 | 1451 | 1448 | 224 | 1177 |
| ce02.marie.hellasgrid.gr | 28279 | 1518 | 1516 | 239 | 1225 |
| ce1.egee.fr.cgg.com | 11088 | failed | | | |
| ce2.egee.unile.it | 253 | failed | | | |
| grid012.ct.infn.it | 136 | 3569 | 3001 | 663 | 2189 |
| grid10.lal.in2p3.fr | 15343 | 1165 | 1161 | 223 | 900 |
| grid8.wdcb.ru | 2369 | aborted | | | |
| gridba2.ba.infn.it | 53571 | 1977 | 1972 | 427 | 1440 |
| gridgate.cs.tcd.ie | 139 | 1947 | 1761 | 338 | 1332 |
| griditce01.na.infn.it | 1434 | 1952 | 1944 | 369 | 1477 |
| helmsley.dur.scotgrid.ac.uk | 50268 | 2396 | 2394 | 457 | 1813 |
| hudson.datagrid.jussieu.fr | 188 | 2641 | 2608 | 510 | 1958 |
| lcgce01.gridpp.rl.ac.uk | 216 | 1637 | 1613 | 325 | 1204 |
| mu6.matrix.sara.nl | 153 | 1526 | 1522 | 296 | 1148 |
| polgrid1.in2p3.fr | 34031 | 1546 | 1544 | 232 | 1263 |
| prod-ce-01.pd.infn.it | 665 | 1596 | 1591 | 240 | 1301 |
| scaicl0.scai.fraunhofer.de | 77881 | aborted | | | |
| spaci01.na.infn.it | 224 | failed | | | |
| tbn20.nikhef.nl | 66 | 1519 | 1518 | 225 | 1245 |
| testbed001.grid.ici.ro | 41631 | 2091 | 2083 | 349 | 1674 |

# 6   Applications

DEM has many applications in various environmental studies, forestry and wild life protection, human health preservation, agricultural economics, global climate changes study, etc. . Some of them are illustrated by the plots below, based on some of the output results of the model.

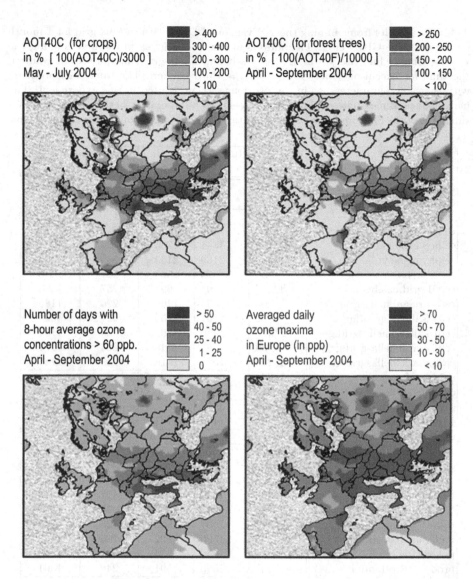

The levels of AOT40 for crops and forests respectively are given in the first two plots for year 2004. This special characteristics are used to evaluate the negative effect of the high ozone concentrations on the vegetation of plants (see [2,3,11,16,17] for more detail). The next two plots are related to the effect of the high ozone levels on the human health.

## 7  Conclusions and Plans for Future Work

- The Danish Eulerian Model is a complicated large-scale air pollution model. Its numerical treatment requires significant computational power, provided either by a high-performance supercomputer or the emerging GRID technology.

- By using splitting of the original PDE system and applying special parallelization techniques to each of the submodels we have created efficient, scalable
  and highly portable parallel implementation of DEM. Important results from
  various application areas can be obtained within a reasonable time.
- The powerful GRID technology allows us to achieve similar or better time results on distributed low-cost resources of the GRID, on the price of somehow
  lower reliability. This is in comparison with relatively bisy supercomputers
  and ordinary (not high-priority) queues.
- The portability of the parallel code, achieved by using only MPI standard
  library, is essential also for its parallel GRID implementation. This is one of
  our tasks for future work. Our preliminary expectations are for decreasing
  sigificantly the run time and increasig the waiting time for free PE-s, but
  in general best results for moderate degree of parallelism (4 or 8 PE-s).
  Targeting the fastest result, the optimal trade-off would be obtenened by
  experiments.

## Acknowledgments

This research was supported in part by the Bulgarian IST Centre of Competence
in 21 Century – BIS-21++ (contract # INCO-CT-2005-016639), by the EGEE-II
project (INFSO-RI-031688) and by NATO grant NATO ARW "Impact of future
climate changes on pollution levels in Europe". A grant from the Danish Natural
Sciences Research Council gave us access to all Danish supercomputers.

## References

1. V. Alexandrov, A. Sameh, Y. Siddique and Z. Zlatev, Numerical integration of
   chemical ODE problems arising in air pollution models, Env. Modeling and Assessment, 2 (1997) 365–377.
2. A. Bastrup-Birk, J. Brandt, I. Uria and Z. Zlatev, Studying cumulative ozone
   exposures in Europe during a 7-year period, Journal of Geophysical Research, 102
   (1997), pp. 23,917–23,935.
3. I. Dimov, Tz. Ostromsky, I. Tzvetanov, Z. Zlatev, Economical Estimation of the
   Losses of Crops Due to High Ozone Levels, Large-Scale Scientific Computations of
   Engineering and Environmental Problems II (M. Griebel, S. Margenov, P. Yalamov,
   eds.), NNFM, Vol.73, Vieweg (2000), pp. 275–282.
4. M. W. Gery, G. Z. Whitten, J. P. Killus and M. C. Dodge, A photochemical
   kinetics mechanism for urban and regional modeling, J. Geophys. Res. 94 (1989),
   pp. 12925–12956.
5. W. Gropp, E. Lusk and A. Skjellum, Using MPI: Portable programming with the
   message passing interface, MIT Press (1994), Cambridge, Massachusetts.
6. E. Hesstvedt, Ø. Hov and I. A. Isaksen, Quasi-steady-state approximations in air
   pollution modeling: comparison of two numerical schemes for oxidant prediction,
   Int. Journal of Chemical Kinetics 10 (1978), pp. 971–994.
7. Ø. Hov, Z. Zlatev, R. Berkowicz, A. Eliassen and L. P. Prahm, Comparison of
   numerical techniques for use in air pollution models with non-linear chemical reactions, Atmospheric Environment 23 (1988), pp. 967–983.

8. G. I. Marchuk, Mathematical modeling for the problem of the environment, Studies in Mathematics and Applications, No. 16 (1985), North-Holland, Amsterdam.

9. G. J. McRae, W. R. Goodin and J. H. Seinfeld, Numerical solution of the atmospheric diffusion equations for chemically reacting flows, J. Comp. Physics 45 (1984), pp. 1–42.

10. Tz. Ostromsky, W. Owczarz, Z. Zlatev, Computational Challenges in Large-scale Air Pollution Modelling, Proc. 2001 International Conference on Supercomputing in Sorrento, ACM Press (2001), pp. 407–418.

11. Tz. Ostromsky, I. Dimov, I. Tzvetanov, Z. Zlatev, Estimation of the Wheat Losses Caused by the Tropospheric Ozone in Bulgaria and Denmark, Numerical Analysis and Its Applications (L. Vulkov, J. Wasniewski, P. Yalamov, Eds.), LNCS-1988, Springer (2001), pp. 636–643.

12. Tz. Ostromsky, Z. Zlatev, Parallel Implementation of a Large-scale 3-D Air Pollution Model, Large Scale Scientific Computing (S. Margenov, J. Wasniewski, P. Yalamov, Eds.), LNCS-2179, Springer, 2001, pp. 309–316.

13. Tz. Ostromsky, Z. Zlatev, Flexible Two-level Parallel Implementations of a Large Air Pollution Model, Numerical Methods and Applications (I.Dimov, I.Lirkov, S. Margenov, Z. Zlatev - eds.), LNCS-2542, Springer (2002), pp. 545–554.

14. WEB-site of the Danish Eulerian Model, available at: *http://www.dmu.dk/AtmosphericEnvironment/DEM*

15. Z. Zlatev, Computer treatment of large air pollution models, Kluwer (1995).

16. Z. Zlatev, I. Dimov, Tz. Ostromsky, G. Geernaert, I. Tzvetanov, A. Bastrup-Birk, Calculating Losses of Crops in Denmark Caused by High Ozone Levels, Environmental Modelling and Assessment, Vol. 6, Kluwer (2001), pp. 35–55.

17. Z. Zlatev, G. Geernaert and H. Skov, A Study of ozone critical levels in Denmark, EUROSAP Newsletter 36 (1999), pp. 1–9.

# Simulation of an Extreme Air Pollution Episode in the City of Stara Zagora, Bulgaria

Maria Prodanova[1], Juan Perez[2], Dimiter Syrakov[1], Roberto San Jose[2],
Kostadin Ganev[3], Nikolai Miloshev[3], and Stefan Roglev[3]

[1] National Institute Meteorology and Hydrology (NIMH), Sofia 1784, Bulgaria
[2] Technical University of Madrid, Computer Science School, 28660 Madrid, Spain
[3] Geophysical Institute (GPhI), Bulgarian Academy of Sciences, Sofia 1111, Bulgaria

## 1 Introduction

Several industrial hot spots exist in Bulgaria and detail study of every one is worth to be done, but often incidents with high pollution levels over usually relatively clean populated areas cause big political concern. Stara Zagora is one of the biggest towns in Bulgaria (300 000 inhabitants) located in the middle of the country. In the summer of 2004, two high level $SO_2$ pollution events happened causing big public complains and even political and judicial consequences. Analogous events happened in 2005, too. All this requires appropriate measures to be taken by local authorities, first of all a suitable monitoring and forecasting system to be set. For the moment such system does not exist; only ambient air concentrations are measured in several points. This is not sufficient neither to predict nor even to explain the cases. As far as the mathematical modeling is alternative and supplementing tool according to the EU Framework Directive on Air Quality (96/62/ES) and its daughter directives (see, [5,6,7,16]), an attempt to simulate one of these events was done applying one of the most comprehensive and up-to-science modeling tools, mainly the US EPA Model-3 system.

## 2 Description of the Pollution Episode

Four days, one after another, very high $SO_2$ concentrations were observed in the afternoon hours over Stara Zagora, leading to appearance of mist. The concentrations were over the alert threshold of 350 $\mu g/m^3$. Keeping in mind that the pollution covered an area of several squared kilometers, it is easy to estimate that tones of sulphur were released over the town. As far as all this happened in summertime and the domestic heating can not be the reason for such pollution, it was supposed that a possible source can be the three thermal power plants (TPPs) disposed at 40 km southeast of the town.

"Maritza-Iztok" TPPs were built in early 70s around a big lignite coal field exploited by open-pit mining. The coals have high sulphur content so these TPPs are the main sulphur polluters not only in Bulgaria but in all south-east Europe. The parameters of these plants are given in Table 1, where TPP Maritza–Iztok 1, 2 and 3 are denoted with MI-1, MI-2 and MI-3. The anual $SO_x$ emission, is

T. Boyanov et al. (Eds.): NMA 2006, LNCS 4310, pp. 483–491, 2007.

**Table 1.** Emission parameters of "Maritza-Iztok" TPPs

| TPP | Latitude | Longitude | Stack | | | | | $SO_x$ anual emission (Mg) | % |
| | | | Nr | Height (m) | Diame-ter (m) | Tempera-ture (C°) | Flow (Nm³/h) | | |
|---|---|---|---|---|---|---|---|---|---|
| MI-1 | 25.91 | 42.16 | 1 | 150 | 6 | 192 | 2116000 | 60139 | 6 |
| MI-2 | 26.08 | 42.23 | 1 | 325 | 12 | 192 | 5400000 | 310714 | 50 |
| MI-2 | 26.08 | 42.23 | 2 | 325 | 10 | 178 | 2900000 | 173572 | 50 |
| MI-3 | 26.01 | 42.14 | 1 | 325 | 12 | 192 | 5150000 | 156938 | 16 |
| | | | | | | | Total emitted | **701363** | **72** |

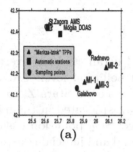
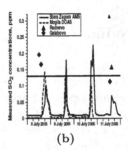

(a)    (b)

**Fig. 1. (a)** Configuration of polluters and receivers in the region of Stara Zagora; **(b)** Measured SO2 concentrations during the episod

about 700 000 tones, i.e. about 2000 tones daily. In the last column of Table 1, the percentage from the total Bulgarian $SO_x$ emission is displayed.

The configuration of Stara Zagora and "Maritza-Iztok" TPPs is given in Fig. 1 (a), the position of the measuring stations shown as well. In the town itself gas analyzers are operating continuously. Another DOAS installation is operating in Mogila village. Other two sampling points are operating in Radnevo and Galabovo – small towns near TPPs.

In Fig. 1 (b), the measured $SO_2$ concentrations are displayed. The straight line shows the alert threshold for $SO_2$ ($350\ \mu g/m^3 = 0.131 ppm$). In the sampling points, only observed data above and near these values are shown. It is clearly seen that during all period, on a background of low $SO_2$ concentration (under 10 ppb) a sharp rises in concentration appear in the afternoon hours. Three possible mechanisms can explain this behavior, provided it is due to the TPPs:

a. In a stable PBL during night and morning hours the plume from high stacks is keeping high $SO_2$ concentrations aloft some tens of kilometers from the sources. If a steady flow exists from southeast such concentrations will exist over the town. The development of convective turbulence in the afternoon hours would drag this pollution to the ground (fumigation). In the evening, the stratification gets stable and pollution does not influence the surface (see, [9,12] and their reference).

b. Meandering of the plume and changes of wind direction.

c. Combination of both mechanisms.

It is quite interesting that, in spite of availability of such powerful $SO_x$ polluters in the vicinity of the town, the pollution episodes are quite rare - several times a year. The reason can be that the synoptic situations favorable for such episodes are rare.

During the period 8-11 July 2005, a high pressure system is centered to the southwest of British Isles at the beginning and moves over the Great Britain at the end of the period, covering the most part of the continent. A depression initially centered over southern France moves eastward and remains blocked in the Balkan region. The pressure field is smooth, relatively low and blocked by the strong highs. The airflow over Bulgaria is southwestern, strong at upper levels (500 hPa), weakening and more southerly at lower levels (700, 850 hPa), and changing to westerly at surface. This means that slow and unstable winds prevailed over the region during the episode.

# 3   US EPA Model-3 System

A big number of models and model systems with different level of complexity were developed in the last twenty years. Many of them can be found in the Model Documentation System of European Environmental Agency or in the respective site of US EPA.

The forthcoming accession of Bulgaria to EU sets necessity of operating a contemporary air quality modeling tool. The choice and implementation of such a tool to different regional and local tasks were the aim of the EC 5thFP project BULAIR (http://www.meteo.bg/bulair) of Bulgarian National Institute of Meteorology and Hydrology (NIMH). The first task of the team was to make an extensive review of the existing models and to choose suitable one/ones. It was done by reviewing papers presenting at the last meetings of the two of the most important events in the field of air pollution modeling. It occurs that the Model-3 system of US EPA is one of the best modeling tools that continues to be developed intensively. This software is free downloadable and can be run on contemporary PCs. It is a modeling tool of large flexibility with a range of options and possibilities to be used for different applications/purposes on range of different regions (nesting). In the frame of BULAIR several tasks of regional ([13,14] and local scale ([15]) was solved making use of this system. The system consists of three parts:

- **MM5** – the 5th generation PSU/NCAR Meso-meteorological Model ([4,8] used as meteorological pre-processor;
- **SMOKE** – the Sparse Matrix Operator Kernel Emissions Modeling System ([3]) used as emission pre–processor, and
- **CMAQ** – the Community Multiscale Air Quality System ([1,2] – the Chemical Transport Model.

## 4   Computational Domains

For this local task meteorological data set from NCEP Global Analysis Data for 2005 is exploited (http://dss.ucar.edu/datasets/ds083.2/). As far as the space resolution of this data is $1° \times 1°$ nesting is applied for downscaling to 1 km step for a domain around Stara Zagora (see Fig.2).

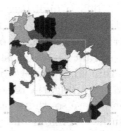

**Fig. 2.** Computational domains

The MM5 program TERRAIN is used to define five domains with 81, 27, 9, 3 and 1 km resolution ($37 \times 37$, $55 \times 55$, $46 \times 55$, $37 \times 37$ and $55 \times 55$ grids, respectively). These domains (referred as $D1 \div D5$) are chosen in such a way that $D5$ is centered in the middle of the distance between TPPs and Stara Zagora. Lambert conformal projection with true latitudes at 30 and 60°N and central point coordinates 42.30N and 25.85E are chosen. Meanwhile, TERRAIN specified topographic, vegetative, and soil type data to all grid points.

## 5   MM5 Simulations

CMAQ needs two kinds of input information - meteorology and emissions. MM5 is used here to provide CMAQ with meteorological fields. As a hole, MM5 solves the non-hydrostatic system of dynamic weather equations in $\sigma$–coordinate system:

$$\frac{\partial u}{\partial t} + \frac{m}{p}\left(\frac{\partial p\prime}{\partial x} - \frac{\sigma}{p^*}\frac{\partial p^*}{\partial x}\frac{\partial p\prime}{\partial \sigma}\right) = -\mathbf{V}\cdot\nabla u + v\left(f + u\frac{\partial m}{\partial y} - v\frac{\partial m}{\partial x}\right)$$
$$- ew\cos\alpha - \frac{uw}{r_{earth}} + D_u$$

$$\frac{\partial v}{\partial t} + \frac{m}{p}\left(\frac{\partial p\prime}{\partial y} - \frac{\sigma}{p^*}\frac{\partial p^*}{\partial y}\frac{\partial p\prime}{\partial \sigma}\right) = -\mathbf{V}\cdot\nabla u + u\left(f + u\frac{\partial m}{\partial y} - v\frac{\partial m}{\partial x}\right)$$
$$- ew\sin\alpha - \frac{vw}{r_{earth}} + D_v$$

$$\frac{\partial w}{\partial t} + \frac{\rho_0}{\rho}\frac{g}{p^*}\frac{\partial p\prime}{\partial \sigma} + \frac{g}{\gamma}\frac{p\prime}{p} = \mathbf{V}\cdot\nabla w + g\frac{p_0}{p}\frac{T\prime}{T_0} - g\frac{p\prime}{p}\frac{R_d}{c_p}$$
$$+ e(u\cos\alpha - v\sin\alpha) - \frac{u^2 + v^2}{r_{earth}} + D_w$$

$$\frac{\partial u}{\partial t} = -\mathbf{V}\cdot\nabla T + \frac{1}{\rho c_p}\left(\frac{\partial p\prime}{\partial t} + \mathbf{V}\cdot\nabla p\prime - \rho_0 g w\right) + \frac{Q}{c_p} + \frac{T_0}{\theta_0}D_0$$

$$\frac{\partial p\prime}{\partial t} - \rho_0 gw + \gamma p \nabla \cdot \mathbf{V} = \mathbf{V} \cdot \nabla p\prime + \frac{\gamma p}{T} \left( \frac{Q}{c_p} + \frac{T_0}{\theta_0} D_0 \right)$$

where $u$, $v$ and $w$ are the wind components, $p$ is the pressure, $T$ – temperature, $\rho$ – air dencity, $f$ – Coriolis parameter, $Q$ – diabatic heating, $D$ – molecular resistances.

Following the CMAQ guidelines MM5 is run on the two rough grids ($D1$ and $D2$) simultaneously with "two-way nesting" mode on first. Then, after extracting initial and boundary conditions from the resulting fields MM5 is run on the finer $D3 - D5$ grids as completely separate simulations with "one-way nesting" mode on. Four-dimensional data assimilation ([17]) is applied to the external domain D1 nudging towards the NCEP data. Specifically, this nudging is performed every 6 hours for the three-dimensional analysis fields aloft.

A 23-level vertical structure is chosen. The first 1 kilometer of the atmosphere, the so called Atmospheric/Planetary Boundary Layer (ABL/PBL), where the dispersion of pollutants occurs, is resolved by 8 levels that seem to be a good presentation.

The MM5 system provides optionally a number of different PBL and other parameterization schemes. The choice of appropriate scheme is usually made by validation. For the purpose, air temperature, humidity, wind data are needed in space and time. In the region of Stara Zagora no validation data are available. The lack of observational data makes any choice of parameterization schemes quite arbitrary, that must be taken into account when analyzing MM5 and CMAQ results. Here, one of the most comprehensive PBL scheme, **MRF**, is applied.

The MM5 simulations started at 12:00 of July 7, 2005, and continue up to 00:00 of July 12, 2005. The first 12 hours were added as to avoid the spin-up effects. The model results are demonstrated in Fig. 3, where the wind and temperature fields at different levels in the PBL are shown. At level $\sigma$=0.995 ($\approx$ 36 m) the wind direction at Stara Zagora is southerly while at "Maritza-Iztok" TPPs it is northerly. The behavior of meteorological parameters in the other days is similar.

The main impression from the analysis is that calm and non-oriented winds prevail during the period. There is a very fast change of wind directions in the different points from the region and at different levels. All this breaks the first hypothetical mechanism able to explain the observed concentration behavior.

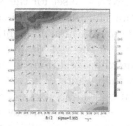 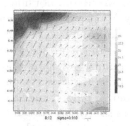

**Fig. 3.** MM5-modeled temperature and wind fields in PBL for 08 July 2005 12:00

## 6  Emission Input to CMAQ

Taking into account the behavior of measured $SO_2$ concentrations, shown in Fig 1 (b), it is decided to neglect all diffuse $SO_x$ sources in the region that create the background concentration, and to consider the three TPPs as the only sources of $SO_x$ provided the emissions in the region are several orders of magnitude less the TPPs ones. This also permits to run CMAQ only over the two inner domains ($D4$ and $D5$).

The emission input files for CMAQ were constructed by using **SMOKE** model. Its purpose is to convert the emission inventory data to the resolution needed by the air quality model. Emission inventories are typically available with annual values for each powerful source and/or for big areas (municipality, region, country). The CTMs however, require emissions data on an hourly basis, for each model grid cell and model layer, and for each model species. So, emission processing involves transforming the emission inventory through temporal allocation, chemical speciation, and spatial allocation, in order to achieve the model's input requirements.

In this case, the role of SMOKE is to produce detailed space/time distribution of the emissions from elevated point sources (the SMOKE module ELEV-POINT), based on the data in Table 1. Some ambient air characteristics (wind speed and temperature) are necessary as well, provided by the MM5 calculations.

## 7  CMAQ Simulations

CMAQ as well as the others CTMs solves the diffusion equation:

$$\frac{\partial c_i}{\partial t} + \frac{\partial(uc_i)}{\partial x} \frac{\partial(vc_i)}{\partial y} + \frac{\partial(wc_i)}{\partial z} = \frac{\partial}{\partial x}\left(K_x \frac{\partial c_i}{\partial x}\right) + \frac{\partial}{\partial y}\left(K_y \frac{\partial c_i}{\partial y}\right) \frac{\partial}{\partial z}\left(K_x \frac{\partial c_i}{\partial z}\right)$$
$$+R_i + S_i + D_i + W_i$$

where $c_i$ is the concentration of $i$-th pollutant, varying in space $(x, y, z)$ and time $(t)$; $u, v, w$ are wind components, $K_x$, $K_y$, $K_z$ - diffusion coefficients; $R_i$ - net rate of production of pollutant $i$ by chemical reactions, $S_i$ - its emission rate, $D_i$ and $W_i$ - its changes due to dry and wet removal processes.

The consideration regarding the measured $SO_2$ concentration behavior allows using zero initial and boundary conditions. They have to be set for the 3-km domain only. By nesting CMAQ produce such conditions for the finer 1-km domain. CMAQ is run from 8 to 11 July 2005 day by day, having the final moment concentration fields of the previous day as initial condition for the next day.

A big number of simulations are made during the execution of this task. Several times MM5 is re-run trying to improve the results. The new sets of meteorological data demand new SMOKE calculations. From its side, SMOKE was run many times to achieve the right parameter setting. And every time, dispersion calculation and visualization of results were done again. In the very beginning of the exercise a relatively rough 6-layer vertical structure of CMAQ

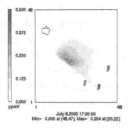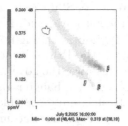

**Fig. 4.** Examples of simulated $SO_2$ concentrations; $D5$ grid

was set. Further it was improved setting up 14 levels, 8 of which in PBL. From the numerous chemical mechanism's options of CMAQ gaseous Carbon Bound IV (CB-4) chemical mechanism was chosen for these calculations.

# 8    CMAQ Model Results and Discussion

The CMAQ calculated concentration fields for different hours of each day of the period were visualized and analyzed. In Fig. 4, graphs for two afternoon hours are shown. The analysis of the results shows that, in spite of the numerous runs, the calculated concentrations do not match the measurements. It must be mentioned, however, that pollution spots near Stara Zagora can be observed every afternoon, but they hardly cover it.

From physical point of view, the calculated $SO_2$ concentration fields have a reasonable behavior. In night hours, in relatively stable PBL, the pollution released from elevated sources (gases with high temperature and release velocity) keeps aloft and the domain is not polluted at the surface. In the day hours the fast development of turbulent mixing drains the pollution to the surface at distances not far from the stacks forming well expressed plumes with very high concentrations. This is quite reasonable for summer time and is discussed by many authors ([12,9,10,11] and others).

This behavior of the calculated fields shows that the second physical hypothesis discussed above is possibly in force, here. A suitable direction of the wind from TPPs to the town of Stara Zagora is observed each afternoon. These flows form pollution spots in different places around the town, but not over it. In some cases, deviation of the wind direction by several degrees or change of the wind speed by several m/s could form spots over the town in the right periods. Small changes in the PBL height and turbulent mixing could lead to the same results. Here, all difficulties faced when trying to model local scale phenomena in complex conditions emerge. According to the author's opinion, the main shortcomings come from the MM5 simulations. In spite of being up-to-science modeling tool, MM5 is **a model** of the reality and as every model has its limitations. The reality can be and often is much more complicated than any model, which is the case, here. Several reasons for the ill-simulation can be identified:

- The region of Stara Zagora is under a shallow low pressure area, the so called nongradient baric field. It is characterized by weak and unstable winds. The lack of synoptic forcing makes every PBL parameterizations very uncertain,
- As far as the PBL schemes present stationary state, its achievement needs some time after the governing parameters has been changed. When the wind characteristics change rapidly there is not enough time for full adaptation.
- Finally, it is quite possible that the used vertical resolution of MM5 is not enough to reproduce accurately the complex character of the local ABL.

The applied MM5/CMAQ model system is quite complex and needs validation for each stage of simulation. As no data are available for the vertical structure of the atmosphere and the surface data are not sufficient, the choice of parameterization schemes is based on literature recommendations. The MRF-scheme is known to predict correctly the PBL height and surface temperature and the surface wind in most of the cases (but evidently not in our case).

## 9    Conclusion

In general, we conclude that the modeling exercise was reasonable. At this stage it could not explain quantitatively this particular $SO_2$ episode in the summer of 2005. For further development validation data for the meteorological model are necessary. In a state of no validation data, numerical experiments can continue in several directions:

- Increase of the number of MM5 and CMAQ vertical levels, especially in PBL.
- Different PBL parameterization schemes must be used.
- Use of other input meteorological data sets - ECMWF, ALADIN and HRM data sets.
- Application of additional emission scenarios, as made in [12].

## Acknowledgements

This study is made under the financial support of European Commission through the 6thFP Network of Excellence ACCENT (Contract Nr. GOCE-CT-2002-500337) and Integrated Project QUANTIFY (Contract Nr. GOGE-003893)

## References

1. Byun, D., Ching, J. (1999): *Science Algorithms of the EPA Models-3 Community Multiscale Air Quality (CMAQ) Modeling System.* EPA Report 600/R-99/030, Washington DC.
   http://www.epa.gov/asmdnerl/models3/doc/science/science.html
2. Byun, D., J. Young, G. Gipson, J. Godowitch, F.S. Binkowski, S. Roselle, B. Benjey, J. Pleim, J. Ching, J. Novak, C. Coats, T. Odman, A. Hanna, K. Alapaty, R. Mathur, J. McHenry, U. Shankar, S. Fine, A. Xiu, and C. Jang, (1998): *Description of the Models-3 Community Multiscale Air Quality (CMAQ) Modeling System,* 10th Joint Conference on the Applications of Air Pollution Meteorology with the A&WMA, 11-16 January 1998, Phoenix, Arizona, 264-268.

3. CEP (2003): *Sparse Matrix Operator Kernel Emission (SMOKE) Modeling System*, University of Carolina, Carolina Environmental Programs, Research Triangle Park, North Carolina.
4. Dudhia, J. (1993): *A non-hydrostatic version of the Penn State/NCAR Mesoscale Model: validation tests and simulation of an Atlantic cyclone and cold front*. Mon. Wea. **Rev. 121**, pp. 1493-1513.
5. EC (1998) *Amended draft of the daughter directive for ozone*, Directorate XI - Environment, Nuclear Safety and Civil Protection, European Commission, Brussels.
6. EC (1999): *Ozone position paper*, Directorate XI - Environment, Nuclear Safety and Civil Protection, European Commission, Brussels.
7. EP (2002): *Directive 2002/3/EC of the European Parliament and the Council of 12 February 2002 relating to ozone in ambient air*. Official Journal of the European Communities, **L67**, 9.3.2002, pp. 14-30.
8. Grell, G.A., J. Dudhia, and D.R. Stauffer, (1994): *A description of thc Fifth Generation Penn State/NCAR Mesoscale Model (MM5)*. NCAR Technical Note, NCAR TN-398-STR, 138 pp.
9. Hurley, P. and Physick, W. (1991): *A lagrangian particle model of fumigation by breakdown of the nocturnal inversion*, Atmospheric Environment, **25A(7)**, pp. 1313-1325.
10. Luhar, A. K. (2002): *The influence of vertical wind direction shear on dispersion in the convective boundary layer, and its incorporation in coastal fumigation models*, Boundary-Layer Meteorology, **102**, pp. 1-38.
11. Luhar, A. K. and Young, S. A. (2002): *Dispersion moments of fumigating plumes- LIDAR estimates and model simulations*. Boundary-Layer Meteorology, **104(3)**, pp. 411-444.
12. Palau, J., Perez-Landa, G., Melia, J., Segarra, D., Millan, M. (2005): *A study of dispersion in complex terrain under winter conditions using high-resolution mesoscale and Lagrangian particle models*, Atmospheric Chemistry and Physics Discussion, **5**, pp. 11965-12030.
13. Prodanova, M., Syrakov, D., Zlatev, Z., Slavov, K , Ganov, K., Miloshev, N., Nikolova L. (2005): *Preliminary results from the use of MM5-CMAQ system for estimation of pollution levels in southeast Europe*, Fist Accent Symposium "The Changing Chemical Climate of the Atmosphere", Urbino, Italy, September 12-16, 2005, p.122. (to be published in a book)
14. Prodanova, M., Perez, J.L., Syrakov, D., San Jose, R. Ganev, K., Miloshev, N. (2006): *Preliminary estimates of US EPA Model-3 system capability for description of photochemical pollution in Southeast Europe*, Proceedings of the 28th ITM on Air Pollution Modelling and Its Application, 15-19 May 2006, Leipzig, Germany.
15. Prodanova, M., Perez, J.L., Syrakov, D., San Jose, R. Ganev, K., Miloshev, N. (2006): *Simulation of some cases of extreme air pollution in the city of Stara Zagora - Bulgaria using US EPA Model-3*, Proceedings of the 28th ITM on Air Pollution Modelling and Its Application, 15-19 May 2006, Leipzig, Germany.
16. Skouloudis, A., (2005): *Preface to the Proceedings of 10th Int. Conf. on Harmonisation within Atmospheric Dispersion Modelling for Regulatory Purposes*, Sissi (Malia) Crete, Greece, 17-20 October, 2005, p.iii.
17. Stauffer, D.R. and N.L. Seaman (1990): *Use of four-dimensional data assimilation in a limited area mesoscale model. Part I: experiments with synoptic data*. Mon. Wea. Rev. **118**, pp. 1250-1277.

# Studying the Properties of Variational Data Assimilation Methods by Applying a Set of Test-Examples

Per Grove Thomsen[1] and Zahari Zlatev[2]

[1] Institute for Informatics and Mathematical Modelling
Technical University of Denmark DK-2800 Lyngby, Denmark
[2] National Environmental Research Institute
Frederiksborgvej 399, P.O. Box 358, DK-4000 Roskilde, Denmark

**Abstract.** The variational data assimilation methods can successfully be used in different fields of science and engineering. An attempt to utilize available sets of observations in the efforts to improve (i) the models used to study different phenomena and/or (ii) the model results is systematically carried out when data assimilation methods are used.

The main idea, on which the variational data assimilation methods are based, is pretty general. A functional is formed by using a weighted inner product of differences of model results and measurements. The value of this functional is to be minimized. Forward and backward computations are carried out by using the model under consideration and its adjoint equations (both the model and its adjoint are defined by systems of differential equations). The major difficulty is caused by the huge increase of both the computational load (normally by a factor more than 100) and the storage needed. This is why it might be appropriate to apply some splitting procedure in the efforts to reduce the computational work.

Five test-examples have been created. Different numerical aspects of the data assimilation methods and the interplay between the major computational parts of any data assimilation method (numerical algorithms for solving differential equations, splitting procedures and optimization algorithms) have been studied by using these tests. The presentation will include results from testing carried out in the study.

## 1 Basic Ideas

Assume that observations are available at time-points $t_p$, $p \in \{0, 1, 2, \ldots, P\}$. These observations can be taken into account in an attempt to improve in some sense the results obtained by a given model. This can be done by minimizing the value of the following functional (see, for example, Lewis and Derber [7]):

$$J\{\bar{c}_0\} = \frac{1}{2} \sum_{p=0}^{P} < W(t_p) \, (\bar{c}_p - \bar{c}_p^{obs}) \, , \, \bar{c}_p - \bar{c}_p^{obs} >, \tag{1}$$

where (a) the functional $J\{\bar{c}_0\}$ is depending on the initial value $\bar{c}_0$ of the vector of the concentrations at time $t_0$ (because the model results $\bar{c}_p$ depend on $\bar{c}_0$),

T. Boyanov et al. (Eds.): NMA 2006, LNCS 4310, pp. 492–499, 2007.

(b) $W(t_p)$ is a matrix containing some weights (it will be assumed here that $W(t_p)$ is the identity matrix, but some weights have to be used in all practical problems) and (c) $<, >$ is an inner product in an appropriately defined Hilbert space (it will be assumed in this paper that the usual vector space is used, i.e. it is assumed that $\bar{c} \in \mathbf{R}^s$ where $s$ is the number of chemical species which are involved in the model).

An optimization algorithm has to be used in order to minimize the functional $J\{\bar{c}_0\}$. Most of the optimization algorithms are based on the application of the gradient of $J\{\bar{c}_0\}$ . The adjoint equation of the model under consideration has to be derived and used in the calculation of the gradient of the functional $J\{\bar{c}_0\}$. Most of the scientific and engineering models are described mathematically by systems of differential equations. Therefore the adjoint equations are also described by systems of differential equations. This short analysis shows clearly that a data assimilation method is a very complicated numerical procedure. The time and storage requirements are the major difficulty. Such a procedure consists of (i) a good optimization algorithm and (ii) good numerical algorithms for solving differential equations. In order to reduce the time and storage requirements it is also necessary (iii) to apply some good splitting technique.

## 2   Need for a Good Set of Test-Examples

The final aim is to apply the data assimilation technique to large-scale air pollution models for studying the transport of harmful air pollutants over Europe ([11]). However, the ideas discussed in this paper are very general and can successfully be applied in connection of many other models which lead (after some kind of semi-discretization) to stiff systems of ordinary differential equations (ODEs).

Before applying a data assimilation method to a given model it is necessary to check carefully (a) the correctness of its modules and (b) the efficiency of the numerical algorithms applied in the different modules. This can successfully be done only if good test-examples are available.

The chemical part of an environmental model is normally the most time consuming part (and the most difficult one because it introduces stiffness in the model). This is why it is especially important to test carefully the correctness and the efficiency of the chemical part.

The chemical part of an environmental model can be represented as a stiff system of ODEs:

$$\frac{d\bar{c}}{dt} = f(t, \bar{c}), \quad \bar{c} \in \mathbf{R}^s, \tag{2}$$

where vector $\bar{c}$ contains $s$ components and function $f$ is in general nonlinear. Five test-examples were devised. We start with a very simple linear system. Then the complexity is gradually increased. The second test-example is a nonlinear but autonomous system. The third test-example is a non-linear and non-autonomous system with a Jacobian matrix which does not depend explicitly on

$t$ . The fourth test-example is a non-linear and non-autonomous system with a Jacobian matrix which depends explicitly on $t$ . The last test-example is a chemical scheme with 56 chemical species, which is really used in many environmental models. It is described by a non-linear and non-autonomous system of ODEs. Both the right-hand-side function and the Jacobian matrix depend on $t$ . It is not possible to express the dependence on $t$ analytically, because some chemical rates depend on some quantities (as, for example, the temperature) which are dependent on the time variable. Analytical solution is not available, but a reference solution has been calculated with a time-stepsize $\Delta t = 10^{-5}$ . The values of this solution were saved at the end of every period of 15 min. The so-found reference solution is used to check the accuracy achieved in different runs.

The first four examples are taken from the book of Lambert ([5]), while the fifth example is similar to to the schemes used in the EMEP models (see Simpson et al., [9], Zlatev, [11]).

## 3   Calculating the Gradient of the Functional

It is convenient to explain the basic ideas that are used when the gradient of the functional $J\{\bar{c}_0\}$ is calculated by the following very simple example. Assume that observations are available only at five time-points: $t_0$, $t_1$, $t_2$, $t_3$ and $t_4$. The gradient of the functional can be calculated in the following way. Assume that some tool, *model*, by which the values of the unknown vectors $\bar{c}(t_0)$, $\bar{c}(t_1)$, $\bar{c}(t_2)$, $\bar{c}(t_3)$ and $\bar{c}(t_4)$ can be calculated, is available. The tool *model* can, for example, be some air pollution model, but in some simpler cases *model* can simply be some ordinary solver for systems of PDEs or ODEs. Under this assumption, the calculations have to be performed, **for this particular example with $P = 4$,** in five steps.

- **Step 1.** Use the *model* to calculate $\bar{c}_1$ (performing integration, in a forward mode, from time-point $t_0$ to time-point $t_1$). Calculate the adjoint variable $\bar{q}_1 = \bar{c}_1 - \bar{c}_1^{obs}$ . Form the adjoint equation (corresponding to the *model* used in the forward mode; adjoint equations will be discussed in Section 5). Perform backward integration (by applying the adjoint equation) from time-point $t_1$ to time-point $t_0$ to calculate the vector $\bar{q}_0^1$, where the lower index shows that $\bar{q}_0^1$ is calculated at time-point $t_0$ , while the upper index shows that $\bar{q}_0^1$ is obtained by using $\bar{q}_1 = \bar{c}_1 - \bar{c}_1^{obs}$ as an initial vector in the backward integration.
- **Step 2 to Step 4.** Perform the same type of calculations, as those in Step 1 to obtain $\bar{q}_0^2$, $\bar{q}_0^3$ and $\bar{q}_0^4$. More precisely, the following operations are to be carried out for $p = 2, 3, 4$:
  - **(a)** use the forward mode to proceed from time-point $t_{p-1}$ to time-point $t_p$,
  - **(b)** form the adjoint variable $\bar{q}_p = \bar{c}_p - \bar{c}_p^{obs}$,
  - **(c)** use the adjoint equation in a backward mode from time-point $t_p$ to time-point $t_0$ to calculate $\bar{c}_0^p$.

– **Step 5.** The sum of the vectors $\bar{q}_0^1$, $\bar{q}_0^2$, $\bar{q}_0^3$, $\bar{q}_0^4$ obtained in Step 1 to Step 4 and vector $\bar{q}_0^0 = \bar{q}_0 = \bar{c}_0 - \bar{c}_0^{obs}$ gives an approximation to the required gradient of the functional $J\{\bar{c}_0\}$.

It is clear that the above procedure can easily be extended for any number $P$ of time-points at which observations are available.

The gradient of the functional $J\{\bar{c}_0\}$ is calculated by performing one forward step from time-point $t_0$ to time-point $t_P$ and $P$ backward steps from time-points $t_p$, $p = 1, 2, \ldots, P$, to time-point $t_0$. This explains the main idea, on which the data assimilation algorithms are based, in a very clear way, but it is expensive when $P$ is large. In fact, the computational work can be reduced, performing only once the backward calculations (see, for example, [1] or [7]).

## 4   Solving the System of ODEs

Six numerical methods for solving stiff systems of ODEs have been used in the experiments. The methods selected by us are listed below:

– the Backward Euler Method,
– the Implicit Mid-point Rule,
– a Second-order Modified Diagonally Implicit Runge-Kutta Method,
– a Fifth-order Three-stage Fully Implicit Runge-Kutta Method,
– a Second-order Two-stage Rosenbrock Method,
– the Trapezoidal Rule.

The Implicit Mid-point Rule and the Trapezoidal Rule are A-stable methods. All the other methods are L-stable. More details about the numerical methods used in this paper can be found in [3], [4], [5] and [10].

## 5   Solving the Adjoint Equations

It is necessary to distinguish between linear models and non-linear models when the adjoint equations are formed and treated numerically. Assume that the model is linear and, furthermore, that the model is written in the following general form:

$$\frac{\partial \bar{c}}{\partial t} = A\bar{c}. \tag{3}$$

Denote by $q$ the adjoint variable. Then the adjoint equation can be written as

$$\frac{\partial \bar{q}}{\partial t} = -A^*\bar{q}, \tag{4}$$

where $A^*$ is the conjugate operator of $A$. If the problem is discretized by using some numerical method, then operator $A$ will be represented by a matrix which is normally also denoted by $A$. If the adjoint equation is discretized, then the transposed matrix $A^T$ will appear in the discretized version of (4).

Consider now a non-linear model:

$$\frac{\partial \bar{c}}{\partial t} = B(\bar{c}). \tag{5}$$

The adjoint equation of the model presented in (5) can be written as

$$\frac{\partial \bar{q}}{\partial t} = - \left[ B'(\bar{c}) \right]^* \bar{q}, \tag{6}$$

where $B'(\bar{c})$ is obtained by differentiation of $B$. In the discrete case, we will have the transposed matrix of the Jacobian of $B$ in (6).

It is seen that the adjoint equations are always linear; compare (4) and (6). However, the right-hand-side in the linear case does not depend on the model variable $\bar{c}$. In the non-linear case this is no more true. The right-hand-side of (6) depends on $\bar{c}$. This fact has serious implications: the values of $\bar{c}$ calculated when the model is treated are to be saved and used when the adjoint equation is handled.

If the chemical scheme (2) is considered, then (6) can be rewritten as

$$\frac{d\bar{q}}{dt} = - \left( \frac{\partial f(t, \bar{c})}{\partial \bar{c}} \right)^T \bar{q}. \tag{7}$$

It is clear now that the numerical methods from the previous section can easily be adapted for the adjoint equation (7) of the chemical scheme (2). For example, the application of the Backward Euler Method in connection with adjoint equation (7) leads to the following formula for the backward computations:

$$\bar{q}_n = \bar{q}_{n+1} - \Delta t \left[ -\frac{\partial f(t_n, \bar{c}_n)}{\partial \bar{c}_n} \right]^T \bar{q}_n. \tag{8}$$

The fact that the adjoint equation is used in the backward mode is taken into account when (8) is derived.

## 6   Application of Splitting Procedures

The application of data assimilation algorithms leads to very time-consuming problems (the computer time may be increased by a factor up to 100 and even more). Therefore splitting, which is commonly used during the treatment of large-scale environmental models, is even more needed when these are used together with data assimilation techniques. The test-examples, which are listed in Section 2, were treated both without splitting and with by four splitting procedures: (i) sequential splitting, (ii) symmetric splitting, (iii) weighted sequential splitting and (iv) weighted symmetric splitting.

The splitting of each of the first four test-examples is not very critical. Let us consider as an example the splitting applied in connection with the second test-example. The operator on the right-hand-side of this example is $f_1 = y_2$, $f_2 = y_2(y_2 - 1)/y_1$ (where $y_1$ and $y_2$ are the components of vector $\bar{c}$ ). It is split into two operators: (a) $f_1^{(1)} = 0$, $f_2^{(1)} = -y_2^{(1)}/y_1^{(1)}$ and (b) $f_1^{(2)} = y_2^{(2)}$, $f_2^{(2)} = (y_2^{(2)})^2/y_1^{(2)}$ . The sum of these two operators is, component-wise, equal to the original operator in the right-hand-side of the second test-example (i.e. $f_1^{(1)} + f_1^{(2)} = f_1$ and $f_2^{(1)} + f_2^{(2)} = f_2$ ).

It is not very obvious how to split the fifth test-example. We grouped in the first sub-model the species which react with ozone. The remaining species formed the second sub-model.

At each time-step during the forward mode the splitting was carried out as usual (see, for example, [11]). At each time-step during the backward mode the splitting operators are applied in reverse order (compared with the order applied in the corresponding forward time-step).

# 7   Minimizing the Functional

The problem of minimizing the functional (1) is in fact an **unconstrained** optimization problem. Therefore, the subroutine E04DGF from the NAG Library, which performs unconstrained optimization, has been used in the beginning. However, we realized quickly that it is better to impose some constraints. There are often physical reasons for this (in the chemical scheme, for example, the concentrations of the chemical species should be kept non-negative). Therefore, the next choice was subroutine E04KDF also from the NAG Library. This is a rather flexible subroutine. It requires simple bounds for the variables of the functional. It is quite reasonable to assume that such bounds could always be derived in real-life problems (by using the physical properties of the studied processes).

The problem with subroutine E04KDF is not the determination of the bounds for the variables, but rather the necessity to scale the model, which is very often a rather difficult task. Unfortunately, such a requirement is, to our knowledge, common for all optimization algorithms.

# 8   Numerical Results

The ability of the data assimilation algorithms to improve the initial values of the solution was tested numerically. This is important for forecasting high pollution levels. However, the data assimilation algorithms can also be used for many other purposes (see, for example, [1], [2], [6], [8]).

A perturbation parameter $\alpha$ was introduced. The values of the initial solution were always perturbed by using ten different values of $\alpha$ (introducing relative errors of 5%, 10%, ..., 50% in the initial values). Data assimilation is used to improve the initial values. The improved initial values are then used to calculate the solution over an increased time-interval. The analytical solution (the reference solution for the fifth test-example) is used to evaluate the relative error, component-wise, at the end of each time-step (each period of 15 min. for the fifth test-example). The max-norm of the vector or relative errors found over the whole time-interval is taken and used in the comparisons of the results from the different runs.

Each test-example has been run with the six numerical methods and the five splittings (including here also the case where no splitting is used). Furthermore, for the first four test-examples we start with a time-stepsize $\Delta t = 0.25$ and carry out successively 18 additional runs (every time reducing the time-stepsize by a factor of two). For the fifth test-example we start with a time-stepsize $\Delta t = 150$ and carry out successively 10 additional runs (reducing again the time-stepsize by a factor of two every time when a new run is started).

The results from the runs show that (i) reducing the time-stepsize leads to a reduction of the error according to the order of the combined method (numerical method + splitting procedure), (ii) if the time-stepsize is sufficiently small then the error obtained with the data assimilation method is practically the same as the error obtained by using exact initial values without data assimilation (which means that the results are optimal in some sense), (iii) the numerical methods that are only A-stable (the Implicit Mid-Point Rule and the Trapezoidal Rule; see [3] and [5]) have difficulties for large time-stepsizes when the stiff chemical scheme is to be handle and (iv) if no splitting is used, then it might be more efficient in some cases to use high-order methods (the Fifth-order Three stage Fully Implicit Runge-Kutta Method performed better, for all five test-examples, than the other methods when no splitting was used).

It should be mentioned here that the stability problems, which were mentioned in (iii), disappear when splitting procedures are used. Since the chemical scheme is a rather general and sufficiently large problem, this fact indicates that the splitting procedures have some stabilizing effect when stiff systems of ODEs are to be handled.

Some results obtained in the efforts to improve the initial value and the accuracy of the ozone component in the chemical scheme with 56 species are given in Table 1. The notation used can be explained as follows: (a) $ERROR\_0\_P$ is giving the relative error in the perturbed initial condition, (b) $ERROR\_0\_I$ is giving the relative error in the improved initial condition, (c) $ERROR\_F\_P$ is giving the global relative error obtained by using the perturbed initial condition, (d) $ERROR\_F\_I$ is giving the global relative error obtained by using the improved initial condition.

**Table 1.** Numerical results obtained when the chemical schemes with 56 compounds is run. The Backward Euler Method is used without splitting. The initial value of the ozone concentration is perturbed by a factor $\alpha = 0.5$.

| Steps | $ERROR\_0\_P$ | $ERROR\_F\_P$ | $ERROR\_0\_I$ | $ERROR\_F\_I$ |
|-------|---------------|---------------|---------------|---------------|
| 1008  | 0.47          | 0.48          | 2.0E-03       | 2.4E-03       |
| 2016  | 0.49          | 0.50          | 1.0E-03       | 1.2E-03       |
| 4032  | 0.47          | 0.47          | 5.0E-04       | 6.1E-04       |
| 8064  | 0.48          | 0.48          | 2.5E-04       | 3.2E-04       |
| 16128 | 0.46          | 0.48          | 1.3E-04       | 1.7E-04       |
| 32256 | 0.49          | 0.50          | 6.3E-05       | 8.8E-05       |

It is seen that reducing the stepsize (i.e. multiplying the number of time-steps by a factor of two) leads to a reduction of both the initial guess and the global error by a factor of two. This is precisely the expected behaviour (because the Backward Euler Method is of order one).

## 9 Conclusions

The results from several thousand runs indicate that the data assimilation modules are able to improve the initial values of the solution if (a) the numerical methods used are sufficiently accurate and (b) the initial perturbations are not very large.

On the other hand, the results indicate also that both the computing time and the storage needed are increased by a factor which is very often greater than 100. Therefore, it is necessary (i) to continue the search for faster but still sufficiently accurate numerical algorithms, (ii) to apply faster computers, (iii) to exploit efficiently the cache memory and (iv) to parallelize the codes.

## References

1. Elbern, H., Schmidt, H.: A four-dimensional variational chemistry data assimilation scheme for Eulerian chemistry transport modelling. Journal of Geophysical Research, **104** (1999) 18583–18598
2. Elbern, H., Schmidt, H., Talagrand, O., Ebel, A.: 4D-variational data assimilation with an adjoint air quality model for emission analysis. Environmental Modelling & Software, **15** (2000) 539–548.
3. Hairer, E., Wanner, G.: Solving Ordinary Differential Equations, II: Stiff and Differential-algebraic Problems. Springer, Berlin-Heidelberg-New York-London, 1991.
4. Hundsdorfer, W., Verwer, J. G.: Numerical solution of time-dependent advection-diffusion-reaction equations. Springer, Berlin, 2003.
5. Lambert, J. D.: Numerical Methods for Ordinary Differential Equations. Wiley, Chichester-New York-Brisbane-Toronto-Singapore, 1991.
6. Le Dimet, F.-X., Navon, I. M., Daescu, D. N.: Second order information in data assimilation. Monthly Weather Review, **130** (2002) 629–648.
7. Lewis, J. M., Derber, J. C.: The use of adjoint equations to solve a variational adjustment problem with advective constraints. Tellus, **37A** (1985) 309–322
8. Sandu, A., Daescu, D. N., Carmichael, G. R., Chai, T.: Adjoint sensitivity analysis of regional air quality models. Journal of Computational Physics, 2005; to appear.
9. Simpson, D., Fagerli, H., Jonson, J. E., Tsyro, S. G., Wind, P., Tuovinen, J-P.: Transboundary Acidification, Eutrophication and Ground Level Ozone in Europe, Part I. Unified EMEP Model Description. EMEP/MSC-W Status Report 1/2003. Norwegian Meteorological Institute, Oslo, Norway, 2003.
10. Zlatev, Z.: Modified diagonally implicit Runge-Kutta methods. SIAM Journal on Scientific and Statistical Computing, **2** (1981) 321–334.
11. Zlatev, Z.: Computer treatment of large air pollution models. Kluwer Academic Publishers, Dordrecht-Boston-London, 1995.

# On the Numerical Solutions of Eigenvalue Problems in Unbounded Domains

Andrey B. Andreev and Milena R. Racheva

Technical University
5300 Gabrovo, Bulgaria
Andreev@tugab.bg, Milena@tugab.bg

**Abstract.** The aim of this study is to propose a procedure for coupling of finite element (FE) and infinite large element (ILE). This FE/ILE method is applied to the second-order self-adjoint eigenvalue problems in the plane. We propose a conforming method for approximation of eigenpairs in unbounded domains. Finally, some numerical results are presented.

## 1   Introduction

Our study is motivated by the great diversity of concepts concerning infinite element approximations. For example, infinite large elements are successfully applied to the approximation of the Helmholz equation [1, 2], parabolic problems [4] and others [3, 5].

In a recent work [6] Han, Zhou and Zheng studied a coupling FE/ILE method applied to the second order eigenvalue problems which is nonconforming. So, our main aim is to propose a new conforming method for connection and coupling of finite and infinite large elements. The considerations are restricted to piecewise bilinear approximations on a bounded domain and we construct corresponding infinite large elements in order to approximate the solution on the unbounded domain. It means that the domain $\Omega$ is divided by two disjoint parts: finite subdomain $\Omega_F$ and infinite one - $\Omega_I$.

In order to avoid some technical details, let $\Omega$ be the exterior of a rectangle:

$$\Omega = \{(x,y) : |x| > r_1 > 0 \vee |y| > r_2 > 0\}.$$

Let the positive numbers $R_1$ and $R_2$ be large enough and such that $R_i > r_i$, $i = 1, 2$. The unbounded subdomain $\Omega_I$ is defined by:

$$\Omega_I = \{(x,y) : |x| > R_1 \vee |y| > R_2\}.$$

Then $\Omega_F = \Omega \setminus \Omega_I$.

A second order eigenvalue problem can be deduced from the corresponding parabolic problem on the unbounded domain $\Omega \times (0, T)$. Many application problems can be modeled by parabolic equations in unbounded domains, such as the heat transfer problem, fluid dynamics problems (see [5, 7]) and recently various option pricing problems in the financial mathematics.

T. Boyanov et al. (Eds.): NMA 2006, LNCS 4310, pp. 500–507, 2007.

Consider the following two-dimensional heat equation:

$$\rho(x,y).\frac{\partial W}{\partial t} = \Delta W, \quad (x,y,t) \in \Omega \times (0,T), \tag{1}$$

where $\rho(x,y) > 0$ is continuous in $\mathbf{R}^2$ and $W(x,y,t)$ is bounded in $\Omega$.

Besides the equation (1) we introduce the following conditions:

$$W_{|_\Gamma} = 0, \qquad 0 < t \le T,$$

$$W_{|_{t=0}} = W_0(x,y). \tag{2}$$

Here $\Gamma$ is the boundary of $\mathbf{R}^2 \setminus \Omega$ and $W_0(x,y)$ is a given function with compact support.

Following the classical steps of the separation of variables method, we look for a solution of (1)-(2) in the form

$$W(x,y,t) = e^{-\lambda t}.u(x,y),$$

where $(\lambda, u(x,y))$ is to be determined. Substituting the last relation into the problem (1)-(2), we arrive at the following eigenvalue problem: Find the number $\lambda$ and the function $u(x,y) \ne 0$ such that

$$-\Delta u = \lambda \rho u, \text{ in } \Omega,$$

$$u = 0, \qquad \text{on } \Gamma. \tag{3}$$

We are interested in the case when the eigenvalues $\lambda$ are real and $u(x,y)$ are real-valued eigenfunctions. This depends on some properties of $\rho(x,y)$ which will be presented below. Since $\Omega$ is unbounded an additional requirement for the eigenfunctions is:

$$\int_\Omega (\nabla u)^2 \, dx + \int_\Omega \rho.u^2 \, dx < +\infty. \tag{4}$$

Here and throughout the following, we use $dx$ instead of $dx \, dy$. We also denote $r = \sqrt{x^2 + y^2}$.

The following two properties concerning $\rho(x,y)$ are supposed [6]:

$$0 < \gamma(R) \le \min_{1 \le r \le R} \rho(x,y), \tag{5}$$

$$\rho(x,y) \le \rho_0(r), \quad 1 \le r < +\infty, \tag{6}$$

where $\gamma(R)$ and $\rho_0(r)$ are given functions.

Let us define the space

$$L_{2,\rho}(\Omega) = \left\{ v : \int_\Omega \rho.v^2 \, dx < \infty \right\}.$$

This space has a natural inner product

$$b(u,v) = \int_\Omega \rho u v \, dx$$

and the corresponding norm is defined as follows:

$$\|u\|_{0,\rho,\Omega} = \sqrt{b(u,u)}.$$

It is obvious that $L_{2,\rho}(\Omega)$ is a Hilbert space. Next we use the Sobolev space $H^1(\Omega)$ and

$$H_0^1(\Omega) = \left\{ u \in H^1(\Omega) : u = 0 \text{ on } \Gamma \right\}.$$

Let

$$a(u,v) = \int_\Omega \nabla u \cdot \nabla v \, dx,$$

then we introduce the space

$$H_0^{1,\rho}(\Omega) = H_0^1(\Omega) \cap L_{2,\rho}(\Omega).$$

The induced norm is defined by

$$\|u\|_{1,\rho,\Omega} = \left( a(u,u) + b(u,u) \right)^{1/2}.$$

The bilinear form $a(u,v)$ is bounded and coercive in $H_0^{1,\rho}(\Omega)$, namely there is a positive constant $\alpha$, such that

$$|a(u,v)| \leq \|u\|_{1,\rho,\Omega} \cdot \|v\|_{1,\rho,\Omega}, \quad \forall u,v \in H_0^{1,\rho}(\Omega),$$

$$a(u,u) \geq \alpha \|u\|_{1,\rho,\Omega}^2, \quad \forall u \in H_0^{1,\rho}(\Omega).$$

The variational formulation of eigenvalue problem (3)-(4) can be written in the following form: Find $\lambda \in \mathbf{C}$ and $u \in H_0^{1,\rho}(\Omega)$, $u \neq 0$ such that

$$a(u,v) = \lambda b(u,v), \quad \forall v \in H_0^{1,\rho}(\Omega). \tag{7}$$

**Lemma 1.** *(see [6], Theorem 2.1) If $\rho(x,y)$ satisfies (5) and (6), then the eigenvalues of problem (7) form an increasing sequence of positive numbers tending to $+\infty$. The corresponding eigenfunctions $\{u_m\}_{m=1}^\infty$ form an orthogonal basis of $L_{2,\rho}(\Omega)$.*

## 2    A Numerical Approximation of the Eigenvalue Problem

We are interested in the numerical approximation of the eigenpairs by finite/ infinite large element method. Our coupling approach differs from those presented in [6] because a conformity between any finite element and the corresponding infinite large element is assured. Moreover, the approximation by ILEs in the "radial" direction does not need the using of polar coordinates.

First, we cover the domain $\Omega_F$ by bilinear finite elements. Toward this end, for each $h$, $0 < h \leq 1$ we let $\tau_h$ be a triangulation of $\overline{\Omega}_F$ by rectangles $K$ of diameter which does not exceed $h$. Next, we also cover $\Omega_I$ by infinite large convex quadrilaterals $K_I$ (see Figure 1). So, the intervals $(-r_i, r_i)$, $i = 1, 2$ are divided into $2n$ parts and the intervals $(-R_i, r_i)$ and $(r_i, R_i)$, $i = 1, 2$ - by $k$ parts. In this way, the number of finite elements covering $\overline{\Omega}_F$ is $4(2kn + k^2)$.

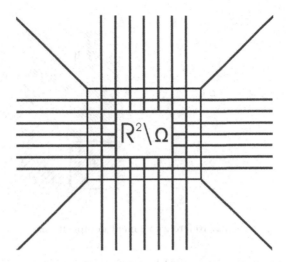

**Fig. 1.** FE/ILE element partition

*Remark 1.* In general, the family $\{\tau_h\}$ is quasi-uniform, i.e. there is a constant $\gamma > 0$ such that

$$\frac{h}{h_K} \leq \gamma \quad \forall K \in \tau_h, \ \forall h,$$

where $h_K$ is the diameter of $K$. It is easy to construct an uniform mesh partition $(h_K = h, \ \forall K \in \tau_h)$ by choosing appropriate values of $R_i$, $i = 1, 2$. Figure 1 illustrates the case $R_i = 2r_i$, $i = 1, 2$.

*Remark 2.* The presented methodology is also applicable to the case where polynomials of higher degree are used. Further analysis on these and related subjects will be provided in forthcoming contributions.

It should be noted that each ILE $K_I$ has exactly one adjacent FE with common side. The ILEs covering the domain $\Omega_I$ form a family $\tau_{h,I}$ related to $\tau_h$. The number of infinite large elements in the case under consideration is $8(k+n)$. The partitions $\tau_h$ and $\tau_{h,I}$ are affine families. It means that for each $K \in \tau_h$ $(K_I \in \tau_{h,I})$ there is an invertible affine map which transforms the reference finite element (reference infinite large element) into the corresponding one.

According to our consideration, in Figure 2 two types of reference infinite large elements are shown, which yield two types of reference element composition. Any element composition consists of an infinite large element and its adjacent finite element.

There are $8(k + n - 1)$ element compositions of type (a) and 4 element compositions of type (b). We shall present the basis functions for both cases. It is reasonable that they will depend on the type of the finite elements we choose.

On the Lagrangian bilinear reference FE the shape functions are determined by the values at the nodes $p_1(1, 1), p_2(0, 1), \ p_3(0, 0)$ and $p_4(1, 0)$.

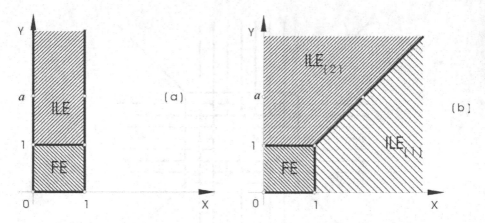

**Fig. 2.** Reference element compositions

Let $\widehat{K}_I = \{(x,y) \in [0,1] \times [1, \infty)\}$ be a reference ILE of type (a). We introduce the following functional space, defined on $\widehat{K}_I$ (see, also Remark 3 and Remark 4):

$$V_{\widehat{K}_I} = \left\{ a_0 + \sum_{i=0}^{1} \sum_{j=1}^{2} a_{ij} \frac{x^i}{y^j} \ : \ a_0, \ a_{ij} \in \mathbf{R}, \ i = 0,1; j = 1,2 \right\}.$$

For any ILE $K_I \in \tau_{h,I}$ we determine the corresponding space $V_{K_I}$ by using an invertible affine transformation. Then the shape functions are chosen to belong to $V_{K_I}$.

Let us consider the ILE of type (a) (see Figure 2). Choosing a number $a$ large enough, such that $a > 1$, we use the nodes $q_1^{(a)}(1,1) \equiv p_1$, $q_2^{(a)}(0,1) \equiv p_2$, $q_3^{(a)}(0,a)$ and $q_4^{(a)}(1,a)$ as well as the node $q_5$ at infinity. In the partition $\tau_{h,I}$ the node at infinity is shared by all infinite large elements. The value of the shape function at the public node is understood in the asymptotical sense. Denoting the corresponding basic functions by $\{\psi_j^{(a)}(x,y)\}_{j=1}^{5}$, where $\psi_j^{(a)}(q_i) = \delta_{ij}$, $i, j = 1, \ldots, 5$, we define $((x,y) \in [0,1] \times [1, +\infty))$:

$$\psi_1^{(a)}(x,y) = \frac{x(a-y)}{(a-1)y^2}, \quad \psi_2^{(a)}(x,y) = \frac{(1-x)(a-y)}{(a-1)y^2},$$

$$\psi_3^{(a)}(x,y) = \frac{a^2(1-x)(y-1)}{(a-1)y^2}, \quad \psi_4^{(a)}(x,y) = \frac{a^2 x(y-1)}{(a-1)y^2},$$

$$\text{and} \quad \psi_5(x,y) = -\frac{(y-1)(a-y)}{y^2} \quad \text{at the public node.}$$

*Remark 3.* Using the same idea, one may easily construct a reference ILE of type (a) in the $x$-direction, i.e. for $(x,y) \in [1, +\infty) \times [0,1]$.

For the reference infinite large element of type (b) (see Figure 2, ILE$_2$) the degrees of freedom are the values of any approximating function at the nodes

$q_1^{(b)}(1,1) \equiv p_1$, $q_2^{(b)}(0,1) \equiv p_2$, $q_3^{(b)}(0,a)$ and $q_4^{(b)}(a,a)$ as well as the node $q_5$ at infinity. Just as previously, $a$ is an arbitrary positive number such that $a > 1$. For $1 \le y < \infty$, $0 \le x \le y$ we define the system of basis functions $\{\psi_j^{(b)}(x,y)\}_{j=1}^5$, $\psi_j^{(b)}(q_i) = \delta_{ij}$, $i,j = 1,\dots,5$ in the following way:

$$\psi_1^{(b)}(x,y) = \frac{x(a-y)}{(a-1)y^3}, \qquad \psi_2^{(b)}(x,y) = \frac{(y-x)(a-y)}{(a-1)y^3},$$

$$\psi_3^{(b)}(x,y) = \frac{a^2(y-x)(y-1)}{(a-1)y^3}, \qquad \psi_4^{(b)}(x,y) = \frac{a^2 x(y-1)}{(a-1)y^3},$$

$$\text{and} \quad \psi_5(x,y) = -\frac{(y-1)(a-y)}{y^2} \quad \text{at the public node.}$$

*Remark 4.* Let us note that the reference infinite large elements in both cases (a) and (b) are equivalent in some sense. It is easily seen, substituting $X = \frac{x}{y}$ in the case (b).

Once the basic functions of the space $V_{\hat{K}_I}$ are established, the following lemma can be proved:

**Lemma 2.** *The basic functions* $\psi_i^{(a)}$, $\psi_i^{(b)}$, $i = 1,\dots,4$ *and* $\psi_5$ *are bounded.*

*Proof.* First we prove the boundedness of the functions $\psi_i^{(a)}$, $i = 1,2$. Let $y \ge 1$ and $0 \le x \le 1$. Since $a > 1$, then

$$0 \le \frac{|a-y|}{(a-1)y} \le 1.$$

On the other hand the nonnegative terms $\frac{x}{y}$ and $\frac{1-x}{y}$ are majorated by 1. Hence

$$|\psi_i^{(a)}(x,y)| \le 1 \quad \text{for } i = 1,2. \tag{8}$$

For the functions $\psi_i^{(a)}$, $i = 3,4$ we use also that $0 \le \frac{y-1}{y^2} \le \frac{1}{4}$. Then

$$|\psi_i^{(a)}(x,y)| \le \frac{a^2}{4(a-1)} \quad \text{for } i = 3,4. \tag{9}$$

In view of Remark 4, substituting $X = \frac{x}{y}$ in $\psi_i^{(b)}$, $i = 1,\dots,4$, from (8) and (9) we also obtain, that

$$|\psi_i^{(b)}(x,y)| \le 1 \quad \text{for } i = 1,2.$$

$$|\psi_i^{(b)}(x,y)| \le \frac{a^2}{4(a-1)} \quad \text{for } i = 3,4.$$

Finally we consider

$$\psi_5(y) = 1 - \frac{a+1}{y} + \frac{a}{y^2}, \quad y \in [1, +\infty).$$

Obviously $\psi_5(1) = 0$ and $\lim_{y\to\infty} \psi_5(y) = 1$. Taking into account that $\psi_5(y)$ has a unique extremum at the point $\dfrac{2a}{a+1} > 1$, one obtains:

$$|\psi_5(y)| \leq \max\left\{1, \frac{(a-1)^2}{4a}\right\}.$$

*Remark 5.* The last inequalities are important from algorithmic point of view. They show that one has to fix the parameter $a$ such that $a \gg 1$ when eigenpairs are calculated in a unbounded domain.

With the partitions $\tau_h$ and $\tau_{h,I}$ we associate the following functional space:

$$V_h = \left\{v_h : v_{h|K} \in Q_1(K), \ \forall K \in \tau_h; \ v_{h|K_I} \in V_{K_I}, \ \forall K_I \in \tau_{h,I}; \ v_{h|\Gamma} = 0\right\},$$

where $Q_1$ is a bilinear polynomial space.

From the definition of the basic functions $\psi_i^{(a)}(x,y)$ and $\psi_i^{(b)}(x,y)$, $i = 1, \ldots, 4$ it is easy to verify the following assertion:

**Theorem 1.** *The functions from the space $V_h$ are continuous, i.e.*

$$V_h \subset C^0(\Omega).$$

According to this theorem, the presented FE/ILE method is conforming.

We introduce an approximate eigenvalue problem, corresponding to (7) : Find $\lambda \in \mathbf{R}$, $u_h \in V_h$ such that

$$a(u_h, v_h) = \lambda_h b(u_h, v_h), \quad \forall v_h \in V_h.$$

## 3    Numerical Results

Consider the problem (3) with

$$\Omega = \{(x,y) : |x| > \sqrt{2}/2, \ |y| > \sqrt{2}/2\}$$

and $\rho(x,y) = \dfrac{x^2 + 2y^2}{(x^2 + y^2)^{2.1}}$.

**Table 1.**

| ILEs | $\lambda_1$ | $\lambda_2$ | $\lambda_3$ |
|------|-------------|-------------|-------------|
| 16 | 0.018272 | 0.413172 | 1.010226 |
| 32 | 0.018090 | 0.411435 | 1.007044 |
| 64 | 0.017811 | 0.410968 | 1.005803 |

Thus $r_i = \sqrt{2}/2$, $i = 1, 2$. The numerical results illustrate the case when $R_i = 2r_i$, $i = 1, 2$.

The interval $(-\sqrt{2}/2, \sqrt{2}/2)$ is divided into $2n$ parts, so the partition consists of $12n^2$ finite elements and $16n$ infinite large elements (Figure 1).In Table 1 we present numerical results, calculated with $a = 3$.

# References

[1] Demkovich, K., Gerdes, K.: Convergence of infinite element methods for the Helmholz equation in separable domains. Numer Math. **79** (1998) 11–42.

[2] Gerdes, K.: A summary of infinite element formulations for ezterior Helmholz problem.: Exterior Problems of Wave Propagation, Special Issue Comput. Methods Appl. Mech. Engrg., 1998, **164**, 95–105.

[3] Bettless, P.: Infinite elements. Internat. J. Numer. Methods Engrg., 1997, **11**, 53–64.

[4] Han, H.D., Huang, Z.Y.: A class of artificial boundary conditions for heat equation in unbounded domains. Comput. Math. Appl. bf 43 (2002), 889–900.

[5] Givoli, D.: Numerical Methods for Problems in Infiniter Domains. Elsevier, Amsterdam (1992).

[6] Han, H., Zhou, Zh. and Zheng, Ch.: Numerical solutions of an eigenvalue problem in unbounded domains. Numer. Math. a Journal of Chinese Universities. Vol. 14, No1 Feb. 2005, 1–13.

[7] Egorov, Y.V., Shubin, M.A.: Foundation of the Classical Theory of Partial Differential Equations. Second Edition. Springer, Berlin (1998).

# Mesh Independent Superlinear Convergence of an Inner-Outer Iterative Method for Semilinear Elliptic Systems

István Antal

Department of Applied Analysis, ELTE University, H-1117 Budapest, Hungary
istvanka@cs.elte.hu

**Abstract.** We propose the damped inexact Newton method, coupled with inner iterations, to solve the finite element discretization of a class of nonlinear elliptic systems. The linearized equations are solved by a preconditioned conjugate gradient (PCG) method. Both the inner and the outer iterations have mesh independent superlinear convergence.

**Keywords:** conjugate gradient method, preconditioning, superlinear convergence, damped inexact Newton method, mesh independence, numerical experiments.

## 1  Introduction

We propose an inner-outer (damped inexact Newton plus PCG) iteration for the finite element discretization of a class of nonlinear elliptic systems. Our aim is to show mesh independent superlinear convergence of the overall iteration. The linearized equations are solved by a preconditioned conjugate gradient method. It is known that the Newton method has quadratic convergence when the exact solution of the linearized equation is given. Instead of this, one may solve the linearized equation in an inexact way, mainly with applying an iteration method, in this paper we consider a preconditioned conjugate gradient method. This way we lose the quadratic convergence of the outer Newton iterations, but we may ensure superlinear convergence as we control the inaccuracy of the inner iteration.

## 2  The PDE System

We consider the class of semilinear PDE-systems described below, which has the short form

$$\begin{cases} -\Delta \underline{u}(x) + f(x, \underline{u}(x)) = \underline{g}(x) \\ \underline{u}\big|_{\partial\Omega} = \underline{0}, \end{cases} \tag{1}$$

where $\underline{u} = (u_1, u_2, \ldots, u_s)^T, \underline{g} = (g_1, g_2, \ldots, g_s)^T$. In this paper all operators, like $\Delta, \nabla, \big|_{\partial\Omega}$, are meant coordinatewise.
We impose the assumptions

T. Boyanov et al. (Eds.): NMA 2006, LNCS 4310, pp. 508–515, 2007.
© Springer-Verlag Berlin Heidelberg 2007

[P1 ] $\partial\Omega \subset \mathbb{R}^d$ ($d = 2$ or $3$) is piecewise $C^2$ and $\Omega$ is locally convex at the corners;

[P2 ] $g_i \in L^2(\Omega)$ ($i = 1, 2, \ldots, s$);

[P3 ] $f : \Omega \times \mathbb{R}^s \to \mathbb{R}^s$, for a.e. $x \in \Omega$ $f(x, \xi)$ has a potential $\psi : \Omega \times \mathbb{R}^s \to \mathbb{R}$, i.e. $f = \partial_\xi \psi$ and it is differentiable w.r.t. $\xi$, and in these points the Jacobians are symmetric positive semidefinite;

[P4 ] for a.e. $x \in \Omega$ the Jacobians $\partial_\xi f(x, \xi)$ are uniformly bounded in $\xi$ by a symmetric matrix $M(x)$, where the eigenvalues $\mu_j(x)$ of $M(x)$ are bounded, $0 \leq \mu_j(x) \leq c_1$, with some constant $c_1 > 0$;

[P4' ] the eigenvalues $\lambda_j^{(f)}(x, \xi)$ ($j = 1, \ldots, s$) of the Jacobians $\partial_\xi f(x, \xi)$ are bounded as follows

$$0 \leq \lambda_j^{(f)}(x, \xi) \leq c_2 + c_3 \sum_{j=1}^{s} |\xi|^{p-2},$$

for some constants $c_2, c_3 > 0$ and $p \geq 2$;

[P5 ] the derivative of $f$ is Lipschitz continuous, that is there exists a constant $C$ that $\|\partial_\xi f(x, \xi_1) - \partial_\xi f(x, \xi_2)\|_2 \leq C\|\xi_1 - \xi_2\|_2$ for a.e. $x \in \Omega$;

[P5' ] the derivative of $f$ is locally Lipschitz continuous, that is there exists a function $C : (0, \infty) \to (0, \infty)$ that $\|\partial_\xi f(x, \xi_1) - \partial_\xi f(x, \xi_2)\|_2 \leq C(r)\|\xi_1 - \xi_2\|_2$ for a.e. $x \in \Omega$ if $\|\xi_1\|, \|\xi_2\| \leq r$.

The weak formulation of problem (1) is: find a function $\underline{u} \in \mathcal{H} = (\mathcal{H}, \langle \cdot, \cdot, \rangle) = (H_0^1(\Omega))^s$ that satisfies

$$\int_\Omega (\nabla \underline{u} \cdot \nabla \underline{v} + f(x, \underline{u}) \cdot \underline{v}) = \int_\Omega \underline{g} \cdot \underline{v}, \ \underline{v} \in (H_0^1(\Omega))^s, \tag{2}$$

Here again the operator $\nabla$ is meant coordinatewise.

*Remark 1.* $\mathcal{H} = (H_0^1(\Omega))^s$ coincides with the energy space of the unbounded operator $S : D(S) \subset (L^2(\Omega))^s \to (L^2(\Omega))^s$, the coordinatewise Laplacian. That is $\mathcal{H}$ is the completion of the space $(D(S), \langle \cdot, \cdot \rangle_S)$ where $\langle \underline{u}, \underline{v} \rangle_S = \int_\Omega S\underline{u} \cdot \underline{v} = \int_\Omega \nabla \underline{u} \cdot \nabla \underline{v}$ is the energy scalar product.

In the following $\langle \cdot, \cdot \rangle$ will always denote the above mentioned scalar product, and $\| \cdot \|$ will denote the induced norm.

The weak formulation of problem (1) has also its (equivalent) variational form: find a function $\underline{u} \in \mathcal{H}$ that minimizes the functional $\phi : \mathcal{H} \to \mathbb{R}$

$$\phi(\underline{u}) := \int_\Omega \left( \frac{1}{2}|\nabla \underline{u}|^2 + \psi(x, \underline{u}) - \underline{g} \cdot \underline{u} \right). \tag{3}$$

By assumption [P3] $\psi$ is a convex function, and therefore $\phi$ is also convex. By (3) we have that $\phi$ is also coercive, therefore the functional $\phi$ has a unique minimum (see e.g. [8]), and hence equation (2) has a unique weak solution.

# 3  Abstract Form

Equation (2) may be considered as an equation on the space $\mathcal{H}$

$$F(\underline{u}) = \underline{b}, \tag{4}$$

where $F(\underline{u})$ and $\underline{b}$ are the Riesz representation vectors defined by the left and right-hand sides of (2) respectively.

**Proposition 1.** *From assumptions [P1-P5] we have that*

1. $F : \mathcal{H} :\longrightarrow \mathcal{H}$ *is differentiable in the Gateaux sense;*
2. $F$ *is regular, and* $\|F'(u)h\| \geq \|h\|$ *independently of* $u, h$;
3. $F$ *has the form* $F = I + N$, *where* $I$ *is the identity operator on* $\mathcal{H}$, $N$ *is also differentiable and for all* $u \in \mathcal{H}$, $N'(u)$ *is a compact self-adjoint operator, further it is Hilbert-Schmidt;*
4. *the operators* $N'(u)$ *are uniformly majorized, that is there exists a compact positive self-adjoint Hilbert-Schmidt operator such that for all* $u \in \mathcal{H}$, $N'(u) \leq K$ *in the sense* $\langle N'(u)h, h \rangle \leq \langle Kh, h \rangle$, $\forall h \in \mathcal{H}$;
5. *if we have [P4'] instead of [P4] we only have the operators* $N'(u)$ *are locally uniformly majorized, that is for all* $r > 0$ *there exists a compact positive self-adjoint Hilbert-Schmidt operator such that* $N'(u) \leq K(r)$ *in the sense* $\langle N'(u)h, h \rangle \leq \langle K(r)h, h \rangle$, $\forall h \in \mathcal{H}$, *for all* $\|u\| \leq r$;
6. $N'$ *is Lipschitz continous with Lipschitz constant* $L$;
7. *if [P5'] holds only instead of [P5] then* $N'$ *is only locally Lipschitz continous, with a function* $L : (0, \infty) \to (0, \infty)$.

*Remark 2.* From 1) and 2) of Proposition 1 we have that $F$ is a homeomorphism, see in e.g. [8], therefore equation (2) has a unique solution.

Now we may state a convergence theorem of the DIN method.

**Theorem 1.** *[5] Let $F$ be the operator defined above, then the following damped inexact Newton method converges. Let $u_0 \in \mathcal{H}$ be arbitrary and let us define the sequence $(u_n) \subset H$ recursively as*

$u_{n+1} = u_n + \tau_n p_n$ $(n \in \mathbb{N})$, *where*

$$\|F'(u_n)p_n + (F(u_n) - b)\| \leq \delta_n \|F(u_n) - b\|, \text{ with } 0 < \delta_n \leq \delta_0 < 1 \text{ and} \tag{5}$$

$$\tau_n = \min\left\{1, \frac{1 - \delta_n}{(1 + \delta_n)^2} \frac{1}{L\|F(u_n) - b\|}\right\}.$$

*Then the sequence $(u_n)$ converges to the exact solution $u^*$ of equation (4)*

$$\|u_n - u^*\| \leq \|F(u_n) - b\| \to 0 \text{ monotonically.}$$

*Further, if $\delta_n \equiv \delta_0$ then we have linear convergence, if $\delta_n \leq const \cdot \|F(u_n) - b\|^\gamma$ $(0 < \gamma \leq 1)$ then the convergence is locally of order $1 + \gamma$, that is the convergence is linear for $n_0$ steps, until $\|F(u_n) - b\| \leq \varepsilon$, where $\varepsilon$ is at most $(1 - \delta_n)\frac{1}{2L}$, and further on $\|u_n - u^*\| \leq d_1 q^{(1+\gamma)^{n-n_0}}$ holds.*

*Remark 3.* The formula (5) gives a bound for the error of the approximate solution of the linearized equation

$$F'(u_n)p_n = -(F(u_n) - b). \tag{6}$$

## 4  FEM Discretization

### 4.1  Discretization of the PDE System

We consider the finite element discretization of the PDE system above. That is we have a finite element subspace $\mathcal{V}_h \subset \mathcal{H}$ with $\mathcal{V}_h = (V_h)^s = \text{span } (\underline{w}_h^i)_{i=1}^m$, where $V_h$ is a finite element subspace in $H_0^1(\Omega)$. Then we seek the element $\underline{u}_h \in \mathcal{V}_h$ that satisfies

$$\int_\Omega (\nabla \underline{u}_h \cdot \nabla \underline{v}_h + f(x, \underline{u}_h) \cdot \underline{v}_h) = \int_\Omega \underline{g} \cdot \underline{v}_h, \text{ for all } \underline{v}_h \in \mathcal{V}_h. \tag{7}$$

This equation could also be understood as an equation on the Hilbert-space $\mathcal{V}_h$ (endowed with the inherited inner product $\langle \cdot, \cdot \rangle$)

$$F_h(\underline{u}) = \underline{b}_h, \tag{8}$$

where $F_h(\underline{u})$ and $\underline{b}_h$ are the Riesz representation vectors defined by the left and right-hand sides of (7) respectively.

**Proposition 2.** $F_h(\underline{u})$ is the projection of $F(\underline{u})$ onto the subspace $\mathcal{V}_h$. It inherits all the analogous properties of $F$ mentioned in Proposition 1.

**Corollary 1.** By Remark 2 we have that (7) also has a unique solution.

Thus we are lead to the problem: find the coefficients $\mathbf{c} = (c_j)_{j=1}^m$ such that $\underline{u}_h = \sum c_j \underline{w}_h^j$ satisfies

$$\int_\Omega \left( \nabla u_h \cdot \nabla \underline{w}_h^j + f(x, \underline{u}_h) \cdot \underline{w}_h^j \right) = \int_\Omega \underline{g} \cdot \underline{w}_h^j, \text{ for } j = 1, \ldots, m.$$

This gives rise to a nonlinear algebraic system of the following form:

$$\mathbf{c} + N_h(\mathbf{c}) = \mathbf{b}. \tag{9}$$

### 4.2  Discretization of the Linearized Equation

From the above formulas we have that the linearization of $F_h$ is

$$\langle F_h'(\underline{u})\underline{p}, \underline{v} \rangle = \int_\Omega (\nabla \underline{p} \cdot \nabla \underline{v} + \partial_{\underline{u}} f(x, \underline{u})\underline{p} \cdot \underline{v}) = \langle \underline{p}, \underline{v} + S^{-1}Q_{\underline{u}}\underline{v} \rangle, \tag{10}$$

for all $\underline{p}, \underline{v} \in \mathcal{V}_h$. We may define the stiffness and mass matrices respectively as

$$\mathbf{S_h} = \left[ \langle \underline{w}_h^i, \underline{w}_h^j \rangle \right]_{i,j=1}^m, \mathbf{D_h}(\underline{u}) = \left[ \langle S^{-1}Q_{\underline{u}}\underline{w}^i{}_h, \underline{w}^j{}_h \rangle \right]_{i,j=1}^m = \left[ \langle Q_{\underline{u}}\underline{w}^i{}_h, \underline{w}^j{}_h \rangle_{(L_2)^s} \right]_{i,j=1}^m. \tag{11}$$

*Remark 4.* It is apparent that $\mathbf{S_h}$ is the $s$-tuple of the discrete Laplacian $-\Delta_h$, and if $\underline{u}_h = \sum c_i \underline{w}_h^i$ and $\underline{v}_h = \sum d_i \underline{w}_h^i$ then $\langle \underline{u}_h, \underline{v}_h \rangle = \mathbf{S_h c \cdot d}$.

From equations (6) and (10) we have that the Newton linearization of (8) leads us to the linear problem: find the element $\underline{p}_h \in \mathcal{V}_h$ that satisfies

$$\int_\Omega \left( \nabla \underline{p}_h \cdot \nabla \underline{v} + \partial_{\underline{u}} f(x, \underline{u}) \underline{p}_h \cdot \underline{v} \right) = - \int_\Omega \left( \nabla \underline{u} \cdot \nabla \underline{v} + f(x, \underline{u}) \underline{v} - \underline{g} \cdot \underline{v} \right), \quad (12)$$

for all $\underline{v} \in \mathcal{V}_h$. We have the following linear equation, with $\underline{p}_h = \sum_j p_j \underline{w}_h^j$ and $\mathbf{p} = (p_j)_{j=1}^m$:

$$(\mathbf{I} + \mathbf{S_h}^{-1} \mathbf{D_h}(\underline{u}))\mathbf{p} = \mathbf{f}, \quad (13)$$

where $\mathbf{f} = \mathbf{S_h}^{-1} (\gamma_1, \ldots, \gamma_m)^T$ with

$$\gamma_j = - \int_\Omega \left( \nabla \underline{u} \cdot \nabla \underline{w}_h^j + f(x, \underline{u}) \underline{w}_h^j - \underline{g} \cdot \underline{w}_h^j \right)$$

### 4.3   Inner CG for the Discretized Equation

In this paper we suggest a preconditioned conjugate gradient method in order to get an inexact solution of (12). By Proposition 1 $F$ is a compact perturbation of the identity. It is then well-known that the CG method applied to (6) has superlinear convergence [7,4,2]. Moreover we have a discretization independent estimate on the convergence:

**Theorem 2 ([4,1]).** *The CG applied to the equation (13) yields the following convergence estimate with the notation* $\mathbf{e}_k = \mathbf{p}_k - \mathbf{p}$:

$$\frac{\|\mathbf{e}_k\|_{\mathbf{S_h}+\mathbf{D_h}(\underline{u})}}{\|\mathbf{e}_0\|_{\mathbf{S_h}+\mathbf{D_h}(\underline{u})}} \leq \left( \frac{3 \, \|\|\mathbf{S_h}^{-1}\mathbf{D_h}(\underline{u})\|\|^2}{2k} \right)^{k/2},$$

*if* $k \in \mathbb{N}$ *is even and* $k \geq \frac{2}{3} \|\|\mathbf{S_h}^{-1}\mathbf{D_h}(\underline{u})\|\|^2$. *This estimate is independent of the subspace* $\mathcal{V}_h$ *used in Galerkin discretization.*

We can combine condition (5) with Theorem 2, thus in the $n$th outer iteration we need to take $k_n$ inner iterations in order to achieve the required estimate (5). This means that for $n \geq 0$ $k_n$ shall satisfy

$$\frac{\|F_h'(u_n)p_n^{k_n} + (F_h(u_n) - \underline{b}_h)\|}{\|F_h(u_n) - \underline{b}_h\|} \leq \delta_n,$$

that is with $p_n^0 = 0$ we have the estimate on $k_n$

$$\left( \frac{3 \, \|\|K\|\|^2}{2k_n} \right)^{k_n/2} \leq \delta_n.$$

## 5 The Algorithm

The DIN algorithm applied to the problem (9) is then

1. we calculate the matrix $\Delta_h$ (since $S_h$ is the $s$-tuple of it) and calculate $\mathbf{b}$ by some fast Poisson solver as a preconditioner, and set the initial guess $\mathbf{c_0} = 0$;

   after $n$ outer iterations

2. we calculate the residual $\mathbf{r_n} = \mathbf{c_n} + N_h(\mathbf{c_n}) - \mathbf{b}$, and its norm $\|\mathbf{r_n}\|_{\mathbf{S_h}}$;
3. we make some inner PCG steps with a stopping criterion $\epsilon = \delta_n$ and calculate $\mathbf{P_n}$:
   (a) we calculate the mass matrix $\mathbf{D}(\underline{u}_\mathbf{h})$ as in (11), and set initial value $\mathbf{p_n^0} = 0$;
   (b) we calculate the residual $\mathbf{e_n^0} = \mathbf{p_n^0} + \mathbf{D}(\underline{u}_\mathbf{h})\mathbf{p_n^0} - \mathbf{f}$, and its norm $\|\mathbf{e_n^0}\|_{\mathbf{S_h}}$, and define $\mathbf{q_n^0} = \mathbf{e_n^0}$;

      after $k$ iterations:

   (c) if $\|\mathbf{e_n^k}\|_{\mathbf{S_h}} \leq \epsilon$ then we set $\mathbf{p_n} = \mathbf{p_n^k}$ and terminate the inner PCG;
   (d) we calculate the constant $\alpha^k$ and then modify $\mathbf{p_n^k}$ and $\mathbf{e_n^k}$ as

   $$\alpha^k = \frac{\mathbf{S_h}\mathbf{e_n^k} \cdot \mathbf{q_n^k}}{(\mathbf{S_h} + \mathbf{D}(\underline{u}_\mathbf{h}))\,\mathbf{e_n^k} \cdot \mathbf{q_n^k}}, \quad \text{and}$$

   $$\mathbf{p_n^{k+1}} = \mathbf{p_n^k} - \alpha^k\mathbf{q_n^k}, \quad \mathbf{e_n^{k+1}} = \mathbf{e_n^k} - \alpha^k(\mathbf{q_n^k} + \mathbf{S_h}^{-1}\mathbf{D}(\underline{u}_\mathbf{h})\mathbf{q_n^k}) \text{ respectively;}$$

   (e) we calculate the constant $\beta^k$ and then modify $\mathbf{q_n^k}$ as

   $$\beta^k = \frac{(\mathbf{S_h} + \mathbf{D}(\underline{u}_\mathbf{h}))\,\mathbf{e_n^{k+1}} \cdot \mathbf{q_n^k}}{(\mathbf{S_h} + \mathbf{D}(\underline{u}_\mathbf{h}))\,\mathbf{e_n^k} \cdot \mathbf{q_n^k}}, \quad \mathbf{q_n^{k+1}} = \mathbf{e_n^{k+1}} - \beta^k\mathbf{q_n^k};$$

   (f) we calculate the residual $\mathbf{e_n^{k+1}} = \mathbf{p_n^{k+1}} + \mathbf{D}(\underline{u}_\mathbf{h})\mathbf{p_n^{k+1}} - \mathbf{f}$, and its norm $\|\mathbf{e_n^{k+1}}\|_{\mathbf{S_h}}$, and step to (c);
4. we calculate the damping parameter $\tau_n$ and let

   $$\mathbf{c_{n+1}} = \mathbf{c_n} + \tau_n\mathbf{p_n},$$

and step to 2.

## 6 Numerical Experiments

We made experiments on some test-problems below:

· the domain was $\bar{\Omega} = [0, 1] \times [0, 1]$,
· we used Courant elements for the FEM discretization using uniform mesh with width $h = 1/N$ where $N$ is the number of subintervals on the interval $[0, 1] \times \{0\}$,
· the coordinates of the exact solutions were chosen among the functions of the form $u(x, y) = C \cdot x(1 - x)y(1 - y)$ and $u(x, y) = C \cdot \sin \pi x \sin \pi y$,
· we had the function $f$ as the potential of the functional $\varphi(\xi) = \|\xi\|^4$,

· the stopping criterion was $\|F_h(\underline{u}_n) - b_h\| \leq 10^{-5}$,
· we used adaptive damping parameters $\tau_n$,
· the code was written in Matlab.

**Proposition 3.** *The above test-problems satisfy assumptions [P1], [P2], [P3], [P4'], [P5'].*

The cases [P1]-[P3] are obvious. By some elementary calculations we have that [P4'] is satisfied with $c_1 = 0, c_2 = 12, p = 4$ and [P5'] is satisfied with $C(r) = 24r$.

Denoting $r_n = F_h(\underline{u}_n) - b_h$, $n_{inn}$ equals the number of inner iterations, we had the following results:

We observe mesh independence for both the outer and inner iterations.

*Remark 5.* The relaxing parameters $\tau_n$ defined in (5) did result linear convergence before the superlinear phase, but then the convergence quotient were so close to 1, that it would needed too much computer time to get to the superlinear phase. That is why we used some adaptive relaxing parameters.

**Table 1.** Results for $s = 2$ and $s = 6$

| $n$ | $N = 25$ | | $N = 55$ | | $N = 85$ | |
|---|---|---|---|---|---|---|
| | $\|r_n\|$ | $n_{inn}$ | $\|r_n\|$ | $n_{inn}$ | $\|r_n\|$ | $n_{inn}$ |
| 1 | 2.4747 | 1 | 2.4804 | 1 | 2.4812 | 1 |
| 2 | 1.8506 | 1 | 1.8547 | 1 | 1.8553 | 1 |
| 3 | 1.1298 | 1 | 1.1319 | 1 | 1.1323 | 1 |
| 4 | 0.4614 | 1 | 0.46195 | 1 | 0.46203 | 1 |
| 5 | $6.2785 \cdot 10^{-2}$ | 2 | $6.2886 \cdot 10^{-2}$ | 2 | $6.2902 \cdot 10^{-2}$ | 2 |
| 6 | $2.85 \cdot 10^{-4}$ | 3 | $2.9349 \cdot 10^{-4}$ | 3 | $2.9479 \cdot 10^{-4}$ | 3 |

| $n$ | $N = 25$ | | $N = 35$ | | $N = 45$ | |
|---|---|---|---|---|---|---|
| | $\|r_n\|$ | $n_{inn}$ | $\|r_n\|$ | $n_{inn}$ | $\|r_n\|$ | $n_{inn}$ |
| 1 | 22.294 | 1 | 22.327 | 1 | 22.341 | 1 |
| 2 | 12.422 | 1 | 12.400 | 1 | 12.390 | 1 |
| 3 | 6.9112 | 1 | 6.9049 | 1 | 6.9023 | 1 |
| 4 | 2.7724 | 1 | 2.7730 | 1 | 2.7732 | 1 |
| 5 | 1.1069 | 2 | 1.1173 | 2 | 1.1217 | 2 |
| 6 | $1.3284 \cdot 10-1$ | 3 | $1.3589 \cdot 10-1$ | 3 | $1.3723 \cdot 10-1$ | 3 |
| 7 | $2.4111 \cdot 10-3$ | 5 | $2.4928 \cdot 10-3$ | 5 | $2.5437 \cdot 10-3$ | 5 |
| 8 | $8.9220 \cdot 10^{-6}$ | 8 | $5.4236 \cdot 10^{-6}$ | 8 | $3.6845 \cdot 10^{-6}$ | 8 |

**Table 2.** Results for $s = 8$

| $N$ | total number of inner PCG iterations | computer time (in $sec$) | the final error $\|\underline{u} - \underline{u}_h\|$ |
|---|---|---|---|
| 15 | 39 | $2.6034 \cdot 10^2$ | $1.4628 \cdot 10^{-1}$ |
| 25 | 28 | $7.6434 \cdot 10^2$ | $4.4973 \cdot 10^{-2}$ |
| 35 | 35 | $2.3816 \cdot 10^3$ | $2.2530 \cdot 10^{-2}$ |
| 45 | 23 | $4.7176 \cdot 10^3$ | $1.3578 \cdot 10^{-2}$ |

# References

1. ANTAL, I. Mesh independent superlinear convergence of the conjugate gradient method for discretized elliptic systems, to appear in *Hung. Electr. J. Sci.*
2. AXELSSON, O., KAPORIN, I., On the sublinear and superlinear rate of convergence of conjugate gradient methods. Mathematical journey through analysis, matrix theory and scientific computation (Kent, OH, 1999), *Numer. Algorithms* 25 (2000), no. 1-4, 1–22.
3. KADLEC, J., On the regularity of the solution of the Poisson problem on a domain with boundary locally similar to the boundary of a convex open set, Czechosl. Math. J., 14(89), 1964, pp. 386,393.
4. KARATSON, J. Mesh independent superlinear convergence of the conjugate gradient method for some equivalent self-adjoint operators, *Appl. Math.* 50 (2005), no. 3, 277–290.
5. FARAGÓ I., KARÁTSON J., *Numerical solution of nonlinear elliptic problems via preconditioning operators: theory and application.* Advances in Computation, Volume 11, *NOVA Science Publishers*, New York, 2002.
6. RIESZ, F., SZ.-NAGY, B., *Vorlesungen über Funktionalanalysis*, Verlag H. Deutsch, 1982. Spriger, 1976.
7. WINTER, R., Some superlinear convergence results for the conjugate gradient method, *SIAM J. Numer. Anal.*, 17 (1980), 14-17.
8. ZEIDLER, E., *Nonlinear functional analysis and its applications*, Springer, 1986

# Solving Second Order Evolution Equations by Internal Schemes of Approximation

Narcisa Apreutesei[1] and Gabriel Dimitriu[2]

[1] Technical University "Gh. Asachi", Department of Mathematics,
700506 Iaşi, Romania
napreut@net89mail.dntis.ro
[2] University of Medicine and Pharmacy "Gr. T. Popa",
Department of Mathematics and Informatics,
700115 Iaşi, Romania
dimitriu@umfiasi.ro

**Abstract.** Numerical approximation for the solution of a second order evolution equation is proposed. An internal scheme of approximation is used. The equation is associated with a maximal monotone operator in a real Hilbert space together with bilocal boundary conditions. A numerical example is investigated.

## 1 Introduction

Let $H$ be a real Hilbert space endowed with the scalar product $(\cdot, \cdot)$ and the associated norm $\| \cdot \|$ and let $V$ be a reflexive Banach space whose norm is denoted by $| \cdot |$. Suppose that $V \subseteq H$ with a continuous injection, i.e. $\exists c_0 > 0$ such that $\|x\| \leq c_0|x|$, $\forall x \in V$ and $V$ is dense in $H$.

Consider a nonlinear operator $A$ from $H$ to $H$ with the properties below:

$(i)$ $A : D(A) = V \subseteq H \to H$ is univoque and hemicontinuous from its domain $D(A) = V$ to $H$, that is $\forall x, y \in V$,

$$\lim_{t \to 0} A(x + ty) = Ax \text{ in the weak topology of } H;$$

$(ii)$ $\|Ax\| \leq c\|x\|$, $\forall x \in D(A) = V$, where $c > 0$ is a constant;
$(iii)$ $\exists \alpha > 0$ such that

$$(Ax - Ay, x - y) \geq \alpha \|x - y\|^2, \ \forall x, y \in D(A) = V.$$

Then the operator $A$ is maximal monotone in $H$. Consider now the bilocal problem

$$\begin{cases} u''(t) = Au(t), & \text{a. e. } 0 < t < T, \\ u(0) = a, \quad u(T) = b, \end{cases} \tag{1}$$

where $a, b \in D(A) = V$. Then problem (1) has a unique solution $u \in W^{2,2}(0, T; H)$ (see for example [4] and [5]).

We give a numerical approximation of the solution $u$ of this problem by an internal scheme of approximation. A similar approximation is described in [6] for first order differential equations.

T. Boyanov et al. (Eds.): NMA 2006, LNCS 4310, pp. 516–524, 2007.
© Springer-Verlag Berlin Heidelberg 2007

Section 2 presents discretizations both for the spaces and the operator $A$. Some boundedness properties and the main approximation result are collected in the third section. An example is given in Section 4.

## 2    The Internal Scheme of Approximation

In order to construct an internal approximation of (1), we associate the following items with the parameter $h$ tending to zero (see [3], [6]):

(a) a finite dimensional linear space $V_h$, for example $V_h = I\!\!R^{n(h)}$, with $n(h)$ finite;

(b) a linear continuous injective mapping $p_h : V_h \to V$ (prolongation);

(c) a linear continuous mapping $r_h : H \to V_h$ (restriction).

Here $h$ is regarded as the step of discretization in space. Assume that:

(iv) $\forall x \in V$, $p_h r_h x \to x$ in $V$ as $h \to 0$ and $\|p_h r_h\|_{L(H,V)} \le c_1$.

We introduce the scalar product $(\cdot, \cdot)_h$ in $V_h$ given by

$$(u_h, v_h)_h = (p_h u_h, p_h v_h), \quad \forall u_h, v_h \in V_h \tag{2}$$

and the associated norm $\|\cdot\|_h$, $\|u_h\|_h = \|p_h u_h\|$, $\forall u_h \in V_h$. We also endow $V_h$ with the norm $|u_h|_h = |p_h u_h|$, $\forall u_h \in V_h$. Obviously, $\|u_h\|_h \le c_0 |u_h|_h$, $\forall u_h \in V_h$ and since $V_h$ is finite dimensional, there exists a positive constant $c(h)$ such that $|u_h|_h \le c(h) \|u_h\|_h$, $\forall u_h \in V_h$.

We now define an internal approximation $A_h$ of $A$, namely

$$A_h : V_h \to V_h, \quad (A_h u_h, v_h)_h = (A p_h u_h, p_h v_h). \tag{3}$$

It is easy to verify that the nonlinear operator $A_h$ is bounded $(\|A_h u_h\|_h \le c\|u_h\|_h)$, strongly monotone on $V_h$, and hemicontinuous on $V_h$ : $\forall u_h, v_h \in V_h$, $\lim_{t \to 0} A_h(u_h + t v_h) = A_h u_h$ in the weak topology of $V_h$.

Let $N$ be a positive integer which will tend to $+\infty$ and let $k = T/(N+1)$ be the step of discretization in time. For fixed $h$ and $k$, consider the problem

$$\begin{cases} \dfrac{1}{k^2}(u_{hk}^{(r+1)} - 2u_{hk}^{(r)} + u_{hk}^{(r-1)}) = A_h u_{hk}^{(r)}, & r = \overline{1, N}, \\ u_{hk}^{(0)} = u_h^0 = r_h a, \quad u_{hk}^{(N+1)} = u_h^T = r_h b. \end{cases} \tag{4}$$

Since $A_h : V_h \to V_h$ is monotone, hemicontinuous, univoque and everywhere defined, then it is maximal monotone in $V_h$. Hence, problem (4) has a unique solution $(u_{hk}^{(r)})_{r=\overline{1,N}}$, with $u_{hk}^{(r)} \in V_h$, $r = \overline{1, N}$ (see [1] and [2]).

One associates the functions $u_h : [0, T] \to V_h$, $z_h : (0, T] \to V_h$,

$$u_h(t) = \begin{cases} u_h^0 (= r_h a), & t = 0, \\ \left[(k-t)u_{hk}^{(0)} + t u_{hk}^{(1)}\right]/k, & 0 < t \le k, \\ \cdots \\ \left[(rk-t)u_{hk}^{(r-1)} + (t-(r-1)k)u_{hk}^{(r)}\right]/k, & (r-1)k < t \le rk, \\ \cdots \\ \left[((N+1)k-t)u_{hk}^{(N)} + (t-Nk)u_{hk}^{(N+1)}\right]/k, & Nk < t \le (N+1)k, \end{cases} \tag{5}$$

$$z_h(t) = u_{hk}^{(r)}, \quad (r-1)k < t \le rk, \qquad r = \overline{1, N+1}. \tag{6}$$

We are going to prove that $p_h u_h \to u$ in $L^2(0, T; H)$, $p_h u_h' \to u'$ in $C([0, T]; H)$ and $Ap_h z_h \rightharpoonup Au$ in $L^2(0, T; H)$.

## 3   The Main Result

First we notice that, in view of $(iv)$, we have

$$p_h u_h(0) = p_h r_h a \to a, \quad p_h u_h(T) = p_h r_h b \to b \quad \text{as} \quad h \to 0. \tag{7}$$

Some boundedness properties are collected in the following lemma.

**Lemma 1.** *Under the above notations,* $p_h u_{hk}^{(r)}$ *is bounded in $H$, for every $r =$* $\overline{0, N+1}$, $p_h u_h'(0)$, $p_h u_h'(T)$ *and* $\sum_{r=1}^{N+1} \|p_h u_{hk}^{(r)} - p_h u_{hk}^{(r-1)}\|^2$ *are bounded in $H$,* $p_h u_h$, $p_h z_h$ *and* $Ap_h z_h$ *are bounded in* $L^\infty(0, T; H)$ *with respect to $h, k$.*

*Proof.* Multiplying (4) by $u_{hk}^{(r)}$ in $V_h$ and using (2), one obtains

$$(p_h u_{hk}^{(r+1)}, p_h u_{hk}^{(r)}) - 2\|p_h u_{hk}^{(r)}\|^2 + (p_h u_{hk}^{(r-1)}, p_h u_{hk}^{(r)}) = k^2(Ap_h u_{hk}^{(r)}, p_h u_{hk}^{(r)}).$$

Then the monotonicity of $A$ and $A0 = 0$ (which follows from $(ii)$) imply that

$$\|p_h u_{hk}^{(r)}\| \le \frac{1}{2}\|p_h u_{hk}^{(r+1)}\| + \frac{1}{2}\|p_h u_{hk}^{(r-1)}\|, \qquad r = \overline{1, N}.$$

Therefore,

$$\|p_h u_{hk}^{(r)}\| \le \max(\|p_h r_h a\|, \|p_h r_h b\|) \le M, \quad \forall h, k, \quad \forall r = \overline{0, N+1}. \tag{8}$$

The same multiplication of (4) by $u_{hk}^{(r)}$ in $V_h$, followed by a summation from $r = 1$ to $r = N$, prove via the monotonicity of $A$ that

$$\sum_{r=1}^{N} \|p_h u_{hk}^{(r)} - p_h u_{hk}^{(r-1)}\|^2 \le -\|p_h u_{hk}^{(N+1)} - p_h u_{hk}^{(N)}\|^2$$

$$+ (p_h u_{hk}^{(N+1)} - p_h u_{hk}^{(N)}, p_h u_{hk}^{(N+1)})$$

$$- (p_h u_{hk}^{(1)} - p_h u_{hk}^{(0)}, p_h u_{hk}^{(0)})$$

and thus (8) yields

$$\sum_{r=1}^{N+1} \|p_h u_{hk}^{(r)} - p_h u_{hk}^{(r-1)}\|^2 \le M_1, \quad \forall h, k. \tag{9}$$

Next, for $(r-1)k < t \le rk$, with $r = \overline{1, N+1}$, we estimate

$$\|p_h u_h(t)\| \le \frac{rk - t}{k}\|p_h u_{hk}^{(r-1)}\| + \frac{t - (r-1)k}{k}\|p_h u_{hk}^{(r)}\| \le M,$$

for all $h, k$. Hence $\|p_h z_h(t)\| = \|p_h u_{hk}^{(r)}\| \leq M$, and by $(ii)$ we have $\|A p_h z_h(t)\| \leq c\|p_h z_h(t)\| \leq M_2$, $\forall h, k$. These inequalities lead to the boundedness of $p_h u_h$, $p_h z_h$ and $A p_h z_h$ in $L^\infty(0, T; H)$.

A simple computation shows that

$$
\begin{cases}
u_h'(0) = \frac{1}{T}\left[ r_h b - r_h a - \sum\limits_{m=1}^{N+1}\sum\limits_{r=1}^{m-1} (u_{hk}^{(r+1)} - 2u_{hk}^{(r)} + u_{hk}^{(r-1)}) \right], \\
u_h'(T) = \frac{1}{T}\left[ r_h b - r_h a + \sum\limits_{m=1}^{N+1}\sum\limits_{r=m}^{N} (u_{hk}^{(r+1)} - 2u_{hk}^{(r)} + u_{hk}^{(r-1)}) \right].
\end{cases}
$$

Applying $p_h$ and using (9), one arrives at the boundedness of $p_h u_h'(0)$ and $p_h u_h'(T)$ in $H$. The lemma is proved.

Now we state the approximation result.

**Theorem 1.** *Under the above hypotheses, if $u$ is the solution of problem* (1) *and $u_h, z_h$ are given by* (5) *and* (6) *respectively, then*

$$p_h u_h \to u \text{ in } L^2(0, T; H), \quad p_h u_h \rightharpoonup u \text{ weakly star in } L^\infty(0, T; H), \tag{10}$$

$$p_h u_h' \to u' \text{ in } C([0, T]; H), \quad A p_h z_h \rightharpoonup Au \text{ in } L^2(0, T; H). \tag{11}$$

*Proof.* By Lemma 1 we can choose subsequences denoted again $p_h u_h$ and $p_h z_h$ such that, for $h, k \to 0$,

$$p_h u_h \rightharpoonup \xi \text{ weakly star in } L^\infty(0, T; H), \tag{12}$$

$$p_h z_h \rightharpoonup \eta \text{ weakly star in } L^\infty(0, T; H), \tag{13}$$

$$A p_h z_h \rightharpoonup \kappa \text{ weakly star in } L^\infty(0, T; H). \tag{14}$$

We prove that $\xi = \eta$. Indeed, by (5) and (6) it follows that, $\forall y \in H$,

$$
\begin{aligned}
\int_0^T (p_h z_h(t) - p_h u_h(t), y)\, dt &= \sum_{r=1}^{N+1} \int_{(r-1)k}^{rk} \left(\frac{rk - t}{k}(p_h u_{hk}^{(r)} - p_h u_{hk}^{(r-1)}), y\right) dt \\
&= \frac{k}{2}\sum_{r=1}^{N+1} (p_h u_{hk}^{(r)} - p_h u_{hk}^{(r-1)}, y).
\end{aligned}
$$

Therefore, for every $y \in H$,

$$\int_0^T (p_h z_h(t) - p_h u_h(t), y)\, dt = \frac{k}{2}(p_h r_h b - p_h r_h a, y).$$

Using (12), (13), and (7), we find that $\xi = \eta$.

We show now that $p_h u_h'$ is strongly convergent in $C([0, T]; H)$. In order to prove this, subtract (4) for $h$ and $l$, and multiply their difference by $p_h u_{hk}^{(r)} - p_l u_{lk}^{(r)}$ in $H$. Summing up from $r = 1$ to $r = N$ and using (5), one obtains

$$
\sum_{r=1}^{N} \|p_h u_{hk}^{(r)} - p_h u_{hk}^{(r-1)} - p_l u_{lk}^{(r)} + p_l u_{lk}^{(r-1)}\|^2 + \alpha k^2 \sum_{r=1}^{N} \|p_h u_{hk}^{(r)} - p_l u_{lk}^{(r)}\|^2
$$

$$
\leq (p_h(u_{hk}^{(N+1)} - u_{hk}^{(N)}) - p_l(u_{lk}^{(N+1)} - u_{lk}^{(N)}), p_h u_{hk}^{(N)} - p_l u_{lk}^{(N)})
$$

$$- (p_h(u_{hk}^{(1)} - u_{hk}^{(0)}) - p_l(u_{lk}^{(1)} - u_{lk}^{(0)}), p_h u_{hk}^{(0)} - p_l u_{lk}^{(0)}). \qquad (15)$$

Remark that

$$k^2 \sum_{r=1}^{N} \|p_h u_{hk}^{(r)} - p_l u_{lk}^{(r)}\|^2 = k \int_0^T \|p_h z_h(s) - p_l z_l(s)\|^2 ds - k^2 \|p_h r_h b - p_l r_l b\|^2. \qquad (16)$$

It is easy to show that (5) leads to

$$\begin{cases} u_h'(0) = \frac{1}{k}(u_{hk}^{(1)} - u_{hk}^{(0)}), \quad u_h'(T) = \frac{1}{k}(u_{hk}^{(N+1)} - u_{hk}^{(N)}) \\ u_h'(t) = \frac{1}{k}(u_{hk}^{(r)} - u_{hk}^{(r-1)}), \quad (r-1)k < t < rk. \end{cases} \qquad (17)$$

Thus, we deduce that

$$\sum_{r=1}^{N} \|p_h u_{hk}^{(r)} - p_h u_{hk}^{(r-1)} - p_l u_{lk}^{(r)} + p_l u_{lk}^{(r-1)}\|^2$$

$$= k^2 \sum_{r=1}^{N+1} \|p_h u_h'(t) - p_l u_l'(t)\|^2 - k^2 \|p_h u_h'(T) - p_l u_l'(T)\|^2. \qquad (18)$$

Introducing (16), (17), and (18) into (15), one arrives at

$$k^2 \sum_{r=1}^{N+1} \|p_h u_h'(t) - p_l u_l'(t)\|^2 + \alpha k \int_0^T \|p_h z_h(s) - p_l z_l(s)\|^2 ds$$

$$\leq \alpha k^2 \|p_h r_h b - p_l r_l b\|^2 + k(p_h u_h'(T) - p_l u_l'(T), p_h r_h b - p_l r_l b)$$

$$- k(p_h u_h'(0) - p_l u_l'(0), p_h r_h a - p_l r_l a).$$

Therefore, using $(N+1)k = T$, we obtain

$$T\|p_h u_h'(t) - p_l u_l'(t)\|^2 + \alpha \int_0^T \|p_h z_h(s) - p_l z_l(s)\|^2 ds \leq \alpha k \|p_h r_h b - p_l r_l b\|^2$$

$$+ (p_h u_h'(T) - p_l u_l'(T), p_h r_h b - p_l r_l b) - (p_h u_h'(0) - p_l u_l'(0), p_h r_h a - p_l r_l a).$$

Lemma 1, together with (7) lead to the convergences

$$p_h u_h' \to \xi' \text{ in } C([0,T]; H), \quad p_h z_h \to \xi \text{ in } L^2(0,T; H) \text{ as } h, k \to 0. \qquad (19)$$

In addition,

$$p_h u_h \to \xi \text{ in } L^2(0,T; H) \text{ as } h, k \to 0. \qquad (20)$$

Indeed, since

$$\int_0^T \|p_h z_h(t) - p_h u_h(t)\|^2 dt = \sum_{r=1}^{N+1} \int_{(r-1)k}^{rk} \|\frac{rk-t}{k}(p_h u_{hk}^{(r)} - p_h u_{hk}^{(r-1)})\|^2 dt,$$

a simple computation leads to

$$\int_0^T \|p_h z_h(t) - p_h u_h(t)\|^2 dt = \frac{k}{3} \sum_{r=1}^{N+1} \|p_h u_{hk}^{(r)} - p_h u_{hk}^{(r-1)}\|^2.$$

Hence, with the aid of Lemma 1 and (19), we obtain (20).

We notice that (14), (19) and the fact that $A$ is demiclosed imply that $\kappa = A\xi$. Therefore

$$A p_h z_h \rightharpoonup A\xi \quad \text{in } L^2(0, T; H). \tag{21}$$

Let us prove that $\xi$ coincides with the solution $u$ of problem (1). To this end, consider $m \in \{0, \dots, N\}$ a fixed integer, and $t \in (mk, (m+1)k]$, $x \in H$ given elements. One multiplies (4) in $V_h$ by $r_h x \in V_h$ and sums up from 1 to $m$. We arrive at

$$(p_h u_{hk}^{(m+1)} - p_h u_{hk}^{(m)}, p_h r_h x) - (p_h u_{hk}^{(1)} - p_h u_{hk}^{(0)}, p_h r_h x)$$

$$= k^2 \sum_{r=1}^{m} (A p_h u_{hk}^{(r)}, p_h r_h x). \tag{22}$$

Denote by $M_l$ and $M_r$, the left-hand side and the right-hand side of this equality, respectively. Using (17) and (6) we find that

$$M_l = k(p_h u_h'(t) - p_h u_h'(0), p_h r_h x),$$

$$M_r = k \sum_{r=1}^{m} \int_{(r-1)k}^{rk} (A p_h z_h(s), p_h r_h x) \, ds = k \int_0^{mk} (A p_h z_h(s), p_h r_h x) \, ds.$$

Going back to (22), one obtains

$$(p_h u_h'(t), p_h r_h x) = (p_h u_h'(0), p_h r_h x) + \int_0^{mk} (A p_h z_h(s), p_h r_h x) \, ds, \tag{23}$$

for all $h, k$. Denoting by $d$, the weak limit of $p_h u_h'(0)$ (when $h, k \to 0$) and using (21), observe that the right-hand side of (23) tends to

$$g(t) = (d, x) + \int_0^t (A\xi(s), x) \, ds. \tag{24}$$

For each continuous function $\varphi : [0, T] \to \mathbb{R}$, we derive via the Lebesgue Theorem that

$$\int_0^t (p_h u_h'(s), p_h r_h x)\varphi(s) \, ds \rightarrow \int_0^t g(s)\varphi(s) \, ds.$$

Since $p_h r_h x \varphi(t) \rightarrow x\varphi(t)$ in $L^1(0, T; H)$, we deduce from (19) that $g(t) = (\xi'(t), x)$, a.e. $t \in [0, T]$. By (24) it follows that

$$(\xi'(t) - d, x) = \int_0^t (A\xi(s), x) \, ds,$$

which leads to $\xi''(t) = A\xi(t)$, a.e. $t \in [0, T]$. Thus, $\xi$ is a solution of problem (1). Since the solution of (1) is unique, we find that $\xi = u$.

Now (10), (11) follow from (12), (19), (20) and (21). The proof is complete.

## 4    Computational Issue

We set $\Omega = (0,1)$, $T = 1$ and consider the following linear bilocal problem: Find $u(x,t)$, $(x,t) \in Q = (0,1) \times (0,1)$, the solution of the equation

$$\frac{\partial^2 u}{\partial t^2} = Au, \tag{25}$$

where the operator $A$ is defined as $A = -\Delta = -\dfrac{\partial^2}{\partial x^2}$, with bilocal conditions: $u(x,0) = sin(\pi x)$, $u(x,1) = e^{\pi} \sin(\pi x)$, $x \in \Omega$ and homogeneous boundary condition: $u(x,t) = 0$, $(x,t) \in \{0,1\} \times (0,1)$.

The analytical solution for (25) is $u(x,t) = e^{\pi t} \sin(\pi x)$. We apply the successive over-relaxation (SOR) method to solve numerically the equation (25). Denoting $\beta = \frac{h}{k}$ and $u_{i,j} = u(ih, jk)$, then using finite differences, the SOR method for (25) is defined by

$$u_{i,j} = (1 - \omega)u_{i,j} + \frac{\omega}{2(1 + \beta^2)}(u_{i,j-1} + u_{i,j+1}) + \frac{\omega\beta^2}{2(1 + \beta^2)}(u_{i-1,j} + u_{i+1,j}).$$

Here $\omega$ represents the sensitivity of the successive over-relaxation factor and it must be inside the interval $(1,2)$. The idea is to choose a value for $\omega$ that will accelerate the rate of convergence of the iterates to the solution.

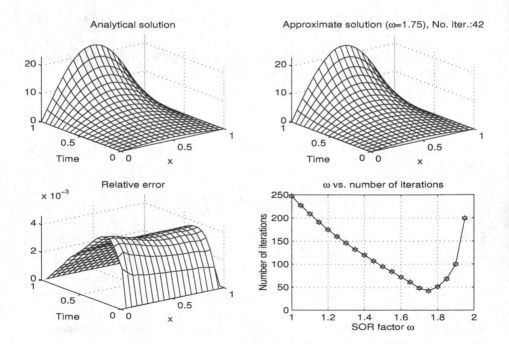

Fig. 1.

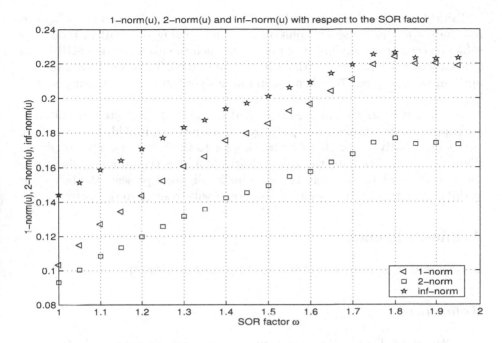

**Fig. 2.**

**Table 1.** Norms of the numerical solution evaluated for different values of $\omega$

| SOR factor $\omega$ | No. iter. | $\|u\|_1$ | $\|u\|_2$ | $\|u\|_\infty$ |
|---|---|---|---|---|
| 1.00 | 247 | 0.1033 | 0.0931 | 0.1442 |
| 1.05 | 227 | 0.1149 | 0.1005 | 0.1513 |
| 1.10 | 209 | 0.1271 | 0.1085 | 0.1587 |
| 1.15 | 191 | 0.1344 | 0.1134 | 0.1640 |
| 1.20 | 175 | 0.1436 | 0.1197 | 0.1708 |
| 1.25 | 160 | 0.1523 | 0.1257 | 0.1771 |
| 1.30 | 146 | 0.1607 | 0.1317 | 0.1832 |
| 1.35 | 132 | 0.1663 | 0.1357 | 0.1874 |
| 1.40 | 120 | 0.1756 | 0.1423 | 0.1939 |
| 1.45 | 107 | 0.1798 | 0.1453 | 0.1972 |
| 1.50 | 95 | 0.1854 | 0.1494 | 0.2012 |
| 1.55 | 84 | 0.1928 | 0.1547 | 0.2063 |
| 1.60 | 72 | 0.1968 | 0.1576 | 0.2093 |
| 1.65 | 61 | 0.2043 | 0.1630 | 0.2144 |
| 1.70 | 48 | 0.2109 | 0.1677 | 0.2194 |
| 1.75 | 42 | 0.2196 | 0.1743 | 0.2253 |
| 1.80 | 51 | 0.2241 | 0.1768 | 0.2265 |
| 1.85 | 68 | 0.2201 | 0.1735 | 0.2233 |
| 1.90 | 100 | 0.2203 | 0.1742 | 0.2229 |
| 1.95 | 199 | 0.2189 | 0.1733 | 0.2233 |

Table 1 contains computed values of the norms $\| \cdot \|_1$, $\| \cdot \|_2$ and $\| \cdot \|_\infty$ corresponding to the numerical solution $u$ when applying the SOR method. The table also contains the number of iterations when the extrapolation SOR factor $\omega$ is varying from 1 to 1.95 with an increment of 0.05. Thus, we remark that an optimal value of $\omega$ is 1.75, when the iterative algorithm converges using only 42 iterations.

The first two subplots in the upper part of Figure 1 illustrate the analytical and approximate solution. The plots appear as indistinguishable because of the very small relative error (the bottom subplot to the left part). The subplot in Figure 1 placed below to the right, clearly indicates the optimal value for $\omega$ assuring the best rate of convergence. Finally, Figure 2 presents the behaviour of different norms of the computed solution $u$ with respect to the factor $\omega$.

## Acknowledgement

This work was supported by the grant CNCSIS 128/2006, Romania.

## References

1. N. Apreutesei. Finite difference schemes with monotone operators, *Adv. Difference Eq.* 1 (2004) 11–22.
2. N. Apreutesei. Some difference equations with monotone boundary conditions, *Nonlin. Functional Anal. Appl.* 10, no. 2 (2005) 213–230.
3. J.P. Aubin. *Approximation of Elliptic Boundary-Value Problems.* John Wiley &Sons, Inc., New York, London, Sydney, 1972.
4. V. Barbu. A class of boundary problems for second order abstract differential equations, *J. Fac. Sci. Univ. Tokyo,* Sect 1, 19 (1972) 295–319.
5. V. Barbu. *Nonlinear Semigroups and Differential Equations in Banach Spaces.* Noordhoff, Leyden, 1976.
6. G. Geymonat and M. Sibony. Approximation des certaines équations paraboliques non linéaires, *Estratto da Calcolo,* 13(1976) 213–256.

# Target Detection and Parameter Estimation Using the Hough Transform*

Vera Behar[1], Lyubka Doukovska[2], and Christo Kabakchiev[2]

[1] Institute for Parallel Processing of Information, Bulgarian Academy of Sciences
"Acad G. Bonchev" St., Bl. 25-A, 1113 Sofia, Bulgaria
behar@bas.bg
[2] Institute of Information Technologies, Bulgarian Academy of Sciences
"Acad G. Bonchev" St., Bl. 2, 1113 Sofia, Bulgaria
ldoukovska@iit.bas.bg, ckabakchiev@iit.bas.bg

**Abstract.** In recent years, the algorithms that extract information for target's behavior through mathematical transformation of the signals reflected from a target, find ever-widening practical application. In this paper, a new two-stage algorithm for target detection and target's radial velocity estimation that exploits the Hough transform is proposed. The effectiveness of the proposed algorithm is formulated in terms of both quality parameters - the probability of detection and the accuracy of velocity estimation. The quality parameters are estimated using the Monte Carlo simulation approach.

## 1 Introduction

Recently, mathematical methods for extraction of useful data about the behavior of observed targets by mathematical transformation of received signals are being widely used for design of new highly effective algorithms for processing radar data. Modern methods for target detection and trajectory parameters estimation, which use mathematical transformation of received signals, allow new highly effective algorithms for radar signal processing to be designed. As a result, extremely precise estimates of moving target parameters can be obtained in conditions of very dynamic radar environment. An approach for linear trajectory target detection by means of Hough transformed coordinates, obtained for few sequential scans of the observation area is considered in papers [1]. According to this approach, the method for target detection uses a limited set of preliminary chosen patterns of a linear target trajectory. The set of target distance measurements is transformed to the pattern space (parameter space) by means of the Hough transform. The association of measurements to a special pattern is done by estimation of the data extracted from the target signals connected to this pattern. Thus the trajectory parameters of the targets, moving in the observation area, are determined through parameters of the corresponding pattern.

---

* This work is supported by IIT - 010059/2004 and Bulgarian NF "SR" with Grants TH - 1305/2003, MI - 1506/2005 and IF - 02-85/2005.

Different new results for statistical analysis of several algorithms, which use the Carlson's approach for target detection in the environment with and without pulse jamming, are obtained in [2,3,4,5,6]. All these papers consider only the detection performance without estimating the target velocity.

In this paper, a new two-stage algorithm for simultaneous target detection and its radial velocity estimation is proposed and tested. At the first stage, the target is detected using the Hough detector. At the second stage, the target radial velocity is found using the estimate of the Hough space parameter, which is found at the former stage. The effectiveness of the algorithm for the combined "detection-estimation" algorithm is formulated in terms of both quality parameters - the probability of detection and the accuracy of velocity estimation. The quality parameters of the detection algorithm are evaluated by means of Monte Carlo simulations.

## 2   Target Detection and Velocity Estimation

Consider radar that provides range, azimuth and elevation as a function of time. Time is sampled by the scan period, but resolution cells sample range, azimuth and elevation. The trajectory of a target that moves in the same "azimuth-elevation" resolution cell, is a straight line specified by several points $(r, t)$ in the range-time data space (r-t space), as it is shown in Fig. 1. Besides, in the (r-t) space, the target trajectory can be specified through the other parameters - the angle $\theta$ of its perpendicular from the data space origin and the distance $\rho$ from the origin to the line along the perpendicular.

The Hough transform maps all points $(r, t)$ of the (r-t) space into curves in the $(\rho - \theta)$ parameter space (Hough parameter space) as follows:

$$\rho = r \cos \theta + t \sin \theta, \tag{1}$$

where $r$ and $t$ are point coordinates in the (r-t) space, $\rho$ and $\theta$ are parameters that specify a straight line in the Hough parameter space.

The mapping of a line into the Hough parameter space can be considered as stepping through $\theta$ from $0°$ to $180°$ and calculating the corresponding $\rho$. The parameter space showing several sinusoids corresponding to different points in the (r-t) space is shown in Fig.2. The trigonometric manipulations of (1) leads to the other form of the Hough transform:

$$\rho = \sqrt{r^2 + t^2} \sin(\theta + \arctan \frac{r}{t}). \tag{2}$$

The mapping by (2) results in a sinusoid with an amplitude and phase dependent on coordinates in the (r-t) space of a point that is mapped. The maximum value for $|\rho|$ is equal to the length of the diagonal across the (r-t) data space. Equation (1) is the simpler version that is actually used for mapping.

Each $(\rho, \theta)$ point in the Hough parameter space corresponds to a single straight line in the (r-t) data space. Any one of the sinusoidal curves in the Hough parameter space corresponds to the set of all possible lines in the data space

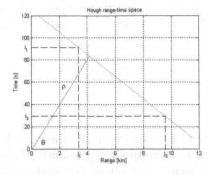

**Fig. 1.** Range-Time Space          **Fig. 2.** Hough Parameter Space

through the corresponding data point. If a straight line exists in the (r-t) space, this line is represented in the Hough parameter space as a point of intersection of all the mapped sinusoids. The slope of the target trajectory presented in Fig.1 is determined by the radial velocity of the target:

$$V = \frac{(j_2 - j_1)\delta R}{(i_2 - i_1)t_{sc}} = \tan\theta \frac{\delta R}{t_{sc}}, \tag{3}$$

where $(i_1\delta R, j_1 t_{sc})$ and $(i_2\delta R, j_2 t_{sc})$ are coordinates of two points in the (r-t) space that belong to the target trajectory, $\delta R$ is the range resolution cell and $t_{sc}$ is the scan period. According to equations (1, 2, 3), the structure of the Hough detector/estimator can be given by Fig.3.

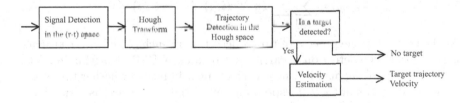

**Fig. 3.** The structure of Hough detector/estimator

In the (r-t) space, a low primary threshold is set, and any range-time cell with a value exceeding this threshold is mapped into the $(\rho - \theta)$ parameter space using (1). The parameter space is sampled in $\rho$ and $\theta$ dimensions. When a primary threshold crossing in any $(r, t)$ cell is mapped into the parameter space, its signal power is added into $(\rho, \theta)$ cells that intersect the corresponding sinusoidal curve in the parameter space. In this way, in the Hough parameter space the accumulator point at the intersection of several sinusoids will reach a high value. A secondary threshold applied to each point in the $(\rho - \theta)$ parameter space can be use to declare detection of a target trajectory. The point $(\hat{\rho}, \hat{\theta})$

where the secondary threshold is exceeded specifies the detected trajectory of a target. According to (3), the estimate of radial velocity of a target can be evaluated as:

$$\hat{V} = \frac{\delta R}{t_{sc}} \tan \hat{\theta}. \tag{4}$$

## 3   Simulation Algorithm

The effectiveness of target detection and velocity estimation provided by the algorithm proposed can be expressed in terms of two quality parameters - the detection probability characteristics and the accuracy of velocity estimation. In order to evaluate statistically these quality parameters, a simulation algorithm for testing of the new detector/estimator is developed:

1. At the first step the (r-t) space is quantized. To do this, the following data are needed - the range resolution cell ($\delta R$), the scan time ($t_{sc}$) and the number of scans ($N_{sc}$). The quantized (r-t) space is of size $[N \times M]$, where $N = N_{sc}$, $M = \frac{R_k - R_n}{\delta R}$, $R_k$ and $R_n$ are the limits of the observation space.
2. At the second step the hypothesis matrix ($IndTr$) is formed as follows:

$$\begin{cases} IndTr(i,j) = 1, \, j = Vt_{sc}i/\delta R, \\ IndTr(i,j) = 0, \, j \neq Vt_{sc}i/\delta R. \end{cases} \tag{5}$$

The number of nonzero elements $K_{target}$ in the hypothesis matrix $IndTr$ equals the number of all target positions in the (r-t) space:

$$K_{target} = \sum_{i=1}^{N} \sum_{j=1}^{N} IndTr(i,j)/IndTr(i,j) \neq 0 \tag{6}$$

3. At the third step the process of target detection in each cell of the (r-t) space is simulated. The detection is carried out by a CA CFAR algorithm [6]. As a result, the following matrix whose each element indicates whether the target is detected or not in the corresponding cell of the (r-t) space, is formed:

$$Det^q(i,j) = \begin{cases} 1, \, target \quad is \quad detected, \\ 0, \, target \quad is \quad not \quad detected, \end{cases} \tag{7}$$

where $q$ is the simulation cycle number.
4. At the fourth step, the ($\rho - \theta$) parameter space is quantized. It is a matrix of size $[K \times L]$. The parameters $K$ and $L$ are determined by the number of discrete values of the $\theta$ parameter, which is sampled in the interval ($\theta_1, \theta_2$) with sampling step $\Delta\theta$, and the size of the (r-t) space.

$$K = 2\sqrt{N^2 + M^2}; \quad L = (\theta_2 - \theta_1)/\Delta\theta. \tag{8}$$

5. At the fifth step all the nonzero elements of the matrix $Det^q$ are performed using the Hough transform. In such a way, the (r-t) space is mapped into the ($\rho - \theta$) parameter space. The resulting matrix is $\{Ht^q\}_{K,L}$.

6. At the next sixth step a target trajectory is detected. This is done by comparing the value of each element of the parameter space, i.e. of matrix $\{Ht^q\}_{K,L}$, with the fixed threshold $T_M$. It means that the decision rule "$T_M$ out of $N_{sc}$" is applied to each element in the parameter space. According to this criterion, the linear target trajectory specified as a point $(\hat{\rho}, \hat{\theta})$ in the Hough parameter space is detected if and only if the value $Ht(\hat{\rho}, \hat{\theta})$ exceeds the threshold $T_M$.

$$DetHo^q(i,j) = \begin{cases} 1, & Ht^q(i,j) > T_M, \\ 0, & otherwise. \end{cases} \tag{9}$$

7. The seventh step is performed in case when a target trajectory is detected at the former step. At this step the target radial velocity is estimated as follows:

$$\hat{V} = \frac{\delta R}{t_{sc}} \tan(\hat{\theta}), \tag{10}$$

where $\hat{\theta}$ is the Hough parameter, where the target trajectory is detected.
8. In order to estimate both the probability characteristics and the accuracy of the velocity estimation, steps 1-7 are repeated $N_q$ times.
The false alarm probability in the (r-t) space is estimated as:

$$\hat{P_{fa}} = \frac{1}{(NM - K_{target})N_q} \sum_{i=1}^{N} \sum_{j=1}^{M} \sum_{l=1}^{N_q} \{Det^q(i,j)/IndTr(i,j) \neq 1\}. \tag{11}$$

The target detection probability in the (r-t) space is estimated as:

$$\hat{P_d} = \frac{1}{K_{target} + N_q} \sum_{i=1}^{N} \sum_{j=1}^{M} \sum_{l=1}^{N_q} \{Det^q(i,j)/IndTr(i,j) = 1\}. \tag{12}$$

The false alarm probability in the $(\rho - \theta)$ space is estimated as:

$$\hat{P}_{FA} = \frac{1}{KLN_q} \sum_{i=1}^{K} \sum_{j=1}^{L} \{DetHo^q(i,j)/i \neq I, j \neq J\}. \tag{13}$$

The probability of trajectory detection in the $(\rho - \theta)$ space is estimated as:

$$\hat{P}_D = max_{I,J} \sum_{l=1}^{N_q} DetHo^q(i,j). \tag{14}$$

## 4   Simulation Results

In this section we apply the simulation algorithm described above to typical surveillance radar. The goal is to analyze statistically the algorithm for target detection and velocity estimation. In order to obtain the statistical estimates of the basic quality parameters (probability characteristics and accuracy of velocity estimation), the following data were used in the simulations: scan period - $t_{sc} = 6s$;

number of scans - $N_{sc} = 20$; range resolution cell - $\delta R = 150m$ and $1500m$; size of the (r-t) space - $128 \times 20$ elements; interval for variation of the $\theta$ parameter - $\theta_1 = 0°, \theta_2 = 180°$; probability of signal detection in the (r-t) space - $P_D = 0, 9$; probability of false alarm per cell in the range-time space - $P_{fa} = 0.0001$; decision rule for trajectory detection in the Hough parameter space - $T_M/N_q = 7/9$ and $T_M/N_q = 7/20$; target velocity - $V_{target} = 333m/s$; the average signal-to-noise ratio - SNR=37dB; number of simulation cycles - 1000.

According to (10), the theoretical accuracy of the velocity estimation ($\Delta V$) can be expressed as:

$$\Delta V_i = V_{i+1} - V_i = \frac{\delta R}{t_{sc}}(\tan(\theta_{i+1}) - \tan(\theta_i)), \qquad (15)$$

where $\theta_{i+1} = i\Delta\theta$, $\theta_i = (i-1)\Delta\theta$, $i = 1, 2, \ldots, L$.

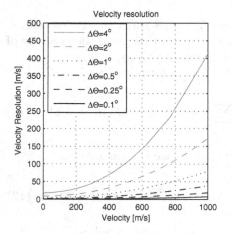

**Fig. 4.** Velocity resolution ($\delta R = 1500m$)

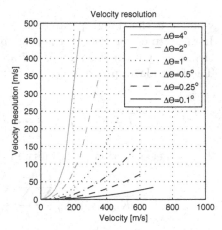

**Fig. 5.** Velocity resolution ($\delta R = 150m$)

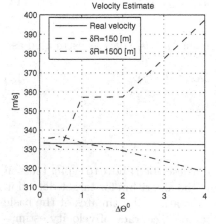

**Fig. 6.** Average velocity estimate

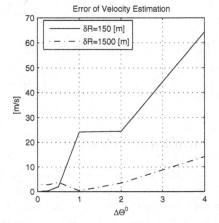

**Fig. 7.** Absolute error of estimation

Therefore, the accuracy of estimation is mainly determined by the sampling rate of the parameter $\theta$ and also depends on the sampling interval of the (r-t) space. It means that for given $\delta R$ and $t_{sc}$, the sampling interval $\Delta\theta$ should be chosen so as to meet the requirements for accuracy of the velocity estimation.

The theoretical accuracy that can be reached depending on the sampling interval of the parameter $\theta$ is presented in Fig. 4 and Fig.5 - for the range resolution cell of $1500m$ and $150m$, respectively. The theoretical accuracy of velocity estimation as a function of the target velocity to be estimated is presented for six different variants of $\Delta\theta$.

The averaged velocity estimates, obtained in simulations for a target velocity are shown in Fig. 6. The velocity estimates are plotted as a function of the sampling interval $\Delta\theta$. For comparison, the averaged velocity estimate is plotted for two values of the range resolution cell - $150m$ and $1500m$. The absolute errors of velocity estimation, calculated for two values of the range resolution cell are shown in Fig. 7. They are also plotted as a function of the sampling interval $\Delta\theta$.

The numerical results that correspond to the graphical results are summarized in Table 1 and Table 2. The velocity estimates calculated for a target moving in straight line with velocity of $333m/s$ are presented for six sampling interval of the parameter $\theta$. In addition, Table 1 and Table 2 contain the estimates of both probability characteristics, the probability of signal detection in the (r-t) space ($P_d$) and the probability of trajectory detection in the Hough parameter space ($P_D$).

For this example the (r-t) space contains 128 range resolution cells. It means that the general number of points specifying the target trajectory in the (r-t) space can be calculated as:

$$N_q = \begin{cases} \frac{128\delta R}{V_{target}t_{sc}}, & if \quad \frac{128\delta R}{V_{target}t_{sc}} < 20, \\ 20, & otherwise. \end{cases} \tag{16}$$

If $t_{sc} = 6s$ and $V_{target} = 333m/s$, then $N_q = 9$ - for $\delta R = 150m$ and $N_q = 20$ - for $\delta R = 1500m$.

Therefore, the decision rule applied to trajectory detection in the Hough parameter space is "7 out of 9" - in case of the range resolution cell of $150m$, and "7 out of 20" - in case of the range resolution cell of $1500m$. For that reason the probability of trajectory detection presented in Table 2 is greater than that presented in Table 1.

**Table 1.** Velocity and probability estimates for value $\delta R = 150m$

| $\Delta\theta^\circ$ | Real velocity - $V_{real}$ Real parameter - $\theta^\circ_{real}$ | $V_{ave}$ [m/s] | $\Delta V$ [m/s] | $P_d$ | $P_D$ |
|---|---|---|---|---|---|
| 0.1 | | 333.1697 | 0.1697 | 0.932 | 0.956 |
| 0.25 | $V_{real} = 333m/s$ | 333.2808 | 0.2808 | 0.909 | 0.940 |
| 0.5 | | 330.9933 | 2.0067 | 0.912 | 0.935 |
| 1.0 | | 357.2151 | 24.2151 | 0.916 | 0.474 |
| 2.0 | $\theta^\circ_{real} = 85.7066^\circ$ | 357.5167 | 24.5167 | 0.914 | 0.485 |
| 4.0 | | 397.3636 | 64.3636 | 0.912 | 0.006 |

Table 2. Velocity and probability estimates for value $\delta R = 1500m$

| $\Delta\theta°$ | Real velocity - $V_{real}$ Real parameter - $\theta°_{real}$ | $V_{ave}$ [m/s] | $\Delta V$ [m/s] | $P_d$ | $P_D$ |
|---|---|---|---|---|---|
| 0.1 | | 335.844 | 2.844 | 0.912 | 1 |
| 0.25 | $V_{real} = 333m/s$ | 335.7617 | 2.7617 | 0.915 | 1 |
| 0.5 | | 336.7357 | 3.7357 | 0.918 | 1 |
| 1.0 | | 333.3181 | 0.3181 | 0.917 | 1 |
| 2.0 | $\theta°_{real} = 53.1026°$ | 329.5216 | 3.4784 | 0.919 | 1 |
| 4.0 | | 318.832 | 14.168 | 0.920 | 1 |

# 5    Conclusions

A new algorithm for detection and velocity estimation of a target moving in straight line in the same azimuth towards or downwards the radar is presented and evaluated in the paper. In order to test and study the new algorithm, the simulation algorithm based on the Monte Carlo approach is developed. The graphical and numerical result show that the quality parameters strongly depend on discretization not only of the (r-t) space but of the Hough parameter space as well. It is also shown that the discretization of both spaces (r-t and Hough) should be optimized in order to meet the requirements for both quality parameters - the probability of target trajectory detection and the accuracy of velocity estimation.

# References

1. Carlson B., Evans E., Wilson S. - Search Radar Detection and Track with the Hough Transform, Parts I, II, III IEEE Trans., vol. AES, pp. 102-124.
2. Behar, V., Chr. Kabakchiev and L. Doukovska - Target Trajectory Detection in Monopulse Radar by Hough Transform, Compt. Rend. Acad. Bulg. Sci., vol. 53, 8, 2000, pp. 45-48.
3. Kabakchiev Chr., I. Garvanov and L. Doukovska - Excision CFAR BI Detector with Hough Transform in Presence of Randomly Arriving Impulse Interference, Proc. of the International Radar Symposium - IRS'05, Berlin, Germany, 2005, pp. 259-264.
4. Kabakchiev Chr., L. Doukovska and I. Garvanov - Hough Radar Detectors in Conditions of Intensive Pulse Jamming, Sensors & Transducers Magazine, Special Issue, "Multisensor Data and Information Processing", 2005, pp. 381-389.
5. Doukovska L. and Chr. Kabakchiev - Performance of Hough Detectors in Presence of Randomly Arriving Impulse Interference, Proc. of the International Radar Symposium - IRS'06, Krakow, Poland, 2006, pp. 473-476.
6. Kabakchiev Chr., L. Doukovska and I. Garvanov - Cell Averaging Constant False Alarm Rate Detector with Hough Transform in Randomly Arriving Impulse Interference, Cybernetics and Information Technologies, vol.6, 1, 2006, pp. 83-89.

# Mechanical Failure in Microstructural Heterogeneous Materials*

Stéphane Bordas[1], Ronald H.W. Hoppe[2,3], and Svetozara I. Petrova[2,4]

[1] Laboratory of structural and continuum mechanics, EPFL, Lausanne, Switzerland
[2] Institute of Mathematics, University of Augsburg, 86159 Augsburg, Germany
[3] Department of Mathematics, University of Houston, TX 77204–3008, USA
[4] Institute for Parallel Processing, BAS, Block 25A, 1113 Sofia, Bulgaria

**Abstract.** Various heterogeneous materials with multiple scales and multiple phases in the microstructure have been produced in the recent years. We consider a mechanical failure due to the initiation and propagation of cracks in places of high pore density in the microstructures. A multi–scale method based on the asymptotic homogenization theory together with the mesh superposition method ($s$-version of FEM) is presented for modeling of cracks. The homogenization approach is used on the global domain excluding the vicinity of the crack where the periodicity of the microstructures is lost and this approach fails. The multiple scale method relies on efficient combination of both macroscopic and microscopic models. The mesh superposition method uses two independent (global and local) finite element meshes and the concept of superposing the local mesh onto the global continuous mesh in such a way that both meshes not necessarily coincide. The homogenized material model is considered on the global mesh while the crack is analyzed in the local domain (patch) which allows to have an arbitrary geometry with respect to the underlying global finite elements. Numerical experiments for biomorphic cellular ceramics with porous microstructures produced from natural wood are presented.

## 1 Global–Local Approach for Heterogeneous Materials

Consider a domain $\Omega \subset \mathcal{R}^d$, $d = 2, 3$, occupied by a heterogeneous material with microstructures of periodically distributed constituents. Suppose that the boundary of $\Omega$, denoted by $\Gamma$, consists of a prescribed displacement boundary $\Gamma_D$ (meas $\Gamma_D > 0$) and a prescribed traction boundary $\Gamma_T$, such that $\Gamma = \Gamma_D \cup \Gamma_T$, $\Gamma_D \cap \Gamma_T = \emptyset$, as shown in Figure 1.

Assume that the periodic cells in the macrostructure are infinitely many but infinitely small and repeated periodically through the medium. The unit microstructure consists of different material constituents and a pore. Both the

---

* This work has been partially supported by the German National Science Foundation (DFG) under Grant No.HO877/5-3. The third author has also been supported in part by the Bulgarian Ministry for Education and Science under Grant I1402/2004.

T. Boyanov et al. (Eds.): NMA 2006, LNCS 4310, pp. 533–541, 2007.

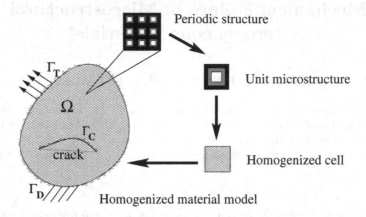

**Fig. 1.** Crack in the macroscopic homogenized material model

macroscopic and microscopic scales are well separated, i.e., the size of the microstructure in the heterogeneous material is much smaller than those of the macroscopic component. The asymptotic homogenization theory [1,2] is applied to find the effective (homogenized) properties of the material and to derive the homogenized macroscopic model. Details are given in the next Section. The main idea for the homogenization of a heterogeneous material with a periodical distribution of microstructures is illustrated in Figure 1.

We allow the domain $\Omega$ to contain discontinuities and consider the crack problem with a crack $\Gamma_C$, see Figure 1. The crack is a multiscale effect which typically appears in regions with microstructures of increasing porosity. The periodicity fails for those microstructures cut by the crack. Therefore, a new finite element analysis has to incorporate microstructural information about the nucleation and growth of micropores. We rely on the mesh superposition method (known as $s$-version of FEM), first developed in [7,8]. The method is based on a finite element approximation by using two independent meshes: the global mesh for the whole domain (also called in the literature a background mesh) and a local mesh in the critical region near the crack (also called a patch mesh). The local mesh is arbitrarily superimposed onto the global mesh without taking care of the matching between nodes in both meshes.

Consider the following governing equations in the domain $\Omega$

$$-\nabla \cdot \boldsymbol{\sigma} = \boldsymbol{b} \qquad \text{in } \Omega \tag{1}$$

$$\boldsymbol{u} = \boldsymbol{g} \qquad \text{on } \Gamma_D \tag{2}$$

$$\boldsymbol{\sigma} \cdot \boldsymbol{n} = \boldsymbol{t} \qquad \text{on } \Gamma_T, \tag{3}$$

where $\boldsymbol{\sigma}$ is the second order symmetric stress tensor, $\boldsymbol{b}$ is the body force, $\boldsymbol{g}$ is the prescribed displacement on $\Gamma_D$, $\boldsymbol{t}$ is the prescribed traction on $\Gamma_T$, and $\boldsymbol{n}$ is the unit normal to the boundary $\Gamma_T$. A traction–free surface $\boldsymbol{\sigma} \cdot \boldsymbol{n} = \boldsymbol{0}$ is assumed on the crack $\Gamma_C$. In case of small strains and displacements, the second order strain tensor $\boldsymbol{e}$ is given by

$$e = e(u) = \left(\nabla u + (\nabla u)^T\right)/2, \tag{4}$$

where $\nabla u$ is the gradient operator. If a linear elasticity is assumed, the constitutive relation is presented by the linearized Hooke law $\sigma = E\!:\!e$, where $E$ is the fourth–order elasticity tensor which depends on the material constants like Young's modulus and Poisson's ratio.

The weak form of the governing equation (1) reads: Find $u \in U$, such that

$$\int_\Omega e(v)\!:\!\sigma(u)\,d\Omega = \int_\Omega b\cdot v\,d\Omega + \int_{\Gamma_T} t\cdot v\,d\Gamma, \qquad \forall\,v \in U_0, \tag{5}$$

where the set of admissible displacement fields is defined by

$$U = \{v\,|\,v \in V,\ v = g \ \text{on}\ \Gamma_D,\ v \ \text{discontinuous on}\ \Gamma_C\} \tag{6}$$

and the test function space is defined by

$$U_0 = \{v\,|\,v \in V,\ v = 0 \ \text{on}\ \Gamma_D,\ v \ \text{discontinuous on}\ \Gamma_C\}. \tag{7}$$

The space $V$ is related to the regularity of the solution in $\Omega$ and allows for discontinuous functions across the crack. At each point $x \in \Omega$ we consider a finite element discretization of (5) with a basis taken from the test function space $U_0$ and nodal shape functions $N(x)$ constructed by the Galerkin method.

We denote the local critical region onto which the local mesh is superimposed by $\Omega^L$, $\Omega^L \subset \Omega$, a subset of $\Omega$, containing the crack. Let $\Omega^G = \Omega \setminus \Omega^L$ be the rest of the domain excluding discontinuities. The domains $\Omega$ and $\Omega^L$ are discretized independently by separate sets of finite elements $\Omega_e^G$ and $\Omega_e^L$, such that $\bigcup \Omega_e^G = \Omega$ and $\bigcup \Omega_e^L = \Omega^L$. Here, the superscript $G$ relates to the global (underlying) mesh and $L$ to the local (superimposed) mesh. $\Gamma^{GL}$ is the boundary between the two meshes excluding external boundaries, i.e., $\Gamma \cap \Gamma^{GL} = \emptyset$.

Let $u^G$ be the global displacement field defined in $\Omega$ and $u^L$ be the local displacement field defined in the local region $\Omega^L$. Note that the superimposed field $u^L$ is in general discontinuous due to a discontinuity across the crack faces. The total displacement field $u$ is constructed by superposition of both displacement fields on the separate meshes and can be written as follows

$$u = \begin{cases} u^G & \text{on}\ \Omega^G,\ \Gamma^{GL} \\ u^G + u^L & \text{on}\ \Omega^L. \end{cases} \tag{8}$$

To insure displacement compatibility between the global and local meshes, we assume homogeneous boundary conditions on the boundary of the patch, i.e., $u^L = 0$ on $\Gamma^{GL}$. Denote by $B^G$ and $B^L$ the discretized gradient operators (also called strain–displacement matrices) for the global and local meshes, respectively. Then, the strain can be expressed as follows

$$e = \begin{cases} e^G = B^G u^G & \text{on}\ \Omega^G \\ e^L = B^G u^G + B^L u^L & \text{on}\ \Omega^L. \end{cases} \tag{9}$$

Based on the Hooke law we get the following constitutive relations

$$\boldsymbol{\sigma} = \boldsymbol{E} : \boldsymbol{e} = \begin{cases} \boldsymbol{\sigma}^G = \boldsymbol{E}^G : (\boldsymbol{B}^G \boldsymbol{u}^G) & \text{on } \Omega^G \\ \boldsymbol{\sigma}^L = \boldsymbol{E}^L : (\boldsymbol{B}^G \boldsymbol{u}^G + \boldsymbol{B}^L \boldsymbol{u}^L) & \text{on } \Omega^L, \end{cases} \tag{10}$$

where $\boldsymbol{E}^G$ and $\boldsymbol{E}^L$ are the elasticity tensors corresponding to the different constitutive laws. By using shape functions $\boldsymbol{N}^G(\boldsymbol{x})$ on the global mesh and shape functions $\boldsymbol{N}^L(\boldsymbol{x})$ on the local mesh, one can get from the standard weak form (5) the following two equations

$$\int_{\Omega^G} \boldsymbol{B}^G(\boldsymbol{x}) : \boldsymbol{\sigma}^G(\boldsymbol{u}) \, d\Omega = \int_{\Omega^G} \boldsymbol{N}^G(\boldsymbol{x}) \cdot \boldsymbol{b} \, d\Omega + \int_{\Gamma_T^G} \boldsymbol{N}^G(\boldsymbol{x}) \cdot \boldsymbol{t} \, d\Gamma, \tag{11}$$

$$\int_{\Omega^L} \boldsymbol{B}^L(\boldsymbol{x}) : \boldsymbol{\sigma}^L(\boldsymbol{u}) \, d\Omega = \int_{\Omega^L} \boldsymbol{N}^L(\boldsymbol{x}) \cdot \boldsymbol{b} \, d\Omega + \int_{\Gamma_T^L} \boldsymbol{N}^L(\boldsymbol{x}) \cdot \boldsymbol{t} \, d\Gamma, \tag{12}$$

where $\Gamma_T^G = \Gamma_T \cap \Omega^G$ and $\Gamma_T^L = \Gamma_T \cap \Omega^L$. Substituting (8)–(10) in equations (11)–(12), we obtain the following discrete system

$$\begin{bmatrix} \boldsymbol{K}^G & \boldsymbol{K}^{GL} \\ (\boldsymbol{K}^{GL})^T & \boldsymbol{K}^L \end{bmatrix} \begin{Bmatrix} \boldsymbol{u}^G \\ \boldsymbol{u}^L \end{Bmatrix} = \begin{Bmatrix} \boldsymbol{f}^G \\ \boldsymbol{f}^L \end{Bmatrix}, \tag{13}$$

where $\boldsymbol{K}^G$ and $\boldsymbol{K}^L$ are the stiffness matrices corresponding to the global and local meshes, respectively, and $\boldsymbol{K}^{GL}$ is the matrix corresponding to the interaction between the two meshes, namely

$$\boldsymbol{K}^G = \int_{\Omega^G} (\boldsymbol{B}^G(\boldsymbol{x}))^T \boldsymbol{E}^G \, \boldsymbol{B}^G(\boldsymbol{x}) \, d\Omega + \int_{\Omega^L} (\boldsymbol{B}^G(\boldsymbol{x}))^T \boldsymbol{E}^L \, \boldsymbol{B}^G(\boldsymbol{x}) \, d\Omega, \tag{14}$$

$$\boldsymbol{K}^{GL} = \int_{\Omega^L} (\boldsymbol{B}^G(\boldsymbol{x}))^T \boldsymbol{E}^L \, \boldsymbol{B}^L(\boldsymbol{x}) \, d\Omega, \quad \boldsymbol{K}^L = \int_{\Omega^L} (\boldsymbol{B}^L(\boldsymbol{x}))^T \boldsymbol{E}^L \, \boldsymbol{B}^L(\boldsymbol{x}) \, d\Omega. \tag{15}$$

The force vectors $\boldsymbol{f}^G$ and $\boldsymbol{f}^L$ are computed from the right–hand sides of (11) and (12), respectively. Note that $\boldsymbol{N}^G(\boldsymbol{x})$ are the shape functions corresponding to finite elements in the global mesh on which continuous displacement field $\boldsymbol{u}^G$ is considered. Furthermore, $\boldsymbol{N}^L(\boldsymbol{x})$ are the discontinuous shape functions of the elements chosen on the local domain to model the crack. The elements of the global and local meshes should not coincide. The cracked mesh is superimposed on the continuous mesh in $\Omega^L$ by using the s-method. The main difficulty of this method is the numerical integration based on Gauss quadratures when solving the system (13). An attractive approach is recently proposed in [10] where only the near–tip crack fields are modeled on a superimposed patch (overlaid mesh) and the rest of the crack is treated within the framework of the eXtended Finite Element Method (XFEM) by introducing additional discontinuous enrichment functions for the elements completely cut by the crack, see [11]. The latter method is also used in our numerical experiments and briefly explained in Section 3.

## 2  Multi–scale Method

In this section, we describe the multi–scale method based on the asymptotic homogenization theory together with the mesh superposition method. For more details we refer the reader to [16]. The homogenization approach is used on the global domain excluding the vicinity of the crack where the periodicity of the microstructures is lost and this approach is not applicable. The crack is considered in the local domain (patch). Two independent (global and local) meshes are generated in the global and local domains, respectively. The patch is allowed to have an arbitrary geometry with respect to the underlying global finite elements. The total displacement field (8) is approximated by adding global (underlying) and local (superimposed) fields and hence, it is discontinuous across the crack.

A double scale asymptotic expansion for the displacement field and a homogenization procedure by taking a zero limit of the scale ratio are applied to come up with computationally feasible macromodels [1,2]. The homogenized macroscopic problem is involved in the governing equation (1). Furthermore, introducing global and local meshes, the global displacement field $\boldsymbol{u}^G$ is expressed by the homogenized displacement, the leading term in the asymptotic expansion form (see, e.g., [9]). The homogenized elasticity problem is transformed to solving the system (13) with a symmetric and usually, sparse, stiffness matrix coupling the integrands on the global and local meshes. Note that the elasticity tensor $\boldsymbol{E}^G$ in the expression (14) is replaced now by the homogenized elasticity tensor $\boldsymbol{E}^H$ with components computed as follows

$$E^H_{ijkl} = \frac{1}{|Y|} \int_Y \left( E_{ijkl}(\boldsymbol{y}) - E_{ijpq}(\boldsymbol{y}) \frac{\partial \xi^{kl}_p}{\partial y_q} \right) dY, \qquad (16)$$

where $E_{ijkl}$ are the elasticity coefficients corresponding to the different material layers in the microstructure. The periodic functions $\boldsymbol{\xi}^{kl}$ (also referred to as characteristic displacements) satisfy the following elasticity equation in the microscopic unit cell

$$\int_Y E_{ijpq}(\boldsymbol{y}) \frac{\partial \xi^{kl}_p}{\partial y_q} \frac{\partial \phi_i}{\partial y_j} dY = \int_Y E_{ijkl}(\boldsymbol{y}) \frac{\partial \phi_i}{\partial y_j} dY, \qquad (17)$$

where $\phi \in \boldsymbol{H}^1(Y)$ is an arbitrary $Y$–periodic variational function. The multi–scale procedure is realized by the following **Multi–Scale Algorithm (MSA)**

> Step 1.  Select a unit microstructure $Y$ in the heterogeneous material.
> Step 2.  Solve (17) to find the characteristic displacement fields $\boldsymbol{\xi}^{kl}$.
> Step 3.  Compute the homogenized coefficients (16) and set $\boldsymbol{E}^G = \boldsymbol{E}^H$.
> Step 4.  Generate a global mesh in $\Omega$ on the macroscopic homogenized model.
> Step 5.  Introduce a local (discontinuous) mesh near the crack $\Gamma_C$.
> Step 6.  Solve (13) to determine the displacements $\boldsymbol{u}^G$ and $\boldsymbol{u}^L$.
> Step 7.  Substitute $\boldsymbol{u}^G$ and $\boldsymbol{u}^L$ in (9) and (10) to find the strains and stresses.

# 3   Crack Modeling with Partition of Unity Enrichment

The extended finite element method [3,11,15] allows treating crack problems without meshing the discontinuity surface. This is possible through enrichment of the standard polynomial finite element space with special functions: discontinuous, to account for the displacement jump, and crack–tip fields to reduce the mesh density required for accurate fracture parameter computations. The method is now becoming quite mature, and has already been applied to industrial fracture mechanics problems [5].

Some elements are split by the crack and others contain the crack tips. Nodes whose support is cut by a crack are in set $\mathcal{N}_{cr}$, while nodes whose support contains one tip are in the set $\mathcal{N}_{tip}$, as in Figure 2.

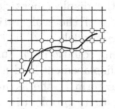
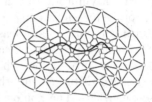

**Fig. 2.** Selection of enriched nodes. Circled nodes (set of nodes $\mathcal{N}_{cr}$) are enriched with the step function whereas the squared nodes (set of nodes $\mathcal{N}_{tip}$) are enriched with the crack tip functions: a) on structured mesh; b) on unstructured mesh.

The XFEM approximation reads

$$\mathbf{u}^h(\mathbf{x}) = \sum_{I \in \mathcal{N}} N_I(\mathbf{x})\mathbf{u}_I + \sum_{J \in \mathcal{N}_{cr}} \widetilde{N}_J(\mathbf{x})(H(\mathbf{x}) - H(\mathbf{x}_J))\mathbf{a}_J$$

$$+ \sum_{K \in \mathcal{N}_{tip}} \widetilde{N}_K(\mathbf{x}) \sum_{\alpha=1}^{4} (B_\alpha(\mathbf{x}) - B_\alpha(\mathbf{x}_K))\mathbf{b}_{\alpha K}, \qquad (18)$$

where $N_I(\mathbf{x})$ and $\widetilde{N}_J(\mathbf{x})$ are finite element shape functions, while $\mathbf{u}_I$, $\mathbf{a}_J$ and $\mathbf{b}_{\alpha K}$ are the displacement and enrichment nodal variables, respectively. Note that the shape functions $\widetilde{N}_J(\mathbf{x})$ associated with the enrichment can differ from the shape functions $N_I(\mathbf{x})$ used for the standard part of the displacement approximation. $H(\mathbf{x})$ is the modified Heaviside function which takes on the value $+1$ above the crack and $-1$ below the crack and $B_\alpha(\mathbf{x})$ is a basis that spans the near tip asymptotic field:

$$\mathbf{B} \equiv [B_1, B_2, B_3, B_4] = \left[\sqrt{r}\sin\frac{\theta}{2}, \sqrt{r}\cos\frac{\theta}{2}, \sqrt{r}\sin\frac{\theta}{2}\cos\theta, \sqrt{r}\cos\frac{\theta}{2}\cos\theta\right] \quad (19)$$

From the enriched approximation (18), Bubnov–Galerkin procedure gives discrete equations of the form $\mathbf{Kd} = \mathbf{f}$. Numerical integration for split elements is done here by partitioning the elements into sub–triangles. Interested readers can refer to [4,13] for details.

# 4    Numerical Experiments

The fracture mechanics computations in this section are performed using the XFEM library OPENXFEM++ [6][1]. Our numerical examples concern biomorphic silicon carbide (SiC) cellular ceramics with porous microstructures produced from natural wood. The open porous system of tracheidal cells which provide the transportation path for water and minerals in the living plants is accessible for infiltration of various liquid or gaseous metals (see, e.g., [14] for the production process). Numerical experiments for the homogenized coefficients in a 2-D material workpiece are given in [9].

Consider a stationary microstructure with a geometrically simple tracheidal periodicity cell $Y = [0,1]^2$ consisting of an outer layer of carbon (C), interior layer of silicon carbide (SiC), and a centered pore channel (P, no material), see Figure 3a). One can also deal with the so-called pure SiC ceramics when enough silicon is infiltrated in the pore channel until the complete reaction between the carbon and silicon, see Figure 3b). The Young modulus $E$ (in GPa) and the Poisson ratio $\nu$ of our two materials are, respectively, $E = 10$, $\nu = 0.22$ for carbon and $E = 410$, $\nu = 0.14$ for SiC.

**Fig. 3.** a) Unit cell $Y = P \cup SiC \cup C$; b) Pure SiC ceramics: $Y = P \cup SiC$

As a first attempt we consider a crack in the macroscopic homogenized model. All computations are performed assuming the theory of Linear Elastic Fracture Mechanics. In Table 1 we report the evaluated data for the homogenized Young modulus $E^H$ (in GPa) and the homogenized Poisson ratio $\nu^H$. In case of SiC ceramics the coefficients are computed for equal widths of the SiC and carbon layers, see Table 1a). The density of the SiC layer and the density of the carbon layer are denoted, respectively, by $\mu_{SiC}$ and $\mu_{carbon}$. Note that the porosity is determined by: meas$(Y) - \mu_{SiC} - \mu_{carbon}$. In case of pure SiC ceramics, see Table 1b), the porosity is given by: meas$(Y) - \mu_{SiC}$.

In all problems, the energy release rate $G = \frac{1}{E'}(K_I^2 + K_{II}^2)$ with $E' = E^H / (1 - (\nu^H)^2)$ is computed for both crack tips. The stress intensity factors (SIFs), $K_I$ and $K_{II}$, for mode I and II, respectively, are determined using the domain form of the interaction integrals [12]. We are interested in the effect of porosity on the energy release rate. Consider a center crack in an infinite plate under remote, unit, uniaxial tension. The geometry of the plate is (2x6), the

---

[1] Available on request.

**Table 1.** Homogenized $E$ and $\nu$: a) SiC ceramics; b) Pure SiC ceramics

| porosity | $\mu_{SiC}$ | $\mu_{carbon}$ | $E^H$ (GPa) | $\nu^H$ |
|---|---|---|---|---|
| 0.81 | 0.0925 | 0.0925 | 89.42 | 0.035 |
| 0.64 | 0.17 | 0.19 | 66.52 | 0.069 |
| 0.36 | 0.28 | 0.36 | 43.47 | 0.142 |
| 0.25 | 0.3125 | 0.4375 | 37.29 | 0.145 |
| 0.16 | 0.33 | 0.51 | 31.31 | 0.137 |
| 0.09 | 0.3325 | 0.5775 | 25.99 | 0.140 |
| 0.04 | 0.32 | 0.64 | 21.69 | 0.153 |
| 0.01 | 0.2925 | 0.6975 | 18.56 | 0.168 |

| porosity | $\mu_{SiC}$ | $E^H$ (GPa) | $\nu^H$ |
|---|---|---|---|
| 0.9025 | 0.0975 | 216.10 | 0.016 |
| 0.7225 | 0.2775 | 231.94 | 0.040 |
| 0.5625 | 0.4375 | 248.93 | 0.062 |
| 0.4225 | 0.5775 | 267.94 | 0.083 |
| 0.3025 | 0.6975 | 289.27 | 0.104 |
| 0.2025 | 0.7975 | 314.16 | 0.124 |
| 0.1225 | 0.8775 | 342.27 | 0.138 |
| 0.0025 | 0.9975 | 409.14 | 0.139 |

crack length is 0.25, and the mesh, completely regular and non–conforming to the crack, contains 2701 four–noded quadrilateral elements. The radius of the circular interaction integral domain is twice the size of the element containing the tip. Note that in this case, the exact SIFs are actually known analytically. The computed SIFs are within one percent of the exact values. The evolution of the energy release rate $G$ as a function of porosity is given in Figure 4b) and shows that, as expected, the energy release rate increases with increasing porosity, i.e. as the pure SiC ceramics become more rigid.

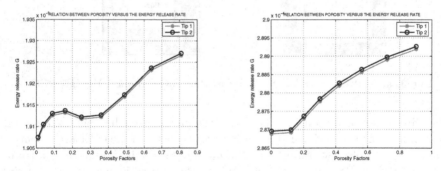

**Fig. 4.** Porosity versus the energy release rate $G$: a) SiC ceramics; b) Pure SiC ceramics

The same problem was solved for a layered SiC ceramics, and results are reported in Figure 4a). We note that the behavior is somewhat more complex in this case, where we observe a decreasing energy release rate for porosities comprised between 0.15 and 0.25.

# References

1. Bensoussan, A., Lions, J.L., Papanicolaou, G.: Asymptotic Analysis for Periodic Structures. North–Holland, Elsevier Science Publishers, Amsterdam, (1978).
2. Jikov, V.V., Kozlov, S.M., Oleinik, O.A.: Homogenization of Differential Operators and Integral Functionals. Springer, (1994).

3. Belytschko T., Moës N., Usui S., Parimi C.: Arbitrary discontinuities in finite elements. Int. J. Numer. Methods Eng., 50 (4) (2001) 993–1013.
4. Bordas S., Legay A.: Enriched finite element short course: class notes. Organized by S. Bordas and A. Legay through the EPFL school of continuing education, Lausanne, Switzerland, December 7–9, 2005.
5. Bordas S., Moran B.: Extended finite element and level set method for damage tolerance assessment of complex structures. Eng. Fract. Mech., 73 (9) (2006) 1176–1201.
6. Bordas S., Nguyen V.P., Dunant C., Nguyen-Dang H., Guidoum A.: An object–oriented extended finite element library. Int. J. Numer. Methods Eng., (2006) (to appear).
7. Fish J.: The s-version of the finite element method. Comput. Struct., 43 (3) (1992) 539–547.
8. Fish J., Guttal R.: The s-version of finite element method for laminated composites. Int. J. Numer. Methods Eng., 39 (21) (1996) 3641–3662.
9. Hoppe R.H.W., Petrova S.I.: Optimal shape design in biomimetics based on homogenization and adaptivity. Math. Comput. Simul., 65 (3) (2004) 257–272.
10. Lee S.-H., Song J.-H., Yoon Y.-C., Zi G., Belytschko T.: Combined extended and superimposed finite element method for cracks. Int. J. Numer. Methods Eng., 59 (8) (2004) 1119–1136.
11. Moës N., Dolbow J., Belytschko T.: A finite element method for crack growth without remeshing. Int. J. Numer. Methods Eng., 46 (1) (1999) 131–150.
12. Moran, B., Shih, C.F.: Crack tip and associated domain integrals from momentum and energy balance. Eng. Fract. Mech., 27 (1987) 615–641.
13. Nguyen V.P.: An object oriented approach to the X-FEM with applications to fracture mechanics. Master's thesis, EMMC–Hochiminh University of Technology, Vietnam, November (2005).
14. Ota T., Takahashi M., Hibi T., Ozawa M., Suzuki S., Hikichi Y., Suzuki H.: Biomimetic process for producing SiC wood. J. Amer. Ceram. Soc., 78 (1995) 3409–3411.
15. Sukumar N., Prévost J.-H.: Modeling quasi-static crack growth with the extended finite element method. Part I: Computer implementation. Int. J. Solids Struct., 40 (2003) 7513–7537.
16. Takano N., Zako M., Okuno Y.: Multi–scale finite element analysis of porous materials and components by asymptotic homogenization theory and enhanced mesh superposition method. Modelling Simul. Mater. Sci. Eng., 11 (2003) 137–156.

# Round-Trip Operator Technique Applied for Optical Resonators with Dispersion Elements

Nikolay N. Elkin, Anatoly P. Napartovich,
Dmitry V. Vysotsky, and Vera N. Troshchieva

State Science Center Troitsk Institute for Innovation and Fusion Research
(TRINITI), 142190, Troitsk Moscow Region, Russia
elkin@triniti.ru

**Abstract.** The round-trip operator technique is widely used for dispersionless optical resonators beginning from pioneering studies of Fox and Li. The resonator modes are determined as eigenfunctions of the round-trip operator and may be calculated by means of numerical linear algebra. Corresponding complex eigenvalues determine the wavelength shifts relative to reference value and threshold gains. Dispersion elements, for example, Bragg mirrors in a vertical cavity surface emitting laser (VCSEL) cause a dependence of the propagation operator on the wavelength and threshold gain. We can determine the round-trip operator in this case also, but the unknown values of the wavelength and threshold gain enter into the operator in a complicated manner. Trial-and-error method for determination of the wavelength shifts and the threshold gains is possible but it is rather time consuming method. The proposed approximate numerical method for calculation of resonator modes is based on the solution of linear eigenvalue problem for the round-trip operator with reference wavelength and zero attenuation. The wavelength shifts and threshold gains can be calculated by simple formulae using the eigenvalues obtained and the computed effective length of the resonator. Calculations for a cylindrical antiresonant-reflecting optical waveguide (AR-ROW) VCSEL are performed for verification of the model.

## 1 Introduction

The traditional approach for numerical modeling of optical resonators [1] is based on the assumption that the wave refraction and gain don't depend on the wavelength. The same property is assumed relatively to the reflectivity of mirrors. This approach is not valid for dispersion optical resonators because light round-trip may depend on the wavelength and attenuation in a complicated manner.

The three-mirror resonator as the simplest example of a dispersion resonator has been studied in [2]. Third mirror is placed outside the usually two-mirror resonator. Feedback field reflected from the third mirror returns to the two-mirror resonator with some phase $\psi$ depending on the exact position of the third mirror. This phase is different for various lateral modes and we cannot calculate the mode spectrum for fixed position of the third mirror solving an eigenvalue

T. Boyanov et al. (Eds.): NMA 2006, LNCS 4310, pp. 542–549, 2007.

problem for a round-trip operator. But we can calculate the mode spectrum for fixed $\psi$ . Each of the calculated modes corresponds to certain position of the third mirror. Varying $\psi$ in the range $-\pi \leq \psi < \pi$ and calculating the mode spectrum for all $\psi$ we may reconstruct the set of modes for fixed position of the third mirror.

Our paper has for an object to demonstrate the applicability of the round-trip operator technique with some correction in the cases when dispersion effects are considerable. The modified algorithm for determination of the exact wavelength and attenuation uses the previously computed effective length of the resonator.

## 2 Round-Trip Operator Technique for Dispersionless Optical Resonator

We start from Maxwell equations and assume that the polarization effects can be neglected. Laser modes have a time dependence of the form $E(r, \varphi, z, t) = U(r, \varphi, z) \exp(-i\Omega t)$, $\Omega = \omega_0 + \Delta\omega - i\delta$, where $\omega_0$ is the reference frequency, $\Delta\omega = \omega - \omega_0$ is the frequency shift and $\delta$ is the attenuation factor. We use cylindrical coordinates for convenience, but the use of Cartesian coordinates is possible also. The reference frequency $\omega_0$ , in other words, the frequency guess must be chosen close to the working frequency of a laser. The reference wavenumber and reference wavelength are defined by standard relations: $\omega_0 = k_0 c$ , $k_0 = 2\pi/\lambda_0$ .

Introducing new variables $g_t = 2\delta/c$, $\Delta k = \Delta\omega/c$, $\beta = g_t + i2\Delta k$, and neglecting small quantities of higher order we have obtained the Helmholtz equation

$$\frac{\partial^2 U}{\partial z^2} + \Delta_\perp U + (k_0^2 n^2 - ik_0 g)U - ik_0 n^2 \beta U = 0, \quad \Delta_\perp U = \frac{1}{r}\frac{\partial}{\partial r}\left(r\frac{\partial U}{\partial r}\right) + \frac{1}{r^2}\frac{\partial^2 U}{\partial \varphi^2}, \quad (1)$$

containing complex eigenvalue $\beta$. Here $n$ and $g$ are the index and the gain respectively.

Following [3], we formulate the problem statement as an eigenvalue problem for the round-trip operator and describe numerical methods for its solution. Let us consider optical resonator formed by two mirrors placed at $z = 0$ and $z = L$ and loaded with an active medium. Typically, transverse sizes of mirrors are much less than $L$. The intracavity wave field can be represented as a sum of two counter-propagating waves $U(r, \varphi, z) = E_+(r, \varphi, z) \exp(ik_0 z) + E_-(r, \varphi, z) \exp(-ik_0 z)$. The envelopes $E_\pm(r, \varphi, z)$ satisfy the parabolic equations: $\pm 2ik_0 \partial E_\pm/\partial z + \Delta_\perp E_\pm - ik_0[g + ik_0(n^2 - 1)]E_\pm - ik_0\beta E_\pm = 0$. The propagating wave field is to be specified at any transverse plane, for example, $u(r, \varphi) = E_+(r, \varphi, L)$ as a start field for the round trip. The round-trip operator is a composition of four operators: $\mathbf{P}(g, n, \beta) = \mathbf{P}^+(g, n, \beta)\mathbf{R_1}\mathbf{P}^-(g, n, \beta)\mathbf{R_2}$.

The first one is reflection from the second mirror: $E_-(r, \varphi, L) = \mathbf{R_2}u(r, \varphi)$. The field distribution $E_-(r, \varphi, L)$ is the initial field for a Cauchy problem for the parabolic equation in the interval $L > z > 0$ . The solution to this problem at $z = 0$ is by definition $E_-(r, \varphi, 0) = \mathbf{P}^-(g, n, \beta)E_-(r, \varphi, L)$. The next operator describes reflection from the first mirror: $E_+(r, \varphi, 0) = \mathbf{R_1}E_-(r, \varphi, 0)$. The result of reflection $E_+(r, \varphi, 0)$ is the initial field for a Cauchy problem for the parabolic equation in the interval $0 < z < L$. The solution to this problem at $z = L$ is by definition $E_+(r, \varphi, L) = \mathbf{P}^+(g, n, \beta)E_+(r, \varphi, 0)$. After round trip the same field $u(r, \varphi)$ should be reproduced. Note, that traditionally the round-trip operator is defined at $\beta = 0$. Further, we call $\mathbf{P}(g, n, \beta \neq 0)$ as the modified round-trip operator. Taking into account round-trip condition we have the non-standard eigenvalue problem

$$\mathbf{P}(g, n, \beta)u = u \qquad (2)$$

because the eigenvalue $\beta$ appears in the operator. It is important that for dispersionless resonators $\mathbf{R}_{1,2}$ don't depend on $\beta$ and as a sequence we have the following expression for the modified round-trip operator [3]: $\mathbf{P}(g, n, \beta) = \mathbf{P}(g, n, 0)\exp(\beta L)$. Consequently, we have the standard formulation of an eigenvalue problem

$$\mathbf{P}(g, n, 0)u = \gamma u, \qquad (3)$$

where $\gamma = \exp(-\beta L)$. It is precisely this problem that was solved for dispersionless resonators beginning from works of Fox and Li. There exist two versions of problem (3). If we neglect the saturation of an active medium we have a linear eigenvalue problem, because gain and index distributions are fixed functions of the spatial variables. $\mathbf{P}(g, n, 0)$ is a linear non-hermitian operator in this case, and $\gamma$ is a complex eigenvalue. After appropriate discretization, the problem can be reduced to a linear algebraic eigenvalue problem for a non-hermitian full matrix of high dimension. Krylov subspace methods (e.g. Arnoldi algorithm) are valid for numerical solution of the algebraic eigenvalue problem in such a situation. The second version is a search of steady-state oscillating mode with medium saturation. Gain and index distributions depend on wave field distribution $U(r, \varphi, z)$ according to the model of an active medium and have to be determined self-consistently. We have an eigenvalue problem for the non-linear operator (3) in this case. The supplementary condition $|\gamma| = 1$ is required for the non-linear problem (3), because $\delta = 0$ ($\mathrm{Re}(\beta) = 0$) for steady-state oscillations.

Summarizing, the round-trip operators $\mathbf{P}(g, n, \beta)$ and $\mathbf{P}(g, n, 0)$ can be synthesized starting from an arbitrary transverse plane between the mirrors in any direction, the eigenvalues do not depend on the start position. The threshold gain $g_t$ and the wavenumber shift $\Delta k$ are determined by the formulas $g_t = -\ln(|\gamma|)/L$, $\Delta k = -(\arg(\gamma) + 2\pi l)/(2L)$, where $-\pi \leq \arg(\gamma) < \pi$, the integer $l$ is the number of the longitudinal mode.

## 3  Numerical Model of the ARROW-Type VCSEL

The ARROW-type VCSEL scheme [4] is presented in Fig.1. 1-wave cavity, containing the active layer (black strip) and the high- and low-index ring reflectors formed by thin GaAs-InGaP spacers are located between $p-$ and $n-$ distributed Bragg reflectors (DBR). The metal contact plate with the output window is fixed on the top Bragg reflector. The laser operates at a wavelength of $\lambda_0 = 0.98\,\mu\mathrm{m}$, so the reference wavenumber $k_0 = 2\pi/\lambda_0 = 64114\,\mathrm{cm}^{-1}$.

The boundary condition at the interface between adjoining layers and the boundary condition at the lateral boundaries must be defined. We can specify the condition of continuity for the wave field $U$ and its normal derivative at the interfaces. It is not an easy task to define correctly the boundary condition at the lateral boundaries but the absorbing boundary condition [5] is more adequate to experimental implementation.

The index and absorption distributions are cellwise constant functions of spatial variables for all layers except the active layer. The non-linear 2D diffusion equation [6] must be solved for carrier density in order to find gain and index at the active layer. It is significant that two-dimensional distributions of $n$ and $g$ at the active layer are controlled at a fixed pump by the two-dimensional distribution $I = |U|^2$ of light intensity in the active layer.

Helmholtz equation was solved using combination of bi-directional beam propagation method and a matrix approach for spectral components within plane layers structure [9]. Taking into account the circular symmetry we shall substitute $U = U_m(r, z) \exp(im\varphi)$ into (1) and eliminate the angular dependence. As a result we have equation for $m$−th angular harmonic

$$\frac{\partial^2 U_m}{\partial z^2} + \frac{1}{r}\frac{\partial}{\partial r}\left(r\frac{\partial U_m}{\partial r}\right) - \frac{m^2}{r^2}U_m + (k_0^2 n^2 - ik_0 g)U_m - ik_0 n^2 \beta U_m = 0, \quad (4)$$

To construct the numerical algorithm we use representation of the wave field $U_m(r, z)$ in terms of $m$−th order Hankel transform over $r$. Introducing appropriate numerical meshes $\{r_l, \ l = 0, 1, \ldots, N_r\}$ , $\{\kappa_n, \ n = 0, 1, \ldots, N_r\}$ for the radius $r$ and for the transverse wave number $\kappa$ and going on to discrete approximation of the Hankel transform we may define $\psi_{nm}(z) = \mathbf{H}_m\{U_m(r_l, z)\}$, where $n$ is the number of radial harmonic and $\mathbf{H}_m$ is the discrete $m$−th order Hankel transform operator. The Fast Hankel transform algorithm [7] was used to evaluate the Hankel transform and its inversion $\mathbf{H}_m^{-1}$ .

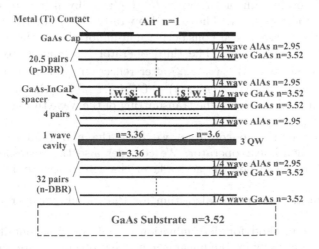

**Fig. 1.** Scheme of the ARROW-type VCSEL (Cross-Section View)

Let us choose the origin of $z$−axis at the plane of the GaAs-InGaP spacers. Let $N$ be the number of layers between the spacers and the metal contact. Assume that $M$ is the number of layers between the substrate and the spacers including the active layer. Thus, we can introduce the array $\{z_j, \ j = -M, \dots, N\}$ of coordinates of the interfaces between the layers so as $z = 0$ to be the GaAs-InGaP spacer position and the interval $z_{j-1} < z < z_j$ to determine the position of the j-th layer $(j = -M + 1, \dots, N)$ .

As a result of Hankel transform, the wave equation (4) in the j-th layer can be expressed in a form: $d^2\psi_n/dz^2 + q_{jn}^2\psi_n = 0, \quad q_{jn}^2 = k_0^2 n_j^2 - ik_0 g_j - ik_0 n_j^2\beta - \kappa_n^2,$ where the index $m$ is omitted. We define $q_{jn} = \sqrt{k_0^2 n_j^2 - ik_0 g_j - ik_0 n_j^2\beta - \kappa_n^2}$ under the condition $\mathbf{Re}(q_{jn}) \geq 0$ . The general solution to the last equation has the form: $\psi_n(z) = C_{jn}\exp(iq_{jn}z) + D_{jn}\exp(-iq_{jn}z), \quad z_{j-1} < z < z_j.$ The coefficients $C_{jn}$ and $D_{jn}$ for different layers are coupled by simple algebraic relations following from the well-known $T$−matrix formalism. Earlier [9], we have applied $T$−matrix formalism for VCSELs with the square index-step lattice.

Let $k$ is the number of layers between the active layer and the GaAs-InGaP spacer. The order number of the active layer will be $(-k)$ in our numeration. We simulate the active layer as a uniform layer containing a non-uniform phase screen with gain and phase incursion according to diffusion equation (see [6], [9]).

Applying the operator $\mathbf{H}^{-1}$ to $\psi_n(z)$ we can represent the wave field in the j-th layer as a sum of upward propagating wave $U_j^+(r_l, z) = \mathbf{H}^{-1}\left(C_{jn}\exp(iq_{jn}z)\right)$ and downward propagating wave $U_j^-(r_l, z) = \mathbf{H}^{-1}\left(D_{jn}\exp(-iq_{jn}z)\right)$ . Similarly to the case of a dispersionless optical resonator, the modified round-trip operator was built up in order to determine the oscillating modes and their losses. We specify as the start field $u(r_l) = U_{-k}^+(r_l, z_{-k})$, that is the upward propagating wave at the upper boundary of the active layer. The required round-trip operator $\mathbf{P}(g, n, \beta)$ may be represented as a composition of four operators. The first one is evaluation of the wave $U_{-k}^-(r_l, z_{-k})$ by means of $u(r_l)$ after reflection from the top DBR. The second operator evaluates transmitting of wave through the phase screen considering non-uniformity of the active layer: $U_{-k}^-(r_l, z_{-k}) \rightarrow W_{-k}^-(r_l, z_{-k})$ . The third operator is evaluation of the wave $W_{-k}^-(r_l, z_{-k})$ by means of $W_{-k}^-(r_l, z_{-k})$ after reflection from the bottom DBR. Lastly, the fourth operator $W_{-k}^+(r_l, z_{-k}) \rightarrow U_{-k}^+(r_l, z_{-k})$ has the same structure as the second one and evaluates the transmitting of the wave through the active layer.

Applying the round-trip condition we have the non-standard eigenvalue problem (2) for non-linear operator. We consider also the corresponding linear eigenvalue problem when the active layer characteristics are fixed and not recalculated.

Details about the numerical algorithm for the round-trip operator can be found in [9].

Trial-and-error method for determination of $\beta$ is possible but it is rather time consuming especially for the linear eigenvalue problem because the complex parameter $\beta$ contains two real parameters.

We can formally define another eigenvalue problem of the form (3). Linear and non-linear problems are formulated similarly. It is reasonable to expect that the oscillating mode fields being found from (2) and (3) differ slightly if dispersion effects are not great. But we cannot apply directly the formula $\gamma = \exp\left(-\beta L\right)$ for resonator with dispersion elements. Moreover, we cannot determine the distance $L$ between the mirrors because the Bragg mirror is a distributed system. Strictly speaking, the eigenvalue $\gamma$ depends on $\beta$ in a complicated way.

Nevertheless, we attempt to develop an effective method for calculation of $\beta$ assuming that it is possible to approximate the dependence of $\gamma$ on $\beta$ by the formula $\gamma \cong \exp\left(-\beta L_e\right)$, where $L_e$ is some effective length to be determined. This guess is based on the fact that the vertical direction of propagation predominates in our case.

If our hypothesis is true we have reasons to calculate $\gamma$ by solving the eigenvalue problem (3) in order to determine the wavelength shift $\Delta k$ and the threshold gain $g_t$. The analytic estimations for $L_e$ ([8], [9]) are known but for more precise determination of the effective length we have to solve at least once a basic eigenvalue problem (2) using trial-and-error method jointly with the problem (3) in order to determine $L_e$ by the equation $\gamma = \exp\left(-\beta L_e\right)$.

# 4    Results and Discussion

Calculations were performed for core diameter $d = 8\,\mu\mathrm{m}$ (see Fig. 1) which has the same value as the diameter of the output window. Low-index reflector width was $w = 2.65\,\mu\mathrm{m}$ , high-index reflector width $s$ was a variable parameter. The phase advance gained on the spacers was taken equal to 0.677.

Layer number $(-k)$ of the start layer of round-trip is assigned as the number of the active layer in the described above algorithm. But in contrast to the case of dispersionless resonator the eigenvalue $\gamma$ may depend on the start position. Varying the start position, we have performed calculations for passive cavity when the active layer has uniform distribution of the index and zero gain. Setting the value $s = 1.25\,\mu\mathrm{m}$ we have calculated the dependence of the eigenvalue $\gamma$ of the 1-st axially symmetric mode ($m = 0$) on the start position $l$ relatively to the active layer. The results are shown in Table 1. The number $l = 0$ corresponds to the active layer. The value $1 - |\gamma|$ is a measure of the attenuation. For a dispersionless resonator this value is proportional to the threshold gain $g_t$ in first order of approximation. The eigenvalue phase $\arg(\gamma)$ is proportional to the wavenumber shift $\Delta k$ for a dispersionless resonator. It follows from Table 1 that values $1 - |\gamma|$ and $\arg(\gamma)$ vary twice over a number layer range from $-7$ to $7$. We have to set a layer number for the round-trip start to determine an effective length $L_e$. It seems most appropriate the active layer ($l = 0$) to be a start position for the round-trip operator. This layer is placed at the center of the cavity and the values $1 - |\gamma|$ and $\arg(\gamma)$ reach minimum when we start from the active layer. The number of mesh nodes is $N_r = 1024$ for all of the presented results. To estimate the error of discretization we have performed some calculations using $N_r = 2048$. For all this the relative change of $1 - |\gamma|$ did not exceed the value

**Table 1.** Eigenvalue module and phase vs start position, $s = 1.25\,\mu m$

| $l$ | -7 | -6 | -5 | -4 | -3 | -2 | -1 | 0 | 1 | 2 | 3 | 4 | 5 | 6 | 7 |
|---|---|---|---|---|---|---|---|---|---|---|---|---|---|---|---|
| $(1 - |\gamma|) \times 10^5$ | 973 | 832 | 711 | 611 | 524 | 469 | 468 | 465 | 468 | 473 | 529 | 617 | 718 | 837 | 975 |
| $\arg(\gamma) \times 10^3$ | 151 | 129 | 111 | 95 | 81 | 73 | 72 | 72 | 72 | 73 | 81 | 95 | 111 | 129 | 151 |

$3 \times 10^{-3}$ and the relative change of $\arg(\gamma)$ did not exceed the value $2 \times 10^{-4}$ what seems to be quite satisfactorily. To check the correctness of our calculations the eigenvalue $\beta = g_t + i2\Delta k$ was calculated by the trial-and-error method for the problem (2) for every start layer of the round-trip. The relative variation of $g_t$ does not exceed $3 \times 10^{-3}$ when the number of a start layer varies from $-7$ to 7. The relative variation of $\Delta k$ does not exceed $2 \times 10^{-4}$ in these conditions. As expected, the eigenvalue $\beta$ does not depend on the start layer within the accuracy of the calculations.

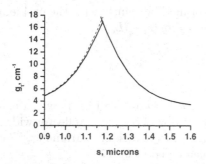

**Fig. 2.** Threshold gain $(g_t)$ vs high-index reflector width $(s)$

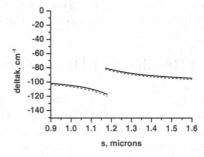

**Fig. 3.** Wavenumber shift $(\Delta k)$ vs high-index reflector width

Two formulae for an effective length determination can be written according to the complex equation $\gamma = \exp(-\beta L_e)$: $L_1 = \ln|\gamma|/g_t$, $L_2 = -\arg(\gamma)/(2\Delta k)$. Here, the eigenvalue $\gamma$ has to be determined as a solution of the linear eigenvalue problem (3), and a pair $(g_t, \delta k)$ is determined as a solution of (2) by trial-and-error method. For the first axially symmetric mode ($m = 0$) we have obtained $L_1 = 4.09\,\mu m$ and $L_2 = 4.18\,\mu m$ using the formulae cited above. Similar calculations for the first mode with angular index $m = 1$ result in $L_1 = 4.14\,\mu m$ and $L_2 = 4.14\,\mu m$. Calculations for the second axially symmetric mode ($m = 0$) give the values $L_1 = 4.26\,\mu m$ and $L_2 = 4.42\,\mu m$.

At last, operating mode was calculated including medium saturation of the active layer according to the model [9]. Because of $|\gamma| = 1$ for the operating mode, only $L_2$ can be determined after simultaneous solution of problems (3) and (2). These calculations give $L_2 = 4.18\,\mu m$.

Let us note, that the effective length depends very weakly on the way of calculations. We have determined the effective length $L_e = 4.1\,\mu m$ for the following calculations.

The calculations were performed when the high-index reflector width $s$ was varied from $0.9\,\mu m$ to $1.6\,\mu m$. The attenuation factor and the wavenumber shift depending on $s$ are presented in Figs. 2 and 3, respectively. Solid lines correspond to values $g_t$ and $\Delta k$ calculated by trial-and-error method for problem (2), dashed lines were obtained by eigenvalue calculations from the problem (3) by using the approximate formulae $g_t \cong -\ln|\gamma|/L_e$, $\Delta k \cong -\arg(\gamma)/(2L_e)$.

We can see that the proposed approximate method possesses quite a good accuracy. Application of this method takes in dozens of times less calculations than the use of trial-and-error method. Singular points in Figs. 2 and 3 are results of the change of mode with the highest $Q$−factor.

## 5  Conclusion

The traditional well-known round-trip operator technique may be successfully applied after some modification to optical resonators with complicated configuration, containing dispersion elements. We believe that the proposed approach to the problem of modeling dispersion resonators has much more wide range of applicability than the specific type of resonator described above.

## References

1. A.G.Fox, T.Li.: Effect of gain saturation on the oscillating modes of optical masers. IEEE Journal of Quantum Electronics, **QE-2**, (1966) 774–783.
2. O. V. Apollonova, N. N. Elkin, M. Yu. Korzhov, et al.: Mathematical simulation of composite optical systems loaded with an active medium. Laser Physics, **2**, (1992) 227 – 232.
3. Elkin N.N., Napartovich A.P.: Numerical study of the stability of single-mode lasing in a Fabry-Perot resonator with an active medium. Appl. Math. Modelling, **18**, (1994) 513 – 521.
4. D. Zhou and L. J. Mawst: High-power single-mode antiresonant reflecting optical waveguide-type vertical-cavity surface-emitting lasers. IEEE J. Quantum Electron., **38**, (2002) 1599 – 1606.
5. R. Kosloff, and D. Kosloff: Absorbing boundaries for wave propagation problem. Journal of Computational Physics, **63**, (1986) 363 – 376.
6. G. R. Hadley.: Modeling of diode laser arrays. Chapter 4 in Diode Laser Arrays, D. Botez and D.R. Scifres, Eds, Cambridge, U.K., Cambridge Univ. Press, (1994) 1-72.
7. A. E. Siegman.: Quasi fast Hankel transform. Optics Letters, **1**, (1977) 13 – 15.
8. S. Riyopoulos, D. Dialetis, J. Ihnman and A. Phillips: Active-cavity vertical-cavity surface-emitting laser eigenmodes with simple analytic representation. JOSA B, **18**(9), (2001) 1268 – 1283.
9. N. N. Elkin, A. P. Napartovich, D. V. Vysotsky, V. N. Troshchieva, L. Bao, N. H. Kim, L. J. Mawst.: Modeling and experiment on vertical cavity surface emitting laser arrays. Laser Physics, **14**(3), (2004) 378 – 389.

# On the Impact of Tangential Grid Refinement on Subgrid-Scale Modelling in Large Eddy Simulation

Jochen Fröhlich[1], Jordan A. Denev[1],
Christof Hinterberger[2], and Henning Bockhorn[1]

[1] Institute for Technical Chemistry and Polymer Chemistry, University of Karlsruhe,
Kaiserstrasse 12, D-76128 Karlsruhe, Germany
{froehlich,denev,bockhorn}@ict.uni-karlsruhe.de
http://www.ict.uni-karlsruhe.de
[2] Institute for Hydromechanics, University of Karlsruhe,
Kaiserstrasse 12, D-76128 Karlsruhe, Germany

**Abstract.** The paper presents Large Eddy Simulations of plane channel flow at a friction Reynolds number of 180 and 395 with a block-structured Finite Volume method. Local grid refinement near the solid wall is employed in order to reduce the computational cost of such simulations or other simulations of wall-bounded flows. Different subgrid-scale models are employed and different expressions for the length scale in these models are investigated. It turns out that the numerical discretization has an non-negligible impact on the computed fluctuations.

## 1 Motivation

The classical approach of large eddy simulation (LES) to turbulence modelling is to fix a computational grid and then to compute all motions of features larger than the step size of the grid $h$. The impact of features smaller than the step size on the resolved flow is accounted for by a so-called subgrid-scale (SGS) model. The SGS model contains a parameter $\Delta$ which is generally set to a multiple of $h$, so that $\Delta$ reduces proportionally when the grid is refined. This approach is, from a principal point of view, well accepted and validated if $h$ is constant in space. For most practical applications, however, a non-constant step size is employed in order to optimize the resolution capacity of a grid with a fixed number of points. With a structured code the step size can vary smoothly in all coordinate directions but may lead to over-refinement in some parts of the domain. A remedy is local grid refinement in an unstructured or block-structured manner. A typical approach for Finite Volume Methods is to locally subdivide cells in regions of refinement. This yields jumps in the step size which impact on the numerical error as well as on the modelling error in a way which is difficult to control.

In the present study we consider local, block-structured grid refinement in both directions perpendicular to the mean flow, which we term "tangential refinement". Such a strategy is interesting for wall-resolving LES, as demonstrated

T. Boyanov et al. (Eds.): NMA 2006, LNCS 4310, pp. 550–557, 2007.
© Springer-Verlag Berlin Heidelberg 2007

in [6]: It allows to resolve the near-wall structures which are finer than the turbulence in the outer flow while using a coarser grid for the latter so that the total number of grid points can be reduced substantially compared to a fully structured grid. An introduction to LES of near-wall flows and LES in general is, e.g., given in [3].

## 2 Numerical Method, Configuration and Subgrid-Scale Models

As a test bench we consider plane channel flow at a nominal Reynolds number of $Re_\tau = U_\tau \delta/\nu = 180$ and 395, where $U_\tau$ is the mean friction velocity and $\delta$ the half-distance between the two plates. Periodic boundary conditions were used in $x$ and $z$ with period length $L_x = 2\pi$ and $L_z = \pi$, respectively, and no-slip conditions were applied at the walls. The runs were conducted by fixing the bulk Reynolds number to $Re_b = U_b\delta/\nu = 2817$ and 6875, respectively, through instantaneous adjustment of a forcing term. The usual coordinate system is employed here with $x$ in streamwise, $y$ in wall-normal and $z$ in transverse direction, respectively. The computations were performed with a block-structured Finite Volume method for the incompressible Navier-Stokes Equations. The discretization scheme is second order central for both, convection and diffusion terms. The code LESOCC2 in its most recent form [5] allows to change the size of the computational cells at block boundaries by integer factors.

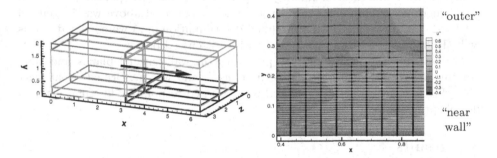

**Fig. 1.** Computational grid. Left: Block structure of the grid in the entire domain. Right Zoom of the grid around the interface between coarse and fine grid. The grey scale represents a snapshot of the instantaneous values $\overline{u}'' = \overline{u} - \langle \overline{u} \rangle$. The dots identify cell centers of the computational grid and the line connections between these (not the finite volume discretization).

The same grid containing 1.22 Mio. cells was used for both Reynolds numbers. Its block structure is displayed in Fig. 1a. In the present runs, refinement was applied in the near-wall blocks with a factor of 2 in both, $\Delta_x$ and $\Delta_z$ as illustrated in Fig. 1b. The dots in this figure represent the cell centers and the lines correspond to the connection between the cell centers (they do not identify the boundaries of the finite volumes used in the discretization). The refinement interface is located at a distance of $0.25\delta$ from the wall. The near-wall part contains 31 internal cells in $y$−direction with geometric stretching of 4% from $\Delta_y = 0.004$ at the wall to $\Delta_y = 0.013$. Beyond the interface in the outer blocks, 16 cells are used until the center of the channel. In the outer blocks $\Delta_y = 0.02$ at the interface with stretching of 15% near the interface and linear decay of the stretching with $y$ to $\Delta_y = 0.07$ at the center. In the near-wall region the cells have a size of $\Delta_x = 0.0491, \Delta_z = 0.0245$ in $x-$ and $z-$direction, respectively, and twice this size in the outer blocks. As the grid for both Reynolds numbers is the same, the step sizes of the grid and the position of the interface change when expressed in wall units. The respective values are provided in Tab. 1. Observe that $y_1^+$ is the wall distance of the center of the wall-adjacent cell with $y_1^+ = \Delta_{y,1}^+/2$. For comparison, Run 395-6 was performed with refining the outer blocks as well yielding uniform $\Delta_x$ and $\Delta_z$ in the entire domain.

The subgrid-scale (SGS) models employed are the following ones. The standard Smagorinsky model [9] was used with a constant of $C_s = 0.1$ (SM01) and van Driest damping near the wall [8]. The dynamic Smagorinsky model (DSM) [4] was employed with a test filter of size $2\Delta_x$ and $2\Delta_z$ in wall-parallel planes, two-dimensional averaging, and clipping of the eddy viscosity to positive values. The dynamic mixed model (DMM) [10] was used with the same parameters for the dynamic procedure. For three-dimensional non-cubic cells, a unique cell size $h$ must be specified. Here, we used

$$h = \left(\Delta_x \, \Delta_y \, \Delta_z\right)^{1/3} \quad , \tag{1}$$

which is one of the most commonly employed formulas. The parameter $\Delta$ used as a length scale parameter in all the models mentioned above was in general set to $\Delta = h$. Two runs with a modified determination of $\Delta$ were carried out as well (Run 395-4 and Run 395-5) and will be discussed below.

For all runs the averaging was performed in wall-parallel planes and time starting after 40 dimensionless time units $\delta/U_b$ and was performed for at least $500\delta/U_b$. Average values are denoted by $\langle\cdot\rangle$ and corresponding fluctuations by a double prime.

## 3 Results

A global assessment of the result of each run is possible by the mean friction velocity $U_\tau = \sqrt{\langle\tau_w\rangle/\rho}$ reported in Table 1. The rightmost column contains the relative error with respect to the DNS value in percent, $E_{rel} = 100(U_\tau - U_{\tau,DNS})/U_{\tau,DNS}$. The reference values $U_{\tau,DNS} = 0.06389$ and $0.05745$ [1] have been used for for $Re_\tau = 180$ and $Re_\tau = 395$, respectively. $U_\tau$ is underpredicted with SM01 and DSM, stronger with the latter, while being overpredicted with DMM.

For the lower Reynolds number $Re_\tau = 180$ all models exhibit an error below 3% in $E_{rel}$ while the error is about twice as large for each model with $Re_\tau = 395$. The best match with the DNS value is obtained with the $DSM$, for the lower and with the $DMM$ for the higher Reynolds number.

**Table 1.** Overview over the runs discussed

| Run | $Re_b$ | $\Delta_x^+, y_1^+, \Delta_z^+$ | refinement | SGS model | $U_\tau$ | $E_{rel}$ |
|---|---|---|---|---|---|---|
| 180-1 | 2817 | 8.8, 0.36, 4.4 | $y^+ < 45$ | DSM | 0.06199 | -2.98 |
| 180-2 | 2817 | 8.8, 0.36, 4.4 | $y^+ < 45$ | DMM | 0.06511 | 1.91 |
| 180-3 | 2817 | 8.8, 0.36, 4.4 | $y^+ < 45$ | SM01 | 0.06276 | -1.76 |
| 395-1 | 6875 | 19.4, 0.79, 9.7 | $y^+ < 99$ | DSM | 0.05366 | -6.59 |
| 395-2 | 6875 | 19.4, 0.79, 9.7 | $y^+ < 99$ | DMM | 0.05938 | 3.36 |
| 395-3 | 6875 | 19.4, 0.79, 9.7 | $y^+ < 99$ | SM01 | 0.05527 | -3.80 |
| 395-4 | 6875 | 19.4, 0.79, 9.7 | $y^+ < 99$ | SM01-double | 0.05321 | -7.39 |
| 395-5 | 6875 | 19.4, 0.79, 9.7 | $y^+ < 99$ | SM01-smooth | 0.05569 | -3.06 |
| 395-6 | 6875 | 19.4, 0.79, 9.7 | all blocks | SM01-fine | 0.05603 | -2.47 |

For lack of space we show plots only for the higher Reynolds number. Similar conclusions as drawn below hold for $Re_\tau = 180$, although the results are generally better in this case since the resolution is better. In the following figures, the average velocity $\langle u \rangle^+$ is reported in the left column and the non-vanishing Reynolds stresses in the right column. Symbols represent values at grid points, thus visualizing the refinement. For normalization the computed value of $U_\tau$ from the same run was used. The fluctuations reported are the resolved ones only. Comparison is performed with the DNS data of [7], labeled "MKM" in the figures.

Fig. 2 shows results obtained with different SGS models and no further adjustment to the jump in grid step size $h$. The profile of $\langle u \rangle^+$ is predicted without any kink or jump at the refinement interface. The DMM yields a very good prediction of this quantity up to the interface but slightly too low values beyond. With the SM01 and the DSM, the shape of the profile is well predicted, but above the DNS curve. Note, however, that this type of semi-logarithmic profile is quite sensitive to the value of $U_\tau$ used for normalization.

The fluctuations show the presence of the refinement by a small kink in the normal stresses. The level of the curves is such that $v-$ and $w-$ fluctuations tend to be below the DNS values while the $u-$fluctuations tend to be larger. In the current plots, the value of $U_\tau$ used for normalization impacts on the height and the horizontal coordinate. Although the DMM yields the best result for $U_\tau$, the best fit with the DNS curves is obtained with SM01. In all cases $\langle \overline{u}'' \overline{u}'' \rangle$ and $\langle \overline{w}'' \overline{w}'' \rangle$ exhibit similar jumps at the interface. These would naturally be attributed to the different resolution capacities of the finer and the coarser grid. Further investigation, however, shows that this is not the case. For Run 395-3 with the SM, e.g., the unresolved contributions to the fluctuations have been determined using [2]

$$\langle u_i'' u_j'' \rangle = \langle \overline{u}_i'' \overline{u}_j'' \rangle + \langle \tau_{ij}^{SM} \rangle + \frac{1}{3} \langle \tau_{kk}^{mod} \rangle \delta_{ij} \tag{2}$$

$$\frac{1}{3} \langle \tau_{kk}^{mod} \rangle = \frac{2}{3} \frac{\langle \nu_t \rangle^2}{(C_s \Delta)^2} \frac{C_s^2}{C_1^2} \quad , \qquad C_1 = 0.094 \quad . \tag{3}$$

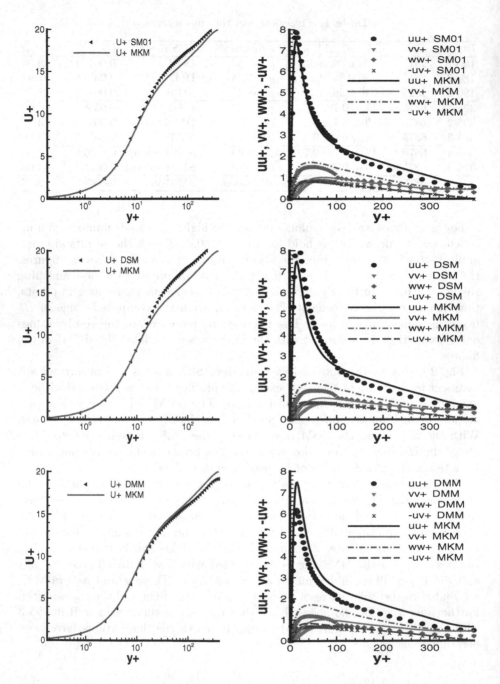

**Fig. 2.** Results for $Re_\tau = 395$ obtained with the three subgrid models SM01, DSM and DMM. Left: dimensionless average velocity $\langle \overline{u} \rangle^+$ (curves with SM01 and DSM extend beyond the upper limit of the plotted range). Right: resolved turbulent stresses $\langle \overline{u''u''} \rangle^+$, $\langle \overline{v''v''} \rangle^+$, $\langle \overline{w''w''} \rangle^+$ and $-\langle \overline{u''v''} \rangle^+$.

Here, $\langle \overline{u}_i'' \overline{u}_j'' \rangle$ are the fluctuations resolved by the LES, $\tau_{ij}^{SM}$ the (traceless) Smagorinsky SGS term, and $\langle \tau_{kk}^{mod} \rangle$ the separately modelled trace of the SGS fluctuations. For technical reasons $C_s \Delta$ was replaced by $C_s \Delta f_D(y^+)$ as used in the computation with $f_D$ the van Driest damping function from [8]. With the current grid, all unresolved fluctuations remain below 3% of the resolved ones beyond the region influenced by the damping. They indeed exhibit a small jump with slightly larger values in the outer blocks, but their magnitude is too small to modify the total fluctuations appreciably.

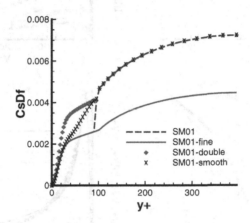

**Fig. 3.** Length scale $l = C_s \Delta f_D$ used for SGS modelling with the SM in different cases. $- - - -$ : original grid without modification (Run 395-3), ————— : grid with refinement of outer block as well (Run 395-6), $\diamond$ : multiplication of $l$ with $4^{1/3}$ in the near-wall block. $\times$ : smooth transition of $l$ from the fine to the coarse grid.

The jump in length scale from the near wall to the outer blocks is visualized in Fig. 3. The quantity depicted is the length scale $l = C_s \Delta f_D$ used in the SGS model. The jump in $l$ leads to the question of whether it impacts on the result of the simulation, in particular the determination of the Reynolds stresses. For this reason, Run 395-4 was conducted with $l$ multiplied by $4^{1/3}$ in the near-wall blocks which compensates for the refinement and yields a pre-factor equal to the one obtained without the refinement near the wall, $\Delta = (2\Delta_x \, \Delta_y \, 2\Delta_z)^{1/3}$. The resolved flow should then be smoother with the discretization still on the finer grid, so that the ratio between numerical and modelling error is improved, here, however, by increasing the modelling error. Indeed, the result (SM01-double in Fig. 4) is less satisfactory than without this change showing that the resolution capacity of the grid is more important than to avoid the jump in the SGS modelling.

Another modification was to introduce a smooth transition in $l$ from the coarse to the fine grid by a corresponding smooth multiplier in Run 395-5. Here, indeed, the result is improved a little (SM01-smooth in Fig. 4). The curves fit better and $U_\tau$ is slightly closer to the DNS value.

As a reference case, Run 395-6 was conducted with the outer blocks refined as well. As expected, it shows the best result but at higher computational cost.

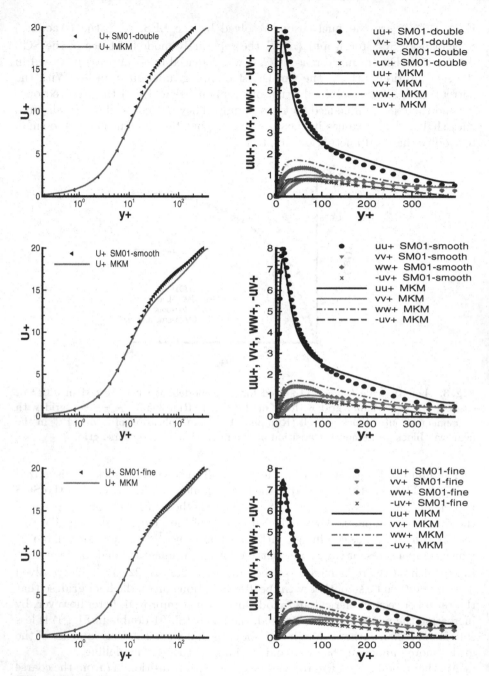

**Fig. 4.** Results for $Re_\tau = 395$ obtained with the SM01 SGS–model with different modifications of the ratio $\Delta/h$, as defined in the text. Left: dimensionless average velocity $\langle \overline{u} \rangle^+$. Right: resolved turbulent stresses $\langle \overline{u''u''} \rangle^+$, $\langle \overline{v''v''} \rangle^+$, $\langle \overline{w''w''} \rangle^+$ and $-\langle \overline{u''v''} \rangle^+$.

# 4   Conclusions

Several computations with different SGS models and different length scales were performed in the presence of tangential refinement by a factor 2 in both directions. Small jumps at the refinement interface persist even with a smooth choice of the length scale in the SGS model. We hence conclude that in the present setting with $\Delta = h$ the numerical scheme impacts on the computed fluctuations and in fact is a major source for the jump observed in these quantities. (This might be different with $\Delta = 2h$ but does not warrant improved results as shown by Run 395-4.) On the other hand, the choice of the SGS model does have an influence as reflected by the different values of $U_\tau$ obtained when altering the model. With a tangentially refined grid the best result among the runs performed was obtained with the SM and a smooth transition in length scale from the fine to the coarse grid. Further computations on coarser grids are under way to investigate the impact of tangential refinement in situations where the SGS model has a larger impact on the momentum balance.

## Acknowledgments

This work was supported by the Priority Programme 1141 of the German Research Foundation and SFB 606.

## References

1. AGARD. A selection of test cases for the validation of Large Eddy Simulations of turbulent flows, AGARD–AR–345, 1998.
2. J.W. Deardorff. A numerical study of three–dimensional turbulent channel flow at large Reynolds numbers. *J. Fluid Mech.*, 41:453–480, 1970.
3. J. Fröhlich and W Rodi. Introduction to Large–Eddy Simulation of turbulent flows. In B. Launder and N. Sandham, editors, *Closure Strategies for Turbulent und Transitional Flows*, chapter 8, pages 267–298. Cambridge University Press, 2002.
4. M. Germano, U. Piomelli, P. Moin, and W.H. Cabot. A dynamic subgrid–scale eddy viscosity model. *Phys. Fluids A*, 3:1760–1765, 1991.
5. C. Hinterberger. *Dreidimensionale und tiefengemittelte Large–Eddy–Simulation von Flachwasserströmungen.* PhD thesis, Institute for Hydromechanics, University of Karlsruhe, 2004. http://www.uvka.de/univerlag/volltexte/2004/25/ .
6. A.G. Kravchenko, P. Moin, and R. Moser. Zonal embedded grids for numerical simulations of wall–bounded turbulent flows. *J. Comp. Phys.*, 127:412–423, 1996.
7. R.D. Moser, J. Kim, and N.N. Mansour. Direct numerical simulation of turbulent channel flow up to $Re_\tau = 590$. *Phys. Fluids*, 11:943–945, 1999.
8. U. Piomelli, P. Moin, and J. H. Ferziger. Models for large eddy simulations of turbulent channel flows including transpiration. *AIAA J. Thermophys. Heat Transf.*, 5:124–128, 1991. also as AIAA Paper 89-0375, 1989.
9. J.S. Smagorinsky. General circulation experiments with the primitive equations, I, the basic experiment. *Mon. Weather Rev.*, 91:99–164, 1963.
10. Y. Zang, R.L. Street, and J.R. Koseff. A dynamic mixed subgrid–scale model and its application to turbulent recirculating flows. *Phys. Fluids*, 5(12):3186–3196, 1993.

# Detection Acceleration in Hough Parameter Space by K-Stage Detector*

Ivan Garvanov[1], Christo Kabakchiev[1], and Hermann Rohling[2]

[1] Institute of Information Technologies, Bulgarian Academy of Sciences
"Acad G. Bonchev" St., Bl. 2, 1113 Sofia, Bulgaria
igarvanov@iit.bas.bg, ckabakchiev@iit.bas.bg
[2] Technical University Hamburg-Harburg,
Eibendorfer Str.40, D-21073 Hamburg Germany
rohling@tu-harburg.de

**Abstract.** The sequential algorithm for target detection in impulse environment by K-stage processing in polar Hough (PH) space is investigated. This algorithm permits the search radar to minimize the time of target detection, because the sequential detector used minimizes the number of the necessary radar antenna scans, while the conventional detector uses a fix scan number. The aim of our study is to minimize the detection time by minimizing the number of radar scans, when the detector's parameters and characteristics are fixed. The average estimates of the minimal number of scans are obtained using the Monte Carlo simulation approach when the detector's parameters are fixed. The proposed algorithm is simulated in MATLAB environment. The efficiency factor of the investigated detector for radar signal detection and track determination in conditions of binomial distributed impulse interference is obtained by using Monte-Carlo approach. The prepositional detector, compared to the conventional detector, accelerates the detection procedure several times.

## 1 Introduction

The standard Hough transform and the related Radon transform have received much attention in the recent years. Using them makes possible the transformation of two-dimensional images with lines into a domain of possible line parameters, where each image line corresponds to a peak, positioned at the respective line parameters. For these reasons, many line detection applications appeared within the image processing, computer vision, and seismic research areas. The use of the standard Hough transform (SHT) for target detection and track determination in white Gaussian noise is introduced by Carlson in [1]. An approach for constant false alarm rate (CFAR) detection by means of SHT for track and target detection in conditions of non-homogeneous background is considered by Behar in [2,3,4]. The Hough detection scheme includes a CFAR signal detector in

---

* This work is supported by IIT - 010059/2004, MPS Ltd. - Grant No IF-02-85/2005 and Bulgarian NF "SR" Grant No TH - 1305/2003.

T. Boyanov et al. (Eds.): NMA 2006, LNCS 4310, pp. 558–565, 2007.

the area of observation, HT of the target distance measurements from the observation area into the parameter space, binary integration of data in the parameter space and linear trajectory detection. These CFAR Hough detectors have been studied in cases when the target flies in one azimuth and the speed is constant. As continuation of this research work, the use of a polar Hough (PH) transform is proposed in [5], which is also suitable for search radar. This transform is analogous to the standard Hough transform, where the input parameters are target range and azimuth obtained from the search radar. The technique used combines data from previous search scans into one large multi-dimensional polar data map. This transform is very comfortable for use in radar detection and track determination, when the target changes its speed and flies at different azimuths. The principle possibility of minimizing the time of radar signal (target) detection by using sequential analysis in conditions of constant false alarm probability and detection probability, is discussed in [6,7]. The advantage of the sequential detector, if compared to the conventional detector, is in reducing of the radar energy at the radar target detection. An important step in the development of an efficient approach for optimization and analysis of truncated sequential procedures is made by Sosulin in [6], within the study of K-stage procedures of statistical hypotheses test. Possible applications of such procedures, as well as optimization methods for various binary detection problems and multi-alternative detection problems are investigated in [6,8,9]. These papers illustrate the universality and efficiency of the developed approach to solutions of various truncated sequential detection problems and reveal the possibility to design K-stage signal detectors that significantly surpass in efficiency wide-spread detectors with fixed sample size. The effectiveness of the multi-channel processing is defined through Monte Carlo computer simulation. The performances of some sequential CFAR detectors are proposed and evaluated in [7,10]. The observations are passed through a dead-zone limited. Numerical results show a significant reduction of the average number of observations needed to achieve the same false alarm and detection probability as compared to a fixed-sample-size CFAR detector using the same kind of test statistic. in this paper In this paper we propose a sequential algorithm for target detection in polar Hough space. The radar efficiency factor of a sequential detector is defined as a ratio between the fixed sample number in the conventional Hough procedure and the average sample number in the K-stage polar Hough procedure. The radar efficiency factor of a K-stage detector in conditions of binomial distributed impulse interference is achieved by using Monte-Carlo simulation. We assume that the noise in the test cell is Rayleigh envelope distributed and the target returns are fluctuating according to Swelling II model, as in [3,4,11]. The efficiency of the researched detector increases with the growth of the signal-to-noise ratio but it decreases with the decrease of the false alarm probability and the increase of the detection threshold $M$. The proposed algorithm may also be applied in other popular CFAR processors using standard or polar Hough transform. The acceleration of the proposed detector is several times higher than that of a conventional detector with constant detection threshold. The research is performed in MATLAB environment.

## 2   Signal Model

We assume that the target in the test cell is fluctuating according to Swerling II model. We further assume that the total background environment includes the binomial distribution of impulse interference-plus-noise situation, which may appear at the output of the receiver with the probability $2e(1-e)$, interference-plus-noise situation with the probability $e^2$ and the noise only situation with the probability $(1-e)^2$, where $e = 1 - (1 - t_c F)^{1/2}$, $F$ is the average repetition frequency of impulse interference and $t_c$ is the length of impulse transmission. The probability density function (pdf) of this signal is given by [12] and used in [9,11]:

$$f(x) = f_1 + f_2 + f_3; \tag{1}$$

$$f_1 = \frac{(1-e)^2}{\lambda(1+S)} \exp \frac{-x}{\lambda(1+S)},$$
$$f_2 = \frac{2e(1-e)}{\lambda(1+S+I)} \exp \frac{-x}{\lambda(1+S+I)},$$
$$f_3 = \frac{e^2}{\lambda(1+S+2I)} \exp \frac{-x}{\lambda(1+S+2I)},$$

where $\lambda$ is the average power of the receiver noise, $I$ is the average interference-to-noise ratio (INR) of impulse interference, $S$ is the average signal-to-noise ratio (SNR). This equation is used for simulation of the input signal model and the interference of the sell average (CA) CFAR processor.

## 3   Detection Acceleration in Hough Parameter Space by Sequential K - Stage Detector

The efficiency of the sequential K-stage detector, which works in Hough parameter space, is investigated. This algorithm includes a CA CFAR processor, which works in the radar area of observation, polar Hough transform of the target distance measurements into the parameter space, binary integration of data and sequential detection of linear trajectory in the parameter space. The algorithm for search radar detection and track with the Hough transform, proposed by Carlson, Evans and Wilson in [1], is improved in our paper. According to the Carlson's approach, only the targets moving radially in the direction of radar (in the same azimuth) are considered. In this case, the target trajectory consists of points with coordinates (range, time) specified in the Cartesian coordinate system "range-time" (range-time data space). The distance in range between points is a function of the radar scans time. Unlike the Carlson approach, in our study the linear target trajectory is specified in the range-azimuth space and consists of points with coordinates (range, azimuth) obtained for each scan time. Therefore, the position of each point of the target trajectory in the coordinate system "range-azimuth" is an indirect time-varying function.

In order to keep the false alarm constant in conditions of impulse interference, we replace the fixed detection threshold in range-time space with a CA CFAR processor. For determination of the non-radial trajectory, we replace the standard transform with polar Hough transform [5]. For detection acceleration we replace the fixed detection threshold in Hough parameter space with a sequential K-stage detector [6,7,8,9].

1. Cell Average CFAR processor

The CA CFAR processor is a detector, which in the process of target detection maintains the false alarm rate constant. Target detection is declared if the signal value $x_0$ exceeds a preliminary determined adaptive threshold $H_D$. The threshold $H_D = VT$ is a multiplication of the noise level in the reference window $V$ and the scale factor $T$. Averaging the outputs $x_i$ of the reference cells, surrounding the test cell, forms the estimate $V = \sum_{i=1}^{N} x_i$ [11].

2. Polar Hough transform

The output data after every radar scan form the polar data space (range-azimuth). There are two approaches for Hough transformation of data - standard and polar Hough transform (SHT and PHT). The SHT is more suitable for image transformation, while the PHT is very convenient for use in radar because the output parameters of the radar (range and azimuth) are input parameters of the transform. One trajectory is formed in a polar data map (range-azimuth) after $N_s$ radar scans. The point's coordinates in $(r - azimuth)$ space form the polar parameter space. The PHT maps points (targets) from the observation space (polar data map) into curves in polar Hough parameter space, termed i.e. $(rho - theta)$ space, by:

$$rho = r \cos(azimuth - theta),\ 0 < (azimuth - theta) \leq \pi; \qquad (2)$$

where $r$ and $azimuth$ are the target range and azimuth, $theta$ is the angle and $rho$ is the smallest distance to the origin of polar coordinate system. The mapping can be viewed as stepping through $theta$ from 0 to 180 and calculating the corresponding $rho$. The result of transformation is a sinusoid with magnitude unites. Each of the points in polar Hough parameter space corresponds to one line in polar data space with parameters $rho$ and $theta$. A single $rho - theta$ point in the parameter space corresponds to a single straight line in the $r - azimuth$ data space with the same $rho$ and $theta$ values. Each cell from the polar parameter space is intersected by a limited set of sinusoids obtained by PHT. Every sinusoid corresponds to a set of possible lines through the point. If a line exists in the polar data space, by means of PHT it is represented as a point of intersection of sinusoids defined by PHT. The polar data space is divided into cells, whose coordinates are equal to range resolution cell number - in range and to the scan number in the history - in time. The parameters $rho$ and $theta$ have linear trajectory in polar Hough parameter space and could be transformed back to polar data space showing the current distance to the target. If the number of binary integration (BI) of data in polar Hough parameter space (of intersections in any of the cells in the parameter space) exceeds the detection threshold, target detection and linear trajectory detection are indicated. Target and linear trajectory detection are carried out for all cells of the polar Hough parameter space.

3. Sequential K-stage detector

In our previous work we investigated different CFAR Hough processors for detection in pulse environment, where we have used fixed value of binary integration procedure in Hough space [2,3,4]. For minimization of the detection time we proposed a sequential detector to be used. The polar data space (*range − azimuth*) after every radar scan is transformed by PHT according to (2) in the parameter space. In each cell of the Hough parameter space two operations are performed - binary integration and comparison with the sequential detection threshold. The sequential detector compares the test statistic (binary integration of intersection of sinusoids in polar Hough space) with the threshold M. The test for target detection is as follows. At the $k$−th observation one of the following three variants is chosen:

$$if \; k < M \; and \; k < N_s \; then \; S_t = \sum_{j=1}^{k} \Phi_j, \tag{3}$$

where $S_t$ is the test statistic, $\Phi_j$ is 1 or 0 respectively if there is or not a sinusoid in a given cell from polar Hough space in the $j$ scan;

$$if \; S_t > M \; and \; k < N_s \; then \; say \; H_1 \tag{4}$$

$$if \; S_t < M \; and \; k = N_s \; then \; say \; H_0 \tag{5}$$

where $H_1$ is hypothesis of target presence and $H_0$ is hypothesis of target absence.

4. Estimation the efficiency of the sequential K-stage detector

The efficiency of the K-stage CA CFAR PH detector is estimated toward the conventional CFAR PH detector with a full number of stages. The efficiency factor of the sequential procedure has the following form:

$$\mu = \frac{N_s}{\bar{k}} \tag{6}$$

where $N_s$ is the full number of radar scans in the BI procedure, $\bar{k}$ is the average sample number in the K-stage procedure.

## 4   Experiment Description

The input of the CA CFAR processor is simulated by using equation (1). The test cell includes signal, noise and impulse interference while the cells of the reference window include noise and impulse interference only ($S = 0$). The number of cells in the reference window of the CFAR processor is chosen to be $N = 16$. The average power of the receiver noise is $\lambda = 1$ , the probability for the appearance of impulse interference is 0.5 and the average interference-to-noise ratio (INR) is $I = 30dB$ (for Fig.4 and Fig.5). For given parameters of both signal and interference, after setting the scale factor of the CFAR detector using the

Monte Carlo simulation approach, we study the efficiency of the Hough detector in relation to the thresholding parameters of the K-stage detector. When the input parameters of both signal and interference are updated, the scale factor of the CFAR detector is additionally adjusted in order to maintain the constant false alarm rate at the output of the Hough detector. The output of the

**Fig. 1.** Block diagram of the experiment

CA CFAR processor is 1 or 0 if target signal is detect or not. This information fills the range-azimuth space and it is the input for the polar Hough transform. The sinusoids obtained from the transform are integrated in Hough parameter space after each antenna scan of the radar. In each cell of the Hough parameter space two operations are performed - a binary integration and comparison with the sequential detection threshold. The sequential detector compares the binary integration of the intersection of sinusoids in polar Hough space with the detection threshold of the K-stage detector. When this threshold is exceeded then the detection procedure is stopped. The efficiency of the researched detector is estimated toward the conventional CFAR PH detector with a full number of stages. We use Monte Carlo approach to obtain statistical stability estimation for the efficiency of the algorithm. The number of runs is 10000.

## 5   Numerical Results

We investigate the influence of the effectiveness of the K-stage procedure in conditions of noise and impulse interference. The experimental results are obtained for the following parameters: average power of the receiver noise $\lambda = 1$; probability for the appearance of impulse interference 0.5; average interference-to-noise ratio (INR) $I = 30dB$; number of cells in the CFAR processor reference window $N = 16$, false alarm probability of the researched Hough detector $P_{fa} = 10^{-5}$, the number of radar scans $N_s = 20$; detection threshold in Hough parameter space $M = 5, 7, 10, 15, 18$.

The effectiveness of the K-stage detector as a function of the SNR and for different binary rules $M/N_s$ is shown on Fig.2. The effectiveness increases with the growth of the signal-to-noise ratio but decreases with the increasing of the detection threshold $M$. When the SNR is constant ($S = 5, 10dB$) and the false alarm probability decreases, then the effectiveness of this algorithm decreases as well. In order to keep the detection probability $P_D = 0.5(P_D = 0.9)$ constant, the signal-to-noise ratio must be increased when the false alarm probability decreases. In this case, the effectiveness is equal for the different values of $P_{fa}$. The experimental results presented on Fig.2 and Fig.3 are obtained in conditions of noise. The experimental results presented on Fig.4 and Fig.5 reveal

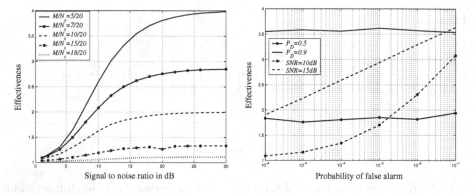

**Fig. 2.** Effectiveness of a K-stage detector as a function of SNR

**Fig. 3.** Effectiveness of a K-stage detector as a function of $P_{fa}$

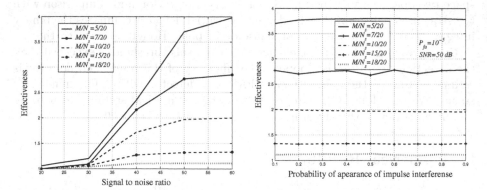

**Fig. 4.** Effectiveness of K-stage detector as a function of SNR

**Fig. 5.** Effectiveness of K-stage detector as a function of probability of appearance of impulse interference

that the probability for the appearance of impulse interference influences the effectiveness of the proposed algorithm. The effectiveness of the K-stage detector increases with the decrease of the detection threshold $M$ and with the growth of the signal-to-noise ratio (Fig.4). The results given on Fig.4 are obtained for the probability of the appearance of impulse interference 0.5 and $P_{fa} = 10^{-5}$. When the probability for the appearance of impulse interference increases, the false alarm is constant and the SNR is 50dB, then the effectiveness is constant (Fig.5).

# 6   Conclusions

The detection acceleration of a sequential polar Hough detector in conditions of impulse interference is investigated. The developed algorithm includes a CA CFAR detector, polar Hough transform, binary integration and sequential detection in parameter space. The efficiency of the researched detector increases when the signal-to-noise ratio grows; the detection threshold $M$ decreases, and the false alarm probability increases. The results are achieved by using Monte-Carlo simulation. The suggested algorithm may also be applied in other popular CFAR processors using standard or polar Hough transform. The acceleration achieved by using the proposed detector is several times higher than that achieved by using a conventional detector with constant detection threshold.

# References

1. Carlson B., E. Evans and S. Wilson, "Search Radar Detection and Track with the Hough Transform", IEEE Trans., vol. AES - 30.1.1994, Part I, pp. 102-108; Part II, pp. 109-115; Part III, pp. 116-124.
2. Behar V., Chr. Kabakchiev and L. Doukovska, "Target Trajectory Detection in Monopulse Radar by Hough Transform", Compt. Rend. Acad. Bulg. Sci., vol. 53, 8, 2000, pp. 45-48.
3. Behar V., B. Vassileva, Chr. Kabakchiev, "Adaptive Hough Detector with Binary Integration in Pulse Jamming", Proc.ECCTD'97, Budapest, 1997, pp. 885 - 889.
4. Behar V. and Chr. Kabakchiev, "Hough Detector with Adaptive Non-coherent Integration for Target Detection in Pulse Jamming", Proc. ISSSTA'98, South Africa, 1998, pp. 1003 - 1008.
5. Garvanov, I. and Chr. Kabakchiev, "Radar Detection and Track Determination with a Transform Analogous to the Hough Transform", Proc. of the International Radar Symposium - IRS 2006, Krakow, Poland, 2006.
6. Sosulin Y. "K-stage Radar CFAR Detection", Proc. of IEEE International Radar Conference -2000, 375-380, 2000.
7. Tantaratana S., "Sequential CFAR Detectors Using a Dead-Zone Limiter", IEEE Trans. Communic., vol. 38, 9, 1375-1383, 1990.
8. Sosulin Y., K. Gavrilov "K-stage Signal Detection", Journal of Communications Technology and Electronics, vol. 43, 7, pp. 835-850, 1998, (in Russian).
9. Garvanov, I. and Chr. Kabakchiev, "K-stage CFAR Detection in Binomial Distribution Pulse Jamming", Proc. of the International Radar Symposium - IRS 2003, Dresden, Germany, pp. 369-375, 2003.
10. Lazarov, A., Ch. Minchev, "ISAR Image Reconstruction Technique with Stepped Frequency Modulation and Multiple Receivers", Proc. DASC'05, Washington, CD-14E2-115, 2005.
11. Garvanov, I., "CFAR BI Detector in Binomial Distribution Pulse Jamming", Comptes Rendus de l'Academie Bulgare des Sciences, vol. 56, 10, pp. 37-44, 2003.
12. Akimov, P., Evstratov, F., Zaharov, S.: Radio Signal Detection, Moscow, Radio and Communication, pp. 195-203, 1989, (in Russian).

# A Method for Calculating Active Feedback System to Control Vertical Position of Plasma in a Tokamak

Nizami Gasilov

Baskent University, Faculty of Engineering, Baglica kampusu, 06530 Ankara, Turkey

**Abstract.** In designing tokamaks, the maintenance of vertical stability of plasma is one of the most important problems. For this purpose systems of passive and active feedbacks are applied. The role of passive system consisting of a vacuum vessel and passive stabilizer plates is to suppress fast MHD (magnetohydrodynamic) instabilities. The active feedback system is applied to control slow motions of plasma. The objective of this paper is to investigate three successive problems the solution of which will allow to determine the possibility to control plasma motions. The first problem is the vertical stability problem under the assumption of ideal conductivity of plasma and passive stabilizing elements. The problem is solved analytically and on the basis of the obtained solution a criterion of MHD-stability is formulated.

The second problem is the vertical stability when finite conductivity of stabilizing elements is taken into account. The dispersion equation relative to instability growth rate is obtained and analyzed. For practical values of the parameters it is shown that there is a unique root with positive real part, which presents the growth rate of only unstable mode.

The third problem is connected with the control of plasma vertical position with application of active feedback system. The problem of calculation of feedback control parameters is formulated as an optimization problem and its approximate solving method is suggested.

## 1 Introduction

The problems of stability and controllability of plasma vertical motions are of special importance for tokamaks with divertor. Necessary condition for the controllability is the stability of plasma column under the assumption of ideal conductivity of passive stabilizing elements. If under this assumption the stability does not take place, the plasma deviates from equilibrium position on Alfven time scales ($\sim 10^{-6}$ sec). The last circumstance makes constructing an effective feedback control system impossible. The problem of plasma vertical stability is well studied [1-7]. However, simple stability criteria are necessary. Therefore one of the purposes of the paper is the analytical solution of the problem, on the basis of which it is possible to formulate a simple enough stability criterion.

The instability growth time determines the restrictions on response time and power of active feedbacks. Analytical investigation of vertical instability problem in the case of finite conductive stabilizers is the second purpose of the paper.

T. Boyanov et al. (Eds.): NMA 2006, LNCS 4310, pp. 566–573, 2007.

The passive system, even if it is constructed successfully, can suppress only fast instabilities. Control of slow instabilities, development time of which is proportional to the characteristic decay time of the eddy currents in passive conductors ($\sim 10^{-3}$ sec), can be ensured using an active feedback system. In designing an active system, the problem of calculation of feedback control parameters arises, research of which is the third purpose of this paper.

## 2   General Set of Equations

The rigid-shift model will be applied to investigate plasma vertical stability. The model assumes that the entire plasma column moves in vertical direction as a solid body. The equilibrium distribution of the current in plasma is taken into account. But it is supposed that this distribution (therefore, the plasma total current $I_p$), and also the shape of plasma remain unchanged as plasma moves. Thus, it is supposed that only the eddy currents, induced in the passive structure by the plasma motion, are changing. The vacuum vessel and passive stabilizer plates are modeled as sets of elementary toroidal coils.

Mathematically, the model is described by plasma motion equation and circuit equations, and is represented by a system of linear differential equations. Unknowns are vertical displacement of plasma and eddy currents in passive conductors. The details of the model used are explained in papers [8, 9].

We first consider the situation without active feedback. Let $\tilde{\xi}(t)$ be a small vertical displacement of the plasma column from a reference equilibrium position. Then under the assumptions made the equation of plasma motion is as follows:

$$M\frac{d^2\tilde{\xi}}{dt^2} = w_0\tilde{\xi}(t) + \sum_{j=1}^{N} w_j\tilde{I}_j(t) \ . \tag{1}$$

Here $M$ is the mass of plasma, $w_0\tilde{\xi}$ represents the reverting Lorentz force on moving plasma due to the external field gradient, $N$ is the number of passive stabilizing coils, $\tilde{I}_j(t)$ is the eddy current in $j$-th passive coil and $w_j\tilde{I}_j$ represents the returning force on the plasma due to this eddy current.

The currents in passive coils are described by Kirchhoff's equations:

$$\sum_{j=1}^{N} L_{ij}\frac{d\tilde{I}_j}{dt} + R_i\tilde{I}_i(t) = -\Phi_i\frac{d\tilde{\xi}}{dt} = -I_p\frac{dL_{ip}}{d\tilde{\xi}}\frac{d\tilde{\xi}}{dt} \ , \qquad i = 1, \ \ldots, \ N, \tag{2}$$

where $R_i$ is ohmic resistance of $i$-th coil, $L_{ii}$ and $L_{ij}$ are self and mutual inductances of the coils, $L_{ip}(\tilde{\xi})$ is the mutual inductance between $i$-th coil and plasma (The value of $L_{ip}$ is calculated by taking the non-homogeneous distribution of plasma current into consideration). It can be shown that, under the assumptions made, the relation $w_i = \Phi_i$ holds.

Let's represent solutions of the problem (1) - (2) in exponential form:

$$\tilde{\xi}(t) = \xi \ e^{\gamma t}, \quad \tilde{I}_j(t) = I_j \ e^{\gamma t}, \quad j = 1, \ \cdots, N. \tag{3}$$

Then the set of equations (1) - (2) is reduced to an algebraic eigenvalue problem:

$$\gamma \left\{ \sum_{j=1}^{N} L_{ij} I_j + w_i \xi \right\} = -R_i I_i , \qquad i = 1, \ldots , N ,$$

$$\gamma^2 M \xi = \sum_{j=1}^{N} w_j I_j + w_0 \xi .$$

(4)

If we designate $\begin{aligned} \mathbf{I} &= (I_1, \ldots , I_N)^T, \\ \mathbf{w} &= (w_1, \ldots , w_N)^T, \end{aligned}$ $L = \begin{bmatrix} L_{11} & \cdots & L_{1N} \\ \cdots & \cdots & \cdots \\ L_{N1} & \cdots & L_{NN} \end{bmatrix}, R = \begin{bmatrix} R_1 & \cdots & 0 \\ \cdots & \ddots & \cdots \\ 0 & \cdots & R_N \end{bmatrix}$

then the problem (4) can be rewritten in the form:

$$R\mathbf{I} + \gamma \, L\mathbf{I} = -\gamma \, \mathbf{w}\xi ,$$

$$\gamma^2 M \xi = ( \mathbf{w}, \ \mathbf{I}) + w_0 \xi .$$

(5)

In particular, it can be found from (5), that $w_0 = M\gamma_0^2$, where $\gamma_0$ is Alfven growth rate in absence of stabilizing elements. Note, that $R$ (matrix of resistances) is a diagonal matrix and $L$ (matrix of inductances) is a symmetric one. In further considerations we suppose that the following natural conditions are satisfied:

$$R \geq 0, \ L > 0, \ w_0 = M\gamma_0^2 > 0, \ \mathbf{w} \neq \mathbf{0}.$$

(6)

## 3    Stability in the Approach of Ideal Conductivity of Stabilizing Elements

To control plasma displacements is possible, only if plasma is stable under the assumption of ideal conductivity of passive stabilizing elements. In this case $R = 0$. For $\gamma = 0$ the plasma is stable, therefore this case is not interesting for us and in the following part we investigate the case $\gamma \neq 0$. Then, if we express $\mathbf{I}$ from the first equation of the system (5) and put it in the second equation, we obtain next formula for the growth rate $\gamma$ in the ideal case:

$$\gamma_{ideal}^2 = \frac{w_0}{M} - \frac{1}{M}(L^{-1}\mathbf{w}, \ \mathbf{w}) .$$

(7)

The ideal conductors stabilize plasma, if $\gamma_{ideal}^2 < 0$. Then the below criterion of MHD-stability, or necessary condition of stabilization, follows from (7):

$$(L^{-1}\mathbf{w}, \ \mathbf{w}) > w_0$$

(8)

## 4    Stability When Finite Conductivity of Stabilizing Elements Is Taken into Account

In this case $R > 0$. We suppose that the other conditions (6) are satisfied also. For system (5) we investigate all solutions with $Re\gamma \geq 0$ (unstable modes). In this case it is not difficult to obtain from (5) the following equation for $\gamma$:

$$M\gamma^2 = M\gamma_0^2 - \gamma \left( (R + \gamma L)^{-1} \mathbf{w}, \ \mathbf{w} \right) .$$

(9)

It can be noted at once that $\gamma = 0$ is not a root of the equation.

At first we consider positive real roots of the equation (9). We rewrite (9) as

$$f(\gamma) = M\gamma^2 + \gamma \left( (R + \gamma L)^{-1} \mathbf{w}, \mathbf{w} \right) - M\gamma_0^2 = 0 .$$

$f(\gamma)$ is a continuous function at $\gamma \geq 0$, and $f(0) = -M\gamma_0^2 < 0$, $f(+\infty) = +\infty$. Consequently the equation (9) has at least one positive root. As $f'(\gamma) = 2M\gamma + \left( (R + \gamma L)^{-1} R (R + \gamma L)^{-1} \mathbf{w}, \mathbf{w} \right) > 0$, this positive root is unique.

Now we study the question whether (9) has complex roots with non-negative real part. Let $\gamma = \alpha + i\beta$ $(\beta \neq 0)$. Then $\frac{1}{\gamma} = x + iy = \frac{\alpha}{\alpha^2 + \beta^2} + i \left( -\frac{\beta}{\alpha^2 + \beta^2} \right)$. Using the formula $(A + iB)^{-1} = (A + BA^{-1}B)^{-1} - i(AB^{-1}A + B)^{-1}$ the following system of equations with respect to $\alpha$ and $\beta$ can be obtained from (9):

$$\gamma_0^2 + \beta^2 - \alpha^2 = \frac{1}{M} \left( \left( L + xR + y^2 R(L + xR)^{-1} R \right)^{-1} \mathbf{w}, \mathbf{w} \right) ,$$

$$\alpha = -\frac{1}{2M(\alpha^2 + \beta^2)} \left( \left( (L + xR)R^{-1}(L + xR) + y^2 R \right)^{-1} \mathbf{w}, \mathbf{w} \right) .$$

From the second equation of the system it follows that $\alpha < 0$. Consequently, under the conditions accepted the equation (9) can not have a complex root with non-negative real part.

The results mentioned above can be summarized in the following theorem.

*Theorem.* If $M > 0$, $w_0 = M\gamma_0^2 > 0$, $\mathbf{w} \neq \mathbf{0}$, and matrices $R > 0$ and $L > 0$ are symmetric real matrices, then the problem (5) has a unique positive real eigenvalue $\gamma$ (unstable mode). Other eigenvalues are either negative real roots, or complex roots with negative real part (stable modes).

In the general case, it is difficult to obtain a formula for positive eigenvalue mentioned in the theorem. But, if the MHD-stability criterion (8) takes place, then ignoring second and higher powers of $\gamma$, and using the relation $(R + \gamma L)^{-1} \approx R^{-1} - \gamma R^{-1} L R^{-1}$, from (9) one can obtain the following approximate estimation for growth rate of slow instability: $\gamma \approx \frac{M\gamma_0^2}{(R^{-1}\mathbf{w}, \mathbf{w})}$.

## 5    Stabilization by Applying Active Feedbacks

We consider active feedback systems consisting of resistive toroidal coils. We suppose that each time moment given the displacement of plasma $(\tilde{\xi}(t))$, and its velocity $(\tilde{\xi}'(t))$ can be measured. According to these values, feedbacks produce a voltage $U_i(\tilde{\xi}(t), \tilde{\xi}'(t))$ in the $i$-th active coil. We assume that control coils are of two kinds: coils reacting to displacement and coils reacting to velocity of displacement. We suppose that the feedbacks allow generating voltage in the first case equal to $a_i\tilde{\xi}(t)$, and in the second case equal to $b_i\tilde{\xi}'(t)$. Note that, in general, a single coil can be of both kinds at the same time. Then $U_i = a_i\tilde{\xi}(t) + b_i\tilde{\xi}'(t)$. Thus, we have the following equation for the $i$-th active coil instead of (2):

$$\sum_{j=1}^{N} L_{ij} \frac{d\tilde{I}_j}{dt} + R_i \tilde{I}_i(t) = -\Phi_i \frac{d\tilde{\xi}}{dt} - a_i\tilde{\xi}(t) - b_i \frac{d\tilde{\xi}}{dt}. \tag{10}$$

Note that (10) is valid for all coils: both for active and for passive, if it is accepted that $a_i = 0$ and $b_i = 0$ for passive coils.

We consider that positions of active coils are given. Then the problem of selection of active system leads to the determination of control parameters $a_i$ and $b_i$, ensuring stabilization of plasma vertical motion. If we represent solutions in the form (3), the problem (1), (10) is reduced to the following algebraic problem:

$$(R + \gamma L)\mathbf{I} = -\gamma (\mathbf{w} + \mathbf{b})\xi - \mathbf{a}\xi, \qquad (11)$$
$$\gamma^2 M \xi = (\mathbf{w}, \ \mathbf{I}) + w_0 \xi,$$

where $\mathbf{a} = (a_1, \ \dots \ , a_N)^T$, $\mathbf{b} = (b_1, \ \dots \ , b_N)^T$. If we express $\mathbf{I}$ from the first equation of the system (11) and put it in the second equation, we obtain:

$$\gamma^2 M - w_0 + \gamma((R + \gamma L)^{-1}(\mathbf{w} + \mathbf{b}), \ \mathbf{w}) + ((R + \gamma L)^{-1}\mathbf{a}, \ \mathbf{w}) = 0 \ .$$

Using the formula $(A + B)^{-1} = A^{-1} - (A + B)^{-1}BA^{-1}$ we have:

$$\varphi(\gamma) = \left[ (R^{-1}\mathbf{w}, \mathbf{a}) - w_0 \right] + \left[ \gamma^2 M + \gamma((R + \gamma L)^{-1}\mathbf{w}, \mathbf{w} + \mathbf{b} - LR^{-1}\mathbf{a}) \right] = 0.$$
$$(12)$$

The root $\gamma$ of the equation $\varphi(\gamma) = 0$ with the greatest real part determines the most unstable mode. Note that $\gamma$ is a function of $\mathbf{a}$ and $\mathbf{b}$: $\gamma = \gamma(\mathbf{a}, \ \mathbf{b})$.

Under the assumptions made the function $\varphi(\gamma)$ is continuous at $\gamma \geq 0$ and it is not difficult to see that $\varphi(+\infty) = +\infty$. Therefore, for the equation $\varphi(\gamma) = 0$ to have no positive root, the fulfillment of the following condition is necessary:

$$\varphi(0) = (R^{-1}\mathbf{w}, \ \mathbf{a}) - w_0 \geq 0. \qquad (13)$$

We will call (13) as the necessary condition for vertical position control.

## 6    Problem of Selection of Active Feedback System

If the coordinates of active coils have been defined, the problem of selection of the feedback system consists of the following: Find control vectors $\mathbf{a}$ and $\mathbf{b}$ ensuring stabilization of plasma vertical motion and requiring minimal power.

Let's calculate total power necessary for realization of active feedbacks. The function $U_i = a_i \tilde{\xi}(t) + b_i \tilde{\xi}'(t)$, expressing feedback, has a meaning of voltage. According to the formula $P = U^2/R$, for total power requirement we have:

$$P(\mathbf{a}, \ \mathbf{b}) = \sum \frac{a_i^2 + b_i^2 \gamma^2(\mathbf{a}, \ \mathbf{b})}{R_i} \xi_m^2,$$

where $\xi_m$ is the parameter of the problem and expresses the maximum amplitude of displacements, for which the system is designed. Hereinafter summation is performed on active coils, though it can be accepted also as summation on all coils with allowance for $a_i = 0$ and $b_i = 0$ for passive coils.

The problem about selection of an active feedback system, in the general form, can be formulated as an optimization problem: Find values of parameters $\mathbf{a}$ and $\mathbf{b}$ satisfying the constraint $\gamma(\mathbf{a}, \ \mathbf{b}) \leq 0$ and minimizing function $P(\mathbf{a}, \ \mathbf{b})$.

In general case, it is difficult to find analytical expression for roots of (12). Let's find an approximate estimation for $\gamma$. If the necessary condition of stabilization is satisfied, i.e. if the plasma is stable in the ideal approach, 2nd and higher powers of $\gamma$ can be neglected. Then from (12) using relation $(R + \gamma L)^{-1} \approx R^{-1} - \gamma R^{-1}LR^{-1}$ an approximate estimation for the growth rate is obtained:

$$\gamma = -\frac{(\mathbf{a},\ R^{-1}\mathbf{w}) - w_0}{(R^{-1}\mathbf{w},\ \mathbf{w}) + (\mathbf{b},\ R^{-1}\mathbf{w}) - (\mathbf{a},\ R^{-1}LR^{-1}\mathbf{w})}. \tag{14}$$

Let's note that $w_0$ and $s = (R^{-1}\mathbf{w},\ \mathbf{w}) > 0$ are determined by input data and consequently they can be considered as known parameters. For the active system to stabilize plasma, condition $\gamma(\mathbf{a},\ \mathbf{b}) \leq 0$ must be satisfied. According to the necessary condition (13), the numerator of fraction in (14) must be non-negative. Therefore for stability it is necessary for the denominator to be positive. Thus, the parameters $\mathbf{a}$ and $\mathbf{b}$ must satisfy the restrictions

$$(\mathbf{a},\ R^{-1}\mathbf{w}) \geq w_0; \qquad (\mathbf{b},\ R^{-1}\mathbf{w}) > (\mathbf{a},\ R^{-1}LR^{-1}\mathbf{w}) - s.$$

In order to prevent errors, concerned with the determination of input data, the fulfillment of these conditions with some reserves $\varepsilon_1$ and $\varepsilon_2$ ($\varepsilon_2 > \varepsilon_1$) will be required. I.e. it will be required that in (14) the numerator be not less than $\varepsilon_1$, and the denominator be not less than $\varepsilon_2$. Then the new restrictions are:

$$(\mathbf{a},\ R^{-1}\mathbf{w}) \geq w_0 + \varepsilon_1 = \overset{\wedge}{w}_0,$$
$$(\mathbf{b},\ R^{-1}\mathbf{w}) \geq (\mathbf{a},\ R^{-1}LR^{-1}\mathbf{w}) - (s - \varepsilon_2) = (\mathbf{a},\ R^{-1}LR^{-1}\mathbf{w}) - \overset{\vee}{s}.$$

In scalar products involving vector $\mathbf{a}$, actually only those components of vectors the numbers of which coincide with numbers of active coils of 1st kind are used. Therefore it is convenient to use the corresponding projections of vectors. Let's designate the subspace spanned on basis vectors the numbers of which correspond to the active coils of 1st kind as $A$. Let's denote $\mathbf{u} - proj_A R^{-1}\mathbf{w}$ and $\mathbf{v} = proj_A R^{-1}LR^{-1}\mathbf{w}$. Then we have: $(\mathbf{a},\ R^{-1}\mathbf{w}) = (\mathbf{a},\ \mathbf{u})$ and $(\mathbf{a},\ R^{-1}LR^{-1}\mathbf{w}) = (\mathbf{a},\ \mathbf{v})$.

Similar things can be done also for $\mathbf{b}$. We designate the corresponding subspace as $B$. If we denote $\mathbf{p} = proj_B R^{-1}\mathbf{w}$, then $(\mathbf{b},\ R^{-1}\mathbf{w}) = (\mathbf{b},\ \mathbf{p})$.

Let's note that the vectors $\mathbf{u}$, $\mathbf{v}$ and $\mathbf{p}$ are represented by input data of the problem and consequently we can consider them as known parameters.

Using the new notations, we can reformulate the problem of selection of active feedbacks: Find values of parameters $\mathbf{a}$ and $\mathbf{b}$, satisfying the restrictions

$$(\mathbf{a},\ \mathbf{u}) \geq \overset{\wedge}{w}_0,$$
$$(\mathbf{b},\ \mathbf{p}) \geq (\mathbf{a},\ \mathbf{v}) - \overset{\vee}{s}. \tag{15}$$

and minimizing the function

$$P(\mathbf{a},\ \mathbf{b}) = P_1(\mathbf{a}) + P_2(\mathbf{a},\ \mathbf{b}) = \left\| \sqrt{R^{-1}}\,\mathbf{a} \right\|^2 \xi_m^2 + \gamma^2 \left\| \sqrt{R^{-1}}\,\mathbf{b} \right\|^2 \xi_m^2, \tag{16}$$

where $\gamma = -\frac{(\mathbf{a},\mathbf{u}) - w_0}{s + (\mathbf{b},\mathbf{p}) - (\mathbf{a},\mathbf{v})}$. Above mentioned $w_0$, $\overset{\wedge}{w}_0$, $s$, $\overset{\vee}{s}$, $\mathbf{u}$, $\mathbf{v}$, $\mathbf{p}$, $\xi_m$ and $R$ are given parameters.

We carry out the selection of a feedback system in two stages, sequentially solving two optimization problems. Actually it means a certain separation of roles between coils of 1st and 2nd kinds. At the first stage by using 1st kind coils we minimize first addend $P_1$ in (16) under first restriction of (15). At the second stage we use 2nd kind coils to satisfy second restriction and minimize $P_2$.

The first problem: Find a value of $\mathbf{a}$, satisfying restriction $(\mathbf{a}, \mathbf{u}) = \overset{\wedge}{w}_0$ and minimizing the function $P_1(\mathbf{a}) = \xi_m^2 \sum \frac{a_i^2}{R_i}$.

Using Lagrange multipliers it can be seen that the solution of this problem is the vector $\mathbf{a}^*$ with coordinates $a_i^* = \frac{\overset{\wedge}{w}_0}{\sum R_i u_i^2} R_i u_i$. Thus $P_1(\mathbf{a}^*) = \xi_m^2 \frac{\overset{\wedge}{w}_0^2}{\sum R_i u_i^2}$.

If the vector $\mathbf{a}^*$ also meets condition $\overset{\vee}{s} - (\mathbf{a}^*, \mathbf{v}) \geq 0$, the pair of vectors $\mathbf{a}^*$ and $\mathbf{b} = \mathbf{0}$ satisfies both restrictions of the general problem (15) - (16) and, as it can be easily checked, it is an optimal solution.

If $\mathbf{a}^*$ satisfies the condition $\overset{\vee}{s} - (\mathbf{a}^*, \mathbf{v}) < 0$, the vector $\mathbf{b}$ can be selected by solving the problem formulated below. Let's designate $D = (\mathbf{a}^*, \mathbf{v}) - s > -\varepsilon_2$. Note that $D$ is easily calculated by input parameters and can be considered as a given parameter also. Let's note that $(\mathbf{a}^*, \mathbf{u}) - w_0 = \varepsilon_1$. Then $\gamma(\mathbf{a}^*, \mathbf{b}) = -\frac{\varepsilon_1}{(\mathbf{b}, \mathbf{p}) - D}$.

The second problem: Find a value of $\mathbf{b}$, satisfying restriction $(\mathbf{b}, \mathbf{p}) = D + E$ (where $E \geq \varepsilon_2$ is a parameter of the problem) and minimizing the function $P_2(\mathbf{b}) = \xi_m^2 \sum \frac{b_i^2}{R_i} \gamma^2$, where $\gamma = -\frac{\varepsilon_1}{E}$.

This problem is similar to the first one and its solution is the vector $\mathbf{b}^*$ with coordinates $b_i^* = \frac{D+E}{\sum R_i p_i^2} R_i p_i$. Note that $\mathbf{b}^*$ depends on the parameter $E$. It can be seen that $P_2(\mathbf{b}^*) = \xi_m^2 \left(1 + \frac{D}{E}\right)^2 \frac{\varepsilon_1^2}{\sum R_i p_i^2}$. Therefore, an increase in $E$ results in a decrease in the required power. However an increase in $E$ also results in an increase in the absolute values of $b_i^*$. The feedbacks are represented by products $b_i \tilde{\xi}'(t)$, second factor $(\tilde{\xi}'(t))$ of which is determined through measurements, so with some error. Therefore, the sharp increase in $b_i^*$ can also lead to increase in the error of the right hand side of equation (10). As a result, the work of feedbacks can be disturbed. Therefore, in the selection of an optional value of $E$ it is necessary to make a decision taking into consideration the power requirement and levels of measurement errors.

The pair of vectors $\mathbf{a}^*$ and $\mathbf{b}^*$ satisfies both restrictions of the general problem (15) - (16), thus, condition $\gamma(\mathbf{a}, \mathbf{b}) \leq 0$. Therefore, the active feedback system with parameters $\mathbf{a}^*$ and $\mathbf{b}^*$ provides stabilization of vertical motion. Let's note that this solution may not be optimum for the general problem. On the other hand, such feedback control means simple separation of roles between coils of 1st and 2nd kinds, which is preferable for technical realization.

# 7    Conclusions

Three problems arising in the study of stabilization of plasma vertical motions in a tokamak were considered. The results can be expressed in the following form.

On the basis of the rigid shift model of plasma the vertical stability problem was analyzed. The problem was solved analytically under assumption of ideal conductivity of plasma and passive stabilizing elements. A MHD stability criterion was obtained, which was expressed in compact form. A system of passive conductors must satisfy this criterion, which is necessary for operating the active feedback system effectively.

The second problem is the vertical stability problem when finite conductivity of stabilizing elements is taken into account. For practical values of the parameters it was shown that stability problem has unique unstable mode with positive real growth rate, other modes are stable.

The question about control of plasma vertical motions with application of an active feedback system was studied. The problem on calculation of active feedbacks control parameters was formulated as an optimization problem and a method of its approximate solution was proposed.

# References

1. Wesson, J.: Tokamaks. Clarendon Press, Oxford (1982)
2. Zakharov, L.E., Shafranov, V.D.: Equilibrium of plasma with current in toroidal systems. In: Problems of the Theory of Plasma, 11th Issue. Energoizdat, Moscow (1982) 118-235
3. Jardin, S.C., Larrabee, D.A.: Feedback stabilization of rigid axisymmetric modes in tokamaks. Nuclear Fusion **22** (1982) 1095-1098
4. Dnestrovskij, Yu.N., Kostomarov, D.P., Pistunovich, V.I., Popov, A.M., Sychugov, D.Yu.: Stabilization of vertical instability in a tokamak with poloidal divertor. Plasma Physics **10** (1984) 688-694
5. Dnestrovskij, Yu.N., Kostomarov, D.P.: Numerical simulation of plasmas. Springer-Verlag, New York (1986)
6. Lazarus, E.A., Lister, J.B., Neilson, G.H.: Control of vertical instability in tokamaks. Nuclear Fusion **30** (1990) 111-141
7. Ward, D.J., Bondeson, A., Hofman, F.: Pressure and inductance effects on the vertical stability of shaped tokamaks. Nuclear Fusion **33** (1993) 821-828
8. Sychugov, D.Yu., Amelin, V.V., Gasilov, N.A., Tsaun, S.V.: Numerical investigation of vertical instability of plasma in tokamak with finite conductive stabilization elements. Moscow State University Herald, Series 15: Computational mathematics and cybernetics **4** (2004) 27-32
9. Bondarchuk, E.N., Dnestrovskij, Yu. N., Leonov, V.M., Maksimova, I.I., Sychugov, D.Yu., Tsaun, S.V., Voznesensky, V.A.: Vertical MHD stability of the T-15M tokamak plasma. Plasma Devices and Operations **11** (2003) 219-227

# Numerical Algorithm for Simultaneously Determining of Unknown Coefficients in a Parabolic Equation

Nizami Gasilov, Afet Golayoglu Fatullayev, and Krassimir Iankov

Baskent University, Eskisehir Yolu 20. km, Baglica kampusu, 06530 Ankara, Turkey

**Abstract.** A numerical algorithm for an inverse problem of simultaneously determining unknown coefficients in a linear parabolic equation subject to the specifications of the solution at internal points along with the usual initial and boundary conditions is proposed. The approach based on TTF (Trace Type Functional) formulation of the problem is used. To avoid instability in this approach the Tikhonov regularization method is applied. Some numerical examples using the proposed algorithm are presented.

## 1 Introduction

In this paper we consider an inverse problem of simultaneously finding unknown coefficients $p(t)$, $q(t)$ and function $u(x,t)$ that satisfy the equation

$$u_t = u_{xx} + q(t)u_x + p(t)u + f(x,t), \quad x \in (0,1), \ t \in (0,T] \tag{1}$$

with the initial-boundary conditions

$$u(x,0) = \varphi(x), \qquad x \in (0,1), \tag{2}$$

$$u(0,t) = g_1(t), \qquad u(1,t) = g_2(t), \qquad t \in (0,T] \tag{3}$$

and the additional specifications

$$u(x_*,t) = e_1(t), \ u(x_{**},t) = e_2(t), \qquad x_*, \ x_{**} \in (0,1), \quad t \in (0,T], \tag{4}$$

where $f(x,t)$, $\varphi(x)$, $g_1(t)$, $g_2(t)$, $e_1(t)$ and $e_2(t)$ are given input functions.

If $u$ represents the concentration then equation (1) models the transport, dispersion and decay of a chemical solute (a tracer) with concentration $u$ moving through a porous medium (an aquifer), $q(t)$ is the average velocity (the drift velocity), $p(t)$ represents the magnitude of the decay [1]. If $u$ is a temperature then the problem (1)-(4) can be considered as a control problem of finding the control $p = p(t)$ and $q = q(t)$ such that the internal constraints (4) are satisfied. Similar inverse problems have been studied in [2-7].

Looking at the problem (1)-(4) we see that if the functions $p(t)$, $q(t)$ and $u(x,t)$ solve the inverse problem, then it follows that

$$e_1'(t) = u_{xx}|_{x=x_*} + q(t)u_x|_{x=x_*} + p(t)u|_{x=x_*} + f(x_*,t),$$

T. Boyanov et al. (Eds.): NMA 2006, LNCS 4310, pp. 574–581, 2007.

$$e_2'(t) = u_{xx}|_{x=x_{**}} + q(t)u_x|_{x=x_{**}} + p(t)u|_{x=x_{**}} + f(x_{**}, t).$$

Thus unknowns $q(t)$ and $p(t)$ are solutions of the following system of two linear algebraic equations

$$\begin{bmatrix} u_x|_{x=x_*} & u|_{x=x_*} \\ u_x|_{x=x_{**}} & u|_{x=x_{**}} \end{bmatrix} \begin{bmatrix} q(t) \\ p(t) \end{bmatrix} = \begin{bmatrix} E_1(t) \\ E_2(t) \end{bmatrix}, \tag{5}$$

where $E_1(t) = e_1'(t) - u_{xx}|_{x=x_*} - f(x_*, t)$, $E_2(t) = e_2'(t) - u_{xx}|_{x=x_{**}} - f(x_{**}, t)$.
For further purposes we represent the system (5) in matrix form

$$Az = b.$$

From (5) we can define $q(t)$ and $p(t)$ provided that the determinant of the system does not vanish. Eliminating these unknown functions from the inverse problem we can obtain a reformulated initial-boundary value problem in which coefficients are functionals of the unknown solution. Such approach is called as TTF (Trace Type Functional) formulation of a problem [5]. However the reformulated problem becomes more unstable. The instability is reflected in the fact that the terms $u_{xx}|_{x=x_*}$ and $u_{xx}|_{x=x_{**}}$ in this case make the coefficients extremely sensitive to variations in the solution which is recursively dependent on itself through these second order derivatives on fixed points. The aim of this paper is to study the possibility of using the Tikhonov regularization method to avoid the instability in this approach. The proposed algorithm will be explained in details in the next section.

## 2  Numerical Algorithm

The system (1)-(4) can be approximated by finite difference method. Let $\tau = \Delta t > 0$ and $h = \Delta x > 0$ be step-lengths in time and space coordinate, $0 = t_0 < t_1 < ... < t_M = T$ and $0 = x_0 < x_1 < ... < x_N = 1$, where $t_j = j\tau$, $x_i = ih$, denote partitions of the $[0, T]$ and $[0, 1]$, respectively. Let also $u_i^j$, $q^j$ and $p^j$ be approximations to $u(x_i, t_j)$, $q(t_j)$ and $p(t_j)$ respectively. Then implicit finite difference scheme for (1)-(4) can be written as follows:

$$\frac{u_i^j - u_i^{j-1}}{\tau} = \frac{u_{i+1}^j - 2u_i^j + u_{i-1}^j}{h^2} + q^j \frac{u_{i+1}^j - u_{i-1}^j}{2h} + p^j u_i^j + f_{ij}, \tag{6}$$

$$u_i^0 = \varphi(x_i), \tag{7}$$

$$u_0^j = g_1(t_j), \qquad u_N^j = g_2(t_j), \tag{8}$$

$$u_{i_*}^j = e_1(t_j), \qquad u_{i_{**}}^j = e_2(t_j), \tag{9}$$

where $f_{ij} = f(x_i, t_j)$, $j = \overline{1, M}$, $i = \overline{1, N-1}$.
We apply the following algorithm to solve system (6)-(9).
At the initial time step $u_i^0 = \varphi(x_i)$. To find initial values $q^0$ and $p^0$ we solve the finite difference approximation of (5) at $t = 0$.

Let at $(j-1)$-th time step we have calculated $u_i^{j-1}$, $q^{j-1}$ and $p^{j-1}$. Then at $j$-th time step to calculate $u_i^j$, $q^j$ and $p^j$ we solve the system (6), (8)-(9) applying the following iterative procedure.

We denote the value of $u_i^j$ at $s$-th iteration step as $u_i^{j,s}$.

At the initial step of iterations ($s = 0$) we put: $u_i^{j,0} = u_i^{j-1}$.

At each $(s+1)$-th iteration step we firstly determine $q^{j,s}$ and $p^{j,s}$. If $s = 0$ then we put $q^{j,0} = q^{j-1}$ and $p^{j,0} = p^{j-1}$. If $s \geq 1$ then we solve the finite difference approximation of the system (5)

$$\begin{bmatrix} \dfrac{u_{i_*+1}^{j,s}-u_{i_*-1}^{j,s}}{2h} & u_{i_*}^{j,s} \\[2mm] \dfrac{u_{i_{**}+1}^{j,s}-u_{i_{**}-1}^{j,s}}{2h} & u_{i_{**}}^{j,s} \end{bmatrix} \begin{bmatrix} q^{j,s} \\ p^{j,s} \end{bmatrix} = \begin{bmatrix} \dfrac{e_1^j-e_1^{j-1}}{\tau} - \dfrac{u_{i_*+1}^{j,s}-2u_{i_*}^{j,s}+u_{i_*-1}^{j,s}}{h^2} - f_{i_* j} \\[2mm] \dfrac{e_2^j-e_2^{j-1}}{\tau} - \dfrac{u_{i_{**}+1}^{j,s}-2u_{i_{**}}^{j,s}+u_{i_{**}-1}^{j,s}}{h^2} - f_{i_{**} j} \end{bmatrix} \quad (10)$$

Then we put the values of $q^{j,s}$ and $p^{j,s}$ into equation (6) and obtain next system which consists of the equations

$$\left[\frac{1}{h^2} - \frac{q^{j,s}}{2h}\right] u_{i-1}^{j,s+1} - \left[\frac{2}{h^2} + \frac{1}{\tau} - p^{j,s}\right] u_i^{j,s+1} + \left[\frac{1}{h^2} + \frac{q^{j,s}}{2h}\right] u_{i+1}^{j,s+1} = -f_{ij} - \frac{u_i^{j-1}}{\tau} \quad (11)$$

and boundary conditions

$$u_0^{j,s+1} = g_1(t_j), \qquad u_N^{j,s+1} = g_2(t_j), \quad (12)$$

where $i = \overline{1, N-1}$, $j = \overline{1, M}$.

We solve the system (11)-(12) by the TDMA (Three Diagonal Matrix Algorithm) method and determine $u_i^{j,s+1}$. If the difference of the values on two successive iteration steps is small enough we stop iterations and we obtain $u_i^j$ on $j$-th time step. Note that by appropriate choice of $\tau$ at each time step we can achieve that the main diagonal of the three-diagonal system (11) be dominant.

The coefficients of the system (10) depend on the solution $u$ and $u_x$, and the right hand side depends on $u_{xx}$. Since $q^j$ and $p^j$ are a solution of (10) these quantities are calculated with some errors. Therefore $q^j$ and $p^j$ are very sensitive to errors, especially if the determinant of the system is close to zero. To avoid this difficulty we apply the Tikhonov regularization method [8]. Let's shortly describe it for the linear algebraic system in general form:

$$Az = b \quad (13)$$

Note that $A$ is not necessarily a square matrix, in general case, when Tikhonov regularization is considered. Generally, the system (13) may have no solution in classical sense. However, we can speak about the normal solution relative to some given vector $z_0$ (Note that $z_0$ expresses guessed solution and is determined from physical considerations). The normal solution for any linear system exists and is unique [8]. The problem of finding the normal solution is ill-posed, i.e. arbitrarily small changes in the input data (i.e. $b$) can cause arbitrarily large changes in the solution. The Tikhonov regularization method is applied for finding normal solution which is stable relative to small perturbations of the right-hand side of the system (experimental data). It is supposed that instead of true data we know their approximate values, i.e. instead of vector $b$ we have a vector $\bar{b}$ such that $\|b - \bar{b}\| \leq \delta$, where $\delta$ is the error of the measurements. The problem is to find for each value of $\delta$ an approximate solution $z_\delta$ which convergence to the exact normal solution $z^*$ as $\delta \to 0$. In Tikhonov regularization method $z_\delta$ is defined from the minimization of the Lagrange functional

$$\left\| Az - \bar{b} \right\|^2 + \alpha \left\| z - z_0 \right\|^2, \tag{14}$$

where $\alpha > 0$ is regularization parameter. The value of $\alpha$ is determined from condition $\left\| Az_\delta - \bar{b} \right\| = \hat{\delta}$, where $\hat{\delta} = \min_z \left\| Az - \bar{b} \right\| + 2\delta$.

Minimization problem (14) is equivalent to solving of the Euler equation

$$\left( A^T A + \alpha I \right) z = A^T \bar{b} + \alpha z_0.$$

Note that as a vector $z_0$ we take the solution at the preceding time step when we apply Tikhonov regularization method for solving the linear algebraic system (10) in our presented numerical algorithm.

Numerical experiments have been performed to test the effectiveness of the proposed algorithm, results of which are discussed in the following section.

## 3   Numerical Examples

In this section we present some results of our numerical calculations using the algorithm described in the previous section to solve the following examples.

**Example 1.** If we take a solution $u(x, t)$, coefficients $q(t)$, $p(t)$ and points $x_*$, $x_{**}$ as $u(x, t) = t \sin x + kx^2$ (where $k$ is a constant parameter), $q(t) = (t-2)^2$, $p(t) = 5 + 2t - t^2$, $x_* = 0.4$, $x_{**} = 0.6$ then substituting in (1)-(4), it can be seen that the input functions are as follows $f(x, t) = (t^3 - 2t^2 - 4t + 1) \sin x - t(t-2)^2 \cos x + k \left[ x^2(t^2 - 2t - 5) - 2x(t-2)^2 - 2 \right]$, $\varphi(x) = u(x, 0) = kx^2$, $g_1(t) = u(0, t) = 0$, $g_2(t) = u(1, t) = t \sin 1 + k$, $e_1(t) = t \sin 0.4 + 0.16k$ and $e_2(t) = t \sin 0.6 + 0.36k$. In calculations we take a grid with size $M \times N = 100 \times 200$, $T = 4$ and $k = 10^{-5}$.

In this example the matrix $A$ is well-conditioned on all grid nodes, and we can use any classical method (without regularization) for solving (10). As seen from the Fig. 1$a$, in this case there are no visual differences between the numerical and exact solutions.

The next task have been performed to test the sensitivity of the algorithm to the errors. Artificial random errors were introduced into the additional specification data by defining functions $\tilde{e}_1(t) = e_1(t)(1 + d_1^\varepsilon(t))$ and $\tilde{e}_2(t) = e_2(t)(1 + d_2^\varepsilon(t))$. Here $d_1^\varepsilon(t)$ and $d_2^\varepsilon(t)$ are random functions of $t$ uniformly distributed on $(-\varepsilon, \varepsilon)$. These functions represent the level of relative errors in the corresponding piece of data. In Fig. 1$b$ we present results for $\varepsilon = 0.005$. It can be seen from the figure that the algorithm is stable with respect to random errors if matrix $A$ remains well-conditioned during the calculation process.

**Example 2.** In this example we consider a situation when matrix $A$ becomes ill-conditioned at some grid points during the calculation process. For this purpose,

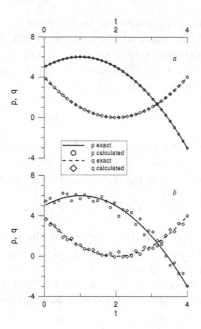

**Fig. 1.** The exact and calculated values of $p(t)$ and $q(t)$ when $A$ is well-conditioned matrix at all grid points. *a*) Additional specifications $e_1(t)$ and $e_2(t)$ are given exactly; *b*) Artificial random errors added to $e_1(t)$ and $e_2(t)$.

we use solution function as $u(x,t) = \sin(2\pi t)\sin x + kx^2$. Coefficients $q(t)$, $p(t)$ and the internal points $x_*$, $x_{**}$ are the same as in the previous example: $q(t) = (t-2)^2$, $p(t) = 5 + 2t - t^2$, $x_* = 0.4$, $x_{**} = 0.6$. The corresponding input functions are:
$f(x,t) = 2\pi \cos(2\pi t)\sin x + \sin(2\pi t)\left[(t^2 - 2t - 4)\sin x - (t-2)^2 \cos x\right] + k\left[x^2(t^2 - 2t - 5) - 2x(t-2)^2 - 2\right]$, $\varphi(x) = kx^2$, $g_1(t) = 0$, $g_2(t) = \sin(2\pi t)$
$\sin 1 + k$, $e_1(t) = \sin(2\pi t)\sin 0.4 + 0.16k$, $e_2(t) = \sin(2\pi t)\sin 0.6 + 0.36k$.
In the calculations we take $k = 5 \cdot 10^{-3}$ and grid with $M \times N = 200 \times 300$.

In Fig. 2*a* we present results of calculations when a classical method to solve (10) is used . As seen from the figure, at the grid points where the matrix $A$ becomes ill-conditioned, the differences between the calculated values and the exact solution are large, i.e. there is no approximation at these points. In Fig. 2*b* the results of calculations by using Tikhonov regularization method for solving (10) are presented. Since the condition number of the matrix $A$ depends on its determinant, it could be reasonable to choose $\delta$ depending on $\det A$. We take

$$\delta = \delta(\det A) = \begin{cases} \widetilde{\delta}, & \text{if } |\det A| > 10^{-2}, \\ (1 - 2\ln|\det A|)\widetilde{\delta}, & \text{if } 10^{-5} \leq |\det A| \leq 10^{-2}, \\ (1 - 2\ln 10^{-5})\widetilde{\delta}, & \text{if } |\det A| < 10^{-5}. \end{cases}$$

In calculations we put $\widetilde{\delta} = 0.001$. As seen from Fig. 2*b*, there is sufficiently good agreement between the exact and the approximate values.

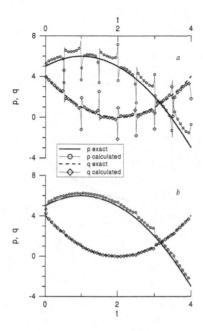

**Fig. 2.** The exact and calculated values of $p(t)$ and $q(t)$ when $A$ is ill-conditioned matrix at some grid points. Results of calculations $a)$ using classical methods; $b)$ using Tikhonov regularization method to solve (10).

Thus, the following procedure can be recommended for solving system (10). During the calculations the matrix $A$ of this system should be investigated. If $A$ is well-conditioned matrix then the system can be solved by usual methods (as Gauss-Jordan elimination). If $A$ is singular or ill-conditioned matrix then Tikhonov regularization method can be applied.

**Example 3.** Here we examine a modification of the proposed algorithm for problem of simultaneously determining of three unknowns in parabolic equation. Namely, we consider the problem (1)-(4) where the source function depends on the time variable only (i.e. $f = f(t)$) and must be determined. For this, in addition to the two measurements (4) we consider the third additional specification $u(x_{***}, t) = e_3(t)$.

For calculations we take a solution $u(x, t)$, coefficients $q(t)$, $p(t)$, function $f(t)$, points $x_*$, $x_{**}$, $x_{***}$ and value of $T$ as: $u(x, t) = e^{t^3-t} \left[e^{2x} - 2e^2 x\right] + \sin t$, $q(t) = -2$, $p(t) = 3t^2 - 1$, $f(t) = \cos t - (3t^2 - 1)\sin t - 4e^{t^3-t+2}$, $x_* = 0.4$, $x_{**} = 0.5$, $x_{***} = 0.6$ and $T = 2$. Then substituting in (1)-(4), it can be seen that the input functions are as follows $\varphi(x) = e^{2x} - 2e^2 x$, $g_1(t) = e^{t^3-t} + \sin t$, $g_2(t) = -e^2 e^{t^3-t} + \sin t$, $e_1(t) = e^{t^3-t} \left[e^{0.8} - 0.8e^2\right] + \sin t$, $e_2(t) = e^{t^3-t} \left[e - e^2\right] + \sin t$ and $e_3(t) = e^{t^3-t} \left[e^{1.2} - 1.2e^2\right] + \sin t$. In calculations we take a grid with size $M \times N = 100 \times 100$. The results of calculations are shown in Fig. 3 and it is seen from this figure that there is sufficiently good agreement between calculated

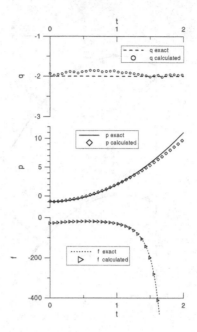

**Fig. 3.** The exact and calculated values of $q(t)$, $p(t)$ and $f(t)$ when $A$ is well-conditioned matrix at all grid points

and exact solutions. Thus, proposed numerical algorithm can be applied also for simultaneously determining of three unknowns in parabolic equation.

## 4    Conclusions

An inverse problem of simultaneously determining unknown coefficients in a linear parabolic equation subject to specifications of the solution at internal points along with the usual initial and boundary conditions is considered. An algorithm based on the TTF (Trace Type Functional) formulation of the problem is applied to solve problem numerically. One of stages of the algorithm consists of solving a linear algebraic system which can be ill-posed. To avoid instabilities the Tikhonov regularization method is used. Several test examples were investigated to study different situations with well and ill-conditioned matrices. The results of calculations show that in the case when the matrix of linear system is well-conditioned during the calculation process, usual methods such as Gauss-Jordan elimination can be used. In the case when the matrix becomes singular or ill-conditioned, using Tikhonov regularization gives sufficiently good approximation to the exact solution. Numerical experiments show effectiveness of the presented algorithm for determining a good approximation of the unknown coefficients.

# Acknowledgments

This research was supported by The Scientific and Technical Research Council of Turkey, TUBITAK, project number 104T137.

# References

1. Marsily, G.: Quantitative Hydrogeology. Academic Press, New York, 1986
2. Cannon, J.R., Lin, Y.: An inverse problem of finding a parameter in a semi-linear heat equation. J. Math. Anal. Appl. **145** (1990) 470-484
3. Lin, Y.: An inverse problem for a class of quasilinear parabolic equations. SIAM J. Math. Anal. **22** (1991) 146-156
4. Fatullayev, A.G.: Numerical procedure for the simultaneous determination of unknown coefficients in a parabolic equation. Appl. Math. and. Comp. **162** (2005) 1367-1375
5. Colton, D., Ewing, R., and Rundell, W.: Inverse problems in partial differential equation. SIAM Press, Philadelphia, 1990.
6. Cannon, J.R., Lin, Y.: Determination of a parameter $p(t)$ in some quasi-linear parabolic differential equations. Inverse Problems **4** (1998) 35-45
7. Cannon, J.R., Lin, Y., Wang, S.: Determination of source parameter in parabolic equations. Meccanica **27** (1992) 85-94
8. Tikhonov, A.N., Arsenin, V.Y.: Solutions of Ill-Posed Problems. Winston and Sons, Washington, 1977

# Data Compression of Color Images Using a Probabilistic Linear Transform Approach

Evgeny Gershikov[1] and Moshe Porat[2]

[1] Technion - Israel Institute of Technology, Haifa 32000, Israel
eugeny@tx.technion.ac.il
http://visl.technion.ac.il/~eugeny
[2] Technion - Israel Institute of Technology, Haifa 32000, Israel
mp@ee.technion.ac.il
http://visl.technion.ac.il/mp

**Abstract.** In this work, we design an efficient algorithm for color image compression using a model for the rate-distortion connection. This model allows the derivation of an optimal color components transform, which can be used to transform the RGB primaries or matrices into a new color space more suitable for compression. Sub-optimal solutions are also proposed and examined. The model can also be used to derive optimal bits allocation for the transformed subbands. An iterative algorithm for the calculation of optimal quantization steps is introduced using the subband rates (entropies). We show that the rates can be approximated based on a probabilistic model for subband transform coefficients to reduce the algorithm's complexity. This is demonstrated for the Discrete Cosine Transform (DCT) as the operator for the subband transform and the Laplacian distribution assumption for its coefficients. The distortion measure considered is the MSE (Mean Square Error) with possible generalization to WMSE (Weighted MSE). Experimental results of compressed images are presented and discussed for two versions of the new compression algorithm.

## 1 Introduction

Image compression has become a common mathematical procedure, where a matrix representing an image is replaced by data of lower bit-rate capacity. No information is lost in the so-called lossless algorithms, however, it is the lossy cases that have become the more acceptable procedure. Although perfect reconstruction is not possible in such cases, the loss of information, or distortion, can be of limited effect on the viewer, making the error negligible in the sense of visual perception. The main tool for evaluating the compression is a bit-rate vs. distortion curve. Such a curve predicts the distortion of coders for a given rate in bits per sample for color images, based on 3 matrices each: Red, Green and Blue (RGB). In this work we present a model for the rate - distortion dependency of subband transform coders. First the subband transform coders are presented in Section 2 and then the rate-distortion connection is derived in Section 3. Section

T. Boyanov et al. (Eds.): NMA 2006, LNCS 4310, pp. 582–589, 2007.
© Springer-Verlag Berlin Heidelberg 2007

4 concentrates on minimization of the compression distortion based on the model and Section 5 introduces a compression algorithm based on the results of this optimization problem. Finally, conclusions are given in Section 6.

## 2    Subband Transform Coders

Subband transforms are vastly used for signal and image compression. Two examples are JPEG [9] and JPEG 2000 [6], [8]. The most familiar representation of subband transforms is perhaps the filter bank scheme, displayed in Figure 1 for 1D (one dimensional) signals and a dyadic filter bank. The input signal $x[n]$ is decomposed by passing through a set of $m$ analysis filters $h_b[n](b \in [0, B-1])$ and down-sampling by a factor $m$. The signal can be then reconstructed by up-sampling the transform subbands $y_b[n]$ by a factor $m$ and filtering them through a synthesis filter-bank of $m$ filters $g_b[n]$. Mathematically, the decomposition is according to:

$$y_b[n] = \sum_l h_b[l] \, x[mn - l] = (x * h_b)[mn], \tag{1}$$

and the reconstruction is by:

$$x[n] = \sum_{b=0}^{m-1} \sum_i g_b[n - mi]y_b[i]. \tag{2}$$

A subband transform encoder is composed of a subband transform operator applied to the input signal $x$, then coding the subband coefficients. Usually this coding consists of quantization followed by some lossless algorithm, based on entropy coding. The decoder reconstructs the signal according to the subband coefficients after inverse quantization introducing a distortion in the reconstructed signal relative to the original one. The benefit is less bits needed for the storage or transmission of the encoded signal. The generalization of subband transform coding to 2D is straightforward, when the subband transform is separable (e.g. in JPEG [9] and JPEG2000 [6]). Also the generalization to multi-scale subband transform coders (e.g. JPEG2000 [6]) is straightforward by applying the subband transform iteratively to the low frequencies subband.

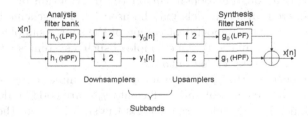

**Fig. 1.** Dyadic (m=2) filter-bank decomposition: analysis and synthesis

## 2.1    Vector Space Representation of Subband Transforms

Equation (1) can be rewritten in the form of the vector product of $\mathcal{C}^n$:

$$
\begin{aligned}
y_b[n] &= \sum_l h_b[l]\, x[mn - l] = \sum_l h_b[-l]\, x[mn + l] \\
&= \sum_l h_b[mn - l]\, x[l] = \sum_l \alpha_b^*[l - mn]\, x[l] = <\mathbf{x}, \alpha_b^{(n)}>,
\end{aligned}
\tag{3}
$$

where $\alpha_b[k] = h_b^*[-k]$ are sequences derived from the subband analysis filters and $\alpha_b^{(n)}$ is the sequence $\alpha_b[k]$ delayed by $mn$ samples in vector form. Thus (3) suggests that the transform coefficients $y_b[n]$ can be regarded as the result of an inner product of the input vector $\mathbf{x}$ and the analysis vectors $\alpha_b^{(n)}$. In a similar way (2), describing the filter bank synthesis operation, becomes:

$$
x[k] = \sum_{b=0}^{m-1} \sum_i g_b[k - mi] y_b[i] = \sum_{b=0}^{m-1} \sum_i y_b[i] s_b^{(i)}[k],
\tag{4}
$$

with $s_b^{(i)}[k]$ denoting the sequences $s_b[k] = g_b[k]$ delayed by $mi$ samples ($s_b^{(i)}[k] = g_b[k - mi]$). Thus in vector form:

$$
\mathbf{x} = \sum_{b=0}^{m-1} \sum_i y_b[i] \mathbf{s}_b^{(i)},
\tag{5}
$$

where $\mathbf{s}_b^{(i)}$ are the synthesis vectors.

# 3    Rate-Distortion of Subband Transform Coders

The rate-distortion performance of a scalar quantizer with independently coded samples for a stochastic source $x$ with variance $\sigma_x^2$ can be modelled for a large enough rate as [2], [8]:

$$
d(R) = \varepsilon^2 \sigma_x^2 2^{-2R},
\tag{6}
$$

where $d()$ is the MSE (Mean Square Error) distortion, $R$ is the rate in bits per sample and $\varepsilon^2$ is a constant dependent on the distribution of $x$. If low image rates are considered, other models may be used [5]. The scheme that performs scalar quantization with independent coding of the source samples is the known PCM (Pulse Code Modulation). An example of such a system is a uniform scalar quantizer with an entropy coded output.

Consider an encoder that first transforms an $N$-sample source signal $x$ into a set of subbands then each subband coefficients $y_b[i]$ are coded independently by the PCM scheme. The decoder reconstructs the signal $\hat{x}$ from the dequantized transform coefficients $\hat{y}_b[i]$. According to (5): $\mathbf{x} = \sum_{b=0}^{B-1} \sum_i y_b[i] \mathbf{s}_b^{(i)}$ and

$\hat{\mathbf{x}} = \sum_{b=0}^{B-1} \sum_i \hat{y}_b[i] \mathbf{s}_b^{(i)}$, where $B$ is the number of subbands. Thus the MSE of the coder for the signal $x$ can be expressed as [8]:

$$d_x = E\left[\frac{1}{N} \sum_k (\mathbf{x}[k] - \hat{\mathbf{x}}[k])^2\right] = \sum_{b=0}^{B-1} \eta_b G_b \sigma_b^2 \varepsilon^2 2^{-2R_b}, \qquad (7)$$

where $\eta_b$ denotes the ratio between the number of coefficients in subband $b$ and the total number of samples in the source $N$, $G_b = \left\|\mathbf{s}_b^{(i)}\right\|^2$ is the energy gain, $\sigma_b^2$ is the variance of subband $b$ and $R_b$ is the rate allocated to it. Equation (7) is valid for orthonormal as well as non-orthonormal transforms.

### 3.1 Extension of the Rate-Distortion to a Color Image

Denote each pixel in a color image in the RGB domain by a 3x1 vector $\mathbf{x} = [R\ G\ B]^T$. We first apply a color component transform (CCT) to the image, denoted by a matrix $\mathbf{M}$ to obtain at each pixel a new vector of 3 components $C1, C2, C3$, denoted $\widetilde{\mathbf{x}} = [C1\ C2\ C3]^T$, related to $\mathbf{x}$ by $\widetilde{\mathbf{x}} = \mathbf{Mx}$. Then each component in the new color space is subband transform coded. The average MSE between the original and the reconstructed images in the RGB domain is [1]:

$$MSE = \frac{1}{3} \sum_{i=1}^{3} \sum_{b=0}^{B-1} \eta_b G_b \sigma_{bi}^2 \varepsilon_i^2\ e^{-aR_{bi}} \left((\mathbf{MM}^T)^{-1}\right)_{ii}, \qquad (8)$$

where $R_{bi}$ stands for the rate allocated for the subband $b$ of color component $i$ and $\sigma_{bi}^2$ is this subband's variance. $a \triangleq 2ln2$.

Note that (8) can be extended to WMSE by introducing subband visual weights $W_{bi}$:

$$WMSE = \frac{1}{3} \sum_{i=1}^{3} \sum_{b=0}^{B-1} \eta_b W_{bi} G_b \sigma_{bi}^2 \varepsilon_i^2 e^{-aR_{bi}}. \qquad (9)$$

Here the WMSE is measured in the C1C2C3 and not in the RGB domain.

## 4   Minimization of the MSE

We would like to minimize the MSE expression of (8) with a constraint [7] on the total image rate: $\sum_{i=1}^{3} \alpha_i \sum_{b=0}^{B-1} \eta_b R_{bi} = R$ and constraints on the subband rates: $R_{bi} \geq 0$. $\alpha_i$ are optional down-sampling factors ($\alpha_i = 1$ means no down-sampling). This optimization problem can be solved analytically for the rates:

$$R_{bi}^* = \frac{1}{a} ln\left[\frac{\frac{\varepsilon_i^2 G_b \sigma_{bi}^2 \left((\mathbf{MM}^T)^{-1}\right)_{ii}}{\alpha_i}}{\prod_{k=1}^{3} \left(\frac{\left((\mathbf{MM}^T)^{-1}\right)_{kk} \varepsilon_k^2 GMA_k}{\alpha_k}\right)^{\frac{\alpha_k \xi_k}{\sum\limits_{j=1}^{3} \alpha_j \xi_j}}}\right] + \frac{R}{\sum\limits_{j=1}^{3} \alpha_j \xi_j}, \qquad (10)$$

for non-zero rates [1]. If we define the set of non-zero rate subbands as $Act_i \triangleq \{b \in [0, B-1] | R_{bi} > 0\}$, then $\xi_i$ and $GMA_i$ in (10) are:

$$\xi_i \triangleq \sum_{b \in Act_i} \eta_b, \quad GMA_i \triangleq \prod_{b \in Act_i} (G_b \sigma_{bi}^2)^{\frac{\eta_b}{\xi_i}}. \tag{11}$$

Finding the sets $Act_i$ can be done iteratively according to the following scheme:

1. Assume that all the subbands are active and calculate the rates.
2. While some $R_{bi} < 0$

   − Set $Act_i = \{b \in [0, B-1] | R_{bi} > 0\}$
   − Calculate new rates.

The optimal CCT matrix $M$ is not derived here analytically, however, it can be calculated by numerical methods of minimizing the target function [1]:

$$f(\mathbf{M}) = \prod_{k=1}^{3} ((\mathbf{M}\mathbf{M}^T)^{-1})_{kk} \prod_{b=0}^{B-1} (\sigma_{bk}^2)^{\eta_b}. \tag{12}$$

An approximate solution can also be proposed, solving the gradient equations

$$\nabla_{\mathbf{m_i}} log(f(\mathbf{M})) = -2\mathbf{m_i} + 2 \sum_{b=0}^{B-1} \eta_b \frac{\Lambda_b \mathbf{m_i}}{\mathbf{m_i}^T \Lambda_b \mathbf{m_i}} = 0 \tag{13}$$

for an orthogonal matrix $\mathbf{M}$. Here, $\Lambda_b$ is the covariance matrix of subband $b$ in the RGB domain ($\mathbf{Y_b} = [y_{bR} \quad y_{bG} \quad y_{bB}]^T$ is a vector of R,G,B transform coefficients):

$$\Lambda_b \triangleq E\left[\left(\mathbf{Y_b} - \mu_{\mathbf{Y_b}}\right)\left(\mathbf{Y_b} - \mu_{\mathbf{Y_b}}\right)^T\right] \quad \mu_{\mathbf{Y_b}} \triangleq E\left[\mathbf{Y_b}\right]. \tag{14}$$

This solution, named the GKLT (Generalized Karhunen-Loeve Transform) can be found by the following numerical algorithm:

1. Take a random 3x1 vector $\mathbf{v}$. Given the $B$ subband covariance matrices $\Lambda_b$, calculate the $\sum_{b=0}^{B-1} \eta_b \frac{\Lambda_b}{\mathbf{v}^T \Lambda_b \mathbf{v}}$ matrix.
2. Find the eigen values $\lambda_i$ and eigen vectors of this matrix.
3. For the $i^{th}$ eigen vector $\mathbf{v_i}$, $i \in \{1, 2, 3\}$ do:
   while($|\lambda_i - 1| \geq \varepsilon$ )

   (a) Calculate the $\sum_{b=0}^{B-1} \eta_b \frac{\Lambda_b}{\mathbf{v_i}^T \Lambda_b \mathbf{v_i}}$ matrix.
   (b) Find its eigen values and eigen vectors decomposition.
   (c) Take the $i^{th}$ eigen vector and eigen value as the new $\mathbf{v_i}$ and $\lambda_i$.

Here the $\varepsilon$ parameter is the threshold, defining how close $\mathbf{v_i}$ will get to solving (13). When the algorithm converges, the GKLT is the matrix with $\mathbf{v_i}$ as its rows.

## 4.1 An Image-Independent Sub-optimal Solution

The one-dimensional DCT (Discrete Cosine Transform) matrix
$$\begin{pmatrix} 0.5774 & 0.5774 & 0.5774 \\ 0.7071 & 0.0000 & -0.7071 \\ 0.4082 & -0.8165 & 0.4082 \end{pmatrix}$$ was applied to images as CCT and found to have
the best performance compared to other CCTs in [3]. It turns out that the DCT
is also a good choice for a sub-optimal transform based on our rate-distortion
model. Table 1 presents the values of the target function of (12) for several
known transforms, such as the RGB to YUV transform and KLT (Karhunen-
Loeve Transform) for a test image. It can be seen that the DCT is very close to
the optimal transform although GKLT is closer. Unlike the GKLT, however, the
DCT is image independent.

**Table 1.** Values of the target function for various CCTs. Identity is the RGB to RGB
transform and Opt. is the optimal transform.

| Identity | RGB to YUV | DCT | KLT | GKLT | Opt. |
|----------|-----------|---------|--------|---------|---------|
| 6.836e4 | 153.344 | 113.240 | 1.860e3 | 110.195 | 109.809 |

# 5 Color Image Compression Algorithm

Based on the results of the previous sections the following coding scheme can be
used based on the DCT subband transform:

1. Apply the DCT as a CCT to the RGB color components of a given image
   to obtain new color components $C1, C2, C3$.
2. Apply the two-dimensional block DCT to each color component $Ci$.
3. Quantize each subband of each color component independently using uniform
   scalar quantizers. The quantization step sizes are chosen to achieve optimal
   subband rates $R_{bi}^*$. The algorithm for the optimal steps calculation follows:

   (a) Set some initial quantization steps $\Delta_{bi}$ and calculate the rates $R_{bi}$.
   (b) Update the quantization steps according to: $\Delta_{bi}^{new} = \Delta_{bi} 2^{-(R_{bi}^* - R_{bi})}$

   until the optimal rates $R_{bi}^*$ are sufficiently close, i.e., $E\left(|R_{bi}^* - R_{bi}|\right) < \varepsilon$ for
   some small constant $\varepsilon$.
4. Apply lossless coding to the quantized DCT coefficients. Use JPEG's post-
   quantization coding [9].

## 5.1 Using Laplacian Distribution for DCT Coefficients

The compression algorithm can be improved both from the complexity and performance points of view if we model the probability distribution of the DCT

subband coefficients as Laplacian [4]. Then the subband rates (entropies) $R_{bi}$ can be approximately calculated as follows, where $k_{bi} \triangleq e^{0.5\mu_{bi}\Delta_{bi}} - e^{-0.5\mu_{bi}\Delta_{bi}}$:

$$R_{bi} = -\left(1 - e^{-0.5\mu_{bi}\Delta_{bi}}\right) \log_2 \left(1 - e^{-0.5\mu_{bi}\Delta_{bi}}\right) - e^{-0.5\mu_{bi}\Delta_{bi}}$$
$$\cdot (\log_2 k_{bi} - 1) + \frac{\mu_{bi}\Delta_{bi}}{k_{bi}} \log_2(e). \tag{15}$$

Here $\mu_{bi}$ is the Laplacian distribution parameter for the subband $b$ of color component $i$ that can be calculated as $\mu_{bi} = \sqrt{\frac{2}{\sigma_{bi}^2}}$. Thus we do not need to calculate the histograms of the subbands but only their variances $\sigma_{bi}^2$.

## 5.2   Results

The algorithm has been implemented and tested on several images. In addition to the popular quality measure of the PSNR (Peak signal to Noise Ratio): $PSNR = 10log_{10}\left(\frac{255^2}{MSE}\right)$, we define the PSPNR (Peak Signal to Perceptible Noise Ratio) as: $PSPNR = 10\log_{10}\frac{255^2}{WMSE}$, where $WMSE$ for each color component is calculated according to (9) and as proposed in [8]. The results for several test images are presented in Table 2. It can be seen that the new algorithm outperforms JPEG for the images tested with a gain of 1.234dB PSNR and 1.484dB PSPNR on average. The gain is even greater when using estimated rates according to (15).

**Table 2.** PSNR and PSPNR results for the new algorithm (New Alg.) and its improved version according to (15) denoted (Est.) compared to JPEG at the same compression ratio (CR)

| Image | PSNR | | | PSPNR | | | |
|---|---|---|---|---|---|---|---|
| | New Alg. | New Alg. (Est.) | JPEG | New Alg. | New Alg. (Est.) | JPEG | CR |
| Lena | 29.765 | 30.011 | 29.785 | 38.671 | 39.038 | 37.132 | 45.07 |
| Peppers | 29.770 | 29.971 | 28.640 | 37.489 | 37.818 | 35.440 | 33.77 |
| Baboon | 28.056 | 30.024 | 26.370 | 37.859 | 38.273 | 36.023 | 16.92 |
| Cat | 29.172 | 30.019 | 28.736 | 39.603 | 40.066 | 38.733 | 21.97 |
| Sails | 29.923 | 29.990 | 28.377 | 39.012 | 39.018 | 37.004 | 14.61 |
| Monarch | 29.721 | 29.975 | 28.665 | 37.871 | 38.221 | 36.690 | 27.08 |
| Goldhill | 29.928 | 29.999 | 27.919 | 40.499 | 40.519 | 39.594 | 13.23 |
| **Mean** | **29.476** | **29.998** | **28.243** | **38.715** | **38.993** | **37.231** | |

## 6   Conclusions

In this work we have introduced a new algorithm for color image compression based on a theoretical model for the rate-distortion connection of subband transform coders. The distortion measures considered are MSE and WMSE. The rate-distortion model transforms the problem of selecting a CCT and subband rates into an optimization problem that can be solved analytically for the rates and by numerical optimization methods for the CCT. This optimization process is rather complex, however. An approximate solution for the CCT has been thus proposed, found by numerical methods with lower complexity, namely the GKLT. An additional image-independent transform has been proposed - the DCT. Both transforms have been shown to be close to the optimal transform. The results for the CCT and subband rates have been utilized in the new compression scheme based on the DCT subband transform. In particular, an iterative algorithm for the design of quantization steps has been introduced. The compression scheme can be made even more effective considering both complexity and compression quality (higher PSNR and PSPNR for the same image rate) using knowledge of the probability distribution of the subband transform coefficients. This has been demonstrated assuming Laplacian distribution of the DCT subband coefficients and modifying the compression algorithm accordingly, so that the subband rates are calculated without the use of histograms. Experimental results of compressed images have been presented and discussed. Our conclusion is that by numerical methods, presently available compression systems such as JPEG could be improved in the sense of lower reconstruction error for the same bit-rate.

## Acknowledgement

This research was supported in part by the HASSIP Research Program HPRN-CT-2002-00285 of the European Commission, and by the Ollendorff Minerva Center. Minerva is funded through the BMBF.

## References

1. Gershikov E., Porat M.: On Subband Transform Coding for Color Image Compression. CCIT report No. 571 (Jan. 2006) Technion, submitted for publication.
2. Gersho A. and Gray R. M.: Vector Quantization and Signal Compression. Boston, MA: Kluwer (1992) ch. 2.
3. Hao P. and Shi Q.: Comparative Study of Color Transforms for Image Coding and Derivation of Integer Reversible Color Transform. ICPR'00 **3** (2000) 224-227.
4. Lam E. Y., Goodman J. W.: A mathematical analysis of the DCT coefficient distributions for images. IEEE Trans. on Image Processing **9** (2000) 1661-1666.
5. Mallat S. and Falzon F.: Analysis of Low Bit Rate Image Transform Coding. IEEE Trans. on Signal Processing **46** (1998) 1027-1042.
6. Rabbani M., Joshi R.: An overview of the JPEG2000 still image compression standard. Signal Processing: Image Communication **17** (2002) 3-48.
7. Porat M. and Shachor G.: Signal Representation in the combined Phase - Spatial space: Reconstruction and Criteria for Uniqueness. IEEE Trans. on Signal Processing **47** (1999) 1701-1707.
8. Taubman D. S. and Marcellin M. W.: JPEG2000: image compression, fundamentals, standards and practice. Kluwer Academic Publishers (2002).
9. Wallace G. K.: The JPEG Still Picture Compression Standard. IEEE Trans. on Consumer Electronics **38** (1992) xviii-xxxiv.

# On a Local Refinement Solver for Coupled Flow in Plain and Porous Media

Oleg Iliev[1] and Daniela Vasileva[2]

[1] ITWM - Fraunhofer Institute for Industrial Mathematics,
Fraunhofer-Platz 1, D-67663 Kaiserslautern, Germany
`iliev@itwm.fhg.de`
[2] Institute of Mathematics and Informatics, Bulgarian Academy of Sciences,
Acad. G. Bonchev str., bl. 8, BG-1113 Sofia, Bulgaria
`vasileva@math.bas.bg`

**Abstract.** A local refinement algorithm for computer simulation of flow through oil filters is presented. The mathematical model is based on laminar incompressible Navier-Stokes-Brinkman equations for flow in pure liquid and in porous zones. A finite volume method based discretization on cell-centered, collocated, locally refined grids is used. Special attention is paid to the conservation of the mass on the interface between the coarse and the fine grid. A projection method, SIMPLEC, is used to decouple momentum and continuity equations. The corresponding software is implemented in a flexible way for arbitrary 3D geometries, approximated by an union of parallelepipeds with different sizes. Results from numerical experiments show that the solver could be successfully used for simulation of coupled flow in plain and in porous media.

**Keywords:** oil filter, numerical simulation, coupled flow in plain and porous media, Navier-Stokes, Brinkman, local refinement.

## 1 Introduction

This paper describes a numerical algorithm for computer simulation of flow through oil filters. The purpose of oil filters is to filter out small dirty particles from the oil. An oil filter consists of a filter box with inlet for dirty oil and outlet for cleaned oil (see Fig. 1). The inlet and outlet are separated by a filtering porous medium. The optimal shape design for filter boxes, achieving optimal pressure drop – flow rate ratio, etc., require detailed knowledge about the flow field through the filters. The numerical algorithm, developed here, provides an efficient way to obtain accurate information about the velocity and pressure distributions by 3D computer simulation of the fully coupled flow in fluid and porous media regions inside the filter box. The local refinement, implemented in the algorithm, ensures detailed resolution of flow peculiarities.

The paper is organized as follows. Section 2 concerns the mathematical model. The discretization of the governing equations is presented in the third section and the last section contains results from numerical experiments.

T. Boyanov et al. (Eds.): NMA 2006, LNCS 4310, pp. 590–598, 2007.
© Springer-Verlag Berlin Heidelberg 2007

**Fig. 1.** A schematic drawing of a vertical cross-section of the filter

## 2  Mathematical Model

Laminar incompressible isothermal Navier-Stokes equations (see, e.g. [1]) are used to model the flow in the pure liquid zones $\Omega_f$

$$-\mu \triangle \mathbf{u} + (\rho\mathbf{u}, \nabla)\mathbf{u} + \nabla p = \mathbf{f}_{\mathrm{NS}},$$
$$\nabla \cdot \mathbf{u} = 0,$$

where $\triangle\mathbf{u} = (\triangle u^1, \ldots \triangle u^N)^T$, $(\mathbf{u}, \nabla)\mathbf{u} = ((\mathbf{u}, \nabla u^1), \ldots (\mathbf{u}, \nabla u^N))^T$, $\mathbf{u} = (u^1, \ldots, u^N)^T$, $(N = 2, 3)$ is the velocity vector, $p$ – the pressure, $\mu$ – the viscosity and $\rho$ – the density. The oil filters are designed to be used for various temperatures and flow rates and although Stokes equations may be used to model the flow for low temperatures (i.e., for high viscosities) and slow flows, Navier-Stokes equations describe more adequately the flow in the liquid zones.

Brinkman extension to Darcy model [2,3] is used to model the flow through the filtering (porous) medium ($\Omega_p$)

$$\nabla \cdot (\mu_{\mathrm{eff}} \nabla\mathbf{u}) + \mu\mathbf{K}^{-1}\mathbf{u} = \mathbf{f}_{\mathrm{B}} - \nabla p,$$
$$\nabla \cdot \mathbf{u} = 0.$$

Here $\mathbf{K}$ is the permeability tensor, $\mu_{\mathrm{eff}}$ is some effective viscosity. The Darcy model needs to be extended by Brinkman term in order to describe more adequately the flow in the case of high porosity. Usually, the porosity of filtering media is more than 90%. As it can be seen, Brinkman model is a Stokes type PDE system. Viscous terms, appearing in the momentum equations, allow for proper setting of some kinds of boundary conditions (for example, no-slip conditions on solid walls).

Interface conditions between plain and porous media are subject of extensive research carried out by physicists, engineers and mathematicians. Here we use continuous velocity – continuous stress tensor interface conditions (see, for example [4] and references therein)

$$\mathbf{u}|_{S_p} - \mathbf{u}|_{S_f} = \mathbf{0}, \quad \mathbf{n} \cdot (\mu_{\mathrm{eff}}\nabla\mathbf{u} - p\mathbf{I})|_{S_p} - \mathbf{n} \cdot (\mu\nabla\mathbf{u} - p\mathbf{I})|_{S_f} = \mathbf{0},$$

where $\mathbf{I}$ is the identity matrix, $S_p$ and $S_f$ means the same interface $S$ seen from porous and fluid parts, respectively.

In our numerical algorithm we use a unique system of equations in the whole domain. The coefficients of the equations vary so that our reformulated problem reduces to the time-dependent Navier-Stokes equations in the liquid zones, and to Brinkman-like model in the porous media. For more details about this approach see, for example, [4,5]. Thus, the Navier-Stokes-Brinkman-type system in the whole domain is

$$\rho\frac{\partial \mathbf{u}}{\partial t} - \nabla \cdot (\tilde{\mu}\nabla\mathbf{u}) + (\rho\mathbf{u}, \nabla)\mathbf{u} + \underbrace{\overbrace{\mu\tilde{\mathbf{K}}^{-1}\mathbf{u} + \nabla p}^{\text{Darcy law}} = \tilde{\mathbf{f}}}_{\text{Navier-Stokes}},$$

$$\nabla \cdot \mathbf{u} = 0,$$

where

$$\tilde{\mu} = \begin{cases} \mu & \text{in } \Omega_f, \\ \mu_{\text{eff}} & \text{in } \Omega_p, \end{cases} \qquad \tilde{\mathbf{f}} = \begin{cases} \mathbf{f}_{\text{NS}} & \text{in } \Omega_f, \\ \mathbf{f}_{\text{B}} & \text{in } \Omega_p, \end{cases} \qquad \tilde{\mathbf{K}}^{-1} = \begin{cases} \mathbf{0} & \text{in } \Omega_f, \\ \mathbf{K}^{-1} & \text{in } \Omega_p. \end{cases}$$

Note, we seek a steady state solution of the above system.

## 3    Numerical Algorithm

### 3.1    Discretization on a Single (Not Locally Refined) Grid

The governing equations are discretized by the Finite Volume Method (FVM) on a cell-centered nonuniform Cartesian grid (3D or 2D). We suppose that each computational volume is completely filled with only one substance – fluid or porous medium. Our single grid algorithm is close to the algorithm, developed in [5], where FVM on a staggered nonuniform Cartesian grid is used. But here we considere collocated arrangements of all the velocity components and the pressure on a cell-centered grid. The main steps in the single grid algorithm are as follows.

**Discretization of the momentum equations.** Let us integrate the $i$-th momentum equation $(i = 1, 2, \ldots, N)$ over the computational volume $V$

$$-\int_{\partial V} (\tilde{\mu}\nabla u^i - \rho u^i \mathbf{u} - p\mathbf{e}^i) \cdot \mathbf{n} + \int_V (\rho\frac{\partial u^i}{\partial t} + \mu\mathbf{k}^i\mathbf{u} - f^i) = 0,$$

where $\mathbf{e}^i$ is the i-th unit vector in $\mathbb{R}^N$, $\mathbf{k}^i$ is the $i$-th row of $\mathbf{K}^{-1}$. Then we may obtain the discretized equation for $u^i$ in $V$, using

- Harmonic averaging of $\tilde{\mu}$ on the fluid-porous walls;
- Linearization of the momentum equation via Picard method, i.e., the coefficients in the convective terms are taken from the previous iteration;
- Central difference, first order upwind or deferred correction scheme (see, for example, [6]) for the convective terms;
- $p$ is taken from the previous iteration;

- $\frac{\partial u^i}{\partial t}$ is discretized using backward differences;
- The diagonal part of the permeability tensor is treated implicitly, while the off-diagonal terms are taken from the previous iteration and added to the right-hand side.

The boundary conditions are specified and discretized in the classical way. Then we solve the corresponding systems of linear equations using ILU preconditioned BiCGStab method and obtain the new pseudo-velocities $u^{*i}$, $i = 1, 2, \ldots, N)$. In order to obtain a new value for the pressure, as well as to correct the pseudo-velocities, we use

**Decoupling via a Projection Method.** The projection method used here can be derived as a fractional time step discretization, or equivalently, as an iterative method for decoupling the momentum and continuity equations. In the first case it is known as Chorin projection method (see, e.g., [1,7]), in the second case it is known as SIMPLE-C (see, e.g., [1,6]). The algorithm consists from a prediction step, where a prediction to the velocity ($\mathbf{u}^*$) is computed using the momentum equations, and from a correction step, where a pressure correction equation is solved in order to correct the pressure and the predicted velocity. The fractional time step discretization on a staggered grid (as well as SIMPLE-C algorithm, see [1,6,7]) leads to a special connection between the velocity and pressure corrections ($\mathbf{u}'$, $p'$) at each time step (at each iteration). For example, at the east wall of $V$ (i.e., the right wall of $V$ in the $x_1$ direction) we consider

$$u'^1_e = (p' - p'_E)/c^1_e, \tag{1}$$

where $u'^1_e$ is the desired correction for $u^{*1}$ on the east wall of $V$, $p'$ and $p'_E$ are the desired corrections for $p$ in the centers of $V$ and its east neighbor $V_E$, respectively, $c^1_e > 0$ is a coefficient, depending on the coefficients in the discretization of the first momentum equation (for more details see, e.g. [1,7,6]). The same relation is used on collocated grids, subject to an additional interpolation procedure (see the above references). A specifics in our case is that we deal with discontinuous coefficients and therefore we use harmonic averaging.

In order to obtain the steady-state solution we perform time steps until the time derivative of $\mathbf{u}$ becomes less than a given tolerance and the original discretization (for $\mathbf{u}$) of the continuity equation is satisfied with the same tolerance.

### 3.2   Local Refinement Algorithm

The initial computations in the algorithm are performed on a single coarse grid, but we also include in the grid some non-computational solid volumes, which are close to the solid boundaries and may become filled with fluid or porous medium during the local refinement process.

After obtaining an initial guess, the grid is refined in regions, where more accurate solution has to be sought. Usually the porous media is not well represented on the coarsest grid, thus the porous region and several cells around it are usually refined. Additional refinement may be performed near the inlet and the outlet, near the walls of the filter box, and in regions prescribed by the user. In this way a *composite grid* is obtained, with fine and coarse parts. The discretization has to be performed on the composite grid, as the solution in the refined part strongly influences the solution in the coarse part. Several stages of refinement may be performed, i.e., after obtaining the new solution on the composite grid, some regions may be chosen for further refinement. An initial guess in the new refined volumes is obtained using linear interpolation from the coarser grid.

Let us consider an interface wall between the coarse and the fine grid in the two-dimensional case. Let, for example, this wall be the east wall of the coarse volume $V$, and west wall for the fine grid volumes $V_1$ and $V_2$ (see Fig. 2). The fine grid volumes $V_1$ and $V_2$ are *irregular*, i.e., they have interface walls with the remaining coarser part of the grid. Then the missing neighbors on the fine grid are added as *auxiliary* volumes (as $V_{1,W}$ and $V_{2,W}$ in Fig. 2) and the discretization in $V_1$ and $V_2$ is performed in the usual way. The discretization on the coarser part of the grid may be performed in a similar way, defining the fathers of the irregular refined volumes as auxiliary ones for the coarse grid part (as $V_E$ in Fig. 2). As the auxiliary volumes produce additional unknowns in the corresponding linear systems for the momentums and the pressure-correction, additional equations are added in the systems. They represent the connections between the values of **u** or $p$ in the coarse volumes and their children volumes. It is supposed that the value of **u** or $p$ in a coarse grid auxiliary volume may be obtained as a mean value of the same variable values in the children's volumes

$$u^i(V_E) = (u^i(V_1) + u^i(V_2) + u^i(V_{1,E}) + u^i(V_{2,E}))/4, \quad i = 1, 2,$$
$$p(V_E) = (p(V_1) + p(V_2) + p(V_{1,E}) + p(V_{2,E}))/4,$$

where $V_{1,E}$ and $V_{2,E}$ are the east neighbors of $V_1$ and $V_2$. The value in a fine grid auxiliary volume may be obtained by linear interpolation from the coarse grid (including the auxiliary volumes)

$$u^i(V_{1,W}) = (9u^i(V) + 3u^i(V_E) + 3u^i(V_S) + u^i(V_{ES}))/16, \quad i = 1, 2,$$
$$p(V_{1,W}) = (9p(V) + 3p(V_E) + 3p(V_S) + p(V_{ES}))/16,$$

where $V_S$ and $V_{ES}$ are the south neighbors of $V$ and $V_E$, respectively. Near the boundaries the geometric factors in the interpolation are changed in accordance with the corresponding distances. Note, we discretize on all grids simultaneously, and the auxiliary volumes are used only in order to simplify the discretization formulas near the coarse-fine interface. They also allow non-neighboring levels of refinement to be treated in the standard way. But a disadvantage of this approach is that we loose the mass (and momentum) conservation on the fine/coarse grid interface. Further we will call this type of discretization *non-conservative*.

An approach to achieve a *mass conservative discretization* for the Navier-Stokes equations on locally refined grids is proposed for example in [8]. Here we follow a similar, but a bit simpler strategy.

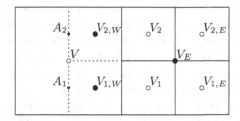

**Fig. 2.** An example of a coarse-fine interface, the centers of the auxiliary volumes are marked with "•"

In order to ensure mass conservation on the interface between $V$ and $V_1 \cup V_2$, the following relation should be satisfied

$$u_e^{*1}(V) + u'^1_e(V) = 0.5 \left( u_w^{*1}(V_1) + u'^1_w(V_1) + u_w^{*1}(V_2) + u'^1_w(V_2) \right).$$

Having in mind (1), we obtain a relation between $p'(V)$ and $p'(V_E)$

$$p'(V) - p'(V_E) = C_1 \left( p'(V_{1,W}) - p'(V_1) \right) + C_2 \left( p'(V_{2,W}) - p'(V_2) \right) + RHS,$$

where $C_1 > 0$, $C_2 > 0$ and $RHS$ are known. That is why here the discretization should be modified - instead of using the unknowns $p'(V)$ and $p'(V_E)$, we have to use $p'(V_1)$, $p'(V_2)$, $p'(V_{1,W})$, and $p'(V_{2,W})$. But if linear interpolation is used for $p'(V_{1,W})$ and $p'(V_{2,W})$, the resulting matrix of the linear system will not be positive definite. In [8] some off-diagonal terms are taken from the previous iteration and moved to the right-hand side, but this turned to be not efficient for our problem, as here pressure corrections may be very high and fast varying. That is why we preferred to use constant interpolation for $p'(V_{1,W})$ and $p'(V_{2,W})$, i.e., we suppose that

$$p'(V_{1,W}) = p'(V_{2,W}) = p'(V)$$

and thus increase the coefficient before $p'(V)$, instead of including $p'(V_{1,W})$ and $p'(V_{2,W})$ in the linear system equation for $p'(V)$. This approach has poor approximation properties, but the pressure-correction equation is used only to predict new corrections, and the stopping criteria are based on the original discretization. Let us also note, that a similar constant interpolation is considered in [9] for elliptic problems and $O(h^{1/2})$ rate of convergence is proved in energy norm. The scheme there is based on auxiliary points like $A_1$ and $A_2$ (see Fig. 2) instead of the centers of $V_{1,W}$ and $V_{2,W}$. We also tested this choice of auxiliary points and the numerical results were very similar. More accurate nonsymmetric and symmetric approximations are also proposed and studied in [9], but we have not implemented them yet.

## 4    Numerical Results

In this paper we present only 2D experiments, which test the performance of the local refinement algorithm. Results from 3D experiments with real filters and a comparison with experimental data may be found in [10].

Our test example uses initial data (density, viscosity, permeability, inflow rate, etc.), close to these of a real oil filter. First we will consider the globally refined solution in the filter geometry, shown in Fig. 1. We started with a very coarse grid, where the porous media has very bad representation. After solving (with tolerance 0.001) the problem on the coarsest grid, we refine it globally, solve on the refined grid and repeat this procedure 4 times, i.e., we perform a mesh-sequencing (MS) algorithm on 5 grids. Some results are shown in Table 1, where: *stage* denotes the corresponding stage in the MS algorithm, i.e., the number of the grid, with 0 for the coarsest grid; $P_I$ denotes the mean pressure, numerically obtained on the inlet ($P_O = 0$ is prescribed on the outlet); $Q_I$ is the flow rate on the inlet and $Q_O$ is the flow rate on the outlet; CVs is the number of computational volumes (including the auxiliary ones); LinIt is the number of linear iterations, performed in order to solve all linear systems, obtained after the discretization. Thus, the amount of computational work performed on each stage may be estimated by the multiplication of CVs×LinIt. As seen, $Q_I = Q_O$, i.e., the MS algorithm is fully mass conservative. The mean inlet pressure on the coarsest grid differs essentially from these, obtained on finer grids, as the porous media is very bad represented on this grid.

**Table 1.** MS solution

| stage | $P_I$ | $Q_I$ | $Q_O$ | CVs | LinIt |
|---|---|---|---|---|---|
| 0 | 8.13e+0 | 2.67e+4 | 2.67e+4 | 184 | 1532 |
| 1 | 2.47e+3 | 2.67e+4 | 2.67e+4 | 734 | 2576 |
| 2 | 3.21e+3 | 2.67e+4 | 2.67e+4 | 2919 | 4117 |
| 3 | 3.58e+3 | 2.67e+4 | 2.67e+4 | 11669 | 8369 |
| 4 | 3.78e+3 | 2.67e+4 | 2.67e+4 | 46583 | 12922 |

The pressure distribution on the finest grid is presented in Fig. 3. As seen, the pressure changes significantly in the porous media region.

The next experiment uses the local refinement (LR) solver, starting on the same coarsest grid and refining on each further stage only the porous media and 2 CVs

**Fig. 3.** Global refinement, the pressure distribution on the last stage

**Table 2.** LR solution, non-conservative discretization

| stage | $P_I$ | $Q_I$ | $Q_O$ | CVs | LinIt |
|---|---|---|---|---|---|
| 1 | 2.49e+3 | 2.67e+4 | 2.67e+4 | 682 | 1874 |
| 2 | 3.21e+3 | 2.67e+4 | 2.68e+4 | 1837 | 2014 |
| 3 | 3.54e+3 | 2.67e+4 | 2.66e+4 | 5114 | 1966 |
| 4 | 3.72e+3 | 2.67e+4 | 2.66e+4 | 14057 | 2509 |

around it. First, the non-conservative discretization is applied on the fine/coarse interface and the corresponding results are shown in Table 2. As seen, the values of $P_I$ and $Q_O$ are almost the same, as for the MS algorithm, but the amount of the computational work is much less. Note, not only the number of CVs, but also the number of linear iterations is much less (especially on the fine grids). A possible explanation of the last observation is that the corresponding linear systems have less unknowns as well as that some peculiarities outside the refined region are not resolved in details on the remaining coarse part of the grid. But we also have to note, that if the stopping criterion is much stronger, i.e., much more SIMPLEC iterations are performed, the mass conservation is not so good and $P_I$ continues to change, i.e., each additional iteration slightly "spoils" the solution.

The pressure distribution, obtained on the last composite grid, is presented in Fig. 4. As seen, the filter geometry is represented very roughly.

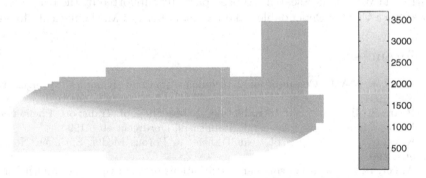

**Fig. 4.** Local refinement near the porous media, the pressure distribution on the last stage

Further we apply the mass conservative discretization on the coarse-fine grid interface. The corresponding results are shown in Table 3. As seen, $P_I$ is the same, as for the MS algorithm, $Q_I - Q_O = 0$, i.e., the mass is really conserved. The amount of computational work is much less, than for MS, but significantly more, compared with the non-conservative discretization.

At the end, we performed local refinement not only in and around the porous media, but also near the walls (2 CVs), the inlet and the outlet (fixed regions). The flow characteristics $P_I$ and $Q_O$ are almost the same, as for the previous local refinement cases. The non-conservative discretization provides faster convergence than the conservative one, and both are much faster than MS.

**Table 3.** LR solution, conservative discretization

| stage | $P_I$ | $Q_I$ | $Q_O$ | CVs | LinIt |
|-------|---------|---------|---------|-------|-------|
| 1 | 2.47e+3 | 2.67e+4 | 2.67e+4 | 682 | 2171 |
| 2 | 3.21e+3 | 2.67e+4 | 2.67e+4 | 1837 | 3780 |
| 3 | 3.58e+3 | 2.67e+4 | 2.67e+4 | 5114 | 6176 |
| 4 | 3.78e+3 | 2.67e+4 | 2.67e+4 | 14057 | 8628 |

## 5    Conclusions

The numerical experiments, presented here and in [10], show that the solver could be successfully used for simulation of coupled flow in plain and porous media. The local refinement ensures significant acceleration of the computations and saving of memory, preserving the accuracy of the globally refined solution. It seems that for practical purposes the non-conservative approach provides faster and reliable way to obtain the locally refined solution, compared with the conservative approach. But one has to keep in mind that the mass conservation must be controlled, i.e., the non-conservative approach is reliable till the mass is really conserved.

*Acknowledgments.* D. Vasileva acknowledges the fellowship from the European Consortium for Informatics and Mathematics (ERCIM) for her nine months work at ITWM. This research has been partially supported by the Kaiserslautern Excellence Cluster Dependable Adaptive Systems and Mathematical Modelling.

## References

1. Fletcher, C.A.J.: Computational techniques for fluid dynamics. Springer, Berlin etc., 1991.
2. Bear, J., Bachmat, Y.: Introduction to Modeling of Transport Phenomena in Porous Media. Kluwer Academic Publishers, Dordrecht etc., 1990.
3. Kaviany, M.: Principles of Heat Transfer in Porous Media. Springer, New York etc., 1991.
4. Angot, Ph.: Analysis of singular perturbations on the Brinkman problem for fictitious domain models of viscous flows. Math. Meth. Appl. Sci., **22** (1999) 1395–1412.
5. Iliev, O., Laptev, V.: On numerical simulation of flow through oil filters. Comput. Vis. Sci., **6** (2004) 139–146.
6. Wesseling, P.: Principles of Computational Fluid Dynamics. Springer, Berlin etc., 2000.
7. Gresho, P.M., Sani, R.L.: Incompressible Flow and the Finite Element Method. Vol. 2: Isothermal Laminar Flow. Wiley, Chichester.
8. Coelho, P., Pereira, C.F., Carvalho, M.G.: Calculation of laminar recirculating flows using a local non-staggered grid refinement system. Int. J. Num. Meth. Fluids, **12** (1991) 535–557.
9. Ewing, R., Lazarov, R., Vassilevski, P.: Local refinement techniques for elliptic problems on cell-centered grids I. Error analysis. Math. Comp., **56** (1991) 437–461.
10. Iliev, O., Laptev, V., Vassileva, D.: Algorithms and software for computer simulation of flow through oil filters. Proc. FILTECH Europa, 2003, Düsseldorf, 327–334.

# Properties of the Lyapunov Iteration for Coupled Riccati Equations in Jump Linear Systems*

Ivan Ganchev Ivanov

Faculty of Economics and Business Administration,
Sofia University "St. Kliment Ohridsky", Sofia 1113, Bulgaria
i_ivanov@feb.uni-sofia.bg

**Abstract.** We analyze the solution of the system of coupled algebraic Riccati equations of the optimal control problem of jump linear system. We prove that the Lyapunov iterations converge to a positive semidefinite stabilizing solution under mild conditions.

## 1 Solution the Coupled Riccati Equations

We consider the solution of the coupled algebraic Riccati equations of the optimal control problem for jump linear systems. This investigation can be considered as an extension of the iterative methods and its properties studied in [1,2,3] Many authors have investigated this problem and they have analyzed different methods for its iterative solution. Val, Geromel and Costa [5] have considered a special model which leads to finding of a positive semidefinite solution of the following coupled algebraic Riccati equations:

$$
\begin{aligned}
& A_k^\top X_k + X_k A_k + C_k^\top C_k + \sum_{j=1}^N \lambda_{kj} X_j \\
& - (C_k^\top D_k + X_k B_k)(D_k^\top D_k)^{-1}(D_k^\top C_k + B_k^\top X_k) = 0, \quad k = 1, \ldots, N,
\end{aligned}
\tag{1}
$$

where $R_k = D_k^\top D_k$ $(k = 1, \ldots, N)$ are positive definite matrices, the numbers $\lambda_{ij}$ are nonnegative for $i \neq j$, and $\lambda_{ii} = -\sum_{j \neq i} \lambda_{ij}$.

The system (1) can be re-written in the following equivalent form:

$$
\begin{aligned}
& \bar{A}_k^\top X_k + X_k \bar{A}_k + \sum_{j=1}^N \lambda_{kj} X_j + \bar{C}_k^\top \bar{C}_k \\
& - X_k B_k (D_k^\top D_k)^{-1} B_k^\top X_k = 0, \quad k = 1, \ldots, N,
\end{aligned}
\tag{2}
$$

with $\bar{A}_k = A_k - B_k(D_k^\top D_k)^{-1}D_k^\top C_k$ and $\bar{C}_k = (I - D_k(D_k^\top D_k)^{-1}D_k^\top)C_k$ for $k = 1, \ldots, N$.

---

* This work was partially supported by the Sofia University under the research project 12/2006.

T. Boyanov et al. (Eds.): NMA 2006, LNCS 4310, pp. 599–606, 2007.
© Springer-Verlag Berlin Heidelberg 2007

For any Hermitian matrices $X, Y$, we write $X > Y$ or $X \geq Y$ if $X - Y$ is positive definite or $X - Y$ is positive semidefinite. A matrix $A$ is said to be c-stable if all eigenvalues of $A$ lie in the open left half plane and $A$ is said to be almost c-stable if all eigenvalues of $A$ lie in the closed left half plane.

Very important for applications is to find a positive semidefinite solution $X = (X_1, \ldots, X_N)$ of (2) with the property that $\bar{A}_k - B_k (D_k^T D_k)^{-1} B_k^T X_k$ is a c-stable matrix. Such solution $X$ is called a stabilizing one. Some assumptions are known (e.g., [1, Assumption 1] in order to exists a unique positive semidefinite stabilizing solution of (2). Here we are interested in a positive semidefinite stabilizing solution of (1). We shall show that a positive semidefinite stabilizing solution of (1) can be found with the same iterative methods under some mild initial conditions.

We denote the matrices $\mathbf{X} = (X_1, \ldots, X_N)$ and $\mathbf{X_k} \| Y = (X_1, \ldots, X_{k-1}, Y, X_{k+1}, \ldots, X_N)$. Let us define the functions

$$
\begin{aligned}
\mathcal{R}_k(\mathbf{X}) = &\hat{A}_k^\top X_k + X_k \hat{A}_k + C_k^\top C_k + \mathcal{E}_k(\mathbf{X}) \\
&- (C_k^\top D_k + X_k B_k)(D_k^\top D_k)^{-1}(D_k^\top C_k + B_k^\top X_k),
\end{aligned}
\tag{3}
$$

where $\hat{A}_k = A_k + 0.5\lambda_{kk}$ and $\mathcal{E}_k(\mathbf{X}) = \displaystyle\sum_{j=1, j \neq k}^{N} \lambda_{kj} X_j$. The system (1) is equivalent to the system of Riccati equations $\mathcal{R}_k(\mathbf{X}) = 0$ for $k = 1, \ldots, N$. We consider two iterative methods for solving the last system. The methods are well known in the literature. The monotonicity and boundedness of the matrix sequences defined by the corresponding iterations have been proved in [5]. However, these properties are proved assuming mean square stabilizable properties for the matrix coefficient in (1) (see [5, Proposition 2.2]). These assumptions guarantee the existence and uniqueness of a positive semidefinite mean square stabilizing solution of [5].

We begin with several properties of $\mathcal{R}_k(\mathbf{X})$. We use the notation $\mathcal{E}_k(\mathbf{X}) = \mathcal{E}_{k1}(\mathbf{X}) + \mathcal{E}_{k2}(\mathbf{X})$ with $\mathcal{E}_{k1}(\mathbf{X}) = \displaystyle\sum_{j=1}^{k-1} \lambda_{kj} X_j$ and $\mathcal{E}_{k2}(\mathbf{X}) = \displaystyle\sum_{j=k+1}^{N} \lambda_{kj} X_j$. In our analysis we also need some well known properties of the solution of the standard Lyapunov equation (see ([4]).

We define matrices $F_k = -R_k^{-1}(B_k^\top X_k + D_k^\top C_k)$, $\tilde{A}_k = \hat{A}_k + B_k F_k$ and $\tilde{C}_k = C_k + D_k F_k$. If we take any symmetric matrix $Z$ instead of $X_k$ in $F_k$, then we shall write $F_Z$, $\tilde{A}_Z$ and $\tilde{C}_Z$.

We quote without proof the following lemma.

**Lemma 1.** *Let $Y$ and $Z$ be symmetric matrices. The functions $\mathcal{R}_k(\mathbf{X})$ $(k = 1, \ldots, N)$ satisfy the following properties:*

$$
\mathcal{R}_k(\mathbf{X}) = (\tilde{A}_k)^\top X_k + X_k(\tilde{A}_k) + \mathcal{E}_k(\mathbf{X}) + (\tilde{C}_k)^\top(\tilde{C}_k),
\tag{4}
$$

$$
\mathcal{R}_k(\mathbf{X_k} \| Y) = (\tilde{A}_Z)^\top Y + Y(\tilde{A}_Z) + \mathcal{E}_k(\mathbf{X}) + (\tilde{C}_Z)^\top(\tilde{C}_Z)
\tag{5}
$$

$$
- (F_Z - F_Y)^\top R_k(F_Z - F_Y),
$$

$$\mathcal{R}_k(\mathbf{X_k}\|Y) - R_k(\mathbf{X_k}\|Z) = (\tilde{A}_Z)^\top (Y - Z) + (Y - Z)(\tilde{A}_Z)^\top \qquad (6)$$
$$-(F_Z - F_Y)^\top R_k (F_Z - F_Y).$$

Using expression (4) and notations $F_k^{(i-1)} = -R_k^{-1}(B_k^\top X_k^{(i-1)} + D_k^\top C_k)$, $\tilde{A}_k^{(i-1)} = \hat{A}_k + B_k F_k^{(i-1)}$ and $\tilde{C}_k^{(i-1)} = C_k + D_k F_k^{(i-1)}$, we derive a recurrence equation whose solution is the unknown matrix $X_k^{(i)}$:

$$\left(\tilde{A}_k^{(i-1)}\right)^\top X_k^{(i)} + X_k^{(i)} \tilde{A}_k^{(i-1)} + \mathcal{E}_k(\mathbf{X}^{(i-1)}) + (\tilde{C}_k^{(i-1)})^\top \tilde{C}_k^{(i-1)} = 0, \qquad (7)$$

for $k = 1,\ldots,N$, and the initial matrices $X_k^{(0)}$ are properly chosen. We call this iteration the Lyapunov method. A natural generalization of the last iteration is the following

$$\left(\tilde{A}_k^{(i-1)}\right)^\top X_k^{(i)} + X_k^{(i)} \left(\tilde{A}_k^{(i-1)}\right) + \mathcal{E}_{k1}(\mathbf{X}^{(i)}) + \mathcal{E}_{k2}(\mathbf{X}^{(i-1)})$$
$$+(\tilde{C}_k^{(i-1)})^\top \tilde{C}_k^{(i-1)} = 0 \quad \text{for } k = 1,\ldots,N. \qquad (8)$$

We call the last iterative formula the accelerated Lyapunov method. These two iterations we use for numerical solution of the coupled algebraic Riccati equations (1). We show that the positive semidefinite stabilizing solution of (1) can be found by iterations (7) and (8) under some mild assumptions. We also compare the rate of convergence numerically.

## 2    The Main Theorem

We prove several properties of a sequence of Lyapunov algebraic equations (8) whose solutions form monotone decreasing matrix sequences bounded from below. The limits of these sequences provide a stabilizing solution to the coupled Riccati equations (1). The main result of this paper is the following theorem.

**Theorem 1.** *Assume that there are symmetric matrices* $\hat{\mathbf{X}} = (\hat{X}_1,\ldots,\hat{X}_N)$ *and* $\mathbf{X}^{(0)} = (X_1^{(0)},\ldots,X_N^{(0)})$ *such that* $\mathcal{R}_k(\hat{\mathbf{X}}) \geq 0$; $\mathbf{X}^{(0)} \geq \hat{\mathbf{X}}$; $\mathcal{R}_k(\mathbf{X}^{(0)}) \leq 0$ *and* $\tilde{A}_k^{(0)}$ *is c-stable for* $k = 1,\ldots,N$. *Then for the matrix sequences* $\{X_1^{(i)}\}_{i=1}^\infty,\ldots,$ $\{X_N^{(i)}\}_{i=1}^\infty$ *defined by (8), the following properties hold:*

(i) *We have* $\mathbf{X}^{(i)} \geq \mathbf{X}^{(i+1)}$, $\mathbf{X}^{(i)} \geq \hat{\mathbf{X}}$ *and* $\mathcal{R}_k(\mathbf{X}^{(i)}) \leq \mathcal{E}_{k1}(\mathbf{X}^{(i)} - \mathbf{X}^{(i+1)})$, *where* $i = 0, 1, 2, \ldots$;

(ii) $\hat{A}_k + B_k F_k^{(i)}$ *is c-stable for* $k = 1,\ldots,N$ *and* $i = 0, 1, 2, \ldots$;

(iii) *The sequences* $\{X_1^{(i)}\},\ldots,\{X_N^{(i)}\}$ *converge to the solution* $\tilde{\mathbf{X}}$ *of the equations* $\mathcal{R}_k(\mathbf{X}) = 0$, *and* $\tilde{\mathbf{X}} \geq \hat{\mathbf{X}}$. *The matrices* $\hat{A}_k + B_k \tilde{F}_k$ ($k = 1,\ldots,N$) *are almost c-stable, where* $\tilde{F}_k = -R_k^{-1}(B_k^\top \tilde{X}_k + D_k^\top C_k)$.

*Proof.* Let $i = 0$. According to the theorem conditions, we have $\mathbf{X}^{(0)} \geq \hat{\mathbf{X}}$, $\tilde{A}_k^{(0)}$ is c-stable and $\mathcal{R}_k(\mathbf{X}^{(0)}) \leq 0$ for $k = 1,\ldots,N$. We shall prove the inequality $\mathbf{X}^{(0)} \geq \mathbf{X}^{(1)}$. From iteration (8) for $i = 1$ we get a matrix Lyapunov equation with the c-stable matrix coefficient $\tilde{A}_k^{(0)}$ for each $k$. Thus, $X_k^{(1)}$ is the unique solution of this equation.

It is easy to derive the equations

$$
\begin{aligned}
\left(\tilde{A}_k^{(0)}\right)^\top \left(X_k^{(1)} - X_k^{(0)}\right) &+ \left(X_k^{(1)} - X_k^{(0)}\right)\left(\tilde{A}_k^{(0)}\right) \\
&= -\mathcal{R}_k(\mathbf{X}^{(0)}) - \mathcal{E}_{k1}(\mathbf{X}^{(1)} - \mathbf{X}^{(0)})
\end{aligned}
\tag{9}
$$

for $k = 1, 2, \ldots, N$. In the last equation for $k = 1$ the right-hand side $-\mathcal{R}_1(\mathbf{X}^{(0)})$ is positive semidefinite and its solution $X_1^{(1)} - X_1^{(0)}$ is negative semidefinite, which means $X_1^{(0)} \geq X_1^{(1)}$. In a similar way, for $k = 2, \ldots, N$, the right-hand side of (9) is positive semidefinite and the matrix $\tilde{A}_2^{(0)} = \hat{A}_2 + B_2 F_2^{(0)}$ is c-stable. Thus $X_k^{(0)} \geq X_k^{(1)}$ for all values of $k$.

Now, assume that there exists a natural number $i = r - 1$ and the matrix sequences $\{X_1^{(i)}\}_0^r, \ldots, \{X_N^{(i)}\}_0^r$ are computed and properties (i) and (ii) are fulfilled, i.e., for $k = 1, \ldots, N$ and $s = 0, \ldots, r-1$ we have $X_k^{(s)} \geq X_k^{(s+1)}$, $X_k^{(s)} \geq \hat{X}_k$, $\mathcal{R}_k(\mathbf{X}^{(s)}) \leq \mathcal{E}_{k1}(\mathbf{X}^{(s)} - \mathbf{X}^{(s+1)})$ and $\tilde{A}_k^{(s)}$ are c-stable. We shall show that for $k = 1, \ldots, N$ the following statements are fulfilled: $X_k^{(s+1)} \geq \hat{X}_k$, $\hat{A}_k + B_k F_k^{(s+1)}$ are c-stable; we shall show how to compute each $X_k^{(s+2)}$, and that the inequalities $X_k^{(s+1)} \geq X_k^{(s+2)}$ hold true. Finally, we shall prove the inequalities $\mathcal{R}_k(\mathbf{X}^{(s+1)}) \leq \mathcal{E}_{k1}(\mathbf{X}^{(s+1)} - \mathbf{X}^{(s+2)})$.

We start with the proof of inequalities $X_k^{(s+1)} \geq \hat{X}_k$ $(k = 1, \ldots, N)$. Using formula (3) for $\mathcal{R}_k(\hat{\mathbf{X}})$, the representation $\hat{F}_k = -R_k^{-1}(B_k^\top \hat{X}_k + D_k^\top C_k)$, and the equality

$$
\begin{aligned}
-(\tilde{C}_k^{(r-1)})^\top \tilde{C}_k^{(r-1)} &- \hat{F}_k^\top R_k \hat{F}_k - (F_k^{(r-1)})^\top B_k^\top \hat{X}_k - \hat{X}_k B_k F_k^{(r-1)} \\
&= \left(F_k^{(r-1)} - \hat{F}_k\right)^\top R_k \left(F_k^{(r-1)} - \hat{F}_k\right) - C_k^\top C_k,
\end{aligned}
$$

we find that the matrix $X_k^{(r)} - \hat{X}_k$ is a solution of the following Lyapunov equation

$$
\begin{aligned}
\left(\tilde{A}_k^{(r-1)}\right)^\top \left(X_k^{(r)} - \hat{X}_k\right) &+ \left(X_k^{(r)} - \hat{X}_k\right)\left(\tilde{A}_k^{(r-1)}\right) \\
&= -\mathcal{R}_k(\hat{\mathbf{X}}) - \mathcal{E}_{k1}(\mathbf{X}^{(r)} - \hat{\mathbf{X}}) - \mathcal{E}_{k2}(\mathbf{X}^{(r-1)} - \hat{\mathbf{X}}) \\
&\quad - \left(F_k^{(r-1)} - \hat{F}_k\right)^\top R_k \left(F_k^{(r-1)} - \hat{F}_k\right).
\end{aligned}
\tag{10}
$$

Thus, the right-hand side of (10) is negative semidefinite for $k = 1$. Hence $X_1^{(r)} - \hat{X}_1 \geq 0$. For the remaining values of $k$ the right-hand side of (10) is negative semidefinite, too. Hence, $X_k^{(r)} - \hat{X}_k \geq 0$ for $k = 3, \ldots, N$.

Now we prove that all matrices $\tilde{A}_k^{(r)} = \hat{A}_k + B_k F_k^{(r)}$ are c-stable. The matrix $X_k^{(r)}$ is a solution to (8). In expression (5) we put $Y = X_k^{(r)}$ and $Z = X_k^{(r-1)}$. For $\mathcal{R}_k(\hat{\mathbf{X}} \| X_k^{(r)})$ we get

$$
\begin{aligned}
\mathcal{R}_k(\hat{\mathbf{X}} \| X_k^{(r)}) &= -\mathcal{E}_{k1}(\mathbf{X}^{(r-1)} - \mathbf{X}^{(r)}) \\
&\quad - (F_k^{(r-1)} - F_k^{(r)})^\top R_k (F_k^{(r-1)} - F_k^{(r)}) \leq 0.
\end{aligned}
$$

On the other hand, we take $Y = \hat{X}_k$ and $Z = X_k^{(r)}$ in (6). Thus

$$
\mathcal{R}_k(\hat{\mathbf{X}}) = -(\tilde{A}_k^{(r)})^\top (X_k^{(r)} - \hat{X}_k) - (X_k^{(r)} - \hat{X}_k)\,\tilde{A}_k^{(r)}
$$
$$
\mathcal{R}_k(\hat{\mathbf{X}}\| X_k^{(r)}) - (F_k^{(r)} - \hat{F}_k)^\top R_k\,(F_k^{(r)} - \hat{F}_k).
$$

Together with the last equality we derive

$$
0 \leq \mathcal{R}_k(\hat{\mathbf{X}}) = -\mathcal{E}_{k1}(\mathbf{X}^{(r-1)} - \mathbf{X}^{(r)}) - (\tilde{A}_k^{(r)})^\top (X_k^{(r)} - \hat{X}_k)
$$
$$
-(F_k^{(r-1)} - F_k^{(r)})^\top R_k\,(F_k^{(r-1)} - F_k^{(r)})
$$
$$
-(X_k^{(r)} - \hat{X}_k)\,\tilde{A}_k^{(r)} - (F_k^{(r)} - \hat{F}_k)^\top R_k\,(F_k^{(r)} - \hat{F}_k),
$$

whence it follows that

$$
(\tilde{A}_k^{(r)})^\top (X_k^{(r)} - \hat{X}_k) + (X_k^{(r)} - \hat{X}_k)\,\tilde{A}_k^{(r)} \leq -(F_k^{(r-1)} - F_k^{(r)})^\top R_k\,(F_k^{(r-1)} - F_k^{(r)}).
$$

Let us assume that there is a number $k$ so that $\tilde{A}_k^{(r)} = \hat{A}_k + B_k F_k^{(r)}$ is not c-stable. Thus there exists an eigenvalue $\lambda$ of $\hat{A}_k + B_k F_k^{(r)}$ with nonnegative real part, $Re(\lambda) \geq 0$ and a nonzero eigenvector $x$ with $(\hat{A}_k + B_k F_k^{(r)})x = \lambda x$. Through the last inequality we get

$$
0 \leq 2Re(\lambda)x^\top (X_k^{(r)} - \hat{X}_k)x
$$
$$
\leq -x^\top (F_k^{(r-1)} - F_k^{(r)})^\top R_k\,(F_k^{(r-1)} - F_k^{(r)})\,x \leq 0.
$$

Hence $F_k^{(r-1)} x = F_k^{(r)} x$. Since

$$
(\hat{A}_k + B_k F_k^{(r-1)})\,x = (\hat{A}_k + B_k F_k^{(r)})\,x = \lambda x,
$$

$\lambda$ must be an eigenvalue of $\hat{A}_k + B_k F_k^{(r-1)}$, which contradicts to the c-stability of this matrix. Our assumption false, hence $\hat{A}_k + B_k F_k^{(r)}$ is c-stable for $k = 1, \ldots, N$.

Further, we compute $X_k^{(r+1)}$ and prove that $X_k^{(r)} \geq X_k^{(r+1)}$ for $k = 1, \ldots N$.

Set $i = r + 1$ in iteration (8). Since $\hat{A}_k + B_k F_k^{(r)}$ is c-stable, $X_k^{(r+1)}$ is the unique solution to (8) for all $k = 1, \ldots, N$.

Using (8), after some matrix calculations we obtain

$$
(\tilde{A}_k^{(r)})^\top (X_k^{(r)} - X_k^{(r+1)}) + (X_k^{(r)} - X_k^{(r+1)})\,\tilde{A}_k^{(r)}
$$
$$
= (\tilde{A}_k^{(r)})^\top X_k^{(r)} + X_k^{(r)} \tilde{A}_k^{(r)} - (\tilde{A}_k^{(r)})^\top X_k^{(r+1)} - X_k^{(r+1)} \tilde{A}_k^{(r)}
$$
$$
\overset{(8)}{=} (\tilde{A}_k^{(r)} \pm B_k F_k^{(r-1)})^\top X_k^{(r)} + X_k^{(r)} (\tilde{A}_k^{(r)} \pm B_k F_k^{(r-1)})
$$
$$
+ \mathcal{E}_{k1}(\mathbf{X}^{(r+1)}) + \mathcal{E}_{k2}(\mathbf{X}^{(r)}) + (\tilde{C}_k^{(r)})^\top (\tilde{C}_k^{(r)})
$$
$$
= -\mathcal{E}_{k1}(\mathbf{X}^{(r)} - \mathbf{X}^{(r+1)}) - \mathcal{E}_{k2}(\mathbf{X}^{(r-1)} - \mathbf{X}^{(r)})
$$
$$
- (F_k^{(r-1)} - F_k^{(r)})^\top R_k\,(F_k^{(r-1)} - F_k^{(r)}).
$$

We obtain the following Lyapunov equation for $\left( X_k^{(r)} - X_k^{(r+1)} \right)$ :

$$
\begin{aligned}
\left( \tilde{A}_k^{(r)} \right)^\top \left( X_k^{(r)} - X_k^{(r+1)} \right) &+ \left( X_k^{(r)} - X_k^{(r+1)} \right) \tilde{A}_k^{(r)} \\
&= -\mathcal{E}_{k1} \left( \mathbf{X^{(r)}} - \mathbf{X^{(r+1)}} \right) - \mathcal{E}_{k2} \left( \mathbf{X^{(r-1)}} - \mathbf{X^{(r)}} \right) \\
&\quad - \left( F_k^{(r-1)} - F_k^{(r)} \right)^\top R_k \left( F_k^{(r-1)} - F_k^{(r)} \right).
\end{aligned} \tag{11}
$$

Let $k = 1$. The right-hand side in (11) is a negative semidefinite matrix. Hence the solution $X_1^{(r)} - X_1^{(r+1)}$ is a positive semidefinite one. Consider (11) for $k = 2$. We know that $X_1^{(r)} - X_1^{(r+1)} \geq 0$. After analogous considerations we get $X_2^{(r)} - X_2^{(r+1)} \geq 0$. In a similar way it is proved that $X_k^{(r)} - X_k^{(r+1)} \geq 0$ for $k = 3, \ldots, N$.

We continue with the proof of the inequalities $\mathcal{R}_k(\mathbf{X^{(r)}}) \leq \mathcal{E}_{k1} \left( \mathbf{X^{(r)}} - \mathbf{X^{(r+1)}} \right)$ for $k = 1, \ldots, N$.

Combining $\mathcal{R}_k(\mathbf{X^{(r)}})$ expressed by (4) with iteration (8), we derive

$$
\begin{aligned}
\left( \tilde{A}_k^{(r)} \right)^\top \left( X_k^{(r)} - X_k^{(r+1)} \right) &+ \left( X_k^{(r)} - X_k^{(r+1)} \right) \tilde{A}_k^{(r)} \\
&= \mathcal{R}_k \left( \mathbf{X^{(r)}} \right) + \mathcal{E}_{k1} \left( \mathbf{X^{(r+1)}} - \mathbf{X^{(r)}} \right).
\end{aligned} \tag{12}
$$

For $k = 1$ the right-hand side in (12) is $\mathcal{R}_1 \left( \mathbf{X^{(r)}} \right)$, and since $\hat{A}_k + B_k F_k^{(r)}$ is c-stable and $X_1^{(r)} - X_1^{(r+1)} \geq 0$, we have $\mathcal{R}_1 \left( \mathbf{X^{(r)}} \right) \leq 0$. Analogously, for $k = 2$ the right-hand side in (12) is $\mathcal{R}_2 \left( \mathbf{X^{(r)}} \right) + \mathcal{E}_{21} \left( \mathbf{X^{(r+1)}} - \mathbf{X^{(r)}} \right)$. It is a negative semidefinite matrix, which means that

$$
\mathcal{R}_2 \left( \mathbf{X^{(r)}} \right) \leq \mathcal{E}_{21} \left( \mathbf{X^{(r)}} - \mathbf{X^{(r+1)}} \right).
$$

In a similar way we prove the inequalities

$$
\mathcal{R}_k \left( \mathbf{X^{(r)}} \right) \leq \mathcal{E}_{k1} \left( \mathbf{X^{(r)}} - \mathbf{X^{(r+1)}} \right), \quad k = 3, \ldots, N.
$$

The induction process for proving (i) and (ii) is now complete.

The matrix sequences $\{X_1^{(i)}\}_0^\infty, \ldots, \{X_N^{(i)}\}_0^\infty$ converge and the limit matrices $\tilde{X}_1, \ldots, \tilde{X}_N$ form a solution to $\mathcal{R}_k(\mathbf{X}) = 0$. Moreover, $\tilde{X}_k \geq \hat{X}_k$ for $k = 1, \ldots, N$. Since all the matrices $\hat{A}_k + B_k F_k^{(i)}$ $(k = 1, \ldots, N, i = 1, 2 \ldots)$ are c-stable, the corresponding "limit" matrices $\hat{A}_k + B_k \tilde{F}_k$ are almost c-stable.    $\square$

In a similar way analogous properties for the matrix sequence defined by (7) can be proved.

**Corollary 1.** *Under the assumptions of Theorem 1, the matrix sequence defined by (7) possesses the same properties* (ii) *and* (iii), *while the property* (i) *becomes* (i) $\mathbf{X^{(i)}} \geq \mathbf{X^{(i+1)}}$, $\mathbf{X^{(i)}} \geq \hat{\mathbf{X}}$ *and* $\mathcal{R}_k(\mathbf{X^{(i)}}) \leq 0$ *for* $i = 0, 1, 2, \ldots$

## 3   Numerical Experiments

We carried out experiments for numerical solving of a special kind of system of stochastic Riccati equations (1) with the introduced iterations: the Lyapunov iteration (7) and the accelerated Lyapunov iteration (8), for different initial points.

Solutions of iterations (7) and (8) can be found in terms of the solutions of $N$ algebraic Lyapunov equations at each step. For this purpose the MATLAB procedure *lyap* is applied and the flops are $N\frac{27}{2}n^3$ per one iteration.

Our experiments are executed in MATLAB on a 900GHz PENTIUM computer. We denote: *tol*- a small positive real number denoting the accuracy of computation; $Error_s = \max_{k=1,...,N}\left\|\mathcal{R}_k(X_1^{(s)},\ldots,X_N^{(s)})\right\|$; *It*- number of iterations for which the inequality $Error_{It} \le tol$ holds. The last inequality is used as a practical stopping criterion. The next example is introduced in [5].

*Example 1.* The coefficient matrices are

$$A_1 = \begin{bmatrix} -2.5 & 0.3 & 0.8 \\ 1 & -3 & 0.2 \\ 0 & 0.5 & -2 \end{bmatrix}, \quad A_2 = \begin{bmatrix} -2.5 & 1.2 & 0.3 \\ -0.5 & 5 & -1 \\ 0.25 & 1.2 & 5 \end{bmatrix}, \quad A_3 = \begin{bmatrix} 2 & 1.5 & -0.4 \\ 2.2 & 3 & 0.7 \\ 1.1 & 0.9 & -2 \end{bmatrix},$$

$$\begin{aligned} B_1 &= diag[0.707; 1; 1]; \\ B_2 &= diag[0.707; 1; 0.707]; \\ B_3 &= diag[0.707; 1; 1], \end{aligned} \quad C_1 = \begin{bmatrix} diag(5\,;1\,;3.31) \\ 0_3 \end{bmatrix},$$

$$C_2 = \begin{bmatrix} diag(6.08\,;8.36\,;5.83) \\ 0_3 \end{bmatrix}, \quad C_3 = \begin{bmatrix} diag(3.16\,;4\,;4.58) \\ 0_3 \end{bmatrix},$$

$$\Pi = \begin{bmatrix} -3 & 0.5 & 2.5 \\ 1 & -2 & 1 \\ 0.7 & 0.3 & -1 \end{bmatrix}, \quad \text{and} \quad D_i = \begin{bmatrix} 0_3 \\ I_3 \end{bmatrix}, \quad i = 1,2,3,$$

where $O_3$ denotes the zero $3 \times 3$ matrix and $I_3$ is the unit $3 \times 3$ matrix. With $tol := 10^{-12}$ we have tested iterations (7) and (8) with initial matrices $X_1^{(0)} = 2I_3, X_2^{(0)} = 4I_3, X_3^{(0)} = 7I_3$ for solving (1). Lyapunov's method requires 17 iterations and achieves the accuracy $Error_{17} = 5.4162 \times 10^{-13}$ for computing a positive semidefinite stabilizing solution. Accelerated Lyapunov's method needs 13 iterations and $Error_{13} = 5.4254 \times 10^{-13}$. We carried out 100 runs with iterations (7) and (8). The time for executing of these iterations are 10.85 and 8.68 seconds, respectively.

*Example 2.* Let $N = 6$. For $k = 1,2,3$ the coefficient matrices from the previous example are used. The remaining coefficient matrices are defined as follows:

$$A_4 = \begin{pmatrix} 2.7 & 0.03 & 1.8 \\ 0.54 & 3 & 0.93 \\ 0.39 & 0.89 & 2.3 \end{pmatrix}, \quad A_5 = \begin{pmatrix} -0.69 & -1.3 & -1.4 \\ -1 & -0.67 & -0.76 \\ -0.4 & -0.04 & -0.66 \end{pmatrix}, \quad A_6 = \begin{pmatrix} 2 & 0.91 & 2.0 \\ 2.1 & 1.1 & 0.9 \\ 1.3 & 0.58 & 2.1 \end{pmatrix},$$

$$\begin{aligned} B_4 &= diag(0.151\,;0.854\,;0.822), \\ B_5 &= diag(0.645\,;0.289\,;0.309), \\ B_6 &= diag(0.838\,;0.546\,;0.795), \end{aligned} \quad C_4 = \begin{pmatrix} diag(0.988\,;0.334\,;0.76) \\ 0_3 \end{pmatrix},$$

$$C_5 = \begin{pmatrix} diag(0.530\,;0.783\,;0.794) \\ 0_3 \end{pmatrix}, \quad C_6 = \begin{pmatrix} diag(0.059\,;0.305\,;0.971) \\ 0_3 \end{pmatrix},$$

$$\Pi = \begin{pmatrix} -3 & 0.5 & 0.5 & 0 & 1 & 1 \\ 1 & -2 & 0 & 1 & 0 & 0 \\ 0.1 & 0.2 & -1 & 0.5 & 0.1 & 0.1 \\ 0 & 1 & 1 & -3 & 0.5 & 0.5 \\ 1 & 0 & 0 & 1 & -2 & 0 \\ 0.5 & 0.1 & 0.1 & 0.1 & 0.2 & -1 \end{pmatrix}, \quad \text{and} \quad D_i = \begin{pmatrix} O_3 \\ I_3 \end{pmatrix}, \quad i = 4, 5, 6.$$

For $tol := 10^{-12}$ we have computed a positive semidefinite stabilizing solution of (1) by using (7) and (8) with initial matrices $X_1^{(0)} = 0, X_2^{(0)} = 4I, X_3^{(0)} = 7I, X_4^{(0)} = 56I, X_5^{(0)} = 3I, X_6^{(0)} = 9I$. Lyapunov's method (7) needs 36 iterations and achieves the accuracy $Error_{36} = 5.5992 \times 10^{-13}$. Accelerated Lyapunov's method (8) requires 24 iterations to achieve the accuracy $Error_{24} = 8.1640 \times 10^{-13}$. We have carried out 100 runs with iterations (7) and (8). The time for executing of these iterations are 41.24 and 28.79 seconds, respectively.

For iterations (8) and (7) the properties of Theorem 1 and Corollary 1 are fulfilled. The numerical experiments indicate that the accelerated iteration (8) is faster than the iteration (7), however we do not have a theoretical proof for this fact.

## References

1. Gajic, Z., Borno, I.: Lyapunov Iterations for Optimal Control of Jump Linear Systems at Steady State. IEEE Transaction on Authomatic Control **40** (1995) 1971–1975
2. Gajic, Z., Losada, R.: Monotonicity of Algebraic Lyapunov Iterations for Optimal Control of Jump Parameter Linear Systems. Systems & Control Letters **41** (2000) 175-181
3. Ivanov, I.: Properties of Stein (Lyapunov) Iterations for Solving a General Riccati Equation. (Accepted for publication in Nonlinear Analysis Series A: Theory, Methods & Applications) .
4. Lancaster, P., Rodman, L.: Algebraic Riccati Equations. Clarendon Press, Oxford (1995)
5. do Val, J. B., Geromel, J. C., Costa, O. L. V.: Solutions for the Linear Quadratic Control Problem of Markov Jump Linear Systems. J. Optimiz. Theory Appl. **103**(2) (1999) 283–311

# Numerical Analysis of Blow-Up Weak Solutions to Semilinear Hyperbolic Equations

Boŝko S. Jovanovic[1], Miglena N. Koleva[2], and Lubin G. Vulkov[3]

[1]University of Belgrade, Faculty of Mathematics, 11000 Belgrade, Serbia
bosko@matf.bg.ac.yu
[2,3] University of Rousse, Department of Mathematics, 7017 Rousse, Bulgaria
mkoleva@ru.acad.bg, vulkov@ami.ru.acad.bg

**Abstract.** We study numerical approximations of weak solutions of hyperbolic problems with discontinuous coefficients and nonlinear source terms in the equation. By a semidiscretization of a Dirichlet problem in the space variable we obtain a system of ordinary differential equations (SODEs), which is expected to be an approximation of the original problem. We show at conditions similar to those for the hyperbolic problem, that the solution of the SODEs blows up. Under certain assumptions, we also prove that the numerical blow-up time converges to the real blow-up time when the mesh size goes to zero. Numerical experiments are analyzed.

## 1  Introduction

This paper is concerned with numerical approximations of weak blow-up solutions of the semilinear hyperbolic equation

$$\frac{\partial^2 u}{\partial t^2} - \frac{\partial}{\partial x}\left(a(x)\frac{\partial u}{\partial x}\right) = f(u), \text{ in } Q_T = \Omega \times (0,T), \quad \Omega = (0,1), \quad (1)$$

$$0 < a_0 \le a(x) \le a_1, \ x \in \Omega. \tag{2}$$

We consider weak solutions of (1), (2) i.e. we do not require $u$ to possess continuous derivatives but we suppose that $u$ is continuous function. Therefore, if it is assumed that $a(x)$ has a jump in a point $\xi$, $0 < \xi < 1$, then

$$[u]_{x=\xi} \equiv u(\xi + 0, t) - u(\xi - 0, t) = 0, \quad \left[a(x)\frac{\partial u}{\partial x}\right]_{x=\xi} = 0. \tag{3}$$

We investigate two phenomena: blow up of the solutions and convergence rate of the finite difference and finite element approximations of such solutions, both depending on the smoothness of the input data, [1,6].

We say that the solution $u(x,t)$ of (1) **blows up** in a finite time $T_b < +\infty$ if: $\lim |||u(\cdot,t)||| = +\infty$, as $t \to T_b - 0$ with respect to appropriately chosen norm $||| \cdot |||$. Among numerous works on blow-up for hyperbolic equations, see [5,10] and references there, we select only the paper [10], where weak solutions are

T. Boyanov et al. (Eds.): NMA 2006, LNCS 4310, pp. 607–614, 2007.

studied. It is shown that the solutions of equation (1) blow-up if the source term satisfies the growth condition (6) and the initial values

$$u(x,0) = u_0(x), \quad \frac{\partial u}{\partial t}(x,0) = v_0(x), \quad x \in \Omega \tag{4}$$

are large enough (12).

In the present paper we are interested in numerical approximations of (1)-(5). The paper is organized as follows. In the next section we present a result for blow-up in (1)-(5). Then, in Section 3 we give an analysis for convergence and blow-up of the numerical solution. In the last section the numerical experiments are discussed. We refer to [1,3,4,8,9,11,2,7] for numerical blow-up results in parabolic problems.

## 2   Blow Up in the Continuous Problem

Our attention is concentrated on the zero Dirichlet problem for equation (1):

$$u(0,t) = 0, \quad u(1,t) = 0. \tag{5}$$

But all the results can be easily extended to the cases of Neumann's and Robin's boundary conditions.

We make the following assumption on the nonlinear source term:

Let $F(s)$ be any indefinite integral of $f(s)$ ($F'(s) = f(s)$). There is a number $s_1$ such that $F(s) - 1/2\lambda s^2$ is non-decreasing on $(s_1, \infty)$ and, for every $\varepsilon > 0$

$$\int_{s_1}^{\infty} \left[ F(s) - \frac{1}{2}\lambda s^2 - (F(s_1) - \frac{1}{2}\lambda s_1^2) + \varepsilon \right]^{-1/2} ds < \infty. \tag{6}$$

Here $\lambda > 0$ (with corresponding spectral function $\psi_1(x)$, which is also positive) is the first eigenvalue of the spectral problem

$$\frac{d}{dx}\left( a(x) \frac{d\psi}{dx} \right) + \lambda\psi = 0, x \in \Omega^- \cup \Omega^+, \Omega^- = (0,\xi), \Omega^+ = (\xi,1), \tag{7}$$

$$[\psi]_{x=\xi} = 0, \quad \left[ a(x)\frac{d\psi}{dx} \right]_{x=\xi} = 0, \tag{8}$$

$$\psi(0) = \psi(1) = 0, \quad \int_0^1 \psi^2(x)\, dx = 1. \tag{9}$$

For the initial functions $u_0, v_0$ we assume $u_0 \in W_2^1(0,1)$, $v_0 \in L^2(0,1)$.

We say $u \in C((0,T), W_2^1(\Omega)) \cap C^1((0,T), L_2(\Omega))$ is a **weak solution** to problem (1)-(5) if

$$\int_0^1 \phi(x,t) \frac{\partial u}{\partial t}(x,t)\, dx = \int_0^1 \phi(x,0) v_0(x)\, dx + \int_0^t \int_0^1 \frac{\partial \phi}{\partial \tau}(x,\tau) \frac{\partial u}{\partial \tau}(x,\tau)\, dx d\tau$$
$$- \int_0^t \int_0^1 a(x) \frac{\partial \phi}{\partial x} \frac{\partial u}{\partial x}\, dx d\tau + \int_0^t \int_0^1 \phi(x,\tau) f(u(x,\tau))\, dx d\tau \tag{10}$$

for all $\phi \in W_2^1((\Omega) \times (0,T))$.

Also, $\psi \in W_2^1(\Omega)$ is a weak solution of (7)-(9) if

$$\int_0^1 a(x) \frac{d\psi}{dx} \frac{\partial v}{\partial x} dx = \lambda \int_0^1 \psi v dx \quad \forall \, v \in W_2^1(\Omega) \quad \text{and} \quad \int_0^1 \psi^2(x) dx = 1. \quad (11)$$

**Theorem 1.** *(Levine [10]) Let u be a weak solution to problem (1)-(5) and let f be convex and satisfies (6). If*

$$G_0 = \int_0^1 \psi(x) u_0(x) \, dx > s_1, \quad G_1 = \int_0^1 \psi(x) v_0(x) dx > 0, \quad (12)$$

*where $\psi$ is a positive solution of (7)-(9) for some real $\lambda$, and if $1 \leq p < \infty$, then*

$$\lim_{t \to T_b - 0} \left( \int_0^1 |u(x,t)|^p \, dx \right)^{1/p} = +\infty, \quad \lim_{t \to T_b - 0} \sup \left( \max_{x \in [0,1]} |u(x,t)| \right) = +\infty, \quad (13)$$

*hold for some $T_b < \infty$, where*

$$T_b \leq \frac{1}{2} \sqrt{2} \int_{G_0}^{\infty} \left[ F(s) - \frac{1}{2} \lambda s^2 - \left( F(s_1) - \frac{1}{2} \lambda s_1^2 \right) + \frac{1}{2} G_1^2 \right]^{-1/2} ds. \quad (14)$$

In fact, it is proved in [10] that $G(t) = \int_0^1 \psi_1(x) u(x,t) \, dx \to \infty$ as $t \to T_b < \infty$.
Further we will use the estimate

$$T_b - t \leq \frac{1}{2} \sqrt{2} \int_{G(t)}^{\infty} \left[ F(s) - \frac{1}{2} \lambda^2 - \left( F(s_1) - \frac{1}{2} \lambda s_1^2 \right) + \frac{1}{2} G_1^2 \right]^{-1/2} ds. \quad (15)$$

# 3   Analysis for Blow-Up and Convergence of the Numerical Solution

We solve (1)-(5) or, in weak formulation, (10), by the difference method of lines. This means that we use semidiscretization in space, and consider time to be continuous. Given a positive integer $N$, $N \geq 3$, let $h = \frac{1}{N-1}$. Then

$$\Omega^h = \{x_i : x_i = ih, \ i = 1, \ldots, N\} \quad \text{is an uniform mesh on } \Omega = [0,1].$$

Integrating equation (1) with respect to the time variable and approximating the space derivative in the usual way by finite differences, we obtain

$$\frac{dU_i}{dt}(t) = v_0(x_i) + \int_0^t D_x^+ \left( \tilde{a} D_x^- U_i(\tau) \right) d\tau + \int_0^t f(U_i(\tau)) \, d\tau, \quad x_i \in \Omega^h, \quad (16)$$
$$U_i(0) = u_0(x_i), \quad U_1(t) = U_N(t) = 0,$$

where, $\tilde{a}(x) = a(x - h/2)$ or $\tilde{a}(x) = [a(x) + a(x-h)]/2$, $D_x^{\pm} U_i = \pm (U_{i \pm 1} - U_i)/h$. The system (16) is a weak formulation of the following SODEs

$$\frac{d^2 U_i}{dt^2} = \frac{\tilde{a}_i U_{i-1} - (\tilde{a}_i + \tilde{a}_{i+1}) U_i + \tilde{a}_{i+1} U_{i+1}}{h^2} + f(U_i), \quad i = 2, \ldots, N-1,$$
$$U_1(t) = U_N(t) = 0, \quad U_i(0) = u_0(x_i), \quad \frac{dU_i}{dt}(0) = v_0(x_i), \quad (17)$$

which is supposed to hold almost everywhere (a.e.) on $[0, T = T_b-\tau]$, $0 < \tau < T$. Further, in the analysis of (16), (respectively (17)) we assume that

$$a(x), \quad u_0(x) \in L_\infty(\Omega) \cap W_2^1(\Omega^-) \cap W_2^1(\Omega^+), \tag{18}$$

which with $\|f\|_{L_2(Q_T)} < \infty$, implies

$$u \in H(Q_T) \equiv C((0,T), W_2^1(\Omega)) \cap C^1((0,T), L_2(\Omega)) \cap C((0,T), W_2^2(\Omega^-))$$
$$\cap C^1((0,T), W_2^1(\Omega^-)) \cap C((0,T), W_2^2(\Omega^+)) \cap C^1((0,T), W_2^1(\Omega^+)).$$

Let $V_h$ be the standard piecewise linear finite element space in $\Omega$ and $\{\varphi_i\}$, $1 \le i \le N$ be the usual Lagrange basis of $V_h$. In view of (10), the scheme (17) is equivalent to

$$\int_0^1 (v(u_h)'')^I dx + \int_0^1 av_x u_{hx} dx = \int_0^1 (vf(u_h))^I dx,$$

$$\int_0^1 (vu_h(0))^I dx = \int_0^1 (vu_0)^I dx, \quad \int_0^1 (v(u_h)')^I dx = \int_0^1 (vv_0)^I dx,$$

a.e. on $[0, T_1]$, $T_1 < T_b$, $\forall v \in V_h$,

where $u_h$ is the FEM solution, $(\cdot)'$ and $(\cdot)_x$ denote the time and space derivatives, while the superindex $I$ denotes linear Lagrange interpolation. Starting from this formulation we prove the following theorem.

**Theorem 2.** *Let $u$ be the solution of a problem like (1)-(5) with $f$ replaced by a globally Lipschitz function $g$ and $U(t)$ is its semidiscrete approximation obtained by semidiscretization of identity (10) (with $f$ replaced by $g$). If $u \in H(Q_{T_1})$ for some $T_1 > 0$ then*

$$\|u - U\|_{L_\infty((0,T_1), L_2(\Omega^h))} \le Ch^2,$$

*where $C$ is a constant depending on $\|u\|_{H(Q_{T_1})}$.*

As a corollary of this theorem one can prove the convergence result.

**Theorem 3.** *Let $u$ be a regular solution of (1)-(5) ($u \in H(Q_{T_b-\tau})$) and $u_h-$ its numerical approximation. Then there exists a constant $C$ depending on $\|u\|_{H(Q_{T_b-\tau})}$ such that*

$$\|u - u_h\|_{L_\infty(\Omega \times (0, T_b-\tau))} \le Ch^{3/2}.$$

**Theorem 4.** *Let $u \in H(Q_T)$ and the assumptions of Theorem 1 are fulfilled. Then, for sufficiently small $h$, the solution of (16) (the weak solution of (17)) blows up in a finite time $T_h$ and*

$$T_h \le \frac{\sqrt{2}}{2} \int_{G_{0h}}^\infty \left[ F(s) - F(s_1) + \frac{\lambda_h(s_1^2 - s^2) + G_{1h}^2}{2} \right]^{-\frac{1}{2}} ds,$$

$$G_{0h} = h \sum_{i=1}^{N-1} \psi_h(x_i) u_0(x_i), \quad G_{1h} = h \sum_{i=1}^{N-1} \psi_h(x_i) v_0(x_i),$$

*where $\psi_h$ is a positive eigenfunction (for some real number $\lambda_h$) of the discrete problem corresponding to (7)-(9).*

The discrete analogues of (13) hold :

$$\lim_{t \to T_h - 0} \|U(t)\|_{L_p(\Omega^h)} = +\infty, \quad (1 \le p < \infty), \quad \lim_{t \to T_h - 0} \|U(t)\|_{L_\infty(\Omega^h)} = +\infty.$$

In fact we prove that $G_h(t) = h \sum_{i=1}^{N-1} \psi_h(x_i)U_i(t) \to \infty$ as $t \to T_h < \infty$,

$$T_h - t \le \frac{1}{2}\sqrt{2} \int_{G_h(t)}^{\infty} \left[ F(s) - \frac{1}{2}\lambda_h^2 - (F(s_1) - \frac{1}{2}\lambda_h s_1^2) + \frac{1}{2}G_1^2 \right]^{-1/2} ds \quad (19)$$

It follows from (15), (19) that the exact and the approximate solutions blow up at close times if $G(t)$ and $G_h(t)$ are large enough. Therefore, if $G(t)$ and $G_h(t)$ are close while $u$ is regular, the solutions will blow up at close times. So, all what we need so as to find accurately the blow up time is the numerical method to provide a good approximation to the exact solution up to large enough values of $u$.

**Theorem 5.** *Let $T_b$ and $T_h$ are the blow up times for problems (1)-(5) and (16) respectively, $f$ is convex and satisfies (6), $a(x)$ and $u_0$ satisfy (18). Then $T_h \to T_b$ when $h \to 0$.*

## 4   Computational Results

In the test example we take $a(x)$ to be a piecewise constant: $a(x) = a^-$ for $x < \xi$ and $a(x) = a^+$ for $x > \xi$, $f(u) = 2(\bar{a}^2 - a^{\pm}(\beta^{\pm})^2(nx - m)^2)u^3 + a^{\pm}\beta^{\pm}nu^2$, $'-'$ for $x < \xi$ and $'+'$ for $x > \xi$, where $a^{\pm}, \beta^{\pm}, \bar{a}, C^{\pm}, m, n$ ($m \le n$- integers) are positive and

$$a^+\beta^+ = a^-\beta^-, \quad \Upsilon := \frac{n}{2}\xi^2 - m\xi + 1, \qquad \begin{cases} m \le \xi n, & a^- < a^+, \\ \xi n \le m \le n, & a^+ < a^-. \end{cases}$$
$$C^- + \beta^-\Upsilon = C^+ + \beta^+\Upsilon,$$

The exact solution is

$$u(x,t) = \begin{cases} (C^- - (\bar{a}t - \beta^-(\frac{n}{2}x^2 - mx + 1)))^{-1}, \, x \le \xi, \\ (C^+ - (\bar{a}t - \beta^+(\frac{n}{2}x^2 - mx + 1)))^{-1}, \, x \ge \xi. \end{cases}$$

It is easy to check that this is a weak solution of problem (1)-(5) in the sense of (10). Now, the initial and boundary conditions can be calculated. The point $(x,t)$, $x \in [0,1]$, $t > 0$ is called a blow-up point of the solution of problem (1)-(5), if there exists a sequence $(x_n, t_n)$, $x_n \in (0,1)$, $t_n > 0$ such that $u(x_n, t_n) \to \infty$, when $x_n \to x$, $t_n \to t$, $t_n < t$ as $n \to \infty$. In this sense, the blow-up point of the exact solution is $(\frac{m}{n}, T_b)$,

$$T_{b(x_i = \frac{m}{n})} = \frac{1}{\bar{a}} \begin{cases} C^- + \beta^- \frac{2n - m^2}{2n}, & a^- < a^+, \\ C^+ + \beta^+ \frac{2n - m^2}{2n}, & a^+ < a^-. \end{cases}$$

The numerical solution is computed, using full discretization. For control of the time $(\triangle t_n)$ and space mesh step size we use CFL condition and the energy conservation law

$$E(t) = \int_0^1 \left[ \frac{1}{2} \left( \frac{\partial u}{\partial t} \right)^2 + \frac{a(x)}{2} \left( \frac{\partial u}{\partial x} \right)^2 - F(u) \right] dx, \quad F(u) = \int_0^u f(s) ds.$$

In the Examples 1,2 we take: $a^- = 1$, $a^+ = 100$, $\beta^- = 100$, $\beta^+ = 1$, $\xi = 0.5$, $C^- = 2$, $\bar{a} = 100$.

*Example 1.* $(\xi \neq \frac{m}{n})$ In this example we take $m = 1$, $n = 4$. Thus $T_b = 0.895$. The solution is computed on nonuniform mesh in time and uniform mesh in space variable. In Table 1, at time $T$ and mesh with $N$ nodes in space, we present: the absolute error $AE_N$ (far away from $T_b$) and the relative error $RE_N$ (for time close to $T_b$) in max norm; the convergence rates $CR_A = \log_2 \frac{AE_N}{AE_{2N}}$ and $CR_R = \log_2 \frac{RE_N}{RE_{2N}}$; the error in $\xi$. The max error is accumulated at or near to the blow-up point. Computations of the solution of problem (1)-(5) with different input data show that the blow-up set is in the subintervals $(0, \xi)$ or $(\xi, 1)$, where the value of $a(x)$ is smaller. Therefore, as in the general case we don't know the exact blow-up set, it is reasonable to refine the mesh (not only in time, but also in space) in the whole region, where the solution could blow-up. In our test example, having in mind that $\frac{m}{n} \in (0, \xi)$, we use a fine mesh $(h_f = \frac{\xi}{N_f - 1})$ on the left of $\xi$ and a coarse mesh $(h_c = \frac{1 - \xi}{N_c - 1})$ on the right. The results (in max norm) are shown in Table 2. When using a fine-coarse mesh the accuracy increases, but the convergence rate and the error in $\xi$ become slightly worse. We may start to refine the mesh (in space) after some time (for example t=0.85 in our case) or when the solution reaches some value. The results are similar to those in Table 2, but the computational costs are smaller.

On Figure 1a we show the graphs of the numerical solution for two different time levels.

*Example 2.* $(\xi \equiv \frac{m}{n})$ Now $m = 1$, $n = 2$ and $T_b = 0.77$. First we compute the solution using uniform mesh in space and nonuniform in time. The results are shown in Table 3 (analogue of Table 1). The computations show that the max error is near to the point $\xi = \frac{m}{n}$, in the interval $(0, \xi)$ when $t$ is far from $T_b$

**Table 1.** Absolute error, relative error, convergence rate, error in $\xi$

| $N$ | $AE_N$ | $CR_A$ | Error in $\xi$ | $RE_N$ | $CR_R$ | Error in $\xi$ |
|-----|--------|--------|-----------------|--------|--------|-----------------|
| | | T=0.85 | | | T=0.89456 | |
| 81 | 6.630663e-4 | | 5.488736e-7 | | | |
| 161 | 1.546649e-4 | 2.1000 | 5.418928e-7 | 2.335503e-1 | | 8.821944e-7 |
| 321 | 3.489349e-5 | 2.1481 | 4.683305e-7 | 4.792738e-2 | 2.2848 | 6.314458e-7 |
| 641 | 7.546011e-6 | 2.2092 | 2.832547e-7 | 1.093085e-2 | 2.1324 | 4.045037e-7 |
| 1281 | 1.692541e-6 | 2.1565 | 1.536091e-7 | 2.515741e-3 | 2.1194 | 2.542696e-7 |
| 2561 | 3.898101e-7 | 2.1183 | 7.988330e-8 | 5.861511e-4 | 2.1016 | 1.401393e0-7 |

**Table 2.** Absolute error or relative error for different meshes

| Total nodes in $[0,1]$ | | 161 | 161 | 161 | 161 | 121 | 101 | 91 |
|---|---|---|---|---|---|---|---|---|
| $T$ | mesh sizes | $h_f = \frac{1}{160}$ $h_c = \frac{1}{160}$ | $h_f = \frac{1}{200}$ $h_c = \frac{1}{120}$ | $h_f = \frac{1}{240}$ $h_c = \frac{1}{80}$ | $h_f = \frac{1}{280}$ $h_c = \frac{1}{40}$ | $h_f = \frac{1}{160}$ $h_c = \frac{1}{80}$ | $h_f = \frac{1}{160}$ $h_c = \frac{1}{40}$ | $h_f = \frac{1}{160}$ $h_c = \frac{1}{20}$ |
| 0.85 | $AE_N$ | 1.5466e-4 | 5.5743e-5 | 3.6666e-5 | 8.9097e-5 | 8.2676e-5 | 7.7945e-5 | 1.7043e-4 |
| 0.89456 | $RE_N$ | 2.3355e-1 | 1.1296e-1 | 5.0354e-2 | 3.5125e-2 | 1.8613e-1 | 1.4757e-1 | 1.9634e-1 |

**Table 3.** Absolute error, relative error, convergence rate, error in $\xi$

| $N$ | T=0.75 | | | T=0.7698 | | |
|---|---|---|---|---|---|---|
| | $AE_N$ | $CR_A$ | Error in $\xi$ | $RE_N$ | $CR_R$ | $RE_N$ in $\xi$ |
| 81 | 2.108088e-3 | | 2.001901e-3 | | | |
| 161 | 4.727657e-4 | **2.1567** | 4.472315e-4 | 3.826691e-1 | | 3.826691e-1 |
| 321 | 1.070448e-4 | **2.1429** | 1.008209e-4 | 7.312934e-2 | **2.3876** | 7.312934e-2 |
| 641 | 2.741007e-5 | **1.9654** | 2.583608e-5 | 1.721622e-2 | **2.0867** | 1.721440e-2 |
| 1281 | 6.688616e-6 | **2.0349** | 6.298093e-6 | 4.235472e-3 | **2.0232** | 4.234052e-3 |
| 2561 | 1.682220e-6 | **1.9913** | 1.584382e-6 | 1.053781e-3 | **2.0069** | 1.053357e-3 |
| 5121 | 4.192929e-7 | **2.0043** | 3.948557e-7 | 2.630462e-4 | **2.0022** | 2.629149e-4 |

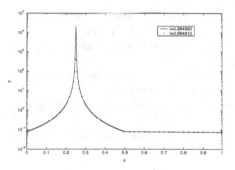

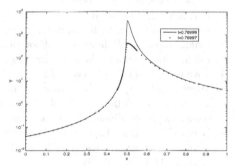

**Fig. 1a.** $N_f = 81$, $N_c = 21$, $\frac{m}{n} = \frac{1}{4}$      **Fig. 1b.** $N_f = 81$, $N_c = 21 + 21$, $\frac{m}{n} = \frac{1}{2}$

and in the interval $(\xi, 1)$, when $t$ approaches $T_b$. Therefore we must refine the mesh also in $x$ direction, around the point $\xi \equiv \frac{m}{n}$. Let for example, in intervals $[0, 0.45]$, $[0.55, 1]$ the mesh is coarse ($N_c = 21$) and in $[0.45, 0.55]$ the mesh is fine ($N_f = 81$). Thus the total number of nodes in $[0, 1]$ is 121. In this experiment we have significant decrease of the relative error (in max norm) - it is $3.9e - 2$ at time $T = 0.7698$.

The numerical solution for two different time levels is plotted on Figure 1b. It is found by using decreasing time step and coarse-fine mesh in space.

## 5    Conclusion

Depending on the initial data (especially on $f(u)$), the solution of problem (1)-(5) may blow up in finite time. Even in this case the approximation, presented above, is very effective. The discrete energy of the scheme is conserved with a good precision. Computations with decreasing time step and fine-coarse mesh improve the results.

**Acknowledgments.** The authors thank to the referee for the suggestions made to improve the results and the exposition.

This investigation is supported by Bulgarian National Fund of Science under Project VU-MI-106/2005.

## References

1. Bandle C., Brunner H. (1998), Blow-up in diffusion equations: a survey, *J. Comp. Appl. Math.* **977**, pp. 3-22.
2. Carpio A., Duro G. (2005), Instability and collapse in discrete wave equations,*Comp. Meth. in Appl. Math.* **3,5**, pp. 223-241.
3. Dimova S., Kaschiev M., Koleva M., Vasileva D. (1998),  Numerical analysis of radially non-symmetric blow-up solutions of a nonlinear parabolic problem, *J. Comp. Appl. Math.* **97**, pp. 81-97.
4. Ferreira R., Groisman P., Rossi J. (2004), Numerical blow-up for the porous medium equation with a source, *Num. Meth. P.D.E.* **20**, **4**, pp. 552-575.
5. Galaktionov V., Pohozaev S., Blow-up, critical exponents and asymptotic spectra for nonlinear hyperbolic equatioms, in press.
6. Jovanovic B. (1993), *Finite difference Method for Boundary Value Problems with Weak Solutions*,  Posebna izdanja Mat. Instituta **16**, Belgrade.
7. Jovanovic B., Vulkov L., Analysis of semidiscrete approximations of blow-up weak solutions to semilinear parabolic equations, submitted.
8. Koleva M., Vulkov L.(2005),  On the blow-up finite difference solutions to the heat-diffusion equation with semilinear dynamical boundary conditions, *Appl. Math. Comput.***161** , pp. 69-91.
9. Koleva M., Vulkov L.,  Blow-Up of continuous and semidiscrete solutions to elliptic equations with semilinear dynamical boundary conditions of parabolic type , *J. Comp. Appl. Math.* , available in online 18 April 2006.
10. Levine H. (1975),  Nonexistence of global weak solutions to some properly and improperly posed problems of mathematical physics: the method of unbounded Fourier coefficients, *Math. Ann.* **214**, pp. 205-220.
11. Nakagawa T. (1976),  Blowing up of a finite difference solution to $u_t = u_{xx} + u^2$, *Appl. Math. & Optimization* **2**, **4**, pp. 337-350.
12. Samarskii A. (2001), *Theory of Difference Schemes*, Marcel Dekker, Inc., New York.

# A Monotone Iterative Method for Numerical Solution of Diffusion Equations with Nonlinear Localized Chemical Reactions

Juri D. Kandilarov

Center of Applied Mathematics and Informatics
University of Rousse, 8 Studentska str., Rousse 7017, Bulgaria
`juri@ami.ru.acad.bg`

**Abstract.** We study the numerical solution of a model two-dimensional problem, where the nonlinear reaction takes place only at some interface curves, due to the present of catalyst. A finite difference algorithm, based on a monotone iterative method and the immersed interface method (IIM), is proposed and analyzed. Our method is efficient with respect to flexibility in dealing with the geometry of the interface curve. The numerical results indicate first order of accuracy.

## 1 Introduction

We consider the following problem

$$- \Delta u(P) = f(P), \quad P \equiv (x, y) \in \Omega \backslash \Gamma, \quad \Omega = (0, 1)^2 \tag{1}$$

$$u(P) = 0, \quad P \in \partial \Omega \tag{2}$$

$$[u(P)]_\Gamma = u^+(P) - u^-(P) = 0, \tag{3}$$

$$\left[ \frac{\partial u}{\partial \mathbf{n}}(P) \right]_\Gamma = F(P, u(P)). \tag{4}$$

Here $\Gamma \subset \Omega$ is a smooth curve which splits the domain $\Omega$ into two separate regions $\Omega^+$, $\Omega^-$: $\Omega = \Omega^- \cup \Omega^+ \cup \Gamma$ and $\mathbf{n}(x)$ is the normal at $P \in \Gamma$. The notation $[u(P)]_\Gamma$ stands for the jump of $u(P)$ across the interface $\Gamma$ in the following sense: for any point $P$ on the interface $\Gamma$,

$$[u(P)]_\Gamma = \lim_{Q \to P, Q \in \Omega^+} u(Q) - \lim_{R \to P, R \in \Omega^-} u(R).$$

The problem describes chemical reaction-diffusion processes in which, due to the effect of catalyst, the reaction takes place only at some local sites. This causes the chemical concentration to be continuous, but the gradient of the concentration to have a jump at these local sites ( $\equiv$ interface $\Gamma$ ). The magnitude of the jump typically depends on the concentration [1,15].

Elliptic and parabolic problems with discontinuous coefficients and different type of concentrated sources (called often interface problems) arise naturally when modeling processes in heat and mass transfer, diffusion in composite media,

T. Boyanov et al. (Eds.): NMA 2006, LNCS 4310, pp. 615–622, 2007.
© Springer-Verlag Berlin Heidelberg 2007

flows in porous media, etc. Various methods have been developed for interface problems [12,18]. The IIM developed by LeVeque and Li [10,12] solves elliptic equations with jump relations on the interface $\Gamma$ (point in $1D$, curve in $2D$ and surface in $3D$ problems),

$$[u] = U(P), \quad [v] = V(P), \quad P \in \Gamma,$$

where $u$ is the solution, $v$ - the flow and $U(\cdot)$, $V(\cdot)$ are known functions, defined on the interface. It has been successfully implemented for $1D$ and $2D$ linear and nonlinear elliptic and parabolic equations [2,10,11,12,13].

$1D$ and $2D$ parabolic problems, respectively with point and line interfaces of the form (3), (4) were solved numerically by Vulkov and Kandilarov [4,8,17], for elliptic equations by Jovanovic, Kandilarov and Vulkov [3] and for $2D$ ones with curvylinear interface by Kandilarov and Vulkov [5,6,9].

The essence of the IIM consists of using uniform or adaptive Cartesian grids and introducing non-zero correction terms in the starting difference approximation near the interfaces. The role of the jump conditions is very important.

In this paper we present an iterative algorithm for numerical solution of problem (1)-(4) which consists in two parts. First, we use an analog of the upper and lower solutions method developed for semilinear elliptic and parabolic problems [13,14] to linearize the problem (1)-(4). Second, we solve numerically by the IIM developed in [9] each linear "upper" and "lower" problem.

The paper is organized as follows. In Section 2, we use the monotone iterative method to show the existence and uniqueness of the solution to the nonlinear problem (1)-(4). In Section 3 we explain the computational algorithm and state a convergence theorem. In Section 4, we consider $1D$ and $2D$ numerical examples and compare our approximate solutions to the corresponding exact ones.

## 2    Monotone Iterative Method

Further, we shall assume that

$$-\frac{\partial F}{\partial u}(P, u(P)) \geq 0, \quad (P, u(P)) \in \overline{\Omega} \times (-\infty, \infty). \tag{5}$$

If $F(P, u(P))$ is sufficiently smooth, then under suitable continuity and compatibility conditions on the data, a unique solution $u(P)$ of (1)-(4) exists. This can be proved by the method of upper and lower solutions and the associated monotone iterations, as for the traditional semilinear elliptic equations [13,14].

We say that $\widetilde{u}(P)$ is an **upper** solution if it satisfies the inequalities:

$$-\Delta \widetilde{u} \geq f(P) \text{ in } \Omega \backslash \Gamma, \quad \widetilde{u}(P) \geq 0 \text{ on } \partial\Omega.$$

$$[\widetilde{u}(P)] = 0, \quad \left[\frac{\partial \widetilde{u}(P)}{\partial \mathbf{n}}\right] \leq -F(P, u(P)), \quad P \in \Gamma.$$

Similarly, $\widehat{u}(P)$ is called a **lower** solution if it satisfies the inverse inequalities. The pair $\widetilde{u}(P)$, $\widehat{u}(P)$ is said to be **ordered** if $\widetilde{u}(P) \geq \widehat{u}(P)$. For a given pair of ordered upper and lower solutions $\widetilde{u}$, $\widehat{u}$ we set

$$D(\widetilde{u}, \widehat{u})\{u \in C(\Omega) : \widehat{u} \leq u \leq \widetilde{u}\}.$$

Let $\gamma(P)$, $P \in \Gamma$ be any nonnegative function satisfying

$$\gamma(P) \geq \max\{-\frac{\partial F}{\partial u}(P, u(P)) : \widehat{u}(P) \leq u(P) \leq \widetilde{u}(P), P \in \Omega\}.$$

Then for any initial guess $u^{(0)}(P)$ we can construct a sequence $\{u^{(m)}\}$ from the linear iteration process:

$$- \Delta u^{(m)} = f(P) \text{ in } \Omega \backslash \Gamma,$$
$$u^{(m)}(P) = 0 \text{ on } \partial\Omega, \tag{6}$$
$$\left[u^{(m)}\right]_\Gamma = 0, \left[\frac{\partial u^{(m)}}{\partial n}\right]_\Gamma = \gamma(P)u^{(m)} - \gamma(P)u^{(m-1)} + F(P, u^{(m-1)}), \ P \in \partial\Gamma.$$

It is obvious that the sequence $\{u^{(m)}\}$ is well defined for each $m = 1, 2, \ldots$. Denote the sequence by $\{\overline{u}^{(m)}\}$ if $u^{(0)} = \widetilde{u}$ is an upper solution, and by $\{\underline{u}^{(m)}\}$ if $u^{(0)} = \widehat{u}$ is an lower solution, and refer to them as maximal and minimal sequences, respectively. The following theorem gives the monotone convergence of these sequences.

**Theorem 1.** Let $F(P, u)$ be a $C^1$ - function in $D(\widetilde{u}, \widehat{u})$, and let $\widetilde{u}, \widehat{u}$ be a pair of ordered upper and lower solutions of (1)-(4). Then the maximal sequence $\{\overline{u}^{(m)}\}$ converges to a maximal solution $\overline{u} = \overline{u}(P)$ of (1)-(4) and the minimal sequence $\{\underline{u}^{(m)}\}$ converges to a minimal solution $\underline{u} = \underline{u}(P)$. Moreover

$$\widehat{u} \leq \underline{u}^{(m)} \leq \underline{u}^{(m+1)} \leq \underline{u} \leq \overline{u} \leq \overline{u}^{(m+1)} \leq \overline{u}^{(m)} \leq \widetilde{u} \text{ on } \Omega$$

and if $F_u \leq 0$ in $D$, then $\overline{u} = \underline{u} = u^*$ is the unique solution of (1)-(4) in $D$.

## 3   The Numerical Method

Let us introduce on $\overline{\Omega}$ the uniform mesh $\overline{\omega}_h = \overline{\omega}_{h_1} \times \overline{\omega}_{h_2}$ [16], where

$$\overline{\omega}_{h_1} = \{x_i = ih_1, \ i = 0, 1, ..., N_1, \ h_1 = 1/N_1\},$$
$$\overline{\omega}_{h_2} = \{y_j = jh_2, \ j = 0, 1, ..., N_2, \ h_2 = 1/N_2\}.$$

Let $\omega_h$ and $\partial\omega_h$ be the sets of mesh points of $\Omega$ and $\partial\Omega$ respectively. Then the difference scheme corresponding to the problem (6) can be written in the form

$$\Delta_h U_{i,j}^{(m)} \equiv \frac{U_{i+1,j}^{(m)} - 2U_{ij}^{(m)} + U_{i-1,j}^{(m)}}{h_1^2} + D_{x,ij}^{(m,m-1)}$$
$$+ \frac{U_{i,j+1}^{(m)} - 2U_{ij}^{(m)} + U_{i,j-1}^{(m)}}{h_2^2} + D_{y,ij}^{(m,m-1)} = -f_{ij}, \quad (x_i, y_j) \in \omega_h \tag{7}$$
$$U_{i,j}^{(m)} = 0, \quad (x_i, y_j) \in \partial\omega_h,$$

where $U_{ij}^{(m)} \approx u^{(m)}(x_i, y_j)$ and $f_{ij} = f(x_i, y_j)$. Here $D_{x,ij}^{(m,m-1)} = D_{xl,ij} + D_{xr,ij}$ and $D_{y,ij}^{(m,m-1)} = D_{yt,ij} + D_{yb,ij}$ are additional terms chosen in order to improve the local truncation error (LTE), see [9]. By $l, r, t, b$, we show the intersection of the interface curve, respectively, with the left, right, top and bottom arm of the standard 5-point stencil for the discrete elliptic operator at $(x_i, y_j)$.

Let us introduce the level set function $\phi(x, y)$ for the curve $\Gamma$, such that $\phi(x, y) = 0$ when $(x, y) \in \Gamma$, $\phi(x, y) < 0$ for $(x, y) \in \Omega^-$ and $\phi(x, y) > 0$ for $(x, y) \in \Omega^+$. The outward normal $\mathbf{n}(n^1, n^2)$ of the curve $\Gamma$ is directed from $\Omega^-$ to $\Omega^+$. We shall call the node $(x_i, y_j)$ **regular**, if $\phi(x_i, y_j)$, $\phi(x_{i-1}, y_j)$, $\phi(x_{i+1}, y_j)$, $\phi(x_i, y_{j+1})$ and $\phi(x_i, y_{j-1})$ are together positive (negative), i.e. the curve $\Gamma$ doesn't intersect the stencil. The rest of nodes we call **irregular**.

At the regular nodes the corrections $D_x^{n+1}$ and $D_y^{n+1}$ are equal to zero. Further at the discretization we need to know the jumps of the derivatives $[u_x]$, $[u_y]$, $[u_{xx}]$ and $[u_{yy}]$, which we find from the conjugation conditions for $[u]$ and $[\partial u/\partial n]$. For this goal we use the idea of Z. Li from [10]. We introduce local coordinate system at each intersection point of the interface curve with the standard 5-point stencil, for example $(\xi_l, y_j)$ of the left arm:

$$\xi = (x - \xi_l) \cos \theta_l + (y - y_j) \sin \theta_l, \quad \eta = -(x - \xi_l) \sin \theta_l + (y - y_j) \cos \theta_l.$$

Here $\theta_l$ is the angle between the axis $Ox$ and the normal vector $\mathbf{n} = (\cos \theta_l, \sin \theta_l)$ at the node $(\xi_l, y_j)$. In the small neighborhood of this point the interface curve is situated near to the tangent with direction vector $\eta = (-\sin \theta_l, \cos \theta_l)$. Then $\Gamma$ can be locally parameterized by $\xi = \chi(\eta)$ and $\eta = \eta$. Note that $\chi(0) = 0$ and for a smooth curve $\chi'(0) = 0$.

If we use the iterative process (6), then for construction of $D_{x,ij}^{(m,m-1)}$, $D_{y,ij}^{(m,m-1)}$ we take the jumps in the form (more detailed derivation of the jumps can be found in [9]):

$$[u_x] = \left(\gamma u^{(m)} + g(u^{(m-1)})\right) \cos \theta_l, \quad [u_y] = \left(\gamma u^{(m)} + g(u^{(m-1)})\right) \sin \theta_l,$$

$$[u_{xx}] = -2 \left(\gamma u^{(m)} + g(u^{(m-1)})\right)_\eta \cos \theta_l \sin \theta_l$$
$$+ \chi'' \left(\gamma u^{(m)} + g(u^{(m-1)})\right) (\cos^2 \theta_l - \sin^2 \theta_l) + [f] \cos^2 \theta_l,$$

$$[u_{yy}] = 2 \left(\gamma u^{(m)} + g(u^{(m-1)})\right)_\eta \cos \theta_l \sin \theta_l$$
$$- \chi'' \left(\gamma u^{(m)} + g(u^{(m-1)})\right) (\cos^2 \theta_l - \sin^2 \theta_l) + [f] \sin^2 \theta_l.$$

where $(.)_\eta$ is the derivative in tangential direction, $[f]$ is the jump of the right hand side, $\chi''$ is the curvature of $\Gamma$ at the intersection point and

$$g(u^{(m-1)}) = -\gamma(P) u^{(m-1)} + F(P, u^{(m-1)}(P)), \quad P = (\xi_l, y_j) \in \Gamma.$$

For approximation of $u^{(m-1)}$ and $u^{(m)}$ at $P(\xi_l, y_j)$ we use Lagrange interpolation with linear polynomials of two variables defined at three points of the stencil.

Using either $U_{i,j}^{(0)} = \tilde{u}(x_i, y_j)$ or $U_{i,j}^{(0)} = \hat{u}(x_i, y_j)$ as initial iteration in (7), we obtain the sequences $\{\overline{U}_{i,j}^{(m)}\}$ and $\{\underline{U}_{i,j}^{(m)}\}$ respectively. The following convergence result holds.

**Theorem 2.** Let (5) be fulfilled. Then the sequences $\{\overline{U}_{i,j}^{(m)}\}$, $\{\underline{U}_{i,j}^{(m)}\}$ converge monotonically in maximum norm to the continuous solution $u(x, y)$ of the problem (1)-(4) when $h_1 + h_2 \to 0$.

# 4   Numerical Experiments

*Example 1.* Let us consider the following 1-D problem with discontinuous coefficients:

$$\beta u_{xx} - u = 0, \qquad x \in (0, \xi) \cup (\xi, 1),$$
$$[u]_\xi = 0, \qquad [\beta u_x]_\xi = -\exp(-u(\xi)),$$

where

$$\beta(x) = \begin{cases} \beta_1 = \alpha_1^2, & 0 \le x \le \xi, \\ \beta_2 = \alpha_2^2, & \xi < x \le 1. \end{cases}$$

The exact solution is

$$u(x) = \begin{cases} A\sinh(x/\alpha_1)/\sinh(\xi/\alpha_1), & 0 \le x \le \xi, \\ A\sinh((1-x)/\alpha_2)/\sinh((1-\xi)/\alpha_2), & \xi \le x \le 1, \end{cases}$$

where $A$ is a solution of the equation

$$A\left(\alpha_2 \coth((1-\xi)/\alpha_2) + \alpha_1 \coth(\xi/\alpha_1)\right) = exp(-A).$$

To find the numerical solution we use the following iterative procedure:

$$\beta U_{xx}^{(m)} - U^{(m)} = 0, \qquad x \in (0, \xi) \cup (\xi, 1),$$
$$\left[U^{(m)}\right]_\xi = 0, \qquad \left[\beta U_x^{(m)}\right]_\xi = \gamma U^{(m)} - (\exp(-U^{(m-1)}(\xi)) + \gamma U^{(m-1)}).$$

We solve this linearized problem using IIM, described in [7]. In Table 1 a mesh-refinement analysis is given for different choices of $\beta_1$, $\beta_2$, and mesh parameter $N$. The tolerance is $10^{-9}$ and the parameter $\gamma = 1$. The difference between the exact solution and the computed solution in maximum norm is

$$\|E_N\|_\infty = \max_{1 \le i \le N} |u(x_i) - U_i|.$$

The rate of convergence $p$ is calculated by the formula

$$p = \left| \frac{log(\|E_{N_1}\|_\infty / \|E_{N_2}\|_\infty)}{log(N_2/N_1)} \right|.$$

The results confirm second order of convergence, when the interface point coincides with a grid one, and first order in the other cases. The number of iterations varies from 7 to 10.

**Table 1.** Numerical test for *Example 1* with $\gamma = 1$, $\xi = 0.5$ and $\beta_i = \alpha_i^2$

| N | $\alpha_1 = 0.1$, $\alpha_2 = 1$ | $\alpha_1 = \alpha_2 = 1$ | N | $\alpha_1 = 0.1$, $\alpha_2 = 1$ | $\alpha_1 = \alpha_2 = 1$ |
|---|---|---|---|---|---|
| | $\|E\|_\infty$ | $\|E\|_\infty$ | | $\|E\|_\infty$ | $\|E\|_\infty$ |
| 15 | 3.5914e-03 | 2.7191e-04 | 16 | 1.6104e-03 | 1.0041e-04 |
| 31 | 1.2536e-03 | 1.2233e-04 | 32 | 4.1372e-04 | 2.5114e-05 |
| 63 | 5.2502e-04 | 5.7809e-05 | 64 | 1.0487e-04 | 6.2793e-06 |
| 127 | 2.5203e-04 | 2.8057e-05 | 128 | 2.6565e-05 | 1.5699e-06 |
| 255 | 1.2461e-04 | 1.3816e-05 | 256 | 6.5723e-06 | 3.8642e-07 |
| p | 1.02 | 1.02 | p | 2.02 | 2.02 |

**Table 2.** Mesh refinement analysis for *Example 2*, where $\Gamma$ is a circle with $a = b = 0.505$ or an ellipse with $a = 0.7505$, $b = 0.2505$

| | | $a = b = 0.505$ | | $a = 0.7505, b = 0.2505$ | |
|---|---|---|---|---|---|
| $N_1$ | $N_2$ | $\|E\|_\infty$ | $m$ | $\|E\|_\infty$ | $m$ |
| 10 | 10 | 6.4712e-02 | 49 | 1.9833e-01 | 51 |
| 20 | 20 | 3.3241e-02 | 55 | 1.1028e-01 | 59 |
| 40 | 40 | 1.7847e-02 | 63 | 5.9011e-02 | 65 |
| 80 | 80 | 9.2173e-03 | 94 | 3.2189e-02 | 93 |
| 160 | 160 | 5.0291e-03 | 123 | 1.7049e-02 | 119 |
| p | | 0.92 | | 0.89 | |

*Example 2.* On the region $\Omega = (-1,1)^2 \setminus \Gamma = \Omega^- \cup \Omega^+$, where $\Gamma :$ $x^2/a^2 + y^2/b^2 = 1$ and $\Omega^\pm = x^2/a^2 + y^2/b^2 - 1 \gtrless 0$, we consider the equation

$$u_{xx} + u_{yy} = f(x,y),$$

with exact solution

$$u(x,y) = \begin{cases} J_0(r), & (x,y) \in \Omega^-, \\ J_0(r)\left(1 + k\left(\frac{x^2}{a^2} + \frac{y^2}{b^2} - 1\right)\right), & (x,y) \in \Omega^+. \end{cases}$$

The boundary condition and the function $f(x,y)$ are found from the exact solution. The solution is continuous, but the jump of the normal derivative on the interface is

$$\left[\frac{\partial u}{\partial \mathbf{n}}\right] = K(x,y)\exp(-u(x,y)),$$

where $K(x,y) = 2kJ_0(r)\exp(J_0(r))\sqrt{x^2/a^4 + y^2/b^4}$, $r = \sqrt{x^2 + y^2}$. We choose $k = 0.1$ and two interfaces: circle with $a = b = 0.505$ and ellipse with $a = 0.7505$, $b = 0.2505$. The mesh refinement analysis is presented in Table 2, when the tolerance is $10^{-3}$. The number of iterations varies from 25 to 120 depending on the initial solution and increases with N. The results show first order of accuracy of the method. In Fig. 1a the error and in Fig. 1b the numerical solution are plotted, when $N_1 = N_2 = 40$ and $a = 0.7505$, $b = 0.2505$. To check the monotonicity of the method, we also control the values of the computed solution

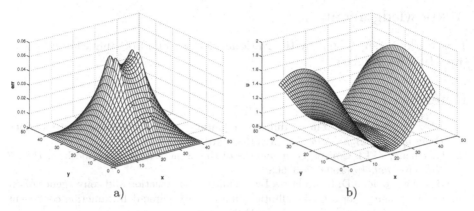

**Fig. 1.** a) The error and b) the numerical solution for *Example 2*, when $a = 0.7505$, $b = 0.2505$, $N_1 = N_2 = 40$, and $k = 0.1$

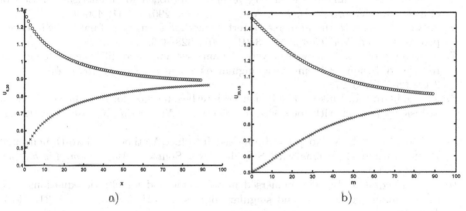

**Fig. 2.** a) The values at point $(6, 20)$ and b) the values at point $(20, 16)$ for *Example 2*, when $a = 0.7505$, $b = 0.2505$, $N_1 = N_2 = 40$, $k = 0.1$, $\gamma = 50$

at the vertexes of the ellipse during the iterations. In Fig. 2a the values of $U_{6,20}$ and in Fig. 2b the values of $U_{20,16}$ are plotted. The values of the parameters are: $N_1 = N_2 = 40$, $a = 0.7505$, $b = 0.2505$, $k = 0.1$, $\gamma = 50$. As initial data for the iteration processes we choose $0.5 \times exact\ solution$ and $1.5 \times exact\ solution$.

## 5  Conclusions

A finite difference method based on the monotone iterative method and the IIM for a diffusion equations with nonlinear own singular sources is developed. The method leads to a sequence which converges monotonically from above or below to the unique solution of the problem. The numerical results indicate first order of accuracy for problems with arbitrary interfaces.

# Acknowledgements

This work was supported by the National Science Fund of Bulgaria under contract HS-MI-106/2005.

# References

1. J.M. Chadam, H.M. Yin, A diffusion equation with localized chemical reactions, *Proc. of Edinburgh Math. Soc.* 37 (1993) 101–118.
2. H. Huang, Z. Li, Convergence analysis of the immersed interface method, *IMA J. Numer. Anal.* 19 (1999) 583–608.
3. B.S. Jovanovic, J.D. Kandilarov, L.G. Vulkov, Construction and convergence of difference schemes for a model elliptic equation with Dirac-delta function coefficient, *Lect. Not. in Comp. Sci.* 1988 (2001) 431–438.
4. J. Kandilarov, The Immersed interface method for a reaction diffusion equation with a moving own concentrated source, *Lect. Not. Comp. Sci.* 2542 (2003) 506–513.
5. J. Kandilarov, Immersed-boundary level set approach for numerical solution of elliptic interface problems, *Lect. Not. Comp. Sci.* 2907 (2004) 456–464.
6. J. Kandilarov, A Rothe-immersed interface method for a class of parabolic interface problems, *Lect. Not. Comp. Sci.* 3401 (2005) 328–336.
7. J. Kandilarov, L. Vulkov, Analysis of immersed interface difference schemes for reaction-diffusion problems with singular own sources, *Comp. Meth. Appl. Math.* 3 (2) (2003) 253–273.
8. J. Kandilarov, L. Vulkov, The immersed interface method for a nonlinear chemical diffusion equation with local sites of reactions, *Numerical Algorithms* 36 (2004) 285–307.
9. J. Kandilarov, L. Vulkov, The Immersed Interface Method for Two-Dimensional Heat-Diffusion Equations with Singular Own Sources, *Appl. Num. Math.*, (accepted, 2006)
10. R.J. LeVeque, Z. Li, The immersed interface method for elliptic equations with discontinuous coefficients and singular sources, *SIAM J. Num. Anal.* 31 (1994) 1019–1044.
11. Z. Li, Immersed interface method for moving interface problems, *Numerical Algorithms* 14 (1997) 269–293.
12. Z. Li, An overview of the immersed interface method and its applications. *Taiwanese J. of Mathematics* 7(1) (2003) 1–49.
13. Z. Li, C. V. Pao, Z. Qiao, A finite difference method and analysis for 2D nonlinear Poisson -Boltzmann equations, *J. Sci. Comput.*, accepted.
14. Pao, C. V., *Nonlinear Parabolic and Elliptic Equations*, Plenum Press, New York, 1992.
15. Pierce, A. P., H. Rabitz, Effect of defect structures on chemically active surfaces: A continium approach, *Phys. Rev. B* 38(3) (1988) 1734–1753.
16. A.A. Samarskii, The Theory of Difference Schemes, Marcel Dekker, Inc. New York, 2001.
17. L.G. Vulkov, J.D. Kandilarov, Construction and implementation of finite-difference schemes for systems of diffusion equations with localized nonlinear chemical reactions, *Comp. Math. Math. Phys.* 40 (2000) 705–717.
18. Y.C. Zhou, S. Zhao, M. Feig, G.W. Wei, High order matched interface and boundary method for elliptic equations with discontinuous coefficients and singular sources, *J. Comput. Phys.* 213 (2005) 1–30.

# Numerical Solution of an Elliptic Problem with a Non-classical Boundary Condition

Natalia T. Kolkovska

Institute of Mathematics and Informatics, Bulgarian Academy of Sciences,
Acad. Bonchev str. bl.8, 1113 Sofia, Bulgaria
natali@math.bas.bg

**Abstract.** We investigate an elliptic problem with a boundary condition given by a sum of normal derivative and an elliptic operator in tangential variables (also known as "Venttsel" boundary condition). The differential problem is discretized by a specific finite difference method. Error estimates of the numerical method in the discrete Sobolev space $W_2^1$ are obtained. The rate of convergence in this space is optimal, i.e. it is $m - 1$ for solutions from $W_2^m$, $1 < m < 2.5$.

## 1  Introduction

The problems for elliptic equations with classical boundary conditions, namely Dirichlet, Neumann, Robin and mixed type, have been studied for many years. In 1959 A. Venttsel introduced [10] a new class of boundary conditions for elliptic equations. The boundary condition in this case is given by an elliptic equation of second order with principle term being an elliptic operator in tangential variables. A simple example of such boundary value problem is the following one:

$$
\begin{aligned}
-\Delta u(r) &= f(r), & r &\in \Omega, \\
\alpha \tfrac{\partial^2 u}{\partial \tau^2}(r) - \beta u(r) - \tfrac{\partial u}{\partial n}(r) &= g(r), & r &\in \partial\Omega.
\end{aligned}
\tag{1}
$$

Here $\Omega$ is a bounded domain in $\Re^2$ with sufficiently smooth boundary $\partial\Omega$, $n$ is the outward normal to $\partial\Omega$, $\tau$ is the tangential direction to $\partial\Omega$, $\alpha$ and $\beta$ are constants. The specific boundary condition in (1) is referred as 'Venttsel' boundary condition.

Elliptic problems with Venttsel type boundary conditions appear in numerous problems in the water waves theory ([5], [9]), in engineering problems of oil wells ([3]), in financial mathematics. In the heat transfer, problem (1) describes the diffusion process in bounded domain combining boundary reflection and diffusion along a surface. This situation arises [2] when the boundary is covered with a thin layer of a material having high permeability.

The specific nature of the problem (1) is determined by the fact, that the boundary condition contains both second order tangential derivative and normal derivative. Thus the boundary equation in (1) is not an autonomous equation on $\partial\Omega$. Moreover one cannot use directly the general theory of elliptic problems to get a-priori estimates.

T. Boyanov et al. (Eds.): NMA 2006, LNCS 4310, pp. 623–627, 2007.

In [5] P. Korman mentioned that the problem is ill-posed in some sense for $\alpha \leq 0$ and $\beta = 0$. However, if $\alpha > 0$ and $\beta > 0$, then a periodic problem with Venttsel type boundary condition has a unique solution in a suitable Sobolev spaces [6] so that the problem is well-posed. We assume in this paper that $\alpha > 0$ and $\beta > 0$.

The unique solvability of the linear problem (1) is established in Hölder spaces $C^{2,\epsilon}(\Omega)$ by Luo and Trudinger in [7]. In [1] Apushkinskaya and Nazarov have considered quasilinear elliptic problems with quasilinear Venttsel type boundary conditions and have proved the existence of solutions in the Sobolev spaces $W_q^2(\Omega) \cap W_{q-1}^2(\partial\Omega)$ for $q > 2$. A survey of results on nonlinear elliptic and parabolic Venttsel problem can be found in [2].

In this paper we propose a numerical method for solving elliptic problems with Venttsel type boundary conditions in rectangular domains. In Section 2 we give some preliminaries. In Section 3 we study a finite difference method for solution of the problem (1). We prove, that if the solution to the differential problem belongs to $W_2^m(\Omega) \cap W_2^m(\Gamma_1)$, $1 < m \leq 3$, then the error of the method in $W_2^1$ mesh-norm is $O(|h|^{m-1})$ for $1 < m < 2.5$ and $O(|h|^{1.5})$ for $2.5 < m \leq 3$.

## 2    Preliminaries

Let $(x, y)$ be the coordinates of a point $r \in \Re^2$. Consider the unit square $\Omega = \{(x, y) : 0 < x < 1, 0 < y < 1\}$ with bottom part of its boundary $\Gamma_1 = \{(x, y) : 0 < x < 1, y = 0\}$.

We use the notations $W_2^m(\Omega)$ and $|\cdot|_{W_2^m(\Omega)}$ for the Sobolev space of functions [4] and for the semi-norm in this space.

We introduce a uniform mesh $\overline{\omega_h}$ in $\Omega$ with mesh sizes $h = (h_1, h_2)$, $h_i = N_i^{-1}$, $N_i \in N$, $i = 1, 2$ and set $\omega_h = \Omega \bigcap \overline{\omega_h}$, $\gamma_h = \overline{\omega_h} \setminus \omega_h$, $\gamma_h^1 = \gamma_h \bigcap \Gamma_1$, $\gamma_h^2 = \{r \in \gamma_h : x = 0\}$.

For functions from $W_2^1(\Omega)$ we define the square of the averaging Steklov's operator in x- direction $T_1$ as

$$T_1 v(r) = \int_{-1}^{1} (1 - |s|) v(x + sh_1, y) ds, \ r \in \omega_h$$

and the operator $T_2^*$ as follows:

$$T_2^* v(r) = 2 \int_0^1 (1 - s) v(x, sh_2) ds, \ r \in \gamma_h^1.$$

The notations of the following operators on discrete functions are taken from [8]. The first finite difference in $y$ direction and the second finite difference in $x$ direction are defined by

$$v_y(x, y) = (v(x, y + h_2) - v(x, y)) h_2^{-1},$$

$$\Lambda_1 u(x, y) = (u(x + h_1, y) - 2u(x, y) + u(x - h_1, y)) / h_1^2.$$

By analogy with $\Lambda_1$ and $T_1$ we define operators $\Lambda_2$ and $T_2$ in y - direction.
In the set of mesh functions, defined on $\overline{\omega_h}$, we consider different scalar products:

$$[u,v]_h^{(1)} = \sum_{r \in \omega_h \cup \gamma_h^2} h_1 h_2 u(r) v(r), \quad [u,v]_h^{(2)} = \sum_{r \in \omega_h \cup \gamma_h^1} h_1 h_2 u(r) v(r),$$

$$(u,v)_h = \sum_{r \in \omega_h} h_1 h_2 u(r) v(r) + 0.5 \sum_{r \in \gamma_h^1} h_1 h_2 u(r) v(r).$$

We denote by $H_h$ the set of discrete functions, defined on $\overline{\omega_h}$, which vanish on $\gamma_h \setminus \gamma_h^1$, with scalar product $(u,v)_h$.

We shall write $C$ for all positive constants, which are independent on the unknown functions and the mesh sizes.

## 3  Finite Difference Method

We consider the following boundary value problem with Venttsel boundary condition on the bottom edge $\Gamma_1$ of $\Omega$ and Dirichlet boundary condition on the other three edges of $\Omega$:

$$\begin{aligned}
-\Delta u(r) &= f(r), & r &\in \Omega, \\
u(r) &= 0, & r &\in \partial\Omega \setminus \Gamma_1, \\
\alpha \tfrac{\partial^2 u}{\partial x^2}(r) - \beta u(r) + \tfrac{\partial u}{\partial y}(r) &= g(r), & r &\in \Gamma_1.
\end{aligned} \tag{2}$$

We assume $\alpha > 0$ and $\beta > 0$.

In the next lemma we summarize some properties of the solution to the problem (2).

**Lemma 1.** Let $f \in L_2(\Omega)$, $g \in L_2(\Gamma_1)$. Then there exists a unique solution $u \in W_2^2(\Omega) \cap W_2^2(\Gamma_1)$ to the problem (2) and the following a-priori estimates are valid:

$$|u|_{W_2^1(\Omega)} + |u|_{L_2(\Gamma_1)} + \left| \frac{\partial u}{\partial x} \right|_{L_2(\Gamma_1)} \leq C \left( \|f\|_{L_2(\Omega)} + \|g\|_{L_2(\Gamma_1)} \right),$$

$$|u|_{W_2^2(\Omega)} + \left| \frac{\partial u}{\partial y} \right|_{L_2(\Gamma_1)} + \left| \frac{\partial^2 u}{\partial x^2} \right|_{L_2(\Gamma_1)} \leq C \left( \|f\|_{L_2(\Omega)} + \|g\|_{L_2(\Gamma_1)} \right).$$

A similar statement for the periodic case is proved in [6].
Using the Laplace equation and the Venttsel boundary condition in (2) we derive the following equation on $\Gamma_1$ for every solution $u$ of (2):

$$T_2^* \Lambda_1 u + \tfrac{2}{h_2} \alpha \Lambda_1 u + \tfrac{2}{h_2}(T_1 u_y - \beta T_1 u) - T_1 T_2^* f + \tfrac{2}{h_2} T_1 g. \tag{3}$$

We shall use (3) in the discretization of (2).
We approximate the problem (2) with the finite difference scheme

$$\begin{aligned}
A_h v \equiv A_1 v(r) + A_2 v(r) &= \varphi(r), & r &\in \omega_h \cup \gamma_h^1, \\
v(r) &= 0, & r &\in \gamma_h \setminus \gamma_h^1,
\end{aligned} \tag{4}$$

where

$$A_1(v)(r) = - \begin{cases} \Lambda_1 v(r), & r \in \omega_h, \\ (\frac{2}{h_2}\alpha + 1)\Lambda_1 v(r), & r \in \gamma_h^1, \end{cases}$$

$$A_2(v)(r) = - \begin{cases} \Lambda_2 v(r), & r \in \omega_h, \\ \frac{2}{h_2}(v_y - \beta v)(r), & r \in \gamma_h^1, \end{cases}$$

$$\varphi(r) \begin{cases} T_1 T_2 f(r), & r \in \omega_h, \\ T_1 T_2^* f(r) - \frac{2}{h_2} T_1 g(r), & r \in \gamma_h^1. \end{cases}$$

**Lemma 2.** *For all $u, v \in H_h$ the discrete operator $A_h$ satisfies the identity*

$$(A_h u, v)_h = [u_x, v_x]_h^{(1)} + [u_y, v_y]_h^{(2)} + \beta \sum_{r \in \gamma_h^1} h_1 u(r) v(r)$$

$$+ (\alpha + 0.5 h_2) \sum_{r \in \gamma_h^1 \cup (0,0)} h_1 u_x(r) v_x(r).$$

*$A_h$ is self-adjoint and positive definite operator on $H_h$.*

The proof of this assertion follows by partial summation and uses the assumptions $\alpha > 0$ and $\beta > 0$. As an immediate consequence of the Lemma 2 it follows, that there exists a unique solution to the difference scheme (4).

To investigate the error of the method, we define the error function $z = u - v$, where $u$ is the solution of (2) and $v$ is the solution of (4). Then the error function $z$ will be a solution to the problem

$$\begin{aligned} A_h z &= -\psi(r), & r \in \omega_h \cup \gamma_h^1, \\ v(r) &= 0, & r \in \gamma_h \setminus \gamma_h^1. \end{aligned} \tag{5}$$

where $\psi$ is given by $\psi = \varphi - A_h u$.

Using the properties of the averaging operators $T_1$, $T_2$ and the equality (3), we represent the function $\psi$ in the form

$$\begin{aligned} \psi(r) &= \Lambda_1 \eta_1 + \Lambda_2 \eta_2, \quad r \in \omega_h; \quad \psi(r) = \Lambda_1 \eta_1 - A_2 \eta_2, \quad r \in \gamma_h^1; \\ \eta_1 &= T_2 u - u, r \in \omega_h; \quad \eta_1 = T_2^* u - u, r \in \gamma_h^1; \quad \eta_2 = T_1 u - u, r \in \omega_h \cup \gamma_h^1. \end{aligned} \tag{6}$$

Then the following statement holds:

**Lemma 3.** *The finite difference scheme (5) with the right-hand side (6) is stable and its solution $z$ satisfies the following inequality:*

$$[z_x, z_x]_h^{(1)} + [z_y, z_y]_h^{(2)} + \sum_{r \in \gamma_h^1 \cup (0,0)} h_1 z_x^2(r) + \sum_{r \in \gamma_h^1} h_1 z^2(r) \tag{7}$$

$$\le C \left( [\eta_{1x}, \eta_{1x}]_h^{(1)} + [\eta_{2y}, \eta_{2y}]_h^{(2)} + \sum_{r \in \gamma_h^1 \cup (0,0)} h_1 \eta_{1x}^2(r) + \sum_{r \in \gamma_h^1} h_1 \eta_{2y}^2(r) \right).$$

The right-hand side terms in (7) can be estimated using the well-known Bramble-Hilbert lemma [4]. In this way we prove convergence of the finite difference scheme.

**Theorem 1.** *Let $u \in W_2^m(\Omega) \cap W_2^m(\Gamma_1)$, $1 < m \leq 3$, be the solution of (2), $v$ be the solution of (4) and $z = u - v$ be the error. Then there exists a positive constant $C$ such that*

$$\left\{ [z_x, z_x]_h^{(1)} + [z_y, z_y]_h^{(2)} + \sum_{r \in \gamma_h^1 \cup (0,0)} h_1 z_x^2(r) + \sum_{r \in \gamma_h^1} h_1 z^2(r) \right\}^{\frac{1}{2}}$$

$$\leq C |h|^{m-1} M_m(h) \, \|u\|_{W_2^m(\Omega) \cap W_2^m(\Gamma_1)},$$

*where*

$$M_m(h) = \begin{cases} 1, & 1 < m < 2.5; \\ |\ln h|, & m = 2.5; \\ h^{2.5-m}, & 2.5 < m \leq 3. \end{cases}$$

The above estimate of the rate of convergence is consistent with the smoothness of the exact solution in the case $1 < m < 2.5$.

In the case $2.5 \leq m \leq 3$ there is a loss of the order of convergence. Note that in the last case the rate of convergence is the same as for the elliptic problem with Robin boundary conditions [8]. A better estimate can be obtained for the scheme with another approximation of the right-hand side similar to the one from [8, p.162].

# References

[1] Apushkinskaya, D., Nazarov, A. : On the quasilinear stationary Ventssel problem. Zap. Nauchn. Sem. S.-Petersburg. Otdel. Mat. Inst. Steklov. (POMI) **221** (1995) 20-29 (in russian)

[2] Apushkinskaya, D., Nazarov, A. : A survey of results on nonlinear Venttsel problems. Applications of Mathematics. **45** (2000) 69–80

[3] Cannon, J., Meyer, G. : On diffusion in a fractured medium. SIAM J. Appl. Math. **20** (1971) 434–448

[4] Ciarlet, P. : The finite element method for elliptic problems. North Holland, New York (1978)

[5] Korman, P. : Existence of periodic solutions for a class of nonlinear problems. Nonlinear Analysis, Theory, Methods, & Applications. **7** (1983) 873–879

[6] Korman, P. : On application of the monotone iteration scheme to noncoercive elliptic and hyperbolic problems. Nonlinear Analysis, Theory, Methods, & Applications. **8** (1984) 97–105

[7] Luo, Y., Trudinger, N. : Linear second order elliptic equations with Venttsel boundary conditions. Proc. Royal Soc. of Edinburgh. **118A** (1991) 193-207

[8] Samarski,A., Lazarov, R., Makarov V. : Finite difference schemes for differential equations with generalized solutions. (1987) (in Russian)

[9] Shinbrot M. : Water waves over periodic bottoms in three dimensions. J. Inst. Maths Applics **25** (1980) 367–385

[10] Ventssel, A. : On boundary conditions for multidimensional diffusion processes. Teor. veroyatn. i primenen. **4** (1959) 172-185 (in Russian)

# Contour Determination in Ultrasound Medical Images Using Interacting Multiple Model Probabilistic Data Association Filter*

Pavlina Konstantinova[1], Dan Adam[2], Donka Angelova[1], and Vera Behar[1]

[1] Institute for Parallel Processing, Bulgarian Academy of Sciences,
25A Acad. G. Bonchev St, 1113 Sofia, Bulgaria
{pavlina,donka,behar}@bas.bg
[2] Department of Biomedical Engineering, Technion-Israel Institute of Technology,
Hajfa 32000, Israel
dan@biomed.technion.ac.il

**Abstract.** The Probabilistic Data Association Filter (PDAF) with Interacting Multiple Model (IMM) approach is applied for contour determination in ultrasound images. The contour of interest is assumed to be a target trajectory which is tracked using IMMPDA filtering. The target movement is assumed to be along a circle and controlled by equally spaced radii from an arbitrary seed point inside the assumed contour. The generalized scores of the candidate points along current radius are determined on the base of two components - the Gaussian probability density function, associated with the assignment of the current point to the trajectory and the edge magnitude. A method for modeling complex contours with known true positions and method for error evaluation are proposed. These methods are used to generate Field II images and to estimate errors of contour determination using IMMPDA algorithm incorporating edge magnitude.

## 1 Introduction

In recent years many algorithms have been developed for tracking targets in clutter. One of the most successful approach used in these algorithms is Probabilistic Data Association (PDA). Instead of one measurement the PDA approach uses all of the validated measurements with different weights (probabilities) [4,7] and improves the algorithm robustness in clutter. Another approach - Interacting Multiple Model (IMM) proved to be suitable for tracking maneuvering targets [4,2]. The combination of the two approaches (IMMPDA) is effectively applied not only for target tracking but also in different areas as robotics, navigation,

---

* This work was partially supported by the Bulgarian Foundation for Scientific Investigations: MI-1506/05 and by Center of Excellence BIS21++, 016639. This work was also partially supported by the Chief Scientist 'Magnet' and 'Magneton' programs, of the Israel Ministry of Industry and Commerce.

T. Boyanov et al. (Eds.): NMA 2006, LNCS 4310, pp. 628–636, 2007.

biomedical imaging and many others. It is also successfully applied in [1] for cavity boundary extraction from ultrasound images. A new approach based on IMMPDA and incorporating edge magnitude is described by the authors. The results are validated through comparison with manual segmentations performed by an expert.

The goal of this work is to develop a method for modeling complex contours with known true positions of the contour points, to generate Field II images with these contours and to estimate errors of contour determination using similar IMMPDA algorithm incorporating edge magnitude.

## 2  Problem Formulation

The common property in medical images is that the cysts and other lesions more often have convex forms. This property allows the following assumption - the contour of interest to be star-shaped i.e. all contour points could be seen from the appropriately selected seed point inside the assumed contour. In this paper we solve contour determination problem of the low quality ultrasound images with unavoidable speckle noise applying the PDA and IMM approaches for low observable maneuvering target tracking in presence of false alarms. *The contour will be treated as target trajectory* and the points of the contour will be defined on equally spaced radii from the selected seed point inside the assumed contour.

The state vector of the system $x$, describing the target dynamic evolves in time according to

$x(k + 1) = F(k)x(k) + \nu(k)$, with measurement vector, given by

$z(k + 1) = H(k)x(k) + w(k)$, where $F$ is the system transition matrix, $H$ is the measurement matrix, $\nu(k)$ and $w(k)$ are zero-mean mutually independent white Gaussian noise sequences with known covariance matrices $Q(k)$ and $R(k)$, respectively, k is index of the time. For more details see [3,7].

### 2.1  Probabilistic Data Association Approach

The Probabilistic Data Association approach considers all points which are near to the predicted point as belonging with some probability to the track (the contour of interest). The area of the valid points, called gate is defined by the following criteria based on Mahalanobis distance:

$$d_j^2(k) = \nu_j'(k) S^{-1} \nu_j(k) \le \gamma, \tag{1}$$

where $\nu_j(k) = \hat{z}(k) - z_j(k)$ is the difference between the distances of the predicted point and the $j^{th}$ validated point, $S$ is the innovation covariance matrix, $\gamma$ is a threshold constant defined from the table of the chi-square distribution [3] p.84.

The weighting probabilities, defining the influence of each point in the gate to the final decision are evaluated using both *kinematic* data and incorporated *edge magnitude* for each of the candidate point. We recall that candidate points at time $k$ are defined to be on the radius from the seed point with angle $\theta_k$. The images are gray level with intensity from 1 to 256. For calculation of the edge magnitude for point $j$ on radius $\theta_k$ the filter, proposed in [1] is used: $Fe_{(j,\theta_k)} = \frac{1}{3}\left\{I_{(j+2,\theta_k)}+I_{(j+1,\theta_k)}+I_{(j,\theta_k)}-I_{(j-1,\theta_k)}-I_{(j-2,\theta_k)}-I_{(j-3,\theta_k)}\right\} \times \left(255 - I_{(j,\theta_k)}\right)^2$.

Then the association probability $\beta_j$ when $j \neq 0$ is the probability for point $j$ to belong to the contour, and $\beta_0$ - the probability none of the validated points to belong to the contour. They are calculated as:

$$\beta_j(k) = \frac{Fe^2_{(j,\theta_k)}e_j(k)}{b(k)+\sum\limits_{i=1}^{m_k}e_iFe^2_{(i,\theta_k)}}; \ j = 1,2,\ldots,m_k; \ \beta_0(k) = \frac{b(k)}{b(k)+\sum\limits_{i=1}^{m_k}e_i(k)Fe^2_{(i,\theta_k)}}, \quad (2)$$

where $e_j(k) = \frac{1}{\sqrt{2\pi|S|}}\exp\left[-\frac{1}{2}\nu_j'(k)S^{-1}\nu_j(k)\right]$, $b(k) = \sqrt{\frac{\pi}{2\gamma}}m_k\frac{1-P_DP_G}{P_D}$, $m_k$ is the number of the points satisfying (1), $P_D$ is detection probability and $P_G$ is factor that accounts for restricting the normal density to the validation gate.

Then combined innovation can be evaluated as [4]:

$$\nu(k) = \sum_{j=1}^{m_k}\beta_j(k)\nu_j(k) \quad (3)$$

and the updated state vector $\hat{x}(k|k)$ and state covariance matrix $P(k|k)$ are:

$$\hat{x}(k|k) = \hat{x}(k|k-1) + W(k)\nu(k), \quad (4)$$

$$P(k|k) = \beta_0 P(k|k-1) + [1-\beta_0]P^c(k|k) + \tilde{P}(k), \quad (5)$$

where $W(k) = P(k|k-1)H(k)'S(k)^{-1}$ is the gain matrix, $S(k) = H(k)P(k|k-1)H(k)' + R(k)$, and using Joseph's form for $P^c$:

$$P^c = [I - W(k)H(k)]P(k|k-1)[I - W(k)H(k)]' + W(k)R(k)W(k)', \quad (6)$$

$$\tilde{P}(k) = W(k)\left[\sum_{j=1}^{m_k}\beta_j(k)\nu_j(k)\nu_j(k)' - \nu(k)\nu(k)'\right]W(k)'. \quad (7)$$

The notation $(k|k-1)$ means that the corresponding value is predicted for time $k$ on the base of the information for time $(k-1)$.

## 2.2   Interacting Multiple Models Approach

To provide more precise tracking of abrupt contour changes the Interacting Multiple Models approach [2,4] with two nested models of the tracking system is applied. The difference between the two models is only in the values of the process noise variance. The filtering of the state vectors for the two models is performed in parallel based on initial combined state vector $\hat{x}^{0j}$ and state covariance matrix $P^{0j}$.

The combining probabilities are:

$$\mu_{i|j}(k-1|k-1) = \frac{1}{\bar{c}_j}p_{ij}\mu_i(k-1) \qquad i,j = 1,2,\cdots,r, \quad (8)$$

where $r$ is the number of models, $p_{ij}$ are the Markovian switching probabilities, $\mu_i(k-1)$ is the probability model $i$ to be true for the time $k-1$ and the normalizing coefficient $\bar{c}_j$ is

$$\bar{c}_j = \sum_{i=1}^{r} p_{ij}\mu_i(k-1) \qquad j = 1, 2, \cdots, r \ . \tag{9}$$

Then combined state vectors and state covariance matrices for each model $j$, where $j = 1, 2, \cdots, r$ are:

$$\hat{x}^{0j}(k-1|k-1) = \sum_{i=1}^{r} \hat{x}^i(k-1|k-1)\mu_{i|j}(k-1|k-1) \tag{10}$$

$$P^{0j}(k-1|k-1) = \sum_{i=1}^{r} \mu_{i|j}(k-1|k-1) \left\{ P^i(k-1|k-1) + \left[\hat{x}^i - \hat{x}^{0j}\right]\left[\hat{x}^i - \hat{x}^{0j}\right]' \right\} \tag{11}$$

The updated model probabilities are:

$$\mu_j(k) = \frac{1}{c}\Lambda_j(k) \sum_{i=1}^{r} p_{ij}\mu_i(k-1), \quad where \quad \Lambda_j(k) = \frac{(Fe_{(j,\theta_k)})^2 \, exp^{-\frac{1}{2}d_j^2(k)}}{\sqrt[2]{2\pi\,|S(k)|}}, \tag{12}$$

$c$ is the normalizing coefficient. As a result the estimation of target state vector for output only (not used in recursive procedure) is:

$$\hat{x}(k|k) = \sum_{j=1}^{r} \hat{x}^j(k|k)\mu_j(k) \tag{13}$$

## 3   Algorithm Description

The system dynamic is forced by equally spaced radii. The schematic presentation of this formulation is shown in Fig.1, where two consecutive radii $R_k$ and $R_{k+1}$ are presented. The estimated point is on the radius $R_k$ and the corresponding predicted point is on the radius $R_{k+1}$. The candidate points around the predicted point with their Gaussian pdf associated with the assignment of the corresponding point to the trajectory of the contour are also presented.

An insight into the image processing during current radius is presented in Fig.2. It can be seen the abrupt change of the pixel intensities along the radius and the corresponding maximum value of edge magnitude. This maximum is in the region of the validated pixels, satisfying condition (1) using Mahalanobis distance. In this case the decision is noted with symbol $*$ on the axis for the points of the current radius.

The state vector $x = \begin{bmatrix} D & \dot{D} \end{bmatrix}'$, where $D$ is the distance from the seed point to the current point and $\dot{D} = \frac{dD}{d\theta}$. The increment of the angle is $\Delta\theta = 2\pi/N_r$, $N_r$ is the number of the radii that corresponds to the evaluated contour points.

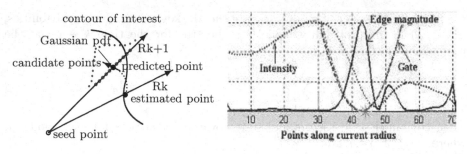

**Fig. 1.** Scheme of the algorithm     **Fig. 2.** An Insight to the Processing

Measurement vector is $Z$, $Z(1) = (D)$, measurement vector dimension $n_z = 1$. System transition matrix $F = \begin{pmatrix} 1 & \Delta\theta \\ 0 & 1 \end{pmatrix}$. The measurement matrix is: $H = [1\ 0]$.

We assume the following Markovian transition probabilities $p = \begin{pmatrix} 0.95 & 0.05 \\ 0.05 & 0.95 \end{pmatrix}$. The measurement noise matrix $R$ is: $R = (\sigma_D^2)$. Process noise matrix $Q$ for model $l$ is $Q_l = \sigma_{\nu l}^2 \begin{pmatrix} \frac{\Delta\theta^4}{4} & \frac{\Delta\theta^3}{2} \\ \frac{\Delta\theta^3}{2} & \Delta\theta^2 \end{pmatrix}$

For degree of freedom $n_z = 1$ and allowable probability to miss true measurement 0.01 the gate threshold is $\gamma = 6.63$.

### Initiating

The contour is initiated by finding the point with maximum of edge magnitude $(x_{maxF}, y_c)$ on the radius, started from the seed point $(x_c, y_c)$ along the positive direction of axes $x$. This point becomes the first point of the contour to be tracked. The state vector $X$ is initialized $X = [D\ \dot{D}]' = [(x_{maxF} - x_c)\ 0]'$, and the predicted state vector is $X(k+1) = F X(k)$. The state vector covariance matrix is $P = \begin{pmatrix} \sigma_D^2 & \frac{\sigma_D^2}{\Delta\theta} \\ \frac{\sigma_D^2}{\Delta\theta} & \frac{2\sigma_D^2}{\Delta\theta^2} \end{pmatrix}$ The predicted point is $Z(1) = X(1)$. The predicted state vector covariance matrix is $P(k+1) = F P(k) F'$ and innovation covariance matrix is $S(k) = H P(k) H' + R$. The angle $\theta = 0$, $\Delta\theta = 2\pi/N_r$.

Using the organization of the computations, proposed in [8] the algorithm can be outlined as follows:

### The algorithm outline

- **Initiating:** $\theta = 0$, $\Delta\theta = 2\pi/N_r$, and matrix preparation as described above.
- **Radius loop:** $\theta = \theta + \Delta\theta$ until $\theta = 2\pi$ :
    * Define validated points, satisfying condition (1) and their edge magnitudes;
    * Find weighting probabilities (2), calculate combined innovation (3) and the matrix, defined by the sum used in the eq. (7);
    * Updating part of Kalman filtering, including equations (4) and (5);
    * IMM part including equations from (8) to (13);

\* Predicting part of Kalman filtering including standard Kalman filter equations:
$$X = FX; \quad P = FPF' + Q; \quad Z(1) = X(1); \quad S = HPH' + R.$$

NB: In this case because of the matrix specifics (for example measurement vector is just a number, and its transpose is the same) the calculations could be simplified.

## 4   Modeling

**Modeling test contours.** Three shapes of contours are used to test the algorithm performance - circle, ellipse and Cassini oval. The aim is to examine different slopes of the curves and curves, which shape not coincide with the assumed circle shape. Simulation program FieldII and the example with cyst phantom [6] are used for the experiments. In modeling function the generated random points are tested if they are inside of the modeled contour.

\* **Circle.** In case of a circle the condition for the random input point $(x_i, y_i)$ to be inside the circle with given center $(x_c, y_c)$ and radius $r$ is trivial:
$$inside = (r^2 \geq ((x_i - x_c)^2 + (y_i - y_c)^2)).$$

\* **Ellipse.** The second modeled contour is ellipse with center point $(x_c, y_c)$, semi-axes $a_x$ and $a_y$ and rotation angle $\alpha_r$. For the ellipse we use a function which evaluates if the input random point $(x_i, y_i)$ is inside the ellipse. We assume that the input point belongs to the family of the concentric ellipses with the same ratio of semi-axes and is on the radius defined by angle $\theta$. At first the polar coordinates of this random point with regard to the center point are calculated and after inverse rotation the following coordinates of a corresponding non rotated ellipse are obtained:
$$x_p = x_i - x_c, \quad y_p = y_i - y_c.$$
$$x_{ir} = x_p \cos(\alpha_r) - y_p \sin(-\alpha_r); \quad y_{ir} = x_p \sin(-\alpha_r) + y_p \cos(-\alpha_r). \quad (14)$$

The angle $\theta$, corresponding to these coordinates and to the parameters of the ellipse is evaluated as: $\Delta x = \frac{x_{ir}}{a_x}; \quad \Delta y = \frac{y_{ir}}{a_y}; \quad \theta = tan\left(\frac{\Delta y}{\Delta x}\right)$. This angle defines the radius, which is the geometric place of the points of the concentric ellipses. The point of the given ellipse located on that radius is:
$$x = a_x \cos(\theta); \quad y = a_y \sin(\theta).$$
The condition the input point to be inside the ellipse is:
$$inside = ((abs(x_{ir}) \leq abs(x)) \& (abs(y_{ir}) \leq abs(y))). \quad (15)$$

\* **Oval Cassini.** The third modeled contour is the oval Cassini [5] with parameters $a$ and $c$ ($a \geq c$). The evaluation of the property *inside* of an input random point $(x_i, y_i)$ with respect to this contour at the beginning is similar to that of the ellipse. The polar coordinates and inverse rotation calculating is the same as for the ellipse (14). The angle, corresponding to the point is $\theta = tan\left(\frac{y_{ir}}{x_{ir}}\right)$. The point $(x, y)$ of the oval Cassini, corresponding to this angle is evaluated as follows:

$$\phi = 2\,\theta; \qquad D = \sqrt{a^4 - c^4 \sin(\phi)^2}; \qquad R_o = \sqrt{c^2 \cos(\phi) + D};$$
$$x = R_o \cos(\theta); \qquad y = R_o \sin(\theta).$$

The condition if the input point is inside the oval Cassini is the same as for the ellipse (15).

### Error evaluation

The *error* evaluation for the modeled shapes is according to the following considerations. The seed point inside the assumed contour in the program is selected manually, using the mouse, or automatically by the program. In both cases this seed point is aimed to be near the ideal modeled center but in the general case does not coincide with it. As a result the evaluated contour points are located irregularly along the estimated contour curve. For that reason we treat them in similar way as input random points in modeling and evaluating function *inside*. But in the case of error evaluation instead of condition estimation we take the difference between the estimated by the tracking algorithm point (treated it as random point ) and the evaluated true position (circle, ellipse or oval point), corresponding to that angle. The evaluated errors are transformed from pixel to world coordinates and are presented in world coordinates.

## 5   Results

The algorithm performance is evaluated by generating images using Field II. Knowing the true positions of the lesion areas (cysts-black or tumors-white) the differences between true positions and estimated by the algorithm are evaluated and presented.

Using Field II and new functions for test *inside* a new cyst phantom is generated with more complicated contours and with known true positions for the contour points. In figure 3 the result contours for modeled ellipse and oval Cassini are illustrated. In figure 4 the errors - absolute differences between estimated and true positions along coordinate axes $x$ and $y$ are presented. It can be seen that the errors along $x$ are greater than along the $y$, because of the specific of the speckle noise which in x direction is more sensible. In figure 5 the IMM probabilities for each estimated point (for each radius) are presented. When the errors are small, the probabilities of the models are in steady state and the first precise model is with bigger probability.

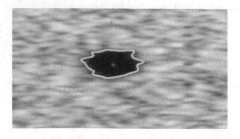 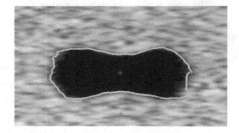

**Fig. 3.** Results - contour determination for ellipse and for oval Cassini

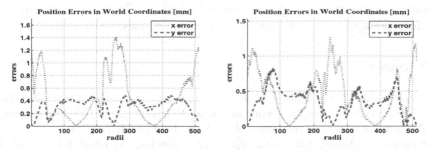

**Fig. 4.** Errors for ellipse (left) and for oval Cassini (right)

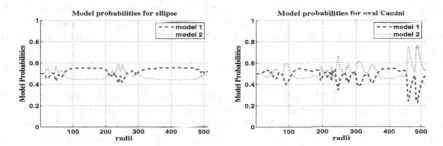

**Fig. 5.** IMM probabilities for ellipse (left) and for oval Cassini (right)

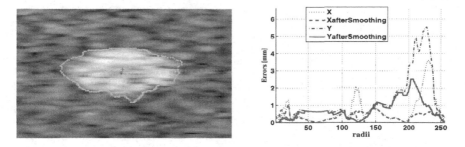

**Fig. 6.** A smoothed image (left) and the errors of the contour before and after smoothing (right)

The contour tracking sometimes goes to wrong directions because of the speckle noise. In such cases the preliminary smoothing improved contour tracking. In figure 6 in the left the smoothed image of an ellipse is shown, and the comparison of the errors before and after smoothing are shown in the right.

## 6   Conclusion

A new method for modeling complex contours with known true positions and evaluation of the errors is proposed. The method is used for error evaluation of contour determination in images generated by Field II simulation program. The contour determination is performed by developed IMMPDA tracking algorithm with incorporated edge magnitude. The results are presented and analysed.

# References

1. Abolmaesumi, P., Sirouspour, M.R.: An Interacting Multiple Model Probabilistic Data Association Filter for Cavity Boundary Extraction from Ultrasound Images, IEEE trans. on Medical Imaging, vol.23, No 6, June (2004) 772-784
2. Angelova, D., Jilkov, V.P., Semerdjiev, Tzv.: Tracking Maneuvering Aircraft by Interacting Multiple Model Filters, Comptes rendus de l'Academie bulgare des Sciences, Tome 49, No.1 (1996) 37-40
3. Bar-Shalom, Y., Li, X.R., Kirubarajan, T.: Estimation with Applications to Tracking and Navigation: Theory, Algorithms and Software. Wiley, New York (2001)
4. Bar-Shalom, Y., Li, X.R.: Multitarget-Multisensor Tracking. Priciples and Techniques, YBS, (1995)
5. Gellert, W., Kestner, H., Neuber, S.: Lexikon der Mathematik, VEB Bibliographisches Institut, Leipzig (1979)
6. Jensen, J.A., Field II Ultrasound Simulation Program, Technical University of Denmark, http://www.es.oersted.dtu.dk/staff/jaj/field/ (visited 10.04.2006)
7. Kirubarajan, T., Bar-Shalom, Y.,: Probabilistic Data Association Techniques for Target Tracking in Clutter, Proc. of the IEEE, vol.92, No 3, march (2004) 536-557
8. Semerdjiev, Tzv., Djerassi, Em., Bojilov, L., Konstantinova, P.: Sensor Data Processing for Target Tracking. Algorithms and Applications, /in Bulgarian/ SoftTrade, (2006)

# Computational Model of 1D Continuum Motion. Case of Textile Yarn Unwinding Without Air Resistance

Yordan Kyosev[1],* and Michail Todorov[2]

[1] Institute for Textile Technology, RWTH Aachen University
Eilfschornsteinstr. 18, 52062 Aachen, Germany
yordan_kyosev@gmx.net
http://www.kyosev.com
[2] Technical University of Sofia
Faculty of Applied Mathematics and Informatics
8 Kl.Ohridski, Blvd., 1000 Sofia, Bulgaria
mtod@tu-sofia.bg

**Abstract.** The textile fibers and yarns are usually modeled as one dimensional continuum of beams or as a system of particles, connected by springs. The governing equation of motion of such systems has many solutions, which switch in between depending on the available energy of the system. Effective modeling of problems with bifurcations like these, applicable for more general geometries is rarely reported. In the present work we demonstrate, that using a system of particles and an appropriate change of variables, a successful time step integration either by the Leap-Frog or the Verlet algorithms is possible.

## 1  Introduction

The textile yarns and fibers are the most significant representative of the one dimensional (1D) continuum. Almost everywhere at the textile processes there are places, where this continuum moves around the 3D space freely, with only one or without any constrains. Examples are the air-jet spinning and weaving, air transport of fibers, etc., where the fibers, resp. yarns are moved by the forces of specially oriented fluid flow. Another example is the yarn unwinding, where the yarn builds large loop, called "balloon". Most authors threat the unwinding of the yarn from bobbin as a boundary value problem. There are several works in the area of modeling of the yarn balloon, mainly in its stable stationary motion during ring spinning [1], [2], [3] and yarn unwinding [4]. The differences between the stationary and transient equations are well described in [5] and the transient solution of the balloon equation was presented in [6] . All these works are based on the solution of the system of partial differential equations with the use of finite differences or relaxation technique and shooting method for boundary value problems. But it is

---

* Alexander von Humboldt research fellow, on leave from Department of Textiles, Technical University of Sofia, Bulgaria.

T. Boyanov et al. (Eds.): NMA 2006, LNCS 4310, pp. 637–645, 2007.

more natural the yarn unwinding to be modeled as initial value problem, because the one end of the yarn is not fixed. In addition, these methods can not be used for more general types of yarn dynamic problems and most of them have numerical difficulties when passing bifurcation points.

Another technique - presentation of the 1D continuum as a mass-spring system is used in [7], [8] and discussed in [9]. Basically the idea is the same as the one in the molecular dynamic and it is used more and more for different kinds of "particle based" models [10]. These models allow implementations for more general geometrical conditions, which is important for industrial applications. In the current work we present a model of the unwinding process, where the yarn is presented as a mass-spring system. Both the Leap-Frog and the Verlet algorithms are used for integration.

## 2    Formulation of the Problem

In this work the textile yarns as 1D continuum are presented. Most commonly used in this case is the Lagrangian description of the equation of motion. The independent curvilinear coordinate $s$ is along the yarn axis (Fig. 1a) and it is moving together with the yarn. In the general case of yarn with linear mass density $\rho$, yarn cross section $A$, and bending rigidity $B$, the governing equation of motion is subjected to the Newton law:

$$\rho A \frac{\partial^2 \mathbf{r}}{\partial t^2} = \frac{\partial}{\partial s}\left(T\frac{\partial \mathbf{r}}{\partial s}\right) - B\frac{\partial^4 \mathbf{r}}{\partial s^4} + \rho A\mathbf{g} + \mathbf{q}, \tag{1}$$

where $t$ is the time, $\mathbf{g}$ the earth acceleration, $T$ is the yarn tension and $\mathbf{q}$ is the resulting vector of the external forces (excluding the earth acceleration). The extensibility condition is given by

$$\frac{\partial \mathbf{r}}{\partial s} \cdot \frac{\partial \mathbf{r}}{\partial s} = (1 + \varepsilon)^2 \tag{2}$$

where $\varepsilon$ is the elongation of the yarn and the dot stands for scalar product. The material properties under tension can be presented as a sum of a linear elastic term and one more general nonlinear term:

$$T = AE\varepsilon + f\left(\varepsilon, \frac{\partial \varepsilon}{\partial t}\right). \tag{3}$$

For given boundary and initial conditions, the yarn coordinates $\mathbf{r}$ and yarn tension $T$ can be obtained by solving the system (1)–(3) at each time step $\Delta t$. If we assume $\varepsilon = 0$ then the force and the elongation can be eliminated and the equations can be solved in the case of inelastic yarns. In this form the above system, however, has a disadvantage – if we want to model some local effects as the influence of yarn guide elements or the contact between the yarn and other machine elements, we should divide the yarn into several parts. Then, system (1)–(3) and appropriate boundary conditions have to be constituted for each part separately and simultaneously to be solved together in all yarn parts. To avoid this shortcoming we use discrete implementation of the problem.

Eq. (1) does not contain the air-drag force and thus, we can assume, that current model represents the yarn motion in vacuum. If the air-drag force is included, it will appear in right-hand side of eq. (1) as $\mathbf{F}_{air} = -p\,|\mathbf{V}_n|\,\mathbf{V}_n$ where $\mathbf{V}_n$ is the air velocity normal to the yarn axis, and the quantity $p$ contains the drag coefficient for a moving circular cylinder in a viscous turbulent flow. In several practical cases (spinning and unwinding balloons) the air resistance force is less then the other forces and can be neglected, i.e., $F_{air} \ll \rho A \frac{\partial^2 \mathbf{r}}{\partial t^2}$. There are areas, for instance, air transport of fibers, air-jet spinning and air-jet weft insertion, where the air-drag force is moving the fibers or yarns. In these cases $F_{air} \gg \frac{\partial}{\partial s}(T\frac{\partial \mathbf{r}}{\partial s}) - B\frac{\partial^4 \mathbf{r}}{\partial s^4} + \rho A\mathbf{g} +$ q. They need additional treatment in order to ensure numerical stability during the integration and can be object of separate work.

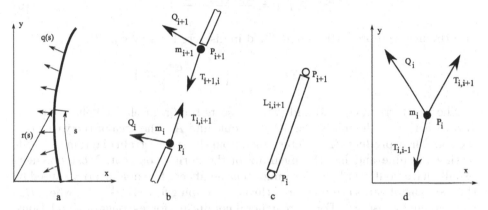

**Fig. 1.** One dimensional continuum (textile yarn): a) original formulation of the problem; b) discretized model; c) linear segment between two particles for description of the material behavior and calculation the tension forces; d) force equilibrium in the particle

## 3   Discrete Model

We can obtain the discrete model presenting the derivatives with regard to curvilinear coordinate $s$ as discrete functions. This has also own mechanical representation (see Fig.1b). Let us divide the yarn into yarn segments with initial equal length $L$ and concentrate all the forces upon particle $P_i$ between the segments. Then each particle has mass $m_i = L\rho$ and the resulting (nodal) force $\mathbf{F}_i$ is applied there.

The equation of motion of $i$-th particle is given by the second Newton law as:

$$m_i \frac{d^2 \mathbf{r}_i}{dt^2} = \mathbf{F}_i \tag{4}$$

where $\mathbf{r}_i \equiv (x_i, y_i, z_i)^T$ is the coordinate vector of the current $i$-th particle. The resulting force $\mathbf{F}_i$ upon $i$-th particle is calculated as a vector sum of the forces acting by neighbor segments $\mathbf{F}_{ei}^{i,i-1}$ and $\mathbf{F}_{ei}^{i,i+1}$ (Fig. 1d):

$$\mathbf{F}_i = \mathbf{T}_{i,i-1} + \mathbf{T}_{i,i+1} + \mathbf{Q}_i, \tag{5}$$

where $i = 1 \div N$ is the particle number, $\mathbf{T}_{i,j}$ are the internal forces in the segments $i$-$j$, and $\mathbf{Q}_i$ are the external forces for the particle. Here $T > 0$, as the yarns can transmit only tension, in this case the initial partial differential equation (PDE) (1) is hyperbolic. If the tension becomes nonpositive, i.e., $T \leq 0$, then PDE type changes and the mechanical instability causes a singularity of the equation of motion. In the case of extensible yarns, the extensibility equation is defined at the segment level by the distance between two particles

$$\varepsilon_{i,i+1} = \frac{|P_i P_{i+1}| - L}{L} \tag{6}$$

and the material properties are defined in discrete form by eq. (3):

$$T_{i,i+1} = AE\varepsilon_{i,i+1} + f\left(\varepsilon_{i,i+1}, \frac{\partial \varepsilon_{i,i+1}}{\partial t}\right). \tag{7}$$

The first term in eqs. (3) and (7) represents the linear elastic behavior of the yarn, where $E$ is the initial elasticity module and $A$ – the area of the yarn cross section. Their product $AE$ is determined from the stress-strain diagram of tensile testing machine and has the meaning of the spring constant. Most polymers usually have nonlinear behavior under tension affect. The nonlinearity caused by the damping of yarns is represented through damping force $AD\frac{\partial \varepsilon_{i,i+1}}{\partial t}$, where $D$ is the damping resistance. For more general nonlinear "force-elongation" relations, additional terms in eq. (7) should be implemented. In most practical applications parabolic functions of both $\varepsilon$ and $\frac{\partial \varepsilon_{i,i+1}}{\partial t}$ satisfy the required accuracy.

Eqs. (4)–(7) and respective boundary conditions at the one end (see below) are considered as a system of $3 \times N$ ordinary differential equations (ODE) for all $N$ particles in 3D space. Theoretically this can be solved by using various numerical methods for ODE, but usually many authors report serious problems of convergence and accuracy [9].

To avoid such kind of problems, we introduce dimensionless variables. Let $\rho$ be the linear density of the yarn, $R$ – the unwinding radius (see Fig.2), $V$ – the yarn velocity, $T$ – the time for unwinding of the yarn in one winding (e.g., with length $2\pi R$), $L$ – the distance between the particles, as we assume an equal distance between each two neighboring particles $j, j+1$. We introduce dimensionless time $\bar{t} = \frac{t}{T}$ where $T = \frac{2\pi R}{V} = \frac{2\pi}{\omega}$ and $\omega$ is the angular velocity of the unwinding point. This dimensionless time differs from those in [6] with $2\pi$ in order to have exact physical sense - i.e., time increment $\Delta\bar{t} = 1$ to correspond to unwinding of one winding. Dimensionless coordinates $\bar{r} = \frac{r}{R}$ and force $\bar{\mathbf{F}} = \frac{\mathbf{F}}{\rho V^2} = \frac{\mathbf{F}}{\rho \omega^2 R^2}$ are according to [6]. Here usually the variable $\xi = \frac{V}{R \cdot \omega}$ is introduced, as for ring, rotor and centrifugal spinning $\xi \neq 1$. In our case of axial unwinding or all other

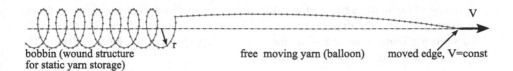

bobbin (wound structure
for static yarn storage)

free moving yarn (balloon)

moved edge, V=const

**Fig. 2.** Initial geometry of the problem

cases, where the bobbin with the yarn is not rotating, one sets $\xi = 1$. The relation of the initial distance between particles $L$ and the unwinding radius $\mathbf{r}$ can be noted as $\lambda = \frac{R}{L}$. By substituting these relations, we obtain the dimensionless equation of motion:

$$\frac{d^2\bar{\mathbf{r}}}{d\bar{t}^2} = 4\pi^2\,\xi\,\lambda\,\bar{\mathbf{F}}. \qquad (8)$$

The dimensionless velocity can be derived as $\bar{v} = v\,\frac{T}{R} = v\,\frac{2\pi}{V}$ and the acting tension force between particles is obtained from eq. (7) in the case of linear elastic material as $\bar{T}_{i,i+1} = \frac{AE}{\rho V^2}\,\varepsilon_{i,i+1}$.

Since the yarn is pulled up with constant velocity at the first particle $i = 1$, we set our boundary conditions:

$$\bar{v}_{1,x} = 0 \qquad \bar{v}_{1,y} = 0 \qquad \bar{v}_{1,z} = 2\pi = \text{const.} \qquad (9)$$

The initial coordinates are calculated from approximate solution of the stationary rotating yarn with special program and the configuration is shown on Fig. 2. Currently we start the iterations with zero initial velocities of all particles, except the first one, where (9) are applied.

## 4   Numerical Scheme and Method of Solution

Equation (8) is an ODE but we have in mind, that force $\bar{F}$ is calculated by spacial derivatives. As it is not explicitly presented in the equations and appears through calculation of the elongation (5) and (6), we have to look for numerical methods for PDE, (not for ODE). Several time-evolutions problems are successfully integrated by either the Leap-frog scheme [10], [11], [12] or Verlet algorithm [13]. We implemented both of these schemes for our problem:

– Leap-Frog algorithm (velocities for time of a half step)

$$\bar{\mathbf{r}}^j = \bar{\mathbf{r}}^{j-1} + \bar{\mathbf{v}}^{j-\frac{1}{2}}\,\Delta\bar{t} \qquad \bar{\mathbf{v}}^{j+\frac{1}{2}} = \bar{\mathbf{v}}^{j-\frac{1}{2}} + \bar{\mathbf{a}}^j\,\Delta\bar{t};$$

– Verlet algorithm [13] (explicit central difference for discretization of the time derivatives):

$$\bar{\mathbf{r}}^{j+1} = \bar{\mathbf{r}}^j + \bar{\mathbf{v}}^j\,\Delta\bar{t} + \tfrac{1}{2}\bar{\mathbf{a}}^j\,(\Delta\bar{t})^2 \qquad \bar{\mathbf{v}}^{j+1} = \bar{\mathbf{v}}^j + \tfrac{1}{2}\left(\bar{\mathbf{a}}^{j+1} + \bar{\mathbf{a}}^j\right)\,\Delta\bar{t}$$

where the superscripts denote the number of the time step. Superscript $j$ corresponds to time $\bar{t} = j\Delta\bar{t}$. The particle indices $i$ are omitted for simplicity. The accelerations $\bar{\mathbf{a}}^j = \frac{d^2\bar{\mathbf{r}}^j}{d\bar{t}^2}$ are calculated according to eq. (8).

For the explicit integration methods the stability limit is defined in terms of maximal time-step $\Delta t$ depending on the highest frequency of the system $\omega_{max}$. This step for linear elastic rod or for mass-spring systems is identical (see [11], [12]). For the terms of the 1D continuum:

$$\Delta t_{stable} = \frac{L}{c} \qquad c = \sqrt{\frac{E}{\rho_m}} \qquad \rho_m = \frac{\rho}{A} \quad \Rightarrow \quad \Delta t_{stable} = L\sqrt{\frac{\rho}{EA}}.$$

More accurate estimate of the maximal time step, which takes into account that 1D continuum is moving in 3D space [14] is the well known Courant-Friedrichs-Lewy stability criterion for 3D problems $\Delta t \leq \frac{L}{\sqrt{3}|V|}$.

## 5   Some Results and Discussion

We calculated the maximal time step size as $\Delta \bar{t}_{max} = 0.00315$, but during the numerical tests the solution is stable for $\Delta \bar{t}_{max} < 0.02$. We used $\Delta \bar{t} = 0.1\Delta \bar{t}_{max}$ in order to ensure enough accuracy of the results. The tension forces in the segment between the first and the second particle, computed for time steps $\Delta t = 0.004$ and $\Delta t = 0.002$ are presented on Fig.3 (left). The relative error

$$\delta \bar{F}^j = \left| \frac{\bar{F}^j_{\Delta \bar{t}=0.004} - \bar{F}^{2j}_{\Delta \bar{t}=0.002}}{\bar{F}^{2j}_{\Delta \bar{t}=0.002}} \right|$$

of these forces is presented on the same figure (right). For the stationary parts of the motion it does not exceed 2% and increases up to 20% when time $\bar{t} > 2.5$. For these times the free yarn length is quite large and an additional reduction of the time step has to be performed, if the relative accuracy needs to remain in

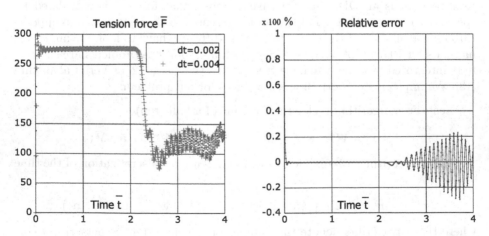

**Fig. 3.** Tension force between the first and second particle of the yarn during the first 4 unwinding cycles, calculated with two time-steps (left) and relative error in [%] (right)

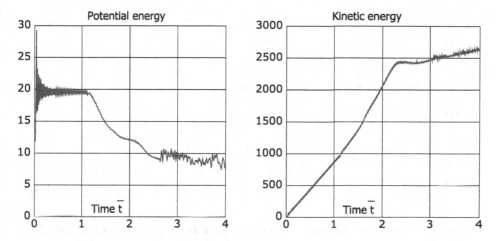

**Fig. 4.** Mean potential (left) and kinetic (right) energy per moving particle of the system

certain limits. The above problem occurs, because the higher order loop balloon forms gain more oscillations and required small integration steps in order the correct yarn position to be followed. The biggest error during the first time steps is a result of the transient process, which is running at the beginning and which is caused by zero initial velocity of the yarn. After this region follows very smooth region with almost constant yarn tension. It is well-known [6] that the equation of motion of yarn has more than one solution and switches between them in the bifurcation points. From mechanical point of view, the solutions can be considered as a "single-loop balloon", "one-and-a half loop balloon", "double-loop balloon", etc. In this way when $\bar{t} = 0 \div 1$ the single-loop balloon is realized. At the next time step, the free yarn length increases and hence more energy is needed to stabilize the single balloon. But the system does not have enough energy and switches itself into double-loop balloon. The last one corresponds to the switching between the different solutions of equation. This qualitative jump can be seen on Fig.4 (left) for time $\bar{t} = 1 \div 2$. On Fig.3 (left) the above mentioned phenomenon is not recognizable because the selected particle is quite away of the region of motion changes. When time $\bar{t} > 2$ there is again an additional switch into a triple-loop balloon, after that the yarn force begins to oscillate by jumping to another solution of higher order. Certain yarn geometries are plotted on Fig.5, and the detailed analysis of the evolution of the solution shows, that all bifurcation jumps are simulated properly. Therefore the presented model can be applied successfully for resolving such problems, which can not be said for other models published till now.

Let us denote that the energy (Fig.4) is computed as a mean energy of the system per moving particle but not as full energy of the whole system. The last is due to the yarn length, which increases linearly in the time for the problem under consideration and hence the investigation of full energy is not appropriate for obtaining useful information about the system. All presented results are

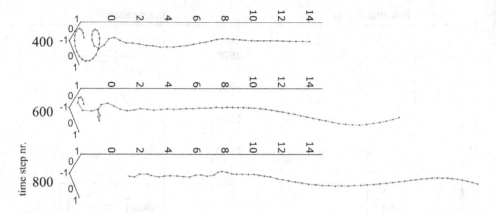

**Fig. 5.** Yarn geometry during the unwinding for time steps $j = 400; 600; 800$. $\Delta \bar{t} = 0.002$, $\rho A = 0.120\,\mathrm{g/m}$, $V = 1240\,\mathrm{m/min}$.

derived with the calculations with Leap-Frog algorithm. Let us denote the same results obtained by the Verlet algorithm though non-presented here, differ in the calculated force with no more than 0.5%.

## 6    Conclusions

Computational model for simulations of the yarn motion, which uses successfully the Leap-Frog and Verlet algorithms for treating Cauchy problems is presented. It is applied for the case of textile yarn unwinding without air resistance. The algorithm simulates the stable states of the yarn motion, as well as transient process if bifurcation occurs. The continuum is discretized in particles connected with segments. This approach allows the algorithm to be directly applied for solving wide range of similar problems.

## Acknowledgments

The authors thank to Alexander von Humboldt Foundation for the research fellowship of Dr. Kyosev in the Institute for textile technology, RWTH Aachen University.

## References

1. Ghosh, T., Batra, S., Murthy, A.: Dynamic analysis of yarn unwinding from cylindrical packages. part i: Parametric studies of the two-region problem. Textile Research Journal **71(9)** (2001) 771–778
2. Batra, S., Ghosh, T, K., Zeidman, M.: Integrated approach to dynamic analysis of the ring spinning process, part 1:without air drag and coriolis acceleration. Textile Research Journal **59** (1989) 309

3. Fraser, W.B.: Ring spinning: Modelling yarn balloons mathematically. Textile Horizons **16(2)** (1996) 37–39
4. Kothari, V., Leaf, G.: Unwinding of yarns from packages.i. the theory of yarn unwinding. Journal of the Textile InstituteUnwinding of Yarns from Packages.I. The Theory of Yarn unwinding **70(3)** (1979) 89
5. Lisini, G., e.a.: A comparison of stationary and non-stationary mathematical models for the ring-spinning process. Journal of the Textile Institute **83(4)** (2001) 550–559
6. Stump, D., Fraser, W.: Transient solutions of the ring-spinnign balloon equations. Journal of APplied Mechanics **June** (1996) 523–528
7. Przybyl, K.: Modelling yarn tension in the process of manufacturing on the ring-spinning machine. Fibres and Textiles in Eastern Europe **6(3)** (1998) 30–33
8. Kyosev, Y., Angelova, Y., Kovar, R.: 3d modelling of plain weft knitted structures from compressible yarn. Research Journal of Textile and Apparel, Hong Kong **9, Feb.**(1) (2005) 88–97
9. Berger, R.: Instationäre bewegung und stabilitätsverhalten eindimensionaler kontinua. In: Fortschrittberichte VDI. Number 189 in 18. VDI Verlag (1996)
10. Krivtzov, A., Krivtzova, N.: Particle method and its application in the mechanics of elastic solids (in russian). Dalnevostochnui matematicheskii journal DVO RAN **3(2)** (2002) 254–276 (original title in russian: Metod chastiz i evo ispolzovanie v mechanike deformiruemogo tverdovo tela).
11. NN.: Getting started with abaqus. Technical report, Abaqus, Inc. (2004)
12. Hallquist, J.: Ls-dyna theoretical manual. Technical report, LSTC Corp. (1998)
13. Verlet, L.: Computer "experiments" on classical fluids. i. thermodynamical properties of lennard-jones molecules. Phys. Rev. **159(1)** (1967) 98–103 Issue 1 5 July 1967.
14. Press, W.H., Teukolsky, S.A., Vetterling, W.T., Flannery, B.P.: Numerical recipes in C: The art of scientific computing. Cambridge University Press, Cambridge New York Port Chester Melbourne Sydney (1992)

# A Numerical Approach to the Dynamic Unilateral Contact Problem of Soil-Pile Interaction Under Instabilizing and Environmental Effects

Asterios Liolios[1], Konstantina Iossifidou[1], Konstantinos Liolios[1], Khairedin Abdalla[2], and Stefan Radev[3]

[1] Democritus University of Thrace, Dept. Civil Engineering & Dept. Environmental Engineering, GR-67100 Xanthi, Greece
liolios@civil.duth.gr
[2] Jordan University of Science and Technology, Dept. Civil Engineering, Irbid 22110, Jordan
abdalla@just.edu.jo
[3] Institute of Mechanics, BAS, 4, Acad. G. Bonchev str., 1113 Sofia, Bulgaria
stradev@imbm.bas.bg

**Abstract.** The paper deals with a numerical approach for the dynamic soil-pile interaction, considered as an inequality problem of structural engineering. So, the unilateral contact conditions due to tensionless and elastoplastic softening/fracturing behavior of the soil as well as due to gapping caused by earthquake excitations are taken into account. Moreover, second-order geometric effects for the pile behavior due to preexisting compressive loads and environmental soil effects causing instabilization are taken also into account. The numerical approach is based on a double discretization and on mathematical programming. First, in space the finite element method (FEM) is used for the simulation of the pipeline and the unilateral contact interface, in combination with the boundary element method (BEM) for the soil simulation. Next, with the aid of Laplace transform, the problem conditions are transformed to convolutional ones involving as unknowns the unilateral quantities only. So the number of unknowns is significantly reduced. Then a marching-time approach is applied and finally a nonconvex linear complementarity problem is solved in each time-step.

## 1 Introduction

Dynamic soil-pile interaction can be considered [1] as one of the so-called inequality problems of structural engineering [2], [3]. As well known [2], the governing conditions of these problems are equalities as well as inequalities. Indeed, for the case of the general dynamic soil-structure interaction, see e.g. [4], [10], [12] – [14] the interaction stresses in the transmitting interface between the structure and the soil are of compressive type only. Moreover, due to in general nonlinear,

T. Boyanov et al. (Eds.): NMA 2006, LNCS 4310, pp. 646–651, 2007.
© Springer-Verlag Berlin Heidelberg 2007

elastoplastic, tensionless, fracturing etc. soil behavior, gaps can be created between the soil and the structure. Thus, during e.g. strong earthquakes, separation and uplift phenomena have often appeared, as the praxis has shown.

The mathematical treatment of the so formulated inequality problems can be obtained by the variational or hemivariational inequality approach [2], [3]. Numerical approaches for some dynamic inequality problems of structural engineering have been also presented, see e.g. [1] – [5].

The present paper deals with a numerical treatment for the inequality dynamic problem of soil-pile interaction where second-order geometric effects for the pile behavior due to preexisting compressive loads are taken also into account. In the problem formulation, the above considerations about gapping as well as soil elastoplastic/softening behavior are taken into account. The proposed numerical method is based on a double discretization and on methods of nonlinear programming. So, in space the finite element method (FEM) coupled with the boundary element method (BEM), and in time a step-by-step method for the treatment of convolutional conditions are used. At each time-step a non-convex linear complementarity problem is solved with reduced number of unknowns. Finally, the presented procedure is applied to an example problem of dynamic pile-soil interaction, and some concluding remarks useful for the Civil Engineering praxis are discussed.

## 2    Method of Analysis

### 2.1    Coupling of FEM and BEM

A spatial discretization is applied for the soil-pile system by coupling the FEM and BEM in the wellknown way, see e.g. Brebbia et al. [6]. For simplicity, the pile is first considered as linearly elastic, and discretized into usual beam/frame finite elements. Each pile node is considered as connected to the soil on both sides through two unilateral constraints (interface soil-elements). Every such interface element consists of an elastoplastic-softening spring and a dashpot, connected in parallel (Figure 1), and appears a compressive force $r(t)$ at the time-moments $t$ only when the pile node comes in contact with the soil. Let $v(t)$ denote the relative retirement displacement between the soil-element end and the pile-node, and $g(t)$ the existing gap. Then the unilateral contact behavior of the soil-pile interaction is expressed in the compact form of the following linear complementarity conditions:

$$v \geq 0, \quad r \geq 0, \quad r.v = 0. \tag{1}$$

The soil-element compressive force is in convolutional form [4]

$$r = S(t) * y(t), \tag{2a}$$
$$y = w - (g + v), \tag{2b}$$

or in form used in Foundation Analysis [11]

$$r = c_s(\partial y / \partial t) + p(y). \tag{3}$$

Here $c_s$ is the soil damping coefficient, $w = w(t)$ the pile node lateral displacement, $y = y(t)$ the shortening deformation of the soil-element, and $p(y)$ the spring force. By $*$ the convolution operation is denoted. $S(t)$ is the dynamic stiffness coefficient for the soil and it can be computed by the BEM [4]. Function $p(y)$ is mathematically defined by the following, in general nonconvex and nonmonotone constitutive relation

$$p(y) \in \Theta P(y), \tag{4}$$

where $\Theta$ is Clarke's generalized gradient and $P(.)$ is the symbol of superpotential nonconvex functions [2]. So, (4) expresses in general the elastoplastic-softening soil behavior, where unloading-reloading, gapping, environmental degrading, fracturing etc. effects are included.

For the herein numerical treatment, $p(y)$ is piece-wise linearized in terms of non-negative multipliers as in plasticity [7], [8]. So the problem conditions for the assembled soil pipeline system are written in matrix form according to the finite element method:

$$\underline{M}\,\ddot{\underline{u}}(t) + \underline{C}\,\dot{\underline{u}}(t) + (\underline{K} + \underline{G})\underline{u}(t)\underline{f}(t) + \underline{A}^T\underline{r}(t) \tag{5}$$

$$\underline{y} = \underline{A}^T\underline{u} - \underline{u}_g - \underline{g} - \underline{B}\,\underline{z}, \tag{6}$$

$$\underline{r} = \underline{S} * \underline{y}, (or\ \underline{r} = \underline{E}\,\underline{y}), \tag{7}$$

$$\underline{\omega} = \underline{B}^T\,\underline{r} - \underline{H}\,\underline{z} - \underline{k}, \qquad \underline{\omega} \leq 0, \quad \underline{z} \geq 0, \quad \underline{z}^T.\underline{\omega} = 0, \tag{8}$$

$$\underline{u}(t = 0) = \underline{u}_0, \qquad \dot{\underline{u}}(t = 0) = \dot{\underline{u}}_0, \qquad \underline{g}(t = 0) = \underline{g}_0. \tag{9}$$

Here (3) is the dynamic equilibrium condition, (4) – (6) include the unilateral and the piece wise linearized constitutive relations and (7) are the initial conditions. As usual, $\underline{M}$, $\underline{C}$ and $\underline{K}$ are the mass, damping and stiffness matrix, respectively; $\underline{G}$ is the geometric stiffness matrix depending linearly on pre-existing stress state [8], [15], [16]; $\underline{u}$, $\underline{f}$ are the displacement and the force vectors, respectively; $\underline{u}_g(t)$ is the vector of (possible) seismic ground displacement; $\underline{A}$, $\underline{B}$ are kinematic transformation matrices; $\underline{z}$, $\underline{k}$ are the nonnegative multiplier and the unilateral capacity vectors; and $\underline{E}$, $\underline{H}$ are the elasticity and unilateral interaction square matrices, symmetric and positive definite the former, positive semidefinite the latter for the elastoplastic soil case. In the case of soil softening, some diagonal entries of $\underline{H}$ are nonpositive [7]. For the case of nonlinear pile behavior, either the linear terms $\underline{C}\,\dot{\underline{u}}$ and $\underline{K}\,\underline{u}$ can be replaced by the nonlinear matrix functions $\underline{C}(\dot{\underline{u}})$ and $\underline{K}(\underline{u})$ , or the local nonlinearities (e.g. elastoplasticity) are included in appropriate internal unilateral constraints [7] - [9].

Thus the so-formulated problem is to find $(\underline{u}, \underline{r}, \underline{g}, \underline{z})$ satisfying (1) – (9) when $(\underline{f}, \underline{u}_g, \underline{u}_0, \dot{\underline{u}}_0, \underline{g}_0)$ are given.

## 2.2   Time Discretization. The Convolutional LCP

Assuming that the unilateral quantities $\underline{z}$ and $\underline{T}$ include all local nonlinearities and unilateral behavior, the procedure of Liolios [9] can be used. So, applying the Laplace transform to (3) – (7), except (8)$_4$, and after suitable elimination of unknowns and back transforming to time domain, we arrive eventually at

$$\underline{w}(t) = \underline{D}(t) * \underline{z}(t) + \underline{d}(t). \tag{10}$$

Thus, at every time-moment the problem of rels. (8)$_{2,3,4}$ and (10) is to be solved. This problem is called here Convolutional Linear Complementarity Problem (CLCP), it has a reduced number of unknowns and it is solved by time discretization [4]. So, for the time moment $t_n n \triangle t$, where $\triangle t$ is the time step, we arrive eventually at a non-convex linear complementarity problem [6]:

$$\underline{w}_n = \underline{D}\,\underline{z}_n + \underline{d}_n, \qquad \underline{z}_n \geq 0, \qquad \underline{w}_n \leq 0, \qquad \underline{z}^T \underline{w}_n = 0. \tag{11}$$

Alternatively, the above inequality problem of rels. (1) - (9) can be solved in time by direct time integration methods as in Liolios [1]. So, some algebraic manipulations and a suitable elimination of unknowns lead to the same discretized LCP (11).

Solving problem (11) by available computer codes of nonlinear mathematical programming [2], [3], [7], we compute which of the unilateral constraints are active and which not in each time-step $\triangle t$. Due to soil softening, matrix $\underline{D}$ is not strictly positive definite in general. But as numerical experiments have shown, in most civil engineering applications of soil-pile interaction this matrix is P-copositive. Thus the existence of a solution is assured [6] - [8].

## 3   Numerical Example

The example problem of Liolios [1] is reconsidered here for comparison reasons. The steel IPB300 H-pile depicted in Figure 1(a) has a length $L = 12m$ and is fully embedded into a clay deposit. The pile has a stiffness $EI = 52857 KN.m^2$, is fixed at the bottom and free at the top. The effects of the over structural framing are approximated by a lumped mass $2KN.m^{-1}.sec^2$ and a rotational inertia $2KN.m.sec^2$. The pile is subjected to a vertical constant top force of $120KN$ and to a dynamic horizontal top force with the time history shown in Figure 2(a). The elastoplastic-softening soil behavior according to eq. (2b) is shown in Figure 2(b) – (diagram p-y [11])- where branch $OA$ has the exponential form $p(x, y)p_u.[1 - exp(-100y)]$, with $p_u = 375[1 - 0.5exp(-0.55x)]$, and for the branch $AB$ holds $p(x, y) = 0.75p_p.(-3\xi^2 + 2\xi^3) + p_p$, with $\xi = (y - 0.02)/0.06$. For unloading-reloading paths the inclination is $100p_u$. Some response results from the ones obtained by applying the herein presented numerical procedure are indicatively reported. So the maximum values of the pile horizontal displacements and the final gaps along the pipeline due to permanent soil deformations are shown in Figures 1(b) and 1(c), respectively. These results are in good agreement with those of Liolios [1].

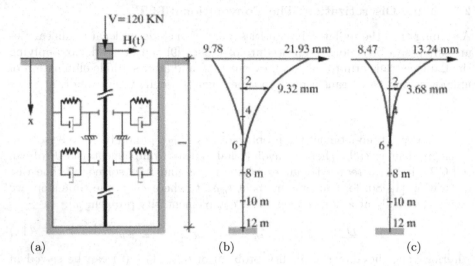

**Fig. 1.** The numerical example: (a) The soil-pile system model, (b) Maximum horizontal pile displacements, (c) Final soil-pile gaps

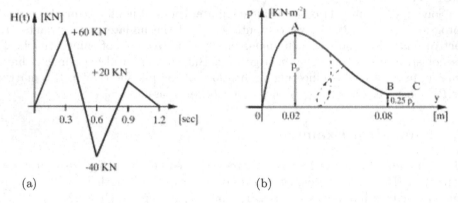

**Fig. 2.** The numerical example: (a) Dynamic loading diagram, (b) Diagram p-y of the soil behavior

## 4   Concluding Remarks

As the above indicative results of the numerical example show, unilateral contact effects due to tensionless soil capacity and to gapping may be significant and have to be taken into account for the dynamic soil-pile interaction. These effects can be numerically estimated by the herein presented procedure, which is realizable on computers by using existent codes of coupling the FEM and BEM as well as optimization algorithms. Thus, the presented approach can be useful in the praxis for the earthquake resistant construction, design and control of piles.

# References

1. Liolios, A.A.: A numerical approach to the dynamic nonconvex unilateral problem of soil pile interaction. In: Dafermos, C.M. Ladas, G. and Papanicolaou, G. (eds.): Differential equations. Marcel Dekker, Inc., Basel (1989) 437-443

2. Panagiotopoulos, P.D.: Hemivariational Inequalities and Applications. Springer Verlag, Berlin (1993)

3. Antes, H. and Panagiotopoulos, P.D.: The Boundary Integral Approach to Static and Dynamic Contact Problems. Equality and Inequality Methods, Birkhauser Verlag, Basel, Boston, Berlin (1992)

4. Wolf, J.P.: Soil-Structure-Interaction Analysis in Time Domain, Prentice-Hall, Englewood Cliffs, N.J. (1988)

5. Brebbia, C.A., Telles, J.C.F. and Wrobel, L.C.: Boundary Element Techniques, Springer-Verlag, Berlin and New-York (1984)

6. Liolios, A.A.: A Linear Complementarity Approach to the Nonconvex Dynamic Problem of Unilateral Contact with Friction between Adjacent Structures. J. Applied Mathematics and Mechanics (ZAMM) **69** (5) (1989) 420-422

7. Maier, G.: Mathematical Programming Methods in Structural Analysis. In: Brebbia, C. and Tottenham, H., (Eds.): Variational Methods in Engineering, Vol.II. Southampton Univ. Press, Southampton (1973) 8/1 – 8/32

8. Maier, G.: Incremental Elastoplastic Analysis in the Presence of Large Displacements and Physical Instabilizing Effects. Int. J. Solids and Structures **7** (1971) 345 – 372

9. Liolios, A.A.: A Boundary Element Convolutional Approach to Unilateral Structural Dynamics. In: Antes, H. (Ed.): IABEM 93-Extended Abstracts. Inst. Appl. Mech., Techn. Univ. Braunschweig (1993)

10. Beskos, D.E.: Application of the Boundary Element Method in Dynamic Soil-Structure Interaction. In: Gulkan, P. and Clough, R.W. (Eds.): Developments in Dynamic Soil-Structure Interaction. Kluwer, London (1993)

11. Scott, R.: Foundation Analysis. Prentice-Hall, London (1981)

12. Schmid, G. and Chouw, N.: Soil-structure interaction effects on structural pounding. In: Proceedings of the 10th World Conference on Earthquake Engineering, Vol. 3. Madrid, Spain, Balkema, Rotterdam (1992) 1651 – 1656

13. Patel, P.N. and Spyrakos, C.C.: Uplifting-sliding response of flexible structures to seismic loads. Engineering Analysis with Boundary Elements **8** (1991) 185 – 191

14. Manolis, G.D. and Beskos, D.E.: Boundary Element Methods in Elastodynamics. Unwin Hyman, London (1988)

15. Chen, W.F. and Lui, E.M.: Structural Stability. Elsevier, New York (1981)

16. Savidis, S.A., Bode, C., Hirschauer, R. and Hornig, J.: Dynamic Soil-Structure Interaction with Partial Uplift. In: Fryba, L. and Naprstek, J. (Eds.): EURODYN '99. Proc. 4th European Conference on Structural Dynamics, Praga, Rotterdam: Balkema (1999) 957 – 962

# Analysis of Soil-Structure Interaction Considering Complicated Soil Profile

Jang Ho Park[1], Jaegyun Park[2], Kwan-Soon Park[3], and Seung-Yong Ok[4]

[1] Ajou University, Suwon, Korea
jangho@ajou.ac.kr
[2] Dankook University, Seoul, Korea
jpark@dku.edu
[3] Dongguk University, Seoul, Korea
kpark@dongguk.edu
[4] Seoul National University, Seoul, Korea
syok@sel.snu.ac.kr

**Abstract.** The effect of soil-structure interaction (SSI) is an important consideration and cannot be neglected in the seismic design of structures on soft soil. Various methods have been developed to consider SSI effects and are currently being used. However, most of the approaches including a general finite element method cannot appropriately consider the properties and characteristics of the sites with complicated soil profiles. To overcome these difficulties, this paper presents soil-structure interaction analysis method, which can consider precisely complicated soil profiles by adopting an unaligned mesh generation approach. This approach has the advantages of rapid generation of structured internal meshes and leads to regular and precise stiffness matrix. The applicability of the proposed method is validated through several numerical examples and the influence of various properties and characteristics of soil sites on the response is investigated.

## 1   Introduction

Over the last few decades, many building structures have been damaged from devastating earthquakes and the concept of seismic resistant design have become worldwide popular. In the seismic resistant design, the soil-structure interaction (SSI) is one of the important considerations for structures on soft soil since the geometry and the properties of the supporting soil exert large influence on their behavior. Various methods have been developed to consider soil-structure interaction effects and are currently being used in the field of earthquake engineering [1], [2], [3]. Most of these methods utilize simplified soil profiles such as horizontally layered soil to avoid the complexity in the modeling of soil. However, these simplified approaches cannot consider accurately complicated soil profile and reduce the reliability of the analysis results. Therefore, a new approach to consider appropriately complicated soil profiles is needed for precise SSI analysis.

One of the viable approaches is the finite element method. However, a general finite element approach can be nearly unpractical for soil sites exhibiting

T. Boyanov et al. (Eds.): NMA 2006, LNCS 4310, pp. 652–659, 2007.
© Springer-Verlag Berlin Heidelberg 2007

complex geometry and material discontinuity since an element is endowed only
with one material property and finite element boundaries coincide with material interfaces. Furthermore, even in case the line of material discontinuity can
be matched with the element boundary, the shape of an element may lose its
convexity and lead to ill-conditioned stiffness matrix. To overcome these difficulties, this paper presents a SSI analysis method, which can consider precisely
complicated soil profiles.

## 2 Soil-Structure Interaction Analysis Method Considering Complicated Soil Profiles

In a structure subjected to ground excitations on soft soil, the structure interacts
with the surrounding soil and the SSI exerts large influence on the dynamic
behavior. Therefore, in this study a precise SSI analysis method is developed
by adopting an unaligned mesh generation approach [4] and by using a direct
method with a modified Lysmer transmitting/absorbing boundary [5].

### 2.1 Numerical Integration of a Finite Element with Discontinuity

In a general finite element analysis, the interface between materials is used as
an element boundary so as to avoid material discontinuity within an element.
However, the modeling is difficult when the interface between materials is complicated - the shape of the element might lose its convexity and might result in
ill-conditioned stiffness matrix [6]. To overcome these shortcomings, this study
adopts an unaligned mesh generation approach which calculates accurately the
stiffness matrix of an element with material discontinuity [4].

The application of an unaligned mesh generation approach leads to uniform
structured finite elements. The numerical integration for an element consisting
of a unique material is trivial, but the integration for an element composed of
two materials is more delicate. The main concept to perform precise numerical
integration for an element with material discontinuity is to use enough number of Gauss quadrature points. Fig.1 shows the concept of an unaligned mesh
generation approach.

The error bound in the integration of a discontinuous function is expressed by
the largest quadrature weight $w_i$ and the maximum distance $\xi_i - \xi_{i+1}$ between
two neighboring Gauss points [4].

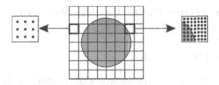

**Fig. 1.** Finite element with material discontinuity

$$\text{Error bound} \leq \max|w_i| \cdot \max|\xi_i - \xi_{i+1}| \approx 5.07N^{-1.82} \qquad (1)$$

where $N$ is the number of Gauss points. The same method can be applied to calculate the stiffness matrix of the finite element. The error bounds in three dimensions are estimated from 0.040 in case of $3 \times 3 \times 3$ to 0.0004 in case of $7 \times 7 \times 7$ Gauss points by equation (1). This shows that the stiffness matrix of an element with discontinuity is calculated with enough precision using an adequate number of quadrature points.

## 2.2   Soil-Structure Interaction Analysis Method Using the Proposed Integration Method

The methods for SSI analysis are mainly classified as direct methods and sub-structure methods according to the way they include the soil in their analysis. This paper adopts a direct method since it is widely used because of its simplicity and accuracy [7], [8]. The soil and the structure are modeled as shown in Fig. 2.

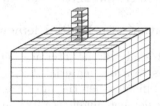

**Fig. 2.** Model for soil-structure interaction analysis

The treatment of seismic input and boundary conditions is also important because those affect significantly the dynamic response. The boundary conditions of the soil media are modeled using a modified Lysmer transmitting/absorbing boundary such that the domain beyond the soil media of interest is replaced with a set of viscous dampers, which are proportional to the wave velocity of soil [5]. At each node on the boundary, the tangential and the normal dampers absorb the energy of the S-wave and the P-wave, respectively.

## 3   Analysis of Soil-Structure Interaction Considering Complicated Soil Profiles

The applicability of the proposed method is validated and the influence of various properties and characteristics of the soil sites on the dynamic response is investigated through numerical examples.

### 3.1   Structure and Soil Profile

The soil-structure interaction analysis was conducted on 6-story structure built on soft soils. The structure was modeled as illustrated in Fig. 2 with the properties summarized in Table 1. The damping ratio of the structure is assumed to

**Table 1.** Properties of a structure model

| Floor | Level(m) | Stiffness(kN/m) | Mass(kg) |
|-------|----------|-----------------|----------|
| 7 | 22.5 | - | 1050000 |
| 6 | 19 | 7820000 | 1050000 |
| 5 | 15.5 | 8540000 | 1150000 |
| 4 | 12 | 9020000 | 1150000 |
| 3 | 8.5 | 9890000 | 1150000 |
| 2 | 4.5 | 10200000 | 1150000 |
| 1 | 0 | 7280000 | 1390000 |
| B1 | -3.5 | $\infty$ | 1750000 |

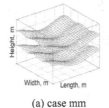 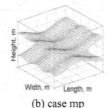 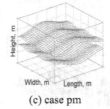 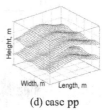

(a) case mm        (b) case mp        (c) case pm        (d) case pp

**Fig. 3.** Four types of soil profile models

be 0.03. Analysis is performed on five different soil profiles (*nn*, *mm*, *mp*, *pm*, *pp*) and three soil types (I, II, III). Soil profile model *nn* stands for conventional soil with horizontal flat layers. The other four soil profiles correspond to realistic three-layered soft soils with uneven soil profiles (Fig. 3). The soil media with dimensions of 100m × 100m × 40m is divided into 20 × 8 × 12 uniform structured finite elements. The material properties of each soil type and layer are listed in Table 2. The number of Gauss points used for element with discontinuity is 7×7×7 and that for element without discontinuity is 3×3×3, which increase the accuracy and the efficiency of the calculation. The artificial earthquake ground acceleration is used as input motion with peak acceleration of 0.2g at the bedrock (Fig. 4).

To validate the proposed method, analysis of free field motion for soil profile case *nn* was performed. One mesh is structured to have discontinuity within an element and the other mesh is aligned with the line of material discontinuity. The response spectra on the ground surface for both cases are in good agreement.

## 3.2   Influence of Properties and Characteristics of Soil

The analysis for the given soil profiles and soil types was performed by the proposed method. Fig. 5 shows the response spectra at the ground surface for model without structure. The differences between cases are distinct in response spectra in contrast to those in the acceleration history such that only spectrum results are presented. The soil type I is stiffer than type II and III such that

**Table 2.** Properties of the soil types and layers

| Soil Type | Layer (m) | S-wave velocity(m/s) | Density (kg/m$^3$) | Poisson's ratio | Damping ratio |
|-----------|-----------|----------------------|--------------------|-----------------|---------------|
| I   | Top(0-13.33)      | 600  | 1900 | 0.35 | 0.10 |
| I   | Middle(-26.67)    | 900  | 2000 | 0.30 | 0.04 |
| I   | Bottom(-40.00)    | 1200 | 2100 | 0.25 | 0.02 |
| II  | Top(0-13.33)      | 400  | 1800 | 0.35 | 0.10 |
| II  | Middle(-26.67)    | 800  | 1900 | 0.30 | 0.04 |
| II  | Bottom(-40.00)    | 1000 | 2000 | 0.25 | 0.02 |
| III | Top(0-13.33)      | 350  | 1800 | 0.35 | 0.10 |
| III | Middle(-26.67)    | 650  | 1900 | 0.30 | 0.04 |
| III | Bottom(-40.00)    | 1000 | 2000 | 0.25 | 0.02 |

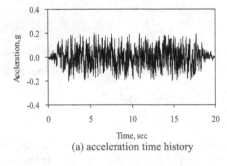
(a) acceleration time history

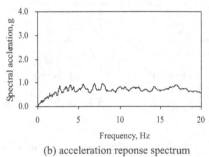
(b) acceleration reponse spectrum

**Fig. 4.** Input ground acceleration

the variation of the response due to the soil profile is relatively small in soil type I. The amplified frequency range in each soil type is changed due to the soil profile. Table 3 presents peak ground accelerations (PGA) in time histories. The variation of PGA due to the soil profile is about 15%, 12% in soil type II and III, but 5% in soil type I. This result shows that the properties and the characteristics of soil exert large influence on the response in soft soil and the precise description of soil layers is important in the analysis.

Fig. 6 presents the response spectra of the structure at the ground level. The spectral acceleration of the structure at the ground level varies according to soil type and the soil profile in Fig 6. The peak value of spectral acceleration occurs at 8.03Hz, 7.92 Hz, and 2.75Hz for the soils I, II, III respectively, which shows that the softer soil is, the lower the frequencies of the dominant modes in the response spectra are. This phenomenon is due to the fact that the frequency of the governing structural mode coupled with soil and amount of amplification vary according to the properties and the characteristics of the soil. The maximum spectral acceleration occurs in case $mm$ (2.88g at 8.03Hz) for soil type I, in case $pp$ (4.08g at 7.92Hz) for soil type II, in case $mm$ (4.33g at 2.75Hz) for soil type III. The minimum peak value occurs in case $pp$ (2.60g at 8.03Hz) for soil type I,

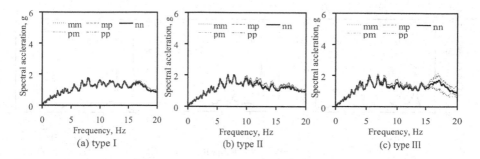

**Fig. 5.** Spectral accelerations at the ground surface without structure for three soil types

**Table 3.** Max. accelerations at the ground surface without structure

| Case | $mm$ | $mp$ | $nn$ | $pm$ | $pp$ |
|---|---|---|---|---|---|
| Soil Type I | 0.359g | 0.349g | 0.349g | 0.349g | 0.342g |
| Soil Type II | 0.422g | 0.417g | 0.366g | 0.385g | 0.372g |
| Soil Type III | 0.397g | 0.396g | 0.356g | 0.390g | 0.381g |

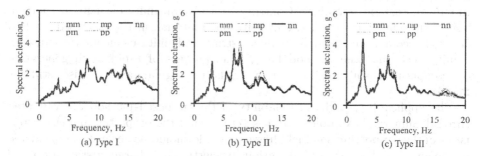

**Fig. 6.** Spectral accelerations at the ground surface with structure for three soil types

in case $mm$ (3.11g at 7.92Hz) for soil type II, in case $pp$ (3.93g at 2.75Hz) for soil type III. These results show that the variation of the peak value changes up to 31% due to the soil profile and the precise description of soil layers is important in the analysis of soil-structure interaction.

Fig. 7 presents the response spectra of the structure at the top of the structure. The variations due to the soil profile are overall smaller than those at the ground level because the frequency of the structural mode differs from the frequency range of amplified soil response. The maximum spectral acceleration occurs in case $mm$ at 3.21Hz for soil type I, in case $pm$ at 3.09Hz for soil type II, in case $nn$ at 2.75Hz for soil type III. The frequency of maximum spectral acceleration decreases as the soil becomes soft.

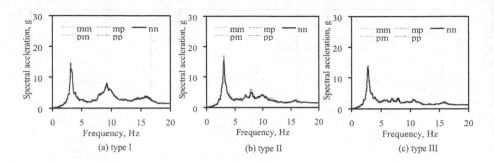

**Fig. 7.** Spectral accelerations at the top of the structure for three soil types

# 4   Conclusion

In a building structure subjected to ground excitations on soft soil, the structure interacts with the surrounding soil and the soil-structure interaction exerts larger influence on the behavior. Therefore, a precise SSI analysis method, which can consider various soil conditions, is necessary. This paper presents an efficient and reliable method for SSI analysis by adopting an unaligned mesh generation approach and using a direct method with modified Lysmer transmitting/absorbing boundary. Contrary to a general finite element method, in which an element is endowed only with one material, in the proposed method the soil media is modeled to uniform structured finite elements with discontinuity such that the method has the advantages of rapid mesh generation of complicated soil media and it leads to regular and precise stiffness matrix.

The applicability and the reliability of the proposed method are validated through numerical examples and it is verified that the properties and characteristics of the supporting soil including soil profile should be accurately considered in the analysis. The influence of various properties and characteristics of the soil on the response is investigated. The softer soil is, the lower the frequencies of the dominant modes in the response are. The variation of the peak value changes largely due to the soil profile and soil properties. In case that the frequency of the dominant structural mode is close to frequency range of dominant soil response, the dynamic behavior may be amplified significantly.

# Acknowledgement

This work is a part of a research project supported by Korea Ministry of Construction and Transportation (MOCT) through Korea Bridge Design and Engineering Research Center at Seoul National University. The authors wish to express their gratitude for the financial support.

# References

1. Luco, J. E.: Linear soil-structure interaction. Lawrence Livermore National Lab. UCRL-15272 (1980)
2. Wang, S.: Coupled boundary and finite elements for dynamic structure-foundation-soil interaction. Computational Structures. **44** (1992) 807–0812
3. Wolf, J. P.: Dynamic soil-structure-interaction in time domain. Prentice-Hall (1988)
4. Zohdi, T. I., and Wriggers, P.:Computational micro-macro material testing. Archives of Computational Methods in Engineering **8** (2001) 132–228
5. Lysmer, J., and Kuhlemeyer, R. L.:Finite dynamic model for infinite media. Journal of Engineering Mechanics, ASCE **95** (1969) 859–877
6. Cook, R. D., Malkus, D. S., and Plesha, M. E.: Concepts and application of finite element method. John Wiley & Sons Inc (1989)
7. Hayashi, Y., Tamura, K., Mora, M., and Takahashi, I.: Simulation analysis of buildings damaged in the 1995 Kobe, Japan, earthquake considering soil-structure interaction. Earthquake Engineering and Structural Dynamics **28** (1999) 371–391
8. Zhang, Y., Yang, Z., Bielak, J., Contel, J. P., and Elgamal, A.: Treatment of seismic input and boundary conditions in nonlinear seismic analysis of a bridge ground system. 16th ASCE Engineering Mechanics Conference (2003)

# Numerical Method for Regional Pole Assignment of Linear Control Systems

Petko Hr. Petkov[1], Nicolai D. Christov[2], and Michail M. Konstantinov[3]

[1] Technical University of Sofia, 1000 Sofia, Bulgaria
[2] Université des Sciences et Technologies de Lille,
59655 Villeneuve d'Ascq, France
[3] University of Architecture, Civil Engineering and
Geodesy, 1046 Sofia, Bulgaria

**Abstract.** A new class of regional pole assignment problems for linear control systems is considered, in which each closed-loop system pole is placed in a desired separate region of the complex plane. A numerically stable method for regional pole assignment is proposed, in which the design freedom is parameterized directly by specific eigenvector (or principal vector) elements and pole location variables that can be chosen arbitrarily. Combined with an appropriate optimization procedure, the proposed method can be used to solve a wide range of optimization problems with pole location constraints, arising in the multi-input control systems design ($H_2/H_\infty$ optimization with pole assignment, robust pole assignment, pole assignment with maximum stability radius, etc.).

## 1 Introduction

Pole assignment is one of the main approaches in the design of state regulators, observers and dynamic compensators for linear control systems. In the multi-input case, the freedom in the pole assignment makes possible to achieve some more design purposes. For instance, several pole assignment algorithms have been proposed which minimize the sensitivity of the closed-loop system poles relative to unstructured and structured perturbations, see [1] − [4].

Along with the standard pole assignment, a regional pole assignment is also used, in which the closed-loop system poles are placed in a specified region of the complex plane, rather than at specified locations [5,6]. The regional pole assignment is applied in various design problems, such as the $H_\infty$ control with pole placement constraints [7], robust regional pole placement [8], etc.

In this paper a new class of pole assignment problems is considered, in which each closed-loop system pole is placed in a specified separate region. A method for regional pole assignment is proposed, in which the design freedom is parameterized directly by specific eigenvector (or principal vector) elements and pole location variables that can be chosen arbitrarily. The method is numerically stable and provides substantially more flexibility in design. Combined with an appropriate optimization procedure, the proposed method can solve a wide range of optimization problems with pole location constraints, arising in the multi-input

T. Boyanov et al. (Eds.): NMA 2006, LNCS 4310, pp. 660–667, 2007.

control systems design ($H_2/H_\infty$ optimization with pole assignment, robust pole assignment, pole assignment with maximum stability radius, etc.) [9].

The following notations are used later on: $\mathcal{F}$ denotes the field of real ($\mathcal{F} = \mathcal{R}$) or complex ($\mathcal{F} = \mathcal{C}$) numbers; $\mathcal{F}^{m \times n}$ – the space of $m \times n$ matrices $A = [a_{ij}]$ over $\mathcal{F}$; $A^T$ – the transposed matrix $A$; $A^H$ – the complex conjugate transpose of $A$; spect($A$) – the spectrum (i.e. the set of eigenvalues counted according to algebraic multiplicity) of the matrix $A$; $\|A\|$ – the norm of the matrix $A$ (we use the spectral norm $\|A\|_2 = \sigma_{\max}(A)$ – the maximum singular value of $A$, and the Frobenius norm $\|A\|_F = (\sum |a_{ij}|^2)^{1/2}$); $I_n$ – the unit $n \times n$ matrix.

## 2 Problem Statement

Consider the linear controllable system

$$\dot{x}(t) = Ax(t) + B_1 w(t) + B_2 u(t) \tag{1}$$

$$z(t) = Cx(t) + Du(t),$$

where $x(t) \in \mathcal{F}^n$, $w(t) \in \mathcal{F}^l$, $u(t) \in \mathcal{F}^m$ and $z(t) \in F^q$ are the state, disturbance, control and performance vectors, respectively, and $A \in \mathcal{F}^{n \times n}$, $B_1 \in \mathcal{F}^{n \times l}$, $B_2 \in \mathcal{F}^{n \times m}$, $C \in \mathcal{F}^{q \times n}$, $D \in \mathcal{F}^{q \times m}$ (we assume that $\mathcal{F} = \mathcal{R}$ or $\mathcal{F} = \mathcal{C}$). Let $(m_1, \ldots, m_p)$ be the set of conjugate Kronecker indexes of the pair $(A, B_2)$:

$$m_1 = m = \text{rank}(B_2), \quad m_i = \text{rank}(P_i) - \text{rank}(P_{i-1}); \; i = 2, \ldots, p,$$

where $p$ is the controllability index of $(A, B_2)$ and

$$P_i = [B_2 \; AB_2 \; \ldots \; A^{i-1}B_2] \in \mathcal{R}^{n \times im}.$$

Applying the state feedback $u(t) = -Kx(t)$, where $K \in \mathcal{F}^{m \times n}$, we obtain the closed-loop system

$$\dot{x}(t) = A_{cl}x(t) + B_{cl}w(t), \quad z(t) = C_{cl}x(t), \tag{2}$$

where $A_{cl} = A - B_2 K$, $B_{cl} = B_1$, $C_{cl} = C - DK$.

Let $\mathcal{L} = \{\Lambda_1, \ldots, \Lambda_n\} \subset \mathcal{C}$ be the prescribed pole region set of (2), where $\Lambda_i(\lambda_{i0}, r_i) = \{\lambda \in \mathcal{C} : (\Re(\lambda) - \Re(\lambda_{i0}))^2 + (\Im(\lambda) - \Im(\lambda_{i0}))^2 \leq r_i^2\}$, $i \in \overline{1, n}$, are the prescribed pole regions and $\lambda_{i0}$ and $r_{i0} \geq 0$ are the regions centers and radii. Then for each pole set $\ell = \{\lambda_1, \ldots, \lambda_n\} \subset \mathcal{L}$, $\lambda_i \in \Lambda_i$, there exists a matrix $K \in \mathcal{F}^{m \times n}$ such that

$$\text{spect}(A - B_2 K) = \ell. \tag{3}$$

The set of matrices $\mathcal{K} \subset \mathcal{F}^{m \times n}$ with the property (3) is $n(m-1)$-dimensional algebraic variety of complicated structure so that its direct parametrization by the free elements of $K$ is a difficult problem except for some low-dimensional cases. A more convenient, although redundant, parametrization of the set $\mathcal{K}$ may be done in the following way [9].

For a given $\ell$ denote by $\mathcal{V}_\ell$ the set of all nonsingular matrices $V \in \mathcal{F}^{n \times n}$ such that

$$\text{spect}(V^{-1}(A - B_2 K)V) = \ell \tag{4}$$

for $K \in \mathcal{K}$. Similarly, let $\mathcal{K}_V \subset \mathcal{F}^{m \times n}$ be the set of all $K \in \mathcal{F}^{m \times n}$ such that (4) holds for some $V$. Note that $V$ may have as its columns not only eigenvectors but generalized eigenvectors of $A - B_2 K$ as well.

The set $\mathcal{V}_\ell$ of all attainable transformation matrices $V$ is an $mn$-dimensional algebraic variety but the set $\mathcal{K}_V$ itself is only $n(m-1)$-dimensional because of the exististence of $n$ non trivial relations given by (3).

The problem for finding $K$ for $\ell \subset \mathcal{L}$ and $V \in \mathcal{V}_\lambda$ is said to be the *region set pole assignment problem* for the system (1). In the next section we present a numerically stable solution of this problem, which utilizes a parametrization of the set $\mathcal{K}_V$ by elements of the eigenvectors and generalized eigenvectors of the matrix $A - B_2 K$, and a parametrization of the set $\mathcal{L}$ by pole location variables.

The freedom in $\mathcal{L}$ and $\mathcal{K}_V$ can be used to optimize some performance index specifying the behavior of the closed-loop system. This makes possible to solve a number of important control systems design problems, e.g. pole assignment with $H_2$ and/or $H_\infty$ optimization, robust pole assignment, design of dynamic output compensators, etc [9].

## 3   Main Result

In this section we first present a computational method which preassigns the closed-loop system poles at a given location $\ell \in \mathcal{L}$ and utilizes in full the freedom in the Schur vectors of the closed-loop system matrix. The method presented is a further extension of the numerically stable pole assignment algorithm described in [10].

The first stage of the method is a reduction of the pair $(A, B_2)$ into orthogonal canonical form [10]

$$(A^0, B_2^0) = (U^T A U, U^T B_2),$$

$$A^0 = \begin{bmatrix} A_{11} & A_{12} & A_{13} & \cdots & A_{1,p-1} & A_{1p} \\ A_{21} & A_{22} & A_{23} & \cdots & A_{2,p-1} & A_{2p} \\ 0 & A_{32} & A_{33} & \cdots & A_{3,p-1} & A_{3p} \\ \vdots & \vdots & \ddots & & \vdots & \vdots \\ 0 & 0 & 0 & \cdots & A_{p,p-1} & A_{pp} \end{bmatrix}, \quad B_2^0 = \begin{bmatrix} B_{10} \\ 0 \\ 0 \\ \vdots \\ 0 \end{bmatrix},$$

where the matrices

$$B_{10} \in \mathcal{R}^{m_1 \times m}, \quad A_{i,i-1} \in \mathcal{R}^{m_i \times m_{i-1}}, \quad i = 2, \ldots, p$$

are of full row rank, and $U$ is an orthogonal $n \times n$ matrix.

This reduction may be done by using singular value decomposition or, more efficiently, using QR decomposition with column pivoting. As it is shown in [11] the corresponding algorithm is numerically stable, the computed pair $(A^0, B_2^0)$ being exact for slightly perturbed data $(A + \Delta A,\ B_2 + \Delta B_2)$.

The second stage of the method is to determine the gain matrix. Since the reduced closed-loop system matrix $A^0 - B_2^0 K^0$, $K^0 = KU$, is in block-Hessenberg form, it is possible to find an eigenvector $v^0$ of this matrix knowing only $A^0$ and $\lambda^0 \in \ell$. In fact, from the equation

$$(A^0 - B_2^0 K^0)v^0 = \lambda^0 v^0,$$

where $v^0 = [v_1\ \ldots\ v_p]^T$, $v_i \in \mathcal{F}^{m_i}$, it follows

$$A_{i,i-1}v_{i-1} = v_i \lambda^0 - \sum_{k=i}^{p} A_{ik}v_k;\ i = p, \ldots, 2, \tag{5}$$

so that by setting $v_p$ it is possible to compute recursively $v_{p-1}, \ldots, v_1$. The elements of $v_p$ are free (except that at least one must be non-zero so that $v_p \neq 0$) and different choices of these elements will lead to different solutions $K^0$.

Apart from the freedom in $v_p$, there is an additional freedom in the solution of equation (5) if $m_i < m_{i-1}$. This freedom may be used by QR or singular value decomposition of $A_{i,i-1}$ when solving (5).

Suppose now a sequence of plane rotations in the corresponding planes is chosen so as to annul successively all elements of $v^0$ except the first one moving from bottom to the top. As a result we obtain

$$Q^H(A^0 - B_2^0 K^0)Q v^1 = v^1 \lambda^0 \tag{6}$$

$$v^1 = Q^H v^0 = [\, v_{10}\ 0\ \ldots\ 0\,]^T,\ v_{10} \neq 0,$$

where $Q$ is the product of plane rotations implemented and the $(n-1) \times (n-1)$ block in the lower right corner of $Q^H A^0 Q$ is again in block-Hessenberg form. This block has the same structure as $A^0$ except that the index $m_p$ has decreased with 1.

From (6) one gets a linear equation for the first column of $K^0 Q$ and the pole $\lambda^0$ is deflated from the problem. It is possible to proceed in the same way at the next step working with a subsystem of order $n - 1$. Thus each pole is prescribed independently which allows to assign multiple poles. In this way, by setting the free elements of the subsystem eigenvectors, it is possible to obtain at each step a specific solution for the corresponding column of the gain matrix.

The determination of the gain matrix $K$ corresponding to specified eigenvector and gain matrix elements is done by the following algorithm which makes use of the matrices $A^0, B_2^0$ of the orthogonal canonical form as well as the transformation matrix $U$:

1 Set $k = 0$, $Q = U$

2 $k \leftarrow k + 1$

3 If $p = 1$ go to 9

4 Set $v_p$

5 For $i = p, p - 1, \ldots, 1$

    5.1 If $i > 1$ compute $v_{i-1}$ from (5) specifying the free elements

    5.2 Determine a product $Q_i$ of plane rotations transforming $v_i$

    5.3 $A^0 \leftarrow Q_i^H A^0 Q_i$

    5.4 If $i = 1$, $B_2^0 \leftarrow Q_i^H B_2^0$

    5.5 $Q \leftarrow Q Q_i$

6 Determine the $k$-th columns of $K^0$ and the upper triangular form of $Q^H (A - B_2 K) Q$ specifying the free elements

7 $m_p \leftarrow m_p - 1$

8 If $m_p = 0$, $p \leftarrow p - 1$

  Go to 2

9 Determine the rest (from $k$-th to $n$-th) columns of $K^0$ and the upper triangular form of $Q^H (A^0 - B_2^0 K^0) Q$ specifying the free elements

10 $K = K^0 Q^H$

In case of real systems the above algorithm is derived as a real arithmetic procedure for both real and complex desired poles.

The numerical properties of the algorithm presented are very favorable. It can be shown that it is backwardly stable, i.e. the computed gain matrix is exact for matrices $A + \Delta A$, $B_2 + \Delta B_2$, where $\|\Delta A\|$, $\|\Delta B_2\|$ are small relative to $\|A\|$, $\|B_2\|$, resp.

Consider finally the parametrization of the prescribed pole region set $\mathcal{L} = \{\Lambda_1, \ldots, \Lambda_n\} \subset \mathcal{C}$. The complex pole regions can be parameterized as

$$\Lambda_i(\lambda_{i0}, r_i) = \{\lambda \in \mathcal{C} : \lambda = \lambda_{i0} + \frac{1 - \exp(-\alpha_i)}{1 + \exp(-\alpha_i)}$$

$$\times r_i \exp\left(j \frac{1 - \exp(-\beta_i)}{1 + \exp(-\beta_i)} \frac{\pi}{2}\right); \ \alpha_i, \beta_i \in \mathcal{R}\},$$

where $j = \sqrt{-1}$. In case of real systems the prescribed pole region set $\mathcal{L}$ and the assigned spectra $\ell \subset \mathcal{L}$ must be self-conjugate, i.e. if $\mathcal{L}$ contains the complex region $\Lambda_i$, it must also contain the conjugate region

$$\bar{\Lambda}_i(\bar{\lambda}_{i0}, r_i) = \{\lambda \in \mathcal{C} : \lambda = \bar{\lambda}_{i0} + \frac{1 - \exp(-\alpha_i)}{1 + \exp(-\alpha_i)}$$

$$\times r_i \exp\left(-j \frac{1 - \exp(-\beta_i)}{1 + \exp(-\beta_i)} \frac{\pi}{2}\right); \ \alpha_i, \beta_i \in \mathcal{R}\},$$

and if $\lambda_i \in \Lambda_i$ belongs to $\ell$, then $\bar{\lambda}_i \in \bar{\Lambda}_i$ must also belong to $\ell$. Thus both $\Lambda_i(\lambda_{i0}, r_i)$ and $\bar{\Lambda}_i(\bar{\lambda}_{i0}, r_i)$ are parameterized by $\alpha_i$, $\beta_i$.

As a particular case, the real pole regions can be parameterized as

$$\Lambda_k(\lambda_{k0}, r_k) = \{\lambda \in \mathcal{R} : \lambda = \lambda_{k0} + \frac{1 - e^{-\gamma_k}}{1 + e^{-\gamma_k}} \, r_k, \, \gamma_k \in \mathcal{R}\}.$$

Changing iteratively the region set parametrization variables and the free eigenvector elements by some optimization technique, the extremum of the corresponding system performance index can be found. For this purpose it is convenient to use direct search methods [12] which make use of the performance index values only and do not attempt to estimate derivatives.

# 4    Applications to Control Systems Optimization

Consider the application of the regional pole assignment to the solution of the following important optimization problems arising in the design of linear control systems.

– **Pole assignment with $H_2$ optimization**

In this case the optimization problem consists of finding the matrix $K$ which assigns the poles of the closed-loop system in the desired region set and minimizing at the same time the 2-norm of the closed-loop transfer function

$$H(s) = C_{cl}(sI_n - A_{cl})^{-1}B_{cl}$$

from $w$ to $z$.

One has that

$$\|H\|_2^2 = \text{trace}[C_{cl}P_cC_{cl}^T] = \text{trace}[B_{cl}^T P_o B_{cl}],$$

where the matrices $P_o, P_c$ are solutions of the Lyapunov matrix equations

$$A_{cl}P_c + P_cA_{cl}^T + B_{cl}B_{cl}^T = 0$$

$$A_{cl}^T P_o + P_oA_{cl} + C_{cl}^T C_{cl} = 0.$$

These equations may be solved in a numerically stable way by the Bartels-Stewart algorithm [13].

– **Pole assignment with $H_\infty$ optimization**

The aim of this design technique is to ensure the desired dynamics of the closed loop system by placing the closed-loop system poles in the desired region set and minimizing at the same time the $H_\infty$ norm of the closed-loop transfer matrix from $w$ to $z$ in order to achieve disturbance attenuation. As it is known [14] the latter will provide also robust stability of the closed-loop system under the presence of unstructured perturbations. The computation of the $H_\infty$ norm can be done by the algorithms from [15,16].

– **Robust pole assignment**

The freedom in the pole region set and in the eigenstructure of the closed-loop system matrix $A_{cl}$ can be used to minimize the condition number $\text{cond}_2(V)$ or $\text{cond}_F(V)$ of the eigenvector matrix $V$ of the matrix $A_{cl}$, which leads to minimization of the overall sensitivity of the closed-loop poles. This objective function was proposed in [1] and it works only if the spectra $\ell \subset \mathcal{L}$ allow a full collection of linearly independent eigenvectors of $A_{cl}$. Instead of minimizing $\text{cond}_F(V)$ it is preferable to minimize $\|V\|_F^2 + \|V^{-1}\|_F^2$ as proposed in [2].

Sometimes it may be desirable to minimize the $p$-norm

$$\|c\|_p = \left( \sum_{i=1}^{n} |c_i|^p \right)^{1/p}, \; p \geq 1$$

of the vector $c = (c_1, \ldots, c_n)^T$ whose elements

$$c_i = \frac{1}{|v_i^H w_i|}; \; A_{cl} v_i = \lambda_i v_i, \; A_{cl}^H w_i = \overline{\lambda_i} w_i; \; \|v_i\| = \|w_i\| = 1$$

are the individual condition numbers of the eigenvalues $\lambda_i$. This approach may also be used only if the spectra $\ell \subset \mathcal{L}$ allow a full collection of eigenvectors of $V$ (i.e. if there are non-defective attainable closed-loop system matrices $F$ with $\text{spect}(F) = \ell$). For this purpose it is appropriate to implement the eigenvalue condition estimators from LAPACK [17].

– **Pole assignment with maximum stability radius**

Minimum sensitivity of the poles does not guarantee that the closed-loop system will remain stable under perturbations in the system matrices of given size. That is why instead of minimization of pole sensitivity one may prefer to maximize the stability radius of the matrix $A_{cl}$, i.e. to maximize the absolute or relative distance from $A_{cl}$ to the set of matrices having eigenvalues with real zero parts. If this distance is greater than a quantity $\Delta$ then the closed-loop system will remain stable for all parametric perturbations whose size is less than $\Delta$. The computation of the complex stability radius may be done by the algorithm proposed in [18].

– **Pole assignment with minimum gain matrix norm**

In some cases for a given pole region set it is desirable to obtain a gain matrix whose elements have magnitudes as small as possible. This may be done, for instance, by minimization of the Frobenius norm $\|K\|_F$ (or its square $\|K\|_F^2$) of the gain matrix $K$. It is important to note that the objective function for this problem is convex and has an unique minimum.

All these optimization problems are relevant to the design of dynamic compensators as well.

It should be stressed that it is also possible to implement the regional pole assignment method along with other objective functions (for instance, the departure from normality of the matrix $A_{cl}$) or to combine some of the objective functions presented above (for instance, a pole assignment with mixed $H_2/H_\infty$ optimization).

The method proposed may be also applied to the design of discrete-time linear control systems.

# References

1. Kautsky, J., N. K. Nichols, N. K., Van Dooren, P.: Robust pole assignment in linear state feedback. Int. J. Control **41** (1985) 1129–1155
2. Byers, R., Nash, S. G.: Approaches to robust pole assignment. Int. J. Control **49** (1989) 97–117
3. Liu, G. P., Patton, R. J.: Eigenstructure Assignment for Control System Design. John Wiley, Chichester (1998)
4. Magni, J. F. : Robust Modal Control with a Toolbox for Use with MATLAB. Kluwer Acad. Publ. Dordrech (2002)
5. Kim, S. B., Furuta, K.: Regulator design with poles in a specified region. Int. J. Control **47** (1988) 143–160
6. Kučera, V., Kraus, F. J.: Regional pole placement. Kybernetika **31** (1995) 541–546
7. Chilali, M., Gahinet, P.: $H_\infty$ design with pole placement constraints: an LMI approach. IEEE Trans. Autom. Control **41** (1996) 358–367
8. Kučera, V., Kraus, F. J.: Robust regional pole placement: an affine approximation. Private communication, 2003
9. Petkov, P. Hr., Christov, N. D., Konstantinov, M. M.: Optimal synthesis of linear multi-input systems with prescribed spectrum. Proc. 25th Spring Conf. of UBM, Kazanlak, 6-9 April 1996, 141–149
10. Petkov, P. Hr., Christov, N. D., Konstantinov, M. M.: Computational Methods for Linear Control Systems. Prentice Hall, N.Y. (1991)
11. Petkov, P. Hr., Christov, N. D., Konstantinov, M. M.: Numerical analysis of the reduction of linear systems into orthogonal canonical form. Systems & Control Lett. **7** (1986) 361–364
12. Higham, N. J.: Optimization by direct search in matrix computations. SIAM J. Matrix Anal. Appl. **14** (1993) 317–333
13. Bartels, R. H., Stewart, G. W.: Algorithm 432: Solution of the matrix equation $AX + XB = C$. Comm. ACM **15** (1972) 820–826
14. Khargonekar, P. P., Rotea, M. A.: Mixed $H_2/H_\infty$ control: A convex optimization approach. IEEE Trans. Autom. Control **36** (1991) 824–837
15. Pandey, P., Kenney, C., Laub, A. J., Packard, A.: A gradient method for computing the optimal $H_\infty$ norm. IEEE Trans. Autom. Control **36** (1991) 887–890
16. Sherer, C.: $H_\infty$-control by state feedback and fast algorithms for the computation of optimal $H_\infty$-norms. IEEE Trans. Autom. Control **35** (1990) 1090–1099
17. Anderson, E., Bai, Z., Bischof, C. H., Demmel, J. M., Dongarra, J. , DuCroz, J. J. J., Greenbaum, A., Hammarling, S. J., McKenney, A., Ostrouchov, S., Sorensen, D. C.: LAPACK Users's Guide. SIAM, Philadelphia (1992)
18. Byers, R.: A bisection method for measuring the distance of a stable matrix to the unstable matrices. SIAM J. Sci. Stat. Comput. **9** (1988) 875–881

# Solution Limiters and Flux Limiters for High Order Discontinuous Galerkin Schemes

Natalia Petrovskaya

School of Mathematics, University of Birmingham,
Edgbaston, B15 2TT, Birmingham, UK
n.b.petrovskaya@bham.ac.uk
http://web.mat.bham.ac.uk/N.B.Petrovskaya

**Abstract.** We analyze a general concept of limiters for a high order DG scheme written for a 1-D problem. The limiters, which are local and do not require extended stencils, are incorporated into the solution reconstruction in order to meet the requirement of monotonicity and avoid spurious solution overshoots. A limiter $\beta$ will be defined based on the solution jumps at grid interfaces. It will be shown that $\beta$ should be $0 < \beta < 1$ for a monotone approximate solution.

## 1 Introduction

Recently a number of new discretization methods have been developed to numerically solve modern problems of science and engineering. One of them is a Discontinuous Galerkin (DG) discretization scheme [3], which affords optimal orders of convergence for smooth problems by using high order approximating spaces. However, the capability of high order DG schemes to resolve solution discontinuities is still an open question. It has been observed many times (e.g. see [2,5,6,7]) that a high order DG discretization may result in oscillations in the vicinity of a shock discontinuity. The study carried out in [8] has shown that high order DG approximations do not provide a monotone solution near the shock even for the simplest linear advection problem.

Since the solution oscillations may have a disastrous impact on the convergence of the approximate solution, a limiting procedure which allows one to obtain a monotone solution near discontinuities should be addressed. A number of authors have contributed to the issue of limiters for the DG scheme in recent years [1,2,7,10]. The examples of limiters implemented in a semi-discrete DG scheme are given in [3]. The approach suggested in [6] is similar to that in ENO schemes and takes data from neighboring grid cells to construct a local solution limiter on a given cell. The discussion of using a limiting algorithm for multi-dimensional problems can be found in [5].

In the present paper we develop an approach to define a limiter on a compact discretization stencil for high order DG schemes. We analyze one-dimensional problems where the definition of a new limiter is straightforward for a high order DG discretization. The limiter $\beta$, which is local and does not require extended

T. Boyanov et al. (Eds.): NMA 2006, LNCS 4310, pp. 668–676, 2007.
© Springer-Verlag Berlin Heidelberg 2007

stencils, is based on the solution jumps at grid interfaces. It will be proved that the values $0 < \beta < 1$ provide a monotone approximate solution over a computational grid. One important feature of the suggested approach is that we also incorporate the evaluation of flux approximation into the limiting procedure, as upwind flux approximation in a high order DG scheme presents another difficulty when steady state solutions are considered.

## 2   The Definition of a Discontinuous Galerkin Scheme

Our concept of limiters in high order DG schemes can be best illustrated by consideration of an ordinary differential equation written for a function $u(x)$ as

$$F_x(x, u) = S(x), \quad x \in \Omega = [a, b]. \tag{1}$$

The function $F(x, u(x))$ is considered as a flux function for a steady state problem (1). The equation above should be augmented with a boundary condition that will be further provided for a given problem under consideration.

We use a discontinuous Galerkin method to obtain a numerical solution to the problem. Let us introduce the element partition $G$ of the region, $G = \bigcup_{i=1}^{N} e_i$, $e_i = [x_i, x_{i+1}], 1 \le i \le N$, where $x_i$ is a nodal coordinate, and $h_i = x_{i+1} - x_i$ is a grid step size. We seek an approximation $u_h(x)$ to the solution $u(x)$ such that $u_h(x)$ is a piecewise polynomial function over $\Omega$. The approximate solution $u_h(x)$ is expanded on each grid cell as

$$u_h(x) = \sum_{k=0}^{K} u_k \phi_k(x), \quad x \in e_i = [x_i, x_{i+1}], \tag{2}$$

where the test functions are $\phi_k(x) = ((x - x_i)/h_i)^k$, $x \in e_i$, $k = 0, 1, \ldots, K$.

Multiplying the equation (1) by the test function $\phi_k(x)$ and integrating over the cell $[x_i, x_{i+1}]$ results in the following weak formulation of the problem,

$$F(x_{i+1}, u(x_{i+1}))\phi_k(x_{i+1}) - F(x_i, u(x_i))\phi_k(x_i) - \int_{x_i}^{x_{i+1}} F(x, u)(d\phi_k(x)/dx)dx = \int_{x_i}^{x_{i+1}} S(x)\phi_k(x)dx, \quad k = 0, 1, \ldots, K \tag{3}$$

where the function $u(x)$ should be further replaced by the approximate solution $u_h(x)$.

Since $u_h(x)$ is discontinuous at the cell interfaces, the above equations considered for the solution $u_h(x)$ require a numerical flux $\tilde{F}(x, u_h)$ consistent with the continuous flux $F(x, u)$ to be defined. Suppose that the flux $\tilde{F}(x, u_h)$ that generally depends on the two values of the approximate solution at any grid interface $x_i$ is chosen for a given problem (see [4] for the discussion of numerical fluxes). Then the DG discretization scheme reads

$$\tilde{F}(u_h(x_{i+1}))\phi_k(x_{i+1}) - \tilde{F}(u_h(x_i))\phi_k(x_i) -$$

$$\int_{x_i}^{x_{i+1}} F(x, u_h(x))(d\phi_k(x)/dx)dx = \int_{x_i}^{x_{i+1}} S(x)\phi_k(x)dx, \quad k = 0, 1, \ldots, K. \quad (4)$$

Although very efficient for smooth problems, a high order DG discretization is not always appropriate when discontinuous functions are concerned. It has been shown in [8] that being applied to a discontinuous problem the DG scheme (4) may generate solution overshoots that do not depend on the grid step size. Below we address the issue of limiters required to eliminate spurious solution oscillations in a high order DG scheme.

## 3    Limiters for a High Order DG Scheme

Let $U(x)$ be the exact solution to the boundary-value problem for the equation (1) and three points $P_i = (x_i, U(x_i))$, $P = (\hat{x}, U(\hat{x}))$, and $P_{i+1} = (x_{i+1}, U(x_{i+1}))$ be chosen at the curve $U(x)$ (see Fig. 1a). Let the distance $x_{i+1} - x_i = h > 0$, and we denote $(\hat{x} - x_i)/(x_{i+1} - x_i) = s_0$, $s_0 \in (0, 1)$. We define the parameter $\theta$ as follows

$$\theta = (U(\hat{x}) - U(x_i))/(U(x_{i+1}) - U(x_i)), \quad U(x_{i+1}) \neq U(x_i), \quad (5)$$

Let us fix $s_0$ and move the points $P$ and $P_{i+1}$ along the curve $U(x)$. The parameter $\theta$ is then considered as a function of the distance $h$.

The behavior of $\theta(h)$ depends on the solution $U(x)$. Let U(x) be a monotone function shown in Fig. 1a. For the monotone solution, the following conditions hold

1. $|U(\hat{x}) - U(x_i)| < |U(x_{i+1}) - U(x_i)|$, if $\hat{x} - x_i < x_{i+1} - x_i$,
2. $sgn(U(\hat{x}) - U(x_i)) = sgn(U(x_{i+1}) - U(x_i))$.

Hence, $\theta(h) > 0$, and $\theta(h)$ is a bounded function over the domain of definition.

We now estimate the value of $\theta$ for $h \rightarrow 0$. The solution is assumed to be a smooth function over the interval $[x_i, x_{i+1}]$. We denote the $k - th$ derivative $d^{(k)}U(x)/dx^k$ taken at the point $x_i$ as $D_k$. The Taylor series expansion of the solution $U(x)$ near the point $x_i$ yields $U(\hat{x}) \approx U(x_i) + D_1 s_0 h$, and

$$1/(U(x_{i+1}) - U(x_i)) \approx 1/(D_1 h + (1/2)D_2 h^2) = 1/(D_1 h(1 + (D_2/2D_1)h)) \approx$$

$$1/(D_1 h(1 - (D_2/2D_1)h)), \quad D_1 \neq 0.$$

Substituting the expansion above into (5), we obtain

$$\theta \approx (D_1 s_0 h(1 - (D_2/2D_1)h))/D_1 h = s_0(1 - O(h)). \quad (6)$$

For the sake of simplicity, let us further consider $x_i = 0$, so that $x_{i+1} = h$. There are three extreme cases of a monotone smooth function $U(x)$ that define the behavior of $\theta(h)$:

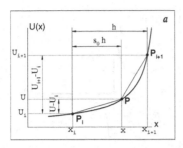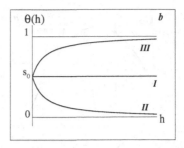

**Fig. 1.** A monotone solution $U(x)$. (a) Geometric interpretation of the function $\theta(h)$ in the $(x, U(x))$ - plane. (b) The function $\theta(h)$ for a monotone solution $U(x)$: (I) a linear function $U(x)$, (II) a concave function $U(x)$, (III) a convex function $U(x)$.

1. $U(x)$ is a linear function. The definition of $\theta$ yields $\theta(h) = const = s_0$. The function $\theta(h)$ is presented by curve $(I)$ in Fig. 1b.
2. $U(x)$ is a concave function, which has a vertical asymptote at the point $x = h^*$: $U(x) \to \infty$, as $x \to h^*$. Since $|U(h) - U(0)| \to \infty$, and $|U(s_0 h) - U(0)| \to \Delta U \neq \infty$, as $x \to h^*$, we have $\theta(h) \to 0$, as $h \to h^*$. If we consider a set of smooth concave functions, then $h^* \to \infty$, and we arrive at $\theta(h) \to 0$, as $h \to \infty$. The function $\theta(h)$ for a concave $U(x)$ is shown as curve $(II)$ in Fig. 1b.
3. $U(x)$ is a convex function, which has a horizontal asymptote: $U(x) \to U_0$, as $x \to \infty$. Since $U(h) - U(0) \to U_0 - U(0)$, and $U(s_0 h) - U(0) \to U_0 - U(0)$, as $x \to \infty$, we have $\theta(h) \to 1$, as $h \to \infty$. The function $\theta(h)$ generated by a convex $U(x)$ is shown as curve $(III)$ in Fig. 1b.

Hence, for a monotone smooth function $U(x)$, the parameter $\theta$ is bounded by $0 < \theta \leq 1$, where $\theta = 1$ for $U(x) \equiv const$ by convention.

We now consider a non-monotone solution $U(x)$ shown in Fig. 2a. Let $P_{ext} = (x_{ext}, U(x_{ext}))$ be an extremum point. For small $h < x_{ext}$, the solution $U(x)$ is a monotone function and we refer to the analysis above, as the function $\theta(h)$ will depend entirely on the derivative $d^2 U(x)/dx^2$. For $h > x_{ext}$, the function $\theta(h)$ is an increasing function which takes the value $\theta(h_0) = 1$ at the point $h = h_0$, where $U(h) = U(s_0 h)$ (see Fig. 2a). The function $\theta(h)$ has a singular point $h = h_d$, defined by the condition $U(h_d) = U(x_i)$. Since $x_i = 0$ and $U(h) - U(0)$ changes the sign at the point $x = h_d$, the asymptotic behavior of $\theta(h)$ is $\theta(h) \to +\infty$ as $h \to h_d - 0$, and $\theta(h) \to -\infty$, as $h \to h_d + 0$. Finally, $\theta(h) \to 0$ (or another constant), as $h \to \infty$, provided there are no other extremum points. The function $\theta(h)$ for a non-monotone solution is shown in Fig. 2b.

The above consideration reveals how a limiting procedure can be defined for an approximate solution in a high order DG scheme. Let $\hat{x}$ be an arbitrary point in the cell $[x_i, x_{i+1}]$. Suppose that the approximate solution $u_h(x)$ coincides with the exact solution $U(x)$ everywhere except for the point $\hat{x}$, so that a local extremum appears at $\hat{x}$ (see Fig. 3a). Since for a monotone function the condition $0 < \theta \leq 1$ always holds, we now compare the approximate solution variation

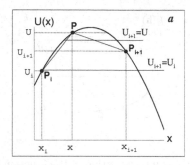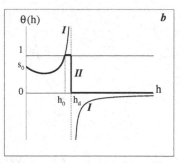

**Fig. 2.** A non-monotone solution $U(x)$. (a) Geometric interpretation of the function $\theta(h)$ in the $(x, U(x))$ - plane. (b) The function $\theta(h)$ for a non-monotone solution: (I) $\theta(h)$ is not bounded near the extremum point, (II) A limiting procedure cuts off $\theta(h)$ near the solution extremum.

$(u_h(\hat{x}) - u_h(x_i))$ with the exact solution variation $(U(x_{i+1}) - U(x_i))$. The value $\theta_i < 0$ or $\theta_i > 1$ of the parameter

$$\theta_i = (u_h(\hat{x}) - u_h(x_i))/(U(x_{i+1}) - U(x_i)) \tag{7}$$

indicates that a local extremum is present in the cell $e_i$. The limiting procedure for a non-monotone function is shown in Fig. 2b in the $(h, \theta)$-plane.

## 4    The Flux Control in the Limiting Procedure

The limiter (7) is not viable, unless an accurate estimate of the exact solution has been given at the points $x_i$ and $x_{i+1}$. Hence, our next purpose is to obtain a reliable solution estimate to be used in the limiter $\theta_i$. Moreover, we also want the evaluation of flux approximation to be incorporated into the limiting procedure, as upwind flux approximation in a high order DG scheme presents another difficulty when steady state solutions are considered. It has been recently shown in [9] that a high order DG scheme is not able to recognize flux extrema that may lead to an underdetermined system of algebraic equations obtained as a result of the discretization. Solving that system of equation will inevitably result in an oscillating numerical solution, so that a flux control procedure should be developed to avoid a divergent solution (see [9] for a further discussion of spurious oscillations arising as a result of incorrect flux approximation).

Our approach to the flux control in the DG scheme (4) is based on the definition of "frame" and "phantom" points on a grid cell. Let $P = (x_i, u_h(x_i))$ be a point in the $(x, u)$ - plane, where $x_i \in G$. Each pair $(x_i, u_h(x_i))$ generates the flux $F(x_i, u_h(x_i))$. We will refer to the point $P$ as a "frame" point and denote it as $P_{\mathcal{F}}$, if $P$ is involved into the definition of the numerical flux, i.e. $\tilde{F}(x_i, u_h)F(P)$. Otherwise, we will refer to the point $P$ as a "phantom" point and will use the notation $P_{\mathcal{P}}$ for it. A solution estimate we use is based on the assumption that a discrete conservation law

$$\tilde{F}(x_{i+1}, u_h) - \tilde{F}(x_i, u_h) = \int\limits_{x_i}^{x_{i+1}} S(x)\phi_k(x)dx, \qquad (8)$$

is consistent with the equation (1), that is $u_{\mathcal{F}}(x_{\mathcal{F}}) \to U(x_{\mathcal{F}}), \quad h \to 0$. In other words, we assume that local extrema which do not vanish on fine grids may only appear at "phantom" points. The above requirement is to guarantee that the correct flux approximation is used in the problem.

Based on the assumption above, we suggest the following approximation to the function (7)

$$\beta_i = \delta u_i / \Delta u, \qquad (9)$$

where $\delta u_i = u_h(x_{i+1}) - u_h(x_i)$ for approximate solution $u_h(x)$ defined at the interval $e_i$. The exact solution variation $\Delta U = U(x_{i+1}) - U(x_i)$ on the cell $e_i$ is replaced in limiter $\beta_i$ with the approximate solution variation $\Delta u = u_{h\mathcal{F}}(x_{i+1}) - u_{h\mathcal{F}}(x_i)$ at the "frame" points.

In order to incorporate the limiter (9) into the DG scheme, the "frame" points should be defined for a given grid cell. Consider an upwind flux approximation that requires one value of the approximate solution at each grid interface for a monotone solution function. Let the "frame" points for the upwind flux be defined as $P_{\mathcal{F}1} = (x_i, u_{i-1} + \delta u_{i-1})$, and $P_{\mathcal{F}2} = (x_{i+1}, u_i + \delta u_i)$ on the cell $e_i$. The solution variation is

$$\Delta u = u_i + \delta u_i - u_{i-1} - \delta u_{i-1} = u_i + \delta u_i - (u_i - [u]_i) = \delta u_i + [u]_i, \quad \beta_i = \frac{\delta u_i}{\delta u_i + [u]_i},$$

where $[u]_i = u_h(x_i + 0) - u_h(x_i - 0) = u_i - (u_{i-1} + \delta u_{i-1})$ is a solution jump at the interface $x_i$.

If the "frame" points are defined as $P_{\mathcal{F}1} = (x_i, u_i)$, and $\Gamma_{\mathcal{F}2} = (x_{i+1}, u_{i+1})$, the solution variation will be

$$\Delta u = u_{i+1} - u_i = (u_i + \delta u_i) + [u]_{i+1} - u_i = \delta u_i + [u]_{i+1}, \quad \beta_i = \frac{\delta u_i}{\delta u_i + [u]_{i+1}},$$

where the jump $[u]_{i+1} u_h(x_{i+1} + 0) - u_h(x_{i+1} - 0) = u_{i+1} - (u_i + \delta u_i)$ is considered at the interface $x_{i+1}$. Generally, the limiter $\beta_i$ can be written as

$$\beta_i = 1 - \frac{[u]_{\mathcal{P}}}{\delta u_i + [u]_{\mathcal{P}}}, \qquad (10)$$

so that $\beta_i$ depends always on the solution jump $[u]_{\mathcal{P}}$ calculated at the cell interface where a "phantom" point presents.

The limiting procedure is illustrated in Fig. 3b for a piecewise linear $(K = 1)$ DG discretization. The "phantom" point $u_i + \delta u_i$ is shown as a white dot, the frame points are shown as black dots. The location $(I)$ of the "phantom" point yields a monotone approximate solution, while the locations $(II)$ and $(III)$ result in a nonphysical local extremum. The limiter (10) detects both cases of the solution overshoot, as $\beta_i < 0$ for the location $(II)$ and $\beta_i > 1$ for the location $(III)$.

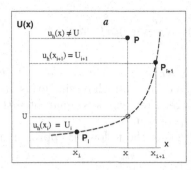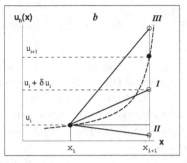

**Fig. 3.** Limiters for a piecewise linear approximate solution. (a) The limiter detects a solution overshoot at the interior point of the domain. (b) The limiter detects a solution overshoot at the cell interface. The "phantom" point (I) yields a monotone approximate solution. The "phantom" point (II) yields an overshoot which is indicated by $\beta_i < 0$. The "phantom" point (III) yields an overshoot which is indicated by $\beta_i > 1$.

We first illustrate the use of limiters by a simple numerical test discussed earlier in [8]. Consider the following linear boundary-value problem

$$u_x = S(x), \quad u(0) = U_0, \quad x \in \Omega = [0, 2], \tag{11}$$

so that flux function $F(x, u) \equiv u$. Let a discontinuous solution U(x) to the equation (11) be given by

$$U(x) = \begin{cases} 1. - \sqrt{0.5 - x} & 0 \leq x < 0.5, \\ 1 & 0.5 < x \leq 1, \\ tanh(200(x - 1.5)) & 1 < x \leq 2. \end{cases} \tag{12}$$

Given the solution (12), the source function $S(x)$ is reconstructed from the equation (11). The boundary condition is $u(0) = U(0)$.

For the advection equation the upwind numerical flux is $\tilde{F}(u_i, u_{i+1}) = u_i$, and we have $[u]_{\mathcal{P}} = [u]_i$ on any grid cell. The approximate solution obtained as a result of the DG discretization with a piecewise linear solution reconstruction is shown in Fig. 4a. The number of grid nodes is $N = 32$.

The DG scheme (4) employed in the problem generates oscillations near the shock. Those oscillations do not vanish on fine grids, so that the limiting is required to eliminate them. Thus, we compute the limiter (10) on each grid cell. In case that $\beta < 0$ or $\beta > 1$ ( a solution overshoot) the solution interpolation between two "frame" points is used to obtain a monotone approximate solution.

After the limiting procedure is applied, the new solution reconstruction is shown in Fig. 4b. It can be seen from the figure that the new solution has no overshoots over the domain.

We now consider the inviscid Burgers' equation

$$\frac{\partial u}{\partial t} + u \frac{\partial u}{\partial x} = 0. \tag{13}$$

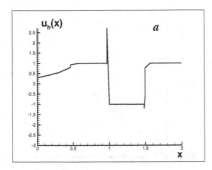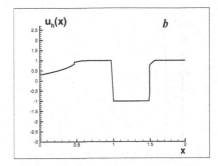

**Fig. 4.** The advection test problem (1) with a shock discontinuity. (a) Oscillations in the vicinity of the shock for the DG $K = 1$ solution. The number of grid nodes $N = 32$. (b) The solution on the same grid after the limiting procedure.

This is a well known example of a nonlinear hyperbolic equation with a quadratic flux function $F(u) = u^2/2$. We solve the equation (13) in the domain $x \in [0,1]$ due to a periodic boundary condition. The initial condition has been chosen as a sine wave function

$$u(x,0) = u_0(x) = 0.25 + 0.5\sin(\pi(2x - 1)). \qquad (14)$$

The exact solution is smooth for any time $t < 1/\pi$, while the solution becomes discontinuous at later times.

We are interested in the numerical solution to the problem (13), (14). A high order DG discretization is implemented, and we apply the Godunov flux $\tilde{F}^G(u_l, u_r)$ in order to discretize the function $F(u)$. The flux approximation is defined as

$$\tilde{F}^G(u_l, u_r) = \begin{cases} \min_{u_l \leq u \leq u_r} F(u), & \text{if } u_l \leq u_r, \\ \max_{u_l \leq u \leq u_r} F(u), & \text{otherwise}, \end{cases}$$

for the left state $u_l$ and right state $u_r$ at a given grid interface.

For numerical solution of the conservation law (13), a DG discretization in space is combined with an implicit time integration scheme. Let us notice that while the limiting procedure for the explicit Runge-Kutta integration has been introduced in the work [2] and further investigated in [5] and other works, limiters for implicit integration schemes have not been intensively discussed in literature. Thus we use a backward Euler time integration scheme in our problem to see how a suggested limiting algorithm will work for time dependent problems.

An approximate solution at $t = 0.47$ is shown in Fig. 5a for a piecewise linear DG discretization on a uniform grid of 128 cells. It can be seen from the figure that the approximate solution oscillates near the shock and we need to apply the limiting procedure (10) at each time step to obtain a non-oscillating solution. A new approximate solution is shown in Fig. 5b. The limiters eliminate spurious oscillations while remaining the solution piecewise linear in the vicinity of the shock. However, further numerical validation of the limiting procedure for

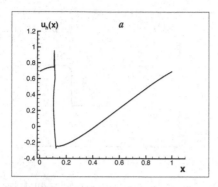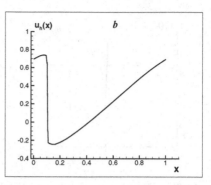

**Fig. 5.** The numerical solution to the inviscid Burgers' equation (13). (a) The approximate DG $K = 1$ solution to the problem (13), (14) oscillates near the shock soon after the shock formation ( time $t = 0.47$). The number of grid nodes $N = 128$. (b) The solution on the same grid after the limiting procedure.

nonlinear equations is required to confirm that the suggested algorithm keeps the order of approximation in a high order DG discretization scheme. That should be considered as a topic for future work.

# References

1. Burbeau, A. *et al.*: A Problem - Independent Limiter for High Order Runge-Kutta Discontinuous Galerkin Methods. J. Comput. Phys. **169** (2001) 111–150
2. B.Cockburn. *Discontinuous Galerkin Methods for Convection - Dominated Problems*, in High-Order Discretization Methods in Computational Physics, T.Barth and H.Deconinck, eds., Lecture Notes in Comput.Sci.Engrg., 9, Springer-Verlag, Heidelberg, 1999, pp.69-224.
3. Cockburn, B., Shu, C.: TVB Runge-Kutta Local Projection Discontinuous Galerkin Finite Element Method for Conservation Laws II: General Framework. Mathemat. Comput. **52** (1989) 411–435
4. Hirsh, C.: Numerical Computation of Internal and External Flows. John Wiley & Sons. (1990)
5. Hoteit, H. *et al.*: New Two-Dimensional Slope Limiters for Discontinuous Galerkin Methods on Arbitrary Meshes. INRIA Rennes. **4491** (2002)
6. Huynh, H.T.: Accurate upwind methods for the Euler Equations. AIAA 95-1737 (1995)
7. Lowrier, R.B.: Compact Higher-Order Numerical Methods for Hyperbolic Conservation Laws. PhD thesis, The University of Michigan. (1996)
8. Petrovskaya, N.B.: Two Types of Solution Overshoots in Discontinuous Galerkin Discretization Schemes. Comm. Math. Sci., **3(2)** (2005) 233–247
9. Petrovskaya, N.B.: On Oscillations in Discontinuous Galerkin Discretization Schemes for Steady State Problems. SIAM J. Sci. Comput. **27(4)** (2006) 1329–1346
10. Tu, S., Aliabadi, S.: A Slope Limiting Procedure in Discontinuous Galerkin Finite Element Method for Gasdynamic Applications. Int. J. Numer. Anal. Model. **2(2)** (2005) 163–178

# Numerical Analysis of the Sinuous Instability of a Viscous Capillary Jet Flowing Down an Immiscible Nonviscous Fluid

Stefan Radev[1], Kalin Nachev[1], Fabrice Onofri[2], Lounes Tadrist[2], and Asterios Liolios[3]

[1]Institute of Mechanics, BAS, 4, Acad. G. Bonchev Str., 1113 Sofia, Bulgaria
[2]Polytech'Marseille,University of Provence, 5, rue E. Fermi, Marseille 13453, France
[3]Dep. Civil Eng., Democritus University of Thrace, Vas. Sofias Str, GR-67 100 Xanthi, Greece

**Abstract.** The sinuous instability of a viscous jet flowing down an inviscid fluid is studied. On the basis of the 3D Navier-Sokes equations for the jet the full dispersion equation of the small disturbances is derived. Numerical results are shown, illustrating both the effect of viscosity and ambient density.

## 1 Introduction

As shown by Rayleigh [1] an isolated liquid jet is unstable to axisymmetrical disturbances only. However when the liquid jet interacts by surrounding immiscible fluid a sinusoidal mode of instability appears, that deflects the jet axis from its rectilinear form. It should be mentioned that from a theoretical point of view the sinuous instability is less analyzed. The first linear analysis of the instability of a nonviscous jet was performed by Weber [2]. Similar solution was proposed by Debye and Daen [3]. Both analyses are based on the Euler equations of motion written in cylindrical coordinates. Using the latter Martinon [4] studied higher non-symmetrical modes of instability. If the viscosity of the jet is to be taken into account the full 3D Navier-Stokes equations should be applied. In [5] the sinuous instability of a viscous jet was studied by using the so-called "quasi-one-dimensional equations" derived for thin jet as a reduced form of Navier-Stokes equations.

The aim of the present paper is to derive a dispersion equation for the sinusoidal disturbances propagating along a viscous jet, flowing down into another inviscid fluid. An asymptotic analysis of this equation is performed for small wave numbers as well as for low viscosity jets. Numerical results are shown, illustrating both the effect of viscosity and ambient density.

## 2 Statement of the Problem

Consider a steady axially symmetric liquid jet of density $\gamma$ and viscosity $\mu$ flowing down an inviscid immiscible fluid of density $\gamma_1$. The steady jet flow (referred

T. Boyanov et al. (Eds.): NMA 2006, LNCS 4310, pp. 677–684, 2007.

further as undisturbed) is assumed to have constant radius $a_0$ and uniform axial velocity $W$ with respect to a cylindrical coordinate system $O\rho\varphi z$, whose $Oz$ axis lies on the jet axis, directed downstream. The corresponding velocity of the outer fluid is denoted by $W_1$.

**Equations of Motion.** In cylindrical coordinates the equations of motion for a non-axisymmetrical jet could be written in the following form:

$$\frac{\partial v}{\partial t} + v \cdot \nabla v + X = -\frac{1}{\gamma}\nabla p + \frac{1}{\gamma}(\nabla \Pi + Y) , \qquad (2.1)$$

where $t$ denotes the time, $v(v, w, u)$ - the radial, transversal and axial component of the velocity, respectively, $p$ - the pressure and $\Pi(p_\rho, p_\varphi, p_z)$ - the well known tensor of the viscous stresses, $\nabla = \rho^0\frac{\partial}{\partial\rho} + \varphi^0\rho^{-1}\frac{\partial}{\partial\varphi} + k\frac{\partial}{\partial z}$ - the gradient operator with $\rho^0$, $\varphi^0$, $k$ as coordinate vectors. The additional symbols $X$ and $Y$ are defined as

$$X_\rho = -\rho^{-1}w^2, \quad X_\varphi = \rho^{-1}vw, \quad X_z = 0 , \qquad (2.2)$$

$$Y_\rho = \rho^{-1}(p_{\rho\rho} - p_{\varphi\varphi}), \quad Y_\varphi = 2\rho^{-1}p_{\rho\varphi}, \quad Y_z = \rho^{-1}p_{\rho z} . \qquad (2.3)$$

The external forces are neglected. The corresponding continuity equation has the form

$$\nabla \cdot v + \rho^{-1}v = 0 . \qquad (2.4)$$

**Interface Geometry.** The interface (jet surface) equation is written into a form allowing non-axisymmetrical disturbances

$$r = R_s = a(\varphi, z, t)\rho^0 , \qquad (2.5)$$

where $a$ denotes the cross-section radius. Let $N$, $t_\varphi$ and $t_z \equiv \psi$ be the outside normal and the tangential vectors to the interface along the corresponding co-ordinate lines.

Further on for convenience a new tangential vector $\phi$ will be used $\phi = \psi \times N$ assuring that the trihedron $(\phi, \psi, N)$ is orthonormal.

**Interface Boundary Conditions.** When written in a scalar form the force balance at the interface is reduced to three boundary conditions for the normal and two tangential stresses:

$$[[-p + (\Pi \cdot N) \cdot N]]_N = -\frac{\sigma}{R_m}, [[(\Pi \cdot N) \cdot \phi]]_N = 0, [[(\Pi \cdot N) \cdot \psi]]_N = 0 , \qquad (2.6)$$

where $\sigma$ denotes the surface tension coefficient, while $R_m$ - the mean curvature radius of the interface. Note that in the above equations the symbol $[[A]]_N$ denotes the (potential) discontinuity of an arbitrary parameter $A$ at the interface in the direction of the outside normal

$$[[A]]_N = A_2^* - A_1^*, \qquad A^* \equiv A(\rho = a) . \qquad (2.7)$$

Due to the immiscibility between the jet and surrounding fluid the zero-mass flux condition should be satisfied at both sides of the interface:

$$a\frac{\partial a}{\partial t} = av^* - \frac{\partial a}{\partial \varphi}w^* - a\frac{\partial a}{\partial z}u^* . \tag{2.8}$$

## 3  Linearized Equations of Motion and Boundary Conditions for the Disturbances

As mentioned above our stability analysis concerns a steady jet of a radius $a_0$ and axial velocity $W$ independent on both the radial and axial coordinates. Further on we will study the evolution of small disturbances (denoted by superscript tilde) imposed on the steady flow:

$$v = \tilde{v}, \quad w = \tilde{w}, \quad u = W + \tilde{u}, \quad p = P + \tilde{p}, \quad a = a_0 + \tilde{a}, \tag{3.1}$$

where $P$ denotes the constant undisturbed pressure.

Substituting the above expressions into eqs. (2.1) and (2.4) and neglecting the products of the disturbed terms results into linearized equations of motion of the following form:

$$B(\tilde{v}) = -\frac{1}{\gamma}\frac{\partial \tilde{p}}{\partial \rho} + \frac{\mu}{\gamma}\left[\Delta\tilde{v} - \rho^{-2}\left(\tilde{v} + 2\frac{\partial \tilde{w}}{\partial \varphi}\right)\right],$$

$$B(\tilde{w}) = -\frac{1}{\gamma}\rho^{-1}\frac{\partial \tilde{p}}{\partial \varphi} + \frac{\mu}{\gamma}\left[\Delta\tilde{w} - \rho^{-2}\left(\tilde{w} - 2\frac{\partial \tilde{v}}{\partial \varphi}\right)\right],$$

$$B(\tilde{u}) = -\frac{1}{\gamma}\frac{\partial \tilde{p}}{\partial z} + \frac{\mu}{\gamma}\Delta\tilde{u}, \quad B \equiv \frac{\partial}{\partial t} + W\frac{\partial}{\partial z}, \tag{3.2}$$

and

$$\rho^{-1}\frac{\partial}{\partial \rho}(\rho\tilde{v}) + \rho^{-1}\frac{\partial \tilde{w}}{\partial \varphi} + \frac{\partial \tilde{u}}{\partial z} = 0, \tag{3.3}$$

where $\mu$ denotes the viscosity, $\Delta$ is Laplace operator in cylindrical coordinates.

Similarly the boundary conditions (2.6) and (2.8) appear in the following linearized form:

$$\tilde{p}^* - \tilde{p}_1^* - 2\mu\left(\frac{\partial \tilde{v}}{\partial \rho}\right)^* = -\frac{\sigma}{a_0}\left(\frac{\tilde{a}}{a_0} + \frac{1}{a_0}\frac{\partial^2 \tilde{a}}{\partial \varphi^2} + a_0\frac{\partial^2 \tilde{a}}{\partial z^2}\right), \tag{3.4}$$

$$\left(\frac{\partial \tilde{w}}{\partial \rho}\right)^* + a_0^{-1}\left[\left(\frac{\partial \tilde{v}}{\partial \varphi}\right)^* - \tilde{w}^*\right] = 0, \quad \left(\frac{\partial \tilde{u}}{\partial \rho}\right)^* + \left(\frac{\partial \tilde{v}}{\partial z}\right)^* = 0, \quad B(\tilde{a}) = \tilde{v}^*, \tag{3.5}$$

where $A^*$ here and below on is used for the value $A(\rho = a_0)$.

## 4  Equations for the Amplitudes of the Disturbances

In the context of the linear theory of instability we will search a solution of the eqs. (3.2) - (3.3) and the corresponding boundary conditions (3.4) - (3.5) in the form of separate modes

$$\tilde{v} = \bar{v}e^{i\xi}, \quad \tilde{w} = \bar{w}e^{i\xi}, \quad \tilde{u} = \bar{u}e^{i\xi}, \quad \tilde{p} = \bar{p}e^{i\xi}, \quad \tilde{a} = \bar{a}e^{i\xi}, \tag{4.1}$$

where the corresponding amplitudes (except $\bar{a}=$const) are assumed unknown functions of the radial coordinate $\rho$. The new independent variable $\xi$ is defined as

$$\xi = \bar{\omega}t - kz + n\varphi, \qquad (4.2)$$

where $k$ is a given wave number, $n$ - a number of the asymmetric mode, when $n > 0$, while $\bar{\omega} = \bar{\omega}_r + i\bar{\omega}_i$ denotes the unknown complex angular frequency of the disturbances. Here and below on only the first asymmetrical (sinuous) mode will be analyzed, hence $n = 1$.

Substituting eqs. (4.1) into equations of motion of the jet (3.2) and continuity equation (3.3) results in the following system of ordinary differential equations:

$$ic_*\bar{v} = -\frac{1}{\gamma}\bar{p}' + \frac{\mu}{\gamma}\left[L_1\bar{v} - \rho^{-2}\left(\bar{v} + 2i\bar{w}\right)\right],$$

$$ic_*\bar{w} = -\frac{i}{\gamma}\rho^{-1}\bar{p} + \frac{\mu}{\gamma}\left[L_1\bar{w} + \rho^{-2}\left(2i\bar{v} - \bar{w}\right)\right], \ ic_*\bar{u} = \frac{ik}{\gamma}\bar{p} + \frac{\mu}{\gamma}L_1\bar{u} \qquad (4.3)$$

and

$$\bar{v}' + \rho^{-1}\left(\bar{v} + i\bar{w}\right) - ik\bar{u} = 0, \qquad (4.4)$$

where the superscript prime denotes a differentiation in respect to $\rho$ and the ordinary differential operator $L_1$ is defined as

$$L_1 \equiv \frac{d^2}{d\rho^2} + \rho^{-1}\frac{d}{d\rho} - \left(k^2 + \rho^{-2}\right) \qquad (4.5)$$

Additionally for convenience the following new unknown parameter is introduced

$$c_* \equiv \bar{\omega} - kW \qquad (4.6)$$

When the disturbances in the surrounding fluid are concerned the continuity equation remains unchanged, while the equations of motion are easily obtained taking into account in eqs. (4.3) that the fluid is nonviscous ($\mu = 0$):

$$ic_{1*}\bar{v}_1 = -\frac{1}{\gamma_1}\bar{p}_1', \ ic_{1*}\bar{w}_1 = -\frac{i}{\gamma_1}\rho^{-1}\bar{p}_1, \ ic_{1*}\bar{u}_1 = \frac{ik}{\gamma_1}, \qquad (4.7)$$

where now

$$c_{1*} = \bar{\omega} - kW_1. \qquad (4.8)$$

In fact the above equations relate the velocity amplitudes $\bar{v}_1$, $\bar{w}_1$ and $\bar{u}_1$ to the pressure amplitude $\bar{p}_1$ in the surrounding fluid. When the continuity equation is used to eliminate these amplitudes we simply obtain an equation for the surrounding pressure:

$$L_1\bar{p}_1 = 0, \qquad (4.9)$$

which corresponds to an irrotational flow in the surrounding fluid. In the next section we will use the fact that eq. (4.9) is easily transformed to a modified Bessel equation.

# 5  Analytical Solution for the Amplitudes. Dispersion Equation

As mentioned above the solution of eq. (4.9) could be written as

$$\bar{p}_1 = A_1 K_1(\zeta), \; \zeta = k\rho, \tag{5.1}$$

where $A_1$ is an integration constant, $K_1$ is a modified Bessel function of a first order and a second kind.

The solution of eqs. (4.3) - (4.4) can be splitted into an irrotational (nonviscous) and viscous part, respectively

$$\bar{v} = \bar{v}_{ir} + \bar{v}_v, \; \bar{w} = \bar{w}_{ir} + \bar{w}_v, \; \bar{u} = \bar{u}_{ir} + \bar{u}_v. \tag{5.2}$$

Similar to the surrounding flow the irrotational part of the jet flow satisfies pressure equation (4.9) in which the subscript "1" should be omitted. Therefore the irrotational velocity modes can again be expressed by the jet pressure:

$$\bar{p} = AI_1(k\rho), \tag{5.3}$$

The equations for the viscous velocity amplitudes are similar to eqs. (4.3) providing that the pressure terms are neglected:

$$ic_*\bar{v}_v = \frac{\mu}{\gamma}\left[L_1\bar{v}_v - \rho^{-2}(\bar{v}_v + 2i\bar{w}_v)\right],$$

$$ic_*\bar{w}_v = \frac{\mu}{\gamma}\left[L_1\bar{w}_v + \rho^{-2}(2i\bar{v}_v - \bar{w}_v)\right], \; ic_*\bar{u}_v = \frac{\mu}{\gamma}L_1\bar{u}_v. \tag{5.4}$$

It should be mentioned that the above system is self consistent: assuming that the velocity amplitudes in the left side of eqs. (5.4) satisfy the continuity equation (4.4) the same is true for the corresponding right sides. As seen the last of equations (5.4) is separated from the remaining equations and can be solved independently. By introducing a modified wave number $k_1$

$$k_1^2 = k^2 + i\frac{\gamma c_*}{\mu}, \tag{5.5}$$

the equation for $\bar{u}_v$ is transformed to a form similar to eq. (4.9), if the wave number $k$ is replaced by $k_1$

$$L_{1,1}\bar{u}_v = 0, \; L_{n,1} \equiv \frac{d^2}{d\rho^2} + \rho^{-1}\frac{d}{d\rho} - \left(k_1^2 + n\rho^{-2}\right). \tag{5.6}$$

Therefore

$$\bar{u}_v = BI_1(k_1\rho). \tag{5.7}$$

By introducing new unknowns $v_{\pm} = \bar{v}_v \pm \bar{w}_v$ the first two of eqs. (4.4) can be splitted into two independent modified Bessel equations of type (5.6) with left hand sides $L_{2,1}v_+$ and $L_{0,1}v_-$, respectively. As a result we obtain

$$\bar{v}_v = CI_2(k_1\rho) + DI_0(k_1\rho), \; \bar{w}_v = CI_2(k_1\rho) - DI_0(k_1\rho). \tag{5.8}$$

Substituting into the continuity equation the following relationship is derived for the constants of integration

$$B = -i\frac{k_1}{k}(C + D).$$    (5.9)

**Dispersion Equation.** By using eq. (5.9) and the interface boundary conditions (3.4) - (3.5), the remaining integration constants $A$, $A_1$, $C$, $D$, can be determined as functions of the (arbitrary) amplitude $\bar{a}$. After rather tedious, but straightforward calculations the dispersion equation can be written in the following dimensionless form:

$$\hat{c}_*^2 \frac{I_1}{qI_1'}\left(1 + \frac{\triangle_1}{\triangle}\hat{I}_2 + \frac{\triangle_2}{\triangle}\hat{I}_0\right) - 2iRe^{-1}\hat{c}_*\frac{qI_1''}{I_1'}\left(1 + \frac{\triangle_1}{\triangle}\hat{I}_2 + \frac{\triangle_2}{\triangle}\hat{I}_0\right)$$

$$+2iRe^{-1}\hat{c}_*q_1\left(\frac{\triangle_1}{\triangle}\hat{I}_2' + \frac{\triangle_2}{\triangle}\hat{I}_1\right) = We^{-1}q^2 + \hat{c}_{1*}^2\hat{\gamma}_1\frac{K_1}{qK_1'}.$$    (5.10)

where $Re = \gamma a_0 W/\mu$ and $We = \gamma a_0 W^2/\sigma$ are the Reynolds and Weber number, respectively, $\hat{\gamma}_1 = \gamma_1/\gamma$ - the density ratio and

$$\hat{c}_* = \omega, \quad \hat{c}_{1*} = \omega + q.$$    (5.11)

with $\omega = a_0\bar{\omega}/W$ -the dimensionless complex angular frequency, $q = a_0 k$ and $q_1 = a_0 k_1$ - the dimensionless wave and modified wave number, respectively. Note that the argument $\zeta = q$ in the above Bessel functions is omitted and denoted by superscript " $\wedge$ " when the argument equals $\zeta_1 = q_1$.

When writing eqs. (5.11) it is assumed that the reference coordinate system $O\rho\varphi z$ is moving with the undisturbed jet flow, while the surrounding fluid is flowing uniformly with a velocity $-W$ parallel to the undisturbed jet axis $Oz$. The symbols $\triangle$, $\triangle_1$, $\triangle_2$ denote second order determinants as follows:

$$\triangle = a_{22}a_{33} - a_{23}a_{32}, \triangle_1 = 2\left(ba_{22} - q^2a_{32}\right), \triangle_2 = 2\left(q^2a_{32} - ba_{23}\right),$$

$$a_{22} \equiv q_1^2\hat{I}_1' + q^2\hat{I}_2, \ a_{23} \equiv q_1^2\hat{I}_1' + q^2\hat{I}_0, \ a_{32} \equiv q_1\hat{I}_2' - 2\hat{I}_2, \ a_{33} \equiv -q_1\hat{I}_1, \ b \equiv \frac{I_1}{qI_1'} - 1.$$

The unknown in eq. (5.10) is the complex frequency $\omega = \omega_r + i\omega_i$, where the imaginary part $\omega_i$ stands for the growth rate of the disturbances and $\omega_r/q$ - for the speed of the propagation of the disturbances. The sinuous instability takes place (in our notations) when the growth rate is negative: $\omega_i < 0$. In the general case the dispersion equation can be solved only numerically. Some possible simplifications will be analyzed in the next section.

# 6    Reduced Forms of the Dispersion Equation

**Nonviscous Jet.** For $\mu = 0$ the dispersion equation is reduced to the form obtained by Debye and Daen [3] and by Martinon [4]:

$$\omega_r = \frac{\hat{\gamma}_1 qb_{11}}{1 - \hat{\gamma}_1 b_{11}}, \ \omega_i^2 = -\frac{q^2}{1 - \hat{\gamma}_1 b_{11}}\left(\frac{\hat{\gamma}_1 b_{11}}{1 - \hat{\gamma}_1 b_{11}} + We^{-1}\frac{qI_1'}{I_1}\right), b_{11} \equiv \frac{I_1'K_1}{I_1K_1'}.$$    (6.1)

**Small Wave Numbers.** Another reduced form of the dispersion equation can be derived if the Bessel functions are substituted by the leading terms in their asymptotic series for small arguments. Then

$$x^2 + 2x(\alpha - i\beta M_1) + \alpha - \delta = 0, \ x = \omega/q, \ M_1 = 2q, \tag{6.2}$$

where the new combinations of dimensionless parameters are defined as:

$$\alpha \equiv \frac{\hat{\gamma}_1}{1 + \hat{\gamma}_1}, \ \beta \equiv \frac{1}{(1 + \hat{\gamma}_1)\,Re}, \ \delta \equiv \frac{1}{(1 + \hat{\gamma}_1)\,We}. \tag{6.3}$$

By solving the quadratic equation (6.2) we get:

$$-\omega_i = \left(\sqrt{s} - \beta M_1\right)q, \ \omega_i = -\alpha\left(1 - \frac{\beta M_1}{\sqrt{s}}\right), \ s \equiv \alpha - \alpha^2 - \delta. \tag{6.4}$$

**Low Viscosity Jets**

$$\omega^2(b+1) - (\omega+q)^2\frac{\hat{\gamma}_1 K_1}{qK_1'} - We^{-1}q^2 = 2iRe^{-1}\omega M(q), \ M(q) = \frac{qI_1''}{I_1'} - (b+1)\left(b - q^2\right). \tag{6.5}$$

# 7   Numerical Results for the Reduced Forms of the Dispersion Equation

The numerical results in Figs. 1 and 2 illustrate the dispersion curves based on eqs. (6.1) and (6.5). The results in Fig. 1a) correspond to the case of water jet into air, when $We = 1000$ and $\gamma_1 \approx 0.001$. In this figure the sinuous mode for the jet assumed nonviscous is compared to the symmetrical mode ($n = 0$) as known after Rayleigh [1]. The effect of the air viscosity ($Re = 1000$) on the sinuous mode is shown in Fig. 2a). The effect of the increased ambient density $\gamma_1 = 0.01$ is shown in Figs. 1b) and 2b) in the conditions similar to that of Figs. 1a) and 2a), respectively.

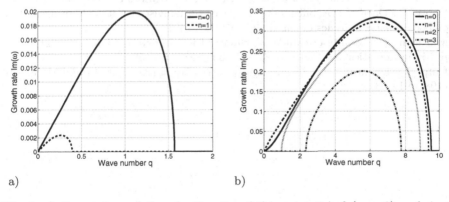

a)                                                                 b)

**Fig. 1.** a) Comparison of the growth rates of the symmetrical ($n = 0$) and sinuous mode ($n = 1$) for water jet in air; b) Effect of the ambient density on the growth rates of the symmetrical, sinuous and higher asymmetrical modes

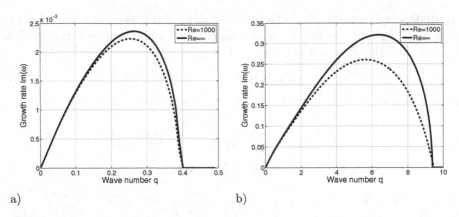

a)                                        b)

**Fig. 2.** a) Effect of the air viscosity on the growth rates of the sinuous mode; b) Combined effect of the viscosity and ambient density on the growth rate of the sinuous mode

## 8   Conclusion

The sinuous instability of a capillary jet appears at high Weber numbers (high injection velocities). It is mainly controlled by the density ratio of the fluids. Both the capillary and viscous forces have a stabilizing effect.

**Acknowledgments.** The present paper is partially supported by the Grant No VU-MI-02/2005 of the National Scientific Fund of Bulgaria.

## References

1. Rayleigh, J.S.W.: On the Instability of Jets. Proc. Lond. Math. Soc. **10** (1878) 4–13
2. Weber, K.: Zum Zerfall eines Flussigkeitsstrahles. ZAMM **11** (1931) 136–154
3. Debye, P., Daen, J.: Stability considerations on nonviscous jets exhibiting surface or body tension. Physics of Fluids **2** (1959) 416–421
4. Martinon, J.: Stability of capillary cylindrical jets in external flow. Journal of Applied Mathematics and Physics (ZAMP) **34** (1983) 575–582.
5. Entov, V.M., Yarin, A.L.: The dynamics of thin liquid jets in air. J.Fluid Mech. **140** (1984) 91–111

# Order Adaptive Quadrature Rule
# for Real Time Holography Applications

Minvydas Ragulskis and Loreta Saunoriene

Kaunas University of Technology
Department of Mathematical Research in Systems
Studentu st. 50-222, Kaunas LT-51638, Lithuania
minvydas.ragulskis@ktu.lt, loreta.saunoriene@ktu.lt

**Abstract.** Order adaptive algorithm for real time holography applications is presented in this paper. The algorithm is based on Master-Worker parallel computation paradigm. Definite integrals required for visualization of fringes are computed using a novel order adaptive quadrature rule with an external detector defining the order of integration in real time mode. The proposed integration technique can be effectively applied in hybrid numerical-experimental techniques for analysis of micro-mechanical components.

**Keywords:** Order adaptability, real time integration, holography.

## 1 Introduction

Holographic interferometry [1] is a powerful experimental technique for analysis of structural vibrations, especially if the amplitudes of those vibrations are in the range of micrometers. Recent advancements in optical measurement technology [2] and development of hybrid numerical-experimental techniques [3] require application of computational algorithms not only for post-processing applications like interpretation of experimental patterns of fringes, but embedding real time algorithms into the measurement process itself [4].

Computation and plotting of patterns of time average holographic fringes in virtual numerical environments involves such tasks as modelling of the optical measurement setup, geometrical and physical characteristics of the investigated structure and the dynamic response of the analysed system [5]. Calculation of intensity of illumination at any point on the hologram plane requires computation of definite integrals over the exposure time. If the analysed structures perform harmonic oscillations that do not impose any complications – there exist even analytical relationships between the intensity of illumination, amplitude of oscillation, laser wavelength, etc. [1]. But if the oscillations of the investigated structures are non-harmonic (what is common when structures are nonlinear) and the formation of patterns of fringes is implemented in real time mode, the calculation of definite integrals becomes rather problematic. The object of this paper is to propose an order adaptive algorithm which could be effectively applicable for calculation of definite integrals in different real time applications.

T. Boyanov et al. (Eds.): NMA 2006, LNCS 4310, pp. 685–692, 2007.
© Springer-Verlag Berlin Heidelberg 2007

## 2  Integration Rule Without Limitation for the Number of Nodes

Higher order Newton-Cotes quadrature formulas [6] require that the number of nodes must be a divisible numeral. For example the second order Newton-Cotes rule already requires that the number of nodes must be odd. Such conditions mean that a significant number of nodes at the end of an experimental time series must be deleted and the integration interval artificially shortened for higher order Newton-Cotes rule, if the number of nodes is not known at the beginning of the experiment. Therefore there exists a definite need for a high order integration rule with a constant time step without any requirement for the number of time steps. Such quadrature formula is proposed in [7]:

$$
\int_{t_0}^{t_0+(k-1)h} f(t)\, dt = \left( \sum_{i=1}^{m} a_i f_i + \sum_{i=1}^{k-2m} f_{m+i} + \sum_{i=1}^{m} a_{m-i+1} f_{k-m+i} \right) h, \qquad (1)
$$

where $a_i$ are the weights and $f_i$ are the discrete values of sampled function $f$ at time moments $t_0 + (i-1) \cdot h$, $i = 1, \ldots, k$. It has been proved that this integration rule is exact when the integrated function is a polynomial of the $m$-th order, if only $m$ is odd [7]. The numerical values of the weights $a_i$ are presented in the Table 1 at different values of $m$. The parameter $p$ in this table denotes the maximum order of exactly integrated polynomials; $l$ is the order of the error term expressed in the form $O(h^l)$.

**Table 1.** Nodal weights of the integration rule

| $m$ | 2 | 3 | 4 | 5 | 6 | 7 |
|---|---|---|---|---|---|---|
| $a_1$ | 0.5 | 0.37500000 | 0.33333333 | 0.32986111 | 0.31875000 | 0.30422454 |
| $a_2$ | 1 | 1.1666667 | 1.2916667 | 1.3208333 | 1.3763889 | 1.4603836 |
| $a_3$ | | 0.95833333 | 0.83333333 | 0.76666667 | 0.65555556 | 0.45346396 |
| $a_4$ | | | 1.0416667 | 1.1013889 | 1.2125000 | 1.4714286 |
| $a_5$ | | | | 0.98125000 | 0.92569444 | 0.73939319 |
| $a_6$ | | | | | 1.0111111 | 1.0824735 |
| $a_7$ | | | | | | 0.98863261 |
| $p$ | 1 | 3 | 3 | 5 | 5 | 7 |
| $l$ | 2 | 4 | 4 | 6 | 6 | 8 |

It can be noted that finite element method was used for the derivation of the proposed quadrature rule which can be interpreted as a new variant of Gregory type formulas [6]. Unfortunately, the proposed quadrature rule (also Gregory type rules) can be used only when the order is predefined before the experiment and does not change over the integration process. This paper proposes a multi-processor parallel algorithm with full order adaptability in real time calculation mode.

## 3    The Basic Real Time Integration Rule

Let's suppose that function $f$ is sampled starting from $t_0$ at equally spaced time steps; the length of a time step is $h$. Due to the real time process the number of nodes is not predefined before the experiment and process continues until the end of the sampling. Let's suppose that the terminal moment of the sampling occurs at $t_0 + 7h$ (8 function values $f_i$, $i = 1, \ldots, 8$ are produced during the sampling process). Order of the integration rule is predetermined to be $m = 3$.

1. The first sum on the right side of eq. (1) is computed:

$$Sum_1 = a_1 f_1 + a_2 f_2 + a_3 f_3, \tag{2}$$

   where $a_1 = 0.375, a_2 = 1.1666667, a_3 = 0.95833333$ (Table 1).
2. Starting from the fourth node, the following sum is computed until the end of the time series:

$$Sum_2 = f_4 + f_5 + f_6 + f_7 + f_8. \tag{3}$$

3. When the sampling is terminated, reverse computation of the third sum of eq. (1) is done:

$$Sum_3 = (a_3 - 1) f_6 + (a_2 - 1) f_7 + (a_1 - 1) f_8. \tag{4}$$

4. Finally, the definite integral $\int_{t_0}^{t_0+7h} f(t) \, dt$ is calculated according to eq. (1):

$$I = (Sum_1 + Sum_2 + Sum_3) h. \tag{5}$$

The process can terminate at any time step, if only $k \geq 2m$, but the last three values of the sampled function must be saved at every time moment in order to calculate $Sum_3$.

Now we will generalize the presented example for $m$-th order integration rule, if only the minimum number of nodes is $2m$. The algorithm is based on Master-Worker paradigm [8]. Schematic graphical representation in Fig. 1 helps to interpret the computation process.

Several notations used in Fig. 1 can be explained in more detail. Order of the integration rule $m$ is predefined before the experiment. Calculation of $Sum_1$ is performed by Master processor (grey right arrow in signal diagram; block $n$ in time diagram and node $n$ in flow chart diagram). After $m$ terms are included into $Sum_1$, the Master processor continues summation of nodal values of the integrand until the sampling process is terminated (white right arrow in signal diagram; block $n^{(1)}$ in time diagram and node $n^{(1)}$ in flow chart diagram). When the sampling is over, Worker processor performs reverse calculation of $Sum_3$ (grey left arrow in signal diagram; block $\bar{n}$ in time diagram and node $\bar{n}$ in flow chart diagram).

It can be noted that the last $m$ values of the sampled function must be remembered at every time node in order to calculate $Sum_3$.

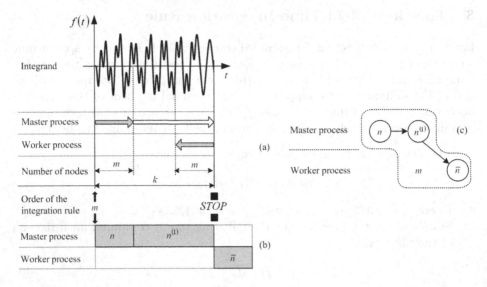

**Fig. 1.** Schematic representation of the basic model: (a) signal diagram; (b) time diagram; (c) flow chart diagram

## 4    Order Adaptive Algorithm for Real Time Applications

The presented basic real time integration rule copes well with integrands which can be approximated by a polynomial of a definite order in the domain of integration. But if the variation of the integrand is fast in some regions and slow in another regions, then order adaptability should be used to increase the accuracy of a definite integral. One can suggest to select very large $m$ at the beginning of the experiment, but then we may face the risk that $k < 2m$.

We assume that there exists a detector which measures the values of the integrand and recommends the order of integration rule at any time moment in the domain of integration. Let's assume that the present order is $m_1$ and the detector recommends order $m_2$. Then two different situations may occur. If the number of sampled nodes since $m_1$ was declared is higher than or equal to $2m_1$, the transition to order $m_2$ can be performed fluently. The Master processor starts calculating $Sum_1$ for order $m_2$, while Worker processor takes care for reverse calculation of $Sum_3$ for terminated $m_1$.

But if the number of sampled nodes since $m_1$ was declared is less than $2m_1$, the Worker processor cannot start reverse calculation of $Sum_3$ without damaging $Sum_1$. Therefore a much more complex transition to order $m_2$ takes place in this situation. If $m_2$ is higher than $m_1$ the Worker processor must return to the point where order $m_1$ was declared and must recalculate $Sum_1$ with order $m_2$. But the simplicity is misleading – the Master processor has already summated $Sum_1$ with order $m_1$ to the total sum! Therefore the Worker processor must evaluate different weighting coefficients for orders $m_1$ and $m_2$. Moreover, the length of the queue where the last function values are stored must be already not $m_i$, but $2m_i$ (here $m_i$ is the current order).

If $m_2$ is lower than $m_1$, but the number of sampled nodes since $m_1$ was declared is less than $2m_1$, the integration with order $m_1$ must be continued until the number of nodes is equal to $2m_1$, and only after that the order $m_2$ can be accepted.

Finally, we may comment what would happen if the sampling process is terminated and the number of sampled nodes since $m_i$ was declared is lower than $2m_i$. Unfortunately, there will be no any possible techniques to preserve order $m_i$ (time step is constant and reverse sampling with smaller time step is impossible in real time mode). The only solution is to select maximum possible order for the available number of time steps (floored half of the number of time steps).

We will illustrate the described situations with the following example (Fig. 2 and Fig. 3).

One Master processor and two Worker processors are necessary for full real time mode. Algorithm control, integrand sampling and summation of sums $Sum_1$ and $Sum_2$ is performed by Master processor. The Worker processors run only when the order is changed. Worker processors send back the results to the Master processor.

As an extreme situation we describe the transition from order $m_3$ to order $m_4$ (Fig. 2) where the second Worker processor is necessary for real time integration. Master processor starts calculating $Sum_1$ (with weights corresponding to order $m_3$) as soon as the order $m_3$ is declared. The number of discrete time nodes

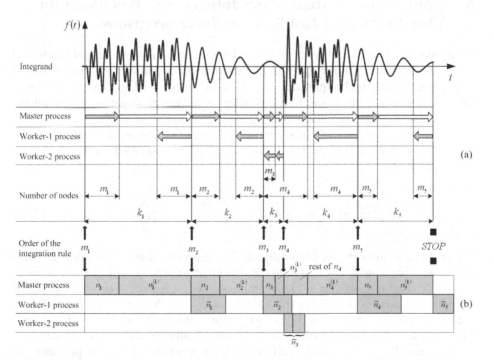

**Fig. 2.** Real time integration, general case: (a) signal diagram; (b) time diagram

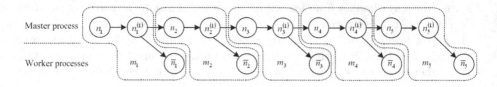

Master process

Worker processes

**Fig. 3.** Real time integration, general case: flow chart diagram

necessary for this procedure is $m_3$. As soon as $Sum_1$ is finished, Master processor starts summing non-weighted discrete function values. This process continues until order $m_4$ is declared. But the order detector has sensed a burst in the digital time series, so $m_4$ is much higher than $m_3$. In this particular situation we have that $m_4$ is even greater than $k_3$ (Fig. 2). Thus, the Worker processor must recalculate both the old $Sum_1$ and the rest non-weighted part ($n_3$ and $n_3^{(1)}$ in time diagram). Moreover, at the same time $Sum_3$ for order $m_2$ must be accomplished ($\bar{n}_2$ in time diagram). Thus Worker-2 processor is unavoidable for real time computation ($\bar{n}_2$ and $\bar{n}_3$ overlap in time diagram). The mathematical formulas for processes $\bar{n}_2$ and $\bar{n}_3$ (consisting from two parts) can be described explicitly in the text of the general algorithm which we omit due to the restrictions for the size of the manuscript.

## 5    Application of Real Time Integration Technique for Visualization of Holographic Interferograms

Computational visualization of holographic interferograms in virtual numerical environments is an important component of hybrid numerical – experimental techniques. These techniques are of crucial importance when the analysed systems perform non-harmonic motions what is a typical situation when micromechanical systems are considered [9].

Whenever a pattern of time average holographic fringes is considered, the intensity of illumination at the hologram plane is described by the following relationship [1]:

$$I(x, y) = \lim_{T \to \infty} \frac{1}{T^2} \left| \int_0^T \exp\left( j \frac{2\pi}{\lambda} \zeta(x, y, t) \right) dt \right|^2, \qquad (6)$$

where $I$ is the intensity of illumination; $T$ – exposure time; $j$ – imaginary unit; $\lambda$ – laser wave length; $\zeta$ – dynamic displacement at point $(x, y)$ at time moment $t$. Usually the function $\zeta(x, y, t)$ is decomposed to a product of time function and coordinate function describing the modal shape. It is clear that accurate computation of definite integral in eq. (6) for finite exposure times is associated with the accuracy of pattern of fringes in the numerically reconstructed hologram.

Dynamic displacements of cantilevered micromechanical bar are presented in Fig. 4(a). Scanning laser is measuring the displacements at the marked nodes

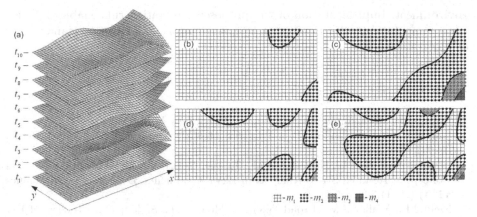

**Fig. 4.** Dynamic displacements of cantilevered micromechanical bar: (a) finite element shapes at different time moments; (b), (c), (d) and (e) – nodal orders of integration at different time moments

**Fig. 5.** Numerically reconstructed pattern of fringes

at discrete time moments. Intensity of illumination in the hologram plane is calculated at every node, so definite integrals are calculated at every node. The system is checking the magnitude of dynamic displacement at every node and generates the recommended order of integration which is based on the absolute value of discrete displacement at appropriate node. Figures 4(b), 4(c), 4(d) and 4(e) present the recommended orders of integration at different time moments; where $m_1 = 3$, $m_2 = 5$, $m_3 = 7$, $m_4 = 9$.

Figure 5 presents the produced time average holographic pattern of fringes.

## 6   Concluding Remarks

The presented procedure for real time calculation of definite integrals can be effectively applied in hybrid numerical-experimental techniques where time average intensities of illumination are reconstructed in virtual computational environment. Implementation of the proposed integration rule enables full real time computations with minimal data queue lengths and effective management of integration order.

# References

1. West, C.M.: Holographic Interferometry. Wiley New York (1979)
2. Kobayashi, A.S.: Handbook on Experimental Mechanics - 2nd ed. SEM Bethel (1993)
3. Holstein A., Salbut L., Kujawinska M., Juptner W.: Hybrid Experimental-numerical Concept of Residual Stress Analysis in Laser Weldments. Experimental Mechanics **41(4)** (2001) 343–350
4. Field J.E., Walley S.M., Proud W.G., Goldrein H.T., Siviour C.R.: Review of Experimental Techniques for High Rate Deformation and Shock Studies. International Journal of Impact Engineering **30(7)** (2004) 725–775
5. Ragulskis M., Palevicius A., Ragulskis L.: Plotting Holographic Interferograms for Visualization of Dynamic Results from Finite-element Calculations. International Journal of Numerical Methods in Engineering **56** (2003) 1647–1659
6. Davis P.J., Rabinowitz P.: Methods of Numerical Integration. Academic Press New York (1984)
7. Ragulskis M., Ragulskis L.: Order Adaptive Integration Rule with Equivalently Weighted Internal Nodes. Engineering Computations **23(4)** (2006) 368–381
8. Mattson T., Sanders B., Massingill B.: Patterns for Parallel Programming. Addison Wesley Professional (2004)
9. Nayfeh A.H., Younis M.I., Abdel-Rahman E.M.: Reduced-order Models for MEMS Applications. Nonlinear Dynamics **41** (2005) 211–236

# FEM Modelling of Resonant Frequencies of In–Plane Parallel Piezoelectric Resonator

Petr Rálek

Department of Modelling of Processes,
Faculty of Mechatronics and Interdisciplinary Engineering Studies,
Technical University of Liberec, Czech Republic
petr.ralek@tul.cz

**Abstract.** In the contribution, we introduce an application of finite element model of the piezoelectric resonator. The model is based on the physical description of piezoelectric materials, using linear piezoelectric state equations. Weak formulation and discretization of the problem lead to a large and sparse linear algebraic system, which results in a generalized eigenvalue problem. Resonant frequencies, the most important parameters of the resonator, are subsequently found by solving this algebraic problem. Depending on the discretization parameters, this problem may become large.

Typically, we are not interested in all eigenvalues (resonant frequencies). For determination of several of them it seems therefore appropriate to consider the Krylov subspace methods (namely the implicitly restarted Arnoldi method implemented in the ARPACK library). For coarser meshes, we compute the complete spectra and we find the frequencies of dominant oscillation modes (the selection is made according to their electromechanical coupling coefficients). Then we focus on the part of the spectra near to the chosen dominant frequency and repeat the computation for refined meshes. From the results, we can also find out intervals between the dominant resonant frequencies (which is other important parameter describing the behavior of the resonator).

The model was tested on the problem of thickness-shear vibration of the in–plane parallel quartz resonator. The results, compared with the measurement, will be given in the contribution.

## 1 Physical Description

We briefly sketch the physical properties of the piezoelectric materials. For more detailed description (including more references), see e.g. [4].

A crystal made of piezoelectric material represents a structure, in which the deformation and the electric field depend on each other. A deformation (impaction) of the crystal induces an electric charge on the crystal's surface. On the other hand, subjecting a crystal to electric field causes its deformation. In the linear theory of piezoelectricity, derived by Tiersten in [5], this process is described by two constitutive equations - the **generalized Hook's law** (1) and the **equation of the direct piezoelectric effect** (2),

T. Boyanov et al. (Eds.): NMA 2006, LNCS 4310, pp. 693–700, 2007.

$$T_{ij} = c_{ijkl}\, S_{kl} - d_{kij}\, E_k, \qquad i, j = 1, 2, 3, \tag{1}$$

$$D_k = d_{kij}\, S_{ij} + \varepsilon_{kj}\, E_j, \qquad k = 1, 2, 3. \tag{2}$$

Here, as in the other similar terms throughout the paper, we use the Einstein's additive rule. The Hook's law (1) describes relation between the symmetric **stress tensor T**, the symmetric **strain tensor S** and the **vector of the intensity of the electric field E**,

$$S_{ij} = \frac{1}{2}\left[\frac{\partial \tilde{u}_i}{\partial x_j} + \frac{\partial \tilde{u}_j}{\partial x_i}\right], \qquad i, j = 1, 2, 3, \qquad E_k = -\frac{\partial \tilde{\varphi}}{\partial x_k}, \qquad k = 1, 2, 3,$$

where $\tilde{u} = (\tilde{u}_1, \tilde{u}_2, \tilde{u}_3)^{\mathrm{T}}$ is the **displacement vector** and $\tilde{\varphi}$ is the **electric potential**. The equation of the direct piezoelectric effect (2) describes dependence between the **vector of the electric displacement D**, the strain and the intensity of the electric field. Quantities $c_{ijkl}$, $d_{kij}$ and $\varepsilon_{ij}$ represent symmetric material tensors, playing role of material constants. Additionally, tensors $c_{ijkl}$ and $\varepsilon_{ij}$ are positive definite.

## 1.1 Oscillation of the Piezoelectric Continuum

Consider a resonator made of piezoelectric material with density $\varrho$, characterized by material tensors. We denote the volume of the resonator as $\Omega$ and its boundary as $\Gamma$. Behavior of the piezoelectric continuum is governed, in some time interval $(0, T)$, by two differential equations: Newton's law of motion (3) and the quasi-static approximation of Maxwell's equation (4) (see, e.g., [3]),

$$\varrho\frac{\partial^2 \tilde{u}_i}{\partial t^2} = \frac{\partial T_{ij}}{\partial x_j} \qquad i = 1, 2, 3, \qquad x \in \Omega, \quad t \in (0, T), \tag{3}$$

$$\nabla \cdot D = \frac{\partial D_j}{\partial x_j} = 0. \tag{4}$$

Replacement of **T**, resp. **D** in (3) and (4) with the expressions (1), resp. (2), gives

$$\varrho\frac{\partial^2 \tilde{u}_i}{\partial t^2}\frac{\partial}{\partial x_j}\left[c_{ijkl}\frac{1}{2}\left(\frac{\partial \tilde{u}_k}{\partial x_l} + \frac{\partial \tilde{u}_l}{\partial x_k}\right) + d_{kij}\frac{\partial \tilde{\varphi}}{\partial x_k}\right] \qquad i = 1, 2, 3, \tag{5}$$

$$0 = \frac{\partial}{\partial x_k}\left[d_{kij}\frac{1}{2}\left(\frac{\partial \tilde{u}_i}{\partial x_j} + \frac{\partial \tilde{u}_j}{\partial x_i}\right) - \varepsilon_{kj}\frac{\partial \tilde{\varphi}}{\partial x_j}\right]. \tag{6}$$

The initial conditions, the Dirichlet boundary conditions and the Neumann boundary conditions are added:

$$\tilde{u}_i(\cdot, 0) = u_i, \qquad x \in \Omega, \tag{7}$$

$$\tilde{u}_i = 0, \qquad i = 1, 2, 3, \quad x \in \Gamma_u,$$

$$T_{ij} n_j = f_i, \qquad i = 1, 2, 3, \quad x \in \Gamma_f,$$

$$\tilde{\varphi}(\cdot, 0) = \varphi,$$

$$\tilde{\varphi} = \varphi_D, \qquad x \in \Gamma_\varphi$$

$$D_k n_k = q, \qquad x \in \Gamma_q,$$

where
$$\Gamma_u \cup \Gamma_f = \Gamma, \ \Gamma_u \cap \Gamma_f = \emptyset, \ \Gamma_\varphi \cup \Gamma_q = \Gamma, \ \Gamma_\varphi \cap \Gamma_q = \emptyset.$$

The right-hand side $f_i$ represents the mechanical excitation by the external mechanical forces, $q$ denotes the electrical excitation by imposing the surface charge (in the case of free oscillations, they are both zero). Equations (5)-(6) define the problem of harmonic oscillation of the piezoelectric continuum under given conditions (7).

## 2   Numerical Solution

### 2.1   Weak Formulation and Discretization

We derive the weak formulation in the standard way, using the Green formula and the boundary conditions. We discretize the problem in spatial variables, using the tetrahedron finite elements with linear base functions. The process of weak formulation and discretization is explained in more details, e.g., in [4]. The system of ordinary differential equations for nodal values of the displacement and the potential results. It has a block structure,

$$M\ddot{U} + KU + P^T\Phi = F, \tag{8}$$
$$PU - E\Phi = Q. \tag{9}$$

After introducing the Dirichlet boundary conditions (see Fig. 1), sub-matrices M, K and E are symmetric and positive definite.

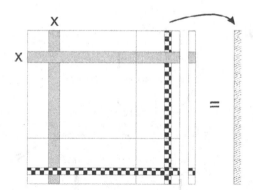

**Fig. 1.** Introduction of the Dirichlet boundary conditions

### 2.2   Generalized Eigenvalue Problem

The core of the behavior of the oscillating piezoelectric continuum lies in its free oscillation. Free oscillations (and computed eigenfrequencies) tell, when the system, under external excitation, can get to resonance. For free harmonic oscillation, the system (8) can be transformed to

$$\begin{pmatrix} K - \omega^2 M & P^T \\ P & -E \end{pmatrix} \begin{pmatrix} U \\ \Phi \end{pmatrix} \begin{pmatrix} 0 \\ 0 \end{pmatrix}, \tag{10}$$

where $\omega$ is the circular frequency of oscillation. Eigenfrequencies can be computed by solving the generalized eigenvalue problem

$$AX = \lambda BX \tag{11}$$

with

$$A = \begin{pmatrix} K & P^T \\ P & -E \end{pmatrix}, \ B = \begin{pmatrix} M & 0 \\ 0 & 0 \end{pmatrix}, \ X = \begin{pmatrix} U \\ \Phi \end{pmatrix}, \ \lambda = \omega^2,$$

where A is symmetric and B is symmetric and positive semi-definite matrix. Resonant frequency $f$ can be computed from the relation $\omega = 2\pi f$. Computed eigenvectors (namely their component U) describe the modes of oscillations. For solving the generalized eigenvalue problem (11), we use implicitly restarted Arnoldi method implemented in ARPACK library [6] (in FORTRAN language). Inner steps in the method make use of direct solver from SKYPACK library

**Table 1.** Comparison between measured and computed dominant resonant frequencies (in the shear vibrational mode)

| Resonance frequency (kHz) | Measured | Computed |
|---|---|---|
| basic ($f_1$) | 4000 | 3696.785 |
| harmonic | 1.017 $f_1$ | 1.014 $f_1$ |
| | 1.03 $f_1$ | 1.04 $f_1$ |

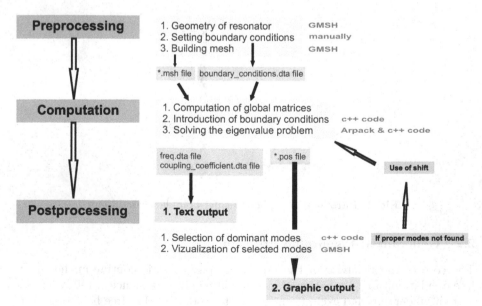

**Fig. 2.** Scheme of computer implementation of the model and its stages

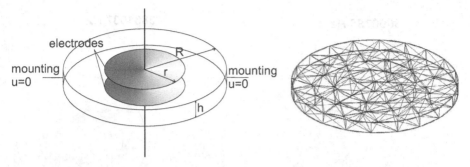

**Fig. 3.** Geometry and discretization of plan-parallel resonator

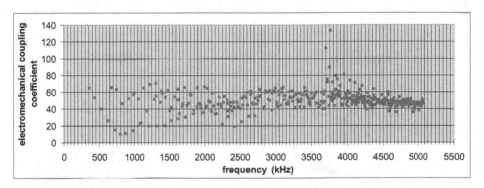

**Fig. 4.** Graph of the electromechanical coupling coefficients

[8] for solving the symmetric indefinite linear systems. The whole method is suitable for solving partial eigenvalue problem (computing of several eigenvalues with high precision) with possibility of the shift and it allows to deal with the sparsity of the matrices. Using of the shift enables to obtain the eigenvalues from the desired part of the spectrum and with good accuracy (if we compute only few eigenvalues from the desired area).

## 3   Practical Problem – Oscillation of the In–Plane Parallel Quartz Resonator

The model was applied on the problem of oscillation of the in–plane parallel quartz resonator $AT_{35°11'30''}$ (Fig. 3) in shear vibration mode in $x$–direction. The dimensional parameters for the resonator are

$$R = 7 \text{ mm}, \ r = 3.5 \text{ mm}, \ h = 0.3355 \text{ mm}.$$

These resonators are manufactured and their behavior is well–known. The objective was to find the dominant resonant frequency and its harmonic frequencies. The comparison between the computed results and the measurements is shown in Table 1. Figure 5 shows the visualization of the computed oscillation modes.

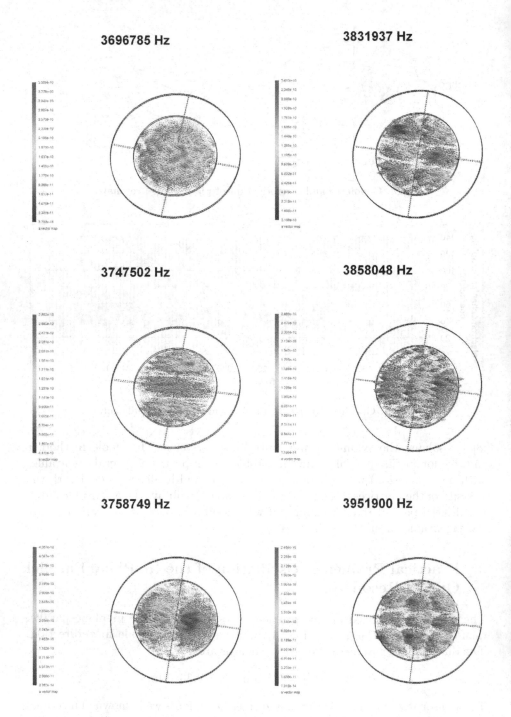

**Fig. 5.** The dominant oscillation modes, lying near to 4 MHz

## 3.1   Computer Implementation

The realization of the model consists of three parts: preprocessing, processing and postprocessing (Fig. 2). In pre- and postprocessing parts, we use the free software GMSH [7] for mesh generation and visualization of the results. For building the global matrices, we developed our own code, written in C++ language. For solving the generalized eigenvalue problem, we use the ARPACK [6] implementation of the implicitly shifted Arnoldi method (all parts are debugged under Windovs XP).

The preprocessing part consists of building the geometry (according to the engineering assignments) and the mesh of the resonator (using GMSH), see Fig. 3. The processing part computes the global matrices and the consecutive eigenvalue problem (using text file with parameters – accuracy, number of computed eigenvalues, shift, etc.). It gives several output files, which are used in the postprocessing. Computed eigenvalues and eigenvectors define the oscillation modes, which are sorted according to their *electromechanical coupling coefficients*. The electromechanical coupling coefficient $k$ is defined by the relation [2]

$$k^2 = \frac{E_m^2}{E_{st} E_d},$$

where

$$E_m = \frac{1}{2} \left( U^T P \Phi \right)$$

is the mutual energy,

$$E_{st} = \frac{1}{2} \left( U^T K U \right)$$

is the elastic energy and

$$E_d = \frac{1}{2} \left( \Phi^T E \Phi \right)$$

is the dielectric energy. The higher the value of $k$ the better the possibility of excitation of the oscillation mode. Figure 4 shows the graph of the coefficients $k$ for a part of the spectra from 0 MHz up to about 5 MHz and the selection of the modes with the highest coefficients $k$.

## 4   Conclusion

The presented mathematical model gives suitable results for the testing problems. It uses methods of numerical linear algebra for solving the partial generalized eigenvalue problem with possibility of shift mode, which allows to compute eigenfrequencies in the neighborhood of the desired value. The restarted Arnoldi methods looks pretty effective for this problem. It is suitable for solving larger problems originated by the discretization of more complicated shapes of resonators. The difference between the calculated and the measured results can be caused by several reasons – mainly in the mathematical formulation, in the use of the simple, linear piezoelectric state equations; in the process of numerical solution, it is the case of rounding errors during the computation (both in discretization and solving the eigenvalue problem of large dimension). The difference between the basic frequencies is rather large (around 9%, which can be caused by the reasons mentioned above), but the relative ratio between the basic frequency and its harmonic frequencies is computed with much small error (about 1%) and after calibration, the model can bring reasonable results.

Nowadays, the next step to do is computation of the graphs of dependance of certain resonance frequency on the geometrical characteristic of the resonator and also the distance of carrier resonance frequency from the spurious frequencies. It still remains as a hard task to improve the postprocessing part of the program for classification of the computed oscillation modes - mainly according to the graphs of amplitudes in several sections of the resonator volume.

# References

1. Allik, H., Hughes, T.J.R.: Finite element method for piezoelectric vibration. International journal for numerical methods in engineering **2** (1970) 151–157
2. Lerch, R.: Simulation of Piezoelectric Devices by Two- and Three-Dimensional Finite Elements. IEEE Trans. on Ultrason., Ferroel. and Frequency Control **37** No. 2 (1990) 233–247
3. Milsom, R.F., Elliot, D.T., Terry Wood, S., Redwood, M.: Analysis and Design of Couple Mode Miniature Bar Resonator and Monolothic Filters. IEEE Trans Son. Ultrason., **30** (1983) 140–155
4. Rálek, P.: Modelling of piezoelectric materials. Proceedings of the IX. PhD. Conference ICS, Academy of Sciences of the Czech Republic (2004)
5. Tiersten, H.F.: Hamilton's principle for linear piezoelectric media. Proceedings of IEEE (1967) 1523–1524
6. Lehoucq, R., Maschhoff, K., Sorensen, D., Yang Ch.:
   www.caam.rice.edu/software/ARPACK/
7. Geuzaine, Ch., Remacle, J.F.: http://www.geuz.org/gmsh/
8. Marques, O.: http://crd.lbl.gov/õsni/#Software

# Numerical Algorithm for Non-linear Systems Identification Based on Wavelet Transform

Elena Şerban

"Gh. Asachi" Technical University of Iaşi,
Faculty of Automation and Computer Science,
Department of Computer Science and Engineering
Bd. Mangeron nr. 53A, 700050 Iaşi, Romania
eserban@cs.tuiasi.ro, eserban27@yahoo.com

**Abstract.** This paper propose a method for identification of complex engineering systems using wavelet transform. This transform is chosen because it can provide a well localization both in time and in frequency. The method is applied to an electrohydraulic system that drives a shaking table. The identification is made using real signals obtained from experimental tests.

**Keywords:** systems identification, wavelet transform, Morlet wavelet, complex systems.

## 1 Introduction

The analysis and the interpretation of signals, obtained from experimental tests by measuring physical quantities with different types of sensors, are important. The extraction of significant information about the phenomenon from an experiment, that simulates the system behavior during its functioning, can be made by using various methods. One of this methods is the identification of the system. This method permits to obtain an experimental model of the dynamical system. The classical approach for systems identification [3], [14] takes into consideration only the time domain or the frequency domain, but not both. This approach may cause losses of important information about the system behavior. An analysis both in time and in frequency can be more accurate and it is possible by using the wavelet transform [2], [6], [7], [11].

Because of their properties, the wavelets were used for system identification. In [8], the authors propose a method for system identification using wavelets networks and three methods for regressor selection. A new class of wavelet network was proposed in [1]. The Morlet wavelet [6], [7] is also used in dynamical system identification because one of its parameters is inversely proportional to the Fourier frequency [10], [13].

In vibration tests, as in seismic engineering, it is important to know at what frequency some phenomena appear. The wavelet transform allows the signal filtering with narrow band filters. In particular, a time-history signal can be extracted for certain frequencies. This paper propose an identification algorithm based on this observation.

T. Boyanov et al. (Eds.): NMA 2006, LNCS 4310, pp. 701–708, 2007.

## 2  Wavelet Transform

The wavelet transform is a linear transform, which realizes the projection of a signal, $x(t)$, which is a time-series, on a two-dimensional space. The two dimensions of the projection space are the scale (or dilatation, associated with the frequency dimension) and the shift (or translation, associated with the time dimension).

The projection of the signal $x(t)$ is possible by means of some fundamental functions [9] using the equation:

$$T_x(a,b) = < x(t), \psi_{a,b}(t) > . \tag{1}$$

In equation (1) $\psi_{a,b}(t)$ is a fundamental function that generates a functions basis, $< \cdot, \cdot >$ means the inner product, $a$ and $b$ are the two dimensions of the plane. The fundamental function choice is the researcher option, but one must take into consideration the purpose of the signal analysis.

The functions basis is generated by functions with some properties, see [2], [8], [11], through dilatations and translations. The functions basis (noted with $\mathcal{W}$) is defined as follows:

$$\mathcal{W} = \left\{ \psi_{a,b}(t) \mid \psi_{a,b} : \mathcal{R} \to \mathcal{C}, \psi_{a,b}(t) = \frac{1}{\sqrt{|a|}} \psi \left( \frac{t-b}{a} \right), a, b \in \mathbb{R}, a \neq 0. \right\} \tag{2}$$

Each function from this basis depends on two parameters: $a$, which is dilatation parameter, and $b$, the translation parameter. The translations (or time shifts) of the original wavelet (also called mother wavelet) permit the extraction of the signal properties in time domain. The frequencies content of the signal can be obtained using dilatations of the mother wavelet.

The mother wavelet must satisfy the following conditions:

$$\int_{-\infty}^{\infty} |\psi(t)|^2 dt < \infty, \tag{3}$$

$$c_\psi = 2\pi \int_{-\infty}^{\infty} \frac{|\Psi(\omega)|^2}{|\omega|} d\omega < \infty, \tag{4}$$

where $\Psi(\omega)$ is the Fourier transform of the function $\psi(t)$.

The original signal can be reconstructed by the formula (5)

$$x(t) = \frac{1}{c_\psi} \int_{-\infty}^{\infty} \int_{-\infty}^{\infty} T_x(a,b) \psi_{a,b}(t) \frac{dadb}{a^2}. \tag{5}$$

Equation (1) denotes the direct wavelet transform, equation (5) denotes the inverse wavelet transform regarding the mother wavelet $\psi(t)$. As the projection is made on a continuous plane, this transform is the continuous wavelet transform.

In order to develop a numerical algorithm, the both directions of the projection plane must be discretized. The discretization is made on a logarithmic scale. The two parameters are discretized accordingly to the following equations:

$$a = a_0^m, \tag{6}$$

$$b = nb_0 a_0^m, \tag{7}$$

where $m, n \in \mathbb{Z}$ and $a_0 > 0$. The choice of the shift parameter $(b)$ depending on the scale parameter $a$ is necessary because the support of the wavelet depends directly on $a$. The $m$ is the discretization parameter of the scale (frequency) axis and $n$ is the discretization parameter of the shift (time) axis. With these relations for $a$ and $b$, the functions of the wavelets basis are defined as follows: in time domain:

$$\psi_{m,n}(t) = \frac{1}{\sqrt{a_0^m}} \psi(a_0^{-m} t - nb_0), \tag{8}$$

in frequency domain:

$$\Psi_{m,n}(\omega) = \sqrt{a_0^m} \Psi(a_0^m \omega) e^{j\omega n b_0}. \tag{9}$$

The discrete wavelet transform is defined, in this case, as follows:

$$T_x(m,n) = \langle x(t), \psi_{m,n}(t) \rangle = \int_{-\infty}^{\infty} x(t) \overline{\psi_{m,n}(t)} dt, \tag{10}$$

where $\overline{\psi_{m,n}(t)}$ is the complex conjugate value of $\psi_{m,n}(t)$. The reconstruction of the signal $x(t)$ is made by means of the equation:

$$x(t) = k_\psi \sum_m \sum_n T_x(m,n) \psi_{m,n}(t), \tag{11}$$

where $k_\psi$ is a constant value for normalization.

The function $\psi_{m,n}(t)$ provides sampling points on the scale (frequency) - shift (time) plane. The most common choice of $a_0$ is:

$$a_0 = 2^{1/v}, \tag{12}$$

where $v$ is an integer value and it defines the *voices* of a signal. The number of voices represents the number of intervals that correspond to an octave (an octave is, as in music, a frequency domain that has the upper bound twice as lower bound).

## 3  Morlet Wavelet

The trigonometric functions (sin and cos) form a function basis for the Fourier transform. This basis is used to analyse the behaviour of the signals in frequency domain because they are well localized in frequency. But, these functions are not well localized in time. In 1946, D. Gabor introduced the Short Time Fourier Transform [4] trying to localize in time the Fourier transform. Due to the Heisenberg uncertainty principle the resolution of this transform is not very good [4].

The basis functions with compact support both in time and in frequency are ideal candidates for the analysis of the signal in time-frequency domain. The wavelet transform uses such a functions basis. J. Morlet [6], [7] has introduced the fundamental function $\psi(t) : \mathcal{R} \rightarrow \mathcal{C}$, defined as

$$\psi(t) = (e^{j\omega_0 t} - e^{-\omega_0^2/2})e^{-t^2/2}. \tag{13}$$

In frequency domain, the Morlet wavelet is a shifted Gaussian adjusted so that it has a zero value for the zero frequency. It is represented by the Fourier transform of the function in equation (13):

$$\Psi(\omega) = e^{-(\omega-\omega_0)^2/2} - e^{-\omega^2/2}e^{-\omega_0^2/2}. \tag{14}$$

This function is known as Morlet wavelet and it is well localized both in time and in frequency, so it can be used for a time-frequency analysis.

Often [11], $\omega_0$ is chosen so that the first two maxima of the wavelet in time domain have a ratio of approximately $1/2$, that is:

$$\omega_0 = (2/\ln 2)^{1/2} = 5.3364... \tag{15}$$

In this paper, the value of $\omega_0$ was chosen to be equal to $2\pi$ for a good correlation between the wavelet transform and the frequency content of the signal. For this value, the second terms in equations (13) and (14) are so small that they can be neglected. As a result, the Morlet wavelet can be considered as a modulated Gaussian function.

The scaling parameter for this wavelet is connected with the Fourier frequency $f$ expressed in [Hz] or $\omega$ expressed in [rad] accordingly to the following relation:

$$a = \frac{f_0}{f} = \frac{\omega_0}{\omega}. \tag{16}$$

## 4    Wavelet Based Identification

System identification is necessary in order to extract information about complex systems. There are many methods [3], [14] to do a system identification, but they work either in time domain or in frequency domain. The key idea of this algorithm is the fact that the wavelet transform realizes a time-frequency analysis and the signal can be decomposed in a such a way that for each frequency, we have a time-series and for each time moment we have instantaneous frequency spectrum.

For identification, the dynamical system schematic form presented in Fig.1(a) will be considered. Here $x(t)$ is the system input signal and $y(t)$ is the system output signal. Accordingly to equation (11) these two signals can be decomposed, using wavelet transform, in $m$ signals, each of them corresponding to an established frequency. If in equation (11) the following notation is considered:

$$x_m(t) = k_\psi \sum_n T_x(m, n)\psi_{m,n}(t) \tag{17}$$

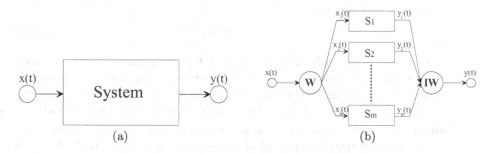

**Fig. 1.** Schematic representation of a system

then the equations for the input and the output signals of the system are:

$$x(t) = \sum_m x_m(t) \tag{18}$$

and

$$y(t) = \sum_m y_m(t). \tag{19}$$

Each $x_m(t)$ and $y_m(t)$ is the input, respectively the output signal of a subsystem of the dynamical system considered, whose schematic form is represented in Fig.1(b).

This way, the identification of a complex system can be reduced to the identification of $m$ systems, each for one frequency considered. The frequencies taken into consideration for the identification are established after a frequency analysis of the system (i.e. a Fourier analysis).

For each of the $m$ subsystems, a classical identification parametric method can be applied [3], [14] (e.g. least square type method: prediction error minimization). The numerical algorithm proposed for system identification is:

***Step 1.*** Determine the Fourier spectrum of the input and output signals. This is necessary for establishing the frequency domain and the frequency values used in the identification. Let us denote with $f_i$ the lower bound of the frequency domain and with $f_s$ the upper bound of the frequency domain.

***Step 2.*** Determine the values of the dilatation parameter $m$ and of the translation parameter $n$ for the discrete wavelet transform.

**Establishing the number of voices.** This is the number of intervals into which an octave is divided. This number is denoted with $v$. It is chosen taking into consideration how fine the wavelet analysis should be.

**Determining the values of the parameter $m$.** These are the points on the frequency axis where the wavelet transform will be calculated. Taking into account equations (6), (12) and (16), the starting value of $m$, denoted by $m_i$, is

$$m_i = v \log_2 \frac{f_0}{f_i} \tag{20}$$

and the ending value is

$$m_s = v \log_2 \frac{f_0}{f_s}.$$ (21)

This way a logarithmic discretization for the parameter $a$ is replaced by a linear discretization of the parameter $m$. For a better correspondence between an actual value frequency and the parameter $m$, in this algorithm the value of $f_0$ is chosen to be 1 Hz ($omega_0 = 6.28$ rad).

**Determining the values of the parameter $n$.** $b_0$ is chosen to be equal to the discretization interval of the signal, denoted by $dt$. If $N$ is the number of samples of each signal, then the starting value for $n$, denoted by $n_i$, is

$$n_i = -\left(\frac{N}{2}\right)$$ (22)

and the ending value for $n$, $n_s$, is

$$n_s = \left(\frac{3N}{2}\right).$$ (23)

**Step 3.** For each value of the parameters $m$ and $n$, the Morlet wavelet $\psi_{m,n}(t)$ is calculated (equation (8)).

**Step 4.** The input and output signals for each of the $m$ subsystems are calculated using wavelet transform (equations (17), (18), (19)).

**Step 5.** A classical identification method is applied for each subsystem (e.g. the least squares method).

**Step 6.** Results validation.

## 5    Numerical Results

The numerical algorithm described in the previous section was used for identification of an electrohydraulic system that drives a shaking table. The shaking

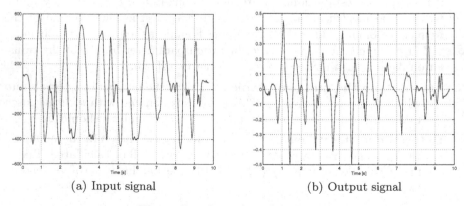

      (a) Input signal                       (b) Output signal

**Fig. 2.** Signals used in identification

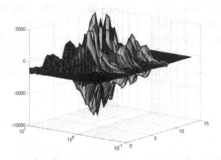

**Fig. 3.** Wavelet transform coefficients for input signal - real part

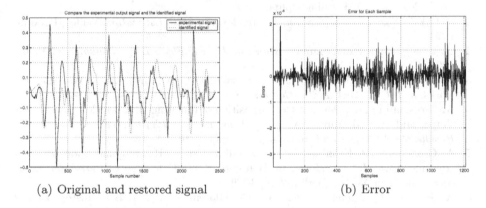

(a) Original and restored signal                    (b) Error

**Fig. 4.** Results of an identification process

tables are used to simulate earthquakes and to test building structures behavior during earthquakes. The output signal of such a system simulates the earthquake and the input signal is called the program signal. Fig. 2(a) presents the experimental input signal used to test the algorithm and in Fig. 2(b) is the corresponding output signal. The sampling interval for these signals ($dt$) is 0.004s.

In the identification process real experimental signals are used. For this reason, the first half of the signals is used for identification and the second half is used for results validation. In Fig. 3 the wavelet coefficients are presented. A comparison between the original signal and the restored signal is presented in Fig. 4(a); the error is presented in Fig. 4(b).

## 6   Conclusions

The paper presents a new method for complex systems identification. This method considers the dynamical system as a sum of many subsystems, one subsystem for each frequency of interest. The method proposes modeling technique able to highlight the influence of some frequencies concerning the system behavior.

The paper presents some facts about the wavelet transform and describes a numerical algorithm able to identify a system using this mathematical tools.

This algorithm was implemented using MATLAB and it permits some developments that can be a basis for a distributed algorithm for system identification.

The algorithm was used for modeling an electrohydraulic system that drives a shaking table.

# References

1. S.A. Billings, and H.L. Wei. A New Class of Wavelet Networks for nonlinear System Identification, in *IEEE Transactions on Neural Networks*, 16(2005), No. 6, 862-874.
2. I. Daubechies. *Ten Lectures on Wavelets*, CBMS-NSF, Regional Conf. Ser. In Appl. Math, No. 61, SIAM, Philadelphia, 1992.
3. P. Eykoff. *System Identification. Parameter and State Estimation.* Willey, London, 1974.
4. D. Gabor. Theory of Communication, in *J. Inst. Elect. Eng. (London)*, 93(1946), 429-457.
5. R.K.H. Galvao, V.M. Becerra, J.M.F. Calado, and P.M. Silva. Linear-Wavelet Networks, in *Int. J. Appl. Math. Comput. Sci.*, Vol. 14 (2004), No. 2, 221-232.
6. A. Grossmann and J. Morlet. Decomposition of functions into wavelets of constant shape, and related transform, in *Mathematics and Physics, Lectures on Recent Results*. World Scientific Publishing, Singapore, 1985.
7. A. Grossmann, R. Kronland-Martinet and J. Morlet. Reading and understanding continuous wavelet transform, in *Wavelets: Time-Frequency Methods and Phase Space*, Springer-Verlag, Berlin, 2-20, 1989.
8. A. Juditsky, Q. Zhang, B. Delyon, P-Y. Glorennec, and A. Benveniste. *Wavelet in identification: wavelets, splines, neurons, fuzzies: how good for identification?* Research Report No. 2315, INRIA, France, 1994.
9. W. Kecs. *Complements of Mathematics with Technical Application.* Ed. Tehnica, Bucureşti, 1981 (in Romanian).
10. T. Kijewski and A. Kareem. Wavelet Transform for System Identification in Civil Engineering, in *Computer-Aided Civil and Infrastructure Engineering*, 18(2003), 339-355.
11. D.T.L. Lee, and A. Yamamoto. Wavelet Analysis: Theory and Application, in *Hewlett-Packard Journal*, 45(1994), No. 6, 44-54.
12. A. Masuda, S. Yamamoto, and A. Sone. New Deconvolution Method for a Time Series Using the Discrete Wavelet Transform, in *JSME International Journal*, C, 40-4(1997), 630-636.
13. E. Şerban. *Automatic Synthesis of Excitation Signals for Seismic Engineering Testing Systems*, Ph. D. Thesis, "Gh. Asachi" Technical University of Iasi, 2003 (in Romanian).
14. M. Tertişco, P. Stoica and T. Popescu. *Computer Aided Systems Identification*, Ed. Tehnica, Bucureşti, 1987 (in Romanian).

# Simulation of Turbulent Thermal Convection Using Finite Volumes

Olga Shishkina and Claus Wagner

Institute for Aerodynamics and Flow Technology,
German Aerospace Center (DLR)
Bunsenstrasse 10, 37073 Göttingen, Germany
{Olga.Shishkina,Claus.Wagner}@dlr.de

**Abstract.** To simulate turbulent Rayleigh–Bénard convection in cylindrical domains an explicit/semi-implicit finite volume method with fourth order approximations in space was developed. Using this method and cylindrical staggered grids of about 11 million nodes clustered in vicinity of the boundary we performed simulations of turbulent Rayleigh–Bénard convection in wide cylindrical containers of the aspect ratios $\Gamma = 5$ and 10 and the Rayleigh number from $10^5$ to $10^8$. In the present paper the method, its numerical stability and mesh generation algorithm are discussed.

## 1 Turbulent Rayleigh–Bénard Convection

The thermally driven turbulent fluid motion between a lower heated horizontal plate and an upper cooled plate, i.e. Rayleigh–Bénard convection (RBC), is one of the classical problems in fluid dynamics. The Rayleigh number $Ra$, which is inversely proportional to the squared diffusion coefficients $\nu - \Gamma^{-3/2}Ra^{-1/2}Pr^{1/2}$ and $\kappa = \Gamma^{-3/2}Ra^{-1/2}Pr^{-1/2}$ in the Navier–Stokes (1) and energy (2) equations,

$$\mathbf{u}_t + \mathbf{u} \cdot \nabla \mathbf{u} + \nabla p = \nu \Delta \mathbf{u} + T\mathbf{z}, \tag{1}$$

$$T_t + \mathbf{u} \cdot \nabla T = \kappa \Delta T, \tag{2}$$

together with the Prandtl number $Pr$ and the aspect ratio $\Gamma$ (the ratio between the linear sizes of the container in the horizontal and vertical directions) are the governing parameters in turbulent RBC. Here $\mathbf{u}$ the velocity vector, $T$ the temperature, $\mathbf{u}_t$ and $T_t$ their time derivatives and $p$ the pressure. The system (1), (2) is closured by the continuity equation $\nabla \cdot \mathbf{u} = 0$. The temperature $T$ varies between $+0.5$ at the bottom plate and $-0.5$ at the top plate. On the adiabatic lateral wall $\partial T/\partial r = 0$ and on all solid walls $\mathbf{n} \cdot \nabla p = 0$ ($\mathbf{n}$ is the normal vector) and the velocity field $\mathbf{u}$ vanishes due to impermeability and no-slip conditions. Note that in cylindrical coordinates $(z, \varphi, r)$ for a vector $\mathbf{a} = (a_z, a_\varphi, a_r)$ one has $\nabla a \equiv \operatorname{grad} a = \left(\frac{\partial a}{\partial z}, \frac{1}{r}\frac{\partial a}{\partial \varphi}, \frac{\partial a}{\partial r}\right)$, $\nabla \cdot \mathbf{a} \equiv \operatorname{div} \mathbf{a} = \frac{\partial a_z}{\partial z} + \frac{1}{r}\left(\frac{\partial a_\varphi}{\partial \varphi} + a_r\right) + \frac{\partial a_r}{\partial r}$ and $\Delta a \equiv \nabla \cdot \nabla a = \frac{\partial^2 a}{\partial z^2} + \frac{1}{r}\frac{\partial a}{\partial r} + \frac{1}{r^2}\frac{\partial^2 a}{\partial \varphi^2} + \frac{\partial^2 a}{\partial r^2}$.

T. Boyanov et al. (Eds.): NMA 2006, LNCS 4310, pp. 709–716, 2007.

To simulate turbulent RBC in cylindrical domains an explicit/semi-implicit finite volume method with fourth order approximations in space was developed. Using this method and cylindrical staggered grids of $110 \times 512 \times 192$ nodes in the vertical, azimuthal and radial directions, respectively, we performed simulations for the cases $\Gamma = 5$ and 10 and $Ra$ from $10^5$ to $10^8$. The considered cases are closely related to many astrophysical, geophysical and meteorological problems, in which $Ra$ varies from $10^5$ to $10^{20}$ and the aspect ratio is large.

In the present paper the numerical method, its von Neumann stability and mesh generation algorithm are discussed. For physical results on turbulent thermal convection, obtained from the simulation data, we refer to Shishkina & Wagner [1].

## 2   Finite Volume Method for Navier–Stokes Equations in Cylindrical Domains

Further we describe the finite volume scheme for the system (1), (2) in cylindrical coordinates by the example of the Navier–Stokes equation (1). For the energy equation (2) the scheme is written analogously.

Consider a finite volume $V = V(z_i, \varphi_j, r_k)$ with the center $(z_i, \varphi_j, r_k)$ and the cell surfaces $A_z^{\pm} = A_z(z_i \pm \frac{\Delta z_i}{2}, \varphi_j, r_k)$, $A_\varphi^{\pm} = A_\varphi(z_i, \varphi_j \pm \frac{\Delta \varphi_j}{2}, r_k)$, $A_r^{\pm} = A_r(z_i, \varphi_j, r_k \pm \frac{\Delta r_k}{2})$, where $\Delta z_i$, $\Delta \varphi_j$, $\Delta r_k$ are the sizes of $V$ in the directions $z$, $\varphi$, $r$, respectively. Each $V$ and its cell surfaces are associated with the coordinates of their own centers. The values $^{\beta\pm}\overline{u_\alpha}$ and $\overline{u_\alpha}$ denote the velocity component $u_\alpha$ averaged over the $A_\beta^{\pm}$-surface and over $V$, respectively, and $^\beta\overline{u_\alpha}$ denotes averaging of $u_\alpha$ over a surface $A_\beta = 0.5(A_\beta^+ + A_\beta^-)$. Using the notations $\Delta V = r_k \Delta r_k \Delta z_i \Delta \varphi_j$, $\Delta A_z = r_k \Delta r_k \Delta \varphi_j$, $\Delta A_\varphi = \Delta r_k \Delta z_i$, $\Delta A_r = r_k \Delta z_i \Delta \varphi_j$ and integrating (1) over $V$ we get the following finite volume scheme

$$\frac{\partial \overline{u_\alpha}}{\partial t} + \sum_{\beta=z,\varphi,r} (K_{\alpha\beta} - D_{\alpha\beta}) + P_\alpha - C_\alpha + T_\alpha = 0, \quad \sum_\alpha (F_{\alpha+} - F_{\alpha-}) = 0, \quad (3)$$

where $\alpha = z, \varphi, r$, $F_{\alpha\pm} = \Delta A_\alpha^{\pm} \, {}^{\alpha\pm}\overline{u_\alpha}$ denotes the momentum flux,

$K_{\alpha\beta} = \frac{1}{\Delta V} \left( F_{\beta+} \, {}^{\beta+}\overline{u_\alpha} - F_{\beta-} \, {}^{\beta-}\overline{u_\alpha} \right)$ the convective term,

$D_{\alpha\beta} = \frac{\nu}{\Delta V} \left( \Delta A_\beta^+ \, {}^{\beta+}\overline{s_{\alpha\beta}} - \Delta A_\beta^- \, {}^{\beta-}\overline{s_{\alpha\beta}} \right)$ the diffusive term,

$P_\alpha = \frac{1}{\Delta V} \left( \Delta A_\alpha^+ \, {}^{\alpha+}\overline{p} - \Delta A_\alpha^- \, {}^{\alpha-}\overline{p} \right)$ the pressure term,

$T_\alpha = \delta_\alpha^z \overline{T}$ the temperature term, $\delta_\alpha^z$ the Kronecker symbol,

$C_\varphi = \frac{\Delta \varphi_j \Delta A_\varphi}{\Delta V} \left( -{}^\varphi\overline{u_\varphi} \, {}^\varphi\overline{u_r} + \nu \, {}^\varphi\overline{s_{\varphi r}} \right)$, $C_r = \frac{\Delta \varphi_j \Delta A_\varphi}{\Delta V} \left( ({}^\varphi\overline{u_\varphi})^2 + {}^\varphi\overline{p} - \nu \, {}^\varphi\overline{s_{\varphi\varphi}} \right)$ and $C_z = 0$ the curvature terms and the surface averaged deformation tensor $^\beta\overline{s_{\alpha\beta}}$ equals

$$\begin{bmatrix} 2\frac{\partial}{\partial z} \, {}^z\overline{u_z} & \frac{1}{r}\frac{\partial}{\partial \varphi} \, {}^\varphi\overline{u_z} + \frac{\partial}{\partial z} \, {}^\varphi\overline{u_\varphi} & \frac{\partial}{\partial r} \, {}^r\overline{u_z} + \frac{\partial}{\partial z} \, {}^r\overline{u_r} \\ \frac{1}{r}\frac{\partial}{\partial \varphi} \, {}^\varphi\overline{u_z} + \frac{\partial}{\partial z} \, {}^\varphi\overline{u_\varphi} & \frac{2}{r}\left( \frac{\partial}{\partial \varphi} \, {}^\varphi\overline{u_\varphi} + {}^\varphi\overline{u_r} \right) & r\frac{\partial}{\partial r}({}^r\overline{u_\varphi}/r) + \frac{1}{r}\frac{\partial}{\partial \varphi} \, {}^r\overline{u_r} \\ \frac{\partial}{\partial r} \, {}^r\overline{u_z} + \frac{\partial}{\partial z} \, {}^r\overline{u_r} & r\frac{\partial}{\partial r}({}^r\overline{u_\varphi}/r) + \frac{1}{r}\frac{\partial}{\partial \varphi} \, {}^r\overline{u_r} & 2\frac{\partial}{\partial r} \, {}^r\overline{u_r} \end{bmatrix}.$$

## 3   Explicit Scheme

The explicit time discretization of the equation (3) can be written as follows

$$\frac{u_\alpha^{n+1} - u_\alpha^{n-1}}{2\Delta t} + \sum_{\beta=z,\varphi,r} \left( K_{\alpha\beta}^{n,n} - D_{\alpha\beta}^{n-1,n-1} \right) + P_\alpha^n - C_\alpha^n + T_\alpha^n = 0, \qquad (4)$$

where $K_{\alpha\beta}^{n,m} = \frac{1}{\Delta V} \left( \Delta A_\beta^+ \, {}^{\beta+}\overline{u_\alpha} \, {}^{\beta+}\overline{u_\beta} - \Delta A_\beta^- \, {}^{\beta-}\overline{u_\alpha} \, {}^{\beta-}\overline{u_\beta} \right)$,

$D_{\alpha\beta}^{n,m} = \frac{\nu}{\Delta V} \left( \Delta A_\beta^+ \, {}^{\beta+}\overline{s_{\alpha\beta}^{n,m}} - \Delta A_\beta^- \, {}^{\beta-}\overline{s_{\alpha\beta}^{n,m}} \right)$,

$P_\alpha^n = \frac{1}{\Delta V} \left( \Delta A_\alpha^+ \, {}^{\alpha+}\overline{p^n} - \Delta A_\alpha^- \, {}^{\alpha-}\overline{p^n} \right)$, $T_\alpha^n = \delta_\alpha^z \, \overline{T^n}$,

$C_z^n = 0$, $\qquad C_\varphi^n = \frac{\Delta\varphi_j \Delta A_\varphi}{\Delta V} \left( -{}^\varphi\overline{u_\varphi^n} \, {}^\varphi\overline{u_r^n} + \nu \, {}^\varphi\overline{s_{\varphi r}^n} \right)$,

$C_r^n = \frac{\Delta\varphi_j \Delta A_\varphi}{\Delta V} \left( \left({}^\varphi\overline{u_\varphi^n}\right)^2 + {}^\varphi\overline{p^n} - \nu \, {}^\varphi\overline{s_{\varphi\varphi}^n} \right)$, and ${}^\beta\overline{s_{\alpha\beta}^{n,m}}$ equals

$$\begin{bmatrix} 2\frac{\partial}{\partial z}{}^z\overline{u_z^n} & \frac{1}{r}\frac{\partial}{\partial\varphi}{}^\varphi\overline{u_z^m} + \frac{\partial}{\partial z}{}^\varphi\overline{u_\varphi^n} & \frac{\partial}{\partial r}{}^r\overline{u_z^n} + \frac{\partial}{\partial z}{}^r\overline{u_r^n} \\ \frac{1}{r}\frac{\partial}{\partial\varphi}{}^\varphi\overline{u_z^m} + \frac{\partial}{\partial z}{}^\varphi\overline{u_\varphi^n} & \frac{1}{r}\left(\frac{\partial}{\partial\varphi}{}^\varphi\overline{u_\varphi^n} + {}^\varphi\overline{u_\varphi^m}\right) + \frac{2}{r}{}^\varphi\overline{u_r^n} & r\frac{\partial}{\partial r}\left({}^r\overline{u_\varphi^n}/r\right) + \frac{1}{r}\frac{\partial}{\partial\varphi}{}^r\overline{u_r^m} \\ \frac{\partial}{\partial r}{}^r\overline{u_z^n} + \frac{\partial}{\partial z}{}^r\overline{u_r^n} & r\frac{\partial}{\partial r}\left({}^r\overline{u_\varphi^n}/r\right) + \frac{1}{r}\frac{\partial}{\partial\varphi}{}^r\overline{u_r^m} & 2\frac{\partial}{\partial r}{}^r\overline{u_r^n} \end{bmatrix},$$

$\Delta t$ is the time step, $n$ is the number of the time step.

The solution of (4) is obtained in three steps using the projection approach by Chorin [2]. First, an approximate velocity field $\mathbf{u}^* = (u_z^*, u_\varphi^*, u_r^*)$ is computed from the equations (5) obtained from (4) by neglecting the pressure term,

$$\frac{u_\alpha^* - u_\alpha^{n-1}}{2\Delta t} + \sum_{\beta=z,\varphi,r} \left( K_{\alpha\beta}^{n,n} - D_{\alpha\beta}^{n-1,n-1} \right) - C_\alpha^n + T_\alpha^n = 0, \qquad \alpha = z, \varphi, r. \quad (5)$$

Then the Poisson equation for an auxiliary function $\overline{\phi^n}$ is solved:

$$\Delta\overline{\phi^n} = \nabla \cdot \overline{\mathbf{u}^*} \equiv \frac{1}{\Delta V} \sum_{\beta=z,\varphi,r} \left( \Delta A_\beta^+ \, {}^{\beta+}\overline{u_{\beta+}^*} - \Delta A_\beta^- \, {}^{\beta-}\overline{u_{\beta-}^*} \right). \qquad (6)$$

On the solid walls $\overline{u_\alpha^*} = 0$, $\alpha = z, \varphi, r$, and $\mathbf{n} \cdot \nabla\overline{\phi^n} = 0$ ($\mathbf{n}$ is the normal vector). The function $\overline{\phi^n}$ and the velocity field $\overline{\mathbf{u}^*}$ are periodic in the $\varphi$-direction. The solution is obtained applying the Fourier transform in the $\varphi$-direction and further any 2D-Poisson solver. Finally, the velocity field is updated as follows

$$\mathbf{u}^{n+1} = \overline{\mathbf{u}^*} - \nabla\overline{\phi^n}. \qquad (7)$$

The correctness of the scheme can be checked as follows. From (4), (5) and $P_\alpha^n = \partial\overline{p^n}/\partial x_\alpha$, $\alpha = z, \varphi, r$, we get

$$\overline{u_\alpha^{n+1}} = \overline{u_\alpha^*} - 2\Delta t \, \partial\overline{p^n}/\partial x_\alpha. \qquad (8)$$

Applying $(\nabla\cdot)$ to this equation and assuming $\nabla\cdot\mathbf{u}^{n+1} = 0$ due to the continuity, from (8) and (6) we get $2\Delta t \Delta\overline{p^n} = \nabla\cdot\overline{\mathbf{u}^*} = \Delta\overline{\phi^n}$. Therefore $\nabla\overline{p^n} = \nabla\overline{\phi^n}/(2\Delta t)$, which together with (8) gives (7).

# 4   Semi-implicit Scheme

Only in a thin subdomain around the cylinder axis the explicit treatment of the viscous term $D_{\alpha\beta}^{n-1,n-1}$ and the convective term $K_{\alpha\beta}^{n,n}$ in the azimuthal direction $\beta = \varphi$ leads to an extremely small time step $\Delta t$ in the scheme (4) due to the numerical stability. In this subdomain we apply the following implicit (in the $\varphi$-direction) scheme

$$\frac{\overline{u_\alpha^{n+1}} - \overline{u_\alpha^{n-1}}}{2\Delta t} + \sum_{\beta=z,r} \left( K_{\alpha\beta}^{n,n} - D_{\alpha\beta}^{n-1,n-1} \right) + K_{\alpha\varphi}^{n+1,n} - D_{\alpha\varphi}^{n-1,n+1}$$

$$+ P_\alpha^n - C_\alpha^n + T_\alpha^n = 0, \quad \alpha = z, \varphi, r,$$

to accelerate the simulations. These equations are solved similarly to (4). First, an approximate velocity field $\mathbf{u}^*$ is computed from the equation without the pressure term using any fast solver for band matrices. The Poisson equation (6) and the equation to update the velocity field (7) remain unchanged, but the pressure is calculated by $\overline{p^n} = \left( \frac{1}{2\Delta t} + \overline{u_\varphi^n} \frac{\partial}{\partial\varphi} + \frac{\partial \overline{u_\varphi^n}}{\partial\varphi} - \nu \frac{\partial^2}{\partial\varphi^2} \right) \overline{\phi^n}$.

# 5   High-Order Discretization

In this section we consider a way to construct high-order schemes to compute the $A_\beta$-surface averaged value $^\beta\overline{u_\alpha}$ and its partial derivatives $\frac{\partial}{\partial\beta}\,^\beta\overline{u_\alpha}$ for each velocity component $u_\alpha$ by the example of a fourth order scheme.

Let $\beta$ be one of the coordinates ($z$, $\varphi$ or $r$) and $V(\beta_i)$ a finite volume bounded by the surfaces $A_\beta(\beta_i \pm \frac{\Delta\beta_i}{2})$, where $\Delta\beta_i$ is the size of $V(\beta_i)$ in the direction $\beta$. The values $\overline{u_\alpha}(\beta_i)$ and $^\beta\overline{u_\alpha}(\beta_i + \frac{\Delta\beta_i}{2})$ denote the $u_\alpha$-component averaged over $V(\beta_i)$ and over $A_\beta(\beta_i + \frac{\Delta\beta_i}{2})$, respectively. Any approximation scheme to compute $^\beta\overline{u_\alpha}(\beta_i + \frac{\Delta\beta_i}{2})$ and $\frac{\partial}{\partial\beta}\,^\beta\overline{u_\alpha}(\beta_i + \frac{\Delta\beta_i}{2})$ involves a certain number of values $\overline{u_\alpha}(\beta_{i\pm k})$, $k \in N$. Everywhere except in the near wall regions we consider central approximation schemes.

## 5.1   Central Discretization in the Directions $\beta = z$ and $\beta = \varphi$

To find the coefficients $\xi_j$ and $\eta_j$, $j = 1, 2, 3, 4$, of the central fourth order approximation schemes

$$^\beta\overline{u_\alpha}(\beta_i + \frac{\Delta\beta_i}{2}) = \sum_{j=1}^{4} \xi_j \overline{u_\alpha}(\beta_{i-2+j}), \qquad \frac{\partial}{\partial\beta}\,^\beta\overline{u_\alpha}(\beta_i + \frac{\Delta\beta_i}{2}) = \sum_{j=1}^{4} \eta_j \overline{u_\alpha}(\beta_{i-2+j}),$$

we assume that the $A_\beta(\beta)$-averaged component $u_\alpha$ equals some polynomial of $\beta$,

$$^\beta\overline{u_\alpha}(\beta) = \sum_{k=1}^{4} \zeta_k \beta^{k-1}, \qquad \frac{\partial}{\partial\beta}\,^\beta\overline{u_\alpha}(\beta) = \sum_{k=1}^{4}(k-1)\zeta_k \beta^{k-2}, \tag{9}$$

with coefficients $\zeta_k$, $k = 1, 2, 3, 4$. From this the $V(\beta_j)$-averaged components $u_\alpha$ can be computed as follows

$$\overline{u_\alpha}(\beta_l) = \frac{1}{\Delta\beta_l} \int_{\beta_l - \frac{\Delta\beta_l}{2}}^{\beta_l + \frac{\Delta\beta_l}{2}} \sum_{k=1}^4 \zeta_k \beta^{k-1} d\beta.$$

Substituting the values $\overline{u_\alpha}(\beta_l)$ for $l = i - 1, i, i + 1, i + 2$ in this equality we get a system of linear equations for the coefficients $\zeta_k$, $k = 1, 2, 3, 4$,

$$\sum_{k=1}^4 (\mathbf{A}_\beta)_{jk} \zeta_k = \overline{u_\alpha}(\beta_{i-2+j}), \quad j = 1, ..., 4, \tag{10}$$

$$(\mathbf{A}_\beta)_{jk} = \frac{1}{k} \sum_{m=0}^{k-1} \left(\beta_{i-2+j} + \frac{\Delta\beta_{i-2+j}}{2}\right)^m \left(\beta_{i-2+j} - \frac{\Delta\beta_{i-2+j}}{2}\right)^{k-m-1}.$$

For the matrix $\mathbf{O}_\beta = \mathbf{A}_\beta^{-1}$ from (9) and (10) it follows

$$^\beta\overline{u_\alpha}(\beta_i + \frac{\Delta\beta_i}{2}) = \sum_{j=1}^4 \overline{u_\alpha}(\beta_{i-2+j}) \sum_{k=1}^4 (\mathbf{O}_\beta)_{kj} \left(\beta_i + \frac{\Delta\beta_i}{2}\right)^{k-1},$$

$$\frac{\partial}{\partial\beta} {}^\beta\overline{u_\alpha}(\beta_i + \frac{\Delta\beta_i}{2}) = \sum_{j=1}^4 \overline{u_\alpha}(\beta_{i-2+j}) \sum_{k=1}^4 (k-1)(\mathbf{O}_\beta)_{kj} \left(\beta_i + \frac{\Delta\beta_i}{2}\right)^{k-2}.$$

In the equidistant case, $\Delta\beta_i = \Delta$, we get the following approximation scheme to compute the values $^\beta\overline{u_\alpha}(\beta_i + \frac{\Delta\beta_i}{2})$ and $\frac{\partial}{\partial\beta} {}^\beta\overline{u_\alpha}(\beta_i + \frac{\Delta\beta_i}{2})$

$$^\beta\overline{u_\alpha}(\beta_i + \frac{\Delta}{2}) = \frac{1}{12}[-\overline{u_\alpha}(\beta_{i-1}) + 7\overline{u_\alpha}(\beta_i) + 7\overline{u_\alpha}(\beta_{i+1}) - \overline{u_\alpha}(\beta_{i|2})], \tag{11}$$

$$\frac{\partial^\beta\overline{u_\alpha}(\beta_i + \frac{\Delta}{2})}{\partial\beta} = \frac{1}{12\Delta}[\overline{u_\alpha}(\beta_{i-1}) - 15\overline{u_\alpha}(\beta_i) + 15\overline{u_\alpha}(\beta_{i+1}) - \overline{u_\alpha}(\beta_{i+2})]. \tag{12}$$

## 5.2  Discretization in the Direction $\beta = r$

For the approximation of the values $^r\overline{u_\alpha}(r)$ and $\frac{\partial}{\partial r} {}^r\overline{u_\alpha}(r)$ we assume that $^r\overline{u_\alpha}(r)$ equals some polynomial of $r$,

$$^r\overline{u_\alpha}(r) = \sum_{k=1}^4 \zeta_k r^{k-1}, \qquad \frac{\partial}{\partial r} {}^r\overline{u_\alpha}(r) = \sum_{k=1}^4 (k-1)\zeta_k r^{k-2}. \tag{13}$$

Therefore the values $\overline{u_\alpha}(r_l)$ for $l = i - 1, ..., i + 2$ equal

$$\overline{u_\alpha}(r_l) \equiv 2\left((r_l + \Delta r_l/2)^2 - (r_l - \Delta r_l/2)^2\right)^{-1} \int_{r_l - \frac{\Delta r_l}{2}}^{r_l + \frac{\Delta r_l}{2}} r \cdot {}^r\overline{u_\alpha}(r) dr$$

$$= 2\left((r_l + \Delta r_l/2)^2 - (r_l - \Delta r_l/2)^2\right)^{-1} \sum_{k=1}^4 \zeta_k \left.\frac{r^{k+1}}{k+1}\right|_{r_l - \frac{\Delta r_l}{2}}^{r_l + \frac{\Delta r_l}{2}}.$$

Substitution of the values $\overline{u_\alpha}(r_l)$ for $l = i-1, ..., i+2$ in this equality gives the following linear system of equations for the coefficients $\zeta_k$, $k = 1, 2, 3, 4$,

$$\sum_{k=1}^{4} (\mathbf{A}_r)_{jk} \zeta_k = \overline{u_\alpha}(r_{i-2+j}), \quad j = 1, ..., 4, \tag{14}$$

where the coefficients $(\mathbf{A}_r)_{jk}$, $j = 1, ..., 4$, $k = 1, ..., 4$, of the matrix $\mathbf{A}_r$ equal

$$(\mathbf{A}_r)_{jk} = \frac{2}{(k+1)r_{i-2+j}} \sum_{m=0}^{k} \left( r_{i-2+j} + \frac{\Delta r_{i-2+j}}{2} \right)^m \left( r_{i-2+j} - \frac{\Delta r_{i-2+j}}{2} \right)^{k-m},$$

$$r_{i-2+j} - \frac{\Delta r_{i-2+j}}{2} \neq 0,$$

$$(\mathbf{A}_r)_{jk} = \frac{2}{k+1} \left( r_{i-2+j} + \frac{\Delta r_{i-2+j}}{2} \right)^{k-1}, \quad r_{i-2+j} - \frac{\Delta r_{i-2+j}}{2} = 0.$$

From this, (13), (14) and $\mathbf{O}_r = \mathbf{A}_r^{-1}$ we get the following formulae

$$^r\overline{u_\alpha}(r) = \sum_{j=1}^{4} \overline{u_\alpha}(r_{i-2+j}) \sum_{k=1}^{4} (\mathbf{O}_r)_{kj}\, r^{k-1},$$

$$\frac{\partial}{\partial r}\, ^r\overline{u_\alpha}(r) = \sum_{j=1}^{4} \overline{u_\alpha}(r_{i-2+j}) \sum_{k=1}^{4} (k-1)(\mathbf{O}_r)_{kj}\, r^{k-2},$$

to approximate the values $^r\overline{u_\alpha}(r)$ and $\frac{\partial}{\partial r}\, ^r\overline{u_\alpha}(r)$ for any $r \in [r_{i-1}, r_{i+2}]$ and, in particular, near the solid wall.

## 6   The von Neumann Stability of the Explicit Scheme

The Leapfrog-Euler scheme (5) remains one of the most popular explicit schemes in turbulent flow simulations, since the scheme does not suffer from false diffusion and is applicable to convection-diffusion problems with large Peclet number $Pe_\alpha = c_\alpha/d_\alpha$, $\alpha = z, \varphi, r$, where $c_\alpha = \frac{U_\alpha \Delta t}{\Delta x_\alpha}$ the Courant number, $d_\alpha = \frac{\nu \Delta t}{\Delta x_\alpha^2}$ the diffusion number, $U_\alpha$ the component of the velocity field, which in the von Neumann stability analysis is supposed to be constant, $\Delta x_\alpha = \Delta \alpha$ for $\alpha = r, z$ and $\Delta x_\varphi = r \Delta \varphi$.

First a sufficient condition for the stability of the Leapfrog-Euler scheme utilizing the second order differences in space was suggested by Schumann [3] in a form of a restriction to the time step $\Delta t$. Further the sufficiency of this condition for the stability was proven by Chan [4] (for the 1D-case) and Wesseling [5] (for the 3D-case). A sufficient condition for the von Neumann stability of the Leapfrog-Euler scheme of any even order on equidistant meshes was derived by

Shishkina & Wagner [6]. This can be resumed as follows. For the Leapfrog-Euler scheme that uses central approximation schemes of the order $2m$, $m \in N$,

$$\beta \overline{u_\alpha}(\beta_i + \frac{\Delta}{2}) = \sum_{j=1}^{m} a_j [\overline{u_\alpha}(\beta_{i+j}) + \overline{u_\alpha}(\beta_{i-j+1})], \tag{15}$$

$$\frac{\partial}{\partial \beta} \beta \overline{u_\alpha}(\beta_i + \frac{\Delta}{2}) = \sum_{j=1}^{m} \frac{b_j}{\Delta x_\alpha} [\overline{u_\alpha}(\beta_{i+j}) - \overline{u_\alpha}(\beta_{i-j+1})], \quad \alpha = z, \varphi, r,$$

the following condition is sufficient for the von Neumann stability of the solution

$$2 \sum_{\alpha=1}^{3} \left\{ d_\alpha b_1 + \sum_{j=1}^{m} (2 - \delta_j^1) \sqrt{c_\alpha^2 a_j^2 + d_\alpha^2 b_j^2} \right\} \leq 1, \tag{16}$$

where $a_j$, $b_j$, $j = 1, ..., m$, are the coefficients of the approximation scheme (15) and $\delta_\beta^\alpha$ is the Kronecker symbol. In particular, for the Leapfrog-Euler scheme (11), (12) of the fourth order in space the sufficient condition (16) can be written as follows

$$\frac{1}{6} \sum_{\alpha=1}^{3} \left( 15 d_\alpha + \sqrt{225 d_\alpha^2 + 49 c_\alpha^2} + 2\sqrt{d_\alpha^2 + c_\alpha^2} \right) \leq 1.$$

We substitute $c_\alpha = \frac{U_\alpha \Delta t}{\Delta x_\alpha}$ and $d_\alpha = \frac{\nu \Delta t}{\Delta x_\alpha^2}$ in this inequality and get the estimation of the critical time step, which guarantees the stability of the calculations

$$\Delta t_{exp}^{crit} < \left( \frac{3}{2} \sum_{\alpha=1}^{3} \frac{U_\alpha}{\Delta x_\alpha} + \frac{16\nu}{3} \sum_{\alpha=1}^{3} \frac{1}{\Delta x_\alpha^2} \right)^{-1},$$

## 7   Mesh Generation Algorithm

In turbulent thermal convection the diffusion coefficients in (1) are very small. In particular, in the considered case $Ra = 10^8$, $Pr = 0.7$, $\Gamma = 10$ we have $\nu \approx 2.6 \times 10^{-6}$ and $\kappa \approx 3.8 \times 10^{-6}$. Therefore the solutions of (1) – both the temperature and the velocity fields – have very thin boundary layers near the horizontal walls. To resolve them some special fine enough meshes must be used in the vicinity of the walls.

Our mesh generation algorithm is based on grid equidistribution approach (see, for example, [7]) and consists of three steps. In the first step a rough solution of the system (1), (2) is found on a mesh, which is equidistant in the vertical $z$-direction. Averaging the temperature in time and also in the $\varphi$- and $r$-directions gives the temperature profile, i.e. a one-dimensional function $\hat{T}(z)$.

In the second step we find the points $\{z_k\}$, $k = 1, ..., N_z$, which equidistribute the monitor function

$$M(z) = \sqrt{1 + \left( \hat{T}'(z) \right)^2},$$

where $\hat{T}'(z)$ - is the derivative of $\hat{T}(z)$. Then the mesh is checked: each cell must be smaller than the Kolmogorov scale $h(Ra) = \pi \Gamma^{-1} Pr^{1/2}(Nu - 1)^{-1/4} Ra^{-1/4}$ [8] to resolve all turbulent scales. Here

$$Nu = \Gamma^{1/2} Ra^{1/2} Pr^{1/2} \langle u_z T \rangle_{t,S} - \Gamma^{-1} \left\langle \frac{\partial T}{\partial z} \right\rangle_{t,S}$$

is the Nusselt number and $\langle \cdot \rangle_{t,S}$ denotes averaging in time and also over any horizontal cross-section $S$. If the constructed mesh is too coarse, the number of nodes $N_z$ is increased and the second step is repeated.

In the third step the hyperbolic tangent algorithm by Thompson [9] is used to make the mesh smoother. This algorithm provides a smooth distribution of the nodes on the interval $[0; 1]$, using the following incoming data: the number of the nodes and the sizes of the first and the last subintervals.

Applying this algorithm we constructed the solution-adapted meshes, which made it possible to resolve the thermal boundary layers in 3D turbulent RBC. For example, 39.2% of the nodes of the adaptive mesh obtained for the case $Ra = 10^5$, $\Gamma = 10$, lie in the thermal boundary layers, while in the equidistant mesh this number equals 24.9%. The constructed meshes are also smooth, since the neighbor intervals differ in size by not more than 7%.

## Acknowledgment

The authors are grateful to the Deutsche Forschungsgemeinschaft (DFG) for supporting this work under the contract WA 1510-1.

## References

1. Shishkina, O., Wagner, C.: Analysis of thermal dissipation rates in turbulent Rayleigh-Bénard convection. J. Fluid Mech. **546** (2006) 51–60
2. Chorin, A.J.: Numerical solution of the Navier-Stokes equations. Math. Comput. **22** (1968) 745–762
3. Schumann, U.: Linear stability of finite difference equations for three-dimensional flow problems. J. Comput. Phys. **18** (1975) 465–470
4. Chan, T.F.: Stability analysis of finite difference schemes for the advection-diffusion equation. SIAM J. Numer. Anal. **21** (1984) 272–284
5. Wesseling, P.: Von Neumann stability conditions for the convection-diffusion equations. IMA J. of Numer. Anal. **16** (1996) 583–598
6. Shishkina, O., Wagner, C.: Stability conditions for the Leapfrog-Euler scheme with central spatial discretization of any order. Appl. Numer. Anal. Comput. Math. **1** (2004) 315–326
7. Qiu, Y., Sloan, D. M.: Analysis of difference approximations to a singularly perturbated two-point boundary value problem on a adaptively generated grid. J. Comput. Appl. Math. **101** (1999) 1–25
8. Grötzbach, G.: Spatial resolution requirements for direct numerical simulation of Rayleigh–Bénard convection. J. Comput. Phys. **49** (1983) 241–264
9. Thompson, J. F.: Numerical grid generation, Elsevier Science Publishing, Amsterdam (1985) 305–310

# Phase-Field Versus Level Set Method for 2D Dendritic Growth

Vladimir Slavov and Stefka Dimova

Faculty of Mathematics and Informatics, University of Sofia,
5 James Bourchier Boulevard, 1164 Sofia, Bulgaria
vladimir-slavov@yahoo.com, dimova@fmi.uni-sofia.bg

**Abstract.** The goal of the paper is to review and compare two of the most popular methods for modeling the dendritic solidification in 2D, that tracks the interface between phases implicitly, e.g. the phase-field method and the level set method. We apply these methods to simulate the dendritic crystallization of a pure melt. Numerical experiments for different anisotropic strengths are presented. The two methods compare favorably and the obtained tip velocities and tip shapes are in good agreement with the microscopic solvability theory.

**Keywords:** phase-field method, level set method, dendritic solidification, finite difference methods.

## 1 Introduction

Various numerical approaches have been proposed to solve the difficult moving boundary problem that governs the dendritic growth. Broadly speaking two approaches for tracking the moving interface between solid and liquid phases can be distinguished - explicit tracking of the interface and implicit tracking of the interface. The most popular methods that use the explicit approach are boundary integral method, front tracking method and immersed interface method. Boundary integral methods are based upon numerical solving an integral equation on the moving boundary. One drawback of boundary integral method is that the necessary parametrization of the boundary makes it hard to extend to higher dimensions. In [1], Juric and Tryggvason presented a numerical method that incorporates front tracking method with the ideas of immersed interface method. Although their method was successful in modeling many physical features of dendritic solidification, special care had to be taken for topological changes such as merging or breaking. The main disadvantages of all the methods, that use the explicit approach, are that topological changes are difficult to be handled and these methods are usually not easily extended to higher dimensions. Implicit representations such as phase-field or level set methods avoid this difficulties by representing the boundary as a level set of a continuous function. Thus the topological changes are easily handled and extension to higher dimensions is straightforward. Moreover, one can easily model additional physics, e.g. material strain, flow past dendrites or dendritic interaction.

T. Boyanov et al. (Eds.): NMA 2006, LNCS 4310, pp. 717–725, 2007.
© Springer-Verlag Berlin Heidelberg 2007

The format of the paper is as follows. In Section 2 we give short description of the sharp interface model of solidification. In Section 3 and 4 we present phase-field and level set models of solidification. Some of the results obtained by using these methods are shown in Section 5.

## 2    Sharp Interface Model of Solidification

For equal and constant in both phases material parameters the sharp interface model of solidification is given by the following system:

$$\frac{\partial u}{\partial t} = D\nabla^2 u , \tag{1}$$

$$V_n = D(\partial_n u|^+ - \partial_n u|^-) , \tag{2}$$

$$u_i = -d(\mathbf{n})\kappa - \beta(\mathbf{n})V_n , \tag{3}$$

where:

- $u = (T - T_m)/(L/c_p)$ is the dimensionless temperature, $T$ and $T_m$ are the temperature and the melting temperature respectively, $L$ is the latent heat of fusion and $c_p$ is the specific heat at constant pressure;
- $D$ is the thermal diffusivity;
- $\mathbf{n}$ is the outer normal to the solid subdomain;
- $V_n$ is the normal velocity of the interface;
- $\partial_n u|^+$, $\partial_n u|^-$ are the normal derivatives of the temperature at the interface for the solid (+) and liquid (-) phases;
- $\kappa$ is the local curvature, $\beta(\mathbf{n})$ is the kinetic coefficient;
- $d(\mathbf{n}) = \gamma(\mathbf{n})T_m c_p/L^2$ is the capillary length and $\gamma(\mathbf{n})$ is the surface tension.

The equation (1) describes the diffusion of the heat inside the bulk solid and liquid phases. Equation (2) gives an expression for the interface velocity, proportional to the discontinuity in the normal derivative of the temperature across the interface. It is needed for conservation of energy and corresponds to release and/or absorption of latent heat. Finally the Gibbs-Thomson condition(3) models the change of melting temperature according to kinetic and capillary effects.

## 3    Phase-Field Model of Solidification

In the phase-field formulation of solidification problems the sharp interface model of solidification is replaced by a pair of non-linear reaction-diffusion type equations. The interface between phases is considered as a diffuse region with small but numerically resolvable thickness and it is given implicitly by the so-called phase-field, i.e., the level set of a scalar function $\phi$ of space and time, called phase-field function. It varies smoothly from -1 in the liquid to +1 in the solid phase. An evolution equation for the phase-field function is solved and the solid-liquid interface is defined by the level set $\phi = 0$. It must be noted that there are many ways to prescribe a smoothing and dynamics of the sharp interface model, so that there is no unique phase-field model. We construct the numerical method on the basis of the phase-field model used in [2], [3] and given by the following system of non-linear equations:

$$\frac{\partial u}{\partial t} = D\triangle u + \frac{1}{2}\frac{\partial \phi}{\partial t} \ , \tag{4}$$

$$\tau\frac{\partial \phi}{\partial t} = \nabla(W^2\nabla\phi) + \partial_x[|\nabla\phi|^2 W\partial_{\phi_x}W] + \partial_y[|\nabla\phi|^2 W\partial_{\phi_y}W] + \phi(1-\phi^2) - \lambda u(1-\phi^2)^2 \tag{5}$$

for the dimensionless temperature $u(x, y, t)$ and for the phase-field function $\phi(x, y, t)$, $(x, y) \in \Omega$, $0 < t \le t_k$. Here:

- $\lambda$ is a dimensionless parameter that controls the coupling between $u$ and $\phi$;
- $W = \delta A_s$, $A_s = (1 - 3\epsilon) + 4\epsilon\frac{\phi_x^4 + \phi_y^4}{|\nabla\phi|^4}$ and $\epsilon$ is the anisotropy strength;
- $\delta$ is the characteristic length;
- $\tau = \tau_0 A_s^2$ and $\tau_0$ is the characteristic time.

The numerical method we have used to solve the phase-field model of solidification is described in details in [4], [5].

The phase-field method was the first method successfully applied to 3D dendritic growth. The main advantages of the phase-field method are: interfacial geometric quantities such as curvature and the outward normal vector do not have to be computed since they are incorporated in the model; because of the implicit representation of the boundary no need to care about topological changes; sidebranching can be obtained by including thermal noise; interaction between several dendrites can be easily simulated. The disadvantages of the phase field method are: it gives only an approximate representation of the front location; the phase-field model requires an asymptotic expansion analysis to be performed with a small parameter proportional to the interface width, $W$. It is important to note that the grid size is proportional to $W$ and only in the limit as $W \to 0$ does the phase-field method converge to the sharp interface model. In that sense, the phase-field method is only a first order accurate approximation to the sharp interface model. In fact it was shown rigorously that if the grid size is not proportional to $W$, the numerical results are generally incorrect.

## 4   Level Set Model of Solidification

Level set method is a numerical technique introduced by Osher and Sethian [6] to track the motion of interfaces. This method is conceptually similar to the phase-field method in that the interface is represented as a zero contour of the level-set function $\phi$. The movement of the interface is tracked implicitly through an advection equation for $\phi$. Unlike the phase-field method, there is no arbitrary interface width introduced in the level set method, the sharp interface equations can be solved directly and thus no asymptotics are required.

## 4.1    Level Set Formulation

Consider a closed moving interface $\Gamma(t)$. Let $\Omega(t)$ be the region that $\Gamma(t)$ encloses. We associate with $\Omega(t)$ a level set function $\phi(x,t)$ that satisfies:

$$\begin{cases} \phi(x,t) < 0, \text{ in } \Omega(t) \\ \phi(x,t) = 0, \text{ on } \Gamma(t) \\ \phi(x,t) > 0, \text{ in } R^n \setminus \overline{\Omega(t)}. \end{cases} \tag{6}$$

The motion of the interface is determined by a velocity field $\boldsymbol{F}$. Thus we obtain the following advection equation for the level set function:

$$\frac{\partial \phi}{\partial t} + \boldsymbol{F}.\nabla\phi = 0. \tag{7}$$

Projecting velocity $\boldsymbol{F}$ normal to the interface, equation (7) becomes:

$$\frac{\partial \phi}{\partial t} + F_n|\nabla\phi| = 0. \tag{8}$$

The outward normal vector $\boldsymbol{n}$ and the curvature $\kappa$ are defined by:

$$\boldsymbol{n} = \frac{\nabla\phi}{|\nabla\phi|} \ , \ \kappa = \nabla.\boldsymbol{n}. \tag{9}$$

Solving equation (8) for one time step results in moving the contours of $\phi$ along the direction normal to the interface according to the normal velocity $F_n$. $F$ is constructed to be an extension of the interface velocity $V_n$, such that $F_n = V_n$ for points on the interface. The velocity extension can be done in different ways: ghost fluid method, local level set method , but we use a PDE based method [8] - every quantity $q$ defined on the interface $\Gamma(t)$ can be extended by finding the steady state solution of the following equation:

$$\frac{\partial q}{\partial t} + sgn(\phi)\frac{\nabla\phi}{|\nabla\phi|}\nabla q = 0. \tag{10}$$

To ensure that the level set function is a smoothly varying function, well suited for accurate computations, it is convenient to initialize $\phi$ to be a signed distance function with $|\nabla\phi| = 1$. Unfortunately, the level set function can quickly cease to be a signed distance function especially for curvature driven flows. Thus reinitialization of the level set function have to be performed after each time step. Reinitialization algorithm maintain the signed distance property by solving to steady state the equation

$$\frac{\partial \phi}{\partial t} + sgn(\phi_0)(|\nabla\phi| - 1) = 0 \tag{11}$$

where

$$sgn(\phi_0) = \frac{\phi_0}{\sqrt{\phi_0^2 + (\Delta x)^2}} \tag{12}$$

Thus the level set method for solidification is given by the following system of equations:

$$\frac{\partial \phi}{\partial t} + F_n |\nabla \phi| = 0, \tag{13}$$

$$\frac{\partial F_n}{\partial t} + sgn(\phi_0)\frac{\nabla \phi}{|\nabla \phi|}\nabla F_n = 0, \tag{14}$$

$$\frac{\partial \phi}{\partial t} + sgn(\phi_0)(|\nabla \phi| - 1) = 0, \tag{15}$$

$$\frac{\partial u}{\partial t} = D\nabla^2 u , \tag{16}$$

$$V_n = D(\partial_n u|^+ - \partial_n u|^-) , \tag{17}$$

$$u_i = -d(\boldsymbol{n})\kappa - \beta(\boldsymbol{n})V_n . \tag{18}$$

## 4.2   Numerical Method

To solve the equation of motion (13) for the level set function and the reinitialization equation (15) we use the method of lines approach - we make finite difference discretization in space and solve the resulting ODE system in time. Let us denote: $\Delta^+\phi_k = \phi_{k+1} - \phi_k$, $\Delta^-\phi_k = \phi_k - \phi_{k-1}$. We use the following fifth order WENO approximations [7] of the first derivatives on the right and left biased stencils respectively:

$$\frac{\partial \phi}{\partial x}^+ \approx \frac{1}{12}\left(-\frac{\Delta^+\phi_{i-2,j}}{\Delta x} + 7\frac{\Delta^+\phi_{i-1,j}}{\Delta x} + 7\frac{\Delta^+\phi_{i,j}}{\Delta x} - \frac{\Delta^+\phi_{i+1,j}}{\Delta x}\right)+$$

$$+\Phi^{WENO}\left(\frac{\Delta^-\Delta^+\phi_{i+2,j}}{\Delta x}, \frac{\Delta^-\Delta^+\phi_{i+1,j}}{\Delta x}, \frac{\Delta^-\Delta^+\phi_{i,j}}{\Delta x}, \frac{\Delta^-\Delta^+\phi_{i-1,j}}{\Delta x}\right),$$

$$\frac{\partial \phi}{\partial x}^- \approx \frac{1}{12}\left(-\frac{\Delta^+\phi_{i-2,j}}{\Delta x} + 7\frac{\Delta^+\phi_{i-1,j}}{\Delta x} + 7\frac{\Delta^+\phi_{i,j}}{\Delta x} - \frac{\Delta^+\phi_{i+1,j}}{\Delta x}\right)-$$

$$-\Phi^{WENO}\left(\frac{\Delta^-\Delta^+\phi_{i-2,j}}{\Delta x}, \frac{\Delta^-\Delta^+\phi_{i-1,j}}{\Delta x}, \frac{\Delta^-\Delta^+\phi_{i,j}}{\Delta x}, \frac{\Delta^-\Delta^+\phi_{i+1,j}}{\Delta x}\right).$$

$$\Phi^{WENO}(a,b,c,d) = \frac{1}{3}\omega_0(a - 2b + c) + \frac{1}{6}(\omega_2 - \frac{1}{2})(b - 2c + d)$$

The weights $\omega_0, \omega_2$ are defined as: $\omega_0 = \frac{v_0}{v_0+v_1+v_2}$ , $\omega_2 = \frac{v_2}{v_0+v_1+v_2}$, $v_0 = \frac{1}{(\epsilon+\beta_0)^2}$ , $v_1 = \frac{6}{(\epsilon+\beta_1)^2}$ , $v_2 = \frac{3}{(\epsilon+\beta_2)^2}$, with smoothness measurements: $\beta_0 = 13(a - b)^2 + 3(a - 3b)^2$, $\beta_1 = 13(b - c)^2 + 3(b + c)^2$, $\beta_2 = 13(c - d)^2 + 3(3c - d)^2$ and $\epsilon$ is a parameter that prevents division by zero.

To finish the space discretization we use the Roe flux with entropy fix [7] for the equation of motion for the level set function (13) and the Godunov flux for the reinitialization equation (15) [8].

To solve the resulting ODE systems in time we use the following third order three stage Strong Stability Preserving Runge-Kutta method [9]:

$$u^{(1)} = u^n + \tau L(u^n) \tag{19}$$

$$u^{(2)} = \frac{3}{4}u^n + \frac{1}{4}u^{(1)} + \frac{1}{4}\tau L(u^{(1)}) \tag{20}$$

$$u^{n+1} = \frac{1}{3}u^n + \frac{2}{3}u^{(2)} + \frac{2}{3}\tau L(u^{(2)}) \tag{21}$$

To solve the velocity extension equation (14) we use again the method of lines approach. For space discretization we use Upwind scheme and for time discretization we use explicit Euler method. The above method reads:

$$F_{n,i,j}^{k+1} = F_{n,i,j}^k - \Delta t((sgn(\phi_{i,j})n_{x,i,j})^+ \frac{F_{i,j}^k - F_{i-1,j}^k}{\Delta x} + (sgn(\phi_{i,j})n_{x,i,j})^- \frac{F_{i+1,j}^k - F_{i,j}^k}{\Delta x}$$

$$+(sgn(\phi_{i,j})n_{y,i,j})^+ \frac{F_{i,j}^k - F_{i,j-1}^k}{\Delta y} + (sgn(\phi_{i,j})n_{y,i,j})^- \frac{F_{i,j+1}^k - F_{i,j}^k}{\Delta y}) \tag{22}$$

It is possible to use WENO scheme for the space discretization of velocity extension equation, but since the approximation of jump condition on the interface is first order accurate, we think that the simple Upwind scheme is a better choice.

To compute components of the outward normal $(n_x, n_y)$ we use central differences.

The spatial discretization of the temperature field equation (16) away from the interface is standard. This discretization is not valid if the interface $\Gamma$ cuts the stencil. For example, suppose that the interface location, $x_I$, falls in between $x_i$ and $x_{i+1}$. Then when discretizing at $x_i$ we define a ghost value $u_{i+1}^G$:

$$u_{i+1,j}^G = \frac{2u_{I,j} + (2\alpha^2 - 2)u_{i,j} + (-\alpha^2 + 1)u_{i-1,j}}{\alpha^2 + \alpha} \tag{23}$$

Here $\alpha = (x_{I,j} - x_{i,j})/\Delta x$. The discretization in y-direction is made by analogy.

For discretization in time we use a second order explicit modification of the Runge-Kutta method [10] with extended region of stability and automatic choice of the time step.

## 4.3  Algorithm Outline

The algorithm of the level set method can be outlined as follows:

1. Initialize $u$ and $\phi$.
2. Compute the extended normal velocity $F_n$, by solving equation (14) for 5 time steps.
3. Solve equation (8) for the level set function for one time step.
4. Reinitialize $\phi$ to be the exact signed distance function by solving equation (15) for 5 time steps.
5. Solve equation (1) for the temperature, taking into account the temperature at the interface, given by Gibbs-Thomson equation (3).
6. Repeat steps 2 through 5 to get the next values of $\phi$ and $u$.

We must note that the main difference between the described level set method and the level set method, used in the literature [8], [11], [12], [13], is that we use an explicit time discretization for the temperature field equation. It is also possible to use other numerical schemes for spatial discretization of the Level set equation (13) and reinitilization equation (15) - third order ENO scheme, third order WENO scheme, etc.. or to use Local Lax-Friedrichs flux or Global Lax-Friedrichs flux for the level set equation, but we found that previously described method gives better results.

## 5    Numerical Experiments

In the first set of experiments we compare the shapes and the tip velocities, obtained by the phase-field and the level set method for two different anisotropy strengths. The computational domain is: $\Omega = \{(x,y) : 0 \le x \le 640, \ 0 \le y \le 640\}$, the seed is placed in the center of the domain, its radius is $r = 4$, the undercooling is $u_0 = -0.55$, the dimensionless grid spacing is $h = 0.4$.

Fig. 1 shows the dendritic shapes for two different values of the anisotropy strength $\epsilon$, the capillary length $d_0$ and the thermal diffusivity $D$. As seen, the shapes almost coincide within the plotting resolution. Fig. 2 shows the dimensionless tip velocities obtained by the phase-field method and by the level set method and predicted by the microscopic solvability theory. We found that the results of the phase-field method and the level set method compare favorably and are in good agreement with the microscopic solvability theory.

In the second set of experiments we examine the effect of grid spacing. The computational domain is as above, the radius is $r = 4$, the undercooling is $u_0 = -0.55$, the anisotropy strength is $\epsilon = 0.05$, the capillary length is $d_0 = 0.139$

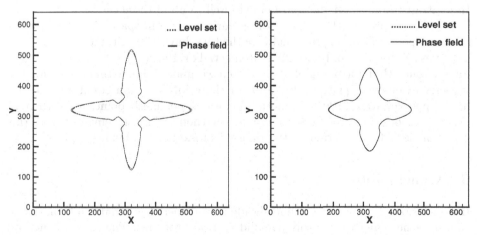

**Fig. 1.** Evolution of spherical seed for $\epsilon = 0.05, d_0 = 0.139, D = 4$ (left) and $\epsilon = 0.03, d_0 = 0.185, \ D = 3$ (right), t = 350

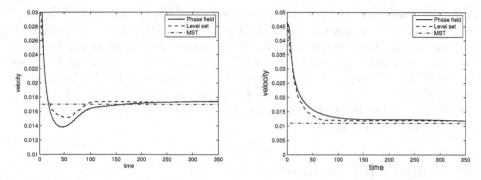

**Fig. 2.** Dimensionless velocity for $\epsilon = 0.05, d_0 = 0.139, D = 4$ (left) and $\epsilon = 0.03$, $d_0 = 0.185, D = 3$ (right)

and the thermal diffusivity is $D = 4$. In the table below we present the tip velocities obtained for three different grid spacings - $h$ is the grid spacing, $m$ is the number of mesh points inside the solid-liquid interface for the phase-field method, column $PhF$ shows the velocities obtained by the phase-field method, column $LS$ shows the velocities obtained by the level set method and column $MST$ shows the velocity predicted by the microscopic solvability theory:

| $h$ | $m$ | $PhF$ | $LS$ | $MST$ |
|---|---|---|---|---|
| 0.2 | $20 - 24$ | 0.01735 | 0.01731 | 0.01700 |
| 0.4 | $10 - 12$ | 0.01740 | 0.01737 | 0.01700 |
| 0.8 | $5 - 6$ | 0.01900 | 0.01770 | 0.01700 |

It can be seen from the table that the phase-field method behaves well when $m \geq 10$ (cases $h = 0.2, h = 0.4$), otherwise it fails to predict the correct velocity. In contrast the level set method behaves well in all of these cases.

Finally, we present the number of operations, used in space discretization for one grid point. For the phase-field method we have 172 arithmetic operations per mesh point and only 1 "if" statement. For the level set method, taking into account that for one global time step we make 5 inner time steps for the velocity extension equation (14) and for the reinitialization equation (15), we have approximately 2100 arithmetic operations per mesh point, more then 15 evaluations of square root function and more then 15 "if" statements. From this point of view the advantage of the phase-field method is obvious.

## 6    Conclusion

In conclusion, the level set method should be considered as a viable alternative to the phase-field method for solving solidification problems. The level set method can handle discontinuous material properties easily, which is currently very difficult with the phase-field approach. On the other hand simulating sidebranching and dendritic interaction is much easier with the phase-field method. Other advantages of the phase-field method are the computational speed and the fact that solving reaction-diffusion equations is numerically more stable process than solving advection equations.

**Acknowledgments.** This research was partially supported by the Sofia University Research Foundation through Contract 86/2006.

# References

1. Juric, D., Tryggvason, G.: A front tracking method for dendritic solidification. J. Comput. Phys. **123** (1996) 127–148.
2. Karma, A., Rappel, W.-J.: Phase-field method for computationally efficient modeling of solidification with arbitrary interface kinetics. Phys. Rev. **E 53** (1996) 3017–3020.
3. Karma, A., Rappel, W.-J.: Quantitative phase-field modeling of dendritic growth in two and three dimensions. Phys. Rev. **E 57** (1998) 4323–4349.
4. Slavov, V., Dimova, S., Iliev, O.: Phase-field method for 2D dendritic growth. Lect. Notes in Comp. Sci. **2902** (2004) 404–411.
5. Slavov, V., Dimova, S.: Phase-field modelling of dendritic interaction in 2D. Mathematics and Education in Mathematics **2902** (2004) 466–471.
6. Osher, S., Sethian, J. A.: Fronts propagating with curvature-dependent speed: Algorithms based on Hamilton-Jacobi Formulations. J. Comput. Phys. **79** (1988) 12–49.
7. Jiang, G.-S., Peng, D.: Weighted ENO schemes for Hamilton-Jacobi equations. SIAM J. Sci. Comput. **21** (2000) 2126–2143.
8. Chen, S., Merriman, B., Osher, Smereka, P.: A simple level set approach for solving Stefan problems. J. Comput. Phys. **135** (1997) 89–112.
9. Gottlieb, S., Shu, C.-W., Tadmor, E.: Strong stability-preserving high-order time discretization methods. SIAM Review **43** (2001) 89–112.
10. Novikov, V., Novikov, E.: Stability control of explicit one-step methods for integration of ordinary differential equations. Dokl. Akad. Nauk SSSR **272** (1984) 1058–1062.
11. Gibou, F., Fedkiw, R., Caflisch, R., Osher, S.: A Level set approach for the numerical simulation of dendritic growth. J. Sci. Comput. **19** (2003) 183–199.
12. Kim, Y., Goldenfeld, N., Dantzig, J.: Computation of dendritic microstructures using a level set method. Phys. Rev. E **. 62** 2000 2471-2474.
13. Peng, D., Merriman, B., Osher, S., Zhao, H., Kang, M.: A PDE-based local Level set method. J. Comput. Phys. **155** (1999) 410–438.

# Author Index

# Lecture Notes in Computer Science

For information about Vols. 1–4295

please contact your bookseller or Springer

Vol. 4345: N. Maglaveras, I. Chouvarda, V. Koutkias, R. Brause (Eds.), Biological and Medical Data Analysis. XIII, 496 pages. 2006. (Sublibrary LNBI).

Vol. 4344: V. Gruhn, F. Oquendo (Eds.), Software Architecture. X, 245 pages. 2006.

Vol. 4342: H. de Swart, E. Orłowska, G. Schmidt, M. Roubens (Eds.), Theory and Applications of Relational Structures as Knowledge Instruments II. X, 373 pages. 2006. (Sublibrary LNAI).

Vol. 4341: P.Q. Nguyen (Ed.), Progress in Cryptology - VIETCRYPT 2006. XI, 385 pages. 2006.

Vol. 4340: R. Prodan, T. Fahringer, Grid Computing. XXIII, 317 pages. 2007.

Vol. 4339: E. Ayguadé, G. Baumgartner, J. Ramanujam, P. Sadayappan (Eds.), Languages and Compilers for Parallel Computing. XI, 476 pages. 2006.

Vol. 4338: P. Kalra, S. Peleg (Eds.), Computer Vision, Graphics and Image Processing. XV, 965 pages. 2006.

Vol. 4337: S. Arun-Kumar, N. Garg (Eds.), FSTTCS 2006: Foundations of Software Technology and Theoretical Computer Science. XIII, 430 pages. 2006.

Vol. 4335: S.A. Brueckner, S. Hassas, M. Jelasity, D. Yamins (Eds.), Engineering Self-Organising Systems. XII, 212 pages. 2007. (Sublibrary LNAI).

Vol. 4334: B. Beckert, R. Hähnle, P.H. Schmitt (Eds.), Verification of Object-Oriented Software. XXIX, 658 pages. 2007. (Sublibrary LNAI).

Vol. 4333: U. Reimer, D. Karagiannis (Eds.), Practical Aspects of Knowledge Management. XII, 338 pages. 2006. (Sublibrary LNAI).

Vol. 4332: A. Bagchi, V. Atluri (Eds.), Information Systems Security. XV, 382 pages. 2006.

Vol. 4331: G. Min, B. Di Martino, L.T. Yang, M. Guo, G. Ruenger (Eds.), Frontiers of High Performance Computing and Networking – ISPA 2006 Workshops. XXXVII, 1141 pages. 2006.

Vol. 4330: M. Guo, L.T. Yang, B. Di Martino, H.P. Zima, J. Dongarra, F. Tang (Eds.), Parallel and Distributed Processing and Applications. XVIII, 953 pages. 2006.

Vol. 4329: R. Barua, T. Lange (Eds.), Progress in Cryptology - INDOCRYPT 2006. X, 454 pages. 2006.

Vol. 4328: D. Penkler, M. Reitenspiess, F. Tam (Eds.), Service Availability. X, 289 pages. 2006.

Vol. 4327: M. Baldoni, U. Endriss (Eds.), Declarative Agent Languages and Technologies IV. VIII, 257 pages. 2006. (Sublibrary LNAI).

Vol. 4326: S. Göbel, R. Malkewitz, I. Iurgel (Eds.), Technologies for Interactive Digital Storytelling and Entertainment. X, 384 pages. 2006.

Vol. 4325: J. Cao, I. Stojmenovic, X. Jia, S.K. Das (Eds.), Mobile Ad-hoc and Sensor Networks. XIX, 887 pages. 2006.

Vol. 4323: G. Doherty, A. Blandford (Eds.), Interactive Systems. XI, 269 pages. 2007.

Vol. 4320: R. Gotzhein, R. Reed (Eds.), System Analysis and Modeling: Language Profiles. X, 229 pages. 2006.

Vol. 4319: L.-W. Chang, W.-N. Lie (Eds.), Advances in Image and Video Technology. XXVI, 1347 pages. 2006.

Vol. 4318: H. Lipmaa, M. Yung, D. Lin (Eds.), Information Security and Cryptology. XI, 305 pages. 2006.

Vol. 4317: S.K. Madria, K.T. Claypool, R. Kannan, P. Uppuluri, M.M. Gore (Eds.), Distributed Computing and Internet Technology. XIX, 466 pages. 2006.

Vol. 4316: M.M. Dalkilic, S. Kim, J. Yang (Eds.), Data Mining and Bioinformatics. VIII, 197 pages. 2006. (Sublibrary LNBI).

Vol. 4314: C. Freksa, M. Kohlhase, K. Schill (Eds.), KI 2006: Advances in Artificial Intelligence. XII, 458 pages. 2007. (Sublibrary LNAI).

Vol. 4313: T. Margaria, B. Steffen (Eds.), Leveraging Applications of Formal Methods. IX, 197 pages. 2006.

Vol. 4312: S. Sugimoto, J. Hunter, A. Rauber, A. Morishima (Eds.), Digital Libraries: Achievements, Challenges and Opportunities. XVIII, 571 pages. 2006.

Vol. 4311: K. Cho, P. Jacquet (Eds.), Technologies for Advanced Heterogeneous Networks II. XI, 253 pages. 2006.

Vol. 4310: T. Boyanov, S. Dimova, K. Georgiev, G. Nikolov (Eds.), Numerical Methods and Applications. XVI, 728 pages. 2007.

Vol. 4309: P. Inverardi, M. Jazayeri (Eds.), Software Engineering Education in the Modern Age. VIII, 207 pages. 2006.

Vol. 4308: S. Chaudhuri, S.R. Das, H.S. Paul, S. Tirthapura (Eds.), Distributed Computing and Networking. XIX, 608 pages. 2006.

Vol. 4307: P. Ning, S. Qing, N. Li (Eds.), Information and Communications Security. XIV, 558 pages. 2006.

Vol. 4306: Y. Avrithis, Y. Kompatsiaris, S. Staab, N.E. O'Connor (Eds.), Semantic Multimedia. XII, 241 pages. 2006.

Vol. 4305: A.A. Shvartsman (Ed.), Principles of Distributed Systems. XIII, 441 pages. 2006.

Vol. 4304: A. Sattar, B.-H. Kang (Eds.), AI 2006: Advances in Artificial Intelligence. XXVII, 1303 pages. 2006. (Sublibrary LNAI).

Vol. 4303: A. Hoffmann, B.-H. Kang, D. Richards, S. Tsumoto (Eds.), Advances in Knowledge Acquisition and Management. XI, 259 pages. 2006. (Sublibrary LNAI).

Vol. 4302: J. Domingo-Ferrer, L. Franconi (Eds.), Privacy in Statistical Databases. XI, 383 pages. 2006.

Vol. 4301: D. Pointcheval, Y. Mu, K. Chen (Eds.), Cryptology and Network Security. XIII, 381 pages. 2006.

Vol. 4300: Y.Q. Shi (Ed.), Transactions on Data Hiding and Multimedia Security I. IX, 139 pages. 2006.

Vol. 4299: S. Renals, S. Bengio, J.G. Fiscus (Eds.), Machine Learning for Multimodal Interaction. XII, 470 pages. 2006.

Vol. 4297: Y. Robert, M. Parashar, R. Badrinath, V.K. Prasanna (Eds.), High Performance Computing - HiPC 2006. XXIV, 642 pages. 2006.

Vol. 4296: M.S. Rhee, B. Lee (Eds.), Information Security and Cryptology – ICISC 2006. XIII, 358 pages. 2006.